American Chemists and Chemical Engineers

American Chemists and Chemical Engineers

Wyndham D. Miles, Ph.D.

Editor

AMERICAN CHEMICAL SOCIETY

WASHINGTON, D. C. 1976

Library of Congress CIP Data

American chemists and chemical engineers.

Includes bibliographical references and index.

1. Chemists—United States—Biography. 2. Chemical engineers—United States—Biography.

I. Miles, Wyndham D., 1916-

QD21.A43 540'.92'2 [B] 76-192
ISBN 0-8412-0278-8

PRINTED IN THE UNITED STATES OF AMERICA

PREFACE

The men and women whose biographies constitute this volume are a small selection of American chemists who ought not to be forgotten. Herein are recorded briefly the lives of researchers, teachers, founders of chemical companies, editors of journals, writers of influential texts and popular books, inventors, presidents of chemical societies, consultants, engineers, interesting characters, and persons who applied chemistry to other professions. They cover a span of 300 years, from alchemists of early Colonial times to chemists who died recently.

This book grew out of my interest in the history of American chemistry. More than a quarter of a century ago I began accumulating information on chemists of this country who, through skill, intelligence, industriousness, and luck accomplished more than the average scientist. Out of several thousand chemists I selected about 500 who are worthy of inclusion in a collection of biographies. My original list underwent modifications for a number of reasons. I added several notable chemists who died during the three years this volume was being prepared. Unable to find writers for about 70 biographies, I replaced some of these chemists with others. Friends with an expert knowledge of certain fields (Guido Stempel, Bernard Schaar, Florence Wall, Robert Hawthorne, Jr., George Dupbernell, Joe Vikin, Aaron J. Ihde, Thomas Chilton) suggested other chemists whom I accepted. Fifty-one biographies I hoped to have in this volume are not here because 26 persons who promised to write them did not do so. This selection of chemists will not please everyone, but then if each reader drew up a list of 500 noteworthy American chemists, each list would be different.

Biographies range in length from less than 200 words to more than 1000, but the majority hover around 700 words. The length of a biography is not necssarily proportional to the importance of a chemist. I did not attempt to list these hundreds of chemists in their order of importance and then assign a specific number of words to each biography.

I would have preferred many of the biographies to have been longer. Would that we could have given each writer all the space he desired. But limited by the planned size of the book—approximately one quarter of a million word—I had to balance the number of biographies against their size. I wrote samples of several lengths and decided that 700 words would provide adequate, although not ideal, coverage, while allowing approximately 500 biographies in the book. The word limit also handicapped many of the contributors in writing as fully as they wished.

References appended to the biographies serve two purposes: they state the source of the writer's data, and they tell the reader where he can obtain additional information. Some chemists in this book, among them Joseph Priestley, Josiah W. Gibbs, Benjamin Silliman, Sr., and Thomas Midgley, Jr., have had so much written about them that it is not practical to refer to every article and book that discusses their work. Furthermore, references can never be complete and up to date because articles and books about chemists appear constantly. "American Men of Science," "Who's Who in America," "Who Was Who in America," and "Chemical Who's Who," have not been included among references because, containing data on many chemists, 20th century chemists in particular, their titles would appear over and over.

The major difficulty in preparing this book was in locating persons interested in history and biography, with time to research and write, and with some knowledge of the field wherein the subject chemist worked. I was not able to find persons to write about a number of chemists on my list although I canvassed members of the American Chemical Society's Division of History and corresponded with companies and chemistry departments that had employed the chemists (chairmen of a score of chemistry departments did not reply), relatives (several did not answer), and colleagues. Hence the biographies of some chemists I desired for this volume are not here.

Thousands of biographies of chemists not mentioned in this volume have appeared elsewhere—in books, pamphlets, periodicals, biographical dictionaries, newsletters, and other publications. There are so many reference works that direct readers to these biographies that I have space to mention only a few of them. *Chemical Abstracts* is the place to start, under the headings of obituaries, biographies, and the chemist's name. If a biography or obituary is not listed in *CA,* it does not mean necssarily that none has been written. One may have appeared in a company or college newsletter, a newspaper, or a literary or historical periodical not covered by *CA.* For biographies in periodicals outside the field of chemistry consult *Biography Index* (1946-), "Poole's Index to Periodical Literature, 1802-1881," with supplements covering the period from 1882 to 1907, *Reader's Guide to Periodical Literature* (1900-), *Current Biography,* monthly and yearbook (1940-), and *American Historical Review* (1895-). For chemists with ties to medicine try *Index Medicus* (1879-), and "Index Catalogue of the Library of the Surgeon General's Office" (five series, 1880-1961). The Royal Society of London's "Catalogue of Scientific Papers, 1800-1900," lists articles of many earlier American chemists. J. C. Poggendorff's "Biographisch-literarisches Handwörterbuch" (1863-) contains biographical information and references to articles by American chemists. Hundreds of city, county, and state biographical dictionaries have been published, but copies are often difficult to locate, and one cannot tell in advance whether or not a biographical dictionary will contain the biography that he is looking for. A search for a biography may be time consuming unless one has access to a university library or library of comparable size and scope where one can follow leads or simply browse.

More and better biographies and obituaries ought to be written about noteworthy chemists who have contributed in some way to the betterment of science and civilization. Not a week goes by but an obituary of a prominent chemist appears in a local American Chemical Society publication, a departmental or company newsletter, or some periodical outside of chemistry; but relatively few explain satisfactorily the chemist's contribution or otherwise do justice to his or her memory.

I do not believe that it is possible to compile a biographical dictionary, within a reasonable length of time and at a reasonable cost, that is free from errors. There are errors in the multivolume "Dictionary of American Biography," prepared with much care and at considerable expense; in "Dictionary of National Biography," the authoratative British work; in "National Cyclopedia of American Biography," now in its 50th volume; in Howard A. Kelly and Walter Burrage's "Dictionary of American Medical Biography,"; and in all other biographical dictionaries that I have consulted.

Some errors creep in because inexperienced researchers do not realize the necessity of seeking out every bit of biographical information about a person and consequently remain unaware of significant material or of discrepancies which, if discovered, might be resolved. Typographical errors can occur among the myriad dates, numbers, and names that are copied and recopied between the initial note taking and the final printing. Errors can be transmitted from incorrect statements in biographies and obituaries consulted by writers. Errors can be copied unwittingly from false data provided to "Who's Who in America," "American Men of Science," and similar reference works by vain persons seeking to further their careers. Errors can have their source in ambiguous or incomplete statements in biographies or obituaries.

We have tried to keep this volume free from errors, but no doubt some are present. Readers who find inaccuracies will do others a favor by sending us corrections so that greater accuracy may be attained in the event that a second edition is published.

It is now 17 years since I began to look for a publisher who was receptive to the idea of bringing out a biographical dictionary. Had the American Chemical Society not accepted the idea, this volume might have remained in the preliminary manuscript stage. I am grateful to the Society and to the head of its Books Department, Robert F. Gould, with whom I have had so many pleasant discussions about the men and women herein.

Gaithersburg, Md. WYNDHAM D. MILES
November 1975

CONTRIBUTORS

Abrahams, Harold J., 45, 188, 271, 473, 490
Adler, Alan D., 405
Anderson, Leigh C., 125, 268, 394, 485
Anderson, Oscar E., 512
Atkins, George M., Jr., 80, 89, 163, 465
Atkinson, Edward R., 67, 100, 148, 189, 191, 248, 436
Bacon, Egbert K., 39, 75, 133, 257, 281, 282, 289, 300, 478, 509
Bailar, John C., Jr., 177, 228
Baker, Dale B., 102
Baker, George W., 301
Baker, Michael H., 264
Bartow, Virginia, 25
Billinger, Robert D., 76, 129, 201, 407, 502
Bishop, Ethel Echternach, 331, 387
Blau, Henry H., 38, 441, 474
Blum, William, 497
Bossert, Roy G., 160
Bowers, Douglas E., 429
Burk, Dean, 97
Butts, Allison, 287
Cady, George H., 61
Caley, Earle R., 311, 369, 517, 520
Castles, Vera Kimball, 59, 273, 525
Chamberlain, Leonard C., 193, 486
Chilton, Thomas H., 470, 531
Clark, Mary C., 93
Cori, Carl F., 95
Corwin, James F., 384
Costa, Albert B., 143, 396, 406, 417
Cox, James C., Jr., 116, 161, 224, 275, 285, 379
Cross, R. James, Jr., 519
Daley, Lawrence R., 74
Dalton, Robert H., 464
Daly, Marie M., 63
Davids, L. Robert, 514
Deal, Dorothy B., 9, 52, 58, 237
Deischer, Claude K., 54, 199, 307
Dinsmore, Ray P., 256
Dowell, Michael B., 145, 146, 215
Dubpernell, George, 4, 17, 141, 154, 424
Eblin, Lawrence P., 26, 150, 210, 223, 329
Eckert, Charles F., 34
Eichelberger, Lillian, 325, 471
Feldman, Martin, 305, 366, 515
Finlay, Gordon R., 410
Finley, K. Thomas, 119, 272, 290, 335
Fisher, Benjamin R., 156
Fitts, Donald D., 277
Florin, Roland E., 491
Forker, Gordon M., 288
Freud, Henrietta Z., 165
Frey, John E., 170
Fruton, Joseph S., 33
Fuchsman, Charles H., 122, 166, 174, 200, 247, 262, 267, 378, 480
Gentieu, Norman P., 400
Gilbert, Arthur D., 175
Gorman, Mel, 8, 165, 349, 431, 475
Gortner, Willis A., 184
Gould, Robert F., 146, 360, 365, 383, 488
Griffin, George R., 376
Grummer, Arnold E., 501
Guthmann, Walter S., 212
Hamer, Walter J., 65
Hanke, Martin E., 279
Hassler, William W., 66
Hawthorne, Robert M., Jr., 44, 62, 87, 91, 98, 108, 118, 124, 132, 207, 217, 229, 249, 261, 274, 309, 315, 332, 389, 403, 419, 432, 461, 493, 495, 503, 518, 526
Hays, Donald R., 131, 182, 233
Healy, John J., Jr., 104
Heindel, Ned D., 93
Heines, Sister Virginia, 114
Herriott, Roger M., 363
Hill, Thomas T., 49, 434
Hitchcock, John, 221
Holran, Bruce G., 426
Horne, Ralph A., 96
Hougen, Olaf A., 467
Hovorka, Frank, 42, 477
Ihde, Aaron J., 7, 15, 55, 112, 144, 260, 320, 420
Jaffee, Bernard, 171
Johnson, Warren C., 198, 428
Jones, Daniel P., 365, 428, 457

American Chemists and Chemical Engineers

A

John Jacob Abel

1857-1938

Abel was born May 19, 1857 on a farm near Cleveland, Ohio the son of George and Mary (Becker) Abel. He graduated with his Ph.B. degree from University of Michigan in 1883. After spending part of a year at Johns Hopkins University, he went to Germany where he studied the medical sciences for about 7 years in various universities. At Strasbourg, where he received his M.D. degree in 1888, he worked in the laboratory of Oswald Schmiedeberg, the most prominent pharmacologist of his day. In 1891, on Schmiedeberg's recommendation, he was appointed lecturer in materia medica and therapeutics at the University of Michigan. Two years later, he became professor of pharmacology at Johns Hopkins, a position which he held until his retirement in 1932. He died May 26, 1938.

Abel has often been called the "father of American pharmacology." The chair he held in pharmacology at Johns Hopkins was the first to be established with that title in the United States. Even at Michigan, however, where his position retained the classical title of materia medica, it was understood that he was being hired to teach modern pharmacology. He brought with him from Europe the tradition, which Schmiedeberg and others were working to establish, of pharmacology as an experimental science and as a separate discipline in its own right. He played an instrumental role in the founding of the American Society for Pharmacology and Therapeutics in 1908 and *Journal of Pharmacology and Experimental Therapeutics,* which he edited until his retirement. Many prominent American pharmacologists, such as Reid Hunt, Carl Voegtlin, Arthur Loevenhart, and Eugene Geiling, received at least part of their training in his laboratory.

His real love was research. He believed firmly in the importance of chemistry for medicine and for pharmacology, and most of his work was in biochemistry. One could describe him as a biochemist as well as a pharmacologist, and he was one of the organizers of *Journal of Biological Chemistry* (1905) and the American Society of Biological Chemistry (1906).

His most important scientific contributions were in the field of hormone research. In 1897 when the study of hormones was still in its infancy, he isolated the substance in extracts from the adrenal medulla which is responsible for raising the blood pressure. While he believed that he had obtained the free hormone, which he named "epinephrine," he had actually isolated the hormone in the form of a benzoyl derivative. In 1900 the Japanese chemist Jokichi Takamine obtained the hormone in pure form, the first isolation of a pure active principle from an endocrine gland.

Abel returned to his interest in hormones again in 1912 when he extracted epinephrine from the parotid secretions of the toad *Bufo agua.* From 1917 until 1924 he was involved in an intensive effort to isolate the active principle of pituitary extracts but did not succeed. His attention next focused on insulin in 1924. Banting and Best had introduced insulin preparations for treating diabetes in 1922 but had not isolated the hormone as a pure chemical entity. Using a series of steps based largely on the principle of isoelectric precipitation from buffered solutions, Abel obtained crystalline insulin in late 1925. Abel's characterization of insulin as a protein was met at first with skepticism because it seemed unlikely that proteins, which many biochemists felt did not possess specific chemical structures, could exhibit the high degree of specific biological activity characteristic of a hormone or an enzyme. At the time, Richard Willstätter's theory that enzymes were actually non-protein molecules which are held in a colloidal state by adsorption on an inert protein carrier was widely accepted. This view was challenged by the isolation and characterization of enzymes such as urease, pepsin, and trypsin in the late 1920's and early 1930's and was gradually abandoned.

In 1913, Abel and his coworkers devised a "vividiffusion" apparatus for removing diffusible substances from the circulating blood of living animals by dialysis. With this apparatus, they were able to demonstrate clearly for the first time the presence of free amino acids in normal blood. Abel also recognized the clinical potential of the apparatus, which is the intellectual forerunner of modern "artificial kidney" machines.

"John Jacob Abel, M.D., Investigator, Teacher, Prophet, 1857-1938: A Collection of Papers by and about the Father of American Pharmacology," Williams and Wilkins (1957), with portrait; Jane Murnaghan and Paul Talalay, *Perspectives Biol. Med.* **10,** 334-380 (1967); Carl Voegtlin, *J. Pharmacol. Exp. Ther.* **67,** 373-406 (1939), with portrait and bibliography; Henry Swain, E. M. K. Geiling and Alexander Heingartner, *Univ. Michigan Med. Bull.* **29,** 1-14 (1963), with portrait; William de B. MacNider, *Biog. Mem. Nat. Acad. Sci.* **24,** 231-257 (1947), with portrait and bibliography; H. H. Dale, *Obit. Not. Fellows Roy. Soc.* **2,** 577-585 (1939), with portrait and bibliography.

JOHN PARASCANDOLA

Edward Goodrich Acheson

1856-1931

Acheson, the son of William and Sarah Diana (Ruple) Acheson, was born in Washington, Pa., Mar. 9, 1856. In his youth he attended Bellefonte Academy, Bellefonte, Pa., but for the most part he was self taught. His inventive tendency exhibited itself as early as 1872; in that year, at age 16, Acheson filed a caveat with the United States Patent Office for a force auger to use in coal-mining operations.

His father's sudden death in June 1873 necessitated the termination of young Acheson's formal education. For the next 7 years he earned his own living in a variety of endeavors and contributed also to the support of his mother and sisters. Surveying, gauging tank capacities in Pennsylvania's oil fields, and mining iron ore in a joint venture with his brother William were representative of the varied nature of his efforts. Concurrently, he was devoting his evenings to his private scientific pursuits, primarily in the field of electricity. One product of his labors was an electric pile, which he brought hopefully to the attention of Thomas Edison. This project led to a visit to Menlo Park in 1880 and to his joining the Edison group there.

He subsequently worked for the Edison interests in England, France, Italy, Belgium, and the Netherlands, as well as in New York. While in Europe he installed the first electric-lighting systems in such public places as Hotel de Ville in Antwerp, Restaurant Krasnapolsky in Amsterdam, Musée du Nord in Brussels, and La Scala Theater in Milan.

He concluded his tenure with Edison in 1884 in order to devote his energies to developing his own inventions. Most notable of the latter were those covering the manufacture of silicon carbide, Carborundum, the preparation of electric-furnace graphite, the synthesis of Siloxicon, and the stabilization of clay in water and of graphite in both water and oil. These were marketed under the trade names Egyptianized Clay, Aquadag, and Oildag.

On June 1, 1886 Acheson received the first of the 70 patents he was to be granted in his lifetime. This was for a "conductor of electricity." He sold the patent to George Westinghouse, who in turn transferred it to Standard Underground Cable Co. for exploitation.

Companies which Acheson established for the commercialization of his more important inventions were The Carborundum Co. (1891), Acheson Graphite Co. (1899), International Acheson Graphite Co. (1900), Acheson Oildag Co. (1908), and British Acheson Oildag Co. (1911).

In 1884 Acheson married Margaret Maher. Five sons and four daughters were born of this union: Veronica Belle, Edward Goodrich, Raymond Maher, Sarah Ruth, George Wilson, John Huyler, Margaret Irene, Jean Ellen, and Howard Archibald.

Acheson received an honorary Sc.D. degree from University of Pittsburgh in 1909. Numbered among his other honors and awards were the Rumford Medal of the American Academy of Arts and Sciences (1908); the Perkin Research Medal (1910); Grand Prix, Paris Exposition (1900) and St. Louis Exposition (1904); and honorary memberships in the Imperial Russian Technical Society and the Swedish Technological Association (1913). He was a president of The Electrochemical Society and vice-president of the American Institute of Chemical Engineers. In 1929 he donated to The Electrochemical Society $25,000 to establish the Edward Goodrich Acheson Prize and Medal to be awarded biennially. In 1930 he was named the first recipient of his own medal by The Electrochemical Society.

Acheson died in New York City, July 6, 1931.

Records of E. G. Acheson; autobiography, "A Pathfinder," (1910); Raymond Szymanowitz, "Edward Goodrich Acheson: Inventor, Scientist, Industrialist," Vantage Press (1971).

RAYMOND SZYMANOWITZ

Isaac Adams, Jr.

1836-1911

Isaac Adams, Jr., pioneer discoverer and promoter of nickel plating, was born in South Boston, Feb. 20, 1836. He was a direct descendant of Joseph Adams, born in Braintree, Mass. in 1654, the grandfather of President John Adams. His father, the Honorable Isaac Adams Sr., is noted as the inventor of the Adams power press for printing, 1828-34, which in time made the father a millionaire.

Isaac Adams, Jr.'s mother died when he was only a few years old. He was educated at Sandwich Academy (1846-48), Chauncey Hall School (1848-54), Bowdoin College (1854-58), Harvard Medical School (M.D., 1862), and École de Médicine in Paris (1862-64). In late 1864 he established himself as a physician in Boston, but attracted to industrial chemistry, he served also as an analytical and consulting chemist for the next few years to his Uncle Seth's Adams Sugar Refinery, to a glass manufacturing business, and to other businesses.

In his testimony in a nickel–plating patent suit against Edward Weston in 1874 Adams stated that he had been successful in producing a nickel electrotype while studying under Josiah Parsons Cooke at Harvard from 1858 to 1860. He was unable to repeat this feat in the winter of 1865-66 while engaged in nickel plating over 100 gross of gas burner tips. Faced with this inconsistency in 1866, he investigated the effect of such impurities as zinc, arsenic, copper and iron on the electrodeposition of nickel. He found that nickel deposited best in neutral or slightly acid solutions, in the range between neutral litmus and congo red paper. Adams also found that cast nickel anodes could be produced which dissolved satisfactorily to maintain the solutions during plating.

Armed with this knowledge, Adams decided in 1868 to devote himself wholly to the commercial introduction of nickel–plating even over his father's objections. He obtained a number of patents in 1869 and early in the year took over the nickel plating business of Wm. H. Remington and Co. of Boston. When the company failed, he formed from it the Boston Nickel Plating Co., which operated for about 100 years. A United Nickel Co. was also incorporated in New York June 14, 1869, with Adams as president. This company owned the patents and after they expired, went out of business in 1890, having filed hundreds of patent suits and collected royalties from more than 1000 licensees.

In the latter part of 1869 and the first half of 1870, Adams went to Europe and visited England, France, and Germany to introduce commercial nickel plating. He was particularly successful in England and France.

Adams also became an expert glass blower and made a large number of Geissler tubes which he sold between 1865 and 1868. In 1865 he developed a vacuum tube carbon "burner" incandescent electric light, about 14 years prior to the Edison-Swann inventions of 1879. Early in 1867 he went to Europe, partly to consider the commercial possibilities of electric lighting with A. Gaiffe of Paris, an instrument maker who later became his

French backer for the sale of nickel plating. At this time electricity was largely of battery origin, and while Adams' introduction of nickel–plating later stimulated the use of dynamo electric machines, in 1867 he decided not to enter the electric light business for lack of a cheap source of electric power.

Adams also made other discoveries of importance from time to time. An historian of Bowdoin College stated that his work on breech–loading rifles was hardly less important than his invention of nickel–plating. He became widely known as an outstanding electrochemical expert and teamed up with Charles F. Chandler of Columbia University in connection with the nickel–plating suits. Some work for Hood Rubber Co. resulted in a patent on the use of copper plate on steel shafts for bonding rubber to the steel for clothes wringer rolls.

Adams married Lucille Emilie Lods in 1869 or 1870. She was in some way related to his French backer, A. Gaiffe, and had two daughters by a previous marriage. The Adams' had two sons, Walter Owen (1877-1926) and Rayne (1880-1931), but no further descendants.

Following the successful outcome of his nickel–plating patent suits in the period from 1871 to 1886, Adams largely retired from public life. He had a special interest in the variations of the magnetic field of the earth and reported his measurements to the U.S. Geodetic Survey. He died at Annisquam, Mass., July 24, 1911.

G. Dubpernell, unpublished paper presented at New York A.C.S. Meeting, Sep. 13, 1954; G. Dubpernell, *Plating* **46,** 599-616 (1959), port. and bibliography.

GEORGE DUBPERNELL

Roger Adams

1889-1971

Roger Adams was born in Boston, Mass., Jan. 2, 1889. He received his early education at Boston Latin School and Cambridge Latin School where he first became interested in chemistry. He received his A.B. degree in 1909, his A.M. in 1910, and his Ph.D. in 1912, all from Harvard. During one summer while in college he tended cattle on a steamer to London and toured England by bicycle.

His thesis work was done under Henry A. Torrey and Theodore W. Richards. He received a college fellowship to study abroad and worked under Otto Diels in Berlin and Richard Willstätter at Kaiser Wilhelm Institute in Dahlem. On returning to Harvard he accepted a postdoctorate fellowship under Charles L. Jackson but quickly transferred to an instructorship which he held for 3 years. In 1916 he accepted an assistant professorship at University of Illinois where he remained for the rest of his career. He became professor in 1919, was department head from 1926 to 1954, research professor from 1954 to 1957 when he retired as professor emeritus. During this time he directed the doctoral theses of 184 candidates and trained many postdoctoral research associates.

His research interests covered a wide range in organic chemistry. He devised new synthetic methods, determined the structures of many natural products, developed a new platinum catalyst for organic hydrogenation reactions, and elucidated the stereochemistry of the ortho substituted biphenyls. The polyhydroxyanthraquinones, chaulmoogric and hydnocarpic acids, gossypol, cannabinol and the various alkaloids of senecio and crotalaria series are examples of the various natural products that he investigated extensively. In 1917 he went to Washington, D.C., as a major in the Chemical Warfare Service to direct a group of research chemists. They synthesized diphenylaminechloroarsine, a sternutator, which was adopted as a standard toxic agent under the name given to it by his associates—Adamsite.

In World War II he was a member of the Office of Scientific Research and Development and the National Defense Research Committee. He was on President Roosevelt's Science Advisory Board, a member of the National Inventors Council, and a member of the board of directors of the National Science Foundation. From the fall of 1945 to the spring of 1946 he was a scientific adviser in Germany to General Lucius Clay. During that time he helped revive Gmelin's "Handbuch der anorganischen Chemie" and the Beilstein "Handbuch der organischen Chemie," which are so valuable to chemical research workers all over the world. In 1947 he headed a seven-man Scientific Advisory Commission to the U. S. Military Government in Japan to advise General Douglas MacArthur.

He was the founder of the important series of publications, "Organic Syntheses," which has had such a marked effect on organic chemistry throughout the world. Later he was one of the founders of the "Organic Reactions Series," and was editor-in-chief for 19 years. The income from these two series is used as the basis for the Roger Adams Award in Organic Chemistry, an international award sponsored by the American Chemical Society through its Division of Organic Chemistry.

Adams was recognized by numerous awards based on the many facets of his research, among them the Nichols Medal, 1927; Willard Gibbs Medal, 1936; Elliott Cresson Medal, 1944; Davy Medal, 1945; Theodore Richards Medal, 1946; Priestley Medal, 1946; Honorary Commander of the Civil Division of the Most Excellent Order of the British Empire, 1948; A. W. Hofmann Medal, 1953; Midwest Medal, 1953; Perkin Medal, 1954; Charles Lathrop Parsons Award, 1958; Franklin Medal, 1960; Gold Medal of the American Institute of Chemists, 1964; National Medal of Science, 1964; John R. Kuebler Award, 1966; Order of Lincoln for Science, 1967; Certificate of Appreciation, Military Planning Division, Office of Quartermaster General, 1945; Certificate of Honor for Meritorious Service, *Chicago Sun,* 1946; Testimonial of Appreciation, University of Illinois, 1950; Certificate of Tribute, *Look* Magazine, 1950; Northwestern University Centennial Award, 1951; Certificate for Meritorious Civilian Service, Department of the Army, 1954; and Gold Plate Award, Academy of Achievement, 1961.

Adams was an active leader in the American Chemical Society, serving as chairman of the Organic Division, president of the Society in 1935, and member of the board of directors and its chairman from 1944 to 1950. He also served as president of the American Association for the Advancement of Science in 1950. He aided the Alfred P. Sloan Foundation and the Robert A. Welch Foundation as an adviser.

Throughout his active career he was always a loyal friend and adviser to his many students and colleagues to whom he was "The Chief." Given a spare moment in his busy life he would start telling his colleagues of his latest work in research which was always his deepest interest.

He married Lucile Wheeler in 1918, and had one daughter, Lucile (Mrs. W. E. Ranz), and four grandchildren. His wife died in 1964; Adams died in Champaign, Ill. on July 6, 1971.

N. J. Leonard, "Roger Adams," *J. Amer. Chem. Soc.* **91,** No. 1, Jan. 1969, with portrait; obituary, *Chem. Eng. News* July 12, 1971, portrait on cover; "McGraw-Hill Modern Men of Science" I, 4-5 (1966); personal recollections.

CARL S. MARVEL

Homer Burton Adkins

1892-1949

As a leader in organic research, Adkins made significant contributions in the field of catalytic hydrogenation during his 30 years at University of Wisconsin. He was born Jan. 16, 1892 near Newport, Ohio, the son of Alvin and Emily (Middleswart) Adkins. With a brother and sister, he grew up on a farm in the Ohio River Valley, attended high school at Newport and in 1915 graduated from Denison University. At Ohio State University he became a student of William Lloyd Evans. His Ph.D. degree was granted in 1918 for studies on the oxidation of organic compounds with alkaline permanganate. Following the completion of his doctorate he served briefly as a chemist in the War Department, taught for a year at Ohio State, held a summer position with Du Pont, and in the fall of 1919 began his career at Wisconsin.

Adkins married a Denison classmate, Louise Spivey, in 1917 while he was still a graduate student. They had three children, Susanne Dorothea, Nance, and Roger.

Outside his professional life, Adkins was an ardent reader particularly interested in the history of the Civil War. He visited several battlefields of that war. Golf was his principal physical relaxation. He was a member of the Congregational Church and served as deacon for several years.

His early research in organic chemistry was aimed toward generalizing the facts by which organic chemistry might be organized. He quickly became disenchanted with the prospect of finding such correlations and became more and more an empiricist seeking to establish the facts of chemical reactions. Very early he began to study catalysis, particularly in hydrogenation reactions. He worked on the catalytic action of metal oxides in the hydrogenation of esters to alcohols, developing copper chromite catalyst in the process. He also did extensive work with Raney nickel. He developed new methods for preparing catalysts by heating various aluminum alkoxides, thus obtaining aluminum oxides with different spacings. His work led to developing equipment for working at steadily higher pressures. Work on hydrogenolysis showed that hydrogen reacts on the catalyst surface to split off two new molecules. A monograph on hydrogenation was published in 1938.

Toward the end of his career Adkins extended his work to include high–pressure reactions of carbon monoxide to convert alcohols to acids. He was always interested in comparative reactivities, particularly the relation of structure, oxidation potential, and other chemical and physical properties to the subject, and was coauthor of the chapter on "Chemical Reactivity" in Henry Gilman's "Organic Chemistry."

Adkins was the author of "Reactions of Hydrogen" (Univ. of Wisconsin Press, 1938) and coauthor, with Samuel M. McElvain, of "Elementary Organic Chemistry" (McGraw-Hill, 1928) and "Practice of Organic Chemistry" (McGraw-Hill, 1928). The latter work went into a second edition, with Michael Klein as an added coauthor. His more than 175 research publications, mostly in collaboration with his graduate students, nearly all appeared in *Journal of the American Chemical Society*.

During World War II Adkins was deeply involved in mission oriented research for the Office of Scientific Research and Development. This led to 38 restricted OSRD reports on work dealing with vesicants, lachrymators, sternutators and detoxifying agents, and especially protective clothing and ointments for use against vesicants. His laboratory at Wisconsin also investigated the synthesis of potential antimalarial agents. The heavy administrative duties and travel associated with his work during these years undoubtedly had adverse effects on his health and led to a heart attack on June 20 which was responsible for his death on Aug. 10, 1949.

Adkins was a no-nonsense scientist who was a good judge of men. A hard worker himself, he expected hard work of his students. At the same time he had a personal charm which made the demands on his students tolerable. In addition he set a personal example of regular work in his own laboratory. He was skeptical of glib generalizations and sought, as he advanced in years, to restrict chemical knowledge to hard cold facts established in the laboratory. Philosophically, he was much influenced by Eric Temple Bell's "The Search for Truth" and P. W. Bridgman's "The Logic of Modern Physics." He loved an argument with his peers and was known to take unexpected positions in order to provoke discussions.

He enjoyed power but preferred to be in a position where he could influence policy rather than administer detail on a routine basis. With his departmental colleague James H. Walton, he exerted a restraining influence on J. Howard Mathews who, though an able chairman, tended to establish policy without departmental sanction.

He took his professional obligations seriously and gave faithful service to university committees, editorial boards, government boards, and ACS matters. At the time of his death he had been nominated for the presidency of the ACS. As a member of a college curriculum committee he had an important role in creating the university's program of integrated liberal studies. His judgment on this committee frequently found him at odds with fellow scientists on the committee while supporting the position of the humanists. Through his wisdom the program had a flexibility and integrity which gave it a permanence so notably lacking in many general education ventures.

Personal recollections; biography by Farrington Daniels in *Biog. Mem. Nat. Acad. Sci.* **27**, 293-317 (1952).

AARON J. IHDE

Carl Lucas Alsberg

1877-1940

Alsberg was born in New York, Apr. 2, 1877 of German-Jewish stock. His father, Meinhard, an organic chemist, had studied under Wöhler, Bunsen, and Genther before emigrating to New York. There the father became chief chemist for the city's Board of Health and later organized his own chemical manufacturing firm. He was one of the founders of the American Chemical Society. Thus, his son was exposed to a background of chemistry.

Carl graduated from Columbia University in 1896 and received his M.A. degree in 1900, at the same time obtaining his M.D. degree from the university's medical school. He then spent almost 3 years in fundamental research with Schmiedeberg in Strassburg and Emil Fischer in Berlin. From 1902 to 1905 he was assistant in physiological chemistry and from 1905 to 1908 instructor and co-chairman in the Department of Biological Chemistry at Harvard Medical School. During the summers of 1906-08, he was a special investigator for the Bureau of Fisheries and studied the value of sea foods. Beginning in 1908 he was engaged by the Bureau of Plant Industry, U. S. Department of Agriculture as a chemical biologist and after 1 year was put in charge of the Poisonous Plant Laboratory.

In 1912 President Taft had the delicate problem of choosing a successor to Harvey W. Wiley as chief of the Bureau of Chemistry, U. S. Department of Agriculture. There were problems within the Bureau, and controversial industrial and public health issues were involved. The selection by the President was based on his desire to choose a man whose personal interests were not at stake and who had all the qualifications to fill the post. Alsberg was recommended by some of the leading American scientists who were aware of his research achievements and humanitarian viewpoint. Although without administrative experience, he had to face the task of making the most formidable political, economical, and scientific decisions almost immediately in matters concerning the administration of the Pure Food and Drug Act. But with tact, justice, and conciliation backed by an extensive store of knowledge and sound judgement, he managed in a relatively short time to blend the scientific activities of the Bureau, the public good, and industrial interests in a fair and efficient manner. After becoming acquainted with the strengths and weaknesses of his subordinates and the channels of communication they had to use, he instituted a series of organizational reforms. The main thrust of the changes involved the fixing of responsibility of the activities of his staff in the field and separating the investigational from the routine regulatory work in the Washington laboratories. In addition to activities germane to the Food and Drug Act, the Bureau was charged with the solution of a variety of problems in agricultural and industrial chemistry. Alsberg, though not disliking administration, was primarily a researcher whose versatility of mind and breadth of information resulted in substantial achievements. Under his direct supervision, the chemists and other scientists of the Bureau produced research reports which appeared in government bulletins and scientific journals and were of such high quality that the Bureau earned the respect of the agricultural, industrial, and academic scientific communities. Among the most admired research projects were those which resulted in improvements in the technology of preparing, preserving, and shipping perishable foods.

When the United States entered World War I the activities of the Bureau were marshalled toward military purposes. Alsberg was on 15 committees. His deep and extensive knowledge of biochemistry, food, and medicine made him of particular value to Herbert Hoover, U. S. Food Administrator. Postwar economics so emasculated the work of the Bureau that Alsberg, unwilling to witness its deterioration, tendered his resignation in 1921.

In the same year Alsberg accepted a co-directorship of the newly founded Food Research Institute at Stanford University. Since this organization was to focus on the economics of food production and consumption, Alsberg veered his mind in this direction and began applying natural science methods to such problems. Initially his attention was directed to wheat and later to fats and oils. With the aid of assistants and graduate students he conducted or directed numerous laboratory studies over a wide range. In due time his administrative ability and inspirational effect on graduate students brought an offer from the authorities at Stanford to become dean of Graduate Studies, a post which he occupied from 1927 to 1933. In 1937 he transferred to University of California, Berkeley where he became professor of agricultural economics and director of the Giannini Foundation of Agricultural Economics.

While at Stanford and California Alsberg gave much time and effort to international relations, with particular concern for food economics, land utilization, and standards of living. He was a member-at-large of the Social Science Research Council and chairman for its Pacific Coast Regional Committee. He held important posts in the Institute of Pacific Relations.

Alsberg died in Berkeley, Calif., Oct. 31, 1940 and was survived by his widow Emma Blount (Peebles) Alsberg.

"National Cyclopedia of American Biography **30,** 31, James T. White & Co. (1943); *Science* **54,** 244 (1921); H. Horton Sheldon, ed., "The Progress of science. A review of 1940," Grolier Society (1941), pp. 28-29; biography by W. W. Skinner in *J. Assoc. Official Ag. Chem.* **26,** no. 2, III-VI (1943); Joseph S. Davis, ed., "Carl Alsberg, Scientist at Large," Stanford Univ. Press (1948), with portrait and partial list of publications; obituaries in *New York Times* and *New York Herald Tribune,* Nov. 2, 1940.

MEL GORMAN

Launcelot Winchester Andrews

1856-1938

Andrews was born in London, Ont., Canada, June 13, 1856, the son of Alfred Augustus and Louisa (Jones) Andrews. He studied at Cornell University for a time and graduated from Yale University with a Ph.B. degree in 1875. From 1876 to 1877 he was science instructor at the Springfield, Mass. High School, practicing as an analytical chemist in the same city for a year. Four years of travel and postgraduate work at the universities of Bonn and Göttingen, Germany followed. Andrews received the degree of Ph.D. at the latter in 1882 and afterwards took a graduate course in history and philosophy at Harvard. He married Anna Lane, daughter of Thomas Ritter of Brooklyn, N.Y. in 1883.

He was professor of chemistry at Iowa Agricultural College for one semester, 1884-85, and head of the Chemistry Department at University of Iowa, 1885-1904. For the next 6 years he was research and consulting chemist to Mallinckrodt Chemical Co., St. Louis, Mo. Then he organized the Andrews Chemical Works in Davenport, Iowa to manufacture oxalic acid, a compound important to the dye industry among others, from corn cobs by a process which he invented in 1909 and patented in 1913. This process involved heating a mixture of ground corn cobs and caustic potash in an atmosphere of superheated steam. Highly satisfactory oxalic acid was produced, but the company lacked capital and suspended operations in 1913.

He became a special investigator of canning for the U. S. Department of Agriculture and in 1915 became research chemist to Victor Chemical Works, Chicago, Ill. which position he held until his retirement in 1921. During this time he was successful in putting a modification of the German process for manufacturing oxalic and formic acids into operation for the first time in the United States. In 1913 he invented and in 1918 patented (U. S. Patent No. 1,283,617) an improvement in the process of manufacturing nitro compounds. He obtained several patents (Nos. 1,165,577 & 1,018,092 & 1,280,622 & 1,281,117) for improvements in the manufacture of formates and oxalates. In addition, with two collaborators he patented in 1920 a fireproofing composite for fabrics, consisting of certain mixtures of phosphates and borates which did not stiffen the fabric but left it soft and pliable.

Andrews also engaged in a number of toxicological investigations. He was the author of "An Introduction to the Study of Qualitative Analysis" (1891) and of many scientific papers on such subjects as the nascent state, the volumetric determination of silver and other elements, and the determination of very small vapor tensions. In addition to his interest in volumetric methods of analysis, he studied the synthesis of formic and oxalic acids and the relation between the densities and optical properties of aqueous solutions and their molecular composition.

Andrews belonged to many scientific societies, including the Chemical Society (London); he was president of the Iowa Academy of Sciences, 1893-94, and chairman of the St. Louis Section of the American Chemical Society, 1908-10. His hobbies were reading, photography, and travel. He died without issue in Belmont, Mass., Apr. 14, 1938 at the age of 81 years.

"The National Cyclopaedia of American Biography" **46,** 398-9, James T. White & Co. (1963); obituary in *New York Times,* Apr. 15, 1938.

DOROTHY B. DEAL

Thomas Antisell

1817-1893

Antisell was born in Dublin, Ireland, Jan. 26, 1817. Educated as a physician he became as much interested in chemistry as in medicine and studied under a number of famous chemists—Dumas, Pelouze, Berzelius, and Kane. He practiced medicine in Dublin, taught scientific subjects, wrote on agricultural chemistry, and left Ireland because of political unrest.

He arrived in New York City in November 1848 and opened a medical office and a chemistry laboratory. Antisell is said to have been the first chemist in New York to give laboratory instruction to women. While living in New York he journeyed north for a few months each year to teach chemistry at Berkshire Medical Institution in Massachusetts and at Vermont Medical College.

In 1854 he went west as geologist with Lt. John G. Parke's Pacific Railroad Survey and made a reconnaissance of sections of Arizona and California. After the survey was completed he settled in Washington, D. C. and in 1856 became chief examiner of the U. S. Patent Office, with responsibility for chemical inventions. When the Civil War broke out he joined the northern army as a surgeon, was in battles, had charge of Harewood Military Hospital in Washington, and was mustered out in 1865 with rank of brevet lieutenant colonel.

The war over, Antisell was appointed chemist of the 3-year old Department of Agriculture. The only government chemist in Washington, he not only ran analyses of minerals and agricultural specimens for his department but at times carried on investigations for other agencies, including an investigation of cancelling ink for the Post Office Department and a study of building stones for the Treasury Department.

In 1871 he accompanied a team headed by Secretary of Agriculture Horace Capron to give technical assistance to the Japanese government. Antisell helped develop inks for paper currency, mucilage for stamps, and other materials. Decorated by the Japanese government, he returned to Washington in 1877 because of Mrs. Antisell's poor health. He again entered the U. S. Patent Office as an examiner and remained there until gradual paralysis forced him to retire in 1891.

Concurrent with his jobs at the Department of Agriculture and Patent Office, Antisell taught a variety of subjects at Georgetown University from 1858 to 1869 and 1880 to 1882, among them hygiene, military surgery, chemistry, toxicology, and physiological chemistry. He also taught chemistry at University of Maryland from 1869 to 1870 and at George Washington University in 1870. He was offered the presidency of Franklin and Marshall College in the 1860's and of a college in Cairo, Egypt in the 1870's.

Versatile and industrious, Antisell wrote for a variety of journals. He published at least six articles on agricultural chemistry, three on geology, four on public health, one on oceanography, two on physiological chemistry, one on biography, five valedictory and inaugural addresses in pamphlet form, and two books, including "The Manufacture of

Photogenic or Hydro-Carbon Oils" in 1859. He edited *New York Journal of Pharmacy* in 1854.

Antisell was one of the founders of Philosophical Society of Washington and a founder and first president of Training School for Nurses. When the Chemical Society of Washington was organized in 1884, he was the most prominent member of the profession in the city and was elected first president of the society.

One of Antisell's colleagues said that "in official life he had the reputation of being reserved and even somewhat brusque, but among his friends he was cordial and even warm-hearted, with an abundant supply of wit and humor." According to another, "he led a very unobtrusive homelife, rarely appearing in public except where his duty called him. He was faithful to duty and conscientious in its performance, unostentatious in manner, and cordial in friendship." He died June 14, 1893 in Washington and was buried in Congressional Cemetery.

One of Antisell's children, Thomas Jr., studied chemistry, was one of the original members of Chemical Society of Washington, and received his M.D. degree from Georgetown University in 1881. He moved to Montana where he died at Warmsprings, Nov. 26, 1893.

Samuel C. Busey, "Personal Reminiscences," pp. 140-41, Medical Society of D.C. (1895); obituary by W. H. Seaman, *Bull. Philos. Soc. Wash.* **13,** 368 (1895-99); obituary, *J. Amer. Med. Ass.* **21,** 93-95 (1893); C. A. Browne, "Lecture and Laboratory Notebooks of Three Early Irish-American Refugee Chemists," *J. Chem. Educ.* **18,** 153-58 (1941); D. Reilly, "Thomas Antisell," *Recorder, Bull. Amer. Irish Hist. Soc.* **16,** 26-29 (1954); W. D. Miles, "CSW's First President, Thomas Antisell," *Capital Chem.* **18,** 7-9 (1968), portrait on cover. Records of Georgetown University and Chemical Society of Washington provide data on Thomas Antisell, Jr.

WYNDHAM D. MILES

John Howard Appleton

1844-1930

Appleton was associated with Brown University as student and teacher for over 70 years—from 1859 to 1930. He graduated Phi Beta Kappa from Brown in 1863, his senior thesis being on the new subject of spectrum analysis. Brown also awarded him an M.A. degree in 1869 and a D.Sc. in 1900.

Although an authority on industrial chemistry, especially synthetic dyes, Appleton was first and foremost a teacher. At the time of his death, it was said that any gathering of prominent Rhode Island citizens was a gathering of his pupils. He became an instructor at Brown in 1864, professor of chemistry and applied arts in 1868, professor of chemistry in 1872 and emeritus professor in 1914. In two decades, 1878-1898, he published 13 textbooks and laboratory manuals covering general chemistry, qualitative and quantitative analysis, organic chemistry, and medicinal chemistry. Several of these went through three editions and one, "The Young Chemist," was published in eight editions.

Perusal by this author of the 1880 edition of his "Handbook of Chemistry," which he called "a popular introduction," shows that he wrote, and undoubtedly lectured, in a style that commanded attention. Anecdotes, the history of chemistry, and economic data are skillfully interwoven with chemical theory and simple laboratory experiments. References to the periodical literature abound at the end of each chapter which the author considered "as among the most important helps offered by the teacher to the student." The book is profusely illustrated with apparatus, flow charts, and portraits of chemists, and includes 14 color plates, unusual for the time.

Appleton was active in the American Chemical Society and served as Vice–President in 1893. A proposal by Appleton and Charles E. Munroe led to the formation of the first local section of the Society in Rhode Island in January 1891. The stated purpose of the section was for "conferences and free discussion rather than the reading of formal papers." Appleton also was a member of the American Association of Advancement of Science, the Society of Chemistry and Industry, and the New York Academy of Sciences. For many years, he was Sealer of Weights and Measures for Rhode Island and Chemist to the State Board of Agriculture. In 1891 he was part of a commission designated by President Harrison to test the coinage of the United States Mint.

Born in Portland, Maine, Feb. 3, 1844 he retained a life-long love of the sea. As a college student, he served with Company I, First Regiment of the Rhode Island Militia but saw only a few days' active duty when marauding Confederate vessels threatened the Providence port. Although the service was brief, he was an inveterate student of the Civil War and especially admired U. S. Grant.

Professor Appleton married Louise Mumford Day in 1874. They had three sons and three daughters, all of whom graduated from Brown. Appleton was a student of Shakespeare and reread all of the plays every few years. A member of the Providence Art Club, he illustrated the novel "John Malcolm" in 1902. Although not a religious man, philosophic discussions with friends were highly prized interludes in his daily routine.

He died Feb. 18, 1930.

Brown Alumni Mag. **14,** 119-120 (Dec. 1913); obituary, *Providence Journal* (Feb. 19, 1930).

HERBERT T. PRATT

Howard Bell Arbuckle

1870-1945

Arbuckle was born near Lewisburg, W. Va., Oct. 5, 1870. He received his B.A. degree, with major work in ancient languages, from Hampden-Sydney College in 1889 and his M.A. in 1890. He then taught for a year in Mississippi. Deciding on a career in teaching, he became professor of ancient languages at the Seminary West of Suwanee, now Florida State University, from 1891 to 1894. During this stay at Tallahassee he decided to become a teacher of chemistry. He became a special student of chemistry for the year, 1894-95, at University of Virginia. He then entered Johns Hopkins University and received his Ph.D. in chemistry in 1898. His research for his dissertation dealt with the determination of atomic weights of zinc and cadmium. In the fall of 1898 he became professor of chemistry at Agnes Scott Institute, now Agnes Scott College. He was appointed professor of chemistry at Davidson College in 1912 and retired in 1937. He was married to Ida Clift Meginnis of Tallahassee, Fla. in 1896, and they had two children: Howard Bell and Adele Taylor.

In addition to his teaching and chemical research, Arbuckle had other strong interests. He had expert knowledge of agriculture and animal husbandry. He was interested in breeding sheep and had a flock of thoroughbred sheep at his summer home in Maxwelton, W. Va. He was a founder of the Continental Dorset Club, for the registry of purebred Dorset sheep, and was a contributing editor of *American Sheep Breeder* for a number of years. He had similar interests in cattle breeding.

Arbuckle's chemical research interests included some work in qualitative analysis but were more particularly concerned with agricultural chemistry and natural products. He did a good deal of work which involved the variation in the protein content of corn in terms of such factors as variety, where grown, and location of grains on the ears of corn. He also did work on the destructive distillation of various forms of cellulose, especially cotton and various kinds of paper. He published a number of papers on these subjects in *Journal of the Elisha Mitchell Scientific Society* of the North Carolina Academy of Science during the years from 1918 through 1934.

One of his interests was membership in the Pi Kappa Alpha fraternity. He served on its national council for many years, and from 1913 to 1933 was Grand Councilor, or national president, of this organization. He was an elder in the Presbyterian Church and it was a matter of pride with him that he influenced young men to enter the ministry, in addition to those whom he turned toward a career in chemistry.

Arbuckle died July 19, 1945 at Morganton, N. C. and was buried at Davidson, N. C.

Archives, Davidson College; *Alumni Bulletin*, Davidson College (Nov. 1945); *The Davidsonian* (May 12, 1937; March 18, 1939); *Charlotte Observer* (June 19, 1927; July 21, 1945; July 25, 1945); *The Shield and Diamond* of Pi Kappa Alpha, 28, 56-57 (Sept. 1968).

THOMAS S. LOGAN
ADELE TAYLOR ARBUCKLE LOGAN

Peter Townsend Austen

1852-1907

Austen was born in Clifton, Staten Island, N.Y., Sept. 10, 1852, son of John H. and Elizabeth (Townsend) Austen. He was a descendant of Peter Townsend, who in 1776 introduced a new process for manufacturing steel and who subsequently forged the chain which stretched across the Hudson River at West Point during the Revolutionary War. Austen was educated at Holden's private school of Clifton and the Columbia School of Mines where he received his Ph.B. degree in 1873. He then went abroad for 3 years to study under August W. Hofmann at University of Berlin and to receive his Ph.D. degree from University of Zurich in 1876.

He returned to America to become an instructor of chemistry at Dartmouth. In 1877 he was appointed to the professorship of general and applied chemistry at Rutgers, a position he kept for 13 years, during which time he also served on the faculty of the New Jersey Science School. In addition to these posts he was appointed to various state and municipal boards as well as state chemist. He married Ellen Monroe in 1878.

He was a pioneer of the university extension work, and his pleasing personality and interesting demonstrations made him a popular lecturer in the principal towns of New Jersey. In 1891 Austen resigned from his position at Rutgers with the intention of becoming a professional consultant. But he found the classroom attraction so great that he accepted the professorship of chemistry at Brooklyn Polytechnic Institute in the fall of 1893 and remained there until 1898. He then gave up educational work completely and established a consulting office and laboratory in New York City.

Austen published several text-books, among them "Kurze Einleitung zu den Nitro-Verbindungen," Leipzig, 1876 and a translation and revision of Pinner's "Introduction to the Study of Organic Chemistry" in 1893. He was the author of some 50 scientific papers, most of which appeared in *American Chemical Journal* and the *Berichte der Deutschen Chemischen Gesellschaft*. He also contributed extensively to such trade publications as *Textile Colorist* and *Druggist's Circular*. He also invented several manufacturing processes used in dyeing and bleaching.

Austen died Dec. 30, 1907 in New York City.

"Appletons' Cyclopaedia of American Biography" **1,** 119-20, D. Appleton & Co. (1888); "National Cyclopedia of American Biography" **13,** 92, James T. White & Co. (1906); "Dictionary of American Biography" **1,** 430-431, Chas. Scribner's Sons (1928); Obituary in *New York Tribune* (Jan. 1, 1908).

CHARLENE J. STEINBERG

B

Earle Jay Babcock

1865-1925

Earle Babcock was born at St. Charles, Winona County, Minn., June 11, 1865, the son of David Loren and Lavinia P. Campbell Babcock. In 1889 he received his B.S. degree from University of Minnesota and went with his bride, the former Lillian G. Cool, to University of North Dakota as instructor in chemistry and English. The university was then 6 years old, and Babcock assumed many tasks in addition to his teaching duties, including monitoring the boys in the main building and serving as the school's first postmaster.

Babcock was convinced early in his career that his research should serve the state of North Dakota and its resources. In 1890 he published a paper on soils and sugar beets and became professor of chemistry and geology. Although he had no formal training in the field, he was appointed the first state geologist in 1895. His pioneering research on the clays and lignites and the water resources of North Dakota were responsible for legislative support of science at the university and the creation of the School of Mines.

When the School of Mines opened in 1897, Babcock was named director and, after a reorganization of the university in 1901, became dean of the College of Mining Engineering. The first geological survey of North Dakota was published in this year, Babcock's laboratory furnishing the chemical analyses. Babcock at this time turned over the survey and the position of state geologist to a professional geologist. In 1909 he became director of the mining experimental station, and in 1915 the U. S. Department of Interior published his "Economic Methods of Utilizing Western Lignites," a summary of his life's work.

During the war Babcock served as chairman of the university branch of the North Dakota Defense Council and director of U. S. vocational and army training.

In 1917 the School of Mines and College of Mining Engineering were consolidated, and Babcock was appointed dean, a position he held until his death. He also served as interim president of the university in 1917-18.

Babcock obtained patents on methods of carbonizing and briquetting lignites and opened and operated a cement plant. He was awarded a Sc.D. degree by University of North Dakota.

A redhead, endowed with energy and imagination, Babcock was described as witty, polished, and cosmopolitan with an interest in art, music, and literature. He suffered a heart attack while swimming at his summer home on Lake Bemidji and died Sept. 3, 1925.

"National Cyclopaedia of American Biography," **25**, 55-6, James T. White & Co. (1936); Louis G. Geiger, "University of the Northern Plains," Univ. of North Dakota Press (1958).

JANE A. MILLER

James Francis Babcock

1844-1897

Babcock was born in Boston, Feb. 23, 1844, the son of Archibald D. and Fannie (Richards) Babcock. He attended Harvard, completing the chemistry course in Lawrence Scientific School in 1862. The following year, although he was only 19 years old, he went into business as an analytical and consulting chemist, and continued as a consultant for the remainder of his life, concurrently carrying on other activities. As with most consulting chemists, we know little of the investigations he undertook in his Boston laboratory since the results were the property of his clients, but it is known that he testified as an expert in patent suits and that he was regarded as an authority on the chemistry of foods. Quite possibly his invention of the Babcock fire extinguisher in 1868 was a by-product of one of his industrial investigations. His interest in the chemistry of foodstuffs led him to write articles on adulteration for consumer-oriented periodicals, and led to his appointments as Massachusetts state assayer of liquor from 1875 to 1885, and Boston milk inspector from 1885 to 1889.

Babcock was professor of chemistry at Massachusetts College of Pharmacy, 1869 to 1874, and at Boston University, 1874 to 1880. He edited volumes 1 and 2, 1874 to 1876, of the short-lived periodical, *The Laboratory,* A monthly journal of the progress of chemistry, pharmacy, recreative science and the useful arts, a popular journal directed toward students, pharmacy apprentices, and young adults interested in science, published by Wm. M. Bartlett & Co., manufacturing chemists.

The compilation of a list of Babcock's writings would be a time-consuming task for they are distributed among official reports, scientific journals, popular magazines, chapters in books, and pamphlets. He contributed the chapter on blood stains to Hamilton's "Legal Medicine," and published one pamphlet entitled, "Laboratory Talks on Infant Foods."

Babcock was married twice, from 1869 to 1890 to Mary P. Crosby, with whom he had five children, and to Maria B. Alden from 1892 until he died in Dorchester, Mass., July 19, 1897.

Catalogs of Massachusetts College of Pharmacy; *The Laboratory* vols. 1-2, 1874-1876; *Science* **6,** 167 (1897); "National Cyclopedia of American Biography" **10,** 445-6, James T. White & Co. (1960); "Dictionary of American Biography" **1,** 457, Chas. Scribner's Sons (1927).

JANE A. MILLER

Stephen Mouton Babcock

1843-1931

Stephen Babcock, remembered primarily for his test for butterfat in milk, was a seminal force in the development of nutritional chemistry. His parents, Pelig and Mary (Scott) Babcock operated a sheep farm near Bridgewater, N. Y. where young Babcock was born Oct. 22, 1843. He earned his A.B. degree at Tufts College in 1866, then undertook engineering studies at Rensselaer Polytechnic Institute. His father's death forced him to return to the family farm. A few years later he became a chemistry assistant at Cornell University where he was made an instructor in 1875. Two years later he entered University of Göttingen where he studied under Hans Hübner, receiving his Ph.D. degree in 1879.

Babcock resumed his instructorship at Cornell but left in 1882 to become a chemist for the New York Agricultural Experiment Station at Geneva. He worked on the analysis of milk and developed a viscosimeter for detecting adulteration of fats and oils. In 1888 he moved to the Wisconsin Station as chief chemist, also holding appointment as professor of agricultural chemistry in the university. He was successor to Henry P. Armsby who had recently moved to Pennsylvania State College.

Because Wisconsin had a large stake in the developing dairy industry, Babcock was assigned the task of developing a simple and rapid test for measuring fat in milk. He published a successful method in 1890. It played an important role in restoring honesty in the dairy business and was an important factor in the improvement of dairy cattle. Babcock took no remuneration for his test but indicated that it should be freely available for use in the dairy world. Despite the honors which the test brought him, Babcock considered his contribution as developmental rather than basic, and he left further application to Wisconsin associates such as Fritz Woll, Edward H. Farrington, and John L. Sammis.

With Harry L. Russell, station bacteriologist, Babcock made studies on the influence of pasteurization on the whippability of cream. Calcium sucrate was found to restore whipping quality. They soon learned that this was unnecessary however, if the cream was not subjected to unnecessarily high temperatures or excessive heating times. Babcock and Russell also studied the changes involved in the curing of cheese and developed the cold-curing process which took on major importance in the production of cheddar cheese.

Babcock's most profound influence grew out of the skepticism with which he viewed chemical analysis as a basis for evaluating quality of feeds. His own exploratory studies in cattle feeding culminated in the single-grain experiments of the Wisconsin Station. Edwin B. Hart, Elmer V. McCollum, Harry Steenbock, and George C. Humphry began four-year studies in 1907. It quickly became apparent that cows fed presumably balanced rations derived from a single plant (wheat, oats, corn) soon showed signs of malnutrition. The study, which Babcock stimulated but which was carried out by his younger associates, triggered the extensive studies on the nutritive importance of vitamins and minerals which were carried out by McCollum, Hart, Steenbock, Elvehjem, and their associates.

Babcock's own work during this period terminated with his studies on metabolic water, showing that creatures such as clothes moths, confused flour beetles, and alfalfa weevils which feed on dry substances with no access to water, obtained their moisture needs from the water formed during metabolic processes. During his retirement years, which began in 1913, Babcock undertook inconclusive studies on the nature of gravity. He also learned to drive an automobile, indulged his passion for raising hollyhocks, and served as a stimulus to younger associates.

Babcock was by nature a cheerful man who enjoyed life and detested all that was pompous. His life with May Crandall, whom he married in 1896, was a long and happy one although there were no children. He preferred watching a baseball game to attending a scientific meeting. He hated writing and published very little, yet he had a profound influence on the development of agricultural science in general and nutrition in particular.

Babcock papers are available in the archives of the State Historical Society of Wisconsin. Also see the *Reports* of the Agricultural Experiment Stations of New York and Wisconsin. For biographies see: Paul De Kruif, "Hunger Fighters," Harcourt, Brace & Co. (1926), ch. 9; A. J. Ihde in Eduard Farber, ed., "Famous Chemists," Interscience (1961), pp. 808–13; A. J. Ihde in Duane H. D. Roller, ed., "Perspectives in the History of Science and Technology," Univ. of Oklahoma Press (1971), pp. 271-282; "Dictionary of American Biography," **21** (supplement 1), 37, Chas. Scribner's Sons (1944).

AARON J. IHDE

Franklin Bache

1792-1864

Bache, a great grandson of Benjamin Franklin, was born in Philadelphia, Oct. 25, 1792. He graduated from University of Pennsylvania in 1810, then studied medicine under Benjamin Rush. When Rush died in 1813, Bache joined the army as a surgeon's-mate. While in the army he managed to attend lectures at the university medical school and received his medical degree in 1814. He served on the frontier as a surgeon until 1816 when he resigned from the army and began to practice medicine in Philadelphia. Although he was a prison physician for a time and an editor of the *North American Medical and Surgical Journal* from 1826 to 1831, it was in chemistry rather than in medicine that he achieved his reputation.

Bache became interested in chemistry in his youth. At the age of 19 he published his first article, an essay "On the Probable Constitution of Muriatic Acid," in *Aurora,* a newspaper founded by his father. He was a vice–president of the Columbian Chemical Society, which existed from 1811 to about 1814, and wrote three articles for the society's single volume of *Memoirs.*

In the 1820's and 1830's he taught chemistry in two or three private schools of medicine and was professor of chemistry at Franklin Institute from 1826 to 1832, at Philadelphia College of Pharmacy and Science from 1831 to 1841, and at Jefferson Medical College from 1841 until his death on March 19, 1864.

Bache wrote "A System of Chemistry for the Use of Students" (1819), he and Robert Hare published an American edition of Andrew Ure's "Dictionary of Chemistry" (1821), he issued a supplementary volume to William Henry's "Elements of Experimental

Chemistry" (1823), edited anonymously James Cutbush's "System of Pyrotechny" (1825), edited the third edition of Robert Hare's "Compendium of the Course of Chemical Instruction" (1836), and saw through the press four American editions of Edward Turner's "Elements of Chemistry" between 1819 and 1841. He was active on committees of revision of the "Pharmacopoeia of the United States" from 1830 to 1860. In 1833 he and George B. Wood published the "Dispensatory of the United States," a reference book that has gone through revisions up to the present time.

Although he was not a researcher, analyst, or inventor Bache was one of the most widely known American teachers of chemistry in his time.

George B. Wood, "Historical and Biographical Memoirs," J. B. Lippincott (1872), port.; Edgar F. Smith, "Franklin Bache, Chemist, 1792-1864" (1922), with portrait, reprinted in *J. Chem. Educ.* **20,** 367-368 (1943); Joseph W. England, "First Century of the Philadelphia College of Pharmacy" (1922).

WYNDHAM D. MILES

Werner Emmanuel Bachmann

1901-1951

Bachmann was born in Detroit, Nov. 13, 1901 the second of four sons of a protestant minister. He attended Wayne State University from 1919 to 1921 and The University of Michigan from 1921 to 1923, receiving his B.S. degree in chemical engineering in the latter year. He later obtained his M.S. and Ph.D. degrees from Michigan, where he was an assistant to Moses Gomberg. During the summer of 1924 he worked in the National Aniline and Chemical Co. laboratory, in the summers of 1926 and 1927 in Eastman Kodak's organic preparation laboratory, and in the summer of 1931 he taught at University of Illinois. From 1925 on he taught at Michigan, becoming Moses Gomberg University Professor of Chemistry in 1947. Fifty-seven students completed their Ph.D. degrees under his direction.

In the area of research Bachmann constantly and intensively expanded his techniques and experience. In 1928-29 he spent a year in Europe on a Rockefeller Foundation Fellowship, working in Paul Karrer's laboratory in Zürich on the chemistry of lycopene, the naturally occurring red colored compound in tomatoes and other fruits and berries. In 1935 he was awarded a Guggenheim fellowship which he utilized for study in Europe. Since Heinrich Wieland of Münich was not available during the summer of 1935, Bachmann arranged to work with J. W. Cook at the Royal Cancer Hospital in London. When he went to Münich in the fall, Wieland declared he had accomplished a year's work in three months on the cancer problem while in London.

Bachmann's friendship with Gomberg resulted in a long collaboration on free radicals and the reducing system magnesium-magnesium iodide. He had a keen appreciation of the experimental approach to physical organic chemistry earlier than most of his contemporaries and made contributions in this area, especially to the pinacol and Beckmann rearrangements and to free radicals. He carried out extensive research on the synthesis of cancer-producing agents, steroids, compounds related to penicillin, and of other compounds. He was a pioneer in developing methods for synthesising steroidal hormones, completing the first such synthesis, that of equilenin in 1939, with John Cole and Alfred Wilds. His total output is partially indicated by his approximately 160 publications.

Bachmann might be said to have been a leader in the introduction of more or less quantitative methods into organic synthesis, where previously low yields in reactions were almost the rule rather than the exception and were taken as a matter of course. He investigated the byproducts and side reactions of each synthesis as much as was necessary and meticulously determined the conditions and manipulative details necessary for a high yield. He did not consider a reaction satisfactorily controlled or understood unless a yield of 90% or more was achieved. His favorite reactions were those with 98-100% yield.

Nowhere was this illustrated more clearly than in his work on the synthesis and production of the explosive RDX in World War

II, an extremely difficult task. When Bachmann received this assignment in November 1940, he recorded that his "heart sank" at the thought that he might be instrumental in bringing harm to his fellow men. However, he set his scruples aside and working long hours was able to anticipate success by January 1941, even though he had no previous experience with explosives.

A wasteful and expensive British process was in use at the time at Woolwich Arsenal, which consisted in the direct nitration of hexamethylenetetramine with 98-100% nitric acid to produce the RDX (cyclotrimethylenetrinitramine). Large excesses of nitric acid were required, and half of the equivalent of formaldehyde was lost. J. H. Ross and R. W. Schiessler at McGill University had obtained RDX from formaldehyde, ammonium nitrate, and acetic anhydride in the absence of nitric acid, but no details were available at the time. Bachmann decided to combine the British and Canadian processes and attempt utilization of the byproducts to obtain two moles of RDX in a single reaction. In this he was successful, working with John C. Sheehan in 1941.

It was found that by portion-wise addition of a mixture of 98% nitric acid and acetic anhydride, and a mixture of hexamethylenetetramine dinitrate and ammonium nitrate, simultaneously and equivalently to a flask at such a rate that the temperature was kept at 75°C, pure RDX was obtained in 60% yield. Even small variations in this procedure produced quite different results. Later, yields were increased to 80% and even 90%.

At first the armed services showed little interest in RDX, but the potential advantages soon became apparent since it is about twice as powerful as TNT on a volume basis. Tennessee Eastman Corp. built the largest munitions plant in the world at Kingsport, Tenn. to produce RDX on 10 continuous production lines. In 1943 when this plant achieved a production of 360 tons of RDX per day, the savings to the government in plant cost alone amounted to over $200 million compared with the British process. This was contrasted to a cost of $45,000 for the research which was done at Michigan, Cornell, and Pennsylvania State. The importance of RDX in the war can scarcely be overestimated. It was used in "blockbuster" rockets and in torpedos.

Bachmann received the Naval Ordnance Award in 1945, the Presidential Certificate of Merit from the United States, the King's Medal from England in 1948, and a posthumous distinguished alumnus award from Wayne State University in 1951.

Bachmann gave a series of lectures in May 1947, in Basel, Zürich, and Geneva, under the auspices of the Swiss-American Foundation for Scientific Exchange. He served on the editorial boards of the *Journal of Organic Chemistry, Organic Reactions,* and *Organic Syntheses.*

At an early stage Bachmann was dubbed "The Chief" by his graduate students, a name which stuck. He had an extremely quiet manner and was very gentle and friendly, with an accompanying delightful, dry sense of humor. He died suddenly of heart failure Mar. 22, 1951 at the age of 49.

A. L. Wilds, *J. Organic Chem.* **19,** 128-130 (1954); biography by R. C. Elderfield, *Biog. Mem. Nat. Acad. Sci.* **34,** 1-30 (1960) (includes bibliography); J. P. Baxter 3rd, "Scientists Against Time," Little Brown (1946); R. K. Schiessler and J. H. Ross, U.S. patent 2,434,230 (Jan. 6, 1948), and *Chem. Abstr.* **42,** 2292; communication from Marie Bachmann Bull (Mrs. H. S. Bull, Ann Arbor, Michigan); obituaries in *Ann Arbor Daily News,* Mar. 22, 1951; *New York Times,* Mar. 23, 1951; *Michigan Alumnus* **57,** 340 (Apr. 14, 1951).

GEORGE DUBPERNELL

Walter Lucius Badger

1886-1958

Badger, an expert in heat transfer and evaporation, was born in Minneapolis, Minn., Feb. 18, 1886 the son of Minor C. and Mary Helen Badger. He received three degrees at University of Minnesota: A.B. 1907, B.S. 1908, and M.S. 1909 and was instructor in chemistry there from 1907 to 1909.

His first industrial experience was as a chemist with Great Western Sugar Co. in Brush, Colo. This was followed by 2 years with the National Bureau of Standards in Washington, analyzing construction materials used in the Panama Canal.

In 1912 he began a 25-year tenure with University of Michigan and was instructor (1912–15), assistant professor (1916–18), and professor of chemical engineering (1918–37). Much of his work at Michigan was concerned with heat transfer and evaporation, and after 1917, he headed a laboratory for research in these fields. He was also director of research for Swenson Evaporator Co. until 1931 when the depression forced Swenson to withdraw its support of the laboratory and cancel its research. This laboratory contributed substantially to the development of evaporators for use in caustic soda manufacture and the concentration of paper mill waste liquor.

In 1937 Badger gave up teaching to become manager of the Consulting Engineering Division of Dow Chemical Co. At Dow, he developed uses for Dowtherm, a high temperature heat transfer material which aided the development of processes for manufacturing synthetic resins, distilling fatty acids, and refining vegetable oils.

In 1944 he established an independent practice as consulting engineer in Ann Arbor, Mich., and in 1957 he formed W. L. Badger Associates. His consulting projects included fatty acid distillation, rayon spin bath recovery, varnish making, recovery of zinc chloride from fiber board waste, and the design of plants to produce salt, chlorine, and caustic soda.

He carried on research for over 40 years in water desalinization and helped to build solar evaporators in 11 countries. The culmination of this work was the Badger long tube vertical evaporator built by the United States Government at Wrightsville Beach, N.C., and placed in operation in 1957.

Badger was author of more than 40 papers on heat transfer and evaporator design and in 1926 he published a monograph "Heat Transfer and Evaporation." He was coauthor of three textbooks: "Inorganic Chemical Technology" (E. M. Baker) 1928; "Elements of Chemical Engineering" (W. L. McCabe) 1931, 1936; and "Introduction to Chemical Engineering" (J. T. Banchero) 1951. Both editions of "Elements of Chemical Engineering" were used worldwide and translated into several languages.

He served three terms as a director of the American Institute of Chemical Engineers and was a member of its committee on chemical engineering education. An abstractor for *Chemical Abstracts,* he also was on the advisory boards of *Industrial and Engineering Chemistry* and *Chemical Engineering News.* In 1940 he was the recipient of the William H. Walker Award of the American Institute of Chemical Engineers for his contribution to the literature of heat transfer.

He was married in 1913 to Helen Franklin of Minneapolis; they had one daughter. His avocations were mountain climbing and collecting rare books on engineering and salt manufacture. Badger died in Ann Arbor, Mich., Nov. 19, 1958.

"National Cyclopedia of American Biography," **47,** page 580, James T. White Co. (1965), with photograph; *Chem. Eng. News* **36,** 89 (Dec. 1, 1958); Sidney D. Kirkpatrick *et al.*, "Twenty Five Years of Chemical Engineering Progress," pp. 277-278, D. Van Nostrand (1933) with partial bibliography of Badger's publications on evaporation; *Trans. Amer. Inst. Chem. Eng.* **36,** B29-30 (June 25, 1940).

HERBERT T. PRATT

Leo Hendrik Baekeland

1863-1944

Baekeland was born in Ghent, Belgium, Nov. 14, 1863 to a family of modest financial means. Although Leo's father wished the boy to become an artisan, his mother recognized his intelligence and encouraged him to receive an academic training. In 1880 he graduated from the École Industrielle and received a scholarship to University of Ghent. A splendid first year record led to his appointment as *prepareteur* in general chemistry. His doctoral degree was awarded *magna cum laude* in 1884 as a result of research taken under Theodore Swarts, an organic chemist who was August Kekulé's successor at Ghent.

The following year Baekeland became Swarts' assistant at the École Normal in nearby Bruges. Here he began doing independent research and won a sum of money for foreign study in an essay competition. In 1889, newly married to his former professor's daughter Celine, he decided to use his prize to see how foreign chemists were being trained for jobs in industrial chemistry. He visited French and British universities and stayed a while in Edinburgh to observe Alexander Crum Brown's methods. Then he went to America where he was most favorably impressed with conditions. He elected to stay in New York City, where Charles F. Chandler had established a program at Columbia University in industrial chemistry.

Because of an interest in photography, Baekeland became a chemist for a photographic supply firm. Later, as an independent consultant, he invented a paper that did not require sunlight for printing, now known as Velox paper. Because this silver chloride paper was warmly received by amateur photographers, George Eastman bought the invention for one million dollars, making Baekeland a wealthy man at the age of 35.

In 1904 Elon H. Hooker engaged Baekeland to solve a problem he had with the asbestos paper diaphragms that were preventing continuous operation of the electrolytic cells in his new caustic soda-chlorine plant at Niagara Falls. The next year he became interested in the formaldehyde-phenol reaction, the tarry products of which had annoyed organic chemists since Adolf von Baeyer first studied them in 1870. Rather than discard the non-crystalline condensation products, Baekeland found value in them, under the name Bakelite plastic. The first of his over 400 patents was issued in 1906. As early as 1910 a Bakelite Geselleschaft was operating in Germany. In 1922 Baekeland got together with his competitors to form the Bakelite Corp. with its main plant in Bound Brook, N.J. In 1939 Union Carbide absorbed the company as its Bakelite Division.

After the merger Baekeland spent much of his time in his second home in Florida, where he raised exotic flowers and fruits as a hobby. He died Feb. 23, 1944 at Beacon, N.Y.

Baekeland was president of the Electrochemical Society in 1909, of the American Institute of Chemical Engineers in 1912, and of the American Chemical Society in 1924.

Biographies by William Haynes in Eduard Farber, ed., "Great Chemists," pp. 1182-1190 (1961), Interscience, portrait; J. Gillis, translated by R. Oesper, *J. Chem. Educ.* **41**, 224-226 (1964) port.; J. Gillis, "Dr. L. H. Baekeland," (Brussels, Belgium, 1965).

SHELDON J. KOPPERL

Edgar Henry Summerfield Bailey

1848-1933

Bailey was born in Middlefield, Conn., Sept. 17, 1848. He did his undergraduate work at Yale where he received his Ph. B. degree in 1873, in that same year serving as an instructor in chemistry there. In 1874 he moved to Lehigh University and held the position of instructor in chemistry intermittently over a period of 9 years. While at Lehigh he married Aravesta Trumbauer. In 1883 he received his Ph.D. degree at Illinois Wesleyan and accepted the position of professor of chemistry and metallurgy at University of Kansas. He spent 2 years of study abroad; 1881 in Strassburg and 1895 in Leipzig. At Lawrence, Kans. he also served on the Kansas State Board of Agriculture and the State Board of Health. In 1900 he became director of the Chemistry Laboratory at the university, serving in this capacity until his death June 1, 1933.

Bailey's energies were devoted to teaching, and he was responsible for the success of many chemistry students during his long career as an educator.

Other than teaching, Bailey's interests were foods, sugars, and the relation of chemical composition to taste and odor. His publications point to his diversity of interests. In 1901 he authored with H. P. Cady "A Laboratory Guide to the Study of Qualitative Analysis," in 1906 "Sanitary and Applied Chemistry," in 1914 "The Source, Chemistry and Use of Food Products," in 1915 "Laboratory Experiments on Food Products," in 1921 "Report on the Dietaries of Some State Institutions under the Care of the Board of Administration," and in 1922 with his son, Herbert Stevens Bailey, "Foods from Afar."

Ind. Eng. Chem., **16**, 316-317 (1927) with portrait

CHARLENE J. STEINBERG

Edwin Myron Baker

1893-1943

Baker was born in Cleveland, Ohio, Feb. 21, 1893. A son of Myron Daniel and Caroline (Maples) Baker he received his early education in New York State, Maryland, and Washington, D.C. He graduated from the course in electrochemical engineering at Pennsylvania State University in 1916 with first honors. He started his professional career with Hooker Electrochemical Co. at Niagara Falls, N.Y. He married Ruth Lewis from nearby Tonawanda, June 14, 1920.

Baker went to University of Michigan in 1918 as instructor in chemical engineering, became assistant professor in 1920, associate professor in 1929, and professor in 1933. He served as acting chairman of the Department of Chemical and Metallurgical Engineering during the second semester of the college year 1937-38.

An outstanding characteristic was his painstaking and conscientious performance of routine duties, and another was the orderly arrangement of his time. It seemed to some of his associates that he considered a week composed of 21 parts for which one should have activities planned, *i.e.* morning, afternoon, and evening of the seven days. He made thorough preparation for his classes and even when pressing demands were at hand, he corrected problems and examinations punctually and carefully. His interest in students was genuine and wholehearted, and he made every effort to be a sympathetic advisor to his students. In this connection, the professional fraternity, Alpha Chi Sigma, owed the maintenance of its chapter house in Ann Arbor through the depression, in large measure, to his business ability and interest.

In the university he chaired the Committee on Coordination and Teaching from its inception in 1938 to 1944. This committee inaugurated surveys in which students recorded their estimate of their instructors. He served as a member of the Senate Advisory Committee of the University for the last 4 years of his life.

Baker's quarter century of research was prominent in three areas: gas absorption, inorganic chemical industry, and the coating (plating) of metal surfaces. He was coauthor with Walter L. Badger of the book, "Inorganic Chemical Technology," first published in 1928 and revised in 1941. Of his some 50 research papers, a goodly number had to do with the absorption of gases in spray towers, theories of gas absorption, including film concepts, entrainment in gas absorption equipment, and the efficiency of bubble plates in fractionating columns. Several papers were involved with nickel and cobalt plating. For his consulting practice in industry, he also carried on continuous laboratory studies on plating and cleaning techniques. The ability to polish and maintain such surfaces was also investigated. At the time his career ended, he was beginning studies on heat transfer, particularly condensation of vapors.

After the United States entered World War II and while he was still carrying on his full share of teaching, Baker devoted much energy to help some of our manufacturers change their equipment and processes to make war products. He often said he wished he could be temporarily relieved of his university work but did not ask for a leave of absence until April 1943. At that time he was asked to take responsible charge of designing a portion of the gaseous diffusion plant for separating uranium. Since he died before information was released on this project, little information is known of his contribution although it is understood that he was a primary participant in developing the sintered metal barrier.

Baker was chairman of the Student Chapter Committee for the American Institute of Chemical Engineers. In the American Society for Testing Materials, he was chairman of the large and important Committee B-8, which was charged with the duty of preparing specifications for electrodeposited coatings and for testing such coatings. He served as president of The Electrochemical Society in 1942.

Baker was closely associated with the Houdaille Hershey Co. and the C & G Spring and Bumper Co. for several years. He not only assisted in the development of their plating and metallurgical operations but served in a capacity akin to chief engineer of their activities. The results of these industrial contacts and many others made him one of the professors who brought to his students a view of the real world which needed the attention of the chemical engineer.

Baker died in the Hotel Biltmore in New York City, May 27, 1943, while engaged in development of the gaseous diffusion plant.

Personal recollections; records of The University of Michigan; brief biography in *Metal Cleaning Finishing* **9,** 1012 (1937).

DONALD L. KATZ

Edward Bancroft

1744-1821

Edward Bancroft remains a man of mystery. His talents in natural history and chemistry of dyeing were well known, but his double spying for the British and the Americans during the Revolution was successfully concealed for more than 60 years after his death. He was a distinguished scientist in his day. He was sponsored by Benjamin Franklin and was a friend of Joseph Priestley. He was a Fellow of the Royal Society, Royal Institution, and American Academy of Arts and Sciences.

Bancroft was born in Westfield, Mass., Jan. 9, 1744. He went to school in Hartford, Conn. where his stepfather owned a tavern called *The Bunch of Grapes.* He was apprenticed to a physician in Killingworth but at 18 broke his contract and ran away to the Barbadoes. Finding no job he journeyed on to Surinam.

Paul Wentworth, a plantation owner originally from New Hampshire, gave him the chance to work while educating himself. Bancroft was employed by the plantation physician and was soon practicing medicine himself. The physician interested Bancroft in the study of tropical plants and animals, and he studied the vegetable dyes and poisons used by the natives. Bancroft remained in Surinam until 1766 when he returned to New England but is said to have returned to Surinam a second time in 1768, a third and last time in 1805.

He wrote long letters to his brother and made these the basis for a book he published in 1769 when he was only 24, "An Essay on the Natural History of Guinea in South America. . . ." The same year he published a political tract: "Remarks on the 'Review of the Controversy between Great Britain and her colonies . . .' To which is subjoined a Proposal for terminating the Present unhappy Dispute with the Colonies . . ."

These works interested the aged Franklin, who from then on was a friend and sponsor of Bancroft. Bancroft and Priestley each wrote tracts concerning the abuse heaped upon Franklin by the Privy Council on Jan. 29, 1774. Franklin obtained for Bancroft a position as political writer on *The Monthly Review* and helped elect him as a Fellow of the Royal Society. As a member of the Royal Institution he associated with such chemists as Davy, Wollaston, Frederick Accum, and Count Rumford.

In 1767 Bancroft went to London and studied medicine at St. Bartholomew's Hospital. He was a founding member of the Medical Society (London) in 1773. He obtained his M.D. degree from University of Aberdeen.

Bancroft voyaged to the Americas in 1770 in search of new natural dyes and inks. The most promising seemed to be the barks of the black oak *(quercus velutina)* whose coloring principle he named *quercitron.* Upon his return to England he obtained a 14-year patent in 1771 which gave him exclusive rights to this dye. It would have made him a rich man but for the outbreak of the Revolution. After peace came in 1783, Bancroft obtained a 14-year extension of his "patent" because of the interruptions of the war but was turned down in 1798 when he asked for another extension. In the meantime many dyers and calico printers had been using quercitron from other sources.

Chevreul was the first chemist to examine quercitron and found it to be composed of a "peculiar" tannin combined with the glucoside of rhamnose. Tree barks were used in America for home dyeing until the arrival of synthetic dyes.

All the while Bancroft carried out experiments and studied colors. In 1784 he published "Experimental Researches Concerning the Philosophy of Permanent Colours and the Best Methods of Producing Them by Dyeing, Calico Printing, etc." This was the first book on dyes in the English language, and it quickly sold out. Second hand copies then sold for six times the original price. It was quoted in C.L. & A.B. Berthollet's "Éléments de l'art de la Teinture," Paris 1804, and in other books.

In 1813 Bancroft published an enlarged edition in two volumes. In 1972 a copy was offered for sale for 225 pounds (about $562). An American edition was published at Philadelphia in 1814. These volumes contained a *vade mecum* of the state of chemistry at the time as well as an exhaustive treatment of natural dyes. Though 100 pages were devoted to inks, there was nothing about the invisible ink he used while a spy during the Revolution. This was his last publication, and he seems to have spent his last days in seclusion. He died at Margate, England, Sept. 8, 1821.

His career as a British spy may have started as early as 1772. In December 1776 he obtained a written agreement to continue spying for 500 pounds down and 400 pounds per annum which was later increased to 1,000 pounds. For this money he gave to the British confidential information about the dealings of the American Commissioners with the French government. At the same time he brought secret news from London to the American Commissioners in Paris, receiving a salary for this from the Continental Congress. He fooled among others, Franklin, Silas Deane, and John Paul Jones. It was not until 1889 that it became known that Bancroft had been a double spy.

John Bigelow, ed., "Autobiography of Benjamin Franklin," J. B. Lippincott & Co. (1868); J. P. Boyd, "Silas Deane: Death by a Kindly Teacher of Treason?," *William and Mary Quarterly* **16,** 165-187 (1959); C. A. Browne, "A Sketch of the Life and Chemical Theories of Dr. Edward Bancroft," *J. Chem. Educ.* **14,** 103-7 (1937); F. Crace-Calvert, "Dyeing and Calico Printing," second edition, pp. 253-261, Simpkin, Marshall & Co., London (1876); S. M. Edelstein, "Historical Notes on the Wet-Processing Industry. VI. The Dual Life of Edward Bancroft," *Amer. Dye. Rep.* **43,** 712-3, 735 (1954); R. J. Popkin, "Medical Intelligence, Doctors afield, Edward Bancroft, M.D., F.R.S., and F.R.C.P. (London)," *New England J. Med.* **268,** 312-3 (1963); Farber, "Great Chemists," Interscience (1961).

DAVID H. WILCOX, JR.

Wilder Dwight Bancroft

1867-1953

Bancroft, born in Middletown, R. I., Oct. 1, 1867 was the grandson of Secretary of the Navy George Bancroft who founded the U.S. Naval Academy at Annapolis. Wilder attended Harvard, played on the football team, assisted in chemistry 1888-89, went to Strasbourg, Berlin, and Leipzig, where he received his Ph.D. degree under Wilhelm Oswald in 1892, and to Amsterdam, where he studied under van't Hoff. Returning home, he taught chemistry at Harvard from 1893 to 1895 and physical chemistry at Cornell from 1895 to 1937.

Bancroft was a pioneer American physical chemist. With graduate students he carried out research in several fields. His early work was on the phase rule, and he wrote a book on the subject simply titled "The Phase Rule" (1897). He then turned to electrochemistry and his research provided basic data on the electrodeposition of metals. Becoming interested in contact catalysis he correlated scattered work that had been done and developed a general theory of the process. He investigated several areas of colloid chemistry, including emulsions, anesthesia, and drug addiction, taught one of the first college courses on colloids in the United States, and wrote a standard book, "Applied Colloid Chemistry: General Theory" (1921). He is said to have coined the term solute.

He founded *Journal of Physical Chemistry* in 1896 because there was no journal in this country at that time which was appropriate for his and his students' papers. He supported the journal financially (he was independently wealthy from an inheritance), coedited it until 1909, and then edited it until 1932.

Bancroft was president of the Electrochemical Society in 1905 and 1919, and of the American Chemical Society in 1910. During World War I he was a lieutenant-colonel in the Chemical Corps, stationed at Edgewood Arsenal, Md., and compiled a manuscript history of chemical warfare operations. Bancroft had a phenomenal memory, was an omnivorous reader, and a voluminous writer. He remained interested in sports all his life, playing golf in later years. In 1937 he was struck by an automobile on Cornell campus, lay in the hospital for months, and thereafter was a semi-invalid until he died in Ithaca, Feb. 7, 1953.

Biography by H. W. Gillett in *Ind. Eng. Chem.* **24,** 1200-1201 (1932) with portrait. Obituaries by C. W. Mason in *J. Amer. Chem. Soc.* **76,** 2601-2602 (1954), with portrait; A. Findlay in *J. Chem. Soc.* 2506-2514 (1953), with portrait, abridged in Eduard Farber's "Great Chemists" pp. 1247-1261, Interscience (1961), with portrait; *Chem Eng. News* **31,** 697 (Feb. 16, 1953), with portrait. In the library, Edgewood Arsenal, is a manuscript history of American chemical warfare operations which Bancroft compiled in World War I.

WYNDHAM D. MILES

Samuel Bard

1742-1821

Bard, an eighteenth century chemistry teacher at Columbia University, was born in Philadelphia, Apr. 1, 1742. He attended a private school in New York City, King's College, and in 1761 went to Great Britain to study at University of Edinburgh. There he learned chemistry from the notable physician-chemist William Cullen. Receiving his M.D. degree in 1765, Bard returned to America and began practice in New York.

In 1767 he and five other physicians founded the Medical School of King's College, Bard becoming professor of theory and practice of physics. In 1770 he moved to the professorship of chemistry. From 1770 to 1773 he may not have lectured on chemistry, the Medical School having so few pupils that the fees would not have been worth his time. In 1774 he gave chemistry lectures in his home. He advertised these lectures in newspapers, inviting citizens to attend upon payment of a fee.

In 1776 he manufactured salt, which was then selling at a high price because of the Revolution, from sea water at Shrewsbury on the New Jersey coast but gave up because the process proved unprofitable.

King's College closed in 1776 because of the Revolution and opened as Columbia in 1784. Bard was reelected professor of chemistry in the Medical School but in 1785 switched to the professorship of natural philosophy in the faculty of arts. He taught natural philosophy for one year then returned to the professorship of chemistry from 1786 until 1787. Thereafter he devoted himself to practicing, teaching, and writing on medicine and, at times, to dabbling in agriculture on his estate at Hyde Park, where he died on May 24, 1821.

John McVickar, "Domestic Narrative of the Life of Samuel Bard," Columbia College (1822); John D. Langstaff, "Doctor Bard of Hyde Park," (1942); B. Stookey, "Samuel Bard's Course on Natural Philosophy and Astronomy, 1785-1786," *J. Med. Educ.* **39,** 397-406 (1964); archives, Columbia University.

WYNDHAM D. MILES

George Frederic Barker

1835-1910

Barker was born in Charlestown, Mass., July 14, 1835. His father was captain of a packet ship plying between Boston and Liverpool. George learned elementary science while attending academies and set up a little chemical laboratory in his bedroom. Between the ages of 16 and 21 he was an apprentice to Joseph Wightman, apparatus maker of Boston. Wightman's shop served as an educational laboratory to Barker; here he learned the techniques of chemistry and physics and became an expert in constructing scientific apparatus. His apprenticeship completed, he attended Yale for 2 years, graduating in 1858. Barker's later eminence in science, based on only 2 years of collegiate education, indicates his considerable talents.

During Barker's second year at Yale he assisted Benjamin Silliman, Jr. He assisted John Bacon at Harvard Medical School during the sessions of 1858-59 and 1860-61, delivered a series of public lectures on science in Pittsburgh during the winter of 1859-60, was professor of natural science at Wheaton College, Illinois from 1861 to 1862, professor of chemistry at Albany Medical School from 1862 to 1864 (here he also studied medicine and received his M.D. degree in 1863), and professor of natural science at University of Pittsburgh 1864-65. He was demonstrator of chemistry, then professor of physiological chemistry and toxicology in Yale Medical School from 1865 to 1873, lecturer on chemistry at Williams College in 1868-69, and professor of physics at University of Pennsylvania from 1873 to 1900.

Barker had the knack of being able to explain scientific phenomena clearly and simply while illustrating his remarks with ingenious, eyecatching demonstrations. This trait made him a popular lecturer in the classroom and on the stages of public halls of Philadelphia and other cities.

Barker was a consultant to industrialists, among them Thomas Edison. His reputation as an expert witness, based on his knowledge and on his ability to explain technical and scientific matters to laymen, was said to have been unexcelled. He testified in important patent cases involving chemical processes, batteries, the telephone, telegraph, and electric light; he also testified in criminal trials, among them the Sherman case in which he proved that Lydia Sherman poisoned three husbands and four children.

For the city of Philadelphia he served on the Board of Education, investigated the wholesomeness of the water supply, the quality of illuminating gas, and the best method of protecting municipal buildings against lightning.

Barker was a delegate to the Paris Electrical Exposition and Electrical Congress in 1881. He also served on the U.S. Commission at the Philadelphia Electrical Congress in 1884 and on the jury of awards of the Columbian Exposition of 1893. He presided over the American Association for the Advancement of Science in 1879 and over the American Chemical Society in 1891.

During his career Barker published more than 100 articles and a text on physics that went through many editions. His chief contribution to chemical education was his "Text Book of Elementary Chemistry," one of the first major texts to use the modern notation and nomenclature that resulted from the Congress of Chemists at Karlsruhe. One of the outstanding American chemistry texts of the 19th century, it sold 10,000 copies in the first 5 years, an impressive number a century ago when fewer persons were studying chemistry, and it was used in hundreds of educational institutions. It was translated into French, Arabic, and Japanese, and through the latter became one of the instruments through which western chemistry was transmitted to the changing Japanese culture.

Barker died in Philadelphia, May 24, 1910.

Biographies in *Pop. Sci. Month.* **15,** 693-697 (1879), with portrait; *Sci. Amer.,* **57,** 231-232 (1887), with portrait; by E. F. Smith, *Amer. J. Sci.* **180,** 225-232 (1910); by E. Thomson, *Proc. Amer. Philo. Soc.* **50,** xiii-xxix (1911), with portrait, list of Barker's writings.

WYNDHAM D. MILES

Edward Bartow

1870-1958

Born Jan. 12, 1870 in the small village of Glenham, N.Y., Edward Bartow attended the village school and graduated in a class of four from Mount Beacon Academy at Fishkill Landing, N.Y. Influenced by the life of President James Garfield, he went to Williams College where he graduated in 1892.

During his junior year, his first course in chemistry determined his future. After an assistantship at Williams, he went to Göttingen to study under Otto Wallach for his Ph.D. degree which he received in 1895. He returned to Williams as an instructor until he was called to University of Kansas in 1897 as an instructor in organic chemistry.

An analysis Bartow made of a spring water in Massachusetts, an investigation of the waters of Kansas for the U.S. Geological Survey, and a severe case of typhoid fever turned his attention to sanitary chemistry. He was invited to University of Illinois in 1905 as associate professor of sanitary chemistry and chief of the State Water Survey. The survey analyzed water samples from within the state free of charge and assisted in any crisis involving potable water. Bartow's 14 annual reports of the survey were standard reference books for many years.

After visiting England to see Gilbert F. Fowler's work on the activated sludge method for sewage treatment, Bartow undertook the first experiments on this subject in the United States. Through the years he was a consultant for many industries which had trade waste problems, including starch manufacturers, canners, meat packers, mines, and oil refineries. The numerous graduate students whom he directed became a nucleus for the sanitary chemical group.

In 1917 Bartow was selected to head a mission of five chemists sent to Europe to help the French in the war effort. All held army commissions with Major Bartow as the leader. He was mustered out July 1919, as lieutenant colonel. Before the war ended, ten analytical laboratories were organized in France and one in England to control the water used by the American Expeditionary Forces. For suggested improvements in the French water supplies, Bartow was awarded the Medaille d'Honneuer des Epidemies d' Argent by the French government.

In 1920 Bartow accepted an invitation from University of Iowa to head the Department of Chemistry and Chemical Engineering. There he remained until he retired in 1940. He was then employed for a year by Johns-Manville as a consultant after which he returned to Iowa on a half-time appointment. He continued to supervise graduate students until he was well over 80. The articles describing his research or related subjects number nearly 200.

Bartow was a member of 41 organizations; he made contributions to almost all of them. He was a member of American Society of Civil Engineers, American Public Health Association, and American Institute of Chemical Engineers. He was Phi Beta Kappa, secretary of the chapter at Illinois, president of the Iowa chapter. He was president of the American Water Works Association ln 1922 and of the American Chemical Society in 1936. His active membership in the latter lasted 62 years. He was founder of two local ACS sections and an editor of the *Chemical Abstracts* water section from 1911 until his death. He had attended four meetings of the Congress of Pure and Applied Chemistry and 11 as delegate to the Union of Pure and Applied Chemistry, including the reorganization meeting in Paris in 1919. From 1934 to 1936 he was a vice-president of the Union. He was active in civic organizations as well as his church. He died in Iowa City, Iowa, Apr. 12, 1958.

Sidney D. Kirkpatrick, *J. Amer. Chem. Soc.* **81,** 5841-5845 (1959), port.; Ralph L. Shriner, *Science* **128,** 289 (1958); personal recollections.

VIRGINIA BARTOW

Charles Baskerville

1870-1922

Baskerville was born in Deer Brook, Noxubee County, Miss., June 18, 1870 the son of Charles and Augusta Louisa (Johnston) Baskerville. He entered University of Mississippi at 16 and graduated from University of Virginia in 1890. He studied at Vanderbilt University in 1891 and received his Ph.D. degree in chemistry at University of North Carolina in 1894. While on leave of absence for part of 1893 he studied under Hofman at University of Berlin.

Baskerville was assistant to the state geologist of North Carolina from 1892 to 1894, assistant chemist, 1894 to 1900, and chemist, 1900 to 1904, with the North Carolina Geological Survey. He was appointed instructor at University of North Carolina in 1894, assistant professor in 1895, associate professor in 1898, and professor of chemistry in 1900. He was head of the Department of Chemistry from 1900 to 1904. From 1904 until his death he was professor of chemistry and director of the chemical laboratories at College of the City of New York. There he designed and supervised the construction of the new laboratory.

Few chemists have been as widely known as was Baskerville. Although he was only 51 when he died, he had written 186 papers and a number of textbooks and monographs. His first paper in *Journal of Analytical and Applied Chemistry* in 1893, was on "Rapid Methods for the Estimation of Phosphorus in Titaniferous Ores." He showed interest in almost every branch of chemistry, and investigated vegetable and mineral oils and their refining; atmospheric pollution and the injury to vegetation caused by noxious gases; oil shales of Canada; chemistry of anesthetics; paper manufacture and the recovery of paper stock; occupational diseases in chemical industries; rare earths; radioactivity; and the utilization of surplus munitions (war gases). His special interest lay in the field of transition and rare earth metals, especially zirconium, thorium, and praseodymium. He thought he had isolated two new elements, which he named carolinium and berzelium, but his claims proved to be erroneous.

Baskerville was general secretary of the American Association for the Advancement of Science in 1900, and vice-president in 1903. He was the United States delegate to the International Congress of Applied Chemistry in London in 1909, serving as chairman of the section on analytical chemistry. He was awarded the Longstreth prize of the Franklin Institute in 1912. He served as president of the North Carolina Academy of Science and as vice-president of the New York Academy of Science. Other offices that he held included: chairman of the North Carolina and New York Sections of the American Chemical Society, secretary of the Chemists' Club, and chairman of the New York section of The Electrochemical Society. He was also a fellow of the Chemical Society (London).

Baskerville wrote "Radium and Radioactive Substances" in 1905. Among the other books that he wrote was a general chemistry textbook, with accompanying laboratory manual, published in 1909. Probably the most widely known of his books was the famous qualitative chemical analysis text written with Louis J. Curtman.

He was chairman of the American Chemical Society committee on occupational diseases in the chemical trades and was a member of the American Institute of Chemical Engineers committee on atmospheric pollution.

Baskerville was a gifted person with attractive personal characteristics and indomitable energy. He died of pneumonia on Jan. 28, 1922 in New York City.

Obituaries: *New York Times*, Jan. 30, 1922; *Chem. Met. Eng.*, **26**, 280 (1922) (portrait); *J. Franklin Inst.*, **193**, 566-67; by F. P. Venable, *J. Ind. Eng. Chem.*, **14**, 247 (1922) (portrait).

LAWRENCE P. EBLIN

Lewis Caleb Beck

1798-1853

Beck, a teacher of chemistry and an investigator with wide interests, was born in Schenectady, N.Y., Oct. 4, 1798. He received his A.B. degree from Union College in 1817, then studied medicine under a preceptor and at College of Physicians and Surgeons in New York City. Although he did not graduate from College of Physicians, he learned sufficient medicine to receive a license to practice in 1818 and he began to do so in his home city.

In 1819 he moved to St. Louis where he conceived the plan of writing a book about the surrounding region. He meandered through the country gathering information on botany, geology, mineralogy, climate, and the life-style of the inhabitants, and in 1823 published "Gazetteer of the States of Illinois and Missouri," regarded today as important Americana.

Giving up thought of settling in the Midwest, he returned to Albany in 1821. He practiced medicine, studied science, and within a few years was regarded highly by scientists of the state. Berkshire Medical Institution engaged him to teach botany in the summer of 1824. He taught botany, mineralogy, and zoology at Rensselaer from 1824 to 1829, botany and chemistry at Vermont Academy of Medicine from 1826 to 1832, chemistry at Middlebury in 1827, botany at Fairfield Medical College in 1827, chemistry at Castleton Medical Academy from 1827 to 1829, chemistry and natural history at Rutgers from 1831 to 1853, chemistry at New York University from 1836 to 1838, and chemistry and pharmacy at Albany Medical College from 1840 to 1853.

He was able to teach at two or three institutions in 1 year because the courses were short (his chemistry course at Rutgers lasted 2 months in 1831; lengthened in later years) and they did not overlap.

During the years Beck was teaching at several institutions, he carried on a number of investigations. For the state of New York he studied the potash industry for the purpose of improving the manufacturing process, assisting producers, and determining standards of purity to protect buyers from accidental and fraudulent adulterants. He worked on the geological survey of New York part time for several years, traveling thousands of miles throughout the state to visit mineral localities and obtain specimens which he analyzed and reported upon in 1842 in a book, "Mineralogy of New York." One of his chief interests was the detection of adulterants in commercial products, foods, and drugs, and in 1846 he published a book "Adulterations of Various Substances Used in Medicine and the Arts, with the Means of Detecting Them." For the federal government he investigated the manufacture of flour and bread, determined the adulterants that were being added to breadstuffs, and ascertained the effects of the sea voyage upon flour shipped to Europe. His "Report on the Breadstuffs of the United States, their Relative Value, and the Injury which they Sustain by Transport," appeared in the annual report of the United States Patent Office in 1848 and was followed by a second report in 1849.

Beck published two texts which were widely used: "Manual of Chemistry," which appeared in 1831 and reached a fourth edition in 1844, and "Botany of the Northern and Middle States," issued in 1833 with editions up to 1848. He also wrote "A Short Series of Elementary Lectures on Chemistry, Electricity, and Magnetism, and the Applications of Science to the Useful Arts," in 1834.

Beck died in Albany, Apr. 20, 1853 at the age of 55.

C. E. Van Cortlandt, *Trans. Med. Soc. County of Albany, N.Y. (Albany Med Annals)* **1,** 292-302 (1864); Alden March in Samuel D. Gross, ed., "Lives of Eminent American Physicians and Surgeons of the Nineteenth Century," pp. 679-696, Lindsay & Blakiston (1861); Dictionary of American Biography" **1,** p. 116, Chas. Scribner's Sons (1929); L. F. Kebler, "A Pioneer in Pure Foods and Drugs; Lewis C. Beck," *Ind. Eng. Chem.* **16,** 968-970 (1924).

WYNDHAM D. MILES

Frederick Mark Becket

1875-1942

Becket was born Jan. 11, 1875 in Montreal, son of Robert Anderson and Ann (Wilson) Becket. Graduating from McGill University in 1895 as an electrical engineer, he was employed briefly by Westinghouse Electric and Manufacturing Co. and by Acker Process Co. before resuming his formal education at Columbia with training in physical chemistry. His master's thesis project at Columbia, completed in 1899, was on electric furnace production of metals and various chemical compounds. This experience was immediately applied by Becket on his return to Acker Process Co. where his employer was trying to develop a commercial process to manufacture caustic soda and bleaching powder by electrolysis of fused sodium chloride. Back at Columbia again for further work in electrochemistry and electrometallurgy, Becket began a Ph.D. thesis on the electrical conductivity of fused salts. Before this object could be attained, however, Becket was persuaded to reenter the industrial field with Ampere Electrochemical Co.

All of this experience was but preliminary to the real beginnings of Becket's career, when in 1903 with two associates from Ampere, he set up the Niagara Research Laboratories to conduct both original and contract research and to offer consulting services in electrochemistry and electrometallurgy. This organization was short-lived as an independent group, for in 1906 it was acquired by Electro-Metallurgical Co., a subsidiary of Union Carbide Co. Thus, Becket began an association with that firm which was to last until his retirement in 1940 when he held the posts of president of Union Carbide and Carbon Research Laboratories Inc., vice-president of Union Carbide Co., Electro-Metallurgical Co. and Haynes-Stellite Co.

In 1900 Becket married Frances Kirby of New York City, who died 3 months later from a ruptured appendix. He married second in 1908, Geraldine McBride of Niagara Falls; they had two daughters, Ethelwynn and Ruth. Becket became a U.S. citizen in 1918 and died in New York City, Dec. 1, 1942.

During his career, Becket amassed a notable group of honors. He received honorary degrees from Columbia and McGill. He was elected to the presidencies of the Electrochemical Society (1926), the American Institute of Mining and Metallurgical Engineers (1933), and The Chemists' Club (1939). He received the Perkin medal of the Society of the Chemical Industry (1924), the Acheson medal of the Electrochemical Society (1937), the Cresson medal of the Franklin Institute (1938), the Howe medal of the AIME (1938) and the National Pioneers Award of the National Association of Manufacturers (1940). A total of 125 patents were registered in his name, the first of which was issued in 1906 and the last shortly before his retirement in 1940.

What were the achievements of this career in industrial research that brought such recognition? The list is long and diversified yet with certain threads tying together the whole. To mention only the main areas, we may cite the commercial development of a variety of ferro alloys via silicon reduction in the electric furnace; the electric furnace production of chromium, manganese, silicon, and calcium carbide; the invention of numerous important alloy steels; vanadium- and molybdenum-bearing steels, high chromium steels for oxidation resistance, columbium-stabilized stainless steel, and chromium-manganese austenitic stainless steel; development of welding rods of tailored composition for joining various materials, contributions to ore dressing; and methods for extracting tungsten from high phosphorus and high tin ores.

Lest it be concluded that all his talents lay in the metallurgical field, it should be pointed out that his earlier successes were with the resolution of electrical problems connected with the design, construction, and operation of electric furnaces of unprecedented size. These problems included those of the unexpectedly low power factor, very high current draws, fragility of large electrodes, and poor life of furnace linings.

The intellectual and personal attributes of the man were direct contributors to the success which Becket attained. Becket contrived to maintain himself at the forefront of both the science and the commerce in which he was concerned, and he adroitly exploited each to the benefit of the other. For example his commercial ventures with the electric furnace were carried on within 10 years of the original experiments of Moissan. His phenomenal powers of mental and physical endurance, perhaps a reflection of his youthful years as an athlete, led to frequent 12-18 hour work days that exhausted his younger colleagues. Fierce determination, unwavering intellectual honesty, keen business acumen, unusual memory, a lively imagination, and extreme thoroughness were among the qualities recalled by his associates. One of the latter once described these characteristics in these words: "First, his ability to select a worthwhile objective, then a persistent drive toward the goal; next an untiring zeal to improve his results, and refusal to be satisfied with past accomplishment; a wide knowledge, and the ability to draw from many fields to aid in the solution of a problem; but most of all, an uncanny sense of the coming needs of an industry before they are realized by the industry itself."

Ind. Eng. Chem. **16,** 197-205 (1924); *Trans. Electrochem. Soc.* **72,** C. E. MacQuigg, 4-7, J. H. Critchett, 7-13, F. M. Becket, 14-24 (1937); J. H. Critchett and H. R. Lee, *Trans. Electrochem. Soc.* **82,** 34-36 (1942).

J. H. WESTBROOK

Clayton Wing Bedford

1885-1922

Bedford was born in New Windsor, Ill., June 13, 1885. He graduated from nearby Galesburg High School in 1903 and taught there a short time before going on to college. He attended Wheaton College and University of Illinois before completing his degree in chemical engineering at The University of Michigan in 1910. He remained at Michigan as teaching assistant for a year and then became instructor of organic chemistry at Case Institute of Technology.

Early in 1912 he joined the technical staff of The Goodyear Tire & Rubber Co. in Akron, Ohio.

At Goodyear Bedford became interested in accelerators for rubber vulcanization. He evolved a theory that the real accelerator was not a particular chemical in the rubber mix but rather a reaction product formed from that chemical and sulfur during vulcanization. About 1915 he was given the assignment of implementing his theory to find better accelerators, and he started his experiments which essentially involved heating various proportions of carbon bisulfide and sulfur with various proportions of known accelerators such as aniline and thiocarbanilide. The resultant resin, known as Dubax, was found to have variable accelerator power. Bedford's 71st batch proved to be very effective; a patent was filed January 1917.

The organic qualitative analysis of Dubax proved to be a stubborn problem. Dubax could be reproduced by following carefully controlled procedures and was used in production for years before the active accelerator ingredient was identified. Indeed, Goodyear employed Samuel P. Mulliken, the eminent authority in the field of organic qualitative analysis, to spend the summer of 1917 analyzing Dubax. Mulliken was able to isolate and identify anilidobenzothiazole, itself inactive, but which was later shown to be a precursor of the active accelerator. The active component of Dubax eluded him. In 1921 Lorin B. Sebrell synthesized the active material and proved it to be 2-mercaptobenzothiazole. Now known as Captax, it remains today one of the most important rubber vulcanization accelerators.

Late in 1921 Bedford resigned from Goodyear to become co-founder of Rubber Service Laboratories, Inc., later purchased by Monsanto. However, as soon as he became a free agent, B. F. Goodrich made him an offer he could not refuse, and he sold his stock in Rubber Service to join Goodrich. Here he continued his accelerator work, was transferred for a short time to tire development, and then became manager of compounding research. In the meantime he collaborated with Herbert Winkelmann in writing "Systematic Survey of Rubber Chemistry," a bibliography covering the entire rubber literature to 1923, and a monumental undertaking.

Bedford was described as "magnetic, affable, dynamic, lusty, and scholarly." He entered enthusiastically into numerous activities besides his work: church service, Boy Scouts, Masonry, and chess. He was also a philatelist of note and was recognized nationally as an authority on United States stamps. He died unexpectedly on June 19, 1922 following a minor operation.

J. M. Ball, *Rubber Chem. Tech.*, **37**, 5, XIX (1964); Bedford and Winkelmann, "Systematic Survey of Rubber Chemistry," Chemical Catalog Co. (1923).

GUIDO H. STEMPEL

Anton Alexander Benedetti-Pichler

1894-1964

"The progress of chemistry is inversely proportional to the size of the sample used in the investigation." If we accept this statement then we must agree that its author, Anton A. Benedetti-Pichler, bears a great deal of responsibility for advancing chemistry with his introduction of microchemistry to the United States, his development of new microtechniques, characterized by elegant simplicity, and his teaching career that spanned 35 years in the United States.

Benedetti-Pichler was born in Vienna, Austria, Apr. 1, 1894. He arrived in the United States in September 1929, accompanied by his wife, the former Jenny Bierbaumer, to become an instructor at the Washington Square College of New York University, where he remained until 1940. He then joined Queens College, later a unit of the City University of New York, where he served until his retirement in 1964. He also held a special appointment in the graduate division of Brooklyn College, 1945-64, where he taught chemical microscopy.

Prior to his coming to the United States Benedetti-Pichler obtained his degree of "Doktor der Technischen Wissenschaften" from the Technische Hochschule, Graz, Austria, where he studied under Friedrich Emich. His doctoral dissertation dealt with "The Microanalysis of Solid Mixtures."

From 1922 to 1927 he was an assistant in his alma mater, and from 1927 to 1929 he was privat dozent.

Benedetti-Pichler received the 11th Annual Anachem award of the Association of Analytical Chemists in 1963, whose citation hailed him as "a soldier, teacher, chemist, student of nature and humanity, and American in the truest sense." The allusion to soldier referred to the fact that Benedetti-Pichler served in the Austrian Army during World War I on the Italian and Russian fronts, reaching the rank of lieutenant on his discharge.

His love of nature was a factor in bringing him to the United States, since Joseph B. Niederl, who secured his appointment at New York University, informed him he could go hiking in Yellowstone National Park. Conveniently, Niederl failed to mention how far Yellowstone Park is from New York City. Regardless of the distance Benedetti-Pichler often went camping there, and later he became a member of the National Park Association. His love of nature perhaps was enhanced by the fact that his wife was the daughter of a forest ranger.

His career as a chemist included the development of microanalytical techniques that were used profitably during the Manhattan Project. Apparently his foreign birth prevented him from taking active part in it. He was one of the first to establish a commercial laboratory using the microchemical techniques developed by Emich. He was a consultant to the Socony Vacuum Oil Co. from 1947 to 1951. He also developed several methods of nondestructive testing that were applied to the investigation of art forgeries.

His career as a teacher included the writing of five books and numerous articles on microchemistry, and the teaching of analytical chemistry at the undergraduate and graduate levels. He was an editor of *Mikrochemie* and in 1936 began to publish an English edition of the journal.

In his teaching Benedetti-Pichler pioneered the use of visual aids, preparing his own slides and illustrations with artistic flair. He also prepared several films to teach microanalytical techniques.

His contributions to the advancement of analytical chemistry included the founding of the American Microchemical Society. He was also an active organizer of the Division of Analytical Chemistry of the American Chemical Society. A biographer has noted his part in raising the analyst from a "cookbook technician" to a true scientist.

The honors Benedetti-Pichler received included the Pregl Prize of the Vienna Academy of Science in 1933, the Emich plaque of the Austrian Microchemical Society in 1955, and the Austrian Honorary Insignia for Arts and Sciences of the President of Austria (First Class) in 1962.

In 1965 the Microchemical Society established the A. A. Benedetti-Pichler award to be given to a chemist "in recognition of his outstanding contribution to microchemistry."

Two of Benedetti-Pichler's outstanding characteristics were his humility and kindness. He treated his students with respect and consideration but without condescension. He paid as much personal attention to undergraduates as he did to graduate students. He had an open, creative, and alert mind and never tried to force his ideas or method on his students and collaborators.

He retired from teaching in January 1964 and went to his farm in Camden, S. C., where he planned to apply his chemical knowledge to improve the fertility of the soil. He did not have a chance to do so as he died Dec. 10, 1964 of a heart attack.

Biography by David E. Sabine, *Chemistry* **42,** 12-15 (1969); obituaries: *Chem. Eng. News* **42** (52), **53** (1964), with portrait; *Microchemical J.* **8,** 448-9 (1964); *Mikrochimica et Ichnoanalytica Acta* 205-6 (1965); *Camden Chronicle* Dec. 11, 1964;Herbert K. Alber, "A. A. Benedetti-Pichler —A Dedication," Eastern Analytical Symposium, New York, Nov. 17, 1965; Frank L. Schneider, "Microchemical Techniques in Research" (A. A. Benedetti-Pichler memorial lecture), *Mikrochimica Acta* 742-9 (1967); personal communication from A. G. Loscalzo.

JOE VIKIN

Francis Gano Benedict

1870-1957

Benedict was born Oct. 3, 1870 in Milwaukee, Wis. When he was 7 years old his family moved to Florida and later to Boston. As a boy of 13 he became fascinated with the wonders of chemistry and set up a laboratory in the cellar of his home. His formal training began with one year at the Massachusetts College of Pharmacy, followed by studies at Harvard University where he received his A.B. degree in 1893 and his A.M. degree in 1894. He continued his graduate studies at Heidelburg under Victor Meyer and received his Ph.D. degree in 1895.

Upon his return to the United States Benedict received an appointment as research assistant in the Department of Chemistry at Wesleyan University under Wilbur O. Atwater, and here he was introduced to the field of physiology, particularly energy metabolism, which became his life work. For 12 years Atwater and Benedict jointly carried out studies using the Atwater-Rosa respiration calorimeter. They dealt with the quantities of nutrients and energy metabolized under different conditions of rest and muscular activity, the relations between external work and the energy and nutrients metabolized in its performance, and the capacity of different classes of nutrients to supply the body with nutrients and energy.

Upon Atwater's death in 1907, Benedict was appointed director of the Nutrition Laboratory in Boston, established by the Carnegie Institution of Washington. Here he continued and greatly expanded the program in which he had been engaged at Wesleyan; he collaborated with clinicians in the hospitals of the city in studies of energy metabolism in various diseases. He devised the Benedict apparatus for measuring the metabolism of patients. Extensive studies were made of the basal metabolism of man, dealing with the conditions required to establish the basal state—the effect of age, sex, body size, race and other factors. The studies resulted in the Harris and Benedict standards for comparing the metabolism of patients with that of normal persons.

Benedict's studies on basal metabolism in man were accompanied and followed by similar studies with a variety of species of animals. The investigations included mammals both domestic and wild, ranging in size from the 8-gram dwarf mouse to a 4000 lb elephant, reptiles, and birds. For several of the species Benedict had to design special equipment. His genius for devising apparatus to meet his research objectives provided tools used by many others and was one of his important contributions to the advancement of scientific research.

The publications of Benedict and his many collaborators number some 400. He received many awards both at home and abroad, honorary memberships in several foreign societies, and three honorary degrees. Benedict's principal hobby was magic. His skill as a magician won him election to the Society of American Magicians. This skill was reflected in an art of showmanship which made his public lectures especially interesting.

A large man with an impressive and dignified bearing, Benedict was regarded as austere by his younger collaborators. Yet he was a friendly person, an interesting conversationalist, and a very entertaining speaker.

Upon his retirement from the Nutrition Laboratory in 1937, Benedict moved to Machiasport, Maine where he became interested in local affairs. He continued an extensive correspondence and went on several lecture tours. He died May 14, 1957.

E. DuBois and O. Riddle, *Biog. Mem. Nat. Acad. Sci.* **32,** 67-78 (1958), port., list of publications; L. A. Maynard, *J. Nutr.* **98,** 1-8 (1969), port.; personal recollections.

LEONARD A. MAYNARD

Stanley Rossiter Benedict

1884-1936

Benedict was born in Cincinnati, Ohio, Mar. 17, 1884. He came from an illustrious family, his father being professor of philosophy and psychology at University of Cincinnati. His maternal grandfather, Asahel Clark Kendrick, was professor of Greek at University of Rochester, and one sister, Mary Kendrick Benedict, was president of Sweet Briar College but later gave up educational work to practice medicine in New Haven, Conn.

Benedict's first interest was in a medical career, but while he was attending University of Cincinnati his interest shifted to teaching and research. He received his B.A. degree from University of Cincinnati in 1906 and his Ph.D. degree from Yale in 1908. He then held the positions of instructor of chemistry at Syracuse University for one year and associate in biological chemistry at Columbia University for one year. In 1910 he was appointed assistant professor of chemical pathology at Cornell University Medical College in New York City, where he stayed for the rest of his life. He was professor of chemistry and chairman of the department from 1913 to 1936.

Benedict's major contributions were in the field of analytical biochemistry as related to the analysis of blood, urine, and other substances related to normal metabolism and the diagnosis of disease. His work in this area, along with that of Otto Folin, made chemical analysis an important tool in the diagnosis and treatment of disease and opened a new era in biological chemistry. One of his first successes was an improved test for sugar in urine; the reagent, known as Benedict's Reagent, is well-known to clinical chemists. Later he was instrumental in developing a simple, rapid, and accurate method for determining blood sugar using blood samples of 2 ml or less. Prior to this time, the large amount of blood necessary for a sugar determination made such a test impractical to use as a routine tool.

Publications from Benedict's laboratory dealt with many topics other than analytical methods; they included various aspects of the metabolism of carbohydrates, uric acid, phenols, creatine, creatinine, and cancer research. His laboratory was credited with the discovery and isolation of two substances, ergothioneine and a uric acid riboside compound, which were not previously known to exist in blood.

From 1920 until his death Benedict was the managing editor of *Journal of Biological Chemistry*. This position made him a strong influence in his field since the journal was the only American publication of research in biological chemistry at that time.

He was president of the American Society of Biological Chemists, 1919-20, and was elected to the National Academy of Sciences. He died of a heart attack at his home in Elmsford, N.Y., Dec. 21, 1936.

New York Times (Dec. 23, 1936); "Dictionary of American Biography," vol. 22, pp. 35-36, Chas. Scribner's Sons (1958); biographies by H. D. Daken, *Science* (Jan. 15, 1937) and E. V. McCollum, *Biog. Mem. Nat. Acad. Sci.* **27** (1952).

CHARLENE J. STEINBERG

Max Bergmann

1886-1944

Bergmann was born Feb. 12, 1886 in Fürth, Germany, the son of a prosperous coal merchant whose family had lived in the town for many generations. After completing his secondary schooling in Fürth, Bergmann studied at University of Münich, where he received his first university degree in 1907. Although originally attracted to botany, his interests soon turned to organic chemistry, and he enrolled in the Chemistry Department of University of Berlin, then headed by Emil Fischer. Working under Ignaz Bloch on acyl polysulfides, Bergmann took his Ph.D. degree in 1911, and thereupon became an assistant to Fischer. After Fischer's death in 1919, Bergmann was appointed in 1920 privatdozent at University of Berlin and head of the Chemistry Department of Kaiser-Wilhelm Institute for Textile Research. In 1921 he became director of the newly-established Kaiser-Wilhelm Institute for Leather Research in Dresden. When Hitler rose to power in Germany, Bergmann came to the United States. In 1934 he was appointed associate member of Rockefeller Institute for Medical Research in New York City, and 3 years later he became a full member, a position he held until his death Nov. 7, 1944.

Bergmann's scientific work before and after coming to the United States showed considerable continuity. In his association with Fischer, he made basic contributions to carbohydrate, lipid, and amino acid chemistry. For example, he elucidated the structure of glucal and developed new methods for preparing α-monoglycerides. While at Dresden he created one of the leading laboratories in protein chemistry. Together with his chief associate, Leonidas Zervas, he made numerous contributions to the chemistry of amino acids and peptides. Among them were studies on the mechanism of amino acid racemization, on the use of oxazolones for peptide synthesis, and on the transfer of the amidine group of arginine to glycine.

In 1932 Bergmann and Zervas devised a new method for synthesizing peptides, which marked a decisive new stage in protein chemistry. Their carbobenzoxy method opened an easy route to the synthesis of peptides which had hitherto been difficult or impossible to prepare.

At Rockefeller Institute Bergmann and his associates applied the carbobenzoxy method to the synthesis of peptides for test as possible substrates for protein-splitting enzymes such as pepsin and trypsin. This work, pursued by Joseph S. Fruton, led to the discovery in 1936-39 of the first synthetic peptide substrates for these enzymes, thus opening the way for study of their specificity. With Carl Niemann, Bergmann proposed in 1938 that the arrangement of amino acids in a protein chain is periodic; although this theory was later shown to be an oversimplification, it stimulated great experimental activity in the protein field. In Bergmann's laboratory, William H. Stein and Stanford Moore began work that later led them to solve the problem of the accurate determination of the amino acid composition of proteins. These researches were suspended after 1941 when Bergmann's laboratory worked on the chemistry of the nitrogen and sulfur mustard gases in connection with the war effort.

H. T. Clarke, *Science* **102,** 168-170 (1945); C. R. Harington, *J. Chem. Soc.* 716-718 (1945); B. Helferich, *Chem. Ber.* **102,** I-XXVI (1969); G. W. Corner, "A History of the Rockefeller Institute, 1901-1953, Origins and Growth" Rockefeller Inst. Press, (1964); personal recollections.

JOSEPH S. FRUTON

Edward Martin Bevilacqua

1920-1968

Bevilacqua was born Nov. 4, 1920 in Philadelphia, Pa. He received his early education in the public schools of Lower Merion Township and Media, Pa. At age 11, the family moved to Staten Island, N. Y., where he attended P.S. 21 and the Port Richmond high school. After graduating first in his high school class, he entered Rensselaer Polytechnic Institute at Troy, N. Y. in the fall of 1937. After majoring in chemistry and graduating from RPI in 1941, he spent the next 3 years as a graduate student and research assistant in the Chemistry Department of University of Wisconsin. He received his Ph.D. degree in 1944.

His research effort as a graduate student with John W. Williams was concerned with the theory and practice of measuring diffusion coefficients of macromolecules in solution. While working on projects sponsored by National Defense Research Committee and the War Department, he co-invented a device for measuring sedimentation rates of finely divided material dispersed in a liquid.

In 1944, Bevilacqua moved to Passaic, N. J. to begin his career as an industrial research chemist in the corporate research laboratories of The United States Rubber Co. (later known as Uniroyal, Inc.). He continued in the employ of Uniroyal and by virtue of his excellent performance became the youngest scientist to reach the top category of professional classification at Uniroyal. His innate curiosity, keen intelligence, and capacity for work resulted in many important contributions to the science and technology of natural and synthetic rubber. His researches resulted in the publication of some 50 papers and 10 United States patents.

His career as an industrial chemist started with an investigation of the preparation and properties of high polymers based on metal-organic coordination chemistry. Later he became interested in the oxidation of natural rubber and through his research publications became recognized as an authority on the subject. He also contributed to the understanding of the complex subject of vulcanization. Finally, he demonstrated his versatility as a physical scientist by publishing a series of papers concerning the traction of pneumatic tires on wet roads.

Bevilacqua's research on the thermal oxidation of natural rubber contributed much to the present knowledge of the processes involved in the oxidation of elastomers. Thermal oxidation of natural rubber results in chain scission and the formation of numerous low-molecular-weight compounds. By identifying the scission products, Bevilacqua concluded that in the course of oxidation, one end of the oxidized rubber molecule is a methyl ketone, the other is presumably aldehyde or acid, and the principal scission products are formic acid and levulinaldehyde which is rapidly oxidized to acetic acid and carbon dioxide. The yield of volatile products increases as the temperature is raised, paralleling increased scission efficiency.

Bevilacqua's research on rubber friction was concerned with the vitally important problem of traction and safety on wet roads. His efforts over a relatively short time culminated in a series of publications dealing with the relation between road safety and tire traction and with the interactions of road surface characteristics with specific rubber properties that affect lubricated friction or skid resistance on wet and icy highways. His studies showed that the principal rubber properties were hardness and resilience, both of which can be measured and controlled by the rubber technologist.

Bevilacqua, from 1965 until his death, was editor of *Rubber Chemistry and Technology*. In tribute to him, the American Chemical Society's Rubber Division sponsored a Memorial Symposium on the Oxidation of Rubber in Cleveland, October 1971. Besides ACS, he was a member of American Association for the Advancement of Science, Sigma Xi, and New York Academy of Sciences.

Outside the laboratory, Bevilacqua participated in a variety of activities concerned with world peace, conservation, public education, and local government, besides being a responsible parent and an avid gardener.

He died at his home in Allendale, N. J., Oct. 27, 1968 survived by his wife, the former Ellen Burtner, and two children, Susan and Jeffry.

Personal recollections; obituaries in *Rubber Age* **100,** No. 12, 119 (1968), and by C. F. Eckert in *Rubber Chem. Tech.* **42,** No. 1, G28 (1969).

CHARLES F. ECKERT

Eugene Cook Bingham

1878-1945

Bingham was born at Cornwall, Vt., Dec. 8, 1878. He received his A.B. degree from Middlebury College in 1899, majoring in chemistry, and his Ph.D. degree at Johns Hopkins in 1905. He then spent a year in Europe. At University of Leipzig he met Wilhelm Ostwald. In Berlin he met Nernst and Van't Hoff, and in Cambridge, England he worked under Sir J. J. Thomson. Apparently in Europe Bingham's interest in viscosity of fluids was born.

From 1906 to 1915 Bingham was professor of chemistry at Richmond (Va.) College. From 1915 to 1916 he was assistant physicist at the U.S. Bureau of Standards. In 1916 he went to Lafayette College as head of the Department of Chemistry and Metallurgy. In 1939 he retired and became research professor at the same college, occupying this position till his death on Nov. 6, 1945.

Almost all of Bingham's research was devoted to viscous and plastic flow. His importance to this field can not be overestimated. He is considered the founder of modern rheology, and introduced the term rheology, itself. He also introduced the concept of "yield value" that is very important in rheology and is credited with naming the unit of viscosity the "poise."

In 1924 a "Plasticity Symposium" sponsored by the American Chemical Society took place at Lafayette College. Many of the contributors to this symposium became prominently active in the Society of Rheology formed on Bingham's initiative in 1929. Bingham was the first editor of *Journal of Rheology* published by this society. He was chairman of the Viscosity Section of the World Petroleum Congress, London in 1933.

Many prominent foreign scientists worked with Bingham at Lafayette. Among them was George W. Scott-Blair from England, to whose book "An Introduction to Industrial Rheology" Bingham wrote a foreword, and M. Reiner from Tel Aviv who contributed very much to the development of rheology. Bingham was in contact with many prominent foreign scientists, among them Wolfgang Ostwald, and T. Erk. Bingham, like Wilhelm Ostwald, was interested in the life and work of his predecessors. He published in the *Journal of Rheology* a biography of Poiseuille by Marcel Brillouin. He also published in "Rheological Memoirs" an English translation of the classic paper by Poiseuille on the flow of water.

According to the testimony of his successor as head of department at Lafayette, Bingham was a stimulating teacher. The department under him increased in number of teachers and number of students. The laboratory facilities were expanded.

He organized the ACS Student Affiliates group at Lafayette, the first in the United States. Bingham was very active in the Lehigh Valley Section of the ACS. He served twice as its chairman and three terms as counselor. In 1918 Bingham started a 4-page leaflet that he called *Octagon,* and that developed into a small journal, to serve as the organ of the Lehigh Valley Section of ACS.

He was a strong supporter of the metric system and fought for its adoption. He experimented with making the highways safer for night driving by imbedding heavy metal plates in roadways. An ardent conservationist he was active in The Blue Mountain Club and helped blaze the Appalachian trail between Wind Gap and Water Gap in Pennsylvania.

On the day of his death while under an oxygen tent, Bingham commented that 160 years ago that day Priestley had discovered oxygen.

The first annual meeting of the Society of Rheology after Bingham's death was named "Bingham Memorial Symposium on Rheology." At this meeting the "Bingham Memorial Award" was established to be given annually to the scientist making the greatest contribution to rheology.

In all, Bingham published two books: "A Laboratory Manual of Inorganic Chemistry" (1911), and "Fluidity and Plasticity" (1922), and 81 papers.

Biographies by A. Nadai, *J. Colloid Sci.* **2,** 1-5 (1947); W. H. Fullweiler, *J. Colloid Sci.* **2,** 5-6 (1947); J. H. Wilson, *Octagon* **29,** 4 (1946); R. D. Billinger, *Octagon* **30,** 72-78 (1947); bibliography by Mary Hertzog, *Octagon* **29,** 8 (1946).

GEORGE SIEMIENCOW
ROBERT YUSZCZUK

Greene Vardeman Black

1836-1915

Black was a self-taught dentist who took up chemistry to develop better dental fillings. He was born Aug. 3, 1836 near Winchester, Ill. Instead of going to school regularly as a boy, he preferred to ramble through the woods and along the river absorbing nature. His parents, sensing perhaps his potential for greater things, did not object, but when he was 17 years old they encouraged him to live with his brother, a physician, in a nearby town. There he became interested in dentistry, watching the local dentist tend his patients. When he was 21, Black decided that he could practice dentistry at least as well as his mentor so he went home and opened an office. From 1862 to 1864 he was in the Union Army, most of the time as a scout.

In 1864 at Jacksonville, Ill., Black resumed his dental practice. By self-instruction in the sciences he added continuously to his meagre educational base. He set up a laboratory next to his office and studied chemistry, hoping to improve the materials used in filling teeth. Becoming enthusiastic over chemistry, he organized chemistry classes among the public school teachers of the area and taught them for several years. He developed an amalgam that was hailed by the dental profession, and his method of preparing amalgams was adopted by practically all dentists. He became too involved in teaching, writing, and practicing dentistry to pursue the study of chemistry extensively, but his excursion into chemistry benefited all who have had their teeth filled. He died Aug. 31, 1915, in Chicago where a monument was erected to him in Lincoln Park.

M. D. K. Bremner, "The Story of Dentistry," Dental Items of Interest Pub. Co. (1946); B. E. Schaar, "Dental Amalgams," *Chemistry* **40,** 21-22 (Sept. 1967); Burton L. Thorpe, "Biographies of Pioneer American Dentists," being vol. 2 of Charles Koch, ed., "History of Dental Surgery," 570-580, The National Art Pub. Co. (1909); "Dictionary of American Biography" **2,** 308-310, Chas. Scribner's Sons (1928).

BERNARD E. SCHAAR

James Blake

1815-1893

Blake was born in Gosport, England, July 14, 1815. He studied medicine in London and Paris, spending some time while in the latter city working with the famous physiologist, Magendie. In 1841 he obtained his M.D. degree from University College, London. During his student years he carried on research in the physiological laboratory of William Sharpey and as early as 1839 published a paper on the effects of direct injection of various salt solutions into the blood stream of dogs. In 1841 he was sufficiently well–known to be chosen a Fellow of the Royal College of Surgeons. He continued his researches for a time in England but in 1847 emigrated to the United States where he became professor of anatomy and surgery in the Medical College of St. Louis. In 1850 no doubt attracted by the California gold rush, he moved to Sacramento and established a medical practice there.

In 1862 he moved his practice to San Francisco. There he edited a medical journal for a short time but gave up this work because there were not enough papers of a quality high enough to satisfy him. He became a member of the faculty of the Toland Medical College, afterward the Medical School of the University of California. His broad scientific interests led to his election as president of the California Academy of Sciences in the years from 1868 to 1872. At about this time he became interested in the open–air treatment of tuberculosis. He established a sanitarium to test his ideas at Calistoga in the Napa Valley north of San Francisco in 1876. However, in 1880 he sold this and moved a short distance north to the small town of Middletown in Lake County. There he remained for the rest of his life except for short trips to Europe to attend scientific meetings. He died Nov. 18, 1893 in Middletown, Calif., and was buried there.

Blake was a man of wide interests. He carried on analytical work on mineral waters and practiced assaying; he was well–known for his geological studies; he investigated wine–making and made studies in zoology. However, his most original and important work was begun during his years in England and was resumed after he settled in Middletown. He began these studies because he was convinced that there was a relation between the physiological action of substances and their chemical structures. Since so little was known of the structure of organic compounds when he began his work, he confined himself to the investigation of inorganic salts. At this time the work of Mitscherlich had brought the phenomenon of isomorphism to the attention of scientists. Blake felt that he could show that substances whose salts were isomorphic with each other would have similar physiological properties when injected into the blood stream. In his later work he related the physiological properties and the toxicities of various salts to the atomic weights and valences of elements with variable valence.

It is of considerable interest that he was able to group the different elements into families on the basis of their physiological actions, and that these families afterwards turned out to be those which appeared in the periodic table. Blake did not recognize their periodicity. Nevertheless his careful studies, accurately carried out under controlled conditions, not only indicated the importance of chemical relationships to physiological behavior but also anticipated later work on the importance of inorganic salts in mammalian metabolism.

Sister Mary Ambrose Deveraux, masters thesis, St. Louis University, 1952; Sister M. A. Deveraux, H. B. Donahoe, and K. K. Kimura, "Physiological Basis for the Grouping of the Elements—James Blake (1815-93)," *J. Chem. Educ.* **33,** 340-43 (1956); W. F. Bynum, "Chemical Structure and Physiological Action," *Bull. Hist. Med.* **44,** 521-24 (1970); J. Parascandola, "Structure-Activity Relationships—The Early Marriage," *Pharmacy in History* **13,** 3-5 (1971); sketch by J. L. Miller, Howard A. Kelly, and Walter Burrage in "Dictionary of American Medical Biography" (1928).

HENRY M. LEICESTER

Albert Victor Bleininger

1872-1946

Bleininger was born in Poling, Germany, July 9, 1872 and attended the grade and intermediate schools in Münich before coming to the United States as a boy of 14 with his family. At first he worked with his father who attempted to establish plants for producing brick, clay tile, and terracotta near Akron, Ohio. Although working and attending school intermittently, he was able at 23 to gain admission to The Ohio State University where he graduated 6 years later with his B.S. degree in chemistry. He continued his association with the university after graduation by working under Edward Orton who was then establishing the first Ceramic Engineering and Science Department in the Western Hemisphere. He served successively as instructor, assistant professor, and associate professor while at the same time conducting a survey of the Portland cement industry as part of his role as a staff member of the Ohio Geological Survey. He assumed the duties of assistant professor of ceramics at University of Illinois in 1907, at which time this new department was being established. He also was serving then as head of the Clay Products Section of the U. S. Geological Survey, for whom he organized and equipped its Ceramic Laboratory in Pittsburgh. In 1910 he was made professor and head of the Department of Ceramic Engineering at University of Illinois. In 1912 he became chief of the Division of Ceramics at the U. S. Bureau of Standards in Pittsburgh where he instituted research in glass, enamelled metals, and porcelain. During World War I he helped develop and establish optical glass factories when the European supply of glass was cut off and also served as a consultant for the U. S. Army's Chemical Warfare Service. He became chief chemist for the Homer Laughlin China Co. of Newell, W. Va. in 1920 and served in that capacity for the remainder of his career. The Ohio State University awarded him the degree of ceramic engineer in 1931 and the following year presented him with its Lamme Medal for outstanding achievement in engineering. In 1933 he received an honorary Sc.D. degree from Alfred University.

Throughout his long career, Bleininger took an active interest in numerous engineering, scientific, and civic organizations but devoted most of his extra-curricular attention to the American Ceramic Society which he helped found and of which he became a charter fellow in 1936. When the society established divisions, he became a member of its White Wares Division and served as its chairman. He was elected a society trustee in 1903, vice president in 1904, and became its president in 1908. He was chairman and member of many society committees. He also was chairman of the National Research Council Subcommittee on Ceramic Chemistry and the U. S. Potters Association Research Committee.

Bleininger brought to bear a rare combination of practical abilities and interests with essentially fundamental scientific approaches to applied problems. These were most extensive in scope and diversity including Portland cement, optical glass, porcelain, spark plugs, catalyzers for ethylene production, new bodies and glazes for the semivitreous china industry, and the elimination of crazing in ceramic products. His 86 publications and his patents covered a wide and diverse range, but his translation of the writings of Herman Seger was a particularly noteworthy contribution in the earlier stages of the development of ceramic industry and education in this country.

As great as his contributions were to science, industry, and technology, his attitude toward his associates, fellow-men and adopted country were even more outstanding and unique. He never neglected opportunities to inspire and encourage the younger workers in his fields nor to be receptive to new concepts and ideas. His career in this country began with the simplest and most menial of occupations in industry but was crowned by attaining the highest of educational and scientific honors. In acknowledging one of the last of these and sketching his own career, he wrote: "Such a story is possible only in our beloved country. Such men as have helped me, exist only in America, the sons of pioneers, teachers, and scholars who founded this country."

Bleininger died of a heart attack in East Liverpool, Ohio, May 19, 1946 at the age of 74 and had been active in industry to the end.

Bull. of the Amer. Ceramic Soc. **20,** 177 (1941); **25,** 241-42 (1946); **27,** 204-09 (1948); **31,** 255 (1952).

HENRY H. BLAU

Rachel L. Bodley

1831-1888

Bodley, first woman member of the American Chemical Society and a teacher of chemistry and toxicology, was born in Cincinnati, Ohio, Dec. 7, 1831. She attended her mother's private school until she was 12 years old, then Wesleyan Female College in Cincinnati. After graduation in 1849 she taught at the college until 1860 when she went to Polytechnic College of Philadelphia, presided over by Alfred Kennedy, to study chemistry and physics.

Returning to Ohio in 1862, Bodley was professor of natural sciences at Cincinnati Female Seminary for 3 years. In 1865 she moved to Philadelphia to teach chemistry and toxicology at Female Medical College of Pennsylvania (later, Medical College of Pennsylvania). She was not the first woman to teach chemistry in a medical school (she had been preceeded by Almira Fowler at Female Medical College), but she seems to have been the first woman to hold the title of professor of chemistry in a medical school and the first to excel as a teacher of the science.

Bodley was appointed dean in 1874 and thereafter acted as chief business officer of the institution as well as professor. Competent, wise, and energetic, she carried on a world-wide correspondence, helped spread the reputation of her college through America, Europe, and Asia, and accelerated the growth of the institution.

Bodley suggested Joseph Priestley's home in Northumberland, Pa. as the meeting place of the Priestley Centennial in 1874 and was elected first vice president of that important gathering of chemists. She was a member of the State Board of Public Charities and the Philadelphia school board. The M.D. degree which is often found after her name was honorary, being bestowed on her by Woman's Medical College in 1879. She died in Philadelphia, June 15, 1888.

Records of the American Chemical Society and of Medical College of Pennsylvania; communication from Harry P. Bodley; *American Chemist* **4, 5** (1873-75); "Papers read at the Memorial Hour Commemorative of the Late Prof. Rachel L. Bodley," Woman's Medical College (1888).

WYNDHAM D. MILES

Marston Taylor Bogert

1868-1954

Of Dutch origin, Bogert's ancestors settled in the New York City area in 1668. He was born in Flushing, N.Y., Apr. 18, 1868, son of Henry A. and Mary B. (Lawrence) Bogert, attended Flushing Institute, and graduated from Columbia in 1890 with his A.B. degree. After spending 4 years as a chemistry student at Columbia School of Mines he received his Ph.B. degree (the Ph.B. degree was often reported as Ph.D.). Later, he received honorary doctor's degrees from Clark University and Columbia. Although he directed the work of many graduate students, he had no earned graduate degree.

After graduating from the School of Mines he became a member of the Columbia faculty and remained at the University until his retirement in 1938 as emeritus professor in residence. In 1904 he was appointed the first professor of organic chemistry. Author of numerous publications in the field of synthetic organic compounds, he worked on perfumes and studied the relationships of odor and molecular structure.

Bogert was active in various organizations. In dress, manner, and diction he was a true gentleman. He had unusual ability in dealing with people. In the classroom he was a skillful lecturer. An ideal person to conduct a meeting, he had great tact, used no notes, and could introduce foreign speakers in the language of their country. One could be assured that any meeting chaired by Bogert would run smoothly. He always presided at the annual award of the Perkin Medal where he wore a dress suit and a mauve tie.

His interests in the international aspects of chemistry resulted in his election as president of the International Union of Pure and Applied Chemistry in 1938. After World War II he helped rebuild the organization and continued as its president until 1947. He was a member of several foreign chemical societies and was president of the Society of Chemical Industry in 1912. In 1927-28 he was visiting Carnegie Professor of Internationl Relations at Charles University in Czechoslovakia. Here he was a medalist and recipient of an honorary degree. He was also a medalist at Comensky University in Bratislava.

As a servant of his country he accepted responsibility in national affairs as an organizer, consultant, and advisor to many government departments and agencies. During the first World War he was a colonel in the Chemical Warfare Service and for the remainder of his life was often addressed as "Colonel Bogert."

Bogert was active in many organizations of his profession. One of the founders of the Chemists' Club, he served as its president in 1908. He was president of the American Chemical Society in 1907 and 1908 and an associate editor of *Journal of the American Chemical Society,* 1909-19 and 1924-29. He held a similar position on *Journal of Organic Chemistry.*

Many honors came to Bogert, among them the Nichols Medal (1905), Medal of the American Institute of Chemists (1935), Priestley Medal (1938), and the Chandler Medal (1949).

He died of pneumonia in a convalescent home on Long Island, Mar. 21, 1954.

J. Soc. Chem. Ind., Special Jubilee Number, 80, July 1931, with portrait; *J. Chem. Ed.* **5,** 378-80 (1928) with portrait; biography by H. L. Fisher in *Ind. Eng. Chem.* **25,** 591-92 (1933); *Chem. and Eng. News* **24,** 2029 (1946), with cover portrait; obituary, *Chem. Eng. News* **32,** 1256-1257 (1954), with portrait; obituary, *New York Times,* Mar. 22, 1954.

Egbert K. Bacon

Elmer Keiser Bolton

1886-1968

Bolton was born in Philadelphia, Pa., June 23, 1886, the son of George G. and Jane E. (Holt) Bolton. He attended Philadelphia Central High School, graduated from Bucknell University in 1908, and received his Ph.D. degree from Harvard University in 1913.

Elmer was awarded the Sheldon Fellowship at Harvard and this enabled him to do postdoctorate research at the Kaiser Wilhelm Institute in Berlin. He worked in the laboratory of Richard Willstätter where he isolated and established the chemical constitution of the pigments of geraniums, scarlet sage, and dark red chrysanthemums. He returned to America in 1915 and went to work at the Du Pont Company Experimental Station in Wilmington, Del. He advanced through positions of increasing responsibility in the company until, in 1930, he became director of Du Pont's Chemical Department. In this capacity he was in charge of the basic and applied chemical research activities of the company.

Bolton was recognized for his leadership in the synthesis and development of the first general-purpose synthetic rubber (neoprene) and for his direction of the development of nylon as a new and revolutionary commercial textile fiber. In 1941 he received the Chemical Industry Medal for valuable application of chemical research to industry. In 1945 he was awarded the Perkin Medal in recognition of his outstanding accomplishments in the field of industrial research. In accepting the Perkin Medal from Marston T. Bogert, Bolton said, "I am deeply conscious of the fact that any credit for research accomplishments with which I have been connected belongs to the organizations of able research chemists with whom it has been my privilege to be associated. As their representative, I am happy to accept this award because, in honoring me, you honor them." In 1954 he was awarded the Willard Gibbs Medal by the Chicago Section of the American Chemical Society in recognition of his contributions to fundamental research. The Chicago Section also honored Bolton by listing him in *The Chemical Bulletin* as one of the top 10 industrial chemists in the United States.

Bolton received honorary Doctor of Science degrees from Bucknell University and University of Delaware. He served as a director of the American Chemical Society from 1936 to 1938 and 1940 to 1943; as a member of the Visiting Committee of the Chemistry Departments at Massachusetts Institute of Technology, and Harvard University; as a trustee of Bucknell for more than 30 years; and as president of Elizabeth Storch Kraemer Memorial Foundation for Cancer Research.

Bolton retired from the Du Pont Co. in June 1951, after 36 years of service. He died July 30, 1968 at the Wilmington Medical Center at the age of 82.

Chem. Eng. News **30,** 4143 (1952); information from Du Pont Co.; *Chemical Bulletin (Chicago); Bucknell Univ. Alumni Magazine.*

LESTER KIEFT

Henry Carrington Bolton

1843-1903

Bolton was born in New York City, Jan. 28, 1843. He learned chemistry at Columbia from Charles Joy, in Paris from Dumas and Wurtz, in Heidelberg from Bunsen, Kirchhoff, and Kopp, in Göttingen where he obtained his Ph.D. degree in 1866 from Wöhler, and in Berlin from Hofmann. Returning to New York after 5 years abroad, he opened a consulting-teaching laboratory, then taught at Columbia University School of Mines from 1872 to 1877, at Woman's Medical College of New York Infirmary from 1875 to 1877, and at Trinity College, Hartford, Conn., from 1877 to 1887.

Independently wealthy, Bolton retired in 1887 and thereafter traveled and worked on history, biography, and other subjects that interested him. He carried out research on musical sands, folk-lore, and uranium compounds (he was the earliest American expert on uranium, dating from his Ph.D. thesis on uranium fluoride in 1866). He traveled in Europe, the Near East, Egypt, and the Hawaiian Islands. He published a genealogy of the Bolton family. His "Student Guide in Quantitative Analysis" passed through three editions. He dashed off notes, letters, and book reviews to scientific journals, literary magazines, and newspapers. But his chief preoccupation was history and bibliography of chemistry. He amassed probably the best private collection of books on alchemy and history of chemistry in the country up to that time. He wrote three books and at least 40 articles dealing with history of chemistry and science.

In 1874 it occurred to Bolton that the centenary of Priestley's discovery of oxygen was approaching. He wrote a letter to the editor of *American Chemist* suggesting that the event be commemorated on August first. Readers approved the idea and Rachel Bodley suggested it be held at Priestley's home in Northumberland, Pa. When the time arrived, about 100 of the foremost chemists in the country met at the Priestley homestead. Discussions among these persons led, 2 years later, to the organization of the American Chemical Society.

In 1892 George Washington University appointed him "non-resident lecturer on the history of chemistry." Bolton, then living in New York, would visit Washington for about a month to deliver the nine or 10 lectures in his course. He was not the first person to lecture on the history of chemistry in the United States, but he was the first to approach the subject equipped as a professional. His course at George Washington was one of the earliest in the country and probably the first of high quality.

At George Washington Bolton was also, from 1894 to 1896, professor of bibliography and bibliology. His work in bibliography took several directions. In 1891 he tried to persuade the American Chemical Society board of directors to start an Index Chemicus similar, apparently, to *Index Medicus.* He tried in 1893 to interest American and European chemical societies in establishing an International Co-operative Index to Chemical Literature. He organized and headed for 20 years an American Association for the Advancement of Science committee on chemical bibliography which stimulated the preparation of bibliographies on many subjects. He compiled two bibliographies still in use: "Catalogue of Scientific and Technical Periodicals (1665-1882)" and "Select Bibliography of Chemistry."

Bolton moved to Washington in 1896, presumably because of the city's splendid library facilities. For some years he was secretary of the Literary Society of Washington, which included Alexander Graham Bell and John Wesley Powell among its members and had President Theodore Roosevelt as an "honorary associate." In 1900 he was president of the Chemical Society of Washington, the American Chemical Society's Washington Section.

Bolton was bald, bearded, and chubby. He was, he said, "blessed with a hearty appetite." He was sociable and friendly with many American scientists of his time. A fluent and charming speaker, he delivered addresses before many groups. "It was said of Dr. Bolton," reported the New York *Times,* "that he belonged to more learned societies than any other living American."

Bolton died in Washington, Nov. 19, 1903. His wife, Henrietta Irving, a great grandniece of Washington Irving, had his body buried in the Irving plot at Tarrytown, N. Y. Mrs. Bolton presented his alchemical and chemical books to Library of Congress.

Biography by Charles A. Browne, *J. Chem. Educ.* **17,** 457-461 (1940), with portrait; W. D. Miles, "Henry Carrington Bolton," *Capital Chemist* **17,** 86-87 (1967), with portrait; Bolton's scrapbooks at Library of Congress; catalogs of George Washington University; "Dictionary of American Biography" **2,** 422-23, Chas. Scribner's Sons (1928).

Wyndham D. Miles

Harold Simmons Booth

1891-1950

Booth, descendant of very early settlers in Connecticut, born in Cleveland, Ohio, Jan. 30, 1891 was a devoted teacher, dramatic lecturer, researcher, and administrator. He achieved many things in the 59 years of his life besides those listed in "Who's Who," professional literature in the field of chemistry, and other biographical sources.

He was a trail blazer, undaunted by unfortunate signals which early appeared in his life. Upon request once he wrote this of his boyhood, "I was interested in everything mechanical and made replicas of roller coasters, sewer digging machines, and electric street cars, and the usual number of play houses, caves, and huts. I lived the life of a boy scout before there were Boy Scouts. My summers in the country were devoted to the usual boyish activities of fishing from morning to night, pilfering the neighboring farms of peaches, muskmelons, and apples. . . .

"My interest in chemistry was first awakened by the high school chemistry teacher at West High in Cleveland, and I went to Adelbert College of Western Reserve University planning to take the combination course with Case School of Applied Science in which the first 3 years were spent at Adelbert and the last two at Case School. Due to financial reverses at the end of my sophomore year, I was forced to find employment and could not come back to school in the fall. That summer I tried to find a position in vain and so, having nothing better to do, I played around with photography." That playing around developed into the Booth Quality Photo Co. about which Booth used to laugh and say his father inherited from him. Its growth—ultimately 125 branches—brought him back to Western Reserve University in 2 years. He had flirted with English and dramatics—which later came in good stead. Also geology intrigued him no end, but chemistry won him. He received his A.B. degree with honors from Adelbert College 1915, M.A. from Western Reserve University 1916, working with Arthur F. O. Germann on the density of air in Cleveland, and his Ph.D. from Cornell University 1919, working with Arthur W. Brown on the atomic weight of nitrogen and silver by thermal decomposition of silver nitride.

However, upon his return to college in 1915 he had to contend with serious illness which medical advisors indicated would cut his life short. This opinion was confirmed in 1917. But Providence, his own extraordinary will to live and intense professional activity—coupled with the ceaseless care of his devoted wife and children—gave him 33 additional years, happy, absorbing, and fruitful. His students remember his graphic lectures, his ability to dramatize the mundane, and his love of life. Few can forget him, rotund figure in chef's outfit, presiding over the hot dogs on picnics in metropolitan parks. How he did enjoy eating outdoors and indoors with his family and friends. Whether he was hiking in the winter snows, hunting the first trillium in spring, or summering in Key Harbor, Ontario, Booth loved life!

Another type of student, the part-time ones, will remember opportunities he helped open up for evening classes in chemistry (1921). The evening courses became one of the corner stones of Cleveland College. He had continued his education in spite of difficulties, and he never forgot there were others who had similar problems. Later in his consulting work he obtained financial help in the form of scholarships and fellowships and, having been a young businessman who knew difficulties, he could sense students' aims and frustrations.

Booth's principal interest was in the field of inorganic chemistry. He taught for many years the three subjects: general chemistry, quantitative analysis, and chemical microscopy. He was very fond of the latter course. He was trained in this work by Emile M. Chamot of Cornell who specialized in this field. He gave about 80 lectures on "Chemistry Through the Microscope" before professional groups. Using his chemical microscopy in crime detection, he also gave before many diversified groups a very interesting lecture on "Crime Pays the Chemist." In his later years he taught advanced inorganic chemistry in place of quantitative analysis.

He thoroughly enjoyed teaching. His lectures always aroused the interest of even the most mediocre freshman. He liked to dramatize his presentations by freely using spectacular demonstrations and well-selected visual aids in the form of lantern slides and movies long before they became routine.

Booth was internationally known for his researches, particularly in the field of the fluoride gases. During his tenure at Western Reserve he published over 100 papers, more than half of which were on fluoride gases. In the course of this work he prepared 24 new fluoride gases. He discovered many new reactions involving fluorine and synthesized many new fluorine compounds. His expertise in fluorine chemistry proved a great asset to our nation during World War II in the development of chemical processes for uranium purification.

Booth was an expert glass blower, an excellent machinist, and a most versatile and talented laboratory worker, qualities which made possible his productive work in his field.

In addition to his research papers, he published (1940) a textbook "Quantitative Analysis" (with Vivian R. Damerell), a monograph (1949) "Boron Trifluoride and Its Derivatives" (with Donald R. Martin), was editor-in-chief of Vol. I of "Inorganic Syntheses" and associate editor of Vol. II and Vol. III.

Booth, in spite of a chronic disease, enjoyed reasonably good health all of his life. In May of 1950 he began to complain of illness. He had a premonition that he would go soon; he mentioned it a number of times. On June 23 his illness did prove fatal. He died at the age of 59.

Archives of Case Western Reserve University; personal recollections; letter written by Booth June 7, 1939 at the request of James Huffkins, Denton, Texas; remarks made by Winfred George Leutner, president emeritus of Western Reserve University, during memorial service for Booth, June 26, 1950; Case Western Reserve chemistry department publication, *The Reserve Chemist* No. 17, Aug. 24, 1950.

FRANK HOVORKA

James Curtis Booth

1810-1888

Booth, founder of the oldest chemical consulting firm in the United States, was born in Philadelphia, July 28, 1810. He graduated from University of Pennsylvania in 1829, studied chemistry and geology under Amos Eaton at Rensselaer for a year, and gave a series of lectures on chemistry in Flushing, Long Island during the winter of 1831–32. Sometime in 1831 or 1832 he may have studied chemistry under Benjamin Silliman at Yale. In late 1832 he went to Europe, studied under Friedrich Wöhler and Gustav Magnus, and learned industrial chemistry processes by visiting factories in Germany, Austria, and England.

Booth returned home late in 1835 or early in 1836 and opened a consulting and teaching laboratory in Philadelphia. He was perhaps the first professional consulting chemist in the United States. He analyzed ores, minerals, fertilizers, sugar, and other materials and he assisted clients in developing or improving industrial processes. He was granted at least one patent for a process to manufacture potassium chromate and bichromate from chrome iron ore. Joseph Henry, around 1840, considered Booth "the most accomplished practical chemist in our country."

To relieve himself of some of the consulting work so that he could devote part of his time to other chemical activities, Booth took partners, several of whom became well known as industrial analysts. Among them were Martin H. Boyé, Thomas H. Garrett, and Andrew A. Blair. A succession of partners continued the firm after Booth's death, and it is in existence at the writing of this book.

In the 1830's students could obtain laboratory chemical instruction in only a few institutions. There were no courses in industrial chemistry. Booth accepted pupils and taught them the techniques of general and analytical chemistry, of chemical research, and of industrial chemistry. Among his students in the 1830's and 1840's were John F. Frazer, Robert E. Rogers, Richard S. McCulloh, Nat R. Davis, and James Van Zandt Blaney, who became professors of chemistry; and Robert B. Potts, Charles Hartshorne, William M. Uhler, Alexander Mucklé, Campbell Morfit, and Clarence Morfit, who became industrial chemists. Approximately 50 young men were, in a sense, apprentices of Booth.

For several months in 1836 Booth was a member of a team making a geological survey of Pennsylvania. In 1837 and 1838 he directed the geological survey of Delaware. He taught chemistry at Franklin Institute from 1836 to 1845, at Central High School in Philadelphia from 1842 to 1845, and at University of Pennsylvania from 1851 to 1855.

Booth published many articles; edited and wrote most of "Encyclopedia of Chemistry," which appeared in 1850 and went through several editions; edited the American edition of Henri V. Regnault's "Elements of Chemistry" in 1852; and with Campbell Morfit wrote "Report on Recent Improvements in the Chemical Arts" in 1851.

In 1849 President Taylor appointed Booth melter and refiner of the U. S. Mint in Philadelphia where more gold and silver were refined than in any other place in the world. Booth held this position until 1887 and made several improvements in the process for separating and refining gold and silver.

A religious man, Booth was active in the Protestant Episcopal Church. He supported several philanthropic organizations. Bucknell and Rensselaer presented him with honorary degrees, and the American Chemical Society elected him president in 1883 and 1884. He died in Philadelphia, Mar. 21, 1888.

Biographies by Patterson DuBois, *Proc. Amer. Phil. Soc.* **25,** 204-211 (1888); S. L. Wiegand and others, *J. Franklin Inst.* **126,** 67-69 (1888); Edgar F. Smith, "James Curtis Booth, Chemist, 1810-1888," (1922), a pamphlet reprinted in *J. Chem. Educ.* **20,** 315-318 (1943); W. D. Miles, *Chymia* **11,** 139-149 (1966); *Pop. Sci. Month.* **40,** 116-123 (1892), with portrait.

WYNDHAM D. MILES

Henry Bower

1833-1896

In the 1850's Bower founded a chemical company in Philadelphia which was to last over a century. Based originally on the recovery of chemicals from the wastes of other industrial processes, his company came to restrict itself to a very small line of high-grade ammonias, prussiates, and dichromates; it was in this form that the company was finally absorbed by Diamond Shamrock Chemical Co. in 1969.

Bower was born in Philadelphia in 1833, son of William Bower, and graduated from Philadelphia College of Pharmacy and Science in 1854. He acted as a chemical broker for a time, but by 1856, at the age of 23, he moved into chemical manufacturing. In 1858 he leased a portion of the land upon which he would later build his chemical works at 29th St. and Gray's Ferry Road, near the Schuylkill River in southwest Philadelphia—

an address unchanged for the lifetime of the firm. Here he set up the first of his recovery projects—the manufacture of ammonium sulfate from ammonia in the effluent of Philadelphia Gas Works. The need for a domestic source of ammonium sulfate of reliable quality made Bower's operation a success from the start. By 1866 he was combining this with other salts to market a "complete manure containing superphosphate of lime, ammonia, and potash," one of the first such mixed fertilizers in this country. The fertilizer venture was dropped after 10 years, however.

Bower's second triumph of by-product utilization was the production of a "pure inodorous glycerine," for which achievement he was awarded the Franklin Institute's Elliott Cresson Medal in 1878. As a drug clerk during his pharmacy school days, Bower had noted that the glycerol available from decomposition of olive oil was "a fetid preparation . . . its use . . . restricted by its quality and exorbitant price." Bower found a way to purify the tons of glycerol that were being poured away in the wastes of soap and stearin candle-makers and was soon recovering glycerol from effluents shipped from all over the country.

In 1867 Bower began manufacture of yellow prussiate of potash, or potassium ferrocyanide, again from by-products: potash from wood ashes, and cyanide (as cyanogen) from animal scrap (leather clippings, waste hoofs and horns). Later, cyanogen was recovered from gas works effluent as well. In 1882 he entered into manufacture of potassium dichromate, forming the Kalion Chemical Co. with Harrison Brothers of Philadelphia. Bower also produced ammonia: aqua–ammonia of strength as great as 15% in 1890 and anhydrous ammonia by 1903. In 1893 he produced stannic chloride by treating waste tin clippings with elemental chlorine generated in what was apparently the first electrolytic chlorine plant in the country.

Bower was concerned with the chemical industry beyond his own manufacturing operations. His firm was one of the early members of the Manufacturing Chemists Association, and he persuaded the U. S. Census Bureau, in the 1880 and 1890 censuses, to begin reporting statistics of the chemical industry as a separate branch of manufacturing. His advice was sought in his last years by federal legislators in drafting tariff regulations.

Bower was married to Lucretia Kirk Elliott, of Philadelphia, who bore him three sons and a daughter. Two of the sons, George Rosengarten (named after Bower's uncle, a Philadelphia chemical manufacturer) and William H., came into the firm in 1885 and 1886 respectively, after graduating from University of Pennsylvania. The depression of 1887 caused serious reverses; the ammonia operations were taken over for a time by another firm, and money was borrowed from William Weightman of yet another Philadelphia chemical company, Weightman and Powers. On Mar. 26, 1896 Henry Bower died, and William and George, now joined by their brother Frank, a mechanical engineer, and their brother-in-law Sydney Thayer, continued the business. By 1906 a number of companies were merged to form Henry Bower Chemical Manufacturing Co., with George as president. In this form, and always with a Bower in some executive position, the firm continued to produce ammonia, dichromates, and ferrocyanides until its absorption by Diamond Shamrock in 1969.

J. M. Wilson, *J. Franklin Inst.* **141,** 387-88 (1896); for history of the company, see W. Haynes, "American Chemical Industry," **VI,** 57-60, 287-90 (1949), as well as scattered references, **II;** and W. Haynes and E. L. Gordy, "Chemical Industry's Contribution to the Nation: 1635-1935," *Chem. Industries,* Supplement, 111-12 (May 1935); Biographies of George R. Bower can be found in "Men of America," 198 (1910); and "National Cyclopaedia of American Biography," **26,** 216-17, James T. White Co. (1937).

ROBERT M. HAWTHORNE JR.

Martin Hans Boyé

1812-1909

Boyé was born in Copenhagen, Denmark, Dec. 6, 1812, son of Mark Boyé, chemist-superintendent of the Royal Porcelain Manufactory and of a large pharmaceutical works. After graduating from University of Copenhagen in 1832, he studied at the Polytechnic School, where his teachers included Oersted, Zeiss, and Forchhammer, from which he graduated with honors in 1835. He emigrated to America in 1836 to follow a career in chemistry. The next year he moved to Philadelphia where he became associated with Robert Hare as laboratory assistant and as auditor of Hare's lectures on chemistry at the School of Medicine of University of Pennsylvania.

With Henry D. Rogers, professor of geology and mineralogy at the same university, he toured the anthracite coal regions of Pennsylvania as assistant geologist and chemist in the first geological survey of the state and analyzed magnetic iron ore and other minerals. Under Robert E. Rogers and James B. Rogers, he further developed his talents in mineral analysis and originated a quantitative separation of calcium from magnesium —by adding dilute sulfuric acid in excess and 40–41% alcohol by volume.

At about that time, he and Henry D. Rogers prepared the compound nitrosoplatinic chloride. This accomplishment led to his election to the American Philosophical Society. With Clark Hare, Robert Hare's son, he prepared the first of the violent explosives, ethyl perchlorate, a substance many times more powerful than gunpowder and manageable only when diluted with alcohol. The discovery of this substance marked him a pioneer in the field of smokeless gunpowder. [Beilstein erroneously ascribed this discovery to Sir Henry E. Roscoe.]

The same year (1840) saw the rise of the American Association of Geologists, organized in Philadelphia by a small number of scientists, one of whom was Boyé. Out of this society grew, in 1848, the American Association for the Advancement of Science. Boyé was a very active member of both societies.

From 1842 until 1844, Boyé attended lectures in the School of Medicine at University of Pennsylvania, where instruction took place from 4 PM until late in the evening. During the day he collaborated with James Curtis Booth in their laboratory for mineral analysis and research. They studied the mineralogy and ores of Pennsylvania and explored other phases of chemistry, biology, and even the aurora borealis. At the 100th anniversary of the American Philosophical Society he read a paper on the conversion of benzoic into hippuric acid. He received his M.D. degree in 1844 but never practiced medicine.

In 1845 Boyé perfected the refining of the viscid, black oil obtained from cottonseed into a bland, colorless cooking and salad oil, which, in addition, yielded a fine soap—equal or superior to the best castile. A sample of this oil, made in 1848, was still of such high quality many years later that it and a sample of recent manufacture were awarded a first premium at the 1876 Centennial Exposition at Philadelphia.

When James Booth retired as teacher of chemistry from Philadelphia's Central High School in 1845, Boyé succeeded to the post and taught there until 1859 as an earnest and enthusiastic educator. One of his students, Thomas M. Drown, later president of Lehigh University, remembered Boyé as the person who inspired in him a love of chemistry.

Boyé published two texts for his students, "Pneumatics or the Physics of Gases," and "Chemistry, or the Physics of Atoms." His zealousness in grounding his students in the fundamentals and principles of science stimulated many of them to become successful chemists and teachers of chemistry.

Boyé translated into English many literary essays and chemical publications, including Berzelius' book, "The Kidney and Urine" in 1843. The University of Pennsylvania conferred upon him an honorary degree of Master of Arts in the same year he earned his M.D. degree. He was author of many of the articles in Booth's "Encyclopoedia of Chemistry," one of which, the lengthy article "Analysis," was eventually published as an excellent, independent volume.

After a most fruitful scientific life, he passed away on Mar. 6, 1909 in Coopersburg, Pa., at the age of 97 years.

Edgar F. Smith, "Martin Hans Boyé, 1812-1909, Chemist," reprinted in *J. Chem. Educ.*, **21,** 7-11 (1944), with portrait; *Science* (March 19, 1909); W. H. Hale, "A Pioneer in Science," *Sci. Amer.*, **75,** 430 (1896).

HAROLD J. ABRAHAMS

Frederick Ernest Breithut

1880-1962

Breithut, dynamic leader of a profession, was born in New York, Aug. 15, 1880, the son of Frederick and Mary (Neuser) Breithut. He attended local schools, including College of the City of New York, from which he received his A.B. degree in 1900.

For a few years he was a journalist, working with the famous editor, Arthur Brisbane, on the *New York Evening Journal.* In 1903 he returned to City College as an instructor in chemistry and worked his way up to a professorship. In 1909 he received his Sc.D. degree from New York University. During this time he also lectured for the Board of Education and served as chemist for the State Factory Commission.

World War I was a serious interruption. Breithut joined the Chemical Warfare Service in 1917 with the rank of major. The next year he became chairman of the chemical division of the War Trade Board; also director of the Bureau of Conservation of the Federal Food Board, and chairman of Personnel of the Chemical Warfare Service (census of available chemists). After the war he was retained as Chief of Salvage and Sales for the chemical industry. His last service for the government at this time was in 1923, when he was sent to Germany as Chemical Trade Commissioner for the U.S. Department of Commerce. The mission was most successful, as everyone soon came to understand that "Breithut knows his stuff."

He was happy to return to teaching and remained at City College until 1930 when Brooklyn College was founded and he moved across the bay to become head of the Department of Chemistry. His many outside contacts had made him interested in far more than straight chemistry, so he offered a course in chemical economics; and, at the request of the Salesmen's Association of the American Chemical Industry, a course on chemistry for salesmen.

As consultant for the Board of Health and the Bureau of Municipal Research in New York, Breithut became acutely conscious of the sad plight of the municipal chemists—rated about with the yard cleaners; and through this he reached the American Institute of Chemists. In 1927, after a year of quiet observation, he was elected chairman of the New York Chapter and immediately started local action for better recognition of, and much better pay for, city chemists. In 1928 he was elected president of the AIC.

Almost his first official act was to create a printed publication for the members to succeed the mimeographed bulletins that had been issued for the first 5 years. This became *The Chemist,* named and first edited by Albert P. Sachs. On awarding the medal of the Institute, he felt that as so many medals are available for deserving chemists, some should go also to non-chemists who help chemistry and chemists. Accordingly, he proposed that the AIC medal for 1929 should go to Mr. and Mrs. Francis P. Garvan, who had established The Chemical Foundation. In 1930 it went to George Eastman, inventor and industrialist; and in 1931, to Andrew W. and Richard B. Mellon, for establishing Mellon Institute.

As president for two terms, Breithut brought the AIC through a most trying but constructive period. He was a brilliant speaker and was always willing to talk to professional societies and to lay audiences about the importance of chemistry and chemists. One of his pet remarks was: "Other societies aim to make a chemist out of a human being; the AIC aims to make a human being out of a chemist." He wrote several papers and booklets on chemists in public service, on engineering, and one that was very special, on "Training for Municipal Work." And he helped considerably on the educational programs of the Chemical Expositions.

Breithut had a delightful personality. Anyone would be quite cheered just to hear his hearty laugh. He was a trustee of The Chemists Club, a member of ACS Chemistry Teachers Club, Society of Chemical Industry, and a fellow of AAAS. He enjoyed golf, was fluent in French and German, and played the piano extremely well.

He retired from teaching in 1938 and moved to California. He died on May 13, 1962, in San Diego, survived by Richard C., son of his first marriage to Edith Commander Breithut.

As a young man Breithut wrote the following philosophy; "The trouble with the scientists is that they pursue their own line of work and seem to forget everybody else. They think that all work outside of science is baby work. I don't believe in science for science's sake, art for art's sake, or literature for its own sake. I believe in all these things for humanity's sake." Breithut's own life and work well fulfilled the implied prophecy.

Personal recollections; *Chemist,* Feb. 1932; obituary in *New York Times* May 15, 1962.

FLORENCE E. WALL

Jonathan Brewster

1593-1659

Jonathan Brewster, an American alchemist of the seventeenth century, was the son of William Brewster, a noted elder in the Pilgrim Church and a leader of the Pilgrims who sailed to America on the *Mayflower.* Jonathan was born in England around 1593 and came to America in 1621. Practically nothing is known of his life other than he was a trader in Connecticut when that state was wilderness, covered by forest, and inhabited by Indians and the first English settlers. For a time he had charge of the Plymouth trading post on the Connecticut River after that post was established in 1633.

He became interested in alchemy, perhaps in Europe before he came to America. He borrowed books on alchemy from John Winthrop, Jr., governor of Connecticut. Presumably with ordinary utensils and native materials, unless he went to the considerable expense and trouble of having materials shipped across the Atlantic, he experimented to find the philosopher's stone. Nothing is known of his experiments or speculations other than his interpretation of obscure passages in alchemical tomes which he remarked upon in letters to Winthrop.

Brewster died in New England around 1659.

R. S. Wilkinson quotes alchemical passages from Brewster's letters to Winthrop on pp. 46-48 of "The alchemical library of John Winthrop, Jr. . . ." *Ambix* **11** (1963).

WYNDHAM D. MILES

Robert Bridges

1806-1882

Bridges, a teacher of chemistry, was born in Philadelphia, Mar. 5, 1806. He attended University of Pennsylvania, received his bachelor's degree from Dickinson in 1824 and his M.D. degree from Pennsylvania in 1828. Thereafter he practiced medicine in Philadelphia for several years but did not build a lucrative practice, probably because he spent too much time studying science rather than concentrating on medicine.

Bridges began to teach chemistry in the 1820's as assistant to Franklin Bache in Thomas Hewson's private medical school. He continued to assist Bache at Franklin Institute from 1826 to 1831, at Philadelphia College of Pharmacy and Science from 1831 to 1841, and at Jefferson Medical College from 1841 to 1864. He was professor of chemistry at the College of Pharmacy from 1842 to 1879, at Philadelphia Association for Medical Instruction from 1842 to 1860, and at Franklin Medical College from 1846 to 1848.

Bridges served on committees for revisions of the 1840, 1860, and 1870 editions of "United States Pharmacopeia," was an assistant editor of *American Journal of Pharmacy* from 1839 to 1845, and an assistant to George B. Wood in bringing out the twelfth (1865), the thirteenth (1870), and the fourteenth (1877) editions of "United States Dispensatory."

Bridges edited American editions of two well-known British texts: Thomas Graham's "Elements of Chemistry" in 1853 and 1858, and George Fownes' "Elementary Chemistry" in 1847, 1850, 1853, 1855, 1862, 1870, 1871, and 1878. His editions of Fownes' text were widely used, particularly in medical and pharmaceutical schools.

Bridges published a number of articles, but he was noteworthy as an amiable, efficient teacher and a learned, industrious compiler, rather than a researcher. He died in Philadelphia, Feb. 20, 1882.

Obituary by W. S. W. Ruschenberger, *Amer. J. Pharm. Allied Sci.* **56,** 241-251 (1884), and *Proc. Amer. Philos. Soc.* **21,** 427-447 (1883-84); "Dictionary of American Biography" **3,** 35-36, Chas. Scribner's Sons (1928); Joseph W. England, ed., "First Century of Philadelphia College of Pharmacy, 1821-1921" 119, 401-2, Phila. Col. of Pharm. and Sci. (1922), with portrait.

WYNDHAM D. MILES

Edgar Clay Britton

1891-1962

Britton was born Oct. 25, 1891 in Rockville, Ind. and died July 31, 1962 in Midland, Mich. He attended Indiana State Normal in Terre Haute, Ind., and taught grade school at Diamond and Montezuma, Ind. before entering Wabash College, where he planned to major in law. It was fortunate for chemistry that an upperclassman at Wabash College persuaded Britton to enroll in a chemistry course taught by a professor who was a campus favorite. This professor, in a single interview, convinced Britton that chemistry should be his life work.

In 1914 Britton entered University of Michigan, where he earned his A.B. degree in 1915 and Ph.D. in organic chemistry in 1918. He later was awarded an honorary Sc.D. degree from University of Michigan in 1952 and one from Wabash College in 1955.

Britton, like many students in his time, was responsible for much of his finances during college. He taught, waited on tables, clerked in a men's clothing store, and worked during summer vacations for his father, a builder of covered bridges. In 1917, during the summer months, he was employed as an analytical chemist in the U.S. Public Health Service. He was an instructor in organic chemistry at University of Michigan from 1918 to 1920. He joined The Dow Chemical Co. in 1920 as an organic chemist, working for William J. Hale. Britton became director of the Organic Research Laboratory in 1932, renamed the Edgar C. Britton Research Laboratory in 1953—the first laboratory which Dow named in honor of an employee. He was vice-chairman of Dow's Executive Research from 1949 to 1952 and played a very important part in organizing the General Research Committee in 1948, serving as its chairman the first year. He was a director of Dow Corning Corp. from its organization in 1943 until his death and served for 18 years as secretary of that corporation.

Britton's research contributions covered a broad spectrum as evidenced by over 300 patents bearing his name. His early work was on the synthesis of phenol from halogenated benzene. The successful commercial production of phenol from chlorobenzene provided a vital raw material for making many advances in agricultural and plastic products. The Dowicide preservatives and dinitrophenols were developed in Britton's laboratory. He became interested in amino acids and developed processes for synthesizing seven of the essential ones; one, methionine, was produced commercially. He felt that amino acids would some day supplement many of the foodstuffs for man. He pioneered production of cellulose methyl and ethyl ethers, ethylene oxide, glycols, polyglycols and glycol ethers. The first tank car of butadiene made in the U.S. was manufactured by a process from his laboratory.

The calcium-nickel phosphate catalyst he developed was widely used in preparing butadiene. He worked with plastics, pharmaceuticals, and a variety of organic chemicals and played a major part in the commercial production of silicone resins. His syntheses of compounds were so numerous that one cannot do justice to all of them.

Britton was known to his friends at University of Michigan as "Hec," and later at Dow everyone knew him as "Doc." He had a remarkable sense of humor and could come up with a suitable story for any occasion. He had an excellent bass voice and enjoyed singing hymns in the Presbyterian choir as well as some lumberjack ballads in a deer hunting camp. He liked people and had many friends. His hobbies included fishing, hunting, golf, and other forms of outdoor life. He was an accomplished woodworker. He served as president of the American Chemical Society during 1952.

Britton was a chemist's chemist. He believed in man and felt that results of a chemist's work should be useful to mankind at some future time. He believed in 100% accountability of reaction products. He had an astounding memory and seemed to know Beilstein almost by heart. His interest in chemistry at the bench persisted up to his death; after retirement he could be found working with a test tube trying to crystallize a new compound prepared by him or one of his co-workers.

When he received the Perkin Medal in 1956, he gave an address entitled "Journeys in Research" in which, typically, he gave the credit to others when he said, "I accept it in behalf of all those with whom I have worked in chemistry."

Obituary by R. P. Perkins: *Chem. Ind.* 2066-2067 (1962), with portrait; Journeys in Research, a talk by Britton; *Chem. Eng. News* **34,** 4638 (Sep. 24, 1956); *The Brinewell,* a Dow publication, August 1956; personal recollections.

EZRA MONROE

Leslie George Scott Brooker

1902-1971

Brooker was an expert in the chemistry of photographic sensitizing dyes. He synthesized cyanine dyes and interpreted the relationship between their color and structure. One of his early cyanine dyes sensitized photographic emulsions to the entire visible spectrum and contributed to the development of Kodachrome film by sensitizing the three separate emulsion layers used in the color film. Other of his dyes sensitized emulsions to red and infra red radiation and thus contributed to the development of astronomical photography and to aerial photography for ecological studies and military uses.

Brooker's interests extended beyond the photographic applications of the compounds that he prepared. At his suggestion, in the 1940's, several cyanine dyes were screened for use in treating disease. One proved to be efficacious against pinworms in humans and was marketed in 1956. Another was found to be useful in studies of blood circulation.

Brooker was born in Gravesend, County Kent, England, Mar. 12, 1902. He was educated at King's College, London, which awarded him his B.Sc. degree in 1923, and at University of London, where he earned his Ph.D. degree in organic chemistry in 1926. After graduation he came to the United States to join Eastman Kodak Research Laboratory in Rochester, N.Y., where he spent his career. He was naturalized in 1942 and at the time of his retirement in 1968 was senior research associate in charge of the sensitizer section of the emulsion research division of the laboratories.

Brooker received more than 140 U.S. patents (as well as Canadian and European equivalent patents) and published outstanding articles in his field. He lectured extensively. He was awarded the Modern Pioneers Award of the National Association of Manufacturers in 1940 and the Henderson Medal of the Royal Photographic Society of Great Britain in 1948.

Brooker died at Rochester, N.Y., Dec. 22, 1971.

Information from Brooker's writings, the archives of Eastman Kodak Co., and from obituaries.

THOMAS T. HILL

Benjamin Talbott Brooks

1885-1962

Brooks, one of the country's best known independent authorities on petroleum, was born in Columbus, Ohio, Dec. 29, 1885, the son of Nathaniel W. and Rae (Saunders) Brooks, and the grandson of the first mayor of Columbus. He attended local schools and graduated with his B.A. degree from Ohio State University in 1906.

After one year as chemist with the Bureau of Standards, he went to the Philippine Bureau of Science, where he remained for 4 years, working principally on essential oils from flowers. His study of champaca is still considered definitive on this oil. While there he married the librarian, Sarah Osgood of Springfield, Mass. They left in 1911, via Europe, and he earned his Ph.D. degree at Göttingen in 1912.

Back in the United States, Brooks worked from 1912 to 1917 at Mellon Institute on petroleum research; and for four of those years he doubled as professor of chemical engineering at University of Pittsburgh. His work was interrupted by military service at the Mexican border in 1916 and later during the war as major in the Chemical Warfare Service. He remained in the reserves for several years.

During a year with Commercial Research, Inc., of New York (1917-1918), Brooks developed processes for the commercial production of ethylene, propylene, their glycols, and other derivatives. After a brief period as a consultant, during which time he was chairman of the committee that organized the American Petroleum Institute (1919), he was chemical engineer with Mathieson Alkali Works at Niagara Falls and then reestablished himself as a consultant.

In 1924 when Brooks was chairman of the Committee for Petroleum Chemistry of the National Research Council, he was commissioned by the American Petroleum Institute to prepare a comprehensive report on the state of the industry. This was duly presented in 1926, but it was ignored until Charles H. Herty, during a meeting of the American Chemical Society upbraided the industry for its inertia and goaded it into a realization of the full potentialities of petroleum as the basis of a whole chemical industry. There had been scattered chemical research, but most of the companies were still content with the distillation of petroleum into salable fractions. Brooks helped in many ways to further the developments that followed.

The first of his considerable and worthy literary output was "The Chemistry of the Non-Benzenoid Hydrocarbons," published in 1922. He was the author also of over 100 technical articles, many on the origins and history of petroleum, and on warnings about dwindling resources, including "Peace, Plenty, and Petroleum" (1944). He was American editor of "Science of Petroleum," four volumes of which were published in 1937 and three supplementary volumes in 1950-52. In 1954-55 he was co-editor of the 3-volume "Chemistry of the Petroleum Hydrocarbons," which was translated into Russian in 1958-59. He was an advisory editor of *Chemical Industries.*

In 1940-41, Brooks was consultant on petroleum to the government of Venezuela, and in 1942 he served on the Petroleum Commission of Mexico. His last major enterprise was as the vice president of Glycerine Corp. of America and Glycerine Corp. of Cuba (1948). He was a consultant for many large companies and was frequently called for expert testimony.

Ben Brooks was an outgoing, pleasant person, with many interests to help him "keep a growing edge." He loved good music, especially opera, and had a wonderful collection of records. His early tennis was dropped for golf. He was a good dancer and an expert at bridge. He was chairman of the American Chemical Society's New York Section in 1926. He was a devoted member of The Chemists Club of New York, for which he served a term as library chairman and which made him an honorary member.

Brooks died at Madison, Wis., Aug. 5, 1962, survived by his wife, a daughter, Mrs. Joseph J. Hickey, and three sons: William B., Benjamin T., and Robert O. Brooks.

Personal recollections; files of the American Chemical Society; obituaries in *J. Inst. Petrol.* **49,** 26-27 (1963) and *New York Times* Aug. 7, 1962.

FLORENCE E. WALL

George Granger Brown

1896-1957

Brown was born Sept. 3, 1896 in New York City, the son of George C. and Emma L. (Tuttle) Brown. He prepared for college at Erasmus High School in Brooklyn and received his bachelor of science degree in chemical engineering from New York University in 1917. He received the degree of chemical engineer from New York University in 1924.

After graduation from college, George worked for Aluminum Co. of America for 1 year, served in the Chemical Warfare Service in 1918, and was production manager for Union Special Machine Co. in Chicago in

1919 and 1920. He joined the University of Michigan faculty as an instructor of chemical engineering in 1920 and started work toward his doctors degree. He was awarded his Ph.D. degree by Michigan in 1924. He remained at the university and was promoted to professor of chemical engineering in 1930.

Brown was appointed chairman of the Department of Chemical and Metallurgical Engineering in 1942 and was appointed Edward deMille Campbell Distinguished Professor of Chemical Engineering in 1947. He was one of eight professors honored when the Board of Regents approved the establishment of distinguished professorships in 1947. He was appointed dean of the College of Engineering in 1951. He also served as chairman of the Engineering Research Council at University of Michigan. His leadership of the College of Engineering marked him as one of the nation's outstanding educational administrators. He instituted educational programs designed to give engineering students greater understanding of the basic sciences to prepare them better to meet increasingly complex problems of modern technology.

Brown's major work involved the application of chemical engineering and thermodynamics to fractional distillation, petroleum, gaseous explosives, and combustion. He investigated the uses of the clay and sand resources of Michigan. He was the author or coauthor of over 140 technical papers in his fields of specialization.

While Brown was at Michigan he also served as director of research for the Natural Gasoline Association of America from 1926 to 1935 and was director of research for National Dairy Products, Inc. from 1945 to 1948. During 1950 he was director of the United States Atomic Energy Commission's Division of Engineering, with responsibility for the chemical engineering phases of the Commission's reactor development program. This involved the establishment of reactors at the Idaho Reactor Testing Station, the development of processes for the recovery of uranium from depleted nuclear fuels, and the treatment of radioactive wastes. He was a consultant to many industrial firms throughout America.

Brown received the William H. Walker Award of the American Institute of Chemical Engineers in 1939 in recognition of his outstanding contributions to chemical engineering and the Hanlon Award of the Natural Gasoline Association of America in 1940 for meritorious service to the natural gasoline industry. He was president of the American Institute of Chemical Engineers in 1944 and chairman of the Institute's educational committee from 1948 to 1954. He also served the Institute as a director and as its treasurer. He was regional chairman of the Engineer's Council for Professional Development.

Brown married Dorothy B. Martin on Dec. 1, 1917. They had three sons, George Martin, Judson Granger, and David Malcolm.

Brown died Aug. 27, 1957 at University Hospital, Ann Arbor, Mich. at the age of 60.

Alumni records of University of Michigan; information from American Institute of Chemical Engineers.

LESTER KIEFT

Arthur Lee Browne

1867-1933

Browne was born in Baltimore County, Md., May 10, 1867. His father was a distinguished scholar and member of the faculty of Johns Hopkins University. The son graduated with an A.B. degree at Johns Hopkins in 1888 and subsequently took special work in chemistry under Ira Remsen. Later he studied medicine at University of Maryland, receiving his M.D. degree in 1906. He was professor of chemistry and toxicology at Baltimore Medical College from 1907 to 1912.

Following his earlier education he was chemist for American Leather Company for a brief period. Then, in 1893, with William B. D. Peniman he organized in Baltimore a firm of consulting chemists and metallurgists which continued until Browne's death. Their outstanding work was with petroleum in which field they devised two systems of distilling and cracking petroleum. His work as an analytical chemist is recorded in the publications of the U.S. Geological Survey and the Maryland and Virginia state surveys.

The examination and compilation of statistics concerning the water supply of Maryland as part of the general work done by the government was made under his direction. He also assisted in developing commercial alloys, various methods of regulating boiler water and water plants, and a bottle seal.

Among the firms that Browne served as a consulting chemist were United Railways and Electric Co., Consolidated Gas Electric Light and Power Co., Standard Lime and Stone Co., Linen Thread Co., Baltimore Copper Paint Co., Samuel Kirk and Sons, Hurricane Zinc Co., and National Lead Co.

Browne died at the age of 66 years at his home in Anne Arundel County, Md., June 17, 1933.

"National Cyclopaedia of American Biography," with portrait.

DOROTHY B. DEAL

Charles Albert Browne

1870-1947

Into the family of Charles Albert and Susan (McCallum) Browne, a son, Charles Albert, was born Aug. 20, 1870 in North Adams, Mass. The son was the eldest of five children in this family.

The father, of colonial ancestry, was a farmer, an inventor and later a manufacturer. He became a prominent explosives chemist, notable for his invention and use of a new type of explosive fuse. The elder Browne experimented with explosives, manufactured some, and was granted several patents. He directed blasting projects for railroads and tunnels. In 1869 he lost his eyesight when some copper fulminate exploded; consequently, he was never able to see his children. His memory, inventive skills, and numerous suggestions were of untold help to his family and associates.

With such a background of a colonial heritage in an agricultural and scientific environment, one need not wonder why the son, Charles Albert, later became an outstanding agricultural chemist, historian, and scientist. Browne pioneered in the field of sugar chemistry and its technology and contributed extensively to the cultural and humanistic aspects of chemistry.

Charles Albert, Jr. was granted his A.B. degree from Williams College in 1892 and his A.M. in 1896. While at Williams, in addition to his studies in the sciences, he became keenly interested in the classical and Germanic languages, which served him well in his later travels and research.

His first position, after graduation in 1892, was as an analytical chemist in the John Sabin Laboratory in New York City. Two years later Browne was appointed an instructor in chemistry at Pennsylvania State College (now Pennsylvania State University) and the following year became an assistant chemist in its Agricultural Experimental Station.

In 1900 Browne decided to attend University of Göttingen to work with Bernhard Tollens in the fields of chemistry, agriculture, plant physiology, and physics, but primarily to specialize in sugar research. The latter remained his chief interest throughout his life. The university granted him his Ph.D. degree in 1902.

Upon his return to America, he stayed briefly at Pennsylvania State. The next 4 years, he was a research chemist at the Louisiana Sugar Experiment Station, where his researches dealt with the chemistry and technology of sugar cane and rice and their products.

In 1906 Browne was elected the U.S. delegate to the Sixth International Congress of Applied Chemistry, held in Rome. On returning to the U.S., he was appointed chief of the Sugar Laboratory, U.S. Bureau of Chemistry. Here he became closely associated with the chief of the bureau, Harvey W. Wiley. For almost 2 years (1906-07) his many researches were concerned with American honeys, various sugars, and starch products.

In 1907 Browne organized and became the chief chemist in charge of N.Y. Sugar Trade Laboratory, Inc. This later became the largest sugar control laboratory in the world. His association with the Sugar Laboratory from 1907 to 1923, his work, skills, and reputation, required that he make many visits to sugar-producing countries. This organization conducted extensive research upon methods of sugar analysis, deterioration of raw sugar, and finding means for its preservation. It was during this period that he began the collection of a library on the history of chemistry.

Browne was assistant editor of *Chemical Abstracts,* 1907-10; assistant editor of *Journal of the American Chemical Society,* 1911-12; assistant editor of *Industrial and Engineering Chemistry,* 1912-16; chairman of the American Chemical Society's Division of Sugar Chemistry, 1919-21; chairman of the Division of History, 1922-23; and editor of the ACS Golden Jubilee volume, "A Half Century of Chemistry in America," (1926).

On Sept. 7, 1920, Browne and Edgar Fahs Smith, attending an ACS convention, chatted under the trees along the lake front at Evanston, Ill. Both were graduates of the University of Göttingen, had a mutual interest in historical chemistry, and in their travels sought to acquire early chemical works, letters, and manuscripts for their personal libraries, Together they brought about the first meeting and organization of the History of Chemistry Division on Apr. 27, 1921.

Fifteen years after leaving the Department of Agriculture, Browne returned in 1923 as chief of the Bureau of Chemistry. From 1927 to 1938 it was known as the Bureau of Chemistry and Soils, and from 1938 to 1943 as the Bureau of Agricultural Chemistry and Engineering. As his years of service continued, much of the administrative work was delegated to others so that he could continue to guide research. Browne retired in 1940.

From 1940 until his death Browne spent his time in study, historical researches, writing, and publishing results of his work—mostly in agricultural chemistry and in the history of chemistry in America. His keen interest in the Edgar Fahs Smith Memorial Library in History of Chemistry at University of Pennsylvania led him to donate, in memory of his parents, some 450 items consisting of books, letters, manuscripts, engravings, and prints—all relating to the history and technology of chemistry. Browne kept detailed diaries of his travels, personalities he met, conversations held, and events of each day. Some of these notes, information, and comments or notes from his experiences were carefully edited and recorded either on slips of paper or in margins and placed at the appropriate places in the volumes which were presented to the collection.

The University appointed Browne as chairman of an Advisory Committee to the Smith Library. He was named editor-in-chief of *Chymia,* an annual publication, sponsored by the University with the objective of publishing scholarly historical research in chemistry. Ill health forced his withdrawal. His death Feb. 3, 1947 occurred before the first volume of *Chymia* was published.

Browne's scientific publications number in the several hundreds. His best known publications dealt with sugar chemistry. Some of his other interests were plant nutrition, presence and effects of trace elements in plants, studies of the action of enzymes, and observations of the loss of sugar content (as a food) in plants when cut and dried. In 1912 Browne published a "Handbook of Sugar Analysis" which appeared in several later editions.

As a historian of chemistry, Browne was equally eminent. Over 100 publications appeared from his pen. His influences and outstanding contributions extended almost three decades. He also wrote many poems and translated others from Latin and Greek. His most highly regarded historical work, published in 1944, was "Source Book of Agricultural Chemistry." Browne had written a major part of "A History of the American Chemical Society" before his death. Mary Elvira Weeks edited and completed it, and it was published in 1952. Many book reviews, biographical articles, and obituaries of well-known chemists also came from his desk.

Browne was a charming and delightful person with tremendous energy and a keen sense of humor. He was a precise and very methodical worker. His wisdom and powers of reasoning were always evident in his discussions, administrative duties, and his writings.

He received honorary degrees of Sc.D. from Williams and Stevens Institute of Technology in 1924. Numerous citations, silver, and gold medals were granted in recognition of his achievements.

Charles married Louise McDanell in 1918. Louise was born in Gallitin County, Ky., Feb. 16, 1883. She obtained her B.S. degree from University of Nashville in 1902, her A.B. from Stanford in 1906, master's degree from University of California in 1912, and Ph.D. in physiological chemistry from Yale in 1917. She taught in California and Georgia high schools, was assistant professor of home economics at State College of Washington from 1912 to 1913, at University of Minnesota from 1913 to 1915, and assistant professor at Goucher from 1917 to 1918. After marrying Charles Browne, she maintained their home in Washington, D.C. The couple had a daughter, Caroline L. Browne.

Louise and her husband were world travelers, both for attending conferences and as tourists. She died in Washington, Feb. 15, 1963.

Diaries and books of Browne in Edgar Fahs Smith Memorial Library, University of Pennsylvania; Charles A. Browne and Mary E. Weeks, "History of the American Chemical Society," ACS (1952), pp. vii-x; personal recollections; biographies by Arnold K. Balls in *J. Assoc. Off. Agri. Chemists* **30,** No. 3, ii-vii (1947), by Claude K. Deischer in *Chymia* **1,** 11-24 (1948), and by Herbert S. Klickstein and Henry M. Leicester in *J. Chem. Educ.* **25,** 315-317, 343 (1948); obituary in *New York Times,* Feb. 4, 1947; obituary of Louise M. Browne in *Washington Post,* Feb. 16, 1963.

CLAUDE K. DEISCHER

Gershom Bulkeley

1636-1713

Gershom Bulkeley is one of the few known seventeenth century American chemists. Born about 1636, perhaps in Concord, Mass., he graduated from Harvard in 1655. During his life he was a clergyman, physician, justice of the peace, and politician. He served as a surgeon with Connecticut troops during King Philip's War and was wounded in an Indian attack near Princeton, Mass. It is said that he was "eminent for his skill in chemistry" and that "even to alchemy . . . he seems to have paid considerable attention." Nothing is known of his experiments. An epitaph published in Boston after he died in Glastonbury, Conn., Dec. 1 or 2, 1713, read in part:

A pure extract and quintessential wrought,
The Caput Mortuum is hereto brought.
Brave chymist death! how noble is thine art?
The spirits thus who from the lees canst part,
By sacred chymistry the spirit must
Ascend, and leave the sediment to dust.

John L. Sibley, "Biographical Sketches of Graduates of Harvard University," **I,** 389-402, Mass. Historical Society (1873), quotes and epitaph are from this book; biography by W. R. Steiner in *Johns Hopkins Hospital Bull.* **17,** 48-52 (1906); G. W. Russell, "Account of Early Medicine and Early Medical Men in Connecticut," *Proc. Connecticut Med. Soc.,* 1892, 89-109; Donald L. Jacobus, "History and Geneology of the Families of Old Fairfield," Eunice Dennie Burr Chapter of D.A.R., **1,** 110 (1930).

WYNDHAM D. MILES

Charles Frederick Burgess

1873-1945

Burgess pioneered in developing the program of chemical engineering at University of Wisconsin and was the founder of Burgess Battery Co., C. F. Burgess Laboratories, and several other companies. He was born in Oshkosh, Wis., Jan. 5, 1873 the first child of Frederick and Anna (Heckman) Burgess, both immigrants from Nova Scotia. The family was poor but hardworking, and their circumstances improved substantially by the time that Charles and his younger brother George reached college age. Their father, a handyman, became county sheriff for several terms and later was chief of police in Oshkosh. Charles and George, both clever with tools, entered University of Wisconsin in 1891 to study electrical and civil engineering, respectively. Charles became a close friend of Dugald C. Jackson, who later became head of the Electrical Engineering Department at Massachusetts Institute of Technology. Jackson was interested in the newly developing electrochemical industries. Following Burgess' graduation in 1895, Jackson arranged a faculty appointment for him in order to set up a course in applied electrochemistry.

Burgess began his new duties enthusiastically, accepting the challenge of creating equipment for an electrochemistry laboratory and creating courses without prototypes anywhere. In 1898 he was granted his E.E. degree in recognition of his original endeavors. Although he was interested in earning his Ph.D. degree, the German language requirement was an obstacle he never mastered. Nevertheless, he advanced rapidly in faculty rank, becoming professor in 1905, the year in which the course in applied electrochemistry was replaced by the course in chemical engineering, administered by a new Department of Chemical Engineering with Burgess as head. In the meantime, his penchant for practical tinkering led to the development of several pieces of apparatus and processes of industrial significance. When the commercialization of his inventions and his consulting activities seriously interfered with his university duties, he resigned his professorship in 1913.

As early as 1906 Burgess had become a consultant for French Battery and Carbon Co., a small Chicago firm making dry cells. It soon moved to Madison, and Burgess became deeply involved in improving its product and its operations. After founding Northern Chemical Engineering Laboratories in 1910, Burgess developed a cell for use in flashlights. It was produced by the Laboratories, which became C. F. Burgess Laboratories in 1915 and marketed by French Battery Co. and by another client, The American Carbon and Battery Co. of St. Louis.

Relations with French Battery Co. gradually deteriorated, and in 1917, at a time when wartime demand for dry cells was high, Burgess Battery Co. was founded as a production and marketing organization. In 1921 Burgess brought suit against French Battery Co. for unethical actions. The lengthy and bitter legal battle was decided in Burgess' favor in 1923. Besides his work on dry cells, Burgess invented a process for purifying iron and iron alloys by electrolytic means, developed a process for manufacturing electrolytic iron, and experimented with iron alloys suitable for use in heating elements and permanent magnets.

During the 1920's Burgess was becoming displeased with Wisconsin's taxation policies. Although the university made him an honorary Sc.D. in 1926, he moved the battery company and other operations to Freeport, Ill. and incorporated the Laboratories under the laws of Delaware. He took up residence in Florida but spent a large amount of time at his business sites. His enterprises continued to grow, and he spread them geographically, even into Canada and Europe. Research at the laboratories spawned new companies. The work of Arlie W. Schorger on cellulose derivatives led to formation of Burgess Cellulose Co. to produce stereotypes, sheet and fiber plastics, and sponges. Studies on zeolite led to Burgess Zeolite Co.

Burgess was hard driving and hard headed. A skilled technician, he had a strong insight into practical innovation coupled with a daring which kept him on the brink of financial disaster for many years. He was admired by many of his associates but was considered a shrewd manipulator by others. He could be charming in a social setting and was known for his loyalty to his family and his colleagues. He was inclined to involve his relatives in his business activities, and a number of his students held responsible positions in his companies. His life with Ida May Jackson, whom he married in 1905, was one of deep devotion. His relations with his children, Betty and Jack, were close despite the time occupied by his many commercial activities. He died in Chicago, Feb. 13, 1945.

Alexander McQueen, "A Romance in Research, The Life of Charles F. Burgess, Student-Teacher-Researcher-Industrialist," Instruments Pub. Co. (1951).

AARON J. IHDE

William Irving Burt

1893-1965

Burt was born in Granger, Ohio, Oct. 15, 1893. He attended Ohio State University from 1911 to 1913 and 1915 to 1917, graduating with a B.S. degree in chemical engineering.

Burt worked as a chemist for several companies, served as a private in the Chemical Warfare Service in 1918, was sales engineer for The Bristol Co., chemical engineer for The Dolomite Product Co., and later for Republic Iron & Steel Co.

In 1927 he joined The B. F. Goodrich Co. at Akron, Ohio, as research chemist and soon became manager of the Chemical Manufacturing Department. He was appointed superintendent of production of chemicals in 1942 and general manager of chemical plants in 1943. Shortly after the formation of The B. F. Goodrich Chemical Co. in 1944, Burt was named vice–president in charge of manufacturing and engineering. Goodrich-Gulf Chemicals, Inc., was incorporated in 1952; Burt served as president of that firm from January 1955 until March 1958, and then chairman of the board until he retired Nov. 1, 1958. From 1955 to 1958 he also served on the board and the executive committee of Neches Butane Products Co., Port Neches, Tex.

Many of the key personnel of B. F. Goodrich Chemical Co. in the production, sales, technical, and engineering departments received their training under Burt when he headed Plant 3 in Akron. He took many processes from the research laboratories and turned them into practical, efficient, production operations. His ability to foresee the engineering problems in transferring laboratory and pilot plant data into full scale production was demonstrated by the fact that all plants built under his direction commenced production on schedule and without difficulty.

He played an important part in building the synthetic rubber industry from the laboratory and pilot plant stage during World War II. Burt was appointed to the key position of chairman of Standard Plant Design Committee, and out of this group under his guidance came the basic design of the synthetic rubber plants which were put into operation successfully within 18 months.

During the war plants were built to produce 705,000 long tons of synthetic rubber annually, and Burt personally directed the construction of 255,000 long tons of this capacity. The B. F. Goodrich Co. was assigned the operation of 165,000 long tons of this capacity, and Burt had the direct responsibility for its operation. In addition, he trained a group of small rubber companies to take over The National Synthetic Co.—a 30,000 long tons per year plant. He also served on the Government Operating Committee for all GR-S plants for many years.

Under Burt's leadership as president of Goodrich-Gulf, facilities for the concentration of high solids latex was installed at the Port Neches plant, and other improvements made which increased the capacity for making rubber more than 20 percent. A steam generating plant was installed at Institute, W. Va., and one line converted to the continuous method for making "cold" rubber, giving that plant a substantial increase in capacity. Processes for black masterbatch and crumb rubber were also implemented during Burt's presidency. At Neches Butane Products Co. major construction was completed in 1958 to give added butadiene capacity.

Goodrich-Gulf's work in research and development to broaden the company's activities resulted in a low pressure polyethylene plant which went into production in 1961 and the construction of facilities to produce *cis*-polybutadiene. The successful duplication of natural rubber (*cis*-polyisoprene) was announced in 1954, and a pilot plant for further development work was built at Avon Lake in 1956. Burt deserved much credit for this petrochemical activity as he was an effective member of the negotiating team which completed the original license agreement with Professor Ziegler of Germany upon which much of Goodrich-Gulf's research was based.

Burt twice served as a director of the American Institute of Chemical Engineers, was vice–president in 1951 and president in 1952. He was also chairman of the Air Pollution Committee of Manufacturing Chemists' Association.

Burt died Nov. 5, 1965.

Records of The B. F. Goodrich Co.; personal recollections.

HARRY B. WARNER

William Meriam Burton

1865-1954

Burton, the son of Erasmus Darwin and Emeline Meriam Burton was born in Cleveland, Ohio, Nov. 17, 1865.

His preliminary education was received in public schools in his native city, and in 1866 he graduated with his B.A. degree at Western Reserve. Afterwards he did graduate work at Johns Hopkins, receiving his Ph.D. degree in 1889.

Shortly afterwards he started his long career with Standard Oil Co. as a chemist with the Ohio branch in Cleveland. The next year he was transferred to the same position in the Indiana branch at its recently organized refinery in Whiting, Ind. where he set up a two-room laboratory on the second floor of an old farmhouse. Later he served as assistant superintendent and in 1895 became superintendent of the refinery. In 1911 he was elected a director of the company and 4 years later a vice-president. In 1918 he was selected president of the Standard Oil Co. of Indiana, which position he held until his retirement in 1927. Thereafter he continued as director and adviser.

Burton's notable contributions to the petroleum industry included his pioneer work in demonstrating the value of laboratory research and testing, his recognition of the need for altering the methods of refining crude oil to produce gasoline for the automobile engine, and his development of the first commercially successful cracking process which more than doubled the potential yield of gasoline from crude oil. Before 1900 the supply of low-boiling fractions of crude oil was much greater than the demand and had limited commercial value. With the increased use of internal combustion engines, gasoline, one of the products of naphtha, a low-boiling fraction of petroleum, came into tremendous demand.

About 1909 Burton began to direct a program designed to find a means of increasing gasoline yield since at that time only about 18% of crude oil had the molecular size suitable for gasoline. His problem was to break up (crack) complicated hydrocarbon molecules of gas oil or fuel oil into fractions suitable for the internal combustion engine. He thus began some of the earliest experiments in an American industrial laboratory toward manufacturing petroleum products by chemical change rather than by simple physical separation. In 1912 Burton received U. S. Patent No. 1,049,667 for his cracking process which consisted of redistilling substances remaining after gasoline and other products had been removed by ordinary methods, the result being to "crack" the stock into its component parts including gasoline. This was achieved by a pressure of 95 pounds per square inch at a temperature of about 750 degrees Fahrenheit. Burton found it necessary to construct stills for this purpose which were also included in the patent. In January 1913 the first battery of twelve Burton stills went into operation, the number increasing to 240 by the end of the year. The Standard Oil Co. of Indiana was operating 500 stills in 1917 and 880 in 1927. During the first 15 years the Burton stills were in use they saved more than one billion barrels of crude oil which would have been required to make the same amount of gasoline using the old methods. During that period, the stills produced more than two hundred thousand barrels of gasoline. The fundamental ideas of Burton's invention were maintained as the original stills became obsolete and were improved. His process enabled oil refining companies to adjust their yield of products to the varying conditions of the market demand. In addition, Standard Oil licensed the Burton process to competitors at a royalty fee, thus beginning the practice of licensing inventions at reasonable royalty fees.

Wide recognition was accorded Burton for his contributions to the petroleum industry. He received the Willard Gibbs Gold Medal of the American Chemical Society in 1918, the Perkin Medal of the American Section of the Society of Chemical Industry in 1921, the Modern Pioneer Award of the National Association of Manufacturers in 1940, and the Gold Medal of the American Petroleum Institute in 1947.

Following his retirement Burton lived in Miami, Fla. where he died Dec. 29, 1954.

Biographical remarks by Charles H. Herty and R. Wiles, and address of acceptance of Perkin Medal by Burton in *Ind. Eng. Chem.* **14,** 159-163 (1922); "National Cyclopedia of American Biography," **C,** 243, James T. White & Co. (1924); obituary in *New York Times,* Dec. 29, 1954.

DOROTHY B. DEAL

Horace Greeley Byers

1872-1956

Byers, a teacher, chemist, and leader in agricultural research, was born near Pulaski, Pa. on Dec. 26, 1872. During his youth "times were hard and money scarce." However, he entered Westminster College, New Wilmington, Pa. After two years lack of funds forced him to leave, and he began teaching in country schools in Kansas. Later he returned to Westminster where, under the guidance of Prof. Thompson, he earned his A.B. and B.S. degrees with honors in 1895, and later received his A.M. degree in 1898. Byers taught chemistry, physics, zoology, botany, and German, at Tarkio College, 1895-1896, and was librarian and secretary to the faculty, at a salary of $600 a year.

He then entered Johns Hopkins University in 1897, earning expenses by instructing part-time at University of Maryland and working one summer as night chemist in a steel plant in Youngstown, Ohio.

In 1899 he received a scholarship and his Ph.D. degree. His thesis, with Harmon Morse, on permanganic acid, foreshadowed the hopcalite gas mask that converts carbon monoxide to dioxide.

That same year, Byers became head of the Chemistry Department of University of Washington. He introduced courses in chemical engineering and brought outstanding chemists to the faculty. His department was the first in the university to award the Ph.D. degree. Chemistry students increased from 75 to 1200.

Byers married Harriett Lane in 1902. They had four daughters.

For several summers, Byers taught chemistry elsewhere. At University of Chicago, he studied perchromic acid and its salts, publishing the results with E. Emmet Reid (1904).

Chemistry was then at its peak development in Europe. Study abroad was essential for an ambitious chemist. Byers had invested hard-earned savings in some corner lots in Seattle. He sold them in 1907 at tripled value to finance a sabbatical year with his family at University of Leipzig, in Germany. There he began a new field of study with Max LeBlanc on the passivity of metals. He visited the leading chemical laboratories in Europe and on his return to University of Washington, he helped to design and equip the new chemical building, Bagley Hall, dedicated in 1909.

In World War I, Byers volunteered for the 30th Regiment of Engineers, a chemical warfare regiment, but was transferred to the Chemical Warfare Service laboratory at American University, Washington, D.C. to take charge of the study of incendiaries. He remained in the Officers' Reserve Corps of the Chemical Warfare Service until his retirement in 1931 as lieutenant colonel.

In 1919 he became head of the Chemistry Department of Cooper Union, New York City. He also was a consultant to firms manufacturing paper, paint, and rubber. He was granted patents on catalysts, electrolytic reduction of carbon tetrachloride, and accelerators and antioxidants for rubber. One patent held jointly with Winfield Scott, claimed 2,4-diaminodiphenylamine as a rubber antioxidant and was soon adopted commercially.

In 1928 Byers joined the U.S. Department of Agriculture as head of the Division of Chemistry and Physics of Soils. He published 76 papers while in the department, at least 41 dealing with selenium. A mysterious sickness known as "alkali disease" had been decimating cattle in Western States. Byers and his associates, including William O. Robinson, Kenneth T. Williams, John T. Miller, and H. W. Lakin, identified the toxic agent as selenium, studied its distribution in soils of the U.S., Canada, and Mexico, and determined which plants concentrate toxic quantities of this element.

Byers wrote textbooks and laboratory manuals on inorganic chemistry and qualitative analysis in addition to a study on "Some Differences in Baking Powders" (1928). His numerous papers in technical journals included series on soil chemistry and on the passive state of metals. In poor health, he retired from government service in 1942 to his home near Arlington, Va.

As a teacher, Byers was highly esteemed by his former students. "None of his achievements outstrip his work as a teacher," Harlan L. Trumbull said. Byers' wife revealed that "his greatest hobby was helping boys and girls who were trying to educate and support themselves. He felt that . . . he was repaying Prof. Thompson of Westminster in some measure for his great kindness to a bewildered boy who was the beneficiary of his sympathetic wisdom and experience."

Byers was a founder and the first president (1923) of The American Institute of Chemists, which awarded him honorary membership in 1954, for "his lifetime of dedicated public service." He died Dec. 2, 1956 in Staunton, Va.

Biography by Harlan L. Trumbull in *The Chemist* (Dec. 1954); *The Chemist*, March (1954); autobiographical data by Byers in *The Washington Common Ion*, University Washington Chemistry Department (May 1916).

VERA KIMBALL CASTLES

C

Hamilton Perkins Cady

1874-1943

When "Hal" Cady was about 14, he watched a neighbor boy in Oberlin, Ohio dissolve a piece of broken silver in nitric acid and then add a solution of salt. Hal was so fascinated that he decided immediately to become a chemist. Before the next day had passed he had read much of a chemistry text that had belonged to his father and had gone far enough in his own "experiments" to have blown up a hydrogen generator. By mowing lawns for a dime he earned enough to buy an old copy of "Elements of Modern Chemistry" by Adolphe Wurtz. He later acquired books by Ira Remsen and by Albert Prescott. He built much of his equipment, including a balance, and carried out many experiments and analyses in his barn laboratory.

Cady was born May 2, 1874 in Skiddy, a village near Council Grove, Kans. When he was 11, his father died; his mother carried on, taking in boarders and giving music lessons, and was able to provide a good education to each of her five children. In 1894 Cady entered University of Kansas. He skipped freshman chemistry and started with quantitative analysis. By the end of his second year, he was put in charge of the course in qualitative analysis. After receiving his bachelor's degree in 1897 he attended Cornell for 2 years on a fellowship, where he worked with Wilder Bancroft. He returned to University of Kansas in 1900 to become assistant professor and married Stella C. Gallup.

For the next 3 years Cady taught and also worked with Edward C. Franklin on electrochemistry in liquid ammonia. In 1903 he received the third Ph.D. degree granted by Kansas University. He also took on the job left by Franklin, who moved to Stanford University. In 1905 he became associate professor and in 1911, professor. From 1920 to 1939 he served as chairman of the Department of Chemistry. He died May 26, 1943.

The studies for which Cady is best known are starting the American work on liquid ammonia as an ionizing solvent (1896) and discovering the existence of helium in natural gas (1905). In the first topic he was greatly encouraged and helped by Franklin. The project soon was expanded and involved the combined efforts of Franklin, Charles Kraus, and Cady. In the second project both Cady and his colleague, David F. McFarland, worked together.

Cady's most important contribution to society was that of a teacher. For years he taught the first semester course in freshman chemistry using his text, "General Chemistry." Through his excellent lectures, fine demonstrations, keen wit, and sound philosophy, hammered home with a demand for high performance, he had a strong positive influence upon thousands of students. Those who took the course always remembered him, not always with pleasure, but always with respect. Because of his initials and drive, many students knew his as "Horse Power" Cady.

His teaching extended over the whole state of Kansas. Kansas University acquired one of the first machines in America for liquefying air. Each year Cady went from the university to a number of places in the state to give a lecture, with demonstrations, on liquid air. The greatest activity of this sort came in the academic year 1916-17 when he gave 37 lectures. He also brought teaching home. As a result, his son, George H. Cady, became a chemist.

Cady had great interest in a variety of natural phenomena. In his pockets one would be likely to find not only a pen and jackknife, but also a small microscope, a file, a compass, a ruler, and an aneroid barometer. He used the latter when hiking in the mountains to tell how much he had climbed, and he kept it with him during severe storms thinking that someday he might have a chance to see what it did when in the heart of a tornado. He never made that observation.

Summer vacations were wonderful. Each year from 1911 to 1917 the family went to Colorado to spend several weeks living in tents. After 1917 a car was used for camping trips. The most extensive trip of this sort was in the summer of 1920 when the family visited all of the national parks in the western states. Cady was interested in Indians and their culture and acquired many items of Indian hand crafts. He included some of these in the western style clothing worn during summer vacations and was highly pleased when a tenderfoot tourist would by mistake take him for an old western scout.

Since his salary as professor left something to be desired, he was from time to time attracted to work as consultant on projects that offered prospects for high return. He never profited much from these, but some of the projects worked out. He used the air liquefier at Kansas University in the late twenties for what was probably the first pilot–plant scale liquefaction of natural gas. He developed a process for removing hydrogen sulfide and other undesirable substances from gas of high carbon dioxide content coming from wells in northern Utah. He developed a process for producing dry ice from combustion gases, but the method was never used commercially. His venture involving a machine for concentrating gold from its ore was also not a success. Even as a student he was involved in one of these extracurricular projects. A local bank had been swindled into buying gold dust which turned out to be counterfeit. Cady's analysis of the material showed it to contain 55% platinum. For his work on this material he received $75, which was nearly enough to pay his expenses for a semester at the university.

Arthur W. Davidson, *Chem. Eng. News* **17,** 660-661 (1939); Wayne E. White, ibid, **19,** 793 (1941); Ray Q. Brewster, *Science* **98,** 190-191 (Aug. 27, 1943); Robert Taft, *J. Chem. Ed.* **10,** 34-39 (1933); Clifford W. Seibel, "Helium, Child of the Sun," Univ. of Kansas Press (1968).

GEORGE H. CADY

George Chapman Caldwell

1834-1907

Caldwell, president successively of the Association of Official Agricultural Chemists and the American Chemical Society, chairman of the Chemistry Department of Cornell University during its great expansion of the 1890's, was one of the leading nineteenth century agricultural chemists in America.

Caldwell was born Aug. 14, 1834 in Framingham, Mass. to Jacob and Mary Ann (Patch) Caldwell and grew up in the Boston area. He attended Lawrence Scientific School of Harvard University and received his B.S. degree in 1854. He did graduate study with Bunsen for a year in Germany and received his Ph.D. degree with Wöhler at Göttingen in 1857. He also attended the College of Agriculture at Cirencester, England during this time.

Caldwell's earliest publications, 1856-57, indicated the direction he was later to follow; they were concerned with peanut and Brazil nut oils and the fatty acids available from plant oils. Thereafter, however, his research talents were to lie fallow for nearly a quarter of a century because of the press of teaching and administrative duties at a variety of colleges and universities. Caldwell was assistant professor of chemistry at Columbia College for 2 years; then professor of chemistry, physics, and botany at the newly organized Antioch College from 1859 to 1862. From 1862 to 1864 he served with the Sanitary Commission in Washington as a Civil War hospital visitor and observer. In 1864 he went to the brand-new (1862) Agricultural College of Pennsylvania (later Pennsylvania State University) as professor of chemistry; in 1867-68 he was acting president of the college. In 1868, once more exercising his penchant for ground-floor appointments, Caldwell was the first faculty member chosen for the newly-formed Cornell College in Ithaca, N.Y. Here he was to stay for the remainder of his life.

Caldwell's initial appointment at Cornell was as professor of agricultural chemistry, and within a year he had distinguished himself in that position by publishing the pioneering "Agricultural Qualitative and Quantitative Chemical Analysis" (1869). The book drew heavily on the work of Emil Wolff and C. Remigius Fresenius, as Caldwell readily acknowledged in the preface; it is said, however, to be the first work of its kind in the English language. From 1875 to 1892 Caldwell was professor of agricultural and analytical chemistry; and from 1892 until his retirement in 1903, professor of general and agricultural chemistry, at Cornell. He wrote three texts: "A Manual of Introductory Chemical Practice," with Abram A. Breneman (1875); "A Manual of Quantitative Chemical Analysis," with Stephen M. Babcock (1882); and "Elements of Qualitative and Quantitative Chemical Analysis" (1892), which subsequently appeared in a number of revised editions as well.

The chemistry facilities at Cornell were strikingly primitive in the college's early years and remained so until the erection of a building to house both chemistry and physics in 1883. Caldwell took charge of the department in 1887. Under his administration, wings were added to the building in 1890 and 1898; on his retirement in 1903 the department was worthy of the major university of which it was a part. Caldwell recommenced his research activities in 1883 producing a series of publications on milk, grain production, cattle foods, and various laboratory devices over the next decade or so. An article of considerable historic interest was his review "The More Notable Events in the Progress in Agricultural Chemistry, since 1870," presented in *Journal of the American Chemical Society,* **14,** 83-111 (1892).

Caldwell was active in his profession outside the university as well as within it. He was one of the founders of the Society for the Promotion of Agricultural Science; was vice-president, 1890, and president, 1891, of the Association of Official Agricultural Chemists; he was president of the American Chemical Society in 1892.

In addition to his professional capabilities, Caldwell's courtesy, fairness, and kindliness in administration and teaching caused his associates and students to hold him in high regard. In 1861 he married Rebecca Wilmarth; they had two children. He died Sept. 7, 1907 at Canandaigua, N.Y.

"National Cyclopedia of American Biography," **26,** 142-3, with portrait, James T. White (1937); L. M. Dennis, *Proc. Am. Chem. Soc.,* (Jan.) *1909,* pp. 7-8; "A Half-Century of Chemistry in America, 1876-1926," *J. Amer. Chem. Soc.* **48** (Special Supplement), p. 184 (1926); "J. C. Poggendorff's Biographisch-literarischer Handworterbuch fur Mathematik, Astronomie, Physik, Chemie und verwandte Wissenschaftsgebiete," 5, 1904-1922, p. 196-7, Verlag Chemie (1926).

ROBERT M. HAWTHORNE JR.

Mary Letitia Caldwell

1890-1972

In those days, it was common for a woman to be addressed as "Miss" even though she had earned a Ph.D. degree. Mary Letitia Caldwell was addressed as "Miss Caldwell" by many of her colleagues, although those who knew her best called her "Mary L." Her students called her "Dr. Caldwell." She was the only female member of the senior faculty of the Department of Chemistry of Columbia University when I was a student there in the late forties.

Mary Caldwell was born Dec. 18, 1890 in Bogota, Colombia, where her father was a Presbyterian minister. Intellectual attainment must have had strong emphasis in her family because all five of the Caldwell children became scholars or educators. Mary attended high school and college in the United States and received her A.B. degree from Western College for Women in 1913. She taught at that college for 4 years and then entered the graduate school of Columbia University. Under the sponsorship of Henry C. Sherman, she conducted research on malt amylase, thus beginning a lifetime research interest in amylases. After she was awarded her Ph.D. degree in 1921, she was appointed university fellow. She gradually rose through the ranks and was appointed professor of chemistry in 1948.

All of the graduate students in the department knew Caldwell. After Sherman retired, she was in charge of the course in the chemistry of food and nutrition. As advisor to graduate students, she worked out a course of study with each student. As secretary of the department, she arranged their assignments as teaching assistants in undergraduate and graduate courses. She was also responsible for the details of financial arrangements for the support of graduate students. Students found her firm but sympathetic.

After half a day or more in the departmental offices, she would go up to her laboratories on the ninth floor of Chandler Hall. At that time the elevators went only to the eighth floor; the ninth floor had to be reached by stairs. Caldwell was afflicted with a progressive muscular disability, which caused her to start using a cane in 1943. The climb to the ninth floor became increasingly difficult. Her students wondered why she could not be given space on another floor. Did she prefer to remain where she was? Was no other suitable space available? Was she too proud to request other space?

Caldwell's research, supported by foundation and industrial grants, was done with the collaboration of numerous associates including 18 graduate students who obtained their Ph.D. degrees under her sponsorship. She inspired her students with respect for technical excellence as well as fine scholarship. Her manners were rather formal; she rarely addressed students by first names and scrupulously changed the "Miss" or "Mr." to "Dr." immediately following a successful thesis defense. Despite her formal manner, she conveyed a sense of concern for a student's personal welfare. She could summon a bright word of encouragement when the work was not progressing fast enough, often ending her comments with a philosophical "Well, child, that's research!"

Many of the principles taken for granted in enzymology today were first applied to amylases in Caldwell's laboratories. She and her collaborators used highly purified enzymes so that the factors influencing their stability and activity could be determined; they were the first to crystallize pancreatic amylase. In a series of distinguished, meticulous studies, they showed that amylases were proteins and demonstrated which chemical groups were necessary for the activity of certain amylases. Amylases of different classes, alpha, beta, and gluc, were known to differ in mechanism of action, but these researchers showed, further, that alpha amylases isolated from different sources varied in mechanism of action.

Caldwell died July 1, 1972, at Fishkill, N.Y. She had retired from active participation in research and teaching several years previously because of illness. During her last years, she must have looked back upon her productive career with a sense of achievement and with the satisfaction of knowing that the quality of her work had been recognized. She was the only woman to achieve the rank of professor in the Department of Chemistry at Columbia. She was awarded an honorary D.Sc. by her alma mater in 1961.

To her, the high point may have been that evening in 1960 when, sitting in a wheelchair, she received the Garvan medal awarded annually by the American Chemical Society to a woman for distinguished service to chemistry.

Personal recollections; biography with portrait, *Chem. Eng. News,* Apr. 18, 1960, p. 86; obituary *New York Times,* July 2, 1972; birthdate from alumni records, Columbia Univ.

MARIE M. DALY

Edward De Mille Campbell

1863-1925

To any scientist, the ability to read the literature in his field and to an experimentalist, the ability to make his own visual observations would seem to be *sine qua non* of a productive life, yet Edward Campbell for all but 6 years of a long and successful professional career as research metallurgist and educator was totally blind. Campbell was born in Detroit, Mich.,.Sept. 9, 1863 the son of James Valentine and Cornelia (Hotchkiss) Campbell. His father was the first dean of University of Michigan Law School and a Michigan Supreme Court Judge for 37 years.

Edward received his B.S. degree in chemistry from University of Michigan in 1886. He spent the next 4 years as an industrial chemist, first with Ohio Iron Co., then with

Sharon Iron Co., and finally Dayton Coal and Iron Co., Dayton, Tenn. In 1890 he was appointed assistant professor of metallurgy at University of Michigan where he remained for the rest of his life, rising through the faculty ranks to professor of chemistry and metallurgy and director of the Chemistry Laboratory.

Campbell's early industrial experience implanted an enduring interest in the microstructure of steel. It was this very research that led to the loss of his sight at the age of 28. Campbell was conducting an experiment on the acid extraction of carbides from steel when a hydrogen explosion occurred, shattering the glass apparatus he was closely observing, blinding him, but scarcely touching his two assistants standing behind him. It was characteristic of Campbell that, despite the trauma of the accident, he lost but 10 days of school. Indeed, upon leaving the hospital, he caused himself to be taken directly back to the chemistry laboratory before going home. He spent some time going about the laboratory, familiarizing himself with it so that, sightless, he could sense the location of rooms, stairs, benches, and storage cupboards. He concluded that loss of his vision need not preclude a successful teaching and research career. The future was to prove him right; for in the remaining 33 years of his life, Campbell carried out research which resulted in 72 published papers (including three posthumous) and trained hundreds of men in his chosen profession. His research contributions were primarily in analytical chemistry, the constitution of steel, and of Portland cement.

Throughout his life, Campbell was active in the affairs of many professional societies and in his university. In 1905 he was made director of the Chemical Laboratory and was responsible for the design of the building. Though blind, his detailed knowledge of plans and construction of the building was phenomenal.

He was distinguished by honorary memberships in the American Society for Steel Treating (now the American Society for Metals), the Society of Detroit Chemists, and the Michigan Gas Association. ASM established a distinguished lectureship in his honor in 1926 to which each year one of the world's leading metallurgical scientists and engineers has since been named.

Campbell was married in 1888 to Jennie Maria Ives by whom he had six children. Athlete, humorist, devoted husband, father, and genial host, he never allowed his handicap to mar his view of life and the world. He died Sept. 18, 1925.

"History of the University of Michigan" (1906), pp. 314-15; *The Michigan Alumnus* **32,** 38 (1925); notes from A. R. Putnam of the American Society for Metals; obituary in *Trans. Am. Soc. Steel Treating* **8,** 444 (1925); Z. Jeffries, *Trans. Am. Soc. Steel Treating* **13,** 369 (1928), (includes a partial bibliography of Campbell's publications).

J. H. WESTBROOK

Henry Smith Carhart

1844-1920

Carhart was a pioneer in physical chemistry many years before it became a recognized scientific discipline. He was a charter member of The Electrochemical Society and active in its organization. He was elected vice president for a 2-year term at the society's organization meeting in Philadelphia, Apr. 3, 1902. The next day he gave a paper before the newly formed society on "A Novel Concentration Cell, namely, $Ni|NiSO_4$(concentrated)$|NiSO_4$(dilute)$|Ni$." This paper was typical of Carhart's work on galvanic cells, a field in which he then had been active for over 20 years. He was elected president of the society in 1904, becoming the second man to hold the office.

Carhart was born in Coeymans, N.Y. Mar. 27, 1844, the son of Daniel Smith and Margaret (Martin) Carhart. In 1865 he entered Wesleyan University from which he received his A.B. in 1869 and his A.M. in 1872. He was a student at Yale in 1871-72, at Harvard in 1876, and at University of Berlin in 1881-82. He received honorary degrees from Wesleyan, from Michigan, and from Northwestern. He married Ellen M. Soule, Aug. 30, 1876.

He was professor of physics and chemistry at Northwestern, 1872 to 1876, and professor of physics at University of Michigan, 1886 to 1909. In 1909 he became professor emeritus at Michigan.

Carhart's main interests were galvanic cells and electrical measurements. While at Berlin he investigated the relation between the electromotive force (emf) of a Daniell cell and the zinc sulfate concentration and observed that cells with lower concentrations had the lower temperature coefficient, an important characteristic for a standard. While at Berlin he also found the emf of the Clark cell to be 1.434 V at 15°C in terms of the Siemens ohm and the silver coulometer. This value agreed with that found by Lord Rayleigh in England the same year. Five years later he described a modified form of Clark cell which became known as the Carhart-Clark cell; it had a temperature coefficient much less than the saturated cell.

Carhart was intensely interested in standard cells. He worked with George A. Hulett at Michigan on cell chemistry. He was chairman of the Committee on the Cadmium Standard Cell of the American Institute of Electrical Engineers in 1904, secretary of the Committee on Standards of Measurements of the American Association for the Advancement of Science in 1900, and served as U.S. delegate to the International Electrical Congress in Chicago in 1893. At that time he was appointed with von Helmholtz and W. E. Ayrton to a committee to work on international specifications for the Clark cell. Carhart also presented several papers to the Electrical Standards Committee of the British Association for the Advancement of Science. One of these carried the proposal that the Weston cell, because of its many advantages over the Clark cell, should replace the latter as an emf standard. Carhart's proposal, made in 1904, was finally accepted internationally in 1909.

Carhart made many Weston cells. On Apr. 15, 1910, two of his cells were added to the group of cells used by the National Bureau of Standards in maintaining the U.S. Legal Volt; they remained a part of this group for approximately 4 years.

His expertise on standard cells led to his appointment to the International Jury of Awards, Paris Exposition of Electricity, 1881; as president of the Board of Judges, Chicago Electrical Exposition, 1893; Jury of Awards, Buffalo Exposition, 1909; U.S. delegate, International Electrical Congress, St. Louis, 1904; British Association for the Advancement of Science, South Africa, 1905; U.S. delegate, London Electrical Congress, 1908; and delegate to the Darwin Centennial Celebration, Cambridge, England, 1909.

He was the author of "University Physics" (1894-96), "Electrical Measurements" (with G. W. Patterson, 1895), "College Physics" (with H. N. Chute, 1910), "High School Physics" (1901, 1910), "First Principles of Physics" (1912), and "Physics with Applications" (1917). These books contained much that is, today, within the realm of physical chemistry.

Carhart retained his keen interest in dissemination of scientific knowledge until his death in Pasadena, Calif. on Feb. 13, 1920.

Trans. Amer. Electrochem. Soc. **1,** 8, 19, 105 (1902); portrait and signature, **5,** frontispiece (1904); **6,** 1, 118 (1904); **7,** 15 (1905); "Appleton's Cyclopedia of American Biography," D. Appleton & Co., **1,** 525 (1886).

WALTER J. HAMER

Wallace Hume Carothers

1896-1937

Carothers, born in Burlington, Iowa, Apr. 27, 1896 to Ira Hume and Mary Evalina (McMullin) Hume, attended Tarkio College where he concurrently taught lower level courses while completing his baccalaureate in chemistry. He earned his doctorate in chemistry at University of Illinois where he received the Carr Fellowship in recognition of the faculty's high regard for him "as one of the most brilliant students who has ever been awarded the doctor's degree."

After teaching organic chemistry briefly at University of Illinois, Carothers joined the chemistry faculty at Harvard where he was associated with James B. Conant who observed in him "that high degree of originality that marked his later work." While at Harvard he began his classic studies on polymerization.

In 1927 Carothers left Harvard to head

the organic section of Du Pont's new fundamental research program in Wilmington, Del. Here he continued his work on the synthesis of macromolecules. His investigation of acetylene type polymers yielded products possessing rubber like elasticity coupled with resistance to organic solvents and sunlight. Du Pont technology translated these developments into a commercial process for the production of neoprene, the first successful synthetic rubber developed in this country. Carothers also synthesized cyclic polymers possessing musk-like aromas which Du Pont marketed for the perfume trade.

During the 1930's Carothers turned his attention to the synthesis of polymers analogous to natural fibers. His interest first centered on the products obtained by reaction of dibasic acids with dihydric alcohols. Although the resulting polymers could be extruded in filament form, the strands were weak, melted at low temperatures, and softened in water.

Carothers then turned his attention to the reaction of dibasic acids with diamines. Just about the time this project was to be abandoned because of disappointing results, his efforts were rewarded by the development of polyamide fibers which Du Pont began marketing in 1940 under the name of Nylon. This first commercial synthetic fiber possessed unique properties not only as a textile but also as a molded plastic for many applications.

On April 29, 1937, two days after his forty-first birthday, Carothers checked into a Philadelphia hotel and committed suicide by drinking cyanide, ending a career which had encompassed several monumental commercial developments coupled with significant contributions to pure science and academe. He published 31 papers, held 50 patents, and served as associate editor of the *Journal of the American Chemical Society* and editor of *Organic Syntheses*.

W. W. Hassler, "How Nylon Was 'Discovered'," *Amer. Hist. Illustrated* **5,** 32-37, (November 1970); biography by R. Adams in *Biog. Mem. Nat. Acad. Sci.* **20** (1939), condensed in Edward Farber, ed., "Great Chemists," Interscience (1961), pp. 1600-1611; "Dictionary of American Biography" **21** (Supplement 2), 96, Chas. Scribner's Sons (1958).

WILLIAM W. HASSLER

Emma Perry Carr

1880-1972

When the Francis Garvan Medal was established by the American Chemical Society to honor outstanding women in American chemistry it was no surprise that the first recipient (1937) was Emma Perry Carr of Mount Holyoke College. During an active career of over 60 years, 36 of which were as head of the Chemistry Department, she won for herself and her college an international reputation for both teaching and scientific research.

Emma Carr was born at Holmesville, Ohio, July 23, 1880. Her undergraduate study was at Ohio State University, Mount Holyoke College and University of Chicago. Her Ph.D. degree was awarded by the latter school following her work with Alexander Smith and Julius Stieglitz. She became a member of the Mount Holyoke faculty in 1910 and began her long service as head of the Chemistry Department in 1913. During her professional career she studied with A. W. Stewart (Belfast, 1919) and Victor Henri (Zürich, 1925; 1929). She was the recipient of four honorary degrees and was an active participant in the national and international activities of many professional organizations. In 1957 the Northeastern Section, American Chemical Society, awarded her the James Flack Norris Award for outstanding achievement in the teaching of chemistry. On that occasion an identical award was made to her colleague, Mary L. Sherrill. Because of Carr's outstanding services to the Mount Holyoke College community, the chemistry laboratory was named in her honor in 1955.

Carr's fame in chemical research was the result of her study of the electronic spectra of aliphatic hydrocarbons, particularly of the simple olefins. Her first major contributions appeared in 1929 and led to the support of her work by the National Research Council and the Rockefeller Foundation. The research program involved contributions by faculty colleagues and a host of students and became a model for group research. The program was particularly noteworthy because of its location in a small liberal arts woman's college devoted primarily to undergraduate instruction. As a result of the work there resulted a more satisfactory theoretical interpretation of the energy relationships involved in ethylenic unsaturation; this was used in later years by Robert S. Mulliken and others in developing extensive theories concerning energy relationships in organic compounds.

Carr's fame as a teacher was based not only on her use of group research but also on her ability to share with each student her enthusiasm for science, politics, pi electrons, baseball, and the circus. Primarily as the result of her presence the Chemistry Department at Mount Holyoke College came to be recognized as the outstanding department among women's colleges. Her active participation in college and town affairs continued long after her retirement. Only the infirmities of advancing age forced her to leave South Hadley for Evanston, Ill., where she died at the age of 92, Jan. 7, 1972.

C. P. Burt, *The Nucleus* **34**, No. 9 (1957) with portraits.

EDWARD R. ATKINSON

Ezra Slocum Carr

1819-1894

Ezra Carr's life and career belong to that phase of nineteenth century natural science which was completely and almost purely American, with little trace of European influence. It lacked essential differentiation and basic specialization, whether in chemistry or any other field. Although born and trained in the East, he migrated westward, ultimately to California. Indeed, Carr became involved in the politics of American education, which perhaps added to his pedagogical importance but detracted considerably from any substantial contribution to science or chemistry.

Carr was born Mar. 9, 1819, in Stephentown, N.Y. He graduated in 1838 from Rensselaer Polytechnic Institute. By 1838 Rensselaer was awarding two degrees then unique in American education, the civil engineer (C.E.) and bachelor of natural science (B.N.S.). Carr earned both of these degrees, but, even more, he was fully imbued with Amos Eaton's zeal for applied science, which he carried west with him.

After graduation Carr worked on the geological survey of New York State, which was to a degree Eaton-inspired and manned by his students. Thereafter Carr took up the study of medicine and received his M.D. from Vermont Medical Academy at Castleton. Apparently he never practiced medicine but became professor of chemistry and pharmacy at Castleton, which was his base of operations for a dozen years. Carr was also professor of medical chemistry for various periods at medical colleges in Philadelphia and in Albany, N.Y. At Albany in 1854 he delivered a course of popular lectures on the application of chemistry to agriculture and the manufacturing arts. They were extremely successful, according to a report in the contemporary *Country Gentleman.*

In 1856 Carr made his first important move westward. He accepted a professorship of agriculture, chemistry, and natural history at University of Wisconsin. It was a young institution, scarcely a half dozen years old, and Carr projected its future course in an inaugural address delivered before the Board of Regents and the legislature. Carr also served as professor of chemistry at Rush Medical College in Chicago. He stayed at Wisconsin for 11 years and became involved in its early political and academic struggles. He was even a regent of the university for a year or two but in the end lost his professorship in 1867.

Carr now moved further west to California, where he was appointed professor of chemistry and agriculture at the newly formed University of California in 1869. Here, as the leading spokesman for a projected college of agriculture to be established at Berkeley, Carr was drawn into a prolonged controversy over the ultimate goals of the university. Carr represented the practical, almost vocational objectives as against the more liberal ones of general education, which were championed by the university president, Daniel C. Gilman. Gilman's views won out although Gilman himself resigned and returned east. In 1874 Carr was dismissed summarily from his professorship and thus lost his campaign for an applied, popular type of education.

Carr's life then assumed a political character. He was associated with the formation of the Grange, or Patrons of Husbandry among the farmers of California and was master of a local grange in Oakland. He addressed himself to them and to the mechanics of the state for support for his educational ideas. He was elected state superintendent of public instruction in 1875 and served for 4 years, apparently as a people's spokesman in education. His wife was deputy superintendent. In 1897 Carr's term of office ended and he retired. He spent the remainder of his life on an estate at Pasadena, where he practiced his theories of farming and horticulture which he had never been able to apply at the University of California and where he died Dec. 4, 1894.

Carr may be described as a generalist in practical education rather than as a specialist in science or chemistry. His writings were of a diverse rather than a particular character. Among them were such titles as "Child Culture," "Claims and Conditions of Industrial Education," "Diseased Moral Conditions" "Medical Education," "Agriculture," and "Genesis of Crime." Most unusual was his production of a substantial volume in 1875 on "The Patrons of Husbandry on the Pacific Coast," which is both a history and an apologetic.

The best public tribute to Carr as a scientific teacher came from John Muir, the great naturalist who was his student at Wisconsin: "I shall not forget the Doctor, who first laid before me the great book of Nature. . . ."

H. B. Nason, "Biographical Record of the Officers and Graduates of Rensselaer Polytechnic Institute" (1887); Aaron J. Ihde, H. A. Schuette, "The Early Days of Chemistry at the University of Wisconsin," *J. Chem. Educ.* **29**, 65-72 (1952), (portrait); Merle Curti, "History of the University of Wisconsin" (1949) **I**, 180; a manuscript biography of Carr, prepared by the Writers' Project of the Works Progress Administration, is preserved at the State Historical Society of Wisconsin; John Muir, "Letters to a Friend, 1866-1879" (1915), contains the quote by Muir.

SAMUEL REZNECK

George Washington Carver

1860-1943

The year of George Carver's (The "Washington" was added later) birth is not known but it must have been in the early 1860's. His mother was a slave of Moses Carver, who owned a farm near Diamond Grove, Mo. George was reared from infancy by the Carvers after his mother disappeared during a raid by anti-abolitionists. After having studied art while at Simpson College in Indianola, Iowa, Carver undertook the study of agricultural science at Iowa Agricultural College, Ames, Iowa. Upon his graduation in 1894, Carver was appointed assistant station botanist at the College. In view of his humble origin and of the prevailing attitude of a populace that could not envision a place for Negroes in higher education, his having attended college and obtained a degree testifies to the intensity of his determination.

Carver left Ames in 1896 to become director of agriculture at Tuskegee Institute where he was to spend the rest of his life. That same year he was named director of the first State Agricultural and Experiment Station in Alabama. Carver's training had been in agriculture and botany. His main avocational interest was painting—in almost any medium. His strong belief that his knowledge should find practical application led him into the realm of chemurgy. On the basis of procedures that were largely empirical, he synthesized a variety of potentially useful products from seemingly mundane sources. Among these were: paving blocks, insulating board, cordage, and paper from cotton; synthetic marble from wood shavings; some two hundred products from peanuts, including beverages, sauces, "coffee," salves, bleach, washing powder, paper, ink, shaving cream, and axle grease; and approximately 100 products from sweet potatoes, including flour, syrup, "tapioca," vinegar, mucilage, and bluing. In their time the development of many of these was a more noteworthy accomplishment than might now be assumed to be the case. Carver never attempted to exploit his discoveries for financial gain.

He was a recognized authority on mycology, plant taxonomy, and several aspects of agricultural science. His counsel was frequently sought and freely given. In 1921 he testified on behalf of the Peanut Association before the Committee of Ways and Means of the U.S. House of Representatives. His ingenuous recital of the protean character of the lowly peanut apparently served to convince the members of the Committee of the advisability of the imposition of an import duty on peanuts. In 1935 Carver was appointed collaborator in the Mycology and Plant Disease Survey of the Bureau of Plant Industry, U.S. Department of Agriculture.

Carver said that he never had time to marry. He lived a somewhat secluded life although he was devoted to his teaching. He died Jan. 5, 1943 and was buried in a small cemetery on the campus of Tuskegee Institute.

Many honors came to Carver during and after his lifetime. In 1916 he was elected to the Royal Society of Arts of Great Britain. The NAACP awarded him the Spingarn medal in 1923. Simpson College conferred the honorary D.Sc. on him in 1928. In 1931 the Tom Houston Peanut Co. presented a bronze plaque to Tuskegee Institute in recognition of Carver's contribution to the peanut industry. In 1939 he was awarded the Roosevelt Medal for distinguished service in science. In 1941 the year the George Washington Carver Museum was dedicated at Tuskegee by Henry Ford, Carver was awarded the honorary D.Sc. by the University of Rochester. In 1943 the Thomas A. Edison Foundation award was designated for Carver, and the G. W. Carver National Monument was established near his birthplace. In 1946 by a joint resolution of the Congress, January 5 was designated as Carver Day. In 1947 a three-cent postage stamp in honor of George Washington Carver was authorized and issued the following year.

Henry Thomas, "George Washington Carver," Putnam (1958); L. Elliott, "George Washington Carver: The Man Who Overcame," Prentice-Hall (1966); Rackham Holt, "George Washington Carver," Doubleday Doran & Co. (1943). There are differences in emphasis in the various biographies as well as minor discrepancies in the accounts of some incidents, but they tend to cover the same ground.

LAWRENCE F. KOONS

Hector Russell Carveth

1873-1942

Carveth was born Jan. 23, 1873 in Port Hope, Ont., Canada, the son of Joseph Lobb and Martha Ann (Butterfield) Carveth. He attended Victoria University from 1892 to 1895, and received his A.B. degree from University of Toronto in 1896. He was a fellow in chemistry at Cornell from 1896 until 1898 when he received his Ph.D. degree. At Cornell, 1898-99, he was Sage Fellow in chemistry; from 1899 to 1900, instructor in general chemistry, and from 1900 to 1905, instructor in physical chemistry. He developed a lasting personal friendship with Wilder D. Bancroft, under whom he was awarded his Ph.D. with a thesis on "Single Potentials." A portion of Carveth's work at Cornell saw light in a series of monographs on various aspects of electrochemistry, the first dealing with sodium and its products. In 1903 he started the first course in chemical engineering at Cornell.

In 1905 Carveth became associated with Edward G. Acheson in the old International Acheson Graphite Co. in Niagara Falls. In 1906 he joined Niagara Electrochemical Co. as works manager. This was the only plant in the United States at that time making metallic sodium by the electrolysis of molten sodium hydroxide.

The Niagara Electrochemical Co. (N.E.C.) had been organized by the firm of Roessler & Hasslacher (R. & H.) of New York with the English Castner-Kellner Alkali Co. and the German Gold und Silber Scheideanstalt (Degussa), each of which in the beginning owned one-third of the stock of N.E.C. Much of the stock originally owned by Degussa was later purchased by R. & H. as was Degussa's stock in R. & H. itself and in the Perth Amboy Chemical Works.

At Niagara Falls, Niagara Electrochemical Co. also made sodium peroxide from the metallic sodium and was making chloroform. R. & H. made sodium cyanide at a small plant in Perth Amboy.

From early in the nineteenth century the possibility of producing metallic sodium directly from cheap, readily available salt had attracted the attention of brilliant chemical minds. At Perth Amboy, R. & H. about 1906 began to support the research of Robert J. McNitt in his efforts to develop a successful cell. In 1907 at Niagara Falls, James C. Downs under the direction of Carveth began work on a molten sodium chloride cell for the production of sodium with chlorine as a co-product. This work stopped in 1914, was resumed in 1922, and resulted in 1924 in the successful Downs Cell (U.S. Patent 1,501,756, July 15, 1924, assigned to R. & H.), the only process now used for producing metallic sodium with chlorine as a coproduct from molten salt in the United States.

In 1907 began Carveth's long personal association with Leo H. Bakeland, later to lead to R. & H.'s relationship with the Bakelite Corp., with R. & H. being an important original stockholder.

On Dec. 22, 1915, Carveth married Josephine McCollum of Lockport, N.Y.

With the U.S. entry into World War I, among the chemical companies in the U.S. seized by the Alien Property Custodian in 1918 was the Roessler & Hasslacher Chemical Co. and its affiliates, Perth Amboy Chemical Works and Niagara Electrochemical Co. The struggle to demonstrate that U.S. citizens were in control of the majority of the stock of R. & H. and its affiliates and controlled their operations was long and difficult but finally successful. In 1918 Carveth became second vice president of the R. & H. Co.

The period after World War I involved Carveth in a great expansion of all R. & H. activities—financing, opening the Pacific Coast operation, the first production of methyl chloride in the U.S. at Niagara Falls, the construction at Niagara Falls of a major air liquefication and separation plant, the entrance of R. & H. into the synthetic ammonia business in Niagara Falls, the beginning of the manufacture of trichloroethylene and of tetrachloroethylene. In 1925, in association with Shawinigan, R. & H. set up Niacet Chemicals Corp. to make such derivatives of acetylene as acetic acid and acetaldehyde, paraldehyde, aldol, and crotonaldehyde.

In 1928 Carveth became president of R. & H. When Roessler & Hasslacher was absorbed by Du Pont in 1931, he became a director of Du Pont, remaining a year and retiring in 1932.

Carveth's interests extended far beyond the bounds of his immediate business. He was a civic leader in Niagara Falls, N.Y. He played a most important part in the introduction of a city manager form of government to the city of Niagara Falls in 1916. He organized the Niagara Falls Boy's Club and was its first president.

In retirement Carveth continued to do electrochemical research in a private laboratory adjacent to his home, 352 Buffalo Avenue, and acquired the Chateau Gay Ltd. winery of Lewiston, N.Y., of which he became president. He was killed Sept. 17, 1942, at the winery when a concrete tank being pressure tested suddenly fractured.

Carveth's eight patents covered the manufacture of sodium perborate, sodium peroxide, black finishing chromium plated articles, apparatus for fusing electrolytes, and other subjects.

H. R. Carveth, manuscript, "Ideas grow—an autobiography"; obituary, *Niagara Falls Gazette* Sep. 18, 1942; archives of Cornell University; patents from Richard C. Woodbridge, U.S. Patent Office.

RICHARD G. WOODBRIDGE, III

Paul Casamajor

1831-1887

Paul Casamajor was one of 35 chemists who attended the meeting on the evening of Apr. 6, 1876 to organize the American Chemical Society. Previously, he had been one of the eight who drafted and signed three letters that were sent out to prospective members. At the organizational meeting he was elected librarian and always held some office in the Society—corresponding secretary, recording secretary, or member of the board of directors.

Casamajor was born at Santiago, Cuba in 1831. Both of his parents were of French descent. His father was born in Cuba and his mother in New Orleans. When he was 14 he came to the United States to study in an academy at Portsmouth, N.H. Later he entered Harvard Scientific School but did not stay long because friends suggested that he go to Paris and enter École Centrale. He graduated as "engenieur chimiste" in 1854. Being quite proud to have earned this degree he was disappointed when the diploma and all of his papers were later burned in a fire at the Brooklyn sugar house of Havemeyer and Elder.

He returned to the United States immediately after graduation and established himself as chemist in New York City. In 1864 the discovery of oil in western Pennsylvania lured him into forming and supervising The Enterprise Mining and Boring Co. Unfortunately, an explosion in 1866 destroyed the entire plant. Casamajor returned to New York where he was employed early in 1867 by Havemeyer and Elder (later American Sugar Co.) as chemist and scientific expert at their sugar refinery. He remained with them the rest of his life.

Casamajor was a frequent and industrious contributor to the *Journal of the American Chemical Society* with original papers, abstracts, and reviews. His first paper appeared in volume I, pages 26-27 (1879) "On the Influence of Variations of Temperature on the Deviation of Polarized Light by Solutions of Inverted Sugar."

His prime concern at the sugar refinery was the easy filtration of turbid sugar liquors. He was successful by the end of his career in finding a practical and economical method which accounted for cleansing about one-half of the 2 million lbs of refined sugar produced daily in the refinery. However, Casamajor maintained a continued interest in other branches of pure and applied chemistry. William Nichols recalled that Casamajor told him of his discovery of a specific cure for whooping cough which he had planned to give to the world.

By 1887 Casamajor's health had deteriorated. He planned to take his family to Europe for a long vacation. However, less than a week after his physician told him that he had heart disease he died. On the day of his death he had gone to Manhattan on personal business and towards evening suffered a heart attack. His friends obtained a carriage to take him home. He died in the cab while crossing the Brooklyn bridge on Nov. 12, 1887.

He left a wife with young children but, fortunately, sufficient estate to take care of them. His son, Louis, became professor of neurology at Columbia University. A grandson, Paul Casamajor, became assistant to the director, Agricultural Experiment Station, University of California, Berkeley.

Notices about Paul Casamajor may be found in the first eight volumes of the *Journal of the American Chemical Society;* obituary by H. Endemann, *ibid,* **9,** 206-8 (1887); Golden Jubilee Number, *ibid,* **48** (August 20, 1926); personal communications from his daughter, Martha Casamajor.

DAVID H. WILCOX, JR.

Hamilton Young Castner

1858-1899

Castner was born in Brooklyn, N.Y., Sept. 11, 1858. Nothing is known of his early education, but he attended Brooklyn Polytechnic Institute for a year, and then transferred to Columbia University School of Mines in 1875, where he studied under Charles Chandler.

While still a student at Columbia, Castner took upon himself the task of analyzing water samples from wells in New York City and found the water so contaminated that Chandler, then president of New York City Board of Health, resolved to keep all of the wells closed permanently.

He left Columbia 3 years later without obtaining his baccalaurate degree, and established himself as a consulting analytical chemist, an endeavor in which he was so successful that "he gathered round him a valuable following that accepted his advice on all chemical matters without question." Within a year he brought his brother into his consulting firm, which allowed Castner to devote his time to devise new manufacturing processes.

His first invention dealt with a method of producing animal charcoal, for which he obtained British patent No. 4507 in 1882. This patent also included "improvements in the manufacture of ammonia." His invention did not become a commercial success because of the depressed financial conditions of the United States at that time.

Undeterred by the commercial failure of his first venture, Castner set out to invent a process for manufacturing sodium. In June 1886 he received British and American patents for "improvements in the manufacture of sodium and potassium," which consisted of reducing caustic soda with iron carbide, as a result of which the price of sodium fell to 25 cents a pound.

Castner expected the main use of sodium would be in the manufacture of aluminum, then produced by the reaction of aluminum chloride with sodium. Using Castner's invention the price of aluminum would fall to about one dollar a pound. He did not attract investors in the United States so he went to England, where the owners of Webster Crown Metal Co. established Aluminum Crown Metal Co., June 1887, with Castner as managing director, to produce sodium needed for the manufacture of aluminum.

His commercial success in England was short–lived, for in 1889 Charles M. Hall in the U.S. and P. L. T. Heroult in France invented an electrolytic process for producing aluminum without the use of sodium. Castner turned this setback to his advantage by devising a method of producing sodium peroxide by burning sodium in an air current.

In addition to his commercial setbacks, Castner suffered personal tragedy by witnessing the death of his two brothers from tuberculosis. He realized his own health problems were from this disease, but instead of seeking rest, he threw himself into his work with greater ardor than ever.

In 1890 Castner developed a method of manufacturing sodium and potassium by the electrolysis of soda and caustic potash. He used mercury at the cathode and rocked this electrode back and forth to remove the sodium amalgam formed. One of the critical points was the purity of caustic soda. Since caustic soda of the required purity was not commercially available, Castner invented a process to produce his own by the electrolysis of sodium chloride. Independently and simultaneously the Austrian chemist Karl Kellner invented a process similar to that of Castner. To avoid costly litigation they reached an agreement that resulted in the founding of the Castner-Kellner Alkali Co.

Castner's process provided sodium for the manufacture of sodium peroxide and ended the dependency of American industry on imported bleaching powder. The Niagara Electrochemical Co. was formed in 1895 to exploit this invention, with Castner as vice-president.

Another of Castner's contributions was the development of a process of converting carbon into graphite, important because carbon electrodes then in use disintegrated very rapidly. The graphitized electrodes invented by Castner lasted longer and were able to resist the corrosive action of electrolysis solutions. The importance of his invention was such that "without graphitized anodes the great electrolytic alkali and chlorine industry of today would not have been possible."

In 1894 Castner developed several processes to produce pure cyanides needed to extract gold from its ores. One of these processes involved the synthesis of sodium cyanide by the reaction of sodium, charcoal, and ammonia. His product found an extensive market in all the gold–producing countries of the world.

Castner returned to the United States in the fall of 1898 when his health, never strong, began to deteriorate. He lived in Florida for a few months and then went to Saranac Lake, N.Y. to be under the care of E. L. Trudeau, the foremost authority on lung diseases.

In spite of Trudeau's efforts and Castner's exposure to the winter air of the Adirondacks that seemed to have helped so many, he succumbed Oct. 11, 1899.

Virginia H. Lord, *J. Chem. Educ.* **19,** 353-6 (1942); Alexander Fleck, *Chem. Ind. (London)* **34,** 515-21 (1947); D. W. F. Hardie, "Hamilton Young Castner (1858-1899) an account of his life and work" (1952); communications from Gordon D. Byrkit; R. B. MacMullin and W. C.

Gardiner, *Trans. Electrochem. Soc.* **86,** 51-68 (1944).

JOE VIKIN

Arthur Douglas Chambers

1870-1961

Chambers, one of the pioneer manufacturing chemists that helped to build the American synthetic organic chemical industry, was born at Woodstock, Ont., May 4, 1870 the son of James Douglas and Josephine (Mollin) Chambers. After attending Woodstock College and graduating from University of Toronto in 1892, Chambers went to Johns Hopkins University and received his Ph.D. degree in chemistry there in 1896.

After working for an oil refinery in Lima, Ohio for a year, he joined the Du Pont Co. in 1897 as assistant superintendent of a gunpowder plant at Ashburn, Mo. He became superintendent of this plant in 1906 but the following year was transferred to Louviers, Colo. to operate a new dynamite plant. The next transfer was to company headquarters in Wilmington, Del. in 1915 as a technical investigator.

About 1910 the Du Pont Co. began to diversify from explosives, which it had made since 1802, and in July 1915, Chambers suggested that the manufacture of dyes and intermediates be investigated. By World War I, Germany supplied 80% of the world's dyes and all of the ingredients for the remaining 20%. American companies had tried to break this monopoly, but there was no protective tariff, and price-cutting by the German dye trusts made it impossible to compete. When the war brought a British blockade of German shipping, the United States found itself short not only of dyes but phenol, potash, cyanide, and mercury. Almost immediately, companies like Dow and General Aniline attacked the problem of dye synthesis, and by 1915 the United States had six producers of crudes, 17 of intermediates, and 12 of dyes.

Chambers convinced his management that if Germany could rapidly convert dye technology to munitions, Du Pont could do the reverse. Research on dye intermediates began in March 1916; a tubful of sulfur black had been produced by January 1917, and in February the company appropriated $600,000 to build a dye plant. When the Dyestuffs Department was created later that year, Chambers was named manager of manufacturing. The new plant came on–stream in 1918 and during that year made 118 different products. America's first colorfast vat dyes were produced in 1919, and the product line expanded in the 1920's to hundreds of dyes in several classes; by 1920 the Dyestuffs Department employed more than 4000 people.

The next two decades brought many new products, among which were: rubber antioxidants and accelerators, tetraethyllead, neoprene synthetic rubber, agricultural chemicals, and a variety of surfactants, finishes, and auxiliaries for the textile industry. In 1931 when the Organic Chemicals Department was created, the Dyestuffs Department became a division of the new department. Chambers became division manager, a post he held until 1943 when he was made production advisor to the department. At the time of his retirement in 1944, he was honored by having the plant site at Deepwater Point, N. J., named for him. That year the plant had more than 9000 employees and shipped 20 million lbs of neoprene, 30 million lbs of intermediates, and 100 million lbs of dyes. Today, the Chambers Works is one of the largest chemical manufacturing complexes in America.

Chambers was married in 1897 to May Fleming; they had three sons. His hobbies were mineralogy and bridge. He died on Aug. 28, 1961. One person said of him, "His slow, quiet manner hid a brilliantly keen, intelligent, and stupendously energetic mind."

Evening Journal, Wilmington, Del., August 29, 1961; *Morning News,* Wilmington, Del., April 4, 1944; Williams Haynes, "American Chemical Industry—A History," D. Van Nostrand (1948), **3,** 242-243—Photograph; Walter J. Smith, "Chambers Works History" (1963), five volumes of unpublished typscript in du Pont's Jackson Laboratory library.

HERBERT T. PRATT

Emile Monnen Chamot

1868-1950

Chamot was born in Buffalo, N. Y. on Mar. 4, 1868. His parents were Alsatian, and he was bilingual from childhood. He matriculated at Cornell University in 1887, receiving the bachelor of science in chemistry in 1891. Chamot's senior thesis under George C. Caldwell proved to be the start of his life's work since it involved the micrography of crystalline phases of lead nitrate and nitrite. His interest in chemical microscopy was further stimulated by an instructorship in quantitative analysis, also under Caldwell.

After receiving his doctorate in 1897, Chamot and his bride Cora (*née* Genung) spent a year in Europe, where he studied with Eugene Mace at Nancy and Robert Otto at Braunschweig. During this time he came into contact with Heinrich Behrens at Delft, who was preparing a new course in microscopical analysis and with whose analysis methods Chamot was already familiar.

On returning to Cornell he began his teaching career in the Department of Chemistry and was at various times instructor in toxicology, instructor and assistant professor in sanitary and toxicological chemistry, and in 1910, professor of chemical microscopy and sanitary chemistry. The chemical microscopy courses of Caldwell, and later of Chamot, were perhaps the first offered in America. In later years, training in microscopy was required of Cornell undergraduates in chemistry and chemical engineering.

"Chammy" was a dedicated and thorough worker. He often entertained his colleagues and students with illustrations from his own experiences with microscopy and with apparently uncanny deductions that were well grounded in fact. He was a lifelong advocate of the use of microscopy in chemistry-related areas and himself coined the term "chemical microscopy."

Chamot's many years in New York's Finger Lakes region nurtured his boyhood love of nature; for many years he maintained an avocational interest in geology, botany, and zoology—studies to which he applied both his habitually careful observation and his microscopical techniques. His examinations of the local water supply near Cornell in 1904 were invaluable in eradicating a typhoid epidemic in Ithaca at that time.

His publications were numerous and included three texts: "Analysis of Water for Household and Municipal Purposes" (1911) with Harry Redfield; "Elementary Microscopy" (1915); and "Handbook of Chemical Microscopy" (1929), the latter with Clyde W. Mason. He also wrote a monograph, "The Microscopy of Small Arms Primers" (1922). Chamot was a founder and first president of the Technical Photographic and Microscopical Society, which suffered a decline during his sabbatical, 1924-25, as exchange professor, lecturing in France. In 1937 he received the Longstreth Medal of the Franklin Institute for his work in microscopy.

After his retirement from teaching duties in 1936, he continued active research in areas of personal interest until his death on July 27, 1950.

G. W. Cavanaugh, *Ind. Eng. Chem.,* **25,** 826 (1933); Clyde W. Mason, *Ind. Eng. Chem., Analytical Edition,* **11,** 341 (1939); communications from Clyde M. Mason.

LAWRENCE R. DALEY

Charles Frederick Chandler

1836-1925

Born Dec. 6, 1836 in Lancaster, Mass. Charles Chandler was an industrial chemist, educator, author and editor, organizer, and public servant. His early days were spent in New Bedford, Mass. where his high school teacher stimulated him to study chemistry and where he experimented in a home laboratory.

He entered Harvard in 1853, studied analytical chemistry under Eben Horsford, and was influenced by Charles Joy to continue his education at University of Göttingen. He studied with Wöhler for a year at Göttingen, assisted in the laboratory of Heinrich Rose at Berlin, and received his doctor's degree from Göttingen without examination. On his return to America in the fall of 1856 he received an invitation from Joy to help develop a chemical laboratory at Union College in Schenectady, N.Y. When Chandler arrived he found that no provision had been made for his salary, so he accepted the $400 budgeted for janitorial services and became "janitor-assistant."

When Joy resigned in the spring of 1857 to go to Columbia, Chandler, who was 21 years old, assumed charge of the Union laboratory as instructor in chemistry. In 1861 he founded the Chemical Society of Union College. While essentially an active student organization, it may be considered as a precursor of the American Chemical Society. Chandler was chairman of the Priestley Centennial at Northumberland, Pa. in 1874 and was a leader in the founding of the American Chemical Society in New York in 1876, the plan of organization following that of the society at Union.

In 1864 Chandler left Union for New York to help organize the School of Mines at Columbia. Later he became dean of the school as well as head of the Columbia Chemistry Department until his retirement in 1911. He taught chemistry at College of Physicians and Surgeons (Columbia's medical school) from 1869 to 1895, at New York College of Pharmacy from 1867 to 1910, and served as president of the College of Pharmacy before it affiliated with Columbia. He was largely responsible for the construction of Havemeyer Hall and the founding of the Chandler Chemical Museum at Columbia.

In the late 1860's Charles edited the American supplement of the British *Chemical News*. When the supplement was discontinued, he and his brother, William, founded, edited, and published *American Chemist,* a news-industrial journal widely read throughout the country during the span of its existence from 1870 to 1877.

Chandler was one of the country's leading industrial chemists during his day. His range of interests included sugar, petroleum, illuminating gas, electrochemistry, sanitation, and the analysis of waters and minerals. He served as president of the American Chemical Society in 1881 and 1889, of the New York City Board of Health from 1873 to 1883, and of the Society of Chemical Industry in 1899. He was a member of many city, state, and United States government commissions, carried out assignments for the Navy, Interior, and Post Office Departments, and was sent by President Arthur as U.S. delegate to the International Medical Congress at Copenhagen. Details of his work may be found in (to use his words) "a great number of reports, analyses, lectures, investigations, articles on chemical subjects . . . far too numerous to be included herein."

It is doubtful that any other American chemist taught as many students as did Chandler between 1857 and 1911, and his influence on chemical research and chemical industry through tens of thousands of students has been incalculable. The Chandler medal and lectureship at Columbia are named in his honor.

He died Aug. 25, 1925 at East Hartford, Conn.

Charles A. Browne and Mary E. Weeks, "A History of the American Chemical Society," ACS (1952); E. K. Bacon, "A Precursor of the American Chemical Society—Chandler and the Society of Union College," *Chymia* **10,** 183-197 (1965); autobiography by Chandler, and sketches by M. C. Whitaker and M. T. Bogert, port., in *Ind. Eng. Chem.* **12,** 183-195 (1920); biographies in *Pop. Sci. Mon.* **16,** 833-841 (1880), port.; M. C. Whitaker, *Ind. Eng. Chem.* **14,** 977 (1922), port.; M. T. Bogert, *Ind. Eng. Chem.* **30,** 117-118 (1938); M. T. Bogert, *Biog. Mem. Nat. Acad. Sci.* **14,** 127-181 (1931), with portrait, list of writings; R. D. Billinger, *J. Chem. Education* **16,** 253-257 (1939), port.; A. W. Hixon, *J. Chem. Education* **32,** 499-506 (1955); obituaries, E. Hendrick, *Ind. Eng. Chem.* **17,** 1090-1091 (1925); *New York Times,* Aug. 27, 1925; the quote is from Chandler's resumé of his professional life in "Directory of Harvard Chemists" (1921).

EGBERT K. BACON

William Henry Chandler

1841-1906

The second son of a merchant, William was born in New Bedford, Mass., Dec. 13, 1841. His ancestors were in New England in 1637. His early training was influenced by his illustrious brother, Charles Frederick. William graduated from Union College in 1862 while his brother Charles was a professor there.

William obtained practical experience in chemistry as an employee of New Bedford Copper Works for 2 years and with Swan Island Guano Co. for 3 years. In 1867 the brothers were reunited at Columbia. William served as instructor and also worked for his A.M. degree which he received in 1871. In 1872 he obtained his Ph.D. degree from Hamilton College.

It was during this period that the brothers began an important literary venture. The Chandlers saw the need for an American chemical journal. The American reprint of the British *Chemical News* had been discontinued. To fill this gap they published *The American Chemist,* forerunner of publications of the American Chemical Society. The journal, the first issue of which appeared in the summer of 1871, was "devoted to theoretical, analytical and technical chemistry," plus news of meetings of American chemists. It was published until April 1877, first in New York, later in Philadelphia. Most interesting was the special edition recounting the Priestley celebration in 1874. This event, held July 31 at Northumberland, Pa. (Priestley's last home), celebrated the one-hundredth anniversary of the discovery of oxygen. Joining this pilgrimage to the banks of the Susquehanna were many chemists who were moved to found the American Chemical Society 2 years later in 1876. All of this was related in the Chandlers' *American Chemist.*

William Chandler was called to the chair of chemistry at Lehigh University in 1871 to succeed Charles Wetherill. Here, while Lehigh was making strides as a rising technical institution, Chandler contributed much to its physical and educational growth. As teacher, author, consultant, and librarian he was a dominant figure on the Lehigh campus from 1871 to 1906. Director of the university library from 1878, he was also acting president of the university in 1895 and 1904-05.

Chandler was especially proud of the large laboratory which he designed and had constructed at Lehigh. This structure, which bears his name, was designed to house departments of chemistry, metallurgy, and mineralogy. It cost $200,000—a huge sum in those days. He thought so well of the laboratory that he wrote a book giving details, pictures, and drawings of it, together with briefer accounts of other laboratories of comparable importance. This was published in a report for the Universal Exposition of 1889 at Paris, to which he was a commissioner. As director of a well-endowed library he was in a position to purchase fine books and built–up the second largest college library in Pennsylvania.

In his later years Chandler undertook the task of editing "Encyclopedia and Epitome of Universal Knowledge." It was a three-volume edition of some 1,700 pages, with many diagrams and colored maps. There were four assistant editors and many experts in special fields, but Chandler was editor-in-chief and organized the material for the 12 major divisions, all this while he directed the main business of the Chemistry Department. He died in Bethlehem, Pa., Nov. 23, 1906.

R. D. Billinger, "The Chandler Influence in American Chemistry," *J. Chem. Educ.* **16,** 253-257 (1939).

ROBERT D. BILLINGER

William Hale Charch

1898-1958

Charch, an authority on the chemistry of textile fibers, was born in Dayton, Ohio, Mar. 20, 1898, the son of a clothing store owner. He attended Stivers High School and was inspired to study chemistry primarily because of his admiration for a cousin, William J. Hale, professor of organic chemistry at University of Michigan. He received his A.B. degree from Miami University in 1920 and M.S. and Ph.D. degrees from Ohio State University in 1921 and 1923, respectively. His first industrial experience as a chemist was with General Motors Corp. where he worked with Thomas Midgely, Jr., on anti-knock compounds for automotive fuels.

In January 1925 Charch applied for work with Du Pont Rayon Co. at Buffalo, N. Y. The chemical director, Ernest B. Benger, was not impressed with the resume, and if his secretary had not insisted, he would not have interviewed Charch. Nevertheless,

Charch was hired and within 7 months was promoted to assistant chemical director. Du Pont had started cellophane production in 1924 and Charch was assigned to the project of finding a means to moisture–proof the film. Over 2000 formulas were tried before one was found that gave a transparent coating that did not change color, "blush with moisture," or crack off; it was a solution of waxes in a solvent with nitrocellulose. The first shipments of moisture-proof film were made in 1927 to a fruit cake baker. This product ultimately created 5000 jobs at Du Pont and 40,000 throughout the industry. Until moisture–proofing, cellophane was largely a decorative material.

For most of 1927 Charch was assigned to the rayon plant in Old Hickory, Tenn., to work on the dry–cake process for viscose. In 1929 he was promoted to assistant director of the Rayon Chemical Division and in 1935 became director of the Pioneering Rayon Research Laboratory. High–tenacity rayon auto tire cord was developed here as well as a number of specialty rayon staple products.

In 1947 Charch transferred to Wilmington, Del., to lay plans for constructing and operating a new Fibers Pioneering Research Laboratory. This laboratory, which Charch headed until his death, examined more than 5000 fiber–forming polymers and was closely associated with the development of Dacron polyester, Orlon acrylic, Lycra spandex, and Teflon fluorocarbon fibers.

For his work in moisture–proofing cellophane and fiber research, Charch received the Charles Frederick Chandler award of Columbia University in 1957 and the Jacob Schoellkopf Gold Medal of the Western New York Section of the American Chemical Society in 1932. Miami University bestowed an honorary degree upon him in 1949. He was not involved in professional societies and did not publish extensively but held more than 50 patents.

Charch married Ruth Hecdrick in 1946. They lived on a farm, travelled extensively throughout the world, collected antiques, and dabbled in archaeology. Charch was hard–driving and an individualist in dress, habit, and thought. He considered himself a physical chemist, not a textile man, and once described pioneering research as the "bootlegging of ideas that sound crazy but not proven so." Commenting on his more than 20 years in fiber research he said, "It's all ahead of us."

Death came on July 24, 1958 as the result of a stroke.

Journal Every Evening, Wilmington, Del., 1 (July 25, 1958); Williams Haynes, "American Chemical Industry—A History," D. Van Nostrand, **4,** 342 (1948); *Chem. Eng. News,* **36,** 89 (Aug. 18, 1958); Robert E. Ellsworth, *Modern Textiles,* **34,** 36, 45, 78, 79 (July 1953); files of Public Affairs Department, E. I. du Pont de Nemours & Co., Inc., Wilmington, Del.

HERBERT T. PRATT

Nicholas Dimitrius Cheronis

1895-1962

Cheronis was born in Greece, June 29, 1895. When he was 16 years old he emigrated to the United States, attended Lane Technical High School in Chicago, and began undergraduate studies at University of Chicago, which the war interrupted. After discharge from the Army's Chemical Warfare Service, he resumed his education. After receiving his bachelor's degree in 1919, he began graduate studies at University of Chicago; to support himself he established a laboratory to prepare chemicals whose major source had been cut off by the war.

He met Irene Hamlin "in the studio of a mutual artist friend," and in 1923 she became Mrs. Cheronis. Six years later he received his Ph.D. in organic chemistry.

After graduation he became an instructor of chemistry at Crane Junior College (now Malcolm X College), and while teaching there he developed micro-methods to allow him to teach organic chemistry at low cost. These methods he described in the book "Macro and Semi-micro Organic Chemistry" published in 1942.

In 1934 Cheronis joined Wright Branch of Chicago City College, where he became chairman of the Physical Science Department. There he pioneered the integrated method of teaching organic chemistry, featuring the use of resonance and mechanisms, concepts that since have become standards in teaching this subject. His interests were not confined to organic chemistry, however. In 1942, in collaboration with James B. Parsons and Conrad E. Ronneberg, he published "The Study of the Physical World," a book that went through three editions, the last in 1958.

At Chicago City College Cheronis gave his students "unknowns" to spark their interest since the endless preparation of compound after compound seemed to stultify rather than activate their minds. As a result, Cheronis and John B. Entrikin wrote "Semimicro Qualitative Organic Analysis," a book chosen to represent the U.S. in the Brussels World Fair, whence the U.S.S.R. published it without the authors' receiving any royalties.

In the fall of 1950 Cheronis joined Brooklyn College as chairman of the Chemistry Department, a move partly motivated by his desire to be closer to his publishers but also to enable him to put into practice many of his teaching innovations.

Cheronis was instrumental in launching *Microchemical Journal,* serving as editor for many years. His contributions to microchemistry were recognized by the Austrian Society for Microchemistry which awarded him the Emich Plakette in 1961.

His researches at Brooklyn College resulted in the book "Organic Functional Group Analysis by Micro and Semicro Methods," published in 1964 and co-authored by Tsu S. Ma. In addition, Cheronis adapted his microtechniques to general chemistry, described in "Semimicro Experiments in General Chemistry and Qualitative Analysis," written with Herman Stein.

Besides his books Cheronis wrote extensively for *Journal of the American Chemical Society, Journal of Chemical Education* and *Microchemical Journal.* He also obtained over 10 patents in fields such as the preparation of amines, treatment of garbage and wastes, and deposition of polymers on leather. He held numerous research contracts and consulting positions with the Pioneer Hi Bread Co., the Air Force, the Army, and the U.S. Department of Agriculture.

Prior to his retirement from Brooklyn College, Cheronis acquired a farm in northwestern Illinois which he named "Halcyon Hill." He established a chemical laboratory there and like his colleague and fellow microchemist, Anton Benedetti-Pichler, he planned to work to restore the soil's fertility. He went into retirement leave in spring 1962, and one day after his retirement became official, Cheronis died in an automobile accident July 2, 1962, leaving his widow and two offspring: Thaleia, a writer and art critic, and Dion, an insurance agent.

When Cheronis announced his retirement he was awarded an honorary membership and a scroll by the Metropolitan Microchemical Society he helped to found in 1957. Brooklyn College awarded him a plaque for "distinctive service as a teacher and administrator." For him, retirement meant "steady work 14 hours or more per day but NO pressures."

His ashes were buried on his farm under a banner that reads: "The village of Kosmos is proud of you." His former pupils simply add: "We are proud to have learned from you."

Information from Mrs. Irene Hamlin Cheronis, Mrs. John B. Entrikin, T. S. Ma and David E. Goldberg; *The Indicator,* p. 35 (Feb. 1959); *Microchemical Journal,* **6,** 313 (1962) with portrait; obituaries: *The New York Times* (July 3, 1962); T. S. Ma, *Microchemical Journal,* **6,** issue 3, IX-X (1962); David B. Sabine, *Microchemical Journal,* **6,** issue 3, I-II (1962) with portrait; John B. Entrikin, *Microchemical Journal,* **6,** issue 3, V-VIII (1962); A. A. Benedetti-Pichler, *Mikrochimica Acta,* 1-5 (1963) with portrait.

JOE VIKIN

Robert Augustus Chesebrough

1837-1933

Chesebrough was born in London, England, Jan. 9, 1837. His parents were American citizens who returned to this country and settled in Brooklyn, N.Y., while Robert was a small boy. It is not certain where he learned chemistry. Instead of seeking a position as a chemist, he set up a laboratory of his own. At that time, whale oil was the chief substance used in lamps for household illumination. The demand was increasing, the price rising, and substitutes were being sought. One was a distillate from oil obtained from cannel coal during the gas-making process. It was sparingly used because of the disagreeable odor. Chesebrough decided he would prepare the distillate, then called kerosene, to tide him over until other business developed.

Chesebrough had hardly started in that venture when, on Aug. 27, 1859, the first oil well was drilled in Titusville, Pa. Soon thereafter, crude petroleum was gushing from hundreds of wells. Chesebrough realized that his little illuminating business would be killed when kerosene that could be distilled from petroleum was marketed. He went to Titusville to try to get in on the ground floor. Strolling leisurely through the district where drilling and pumping were going full–blast, he saw a worker scraping a waxy material from a pump. It was called "rod wax," was causing trouble, and had to be removed periodically. The man told him that when a worker cut or burned himself, application of this material soothed the injury.

Mulling over this information, Chesebrough remembered similar benefits his mother ascribed to goose grease, olive oil, and other like materials. Maybe something medicinal was lurking in rod wax! He obtained a bucket full of rod wax and took it back to Brooklyn, determined to find out what that substance was. He knew that illuminating oil could be distilled from crude petroleum just as he had distilled it from gas-making oil, and that he could continue in this activity. However, he decided to look into rod wax as a possible medicinal substance.

After several years of experimentation, Chesebrough thought he had discovered the medicinally useful part of rod wax. It was the residue left in the still after distilling off the lighter fractions from crude petroleum. After subjecting that to further purification, he obtained a translucent, semisolid, jellylike product without odor. It did not become rancid like animal or vegetable oils and greases. He made a guinea pig of himself to find out how effective the product was. He cut and burned his skin in various ways, applied the grease and found that it soothed and healed better than rod wax.

Still not convinced that his "miracle jelly" would be accepted by others, he sought additional patients by planting himself where construction was going on, and giving samples to workmen. They found it to be just as satisfactory as he had. At this point, he began to manufacture his "jelly." He named it "Vaseline," registered the name, and established a small factory, Chesebrough Manufacturing Co., in Brooklyn. Samples were sent to physicians and druggists, but no orders resulted. He then conceived a striking method of building up a demand. He hired a horse and wagon and drove through the state of New York, giving a small bottle of Vaseline to each person he passed. Soon orders began to flow. He hired a dozen or more "salesmen" with wagons, expanding his activity, giving away thousands of samples. The business increased by leaps and bounds. Chesebrough produced kerosene and lubricating oils as well until 1881 but thereafter turned out only Vaseline.

As sidelines from the late 1870's onward, Chesebrough dealt in real estate and developed tracts of land in New York City. He promoted one of the trolley car lines in the city. He wrote drama, poetry, and technical articles about petroleum. He remained active athletically until he was in his nineties, swimming and diving until he was 94. He died in Spring Lake, N.J., Sept. 8, 1933.

"National Cyclopedia of American Biography," **B,** 515, James T. White & Co. (1927); Mitchell Wilson, "American Science and Invention," Simon & Schuster (1954); information from Chesebrough-Pond's, Inc.; obituary in *New York* Times Sept. 9, 1933.

BERNARD E. SCHAAR

William Mansfield Clark

1884-1964

By the time of his death on Jan. 19, 1964, Clark had received many honors and awards attesting to his success as author, experimenter, lecturer, and teacher. He was awarded the Nichols Medal (1936), the Borden Award (1944), the Passano Award (1957), and the Award of Merit of the Maryland Section, American Chemical Society (1963). He received honorary degrees from

Williams College (1935) and University of Pennsylvania (1940) and was appointed to several honorary lectureships. Clark was elected president of the Society of American Bacteriologists (1933) and of the American Society of Biological Chemists (1933-34).

Clark was born in Tivoli, N.Y., Aug. 17, 1884. He developed an interest in experimentation under the encouraging guidance of his father who operated a school at Tivoli. After his family left Tivoli, Clark attended the Hotchkiss School in Lakeville, Conn. He entered Williams College in 1903. Upon graduation in 1907 he was invited to remain at Williams as an advanced student in chemistry and as assistant to Leverett Mears. His M.A. degree was awarded 1 year later. Clark then entered Johns Hopkins University where he worked toward his doctor's degree under Harmon Morse. Morse assigned to Clark the thesis problem of redetermining the osmotic pressure of cane sugar solutions. Clark left Johns Hopkins with his Ph.D. in 1910.

During his college years, Clark worked several summers at Woods Hole, and it was through his contacts there that he came to be offered the job of chemist in the laboratories of the Dairy Division of the U.S. Department of Agriculture. Given a great amount of freedom in establishing work objectives, Clark researched a variety of topics which interested him—including the holes in Swiss cheese. He grouped certain bacteria from milk, bovine feces, and some grains according to the composition of the gases given off by them. His studies which showed that cows' milk had about the same level of acidity as human milk (instead of a higher level as was previously thought) brought to an end the practice of adding alkali to cows' milk which was to be fed to infants. With Herbert A. Lubs, Clark established a group of some 13 dyes which could be used as indicators over nearly the entire range of pH. Clark and Lubs also described a number of useful phosphate and borate buffer systems.

It was during his years at the Dairy Division that Clark began his first book, "The Determination of Hydrogen Ions," which was published in 1920. The book was well received; it had gone through several printings and was in its third edition by 1928.

In 1920 Clark accepted the position of professor of chemistry in the Hygienic Laboratory of the U.S. Public Health Service. At the Hygienic Laboratory, Clark expanded studies begun at the Dairy Division labs on the oxidation-reduction potentials of dye systems. In response to the large number of requests for his reports on this subject, Clark's reports were combined and reprinted in 1928 as Hygienic Laboratory Bulletin No. 151, a book of 340 pages.

Over his protestations that he knew very little about medical matters, Clark was appointed DeLamar Professor of Physiological Chemistry in the School of Medicine of Johns Hopkins University in 1927. Much of Clark's research activity at Johns Hopkins continued to be in the area of oxidation-reduction behavior of organic compounds—particularly the metalloporphyrins. The dozens of papers published on this subject by Clark and his students report the results of their efforts at meaningful experimental and mathematical treatments of rather complex systems. Often using model compounds in order to make the experiments manageable, Clark developed the concept of an oxidation-reduction acid-base continuum in natural systems.

In 1948 "Topics in Physical Chemistry" was published. This book was written by Clark to be a supplementary chemistry text for medical students. After his retirement in 1952, Clark continued to conduct research on a few topics which interested him, the results of which were never published. His last book, "Oxidation-Reduction Potentials of Organic Systems," appeared in 1960.

Hubert Bradford Vickery, *Biog. Mem. Nat. Acad. Sci.* **39,** 1-36 (1967); William Mansfield Clark, *Ann. Rev. Biochem.* **31,** 1-24 (1962); *Chem. Eng. News* **30,** 2391 (1952); *ibid.* **41,** 114 (Nov. 18, 1963); *ibid.* **42,** 119 (Feb. 3, 1964).

GEORGE M. ATKINS, JR.

William Smith Clark

1826-1886

Clark was born in Ashfield, Mass., July 31, 1826 son of Dr. Atherton and Harriet (Smith) Clark. After graduating from Amherst College in 1848, he taught natural sciences at Williston Seminary where he had received his earlier education. He left in 1850 to study chemistry and botany at University of Göttingen, Germany, where he received his Ph.D. degree in 1852.

When he returned to the United States, he was elected to the Chair of Analytical and Applied Chemistry at Amherst. The following year he was given the Chair of Zoology and that of Botany in 1854. He gave up the latter two in 1858 but held the Chair of Chemistry until 1867. He was a member at large of the Massachusetts State Board of Agriculture from 1859 to 1861 and a member ex-officio from 1867 to 1879.

He was one of a commission of three appointed by Governor Andrew to consider the expediency of a State Military Academy in 1863, presidential elector and secretary of the Electoral College in 1864, representative to the legislature in 1864, 1865, and 1867, and trustee of several academic institutions.

Clark was a dynamic member of the building committee at Amherst. He procured funds, worked with architects, participated in the selection of sites, and aided in the general direction of building.

At the outbreak of the Civil War, Clark enlisted. He was commissioned as a major of the Twenty-first Regiment of Massachusetts Volunteers in 1861, promoted to lieutenant-colonel in February 1862, colonel in May 1862, and was recommended for promotion as brigadier-general in September 1862. He fought in most of the principal battles of the first 2 years of the war. Colonel Clark was reported first as captured and then as killed at the battle of Chantilly. Actually he had hidden within the enemy lines for 3 days after those with him had been shot down and finally reached the Union forces in safety.

After his service in the War, Clark returned to the Chair of Chemistry at Amherst. In 1863 he was appointed a trustee of Massachusetts Agricultural College which was to be established. The following year he was elected to the legislature. He was a forceful and persuasive speaker. Amherst was selected as the seat of Massachusetts Agricultural College owing primarily to Clark's effort and influence in general court and town meeting. His energy decisiveness, and resourcefulness were enormous.

He was elected president of Massachusetts Agricultural College Aug. 7, 1867. The presidency and professorship of botany and horticulture he held until 1879. He was the first president of the functioning college which opened to the students in October 1867. His predecessors, presidents French and Chadbourne, were involved only in the planning and initial building stages.

Clark carried on experiments in 1874 involving the expansive force exerted by the vegetable cell in its growth. He contributed new facts of great value in his results obtained in 1873-76 relative to the movement of sap in trees.

While at MAC Clark was invited by the Japanese to establish and organize the Imperial College of Agriculture at Sapporo, Japan, now Hokkaido Imperial University. During an 8-month period from the summer of 1876 to the spring of 1877, he thoroughly organized the institution. He was its president and taught two classes a day. He laid out a model farm and built the first American barn in Japan. He established the first collegiate military unit in the country. He initiated agricultural research and provided a meteorological observatory. He recorded the new flora and noted those plants which were desirable to introduce in the United States. He sent or brought back to the United States many plants, trees, and seeds. He founded the Convent of Believers in Jesus which eventually built a church in Sapporo on which the name of William S. Clark is carved in stone.

To this day Clark is held in great esteem by the Japanese as one who conferred great benefits upon their country. There is a bust of him at Hokkaido Imperial University. On the coat of arms of the university there are three letters: B B A. They stand for Clark's parting message to the youth of Japan, "Boys, be ambitious."

Since Clark's tenure at Hokkaido, there has been an exchange of professors and experiment station directors between MAC, now University of Massachusetts, and the Imperial University. On one of these exchanges, the soybean was given to University of Massachusetts. From this the U.S. soybean industry developed. In exchange and in gratitude for the soybean, five varieties of cranberries were taken to Hokkaido in 1967 by the head of the University of Massachusetts Cranberry Experiment Station who lectured on them.

The intense enthusiasm and personal magnetism which Clark brought to the lecture room established a bond of sympathy with students that was never broken. In his teaching he always stressed that for the professional botanist, a thorough groundwork of chemistry is not only essential but the more one specializes in botany, the greater his knowledge of chemistry should be.

His best influence was exerted in behalf of scientific agriculture when that science was in its infancy. Clark's claim to scientific recognition rests not so much upon his actual achievements in research as upon his administrative capacity and the manner in which he stimulated and encouraged investigation on the part of others.

Clark published papers in scientific journals and in the annual reports of the Massachusetts State Board of Agriculture. In 1869 he translated Theodor Scheerer's "Blow-pipe Manual" for use in the Agricultural College.

He resigned from the presidency of MAC in 1879 to become president of a "floating college." He directed his energy from 1879 to 1880 to the development of a plan uniting scientific study with a tour around the world. It was abandoned on the sudden death of its originator. Clark subsequently engaged in mining operations. He spent the last 3 years of his life at his home in Amherst with a heart condition. He died in Amherst, Mar. 9, 1886.

William S. Tyler, "History of Amherst College . . . 1821-1891," F. H. Hitchcock (1895); William S. Tyler, "History of Amherst College during its first half century," C. W. Bryan & Co. (1873); Frank P. Rand, "Yesterdays at Massachusetts State College," The Associate Alumni of Massachusetts State College (1933); "Dictionary of American Biography," **4,** 146-7, Chas. Scribner's Sons (1930); "Appletons' Cyclopedia of American Biography," **1,** 632-33, D. Appleton & Co. (1888); information from Chester E. Cross; biographies in *Proc. Amer. Acad. Arts Sci.* **21,** 520-523 (1885-1886) and by D. P. Penhallow in *Science* vol. 27, pp. 172-180.

ANN MARIE LADETTO

Frank Wigglesworth Clarke

1847-1931

Clarke was born in Boston, Mar. 19, 1847 to Henry and Abby (Fisher) Clarke. His father was a hardware dealer, and his mother died when he was 10 days old. In 1865 young Clarke entered the Lawrence Scientific School of Harvard to study chemistry. He remained at Harvard for a year following his graduation in 1867 to study with Wolcott Gibbs and to publish some studies of mineral analyses.

In 1869 he served at Cornell University as assistant to James M. Crafts. He then spent 4 years as professor of chemistry at Boston Dental College. During this time he wrote newspaper and magazine articles to supplement his salary. His articles took the form of reports of scientific meetings and popularizations of scientific topics, especially for young readers. In 1873 he was appointed professor of physics and chemistry at Howard University, and the next year he went to University of Cincinnati as professor of chemistry and physics. He held this position until 1883, when he was appointed chief chemist to the United States Geological Survey in Washington, D.C. and honorary curator of minerals at the United States National Museum. He held these offices until his retirement in 1924. In 1874 he married Mary P. Olmsted, by whom he had three daughters. He died of pneumonia in Chevy Chase, Md., May 23, 1931.

From his early childhood Clarke had been interested in comparing and tabulating collections of things: flowers, stamps, coins, and minerals. His earliest papers on the constants of nature published in the *Smithsonian Miscellaneous Collections* clearly illustrate this trait for compilation. His first paper in the series, "A Table of Specific Gravities, Boiling Points, and Melting Points of Solids and Liquids," appeared in 1874. Earlier papers in *American Journal of Sci-*

ence had dealt with atomic and molecular volumes.

While at Cincinnati Clarke began his series of investigations on the composition of minerals. He was assisted in this work at the Geological Survey by such well-known chemists as William F. Hillebrand, Frank A. Gooch, and Eugene Sullivan, the inventor of Pyrex glass. The most important contribution he made to this field was his book "The Data of Geochemistry," first published in 1908 and republished in 1924. This volume contained Clarke's calculations of the elementary composition of the outer crust of the earth. From 8,600 analyses of rocks Clarke and his associate Henry S. Washington selected nearly 5,200 superior assays. They averaged these to obtain the weight percentage of various oxides (Si, Al, Fe, Mg, Ca, Na, K, Ti, P) in the crust. Since silica is the most prevalent of the oxides, Clarke spent a lot of his time on silicates, and as early as 1895 he published a lengthy paper on "The Constitution of the Silicates" in U.S. Geological Survey *Bulletin 125*.

Chemists probably best remember Clarke for his work on atomic weights. In 1882 he published the first edition of his "Recalculation of Atomic Weights," (*Smithsonian Miscellaneous Collections,* vol. 27), which he began in 1877. A fourth edition appeared in 1920. Annually from 1893 until 1913 he issued reports on atomic weights in the *Journal of the American Chemical Society.* This 20–year period represented an exciting era in atomic weight determinations, largely because of the active investigations of Theodore W. Richards of Harvard. In 1900 Clarke was appointed chairman of the newly organized International Committee on Atomic Weights; in this capacity he issued annual reports until his retirement in 1922 (when he was appointed honorary chairman).

Clarke was among the American Association for the Advancement of Science members who petitioned for a chemistry subsection in 1873. He attended the Priestley meeting in 1874 and was active in the founding of the American Chemical Society; however, he resigned in 1877 within 2 months of his election to the Society, objecting to the dominance of the New York members in Society affairs. In 1889 he and Harvey W. Wiley sent out notices suggesting the formation of a rival National Chemical Society. Fortunately, the ACS agreed to hold general meetings outside of New York and organize local sections; as a result the secessionists returned to the fold. In 1901 Clarke served as ACS president.

The most complete sketch (containing a full bibliography of Clarke's works) is by L. M. Dennis in *Biog. Mem. Nat. Acad. Sci.* **15,** 146-165 (1932). Also helpful is Charles E. Munroe's sketch in *Proc. Amer. Chem. Soc.* **57,** 21-30 (1935), and Charles A. Browne's sketch in "Dictionary of American Biography" **21** (Supplement 1), 177-78, Chas. Scribner's Sons (1944).

SHELDON J. KOPPERL

Parker Cleaveland

1780-1858

Cleaveland, the first professor of chemistry at Bowdoin College and a member of its faculty for 53 years, was born Jan. 15, 1780 the son of a physician of Byfield, Mass. He entered Harvard at the age of 15, graduating in 1799. Undecided as to teaching, law, or the ministry, he taught school for 4 years before accepting an appointment as tutor in mathematics and natural philosophy at Harvard for 2 years. He was called to the newly established professorship of mathematics and natural and experimental philosophy at Bowdoin in 1805.

The curricula of the colleges of that period were essentially classical and sometimes included a senior year lecture course in natural philosophy. Cleaveland based his Bowdoin course upon the Harvard model but by 1808 had broadened the course to include chemistry and mineralogy. His interest in the latter field came about almost by accident when his advice was sought in connection with the discovery of a crystalline deposit at the falls of the Androscoggin River at Brunswick.

Cleaveland identified the crystals as mainly quartz and iron pyrite and sent samples to Aaron Dexter, professor of chemistry at Harvard, who confirmed his identifications and sent him a collection of minerals in exchange. Stimulated by the discovery of the local crystalline minerals, Cleaveland began a period of intensive study and collecting of minerals from a wide range of localities. His work led to the publication of the first American treatise on mineralogy in 1816, the "Elementary Treatise on Mineralogy and Geology." The publications of French and German mineralogists had not been translated, nor had these volumes much information on the minerals to be found in many of the newly opened American localities. The treatise received high praise in the leading scientific journals of the period, and Cleaveland received special commendation for his decision to base the scientific classification of minerals on their chemical composition rather than their external character.

The publication of the treatise led to an extensive correspondence with scientists at home and abroad and exchanges of minerals which brought an outstanding collection to Bowdoin, including a number from Berzelius. Cleaveland was elected to membership in 16 or more literary and scientific societies and received numerous offers of professorships elsewhere, all of which he declined.

Cleaveland published a second and enlarged edition of his treatise in 1822, but circumstances related to his growing interest in chemistry and his heavy duties as secretary of the faculty of the then newly established Medical School of Maine tended to turn his attention away from mineralogy, and he never completed his extensive notes for a projected third edition.

Cleaveland was remembered by his students and others as a gifted and popular lecturer. His lectures, generously illustrated with demonstrations, were presented not only to Bowdoin seniors but also to groups in neighboring towns—this latter a considerable undertaking when it is remembered that in those days an ox team was required to transport his apparatus.

When instruction began in the Medical School, Cleaveland gave the course of lectures on materia medica in addition to the chemistry course for the medical students. His lecture notes and notebooks of several of his students emphasize practical aspects of chemistry. Although he was an industrious student and classifier of scientific information, there is no evidence that he was involved in any extensive research program in chemistry. In addition, Cleaveland gave lectures on conchology and built up a substantial shell collection. For a period of 52 years, he recorded observations of the pressure, temperature, and weather conditions thrice daily at his home in Brunswick.

Cleaveland was an extremely methodical man who was said to have "a large bump of caution" and tended increasingly to restrict his activities to the campus and the town. He served the town for 20 years as commander of the Volunteer Fire Department, managing the hosepipe of the hand-operated pumper engine.

Cleaveland died in Brunswick, Oct. 15, 1858 at the age of 79 and is remembered as a great teacher and the "Father of American Mineralogy."

Leonard Woods' eulogy published in *Collections of the Maine Historical Society* **6** (1859); memoir, Nehemiah Cleaveland, Archives of Bowdoin College; other material, Cleaveland archives.

SAMUEL E. KAMERLING

Thomas Green Clemson

1807-1888

Clemson—chemist, geologist, agronomist, farmer, and diplomat—was born in Philadelphia, Pa., July 1, 1807. The son of a well-to-do merchant, Clemson attended Central High School and developed an interest in chemistry early in life. At age 19, he went to Paris and seems to have lived there for the next 10 years. At the Sorbonne, he attended the chemistry lectures of Louis Thenard, Joseph Louis Gay-Lussac, and Pierre Dulong. During the years 1828-32 he audited classes at the Ecolé de Mines Royales, but there is no record that he received a degree. Nevertheless, his stay in Paris was extremely productive. From 1830 to 1836 he published 15 papers in *American Journal of Science, Journal of the Franklin Institute,* and *Transactions of the Pennsylvania Geological Society.* Of these papers, three were on geology. The other 12 covered the chemistry of piperin, the manufacture of sulfuric acid, and the analysis of minerals, limestone, coal, and iron and copper ores.

When Clemson returned to the U.S. in 1837, he established himself as a consulting chemist in Washington, D.C. Here, he met Anna Calhoun, daughter of United States Senator and former Vice–President, John C. Calhoun of South Carolina. They were married Nov. 13, 1838. From then until 1844, Clemson worked as a mining engineer and farmed a thousand-acre plantation in South Carolina.

Through the efforts of Calhoun, now U.S. Secretary of State, President Tyler appointed Clemson Charge d' Affaires to Belgium in 1844, a post he held until 1851 when he returned to his plantation.

He sold his South Carolina property in 1853 and moved to a farm in Maryland just outside of Washington, D.C. where he lived until 1861. He kept a small chemical laboratory at his home and during this period published six papers on the chemistry of soils and fertilizers. After 1855 he was connected with the Agricultural Division of the U.S. Patent Office and became Superintendent of Agricultural Affairs for the United States in 1860. The impact that science could make on agriculture had fired his imagination, and in two papers published in *American Farmer,* he proposed that an agricultural college be established for the state of Maryland. In the report of the Patent Office for 1860, Clemson proposed the establishment of a U.S. Department of Agriculture, outlined its scope, and emphasized the importance of an efficient chemical laboratory as part of the organization. He was a forceful agitator for any cause that he believed in so his plea found public support. President Lincoln suggested such a department in his message to Congres in December 1861 and the department was created in May 1862. This was followed swiftly with the Land Grant College Act in July.

Threats of war between the Northern and Southern States ended Clemson's connection with the Patent Office in 1861. He was thoroughly Southern in his views and convictions, and following the outbreak of hostilities he offered his services to the Confederate Government. Assigned to the Army of Trans Mississippi, he was involved with the manufacture of raw materials for explosives. At the close of the war in 1865, he was supervisor of Mines and Metal Works.

After the war the Clemson family returned to South Carolina and for awhile lived in Pendleton, where Clemson was president of the Pendleton Farmers Association, which claimed to be the oldest such organization in the country. When "Fort Hill," the John Calhoun Homestead, was sold at auction in 1871, Clemson bought it for $15,000. The same year, Clemson's surviving children died and with no heirs, the Clemsons began plans to make Fort Hill a memorial to their children and to Calhoun that would serve some useful purpose. Clemson outlived his wife and became a recluse bent on saving every possible dollar for his memorial. When he died Apr. 6, 1888, he left the Fort Hill estate and about $60,000 to the state of South Carolina for the establishment of an agricultural college—now Clemson University.

A member of American and European learned societies, Clemson was decorated in Belgium by the Order of Leopold and by France with the Legion of Honor. An amateur violinist and connoisseur of art, during his years in Belgium he acquired a collection of 40 paintings.

R. N. Bracket, "Thomas Green Clemson, LLD, The Chemist," *J. Chem. Educ.* **5,** 433-444, 576-585 (1928).

HERBERT T. PRATT

Alexander Cochrane

1802-1865

Alexander Cochrane, Jr.

1840-1919

Alexander Cochrane, together with his sons Alexander Jr. and Hugh, built one of New England's two largest chemical companies during the industrial expansion of the late nineteenth century. When Cochrane Chemical Co. merged with its major competitor, Merrimac Chemical Co., in 1917, the combination was indisputably the largest; it was this plum which Monsanto plucked in 1929 when it bought Merrimac in its first major move to invade the eastern market.

About the elder Cochrane little is known beyond his connection with the company. Born in 1802 in Scotland, he worked there as an acid maker before coming to America in 1843. He settled his family first at Lodi, N.J., but quickly moved to Massachusetts to work for Lee and Blackburn; he soon left there to take charge of the chemical operations at Talbot Chemical Works in Billerica, north of Boston near Lowell. The old Talbot Dye Works, specialists in logwood, had just expanded its manufacturing operations to include sulfuric and muriatic acids and blue vitriol to supply the burgeoning textile and paper industries in New England. As supervisor for Talbot, Cochrane was invaluable; but recognizing the implications of expansion, he soon left to start his own works in nearby Malden.

From the start Cochrane stuck with heavy chemicals—acids and a few inorganic salts such as blue vitriol (copper sulfate). Very nearly from the start he brought his sons into the business. Alexander Jr., born May 12, 1840 at Bar Head, Scotland, spent 3 years with Talbot, beginning in 1856, then came over to his father's works. Hugh Cochrane, somewhat younger, came in at about the same time; and in 1863 the business, now known as Malden Chemical Works, A. Cochrane & Sons, Proprietors, was made a three-way partnership. Old Alexander Cochrane died at the family's summer home at Newburyport Aug. 11, 1865; and the sons continued with Alexander Jr. acting as front office man and Hugh as plant supervisor.

With business booming all about them, the Cochrane brothers did not long confine themselves to local operation in Malden. Before the elder Cochrane died the firm had already acquired Newton Chemical Co., at Waltham, obtaining in the deal a platinum still which allowed them an edge on their competitors in the strength and purity of their sulfuric acid. Now the brothers branched out into indigo and began a program of acquisition and expansion which by the World War would see the company broad, strong, and diversified.

Their first move, in 1872, was to buy New England Chemical Co., competitors in sulfuric acid manufacture, with buildings and land along the Mystic River in Everett, with direct access to Boston harbor. Over the next few years they added to their land holdings there, and this is today the site of Monsanto's Merrimac Division. In 1879 the "New South" plant was built and in 1881 the "West Works" and an acetic acid and Glauber's salt (sodium sulfate decahydrate) plant. In 1881 also Cambridge Chemical Works was bought, adding sulfuric acid production capacity and introducing a new product, ammonia. More plants were added in 1892 (indigo); 1893 (aluminum sulfate; rebuilt in 1911); 1894 (ammonia); 1895 (acetic acid); 1897 (general manufacture); 1907 (contact-process acid); and 1915 (glue works).

As an interesting sidelight on all this expansion, the brothers at one point accepted stock in a struggling battery works which was slow in its payments; later Alexander bought out his brother's shares. The technical brain in the business was that of Alexander Graham Bell; and it was the original investment, plus his certainty of Bell's worth, which made Cochrane become one of the major investors in the early Bell Telephone Company—a stroke of combined wisdom and luck which made him one of the richest men in New England.

Hugh Cochrane died in January 1904. Alexander, in his sixties by this time and with none of his three sons inclined toward chemical manufacture, turned over much of his management function to his son-in-law, Lindsley Loring. Recognizing the business opportunities opened up by World War I, and foreseeing the profound changes which would overtake the chemical industry after the war, Cochrane knew that the business required reorganization; but he was no longer young enough to handle anything as far-reaching as he envisioned. Accordingly he arranged the merger with Merrimac Chemical in 1917. Both companies were strengthened thereby, with top level managers from both firms retained in strong positions. Alexander Cochrane died in Boston, Apr. 10, 1919. Merrimac was absorbed into Monsanto in 1929.

W. Haynes, "Chemical Pioneers," 57-73, D. Van Nostrand Co. (1939, reprinted 1970), gives the full history of the company and all three Cochranes. For Alexander Cochrane Jr., *see* "National Cyclopaedia of American Biography," **27,** 204, James T. White & Co. (1940); *New York Times,* April 11, 1919.

ROBERT M. HAWTHORNE JR.

Wallace Patten Cohoe

1875-1966

Cohoe was born in Norwich, Ont., Canada, Feb. 22, 1875 a descendant of eight generations of Quaker stock. As a teen–ager, he became a student at McMaster University where he received his B.A. degree in 1896, his M.A. in 1898, and an LL.D. in 1944. He also received an A.M. degree from Harvard in 1944. He served in various capacities at McMaster. After a year as an assistant in the Chemistry Department of Harvard University, he was invited back to McMaster University and remained there for 5 years, leaving with the rank of professor of chemistry in 1912. From 1907 to 1912 he was also president of Chemical Laboratories, Ltd. of Toronto.

Possessed of considerable initiative and ambition—as well as a warm, genial and friendly nature—he left the security of university life to become a consultant in 1912. For this purpose Cohoe left Canada for the United Sates where he hoped to broaden his horizons. This worked out well as he was shortly appointed an advisor to the U.S. Bureau of Mines and quickly expanded his consulting contacts and commitments, particularly in the field of textiles. He became a U.S. citizen in 1913.

Cohoe was a great reader, frequently quoting from the Bible, the classics, or Gilbert and Sullivan. He was an authority on medieval architecture. His hobby was fine machine-work, which he carried out in the well–equipped basement of his Riverdale, N.Y., home. The owner of a handsomely trimmed beard and mustache, as well as a deep, booming voice, he was cordially received by American chemists and their various organizations. In 1904 he became a member of the Society of Chemical Industry, a London-based multinational technical society in Canada and was chairman of the Canadian section in 1912. He became active in the American section of the Society of Chemical Industry and was chairman of the American Section in 1938-40. He became vice-president and later president of the Society in 1942, which involved attending many international meetings in England and other countries of Europe, where many scientific articles of his were published in various technical journals. He received the Messel Medal of the Society of Chemical Industry in 1948. Cohoe was president of The Chemists' Club from 1945 to 1947.

During World War I he collaborated with Frederick Zinsser to put into operation the manufacture of mustard gas using a process cabled from Great Britain by the late Sir William Pope. An early specialty of Cohoe's concerned the hydrolysis of sawdust to fermentable sugars using hydrochloric acid. This work resulted in two patents. This line of research was followed up in Germany, where Friedrich Bergius of Heidelberg further improved it. In the U.S. two chemists, Theodore Wagner and Frederick Kressman, applied a sulfuric acid modification of this to produce alcohol from Louisiana pine sawdust during World War I. The process was further used in World War II when alcohol became an important precursor of synthetic rubber, but the process proved uneconomical in peacetime.

A more important invention of Cohoe's was a synthetic sausage casing of cellulose. Patents covering this were applied for in all the countries of the world important in chemical development, including Germany, Austria, and Russia. In Germany, chemists accepted this with reluctance as they had already spent 40 years of research on the same project. In fact, Cohoe had to go to Berlin in order to demonstrate the viability of his claims before they were accepted. But following this were many accolades and much clicking of heels. The Austrians were likewise skeptical but once convinced by concrete evidence, they ate all of the demonstration sausages!

He put viscose finishes on cotton piece goods to give damask quality, and he improved dyeing processes. He was also active as a consultant in other fields including pigments, dielectric heating, bottle closures, non-curling paper, and electrostatic drying of thin sheets.

Cohoe had one daughter by his wife Edith. He died Nov. 10, 1966, shortly before his 92nd birthday.

Chem. Eng. News **19,** 689 (1941), with portrait; editorial, *Chem. Ind.* (London) **62,** 245 (1943); obituaries, *The Percolator* **42,** No. 1, 21 (Dec. 1966), and *New York Times,* Nov. 12, 1966; personal recollections.

CORNELIA TYLER SNELL

Peter Collier

1835-1896

Peter Collier, son of Jacob and Mary Elizabeth Collier, was born Aug. 17, 1835, in Chittenango, N.Y. He attended local schools until he was 14 years old and then went to work in a drygoods store. He left there after a year and was employed as a salesman for 4 years. He then attended Late's Polytechnic Institute of Chittenango, entered Yale College, and graduated in 1861. He continued graduate work in chemistry under the direction of Samuel W. Johnson. After receiving his Ph.D. degree in 1866, Collier went to University of Vermont in 1867 as professor of chemistry, mineralogy, and metallurgy, and as professor of toxicology and chemistry in the university's Medical School. Collier also studied medicine, received his M.D. degree in 1870, and became dean of the medical faculty. However, Collier's interests turned more and more to agriculture. In 1871, he was elected secretary of the state board of agriculture, mining, and manufacture and soon became active in the farm institute movement, in which educational leaders went to the farmers with lectures and demonstrations in an attempt to assist them with more productive agricultural practices. In 1873 President Grant appointed Collier as one of six scientists to represent the United States at the International Exposition at Vienna. Collier later made a report on the fertilizers at the Exposition.

In 1878, Collier was appointed chemist of the United States Department of Agriculture and began work on grasses and forage crops. At that time, William LeDuc of Minnesota was serving as Commissioner of Agriculture. One of LeDuc's major aims was to make the nation as self-sufficient as possible. He seized upon sugar and tea as two products, largely imported, that should be produced in the United States. Collier was placed in charge of the sugar work and first concentrated on improving the yield of sugar from cane and upon obtaining sugar from beets, corn, and other products. The beet sugar industry was established during this period, but its future possibilities were underestimated in the Department. Instead LeDuc became impressed with the possibilities of sorghum sugar and directed Collier to concentrate his efforts there. A sorghum sugar industry was never established. However, Collier became so confident that success was near he continued to work on it after he was forced, because of a change in administration, to leave the department. In 1884, he published a 570-page volume on sorghum.

Collier in 1887 was named director of the New York Agricultural Experiment Station at Geneva, New York. There he directed the research programs along careful scientific lines upon problems which had direct practical importance to the farmers of New York. As a result, the station became firmly established as a great research institution of outstanding practical bent and value.

In 1895, afflicted with an incurable illness, Collier resigned and moved to Ann Arbor, Mich., where he died on June 29, 1896. He was survived by his wife, Caroline Frances Angell, whom he had married in 1871, and by a daughter. Collier was widely admired as a man, not only of scientific achievement, but of charm, wit, and friendliness.

"Dictionary of American Biography" **4,** 304, Chas. Scribner's Sons (1930); "National Cyclopaedia of American Biography" **8,** 356, James T. White & Co. (1898); United States Department of Agriculture, "Annual Reports of the Commissioner," 1877-83; New York Agricultural Experiment Station, "Annual Reports," 1888-95; sketch by W. H. Jordan in L. H. Bailey, ed., "Cyclopedia of American Agriculture," vol. IV.

WAYNE D. RASMUSSEN

Arthur Messinger Comey

1861-1933

Comey was born Nov. 10, 1861 in Boston, Mass. He attended Harvard University and graduated with his A.B. degree in 1882. He then traveled to Europe to continue his education in chemistry, studying first with Victor Meyer in Zürich and later in the laboratory of Robert W. E. v. Bunsen at Heidelberg. Comey received his Ph.D. degree from Heidelberg in 1885.

Upon his return to the United States, Comey was appointed instructor in chemistry at Harvard. One of his assignments at Harvard was to organize and serve as director of a summer school for chemistry. In 1889 he became professor of chemistry at Tufts College. Four years later, he left Tufts and returned to Boston to open his own analytical and consulting laboratory. Comey devoted much of his time immediately after leaving Tufts to compiling data for the first edition of "A Dictionary of Chemical Solubilities: Inorganic" which he published in 1896. One of the commissions which came to his consulting laboratory was to investigate a process for extracting gold from sea water. While he was successful in obtaining a bead of gold (visible only under a microscope), Comey demonstrated the process to be economically unattractive.

In 1906 Comey was invited by Charles L. Reese to join Du Pont Co. as director of the company's Eastern Laboratory. Reese had founded Eastern Laboratory in 1902 as a site for conducting explosives research. Comey and Reese had first become acquainted while both were students at Heidelberg. The period of Comey's directorship of Eastern Laboratory (1906-19) was one of great activity and included the years leading up to and including the first world war. In addition to his administrative duties, Comey found time to publish several papers dealing with explosives and chemicals used in the manufacture of explosives. His most notable contribution was the development of an explosive made by nitrating a mixture of sugar and glycerine. The material obtained was less expensive and less hazardous than nitroglycerine. In 1919 Comey transferred to the company's headquarters in Wilmington, Del. He retired in 1921.

During the war, Comey was chairman of the Sub-Committee on Explosives of the National Research Council. He published the second edition of "A Dictionary of Chemical Solubilities: Inorganic," co-authored by Dorothy A. Hahn, in 1921.

Comey was an amateur musician and an enthusiastic gardener, woodsman, and hiker. It was to these pursuits that he devoted himself after retirement. For his hobby of keeping weather records he invented a very sensitive recording thermometer. He divided his time between homes in Cambridge, Mass., and Lake Chocorua, N.H. He died Apr. 6, 1933.

C. L. Reese, *Ind. Eng. Chem.* **20,** 438 (1928); J. F. Norris, *Proc. Amer. Acad. Arts Sci.* **68,** 618-9 (1933); A. P. Van Gelder and H. Schlatter, "History of the Explosives Industry in America," Columbia Univ. Press (1927).

GEORGE M. ATKINS, JR.

George Hammell Cook

1818-1889

Cook can be called a chemist more by the uses he made of that science than by his explicit activities in it; he was appointed professor of chemistry and natural sciences at Rutgers College in 1853 but thereafter became state geologist, founder of New Jersey's Agricultural Experiment Station and builder of the university's School of Science. Cook was born Jan. 5, 1818 in Hanover, N.J. the son of John and Sarah Munn Cook; he was of the eighth generation of descendants of Ellis Cook, who settled on Long Island in 1644. His education was somewhat patchy, apparently as a result of his desire to do rather than to study. After leaving the regional school in Hanover he taught himself surveying and helped lay out railroads in New York and New Jersey and roads in Greene and Schoharie Counties, New York. He then enrolled in the Rensselaer Institute, as it was called at the time, and received his C.E. degree in 1839. Later he would earn his B.N.S. (Bachelor of Natural Sciences) and M.S. degrees at the same school.

The next decade and a half was a period of casting about; he taught at Rensselaer after his graduation, becoming adjunct professor and in 1842 senior professor of geology and civil engineering. In 1846 he quit to engage in glass manufacture across the Hudson River in Albany; but in 1848 he returned to teaching, this time as instructor in mathematics and natural philosophy at the Albany Academy, where he became principal in 1851. In 1852 the state of New York sent him to Europe to study salt deposits, to learn ways of exploiting the deposits in Onondaga County. His reports of his findings, both here and abroad, constitute the first of his many technical publications.

In 1853 at the age of 35, Cook finally found a place with scope enough to absorb his prodigious energies. He accepted the chair of chemistry and natural science at Rutgers College and stayed there the rest of his life. His activities were so many and so mixed that it is simpler to take them one by one rather than all at once, chronologically: Academically he remained professor of chemistry until 1864, when, with a land grant under the Morrill Act, Rutgers established a New Jersey state college for the promotion of agricultural and mechanic arts. Land for an experimental farm was obtained, and Cook became a vice-president of the school, with emphasis on agriculture. Early in the school's operation he introduced the Rensselaer system of student laboratory experiment, rather than instructor demonstration. As late as 1878 he was still professor of analytical chemistry, geology, and agriculture; 2 years later he dropped the chemistry duties.

In geology, Cook became assistant state geologist in 1854, under William Kitchell, who was conducting a geological survey of New Jersey. Although the survey was interrupted in 1856, its mapping was of a quality unrivalled for years, and Cook produced papers on the geology of Cape May County and on the subsidence of coastal lands in New Jersey and Long Island. These, together with his later mapping of the terminal moraine across northern New Jersey, a relic of the Wisconsin glacial stage, were virtually the only theoretical papers in his career. His interest in geology was for the most part practical and economic, focussing on the mineral deposits in North Jersey, the agricultural soil types in the south, economically valuable clay deposits, etc. In 1864 the geological survey was reactivated, and Cook was named state geologist. In 1868 his book "Geology of New Jersey" appeared.

After 1864 his interest in geology, although never extinguished (he would attend the international geological congress in Paris in 1878, as a delegate, for example), began to give way to his concern for state agriculture. It was in that year that the new science and agriculture school was established, and by 1870 Cook went to Europe to study agricultural experimentation there. In 1871 he was instrumental in setting up the New Jersey Board of Agriculture; by 1879 he persuaded the state legislature to establish a state agricultural experiment station; and in 1887 he was prominent among the advocates who influenced the federal government to provide funds for similar stations in all states.

In addition to his many other activities Cook found time to begin making regular weather observations in 1857, an activity which culminated in his founding and directing a state weather service in 1886; and to concern himself with water supplies, both for New Brunswick and for many coastal communities which had depended upon contaminated surface sources until Cook showed them how to reach deep aquifer layers for pure water. Cook's lifetime output ran to more than two dozen papers, thirty annual reports of the geological survey, and the book already mentioned.

Cook was awarded an honorary Ph.D. degree by University of New York, as early as 1856; a decade later, in 1865, Union College gave him an LL.D. degree. Cook was married in 1846 to Mary Halsey Thomas. They had six children: Paul, Sarah Cook Williamson, John Willard (who died in infancy), Emma Willard, Anne Bigelow, and Robert Anderson. Cook died of heart failure Sept. 22, 1889.

Biographical material on Cook is plentiful; the fullest biographies are by G. K. Gilbert in *Biog. Mem. Nat. Acad. Sci.* **4,** 135-44 (1902), and John C. Smock in *Am. Geologist* **4,** 321-26 (1889).

In addition see Wm. H. S. Demarest, "A History of Rutgers College" (1924), pp. 368-70, 463-66; Geo. P. Merrill, "The First One Hundred Years of American Geology," 421-22, Hafner Pub. Co. (1924); and *Bull. Geol. Soc. Am.* **1,** 519-20 (1890). Genealogical information may be found in Charles C. Baldwin, "Baldwin Genealogy," 1296, Cleveland Supplement (1889); of the accounts in compilations of biography, those in "Dictionary of American Biography," **4,** 373, Chas. Scribner's Sons (1930) and "Lamb's Biographical Dictionary of the United States," **2,** 162, James H. Lamb Co. (1900) are thorough; that in "National Cyclopedia of American Biography," **6,** 304, James T. White & Co. (1892) is less so, and contains inaccuracies as well.

ROBERT M. HAWTHORNE JR.

Josiah Parsons Cooke

1827-1894

Cooke was born in Boston, Mass., Oct. 12, 1827. His father, also Josiah Parsons Cooke, was an eminent lawyer. His mother, Mary (Pratt) Cooke, died when he was six. Throughout his life he suffered from bad health. A severe affliction eventually led to the loss of one of his eyes. In 1860 he married Mary Hinckley Huntington. They had no children.

As a boy he attended Benjamin Silliman's public lectures on chemistry and was so attracted to the science that he fitted out a home laboratory. He went to Harvard and graduated in 1848. The next year Cooke was appointed tutor in mathematics and the following spring (1850) was appointed instructor in chemistry and mineralogy on the condition "he provide at his own charge the consumable materials necessary in performing chemical experiments." At 23 he was elected to fill the Erving Professorship of Chemistry and Mineralogy, a position he held until his death.

Cooke considered the methods of teaching at Harvard excellent for the classics and mathematics but unsuited for a subject like chemistry. His first step, after his appointment, was to obtain a leave of absence, which he spent in Europe buying chemicals and apparatus, mostly at his own expense.

On his return in 1851 he began to reorganize chemical instruction at Harvard. Cooke recognized the shortcomings of the recitation method and eagerly turned to the laboratory method. It took the college 7 years finally to recognize the laboratory course and to provide accommodations by expanding Boylston Hall.

In 1854 Cooke published his first scientific paper "The Numerical Relation Between the Atomic Weights and Some Thoughts on the Classification of the Chemical Elements." His first large book, "The Elements of Chemical Physics" (1860), ran through three editions. An even more striking success was "The New Chemistry" (1874), containing a popular account of the then current chemical theories. It ran through five editions and was translated into numerous languages including Russian.

In 1876 the London Chemical Society elected Cooke Honorary Member, and 6 years later the University of Cambridge conferred on him the honorary degree Doctor of Laws.

Cooke's experiments on determining atomic weights were described in five papers on antimony. His results were very satisfactory and are close to the modern values.

Shortly after Bunsen and Kirchhoff invented the spectroscope, Cooke constructed the most powerful model of that time and carried out many experiments in spectroscopic analysis.

Cooke also found time to probe the distinctions between science and religion and science and philosophy. His "Religion and Chemistry, or Proof of God's Plan in the Atmosphere and its Elements" (1864) was an attempt to argue natural theology based on scientific data. A similar undertaking marked "The Credentials of Science and The Warrant of Faith" (1888), which he intended for divinity students.

By 1872 Cooke had essentially accomplished the two tasks which confronted him as Erving Professor at Harvard: the "new" method of instruction had been implemented; and equal rights had been won for chemistry and the sciences. Cooke's accomplishments came at a time when Greek and Latin were still considered the most important keys to formal education and instruction in the sciences was either inadequate or confused.

When Cooke died in Newport, R.I., Sept. 3, 1894 one of his friends wrote: "Few men have seen so many of their goals reached." In all, Cooke published eight books, 41 papers on chemistry, and 32 papers on other subjects. He is regarded as the founder of the modern chemical and mineralogical departments at Harvard University.

"Dictionary of American Biography," **4**, 387, Chas. Scribner's Sons (1930); obituaries by T. W. Richards in *Amer. Chem. J.* **16**, 566-568 (1894), by Charles L. Jackson and others in *Proc. Amer. Acad. Arts Sci.* **30**, 513-547 (1894-95), and by Charles L. Jackson in *Biog. Mem. Nat. Acad. Sci.* **4**, 175-183 (1902).

GEORGE SIEMIENCOW AND
ROBERT YUSZCZUK

Thomas Cooper

1759-1839

Chemist, lawyer, theologian, geologist, author, judge, and physician—all these appellations apply with equal validity to Thomas Cooper, a Renaissance Man of the early nineteenth century. Born in Westminster, England, Oct. 22, 1759, he studied law at Oxford, largely to please his father. Yet, because of his own special interests in medicine and science, he attended the lectures in anatomy, veterinary dissection, chemistry, medicine, and theology. He left, however, without securing a degree in anything.

If Cooper were to be judged by history for his experimental discoveries, the record would be bare indeed. About 1790, as part owner of a Manchester dyeworks, he developed an improved synthesis of "oxymuriatic acid" (chlorine) for bleaching. While occupied as a barrister and as chemist for the dyeworks, Cooper became a close friend of Joseph Priestley, with whom he held similar views on religion and science. The Birmingham riots which precipitated Priestley's emigration from England coincided with the financial collapse of Cooper's dyeworks, and the two families sailed for America and settled in Northumberland, Pa.

Upon naturalization in 1795, Cooper was admitted to the bar, and for 9 years he practiced law and medicine, wrote political tracts, edited a local newspaper, and served a 6 months jail term for violation of the Sedition Act (he "libeled" President John Adams).

He was appointed a district judge in 1804, but the harsh penalties, strict decorum, and rigid formalism which characterized his courtroom prompted more than 50 local petitions for his removal. In 1811 the legislature impeached him, and from that turning point until his death he devoted his major attention to science and education rather than to the legal profession. Later that year he accepted a professorship of chemistry at Dickinson College and prepared an introductory lecture for beginning chemists. The lecture was an historical view of chemistry, and its popularity among the students prompted the trustees to publish it in book form. In the volume, Cooper commented at length on Lavoisier's oxidation theory and on Davy's evidence that oxymuriatic acid was an element and not a compound. In both the lecture and in his other writings he continually emphasized the practical aspects of chemistry. He applied his knowledge to tracts on brewing, cooking, bleaching, and gas-lighting of cities. Furthermore, while at Dickinson he edited a popular science periodical called *An Emporium of Arts and Letters.* This, too, dwelt on applications of science to daily life.

In 1817, while serving as professor of chemistry and mineralogy at University of Pennsylvania—a post he received the previous year—he was awarded an honorary M.D. from University of New York. Armed with the degree, he vigorously sought a post on the medical faculty at University of Pennsylvania but was passed over for another. Keenly disappointed, he began to negotiate with Thomas Jefferson about a chair of chemistry, mineralogy, and law at University of Virginia which Jefferson was then organizing. Religious opposition to Cooper's libertarian theology (he was a nominal Unitarian) thwarted consummation of a proffered position; in 1819 he accepted employment at University of South Carolina.

Cooper, well-liked by the Carolina trustees, was offered the professorship of chemistry, and later of geology and of mineralogy; within 2 years was appointed university president, a post he held until 1834. The presidency years were difficult ones, for the South was rife with debate and dissension regarding states' rights, slavery, tariffs, and economic exploitation by the North. As early as 1827 Cooper was one of the first and most outspoken advocates of secession from the Union. When he retired from the presidency in 1834 at the age of 75, his friends in the legislature arranged for him to codify the state statutes, a job which occupied most of his declining years until his death May 11, 1839.

Thomas Cooper is also to be remembered for his progressive contributions to education in early America. He popularized the views that state-supported higher education was indispensable, that a liberal arts education should be a basis for professional training, that the time of study for an M.D. degree be increased, and that a properly trained physician should have a strong background in chemistry. Furthermore, he believed in encouraging experimental investigation at an early stage of schooling. To that end, while Cooper was at Dickinson College, his students engaged in research using Priestley's apparatus, and one managed to obtain publishable results.

A perfect example of a 19th century Renaissance Man, Cooper published with insight and clarity on such diverse topics as witchcraft, political economy, law, medicine, chemistry, geology, cookery, and tariffs, to mention but a few. At the time of his death his library totaled more than 2,500 volumes, not only on the subjects with which we normally associate his efforts but also on such topics as hieroglyphics, Chinese writing, Hebrew lexicons and grammars, Egyptian monuments, world travel, classics, and theology. By his writings and his personal prestige, he helped focus the attention of post-Revolution America on the science of chemistry and its promulgation, not only as a foundation for medicine but as a field in its own right.

Dumas Malone, "The Public Life of Thomas Cooper," Yale University Press (1926); Charles P. Himes, "The Apparatus of Joseph Priestley," Dickinson College Publication, Carlisle, Penna., undated; Eva V. Armstrong, "Thomas Cooper as an Itinerant Chemist," *J. Chem. Educ.* **14**, 153-158 (1937).

Ned D. Heindel
Mary C. Clark

Arthur Clay Cope

1909-1966

Cope was born at Dunreith, Ind., June 27, 1909. He graduated from Butler University in 1929 and received his Ph.D. degree from University of Wisconsin in 1932.

Following graduation from Wisconsin, Cope was a national research fellow at Harvard University, working with Elmer Kohler. In 1934 he accepted a position as associate in chemistry at Bryn Mawr College, was appointed assistant professor in 1935, and associate professor in 1938. While at Bryn Mawr he discovered the reaction now known by his name involving rearrangement of allyl groups in a three-carbon system. In 1941 while holding a Guggenheim fellowship for "Studies of the Phenomenon of Tautomerism," he joined the faculty of Columbia University as an associate professor.

Cope was on leave from Columbia from 1942 to 1944 as a technical aide and section chief in the Division of Chemistry of the National Defense Research Committee. He was responsible for aiding in the administration of various military research programs in such diverse areas as chemical warfare agents and antimalarial drugs.

Cope went to Massachusetts Institute of Technology in 1945 as professor of chemistry and head of the Division of Organic Chemistry. Six months later he became head of the Department of Chemistry and served in that capacity until 1965. He was appointed the first Camille Dreyfus Professor in Chemistry in 1965 in recognition of his outstanding contributions to the department.

Cope was for many years a principal officer of the American Chemical Society. He was elected to the board of directors in 1951 and in 1959 he was elected chairman of the board. He served in this capacity until his death except for one year, 1961, when he was president of the society. He also served on the board of editors of *Journal of the American Chemical Society* and *Journal of Organic Chemistry*. Few members in ACS history devoted more time to the Society than Cope.

He received numerous awards and honors for his research in organic chemistry. In 1944 he received the American Chemical Society Award in Pure Chemistry for his contributions in the fields of synthetic organic chemistry and molecular rearrangements. In 1958 he received the Charles Frederick Chandler Medal from Columbia University for his pioneer work on the chemistry of medium-sized ring compounds and for his recognition of transannular reactions. In 1964 he received the William H. Nichols Medal from the New York Section of the American Chemical Society. In 1965 he received the Roger Adams Award, which recognizes outstanding contributions in organic research. He published more than 200 papers in his field and was a member of the National Academy of Sciences.

Cope married Bernice Mead Abbott in 1930. They were divorced in 1963. He later married Harriet Packard and had one stepson, Gregory Cope.

Cope suffered a heart attack and died while dining with associates of ACS in Washington, D.C., June 4, 1966.

Communications from Butler University Alumni Association, Wisconsin Alumni Association, and Massachusetts Institute of Technology; *Chem. Eng. News* **39,** 72 (1961), **44,** 25-27 (1966), and **42,** 78 (1964).

LESTER KIEFT

Gerty Theresa Cori

1896-1957

Born Aug. 15, 1896 in Prague, Czechoslovakia (then Austria), the daughter of Otto and Martha Radnitz, Gerty Cori received her first education in a school for girls from which she graduated in 1912. Her father was a business man and manager of a sugar refinery, but it was a maternal uncle, a professor of pediatrics at the university, who encouraged her. At that time it was not easy for a woman to enter a university. After passing a special entrance examination which included subjects ranging from literature, history, and Latin to mathematics, physics, and chemistry, she entered the medical school of the German University of Prague in 1914 and received her M.D. degree in 1920. Shortly after graduation she married a fellow student, Carl Ferdinand Cori, and they both moved to Vienna, where she worked as assistant at the Karolinen Children's Hospital. Her work, both clinical and experimental, dealt with studies of the influence of the thyroid on temperature regulation. This was the difficult post-war period when there was little hope of pursuing a scientific career in Central Europe. In 1922 Gerty came to the United States to join her husband at the State Institute for the Study of Malignant Diseases in Buffalo, N.Y. (now the Roswell Park Memorial Institute). Her appointment as assistant pathologist involved routine examination of histological specimens and left her some free time. It was then that the Coris initiated a close collaboration in research on carbohydrate metabolism which was to continue for many years. She became an American citizen in 1928. In 1931 the Coris moved to the Washington University School of Medicine in St. Louis, where she was appointed research associate in the Department of Pharmacology. In 1946 the Coris transferred to the Department of Biochemistry at the same University. Gerty became professor of biochemistry, a post she occupied until her death on Oct. 26, 1957 in St. Louis. She had one son, C. Thomas Cori, born in 1936.

The early work in Buffalo dealt with carbohydrate metabolism on a broad basis, including that of tumors, which paved the way for later developments. A study of the mechanism of conversion of glycogen to glucose in the liver and to lactic acid in muscle, begun in Buffalo and continued in St. Louis, led to the isolation of a new phosphorylated intermediate, α-glucose-1-phosphate, which was shown to be the first product of glycogen degradation. The enzyme which catalyzed this reaction in the presence of inorganic phosphate was named phosphorylase. It split the α-1, 4-glucosidic bonds of glycogen in a reversible manner, leading in the direction of cleavage to the formation of glucose-1-phosphate. Another enzyme, phosphoglucomutase, catalyzed the conversion of glucose-1-phosphate to glucose-6-phosphate, thus joining the glycolytic pathway at the point at which glucose entered it. In the liver, glucose-6-phosphate formed from glycogen was split by a phosphatase, which explained the mechanism of blood glucose formation in this organ. The crystallization of phosphorylase led to the observation that the enzyme occurs in two interconvertible forms, one independently active and the other active only in the presence of adenosine monophosphate (AMP), shown much later to act as an allosteric modifier of the enzyme. Two other enzymes, one concerned with the cleavage and the other with the synthesis of the α-1,6-glucosidic bonds of glycogen, were described. This made possible the first synthesis of a biological macromolecule in the test tube. In 1947 Gerty Cori, jointly with her husband, was awarded the Nobel Prize in Physiology and Medicine.

In the final years of her life she returned to problems in pediatrics. A group of heritable diseases in children, characterized by excessive storage of glycogen in the tissues, was shown to be in reality a group of diseases, each having a specific, single enzyme deficiency. Four such deficiencies were recognized and used as a system for classification. Structure analysis by purified enzymes showed that some of the stored glycogens were of normal structure while others had an abnormal molecular configuration and this was explicable by the particular enzyme deficiency encountered. Gerty Cori's work demonstrated the importance of isolating and characterizing individual enzymes to gain an understanding of normal and abnormal metabolic processes.

B. A. Houssay, *Biochim. Biophys. Acta* **20,** 11-26 (1956); S. Ochoa and H. M. Kalckar, *Science* **128,** 16-17 (1958); C. F. Cori, *Ann. Rev. Biochem.* **38,** 1-20 (1969); Edna Yost, "Women of Modern Science" (1959); "Current Biography Yearbook" (1947).

Carl F. Cori

Charles Dubois Coryell

1912-1971

Coryell was born near Los Angeles, Feb. 21, 1912. He received both his B.S. (1932) and Ph.D. (1935) degrees from California Institute of Technology. His undergraduate and graduate studies were performed under the direction of Don M. Yost and Arthur A. Noyes on the chemistry of vanadium and silver. A bit later he was an American-German Exchange Fellow at Technische Hochschule, München.

For 3 years he was a research associate with Linus Pauling. From this association came a series of magnetic investigations of the nature of the metal bonding in the respiratory pigments. These papers are a landmark in the application of physico-chemical methods to the resolution of problems in molecular biology and still represent one of the most important advances in our far from complete understanding of the chemistry of hemoglobin and related compounds.

No scientist loathed war more than Charles did or was more dedicated to the cause of world peace, yet this did not blind him to the necessity of defending democratic institutions from totalitarianism. He was throughout his life patriotic in the best sense of that term, loyal to his country, to the human race, and to the ideals and aspirations of free men everywhere. In 1942 he left the staff of University of California at Los Angeles to contribute his scientific talents to the war effort. This work at the Metallurgical Laboratory of University of Chicago and later at Oak Ridge centered on the separation and radiochemistry of uranium fission products. The techniques developed in this applied research were later used in filling a final gap in the periodic table, and, with his characteristic generosity, Coryell gave his wife, Grace Mary, the honor of naming element 61 "Promethium," and his young coworkers, Jacob A. Marinsky and Lawrence E. Glendenin, a lion's share of the credit for its discovery. This name, said Coryell, symbolized the hope that man would not be punished, like Prometheus, for developing the atomic bomb and have his vitals torn out by the vultures of war.

In 1946 Coryell came to Massachusetts Institute of Technology, where he remained in the Department of Chemistry and the Laboratory of Nuclear Science until his death Jan. 7, 1971. Here the research conducted by the students and postdoctoral fellows who flocked to his laboratory from all over the world continued to contribute mightily to the physical chemistry of metal-ion complex formation and ion-exchange processes and to nuclear decay schemes. He assigned particular significance to his theory of the nuclear processes responsible for the cosmic creation of the heavy elements, and to the structural relationships between atomic nuclei and their "magic numbers."

In addition to about 100 technical papers, Coryell also co-edited a collection of Manhattan Project papers, "Radiochemical Studies: The Fission Products" in 1951 with Nathan Sugarman.

In 1953-54 Coryell was Louis Lipsky Visiting Professor at Weizmann Institute of Science in Rehovot, Israel, an honor which gave him particular pleasure. Always aware of the Nazi terrors, he held a special place in his heart for the Jewish people and their young nation. He was also the welcomed guest of University of Colorado, University of Notre Dame, Middle Eastern Technical University in Turkey under a Ford Foundation Grant, and a Fulbright Lecturer and Guggenheim Fellow at the Institut du Radium of University of Paris. He was a tireless worker on countless agencies and committees in the cause of world peace and the peaceful uses of atomic energy. In 1960 he received the American Chemical Society Award for Nuclear Applications in Chemistry, and 10 years later, in 1970, he was presented with the Atomic Energy Commission Citation. His many friends and colleagues at the presentation ceremony were deeply touched when, speaking under severe difficulties because of the toll taken by his final illness, he expressed his gratitude to the AEC, not only for its support of a lifetime of research, but for literally giving him and his fellow victims of cancer extra days of life by the techniques of radiotherapy.

Charles suffered more than his share of adversity, but he bore these difficulties with a cheerfulness that baffled and buoyed the spirits of all those around him.

It is said that his scientific achievements might have been even greater if he had troubled himself less with people and their problems. If this was a "fault," it is one for which all his students, co-workers, and friends were deeply grateful. Some teachers only teach, but Professor Coryell inspired. Long after the titles of the more prolific article scribblers are buried in old abstracts, science will continue to enjoy the influence of Charles Coryell in the scientific careers which he encouraged.

Personal recollections; communications from Mrs. Barbara Coryell and Julie Coryell; M.I.T. Chemistry Department files; Atomic Energy Commission citation, 1970; *Chem. Eng. News,* 73, Apr. 18, 1960, 56, Jan. 25, 1971; *New York Times,* Jan. 9, 1971.

RALPH A. HORNE

Frederick Gardner Cottrell

1877-1948

Cottrell was born in Oakland, Calif., Jan. 10, 1877 the son of Henry and Cynthia Cottrell and a descendant of Nicholas Cottrell who came to Rhode Island in 1638. After graduating from University of California at Berkeley in 1896 and following this with a year of graduate work there, "Cot" taught high school in Oakland for the next 3 years; he then proceeded in 1900 to Germany where he worked first with Jacob van't Hoff in Berlin and finally with Wilhelm Ostwald in Leipzig, where he obtained his Ph.D. degree in 1902. He became an instructor in physical chemistry at University of California in Berkeley but resigned in 1911 to become chief physical chemist of the U.S. Bureau of Mines, and in 1914 chief metallurgist there, assistant director in 1916, and director in 1919. In 1921 he became chairman of the Division of Chemistry and Chemical Technology of the National Research Council, and in 1922 director of the Fixed Nitrogen Research Laboratory, which post he held until 1930.

Although Cottrell had an enormous range of interests, he is best known for his development of electrostatic precipitators of suspended particles from gases, now used throughout the world to cleanse industrial smokes, as from power plants and cement kilns, with enormous prevention of air pollution of environment, combined with recovery of materials that would otherwise be lost, *e.g.,* cement, potash, and minerals. These developments were made during Cottrell's university professorship days. Rather than spend the rest of his lifetime in this field, he formed in 1912 the Research Corporation, which took over the patents involved, and in due course became a boon and aid for other inventors who would similarly turn over their inventions and patents to the corporation for full expansion. The profits so derived from the corporation became a source of funds for further research, of which one of the most famous resulted in the development of the cyclotron by Ernest Lawrence. The Research Corporation became a prototype for similar beneficent organizations of philanthropic nature. It was clearly the brain-child of Cottrell's outlook that every man should give of himself as much as he was able for the welfare of all.

As the years moved along, Cottrell played a large role in the development of industrial helium, petroleum dehydration, and nitrogen fixation by a variety of means. In the early 30's he played an important role in the development of Muscle Shoals and the Tennessee Valley Authority from both technical and legislative standpoints. The Research Corporation, with Cottrell's backing, provided the initial funds for the development of rocketry by R. H. Goddard, and, in quite another direction, the production of tasteless, odorless vitamin A by Nicholas Milas of the Massachusetts Institute of Technology.

Other Research Corporation-backed developments include the volume production of vitamin B_1 following upon the scientific foundation laid by R. R. Williams and colleagues, the blood fractionation work on large scale by E. J. Cohn of Harvard, and medical developments by M. Kharasch at the University of Chicago on germicidals such as Merthiolate and Mercurochrome. From all this flowed royalties that were redistributed to scores of institutions for continued research.

Few men have given so much personal inspiration to so many others throughout the United States and indeed the world. Cottrell was wholly dedicated to the advancement of scientific knowledge by himself and others who magnified his own career a thousand fold. He died Nov. 16, 1948 on the very grounds of his alma mater and but a few miles from his birthplace.

Frank T. Cameron, "Cottrell—Samaritan of Science," Doubleday (1952).

DEAN BURK

Alfred Hutchinson Cowles

1858-1929

Cowles, metallurgist and pioneer of the electric furnace, was born in Cleveland, Ohio, Dec. 8, 1858. He was educated in Cleveland's public schools, but beyond this, the Cowles family provided an extraordinarily rich environment for growth. Cowles' father Edwin C. published the Cleveland *Leader*, was a prominent Abolitionist, a champion of religious liberty, a founder of the Republican party, and possessed considerable technical ingenuity as well, with several mechanical patents to his credit. Alfred's aunt was one of the first women in the country to receive a college degree. His brother Eugene showed a similar breadth and ability: he began as an excellent *Leader* reporter but soon joined Charles S. Brush in founding the Brush Electric Light and Power Co., which lit Cleveland's business district electrically in 1881, the first such system in the world. His mother was Elizabeth Hutchinson Cowles. The Cowles family, already extensive and distinguished at the time of Alfred's birth, has been equally prominent throughout the present century.

At 17 Cowles went to Ohio Agricultural and Mechanical College (now Ohio State University) but moved after 2 years to Cornell where he "pursued an eclectic course in science" and distinguished himself at eight- and four-oared crew. In November 1882 he joined his brother Eugene at Santa Fe, N.M. Eugene, who had been sent to the Western mountains to cure his tuberculosis, had become interested in mining and smelting operations. Together the brothers investigated ores from the nearby Pecos River Mine, which consisted of zinc admixed with gold, silver, lead, and copper in so complex and refractory a mixture that conventional smelting methods were useless. For this reason they chose electricity for smelting, both for the tremendous heat of the electric arc and for electroreduction.

They did no important experimentation in New Mexico but were so confident of success that they persuaded their father to buy the Pecos River Mine. On their return to Cleveland in 1884 they swiftly set about building an electric furnace, and by Dec. 24 of that year applied for a patent claiming the "method of smelting or reducing ores . . . which consists in pulverizing the ore and mixing it with pulverized or broken carbon or like material, then introducing the mixed ore and carbon within an electric circuit, of which it forms a continuous part . . . whereby the heat is generated . . .". The original application claimed electroreduction also, but this was removed on Patent Office objection of prior art in European work 30 years earlier.

On this basis the Cowles Electric Smelting and Aluminum Co. produced electrolytic aluminum bronze, and by chemical removal of the alloying metal, pure aluminum, in 1885. A year later Charles Hall, who is customarily credited with originating commercial aluminum production, applied for his basic patent on electroreduction of alumina using molten cryolite (Na_3AlF_6) as solvent. Hall worked briefly for the Cowles brothers but left in 1886 to help set up Pittsburgh Reduction Co.

Within a very few years Pittsburgh Reduction and Electric Smelting and Aluminum (the Cowles name having been dropped) were at each other's throats with patent litigation. The Cowles needed Hall's cryolite-solvent process, and Pittsburgh Reduction needed the internal-heating concept covered by Cowles patents, or those of Charles S. Bradley which were owned by Cowles. At the end of a decade and a half of legal battle the Cowles' position was so weakened—Pittsburgh Reduction (later Aluminum Company of America) having cornered the bauxite market—that they eventually gave up aluminum production. Their electric furnace idea, however, is credited with building several industries and processes which Alfred Cowles controlled after Eugene's death in 1892, among them were those which produced Carborundum, graphite, phosphorus, and calcium carbide for acetylene. The Franklin Institute presented the Cowles brothers with both the Elliott Cresson and the John Scott Legacy medals for this invention in 1886.

Alfred Cowles' later publications showed a continuing interest in recovery of useful chemicals such as inorganic fertilizers from mineral sources. His discovery of sodium metasilicate's cleaning properties while trying to recover aluminum from clay led in 1923 to the founding of the Cowles Detergent Co.

Cowles was married late, in 1906, to Helen J. Wills; they had no children. He died at his home in Sewaren, N.J., Aug. 13, 1929.

"National Cyclopedia of American Biography," **22,** 44-43, James T. White & Co. (1932); Alfred Cowles (a collateral descendant), "The True Story of Aluminum," Regnery (1958); *J. Four Electrique et des Industries Electrochimiques* **38,** 419-20 (1929); in addition the Cowles process for aluminum production is well discussed in Joseph W. Richards, "Aluminum: Its History etc.," 3d ed., Baird (1896).

ROBERT M. HAWTHORNE JR.

John Redman Coxe

1773-1864

Coxe was born in Trenton, N.J., Sept. 16, 1773. His parents, Daniel and Sarah (Redman) Coxe, sent him to Great Britain for his education. In 1790 he returned to America, became a medical pupil of Benjamin Rush, attended the University of Pennsylvania Medical School, and received his degree in 1794. He spent two additional years in Great Britain and France, mainly in extending his medical experience, then returned to Philadelphia where he practiced medicine, was port physician, physician to a hospital and dispensary, and ran a drugstore.

Coxe became interested in chemistry in his youth, perhaps being stimulated by his medical preceptor, Rush. He helped organize the Chemical Society of Philadelphia in 1792 and was a president of the group. Abroad from 1794 to 1796 he attended chemistry lectures in London, Edinburgh, and Paris. In 1809 he was elected professor of chemistry in the University of Pennsylvania Medical School.

From all accounts he was a failure in this position; he did not know sufficient chemistry, had no inclination to carry on laboratory investigations, and did not keep up with the rapid advance in the theories and content of the science. His clumsiness as a lecture demonstrator led Charles Caldwell, who observed him, to remark: "In his manipulations, Dr. Coxe gave proof that a chemist does not, by failing once in an experiment, always learn how to succeed. . . . He failed dozens of times in the same experiment."

Coxe finally realized that he had little aptitude for chemistry, and in 1818 he transferred to the professorship of materia medica and pharmacy. His proposal that the university grant degrees in pharmacy stimulated apothecaries to establish their own institution in 1821, The Philadelphia College of Pharmacy and Science, the first pharmaceutical college in America. In 1835 he was the target of a student riot and was fired by the trustees of the university.

Not suited to be a chemist or a teacher, Coxe nevertheless seems to have been a competent physician. He compiled "Philadelphia Medical Dictionary"; edited *Philadelphia Medical Museum,* a journal in which local chemists occasionally placed articles, from 1804 to 1810; and published "American Dispensatory . . . according to the Principles of Modern Chemistry," which superseded imported European dispensatories and passed through nine editions over a quarter of a century. In 1812 he founded *Emporium of Arts and Sciences* and edited the first two volumes. In 1818 he translated Orfila's "Elemens de Chimie Medicale" into English and published it under the title "Practical Chemistry." His few articles in chemistry were of no consequence.

Coxe was a scholar, able to read and speak nine languages, and an antiquarian bookworm who gathered one of the largest, private libraries in the country. After "never having been sick in his life" he died in Philadelphia, Mar. 23, 1864, at the age of ninety.

Edgar F. Smith, "Chemistry in Old Philadelphia," J. B. Lippencott, 52-61 (1919); Harriot Warner, ed., "Autobiography of Charles Caldwell," Lippincott, Grambo & Co., 306-309 (1855, reprint 1968); "Dictionary of American Biography," **4,** 486, Chas. Scribner's Sons (1930); biographies by "Cato" in *Boston Med. Surg. J.* **41,** 156-159 (1849-50), and M. C. Coxe in *Univ. Pennsylvania Med. Bull.* **20,** 294-301 (1908), with portrait.

WYNDHAM D. MILES

James Mason Crafts

1839-1917

There are few, if any, "name reactions" in chemistry better known than that discovered and developed by Charles Friedel and his student Crafts and since universally known as the "Friedel-Crafts Reaction." However, Crafts' other contributions to American science were significant even though not as well–known.

Crafts was born in Boston, Mass., Mar. 8, 1839 the son of a prominent New England wool merchant and manufacturer. He graduated from Harvard University in 1858 and spent the next 2 years in the study of mining and metallurgy at Harvard and with Plattner at Freiburg. An interest in chemistry developed rapidly. He was an assistant to Bunsen at Heidelberg (1860) and to Wurtz and Friedel at the École des Mines in Paris (1861-1865). At the conclusion of the Civil War he became chairman of the Chemistry Department at the newly-established Cornell University (1867). In 1870 he became chairman of general and analytical chemistry at Massachusetts Institute of Technology, founded a few years previously by William Barton Rogers, a friend of the Crafts family. In 1874 ill health forced Crafts to give up his position, but he was able to return to Paris where he remained until 1891. It was during this period (1877) that Friedel and Crafts became interested in a reaction between alkyl halides and metallic aluminum. When these substances were warmed, hydrogen chloride was evolved vigorously and a significant increase in boiling point of the mixture was observed. In an effort to decrease the vigor of the reaction benzene was used as a diluent and substitution of the benzene nucleus by alkyl groups occurred. It was discovered subsequently that a trace of anhydrous aluminum chloride was the active catalyst for the process. During Crafts' 17 years in Paris over 100 scientific papers were published, most of them describing the development of the Friedel-Crafts reaction.

In 1891 Crafts returned to MIT to direct instruction in organic chemistry and in 1897 he became president of the Institute. During a short tenure (1897-1900) he sought to improve the standards of the school so that it could ultimately become equivalent to the larger European universities. The chores of administrative work and continuing ill health led him to resign the presidency in 1900 and return to research, but the last 6 years preceding his death June 20, 1917 could not be spent in the laboratory; they were devoted to the preparation of papers describing his researches in catalysis and vapor pressure thermometry.

Among his many honors were included the Jecker Prize (1880), the Rumford Medal (1911), and Chevalier, Legion of Honor (1885).

G. A. Olah and R. E. A. Dear in G. A. Olah (Ed.), "Friedel-Crafts and Related Reactions," Interscience-Wiley (1963) Vol. 1, p. 1; obituary, T. W. Richards, *Proc. Amer. Acad. Arts Sci.* **53**, 801-804 (1917-1918); biography, A. Ashdown, *J. Chem. Educ.* **5**, 911-921 (1928), with portrait.

EDWARD R. ATKINSON

David Craig

1905-1964

Craig was born Jan. 12, 1905, on a farm near Palmyra, Iowa. His father, the owner of the farm, was a great reader and largely self-educated. The ideals of honesty, industry, education, and service to others were inculcated in David at an early age and served as guiding principles for all of his life. Craig obtained his education in Iowa and received his Ph.D. degree in 1928 from University of Iowa where he conducted research under George Coleman on certain reactions of nitrogen trichloride and other noxious and hazardous chloramines.

Harlan L. Trumbull, manager of chemical research for the B. F. Goodrich Co., met Craig at an organic symposium in Columbus, Ohio in December 1927 and laid before Craig the chemical challenge of the rubber industry. As a result, after receiving his degree Craig went to B. F. Goodrich to investigate the chemistry associated with the rubber industry. The detailed mechanisms of vulcanization and of deterioration of rubber were unknown, as were the exact chemical composition of many accelerators, antioxidants, and flex-resisters used in rubber compounding. Materials which would prevent the deterioration of rubber as used in tires and other consumer goods were badly needed so Craig started a long program of synthesizing amines, separating complex commercial mixtures, and testing highly purified compounds to determine their effect upon natural rubber compounds. Craig developed a philosophy of utilizing two or more processes, one after the other, as a means for purification. He used such schemes as crystallization from one solvent followed by recrystallization from another, fractional distillation followed by crystallization or selective extraction followed by distillation. Craig used the same principles in setting up his criteria for purity, such as melting point and mixed melting point followed by refractive index of the unknown compared with the compound of known structure and purity. He even purchased an Abbé refractometer for his own use and maintained and calibrated it periodically. Craig made major contributions in explaining the reactions of aromatic amines with aldehydes and ketones. Here his philosophy was stated: "unless you have a 100% material balance you can never really know how a reaction has proceeded."

Craig entered the synthetic rubber program at an early date with the mission of preparing pure monomers and other needed reagents. Here his philosophy of double confirmation paid off handsomely. For instance, styrene of purity stated to be greater than 99% was found to be unsuitable for making high quality copolymers. Styrene prepared by Craig from purified 1,2-dibromoethylbenzene gave excellent results. When the supplier applied his analytical method to Craig's styrene, he found it to be of 101.5% purity! Soon, thereafter, the supplier supplied styrene of suitable quality.

When emulsion polymerization was run using oleic acid from different sources there was great variability in the kinetics of the polymerization. Craig found the USP "oleic acid" which was being used contained only approximately 60% oleic acid, the remainder being fatty acids having two or more nonconjugated double bonds. The allylic hydrogens in the impurities were causing premature termination of the free radical polymerization process.

Craig was highly successful in separating pure 1,3-butadiene from commercial grades of butylene by extractive distillation. He also separated and determined the properties of cis and trans 1,3-pentadienes. But of even greater importance, he determined which impurities might be present in commercial butadiene or isoprene. He prepared these materials in a state of purity so that by adding known amounts to pure monomer, the quantitative effect upon polymerization and rubber quality could be determined. Based upon his work, realistic specifications for monomers to be used in polymerization could be set up.

When asked to find why the process for manufacturing 1,1-dichloroethene was operating poorly, Craig ran a material balance study in the laboratory. One of the by-products was volatile and could only be condensed in a dry ice trap. When Craig poured this by-product into a beaker in order to weigh it, there was a violent detonation, for, as was found subsequently, the by-product was chloroacetylene. Craig was wearing goggles during the experiment, however he did get glass particles and carbon black into his face and arms. Sent to the hospital, Craig said his accident was a God-send for the doctors found he had diabetes and could live a normal life only by the use of insulin.

After his experience in the synthetic rubber program, Craig became interested in some of the many unsolved chemical problems in the rubber industry. He studied the mechanism of the vulcanization of rubber with thiuram disulfides. In order to follow the reaction he used radioactive tagged sulfur atoms. Logically, this led to his interest in deuterated compounds, monomers, plastics, and rubbers. Craig prepared the first hydrogen-free organic rubber, namely the deuterio analog of natural rubber; poly(*cis*-1,3-perdeuterioisoprene).

Craig became editor of *Rubber Chemistry and Technology* in 1957, however, he still spent 8 hours a day in the laboratory performing with his own hands, and with the aid of an assistant, studies of a more general nature on the behavior of rubbery materials. Some of these were the physical flow characteristics such as viscous flow *vs.* turbulent flow as grains of polymers in the Mooney viscosimeter; the breakdown of the molecular structure of rubber by mechanical mastication and the effect of oxygen upon this process; rejoining of free radicals from the breakdown of rubber, giving products of basically different properties; fractionation of rubbers by the so-called coacervation process; and finally the segmental behavior of elastomers.

Craig was an outdoors man. He loved travel, canoeing, the woods, and the mountains, but he did not let this interfere with his scientific work. He died July 16, 1964 while at work in the laboratory, perhaps a fitting location for a man so devoted to his science. He was survived by his wife Catherine and their daughter Martha Coy.

David Craig, *Rubber Chem. Technol.* **37,** XXXII-XXXVIII (1964) and list of publications; personal recollections.

WALDO L. SEMON

Evan J. Crane

1889-1966

Crane was born in Columbus, Ohio, Feb. 14, 1889, graduated from Ohio State University with his B.S. degree in chemistry in June 1911, and immediately joined *Chemical Abstracts* as an associate editor. He became acting editor in 1914 and editor in 1915, a position he held until retirement at the end of October 1958. In 1956 when the organization was renamed Chemical Abstracts Service, Crane became the first director.

In the beginning, chemical nomenclature was vague and confused, almost hopeless. Crane began to organize, with the help of Austin M. Patterson, and to weld chemical information into a discipline. In 1918, he became chairman of the American Chemical Society Committee on Nomenclature, Spelling, and Pronunciation. He served on numerous national and international organic and inorganic nomenclature commissions. In 1948 he became a U.S. State Department delegate to UNESCO's Paris Conference on Scientific Abstracting. In 1958, Crane was the chairman of the first meeting of the National Federation of Science Abstracting and Indexing Services. From 1954 through 1960, he was the U.S. representative in chemistry to the Abstracting Board of the International Council of Scientific Unions. Crane was also chairman of the ACS Division of Chemical Literature in 1950.

Over the years his fellow scientists paid Crane tribute many times. The Ohio State University bestowed an honorary degree of Doctor of Science on him in 1938. The American Section of the Society of Chemical Industry awarded him its Chemical Industry Medal in 1937. In 1946 the ACS Board of Directors dedicated *CA's* Fourth Decennial Index to him for 36 years of service "in grateful appreciation of his success in opening the world's chemical literature to all chemists." Then in 1951, the Society, at its Diamond Jubilee meeting, gave him the highest honor available, the Priestley Medal. In 1953, Crane received the Austin M. Patterson Award for Documentation in Chemistry from the Dayton Section of the ACS. In 1965, Crane was honored by an Ohioana Citation "for prestige he brought to the State capital as the headquarters of the world's leading scientific information service." Finally, in 1971, the ACS created the first national award "to recognize major or specific contributions involving structuring and improvement of chemical information transfer," to be known as the E. J Crane Award in Chemical Information Science.

Crane was president of the Columbus Club of Rotary International, Mayor of the City of Upper Arlington, secretary of the Columbus Torch Club, and secretary of the Kit-Kat Club in Columbus for many years. He served on the board of The Ohio State University Research Foundation. He coauthored "A Guide to the Literature of Chemistry" in 1927 and 1957 and "The Kit-Kat Club of Columbus, Ohio" in 1961, and authored "CA Today—the Production of Chemical Abstracts" in 1959.

Crane remained active in chemical information activities until his death in Columbus, Dec. 30, 1966. He had no middle name, only a middle initial "J," but he picked up the nickname "Jay," which he used.

Profile of Crane, *Chem. Eng. News* **36,** 80 (Aug. 4, 1958), with portrait; obituary, *Chem. Eng. News* **45,** 18 (Jan. 9, 1967), with portrait; personality profile, *Information* **3,** 173-176 (May-June 1971); personal recollections.

DALE B. BAKER

Moses Leverock Crossley

1884-1971

Crossley was one of American Cyanamid's renowned scientists and was research director of the Calco Chemical Division at his retirement. He was born on Saba Island, N.E. West Indies, in the Netherlands Antilles July 3, 1884. Coming to the United States for his education, he graduated from Brown University with a Ph.B. degree in 1909, an M.S. degree in 1910, and a Ph.D. degree in 1911. From 1909 to 1911 he was instructor in chemistry at Brown. In 1940 he was awarded the Brown Bear for outstanding achievements and in 1944 received an honorary D.Sc. degree.

Crossley was associate professor of chemistry at William Jewell College from 1911 to 1913. He left to lecture on organic chemistry at Wesleyan College in 1913 and became associate professor and acting head of the Chemistry Department from 1914 to 1918. Wesleyan conferred on him an honorary Sc.D. degree in 1947. While at Wesleyan he was a consulting chemist to Barrett Co.

In 1918 Crossley became chief chemist for Calco Chemical Co. at Bound Brook, N.J., and rose to the position of director of research in 1936. When Calco became a division of American Cyanamid Co. he became a director of research in both organizations until his retirement Aug. 1, 1949. At that time he was named special consultant to the heads of the Research and Development Departments.

At the end of the European phase of World War II Crossley was a member of a technical team that was sent to Germany to investigate the military uses of dyes and textile auxiliaries.

Upon his retirement from Calco, Crossley was appointed a research specialist in the Bureau of Biological Research at Rutgers University to direct the chemotherapeutic phase of the bureau's cancer research project. Here he became interested in phosphorus as an index of nerve metabolism, in changes in certain constituents of blood during pneumonia, and in the chemotherapy of cancer.

Crossley was chairman of the Connecticut Valley Section of the ACS from 1915 to 1917, president of the American Institute of Chemists in 1924 and 1934, received the institute's Gold Medal in 1947, and was president of New York Academy of Sciences in 1950. He was an American delegate to many world scientific meetings such as the International Union of Pure and Applied Chemistry in 1926 and 1928. He published more than 100 papers on scientific and educational subjects.

He served as trustee of Brown University and Union Junior College and as a member of the advisory board of the research council of Rutgers University.

During his teaching period he became interested in anthraquinone ring synthesis using hydroquinone; for example, 1,4,6-trihydroxyanthraquinone, and the sulfonation of anthraquinone with oleum. Upon entering industry any successful researches of practical value were company property. It may be noted that Crossley was particularly interested in dyes and their intermediates, color and constitution, and pharmaceuticals. He did much of the pioneer work on sulfa drugs. During World War II Crossley and his associates were continually called upon by branches of the armed services for assistance. They solved many problems, and Calco was one of the first firms awarded an Army-Navy E.

Crossley was about five feet nine inches in height, of medium build, and wore glasses. He was shy, and sometimes this showed in his speech. However, he was a well-informed chemist and knew what he wished to say. He was friendly with young chemists and did not try to leave the impression that he was an important person. He was a most gentle man. Crossley died May 29, 1971 at the home of his son, Evan, in Hagerstown, Md.

Chem. Eng. News **24,** 2746 (Oct. 25, 1946); *Cyanamid Bound Brook Diamond* **37,** 1, 3 (June 17, 1971); personal recollections.

DAVID H. WILCOX, JR.

Francis Joseph Curtis

1894-1960

Curtis was born in Cambridge, Mass., Apr. 22, 1894. He attended Cambridge Latin School and Harvard University. Immediately upon receiving his B.A. degree in chemistry in 1915 he started in research at the TNT plant that Merrimac Chemical Co. was running for the Army. Five years later he switched to production. Then in 1925 he joined the company's Sales Department in Boston. In 1929 the company was taken over by Monsanto Chemical Co., and in 1930 Curtis was appointed director of development for the Merrimac Division. In that year he started annual visits to European chemical companies in search of new chemical developments. This was a pioneer idea for an American chemical firm, but it soon became a universal practice for all. In 1935 he was transferred to the St. Louis headquarters where he was assistant director of development for Monsanto. Among his projects were setting up Monsanto's manufacture of styrene and of elemental phosphorus.

In 1942 Curtis and six other members of the American Chemical Society, dissatisfied with the lack of attention within the society toward the marketing of new products, organized the Commercial Chemical Development Association, later known as the Commercial Development Association.

During the spring of 1945 he was a member of a group of technical investigators sent to Germany by the Chemical Warfare Service to evaluate Germany's wartime chemical industry.

In 1949 after serving for 6 years as vice-president and secretary of the executive and finance committees of Monsanto, Curtis became a member of the board of directors. He was appointed Monsanto's director of industrial preparedness in Washington, D.C., serving as liaison between government agencies and Monsanto.

Curtis' interest in improving the professional status of chemists and chemical engineers was reflected in the method of engaging technical staff members that he inaugurated when working with Monsanto in 1936. He was among the first to adopt the view that it was important that chemists should be familiar with the economics of production planning, and that scientists should be able to advance in the industrial hierarchy because of high scientific attainment. Thus, in 1951 he began Monsanto's senior scientist program, which was to be copied throughout the industry.

In 1954 Curtis went to England to study the sulfur situation for the National Production Authority. Upon his return he spent 6 months with this authority as an assistant administrator of the Chemical, Rubber, and Forest Products Bureau. In the same year he received the Commercial Development Association's honor award in recognition of his 36 years of contribution to commercial chemical development in the United States.

Curtis was, at various times, president of the American Institute of Chemical Engineers, vice-president of American Association for the Advancement of Science, chairman of the American Chemical Society's Division of Industrial and Engineering Chemistry, and president of the Society of Chemical Industry, having been previously chairman of the American Section of the society.

Curtis' work was evidence of his concern for his fellow men, with whom his humanity, wit, and active, cultivated mind made him popular. He was not one of the world's great chemists, but he was a master of human relations. He died Apr. 21, 1960.

Personal recollections; information from Monsanto Company Public Relations Dept.; *Chem. Eng. News* Oct. 6, 1947, Feb. 5, 1951, May 2, 1960; obituary: *St. Louis Post Dispatch* Apr. 21, 1960.

JOHN J. HEALY, JR.

Jonathan Peter Cushing

1793-1835

In Ernest Child's "The Tools Of The Chemist," there is pictured an itemized bill of sale dated Nov. 21, 1822 for nearly $300 worth of chemicals and apparatus sold to Jonathan P. Cushing, a chemist so obscure that Child "after diligent search," could not identify him. Even this writer learned only by accident that Cushing, at the time, was president of Hampden-Sydney College.

Cushing was born in Rochester, N. H., on Mar. 12, 1793. Orphaned early, he worked on his guardian's farm until in his early teens when he apprenticed to a saddler. Not resigned to being a tradesman, he bought the remainder of his time and determined to get more education if it cost him "40 years of his life." He entered Exeter, supported himself by his trade, and went on to Dartmouth College where he earned his degree in 1817.

Heading for Charleston, S. C., and a law career, Cushing met a friend from Hampden-Sydney in Richmond who persuaded him to try teaching.

It has not been established just when chemical instruction started at Hampden-Sydney, but the catalog for 1817, the year that Cushing arrived, shows that "one session of chemistry with experiments" was required of all third year students followed by natural philosophy with astronomy. At the time, only four other colleges in the South offered chemistry.

In January 1819, the college trustees established a Chair of Chemistry and Natural Philosophy and appointed Cushing professor. Some time in the early 1820's, John Redman Coxe of Philadelphia, was asked for advice on equipment and since the school had no money, Cushing paid for what was needed out of his own pocket. He became president in the fall of 1821 but continued his teaching duties. Opening of University of Virginia in 1825 thinned Hampden-Sydney's student ranks; nevertheless, Cushing was able to raise $30,000 for expansion.

He married in 1827 and had several children. In 1831 he founded the Virginia Historical and Philosophical Society which is still in existence. His only publication appears to be an address before that organization in 1833 on the "Branches of Philosophy."

Poor health plagued Cushing, and in 1835 he was advised to move to a warmer climate. He died en route at Raleigh, N. C., on April 25, 1835 and was buried in the Episcopal cemetery there. John W. Draper, who 40 years later became the first president of the American Chemical Society, assumed the professorship left vacant by Cushing's death.

Ernest Child, "The Tools Of The Chemist," Reinhold, p. 184 (1940); *Southern Literary Messenger,* **2**; 163-166 (Feb. 1836); A. J. Morrison, "College Of Hamden-Sydney Dictionary Of Biography 1776-1825" (1920); *American Quarterly Register,* **11**, 113 (Nov. 1838).

HERBERT T. PRATT

James Cutbush

1788-1823

Edward Cutbush

1772-1843

James Cutbush was one of the most active American chemists of the early nineteenth century. He wrote at least eight books and pamphlets, 14 journal articles, and 28 newspaper articles. It is largely through these works that we know of him—his personal life is a mystery.

The son of a British stonecutter, Edward Cutbush and his wife Anne Marriat, James was born in Philadelphia in 1788 and left an orphan in 1790. Nothing is known of his early life, but it is a reasonable assumption that he was raised by his oldest brother, Edward, a naval surgeon. It is not known where James learned chemistry, but it may have been under the guidance of Edward. His first appearance as a writer of chemical topics was in a series of 28 articles on the applications of chemistry, in the Philadelphia newspaper *Aurora,* between July 1808 and April 1809. In the winter of 1810 he and Edward presented a series of public lectures on chemistry for citizens of Philadelphia. James presented lectures on chemistry in 1811, on theoretical and practical pharmacy in 1812, and chemistry again in 1813. During these years he seems to have made his living mainly as an apothecary and a dealer in chemical supplies.

In 1812 James was elected professor of chemistry, mineralogy, and natural philosophy at St. John's College, a transient Lutheran institution of Philadelphia. He taught there probably no more than 2 or 3 years, but during this time wrote for his students a two-volume text, "The Philosophy of Experimental Chemistry," which appeared in 1813.

Members of the Columbian Chemical Society, organized in 1811, elected Cutbush president in 1812 and continued to re-elect him each year until the Society disbanded in 1814. During his tenure the organization published *Memoirs of the Columbian Chemical Society* (1813), which was conceived as a periodical but which ceased after volume 1 appeared. At Cutbush's suggestion, Thomas Jefferson was named patron of the society, an honor Jefferson appreciated.

In 1814 Cutbush was appointed assistant apothecary general of the Army. He remained in the Army after the war, going to the U.S. Military Academy in 1820 as post surgeon. In the autumn of 1820 he was made acting professor of chemistry and mineralogy, the first teacher of those subjects at the institution. For the cadets he wrote a small "Synopsis of Chemistry" in 1821. While stationed at the academy, he delivered at least one series of public lectures in the neighborhood in 1823 and published the substance of his remarks in a brochure, "Lectures on the Adulteration of Food."

Cutbush died suddenly in the winter of 1823. The cadets had the following inscription cut into his tomb:

> Sacred to the memory of Dr. James Cutbush, member of the American Philosophical Society, late Surgeon, U.S. Army and Professor of Chemistry at the U.S. Military Academy, West Point, N.Y., who departed this life December 15, 1823, aged 35 years. An honourable tribute of respect from his grateful pupils.

He had prepared a manuscript work on explosives and fireworks, which Franklin Bache edited anonymously after his death and which Clara Cutbush, his widow, published in 1825 under the title, "A System of Pyrotechny."

Of Edward Cutbush as a chemist we know less than of James. Born in Philadelphia, Jan. 5, 1772, Edward received a degree from the University of Pennsylvania Medical School in 1794, was a surgeon with Pennsylvania troops during Washington's expedition against the Whiskey rebels, and was surgeon in the United States Navy from 1799 to 1829.

Somehow, while serving as a naval surgeon, Edward managed to become and remain proficient in chemistry. He was not an investigator but seems to have been a competent teacher. He was a member of the Chemical Society of Philadelphia, an officer in the Columbian Chemical Society, professor of chemistry in the George Washington University Medical School from 1825 to 1827, and professor of chemistry at Geneva College, Geneva, N.Y. from 1830 to 1839 and in the Geneva College Medical School from 1835 to 1839. He died at Geneva, June 23, 1843.

Edgar F. Smith, "James Cutbush, an American Chemist, 1788-1823" (1919), J. B. Lippincott Co., has no references; biography by H. G. Wolfe in *Am. J. Pharm. Educ.* **12,** 89-125 (1948), has references and a list of Cutbush's writings; "Dictionary of American Biography," **5,** 10, Chas. Scribner's Sons (1930); W. D. Miles, "The Columbian Chemical Society," *Chymia* **5,** 145-154 (1959); F. L. Pleadwell, "Edward Cutbush, M.D., The Nestor of the Medical Corps of the Navy," *Ann. Med. Hist.* **5,** 337-386 (1923), with portrait, ill.; Stephen W. Williams, "American Medical Biography" (1845), pp. 118-120, Merriam & Co. (1845).

WYNDHAM D. MILES

D

Charles William Dabney

1859-1945

Dabney, an administrator and educator, began his life work as a chemist.

He was born in Hampden-Sidney, Va., June 19, 1859. He frequently signed his name Charles W. Dabney, Jr., but he was not a junior; his father's name was Robert Lewis Dabney. He attended public schools in Hampden–Sidney and received his bachelor of arts degree from Hampden Sidney College in 1873. He engaged in graduate study at University of Virginia from 1874 to 1877, majoring in chemistry. After teaching chemistry for a year at Emory and Henry College, he went to Germany for further graduate study. At Berlin and Göttingen universities he studied under such men as Wöhler, Huebner, Hoffman, Klein, Helmholtz, DuBois, and Raymond. He received his Ph.D. degree at Göttingen in 1880. His dissertation on organic chemistry was published subsequently in *Annalen der Chemie.*

Upon his return to the United States Dabney became professor of chemistry at University of North Carolina and was soon appointed state chemist and director of the North Carolina State Experiment Station. The position of chemist and director of the Experiment Station had been established in 1877 with an office at University of North Carolina in Chapel Hill. Soon after Dabney arrived, the work was moved to Raleigh, where several years later an experimental farm and industrial institute were established. Under his direction, the work was expanded from chemical testing of fertilizers, soils, and materials for the geological survey and the State Board of Health to seed inspection and exploration for substances needed to improve agriculture such as rock phosphate and field experiments to supplement laboratory work.

In 1887, a year that gave further impetus to scientific research through Federal aid to state experiment stations under the Hatch Act, Dabney took a step that led him further from his involvement in chemical research. He became president of University of Tennessee and director of its experiment station. Six years later, concerned about the expanding research in the state experiment stations across the country, he sought and received the post of assistant secretary of the United States Department of Agriculture. In this position he had general supervision over scientific work in the department. He returned to his work as president of University of Tennessee in 1897. In 1904 he became president of University of Cincinnati, retiring from this position in 1920. He died June 15, 1945.

Charles W. Dabney, "Universal Education in the South," University of North Carolina Press, 2 vols. (1936); papers documenting Dabney's work are in the libraries of the universities of North Carolina and Tennessee and in the National Archives; "National Cyclopaedia of American Biography" **E,** 323, James T. White & Co. (1938); obituary, *School and Society,* 411 (June 23, 1945).

VIVIAN WISER

Henry Drysdale Dakin

1880-1952

Born Mar. 12, 1880, in London, the youngest of eight children, Dakin received his public school education at Leeds Modern School. Before he entered the university he was apprenticed to the Leeds City Analyst, a position he held for 4 years, which greatly influenced his interests in chemistry and his practical approach to it. In 1898 Dakin entered Victoria University at Leeds, where he received his B.Sc. degree in 1901. Later he received his M.Sc., and in 1909 his D.Sc. degree from the same institution, renamed the University of Leeds.

While an undergraduate at Leeds Dakin worked with Julius Cohen, principally on aromatic substitution chemistry. At the same time he published analytical papers resulting from his connection with the city analyst. He stayed on for a year after his degree to work with Cohen and act as his laboratory demonstrator. In 1902 Dakin received an 1851 Exhibition grant which enabled him to work at the Lister Institute for Preventive Medicine, in London, and with Albrecht Kossel in Heidelberg. He shared the discovery of arginase with Kossel, the beginning of a lifelong interest in enzymes.

In 1905 Dakin was invited by Christian A. Herter to work in the latter's private laboratory in New York City, one of the very few in America relating medicine and science through the new field of biochemistry. Dakin remained in the United States for the rest of his life though he never gave up his British citizenship. On Herter's untimely death in 1910 Dakin continued the work of the laboratory at Mrs. Herter's request, taking over Herter's editorship of *Journal of Biological Chemistry* as well. In 1916 Dakin married Herter's widow and 2 years later their residence and laboratory were removed from the city and re-established in Scarborough-on-Hudson, where Dakin worked in something approaching seclusion for the rest of his life.

The range of Dakin's investigations was enormous; a list of the topics covered in his two books and more than 150 papers looks rather like an outline for a basic biochemistry course. At the Lister he demonstrated enzyme stereospecificity and synthesized adrenalin. In Herter's laboratory he worked on biological oxidation, finding that peroxide under mild conditions could serve as an *in vitro* analog of biological oxidants and providing support for the recently advanced theory of beta-oxidation of fatty acids. (The "Dakin reaction"—conversion of an aromatic aldehyde or ketone to a phenol with alkaline peroxide—was an offshoot of this work.) Dakin's interest resulted in his state-of-the-art book "Oxidations and Reductions in the Animal Body," 1912, with a second edition in 1922 to record the further unraveling of the complexities of biochemical oxidation.

Dakin developed an early method of end-group analysis for proteins and used it for structure determinations. With Harold Dudley he discovered glyoxalase, one of many enzymes he investigated. He and Randolph West sought with partial success for the hemopoietic liver factor connected with pernicious anemia. Dakin introduced the technique of separating amino acids in a protein hydrolysate using wet *n*-butyl alcohol, and was the first to use *p*-nitrophenylhydrazine.

Perhaps Dakin's best-known work was his development, with Alexis Carrel, of the borate-buffered hypochlorite disinfectant called "Dakin's solution," widely used as a wound irrigant during the World War and until it was superseded in the thirties. This and other chemical antiseptics were discussed in the extraordinarily useful "A Handbook on Antiseptics," published in 1917 by Dakin and Edward Dunham.

Formal honors came to Dakin throughout his life. He was a fellow of the Institute of Chemistry and of the Royal Society. He was awarded honorary degrees by Yale, Leeds, and Heidelberg. For his World War I work he was made a Chevalier Legion d'Honneur. He received the Conné medal and Davy medal of the Royal Society.

Dakin went on publishing until 1946; his later works were less frequent but of an "unhurried perfection and elegance." The location of the Scarborough laboratory suited his retiring temperament perfectly; Dakin's shyness caused him never to suffer fools gladly, and he could exercise a mordant wit at their expense. The Scarborough establishment was nonetheless frequently filled with guests, and many from the world of biochemistry made their way to Dakin for discussion and advice.

Dakin died at Scarborough on Feb. 10, 1952, surviving his wife by less than a year.

Obituary Notices of Fellows of the Royal Society, **8**, 129-148 (1952-53); "J. C. Poggendorff's biographisch-literarischer Handworterbuch"; Arthur V. Howard, ed., "Chambers' Dictionary of Scientists," Dutton (1951); "Encyclopaedia Britannica," 15th Ed., "Micropaedia" (1974).

ROBERT M. HAWTHORNE, JR.

James Freeman Dana

1793-1827

Dana was born in Amherst, N.H., Sept. 23, 1793. After attending Phillips Academy he entered Harvard in 1809. A scholarly lad, he was particularly interested in natural history and was "much in advance of his classmates" in studying chemistry. At that time it was customary for a senior to be offered the opportunity of assisting the chemistry teacher in preparing and performing demonstrations. John Gorham, adjunct professor of chemistry, asked Dana to assist. Dana proved so competent that the two men became friends. Gorham, a practicing physician, took Dana under his tutelage without charge after Dana graduated in 1813 and began to study medicine.

In the winter of 1813-14 Gorham allowed Dana to use his lecture room to present a public lecture on nitrous oxide. By this time Dana was regarded so highly as a practicing chemist that he received "constant applications for the analysis of minerals and the preparation of particular tests." In 1815 he received the Boylston prize for an essay on tests for arsenic.

For several years before this time Gorham and Aaron Dexter had been asking Harvard to obtain new chemical apparatus. In 1815 the university officials, at the suggestion of Gorham, selected Dana as their purchasing agent and sent him to England to obtain apparatus. While in London Dana studied for several months in the laboratory of Frederick Accum, one of the leading analytical and industrial chemists of Great Britain. On shipboard during the return voyage from England, Dana wrote an essay on the composition of oxymuriatic acid, for which he received another Boylston prize in 1816.

Home again, Dana was appointed to superintend the renovation of Harvard's laboratory. When completed in 1816 it had "a neatness and completeness not surpassed by any laboratory, and in the value and extent of apparatus, not at that time equalled in the United States." Gorham engaged Dana as his assistant from 1816 to 1819, and then Harvard hired him as instructor until 1821.

While assisting Gorham, Dana had completed his medical education and had received his medical degree from Harvard in 1817, his thesis being "on the importance of animal chemistry." While he was teaching at Harvard part of the year, during another part he was lecturer on chemistry at Dartmouth from 1817 to 1820 and professor of chemistry there from 1820 to 1826.

Dana moved to New York in 1826 to become professor of chemistry at College of Physicians and Surgeons. He delivered a series of public lectures on electromagnetism at New York Atheneum, and he was part way through a series of public lectures on chemistry when he died Apr. 15, 1827, at the age of 33.

Dana published about 10 articles, one of which, on the mineralogy and geology of the Boston area, was coauthored by his brother, Samuel Luther Dana, a noteworthy industrial and agricultural chemist. He also published a text, "Epitome of Chymical Philosophy," (1825) used at Columbia, Princeton, and elsewhere.

One of Dana's contemporaries said that "as a lecturer he had few superiors," and another that "as an experimenter he was unrivalled."

The obituary by J. B. B. [John B. Beck] in *New York Med. Physical J.* **6,** 314-318 (1827) contains most of the quotes above; James Thacher, "American Medical Biography," pp. 245-250, DeCapo Press (1828, reprint 1967); I. Bernard Cohen, "Some Early Tools of American Science," Harvard University Press (1950); "Quinquennial Catalogue of . . . Harvard University" (1900).

WYNDHAM D. MILES

Samuel Danforth

1696-1777

Samuel Danforth, Jr.

1740-1827

Samuel Danforth, an American alchemist, was born in Dorchester, Mass., Nov. 12, 1696. After graduating from Harvard in 1715, he taught school in Dorchester and later at Cambridge until 1730. During these years he also surveyed, helped edit *New-England Weekly Journal,* and dabbled in medicine, at least to the extent of inoculating against smallpox. He studied law and became a judge in the Court of Common Pleas for many years. He also entered politics, became a member of the province council, and presided over the council from 1766 until 1774 when opposition from the revolutionaries caused him to resign. He died in Boston, Sept. 26, 1777.

Danforth was attracted to alchemy when he was a young man and believed in the art the remainder of his life. He purchased European alchemical books, of which perhaps 19, with his annotations, have survived. He thought it possible to produce the philosopher's stone, the imaginary substance supposed to have the power of transmuting baser metals into gold and of prolonging life. In 1754 a political opponent ridiculed him as "Madam Chemia (a very philosophical lady), who some years since (as is well known) discover'd that precious stone, of which the Royal Society has been in quest a long time." Writing to Benjamin Franklin in 1773, Danforth apparently told Franklin about his search for the philosopher's stone and promised to give Franklin some, for Franklin replied: "I rejoice . . . in your kind intentions of including me in the benefits of that inestimable stone, which, curing all diseases (even old age itself), will enable us to see the future glorious state of our America. . . . I anticipate the jolly conversation we and twenty more of our friends may have 100 years hence on this subject." A few days after Danforth died, Ezra Stiles, later president of Yale, wrote in his diary: "He was deeply studied in the writings of the adepts, believed the philosopher's stone a reality and perhaps for chemical knowledge might have passed among the chemists for a [adept]."

Samuel Danforth, Junior, born in Cambridge in 1740, graduated from Harvard in 1758, studied medicine, and became a physician. He paid "an ardent attention to the study of chemistry, which was then so little known in this country as to be considered an occult and somewhat mysterious science." Aided by an unidentified French physician-chemist, whom the war brought to America, "he pursued the study far enough to get a glimpse of some of the important facts which soon after broke out with such lustre in the discoveries of Lavoisier and his coadjutors." One of his medical students, Aaron Dexter, became the first professor of chemistry in Harvard's Medical School. His son, completing his medical education in Europe, sent back for him "the most complete chemical apparatus which had been seen in Boston." But Danforth Junior's medical practice eventually became so large he had to abandon chemistry. In 1812 he gave 21 books on alchemy, 19 of which he inherited from his father, to Boston Athenaeum. He died in Boston, Nov. 13, 1827.

A sketch of Samuel Danforth is in: Clifford K. Shipton, "Biographical Sketches of Those Who Attended Harvard College in the Classes of 1713-1721," Massachusetts Historical Society (1942), 80-86, with portrait; the quote regarding Franklin is from this sketch. Franklin B. Dexter, ed., "Literary Diary of Ezra Stiles," Chas. Scribner's Sons (1901), II, 216.

Sketches of Samuel Danforth, Jr., are in: *Boston Med. Surg. J.* **1,** 17-21 (1828-29), reprinted essentially in James Thacher, "American Medical Biography," DeCapo Press (1828, reprint 1967), II, 233-38, portrait, with the quotes given above; Clifford K. Shipton, "Biographical Sketches of Those Who Attended Harvard College in the Classes of 1756-1760," Massachusetts Historical Society (1968), 250-254, with portrait; R. S. Wilkinson, "New England's Last Alchemists," *Ambix* **10,** 128-138 (1962), deals with both Danforths, gives titles of books on alchemy owned by them, and gives the quote, "Madam Chemia."

WYNDHAM D. MILES

Farrington Daniels

1889-1972

Daniels' work in physical chemistry transcended many areas of teaching and research including nitrogen fixation, chemical kinetics, nuclear energy, thermoluminescence, solar energy, and the social implications of science.

He was born Mar. 8, 1889 in Minneapolis, Minn., the son of Frank Burchard and Florence Louise (Farrington) Daniels. He received his B.S. degree at University of Minnesota in 1910, his M.S. a year later. He then joined Theodore W. Richards at Harvard where he made an electrochemical study of thallium amalgams and received his Ph.D. in 1914. Before joining the University of Wisconsin faculty in 1920 he taught at Worcester Polytechnic Institute and served as an electrochemist with the U.S. Nitrogen Fixation Laboratory, 1919 to 1920. His career at Wisconsin did not end when he received emeritus status in 1959, following 7 years as departmental chairman, but he pursued his work on solar energy energetically up to within a few weeks of the hospitalization which preceded his death June 23, 1972.

He married Olive M. Bell Sept. 15, 1917. They had four children. Farrington, Jr., Florence, Mariam, and Dorin. For many years, his cottage in Door County served as a summer retreat attracting children and grandchildren while serving as a spot for continued work on solar energy and other professional objectives.

As a teacher, Daniels had a leading role in the development of physical chemistry in America. His interest in laboratory instruction at Wisconsin led to coauthorship of a laboratory manual published in 1929. His conviction that mathematical preparation of students for physical chemistry was inadequate led to publication of a book aimed toward correction of the deficiency. In 1931 he became coauthor of the fifth edition of Getman's "Outlines of Theoretical Chemistry," a book which had wide acceptance throughout the country. He continued to revise the book substantially in later editions.

His interest in education transcended chemistry, however. He was chairman of a faculty committee which had responsibility for creation of a "Freshman Forum" dealing with problems of contemporary relevance, and stimulated the creation of a Department of the History of Science. The latter department had his lifetime support and he personally gave much encouragement to broadening the program of history of chemistry in his own department. Following World War II he was a leader in creating an interdisciplinary course in "Contemporary Trends." The materials from this course were edited and published as "Challenge of Our Times."

During his active career at Wisconsin he directed the research of more than 60 candidates for the doctorate. His early research dealt with the oxides of nitrogen, leading on the one hand to his contributions to chemical kinetics, on the other to development of a gas fired, magnesia brick, regenerative furnace for producing nitric oxide from air. The latter work attained pilot plant development toward the production of fixed nitrogen for fertilizer but was ultimately abandoned as being non-competitive with nitrogen fixed by the Haber Process.

His other studies dealt with photosynthesis, production and application of radioactive isotopes and the use of thermoluminescence in the study of geological and archeological problems. During the last two decades of his life he was deeply involved in research on solar energy, believing that the sun was a never-ending, non-polluting source of energy available to all members of the world community. He and Mrs. Daniels travelled to many parts of the world in an effort to stimulate practical and inexpensive projects toward utilization of the sun's energy.

During World War II Daniels was actively involved in the atomic energy program, serving as director of the Metallurgical Laboratory in 1945-46 and as first chairman of the board of governors of the Argonne National Laboratory. His design for a nuclear reactor for peacetime production of energy was studied extensively for 2 years at Oak Ridge but was abandoned in the face of a change in governmental policy regarding development of nuclear energy.

Daniels published more than 300 articles and, in addition to the books mentioned previously, wrote "Chemical Kinetics" (1938), and "Direct Use of the Sun's Energy" (1964). He coedited "Solar Energy Research" (1955), and "Photochemistry in the Solid and Liquid State" (1960).

Farrington Daniels was by nature a humble man with a driving enthusiasm for the things which he believed in. A man of small stature, he nevertheless had boundless energy which enabled him to accomplish much more than most men. He had a compassion toward all mankind, and his research was frequently directed toward studies with broad social benefits. He had doggedness in the face of adversity which enabled him to achieve desired goals in the face of apparently insurmountable obstacles. He set high goals in all his activities, and even those ventures which were not fully successful left new knowledge which would have future usefulness. Without seeking honors, he had many bestowed on him. Without seeking responsibility, he was called upon again and again to lead. The American Chemical Society made him its president in 1953, and he held national office for a number of other societies. In his quiet way, he was an inspiration to his students, his associates, and even to those who knew him only by reputation.

Daniels' papers in archives of University of Wisconsin and of State Historical Society of Wisconsin; autobiographical sketch in "McGraw-Hill Modern Men of Science," **1,** 126-7, McGraw-Hill (1966); biographies in *Chem. Eng. News* **31,** 54 (1953), in *Current Biography,* 106-08, H. W. Wilson (1965), and by James C. Spaulding in *Saturday Review,* April 4, 1959, pp. 66-67; personal recollections; a book length biography is being written by Olive Bell Daniels.

AARON J. IHDE

Carroll Campbell Davis

1888-1957

Davis was born in West Roxbury, Mass., Sept. 4, 1888. After receiving his B.S. degree in chemistry in 1911 from Dartmouth College, he studied at Massachusetts Institute of Technology where he was granted his S.B. degree in chemical engineering in 1914. He immediately joined Boston Woven Hose & Rubber Co. where he remained until his retirement as chief chemist in 1956.

In 1924 Davis with John H. Bierer developed an oven in which samples of rubber can be artificially aged by exposure at 70° C. to oxygen at a pressure of 300 p.s.i. This greatly accelerates the aging process, known to be the result largely of reaction with atmospheric oxygen, so that the potential life of the rubber compound in service can be quickly appraised. This greatly reduces the time to optimize a rubber compound for production use, an important economic consideration.

Davis' most important contributions were made in the field of rubber literature. In 1925, after having served as an abstractor since 1919, he became editor of the Rubber Section of *Chemical Abstracts.* In 1928 he was appointed the first editor of *Rubber Chemistry and Technology.* He held both of these positions until 1957, the year of his death. A master of several languages, he was instrumental through *Rubber Chemistry and Technology* in bringing to American chemists translations of important research articles from rubber journals all over the world. Under his editorship *Rubber Chemistry and Technology* became internationally known as one of the most important rubber journals.

In 1937 Davis was editor, with John T. Blake as associate editor, of the American Chemical Society Monograph "Chemistry and Technology of Rubber" which was quickly accepted as a standard reference work in the field of rubber. In 1954, with Raymond F. Dunbrook, he assisted G. Stafford Whitby in editing another ACS Monograph, "Synthetic Rubber." He was also section editor for "Rubber Bibliography" published by *Rubber Age* for the Division of Rubber Chemistry under the editorship of John McGavack. In recognition of his pre-eminence in the field of rubber literature, Davis was awarded the Charles Goodyear Medal by the Division of Rubber Chemistry in 1950.

Davis was married to Caroline Deane Sparrow in Boston, Oct. 25, 1917. They had two sons: Frederic Campbell and Alan. Davis died in Worcester, Mass., Aug. 10, 1957.

"The National Cyclopaedia of American Biography," **44,** p. 369, James T. White, New York, 1962; *Rubber Age,* **81,** 1022 (1957); *Rubber World,* **136,** 873 (1957); *Rubber Chemistry and Technology,* **30** (5), vi (1957).

GUIDO H. STEMPEL

Tenney Lombard Davis

1890-1949

Davis was born Jan. 7, 1890 in Somerville, Mass. From his grandfather, a teacher of mathematics, he gained an interest in science while his mother, a close student of the Bible and poetry, gave him a strong liking for literature and philosophy. Even as a boy he showed great interest in chemistry, setting up his own laboratory at home. During his college years at Dartmouth from 1907 to 1909 and then at the Massachusetts Institute of Technology, from which he received a B.S. degree in 1913, he devoted himself to the study both of chemistry and philosophy. He continued to be interested in these subjects while working for his M.A. and Ph.D. degrees, both of which he received from Harvard. So strong were his philosophical interests that he spent the academic year 1916-17 as Sheldon Traveling Fellow in Philosophy at University of California.

During World War I he served in the Ordinance Corps in France and there began his work on explosives to which he devoted his chemical researches during the rest of his laboratory career. The culmination of his work in this field came with the publication of "The Chemistry of Powder and Explosives" in 1941.

In 1919 Davis was appointed instructor in chemistry at MIT and remained there for the rest of his teaching career. He became a professor in 1938. During this time he continued to be interested in philosophy and published several papers on the philosophy of science. From this work it was an easy step to an interest in history of science, and his studies gradually became more and more historical. He soon established himself as an authority on history of chemistry. He applied to his historical work the same standards of accuracy which he required in laboratory activity.

During the decade of the twenties he wrote chiefly on western alchemy and chemistry. In 1930 his attention was directed by one of his Chinese students to the alchemical treatises contained in the Taoist literature of China. He began to devote himself largely to the almost unstudied field of Chinese alchemy. Working with a large group of collaborators both in the United States and China, he published a number of translations of alchemical works from the writings of the Taoist philosophers.

In 1942 increasing ill health forced his retirement from his teaching position. He retained a consulting position with National Fireworks, Inc., but otherwise he lived quietly at his home in Norwell, Mass. where he devoted himself mostly to historical studies. He was a member of the editorial boards of several journals in history of science and continued his interest in the Division of the History of Chemistry of the American Chemical Society, a division which he had helped to establish. Toward the end of his life he began the study of the history of pyrotechnics, with special attention to Chinese contributions to this subject.

Davis was married to Dorothy Munch in 1923. He was a versatile individual—an ardent naturalist as well as a chemist and historian. He made forestry one of his hobbies and was greatly interested in the life history of spiders. He often wrote poetry and humorous couplets. After his retirement his health grew worse, and he died suddenly Jan. 25, 1949.

Henry M. Leicester and Herbert S. Klickstein in *J. Chem. Educ.* **27,** 222-224 (1950) and in *Chymia* **3,** 1-16 (1950), port., bibliography of works.

HENRY M. LEICESTER

David Talbot Day

1859-1925

Day, son of Willard Gibson Day, a Swedenborgian minister, was born at Rockport (now Lakewood), Ohio, Sept. 10, 1859. His family moved to Baltimore in his early life, and after high school training he entered Johns Hopkins University in 1878. There he received his A.B. degree in 1881 and his Ph.D. in 1884. His major was chemistry under Ira Remsen, whom he served as assistant for several years. Later, as demonstrator in chemistry at University of Maryland from 1884 to 1885, he prepared short articles on manganese, chromium, and tungsten which were published by the United States Geological Survey in the first volume of "Mineral Resources of the United States." Given charge of preparing mining statistics for the Survey in 1886, Day helped to develop a strong organization for collecting a complete set of data covering the calendar year 1889. This was the first actual census of the mining industry and was published in the "Eleventh United States Census."

Day, as the chief of the Division of Mineral Resources of the survey, suggested as early as 1897 that it could be experimentally demonstrated that if limestone is saturated with oil (characteristic of such rock as "Trenton limestone"), slight pressure exerted upon it may cause the oil to flow upward and to be changed in its composition as it passed through finely divided material. Day disclosed results of experimental research and presented samples at the First International Petroleum Congress in Paris on Aug. 20, 1900. Some have called this day "The Birthday of Chromatography." He wished the new method to be considered as a general research tool for resolving the hydrocarbon mixtures in crude petroleum, but much better adsorbents and improved procedures had to be designed to permit complete separation. However, the establishment of spatial separation of substances by selective adsorption should be attributed to him.

Day was vice–president of American Institute of Mining Engineers in 1893 and again in 1900. In 1912 he was president of the fuel section of the International Congress of Applied Chemistry.

A highly efficient organizer of exhibits, Day served as director of the mining department at the exposition of Atlanta, Ga. in 1896, secretary of the jury of awards at the Tennessee Centennial in 1897, director of mining at the Trans-Mississippi Exposition at Omaha in 1898, and was in charge of the petroleum department at the Philadelphia Exposition in 1899 and at the Paris Exposition in 1890. He was in charge of the mining department of the Buffalo Exposition in 1901, honorary chief of the department of mines and metallurgy at the St. Louis Exposition in 1904, and honorary commissioner of mining at the Lewis and Clark Exposition at Portland in 1905 and at Jamestown Exposition in 1907.

In 1907 Day became a consulting chemist and undertook petroleum investigations for the Geological Survey. In 1914 he became consulting chemist to the Bureau of Mines and continued in that capacity until 1920. Thereafter he gave his entire time and expert services to experimental work for oil companies. In 1922 he published his fine "Handbook of the Petroleum Industry" in two volumes.

Day had a theory that when petroleum supplies began to fail it would be necessary to utilize oil shale reservations for the Navy Department. He gave much attention to developing the cracking process that would change the heavier oils into gasoline and had considerable success in perfecting this process. Still actively prosecuting his studies on the cracking process, he died suddenly on Apr. 15, 1925 in Washington, D. C. at the home of a friend whom he was visiting.

Day's chief characteristic was his readiness to aid and encourage all with whom he associated. His ability to gain the confidence and friendship of producers and his astuteness in choosing for his assistants men of high intelligence and efficiency were of great importance in securing data for statistical compilations of mineral production, especially when it had to be taken from confidential records and used with discretion.

Herbert Weil, "Industrial Petroleum Chromatography," *Petroleum,* **14,** 9 (1951); N. H. Darton, "Memorial of David Talbot Day," *Proc. Geol. Soc. America for 1933,* 187 (1934); obituary by M. R. Campbell, *Mining Met.* (June 1925); *Trans. Amer. Inst. Mining Met.,* **71,** 1371 (1925); obituary, *Washington Evening Sun* (April 16, 1925).

SISTER VIRGINIA HEINES

Peter Joseph William Debye

1884-1966

Known as the "Master of the Molecule" because of his pioneering work in molecular structure, Peter J. W. Debye was born in Maastricht, Holland, Mar. 24, 1884, son of Wilhelmus and Maria Reumkens Dibje. He completed the Hoogere Bürger School at Maastricht in 1901 and for the next 4 years studied electrical engineering at the Technische Hochschule in Aachen, graduating in 1905.

While still an undergraduate at Aachen, Debye became an assistant in mechanics to Arnold Sommerfeld and completed a study of the diffraction of light by cylindrical and spherical objects. He became so enraptured with physics and chemistry that he was permitted to use the school's laboratory after hours for experimentation. It was at Aachen that Debye's first original work, a theoretical analysis of Foucault currents in a rectangular conductor, was published in *Zeitschrift für Mathematik und Physik* in 1907.

When Sommerfeld moved to Ludwig-Maximilian University in Munich in 1906 as professor of theoretical physics, he took Debye along as his assistant. Receiving his Ph.D. in physics in 1908 for his thesis on the effects of radiation pressure on spheres of arbitrary electrical properties, Debye continued on at the university as a lecturer (*privatdozent*).

In 1911 Debye succeeded Einstein as professor of theoretical physics at University of Zurich, and it was here that he developed his theories on polar molecules and the specific heat of solids. In 1912 Debye became professor of theoretical physics at University of Utrecht.

He married Mathilde Alberer, Apr. 10, 1913, and two children were born—Peter Paul Ruprecht, later a physicist, and Mathilde Marie Gabriele, later Mrs. Gerhard Saxinger. After 2 years at Utrecht, he accepted the professorship of theoretical and experimental physics at University of Göttingen, where he remained until 1920. The research facilities at Göttingen enabled him to test his theory of permanent dipoles, and in 1916 with Paul Scherrer he published the powder method of x-ray diffraction, now known as the Debye-Scherrer method of identifying crystalline substances by photographing the diffraction pattern of a beam of x-rays directed onto the powdered crystalline material.

Returning to Zurich in 1920 as professor of experimental physics, Debye developed a concept of magnetic cooling and an interionic attraction theory of electrolytes. In 1923 with Hückel he suggested that the deviation of solutions of electrolytes from the laws of ideal solutions is due to interionic attractions.

Becoming director of the Physical Institute of University of Leipzig in 1927, Debye experimented with the measurement of interatomic distances in molecules by x-ray scattering and continued his work on dipoles and electrolytes, publishing a number of books, some of which were translated into English, including "Quantum Theory and Chemistry" (1928), "Polar Molecules" (1929), "The Dipole Moment and Chemical Structure" (1931), "The Interference of Electrons" (1931), "The Structure of Molecules" (1932), "Magnetism" (1933), "The Structure of Matter" (1934), and "Nuclear Physics" (1935). From 1915 to 1940 he also served as an editor of *Physikalische Zeitschrift*.

In 1935 Debye became professor of physics at University of Berlin and director of the Kaiser Wilhelm (now Max Planck) Institute for Physics, with its excellent research facilities. His contribution to an understanding of molecular structure on the basis of dipole moments, x-ray diffraction, and ionization won him the Nobel Prize in Chemistry in 1936.

With the start of World War II in 1939, he was told that he would have to become a German citizen to continue as director of the Institute for Physics. He refused and instead left for the United States, where he had been invited to deliver the Baker Lectures at Cornell University.

From 1940 to 1952 Debye served as head of the Chemistry Department at Cornell, which soon became a leader in solid state research, largely due to his influence. During World War II he became a consultant in the synthetic rubber program. Retiring from Cornell in 1952 as professor emeritus, he continued his researches in the field of high polymers, showing that degree of light scattering is an accurate indication of molecular size.

The number of scientific concepts named after him attests to the originality of his work. These include the Debye theory of specific heat, the Debye-Hückel theory, the Debye-Scherrer method of x-ray diffraction, the Debye-Sears effect in transparent liquids, the Debye theory of wave mechanics, the Debye temperature, the Debye shielding distance, the Debye frequency, and the Debye unit of electric moment. The American Chemical Society also has an award named in his honor.

Debye won many honors in American chemistry, including the Priestley Medal. He was awarded honorary degrees by several universities, and a statue was placed in the city hall of Maastricht in 1939. He became an American citizen in 1946 and died Nov. 2, 1966.

Eduard Farber, "Nobel Prize Winners in Chemistry," 147-151, Abelard-Schuman (1963); *Chem. Eng. News* **27,** 1210 (1949); "Current Biography," 102-04, H. W. Wilson (1963).

JAMES C. COX, JR.

Guillaume Louis Jacques de Chalmot

-1899

de Chalmot, a pioneer in electrothermal chemistry and metallurgy, was born in Holland. Beyond that, nothing is known of his family or personal life. He studied chemistry first at the Realschule and later at Agricultural College in Wageningen. He moved to Germany and received his Ph.D. degree at University of Göttingen in 1891. His two-part dissertation dealt with analytical methods for pentaglycosan and the condensation products of furfural with aromatic bases. Shortly after receiving his degree, he came to America to work with Ira Remsen at Johns Hopkins University. However, after a few months, he accepted the position of assistant chemist in the Virginia State Agricultural Department at Richmond. His work here led to four published papers on plant chemistry, particularly on the pentosans.

In 1894, he joined Willson Aluminum Co. at Spray (now Eden), N. C. where he built the first commercial plant for calcium carbide. This plant went into operation on a round-the-clock basis May 1, 1895. The industrial processes for making calcium carbide and acetylene had been discovered accidentally at Spray in 1892 by Thomas L. Willson during attempts to make calcium metal by reducing lime with carbon in an electric arc furnace. There were no uses for these materials until it was realized that acetylene emitted three times as much light per dollar as water gas; its use as an illuminant grew rapidly.

In the summer of 1895 de Chalmot studied extensively the effects of raw materials, furnace design, and unit operations such as crushing and grinding on power consumption and process efficiency. In the spring of 1896 tests by a team of experts consisting of Edwin Houston, Leonard Kinnicutt, and Arthur E. Kennelly, showed that the plant could produce a ton of carbide per day at a cost of $32 per ton that would yield almost pure acetylene at 78% of the theoretical yield. After the Spray plant burned in May 1896, a new plant was built on the James River near Lynchburg, Va. de Chalmot was manager of these works at the time of his death, Oct. 9, 1899.

During the years 1895–99 de Chalmot's work resulted in 10 patents covering improvements in the electric furnace and electrothermal processes for making silicon, silicon alloys, soluble phosphates, and nitrogen compounds. His 1896 patent for the nitride process for making ammonia clearly preempted the work done by Serpak in France a decade later. High grade ferrochromium, made by processes which he perfected, was used extensively in chrome steels during the Spanish–American War. During this period nine papers were published in *American Chemical Journal.* Six of these dealt with the silicides of calcium, iron, chromium, and copper and brought him recognition as an authority on silicides.

J. T. Morehead, and G. de Chalmot, "The Manufacture of Calcium Carbide," *J. Amer. Chem. Soc.* **18,** 310-331 (1896); *Progressive Age* **18,** 169-181 (Apr. 15, 1896) with photograph.
G. de Chalmot, U. S. Patent 741,396; obituary, *Amer. Chem. J.* **23,** 447 (1900).

HERBERT T. PRATT

Clara Marie de Milt

1891-1953

At the time of her death, May 10, 1953, Clara de Milt was professor of chemistry, head of the department at Newcomb College of Tulane University, and chairman of chemistry for the entire university. Her primary field was organic chemistry, but she was well-known for her work in history of chemistry.

Clara de Milt was born May 8, 1891 in New Orleans. In 1907 she entered Newcomb College where she majored in history and science, being elected to Phi Beta Kappa in 1911. Because of financial pressure, she took a position in McDonough-Jefferson High School, teaching English and science. In 1920 she became instructor in chemistry at Newcomb College and in 1921 obtained her degree of Master of Science from Tulane University. In December 1925 she submitted a doctoral thesis to the faculty of University of Chicago.

On her return to New Orleans she was appointed assistant professor in charge of the Department of Chemistry at Newcomb College, associate professor and head of department in 1928, and professor in 1930. During the years between 1929 and 1940 she directed graduate study in organic chemistry and wrote three laboratory manuals, in addition to teaching, and directing the Newcomb Alumnae Loan Fund. She was appointed chairman of the Department of Chemistry of Tulane University in 1941, a position which she held until 1949 when she resigned in order to have time to devote to research and teaching in history of science, a course which she had instituted in 1923.

Her scientific papers were largely in the field of history of chemistry and primarily on the chemistry of the seventeenth and nineteenth centuries in France, her most important contribution being a study of the work of Auguste Laurent. She was on the editorial board of *Chymia.*

Clara de Milt was fundamentally interested in that interplay of people and ideas which forms the background to discovery. Her handling of biography reflected her sincere interest in people. She was always scrupulously fair, although she did not hesitate to express a forthright opinion on the character or ability of anyone under consideration. Her positive outlook and outspoken and positive expressions of opinion were incisive but were always tempered with a sense of equity.

Obituary, John M. Scott, *J. Chem. Educ.* **31,** 419-420 (1954), list of publications; biography, V. F. McConnell, *Chymia* **9,** 201-213 (1964), with portrait; personal recollections.

VIRGINIA F. MCCONNELL

Martin Dennis

1851-1916

Dennis, leather chemist and founder of the company which bore his name, was born in Newark, N.J., Jan. 8, 1851. He was educated at Newark Academy and graduated in 1873 from Princeton University, where he studied the sciences in preparation for a career as a surgeon and where he distinguished himself as a fine amateur musician and organizer of the first Princeton Glee Club. He won academic honors both at Newark Academy and at Princeton and seemed well on his way into studies leading to medicine, but his father suffered business reverses in the panic of 1873, and Dennis was forced to alter his plans.

He decided to accept a position offered by two of his uncles, as superintendent of their plant in Yonkers, N.Y., which made a variety of fancy leathers. He quickly justified their establishment of the new position by his close attention to details of manufacturing processes, which resulted in more uniform products, greater sales, and a rapidly expanding business. But Martin soon turned his attention to the chemical nature of the processes themselves.

Tanning of hides was at the time (and still is, in large degree) a lengthy, complicated, ill-understood operation, more art than science. Basically it consists of two parts, the removal of superfluous materials from skins, followed by conversion of the remaining fibrous material to leather. In the first part the outer skin and hair is removed by soaking the skins, usually in alkaline solution with calcium hydroxide, followed by mechanical scraping; thereafter the hides are "bated," a process in which the remaining tissues are dissolved out of the inner skin by enzymatic processes, leaving mostly a mat of collagen fibers ready for tanning. Tanning appears to be mainly a preservation of the collagens against further degradation and is accomplished by treating the hide with one of three agents: tannins (high molecular weight, phenolic-type compounds derived from nut galls and barks), salts of chromium (III), or oils such as turkey red, a sulfonated castor oil.

It was the tanning process which Dennis

focussed on first. The existing methods using tannins were incredibly time-consuming, requiring weeks to months for thick hides. Chrome tanning had just come into experimental use but was done in a two-step process in which the hide was first soaked in an acid solution of a dichromate salt, then removed to another solution in which the chromium was reduced *in situ,* usually by bisulfite. Dennis reasoned that this harsh treatment could do little good to the hides, and after much experimentation devised a milder single-bath treatment using chromium (III) chloride as a direct source of chromium for complexation with the protein fibers.

Having perfected and patented his process, Dennis experienced the first of many setbacks from the industry's extraordinary conservatism when his uncles refused to employ the new method. He thereupon set up his own factory in Brooklyn, producing chrome-tanned, glazed kid shoe leather but was forced by lack of capital to close down within a year. In 1893 Dennis decided to market the tanning material to tanners under the trade name "Tanolin," and with a friend from Princeton, Harry E. Richards, formed Martin Dennis Chrome Tannage Co. (later just Martin Dennis Co.), with a single small plant in Newark. Selling the industry his new product was an uphill battle, with England and Europe proving even more resistant to change than America; but eventually customer preference for chrome-tanned shoe leathers, which could be wet and dried without losing flexibility, forced adoption of chrome tanning, and success was on its way.

Dennis next turned his attention to the problem of oiling chrome-tanned leather for flexibility and preservation. As the leather would not accept conventional oils, he made commercially available the emulsified oil called "fat liquor" recently developed by Robert Foerderer in making Vici kid leather. Finally, Dennis considered the bating process, easily the messiest and most disgusting part of the curing of skins, as the enzymes for the bate, or "puer," were obtained by use of chicken, dog, or pigeon manures. He developed a stable enzyme preparation which he marketed as "Puerine," which was the first commercially successful artificial bate.

Dennis died Feb. 6, 1916, and Richards managed the business until 1923, when Dennis's son Harold succeeded Richards as president. In 1948 the firm was absorbed by Diamond Alkali Co. (later Diamond Shamrock Chemical Co.).

W. Haynes, "Chemical Pioneers," Van Nostrand (1939), pp. 197-208; W. Haynes and E. L. Gordy, "Chemical Industry's Contribution to the Nation: 1635-1935," *Chem. Industries,* May 1935 (Supplement), pp. 17-18; obituary, Newark, N.J. *Sunday Call,* Feb. 13, 1916; information from Princeton University archives.

ROBERT M. HAWTHORNE, JR.

Chester Dewey

1784-1867

At the midpoint of the nineteenth century, at the considerable age of 66, Chester Dewey was elected professor of chemistry and natural philosophy at University of Rochester. He accepted the appointment as one of the six founding faculty members with the following observation and promise to John N. Wilder, the university's first president,

> "To secure the prosperity and success of the University, its faculty must be a body of workers. I . . . assure you . . . of the employment of all the powers I can bring to its aid."

The promise he faithfully kept to his university, profession, and students.

The state of scientific development and the needs of the university scientific education in the 1850's called for a natural philosopher. Dewey fitted that role very well. While it is interesting that his official title singled out chemistry for special notice, his assignment required lecturing on botany, geology, physiology, and astronomy as well. In addition to these formal teaching duties Dewey bore the responsibility (rather a common one in colleges and universities of that day) for obtaining collections of mineral and botanical specimens, library materials, laboratory demonstration apparatus, etc. In his 13-year tenure at Rochester, Dewey was able to establish some of these vital facilities, but laboratory instruction was impossible.

He stopped active teaching in 1860 but maintained a nominal connection with the university until his death.

Dewey's teaching and inspirational powers are illustrated by the fact that a notable number of his students entered scientific careers. His public lectures and popular writings on scientific topics, both basic and applied, were numerous and greatly respected by scientists and laymen. One of his meticulous reports to the university board of trustees in February 1852 reveals something of the acceptance his teaching received as well as a very liberated view of education for women,

"Four young ladies from Miss Tracy's Seminary and two from families in the city, have attended the lectures on Chemistry with the Juniors during most of the month, as proposed by Mr. Wilder."

The characteristics described in connection with Dewey's career at Rochester are also found in his earlier 17 years (1810-1827) as professor of mathematics and natural philosophy at his alma mater, Williams College. Between these two positions in higher education he served as principal of the Berkshire Gymnasium in Pittsfield, Mass. (1827-1836) and of the Collegiate Institute in Rochester (1836-1850). These ample credentials illustrate a long and dedicated teaching career.

Chester Dewey's scientific career was also well-suited to his time. His principal scholarly writings were in botany; especially of North American plants. A long series of papers on the sedge grasses were highly respected. Certain California plants bear the genus *Deweya* in his honor. He did publish some original chemical observations including: the analysis of minerals, the conduction of water, and a discussion of Grove's Battery. Also typical of Dewey and his time is the amateur complementing the professional. While he never pretended to be a meteorologist and never mentioned it as a significant part of his instructional activities, he compiled valuable weather observations for decades on behalf of the State Bureau and the Smithsonian Institution.

Dewey was born in Sheffield, Mass., Oct. 25, 1784 the son of Stephen and Elizabeth Dewey. He married twice: to Sarah Dewey in 1810 and to Olivia Dewey in 1825. With the help of these two wives Dewey raised 15 children. Several honorary degrees were conferred upon Dewey; *e.g.*, M.D. (Yale, 1825), D.D. (Union, 1838), and L.L.D.(Williams, 1850). He died in Rochester, Dec. 5, 1867. The picture of Chester Dewey is a master teacher and a competent scholar in the context of his time. While it would be unwise to consider him a chemist as we understand that term, he did play an important role in the early development of scientific education, including chemistry.

Biography by A. Gray, *Amer. J. Sci.*, (2nd series), **45**, 122-123 (1868); "Dictionary of American Biography," **5**, 267, Chas. Scribner's Sons (1930); "National Cyclopedia of American Biography," **6**, 328, James T. White Co. (1892); A. J. May, unpublished History of the University of Rochester, and other documents in University of Rochester Archives.

K. THOMAS FINLEY

Aaron Dexter

1750-1829

Dexter was the first professor of chemistry in Harvard Medical School, from 1783 to 1816. Born in Malden, Mass., Nov. 11, 1750, he attended Harvard where he learned a bit of chemistry from John Winthrop, professor of natural philosophy. After receiving his bachelor's degree in 1776, he studied medicine under Samuel Danforth, Jr., a physician and chemist of Boston. Dexter probably obtained most of his early knowledge of chemistry from Danforth and from reading under Danforth's direction. He did not attend medical school—his M.D. degree was honorary from Harvard in 1786—but after he finished his apprenticeship under Danforth he began to practice in Boston and became a prominent physician.

In 1783 he became professor of chemistry and materia medica in the newly founded Harvard Medical School. Medical School sessions lasted only 3 or 4 months, and Dexter spent the remainer of each year practicing medicine.

Isolated in New England from the centers

of chemistry in Europe, Dexter kept up with the rapid developments in the science by reading imported, up-to-date books. Tradition indicates that he was not skillful in performing lecture demonstrations; perhaps during his studies under Danforth and his teaching career at Harvard he did little experimenting and therefore did not know much about the technique of chemistry.

Dexter taught at Harvard for 33 years and then resigned. He was influential in having his student, John Gorham, elected his successor. Thereafter Dexter practiced medicine until he died in Cambridge, Feb. 28, 1829.

Thomas F. Harrington, "The Harvard Medical School," Lewis Pub. Co. (1905); "Quinquennial Catalogue of . . . Harvard University" (1900); I. Bernard Cohen, "The beginning of chemical instruction in America," *Chymia* **3**, 17-44 (1950) or substantially the same discussion of Dexter in Cohen's book, "Some early tools of American science," Harvard Univ. Press (1950).

WYNDHAM D. MILES

Barnett Fred Dodge

1895-1972

Dodge was born in Akron, Ohio, on Nov. 29, 1895. He received his B.S. degree in chemical engineering from Massachusetts Institute of Technology in 1917 and then spent 3 years working for E. I. du Pont de Nemours and Co., chiefly on explosives, and 2 years working for Lewis Recovery Corp. of Boston. He enrolled for graduate study at Harvard University in 1922 and received his doctor of science degree in 1925. He was appointed to the faculty of Yale University in 1925 as assistant professor, promoted to associate professor in 1930 and to professor in 1935. Dodge served as dean of the School of Engineering from 1960 to 1962, overseeing the conversion of the school to the Department of Engineering and Applied Science. His retirement in 1964 was followed by 8 years of continuing productive scholarship before he died, Mar. 16, 1972.

Dodge's doctoral research on vapor-liquid equilibrium in the oxygen-nitrogen system gave him strong interest in research on cryogenics, thermodynamics, and phase equilibrium at high pressures, and he continued to work on these subjects for many years. In the field of cryogenics, he was an official investigator for the National Defense Research Council during World War II and was involved in developing portable oxygen plants for the Navy. He was also a consultant to Oxygen Process Corp. of New England and to Phillips Petroleum Co.; he had just completed the manuscript of a book on cryogenic engineering at the time of his death.

Much of Dodge's professional work was concerned with thermodynamics. In 1944 he published "Chemical Engineering Thermodynamics," an excellent textbook which, for more than a quarter of a century, inspired students and colleagues by its clarity and depth of presentation of subject matter. His studies in the field of high pressure were also concerned in many cases with thermodynamics, particularly in phase equilibrium. His studies of permeation of metals by hydrogen and of chemical reaction rates of high pressures (ammonia synthesis, polymerization) were fundamental to workers in these fields.

His other research interests included water pollution control, a field which he entered before World War II in studies for the Connecticut State Water Resources Commission. This work was concerned primarily with brass mill wastes. Later studies with the American Electroplating Society concerned plating room wastes. Dodge also aided in the design of a waste-water treatment plant for Oneida, Ltd., 1950-53. Still other interests were in the separation of the isotopes of uranium and in desalination of sea water. During World War II Dodge spent a year at Oak Ridge, Tenn., directing experimental investigations and process control studies concerned with uranium isotope separation.

Dodge was in great demand as a lecturer and devoted considerable time to lecture tours for Sigma Xi and for the American Institute of Chemical Engineers. He was well–known in international circles and taught and lectured at various times at University of Toulouse, University of Lille, University of Barcelona, Universidad Central de Venezuela, and in Iran and Uruguay. A student of languages, he was quite capable of lecturing in French and in Spanish.

It is interesting that this excellent lecturer was not inclined to use the lecture method extensively in teaching. He believed strongly that students had to assume much of the responsibility for their own education and that students learn primarily by doing. His well-formulated, demanding problem assignments taxed the students' knowledge and ability, and there were always helpful comments on the solutions returned to students. The problems and examinations, together with respectful attention to fundamentals at all times and his own vast knowledge of his field, provided students with an excellent atmosphere for learning.

Dodge received honorary degrees from Worcester Polytechnic Institute, the University of Toulouse, and the Universidad Central de Venezuela, as well as the American Institute of Chemical Engineers Founders Award, the Warren K. Lewis Award, and the Walker Award. He served the American Institute of Chemical Engineers as vice-president in 1954 and president in 1955.

Dodge pursued hobbies with the same intensity and pleasure as his professional work. He was a fine photographer of near-professional caliber. During all of his adult life he played tennis, swam, and climbed mountains. At age 55 or so he added skiing and ice-skating to his sports and became quite proficient at both.

He was survived by his widow, Constance Woodbury Dodge, who was a novelist, and by a daughter, Phyllis Dodge Putney, and a son, Richard W. Dodge.

Personal recollections; biography by Charles A. Walker in *Chem. Eng. Educ.,* 150-152 (Fall 1972).

CHARLES A. WALKER

Francis Despard Dodge

1868-1942

Dodge, an American chemist who contributed substantially to contemporary knowledge of the compositions and reactions of essential oils, was born Jan. 14, 1868 in Washington, D.C. He was educated at Columbia University. After receiving his Ph.B. in 1888, he was appointed assistant organic chemist, a position he held during the course of his doctoral studies. In 1889 he isolated the aldehyde, citronellal, from Indian lemongrass oil (oil of citronella) and established its structure. By treating an alcoholic solution of citronellal with sodium amalgam, in the presence of acetic acid, Dodge prepared citronellol, subsequently shown to be a major ingredient of rose oil. Treatment of citronellal with phosphorus pentoxide produced a crystalline substance which Dodge identified, in 1915, as a phosphoric acid derivative of isopulegol, one of the few known crystalline derivatives of that terpene alcohol.

In 1890 Dodge reported the isolation of citral from lemongrass oil. The structure of citral was soon established when Semmler showed it to be a geraniol oxidation product. The year 1890 was an eventful one for Dodge in other ways. Having completed his doctoral dissertation on "Oil of Citronella" earlier in the year, Dodge received his Ph.D. degree and left for University of Heidelberg to work in the laboratories of Victor Meyer. Meyer had arrived in Heidelberg in October 1889 to found a new organic chemistry department.

After a year in the heady atmosphere of German chemical research, Dodge returned to America and joined the family firm of Dodge and Olcott, importers of essential oils.

As chief chemist and factory manager, he soon expanded and transformed the company's manufacturing operations, which until then had been largely limited to oil rectification and methyl salicylate synthesis. Under Dodge's guidance the Bayonne, N.J. plant produced safrole from oil of camphor and imitation sassafras; he isolated in commercial quantities cintronellol, eucalyptol (cineole), and citral. He also contributed to the production of ethyl and amyl acetates for use in flavors.

Dodge was much concerned with analytical procedures—for assaying purposes as well as for detecting adulterants and impurities. In 1904 he became the first proponent of the use of potassium acid phthalate as a reference standard in alkalimetry. He developed a series of analytical methods based on the difference in oxidizability of various organic substances by potassium permanganate. Thus, the cineole assay in eucalyptus oil is based on the resistance of cineole to oxidation in the presence of readily oxidized impurities. The method was adaptable to the estimation of petroleum admixed in turpentine or to that of bornyl acetate in the presence of more oxidizable isomeric esters.

In 1917 Dodge argued that "ursone" (ursolic acid) was a hydroxylactone and held to that view at least as late as 1930, when the hydroxycarboxylic acid structure had become generally accepted.

Although most of his published scientific work after 1891 strongly emphasized analytical methods and techniques, Dodge retained an interest in organic synthesis. One of his last publications was a U.S. Patent (No. 2,211,538, issued Aug. 13, 1941) on a method for synthesizing substituted quinolines.

He remained professionally active until his death, Mar. 14, 1942. He is described by those who knew him as quiet, modest, and gracious. His widow survived him; he left no direct descendants.

William Haynes, "American Chemical Industry," **3,** 331, D. Van Nostrand Co. (1945); *Chem. Abstracts* **6-35;** J. Simonsen, "The Terpenes," 2nd ed. (1941-1957); communications from Val H. Fischer of Dodge & Olcott; communication from Alice H. Bonnell, Columbia University.

CHARLES H. FUCHSMAN

John Van Nostrand Dorr

1872-1962

Dorr was one of the pioneers in chemical engineering both in this country and around the world. Going to work in the laboratory of Thomas A. Edison at the age of 16, Dorr was inspired to study chemistry at Rutgers University where he graduated in 1894. He was one of the first to apply chemical engineering principles to extractive metallurgy, and *Chemical Engineering* magazine in 1949 said editorially, "The contributions which his inventions . . . have made to mankind are probably greater than those of any other chemical or metallurgical engineer of our times."

Dorr was born in Newark, N.J., Jan. 6, 1872, the son of John Van Nostrand and Nancy (Higginson) Dorr. His direct involvement with Edison as a chemical experimenter led him to adopt Edison's slogan, "Try anything once." After working in New York City for a year following graduation he decided that once was enough and headed west.

He settled first in the Black Hills of South Dakota where he worked as a chemist and assayer at various mining operations, operating two of them himself. He also engaged in construction of a dam and power plant in Green River, Utah and in building an apartment house in Denver.

In 1910 he incorporated himself as the Dorr Co., and the major part of the business of this company throughout its life was the development and marketing of three of Dorr's inventions—the Dorr Classifier, Dorr Thickener, and Dorr Agitator. These devices are today the basic tools for the treatment of water, sewage, and industrial wastes. In 1955 the Dorr Co. merged with Oliver United Filters, founded by another distinguished engineer, Edwin L. Oliver, to become an engineering giant, Dorr-Oliver, Inc., of Stamford, Conn., with net sales and other revenues in 1971 of over 90 million dollars.

One of Dorr's sayings was that "the knowledge a man wins goes into a common store, but the inventions he makes go into worldwide use." So he concentrated his research efforts on practical devices for engineering use. His machines and processes came into use all over the world, until today Dorr-Oliver, Inc., operates wholly owned subsidiary companies in 11 countries in addition to the United States. It has representatives in 30 other countries. Dorr established subsidiary companies in London, Paris, and Berlin in the early 1920's, well ahead of the present general trend.

Through the Dorr Foundation, established in 1940, he supported programs and investigations in many fields outside of chemistry and chemical engineering. For example, the Dorr Foundation sponsored work in geriatrics, public education, public safety, and the arts.

He was the author of the book "Cyanidation and Concentration of Gold and Silver Ores," published in 1936 and revised in 1950. He wrote over 65 articles on technical subjects.

Among his many honors were the John Scott Medal of the Franklin Institute and the Perkin Medal of the Society of Chemical Industry. On the occasion of the award of the Perkin Medal in 1941, his brother said of him: "The prime characteristic of his professional and business life is his perfectionism. Every angle must be explored . . . the engineering must be perfect. . . . He has a genius for attracting and inspiring younger men and holding them in his organization. That is the way he has built up his staff the world over. That staff, he will tell you, is the achievement he really takes most pride in."

Dorr died June 29, 1962 in his ninety-first year, survived by his wife, Virginia Nell Dorr, and a daughter and his brother.

Biography by G. H. Dorr and Perkin Medal Address by J. V. N. Dorr, *Ind. Eng. Chem.* **33,** 366-367 (1941); J. V. N. Dorr, "Dorrco World-wide—1919-1954," Dorr Co. (1954); *Chem. Eng. News,* p. 100, July 7, 1958; obituary issued by Dorr-Oliver Co.

JOSEPH A. SCHUFLE

James Douglas

1837-1918

Douglas was a man of such remarkably varied talents that only accident forced him into the career for which he is best remembered—that of manager and prolific innovator in mining and smelting, and sometime president of Phelps, Dodge and Co.

Douglas was born Nov. 4, 1837 in Quebec City, Canada to Dr. James and Elizabeth (Ferguson) Douglas and early showed his characteristic productivity; the first publication recorded under his name was a "Report of a canoe expedition along the east coast of Vancouver Island," in the *Geographical Society Journal* of 1854.

Douglas's education was marked by many changes of direction. He studied medicine, theology, and chemistry, and at one point earned a prize in English literature at Edinburgh University, where he began his college studies. His only earned degree was an A.B. from Queens University, Ontario in 1858. After this he travelled with his father gathering archeological material in Europe and Africa, studied further at Laval University and at Edinburgh, married Naomi Douglas in Frankfurt, Germany in 1860, took charge of his father's Quebec Lunatic Asylum for a time, served as assistant minister at St. Andrew's Presbyterian Church, Quebec, and taught chemistry for 3 years at Morrin College in Quebec.

In 1871 when he was 33, Douglas took up mining and metallurgy, almost by accident. His father had invested heavily in gold and copper mining in Canada but found that the 2% ores of the Harvey Hill mines in Quebec Province could not be refined by existing methods. Douglas and Thomas Sterry Hunt, of Laval University (later of McGill and MIT), developed a process in which even small quantities of cupric oxide could be dissolved with partial reduction in hot aqueous solution of ferrous chloride and sodium chloride. Ferric oxide and inert materials were filtered off, with the mixed copper (I) and (II) held in the filtrate as the tetrachloro complexes, to be reduced to metallic copper by iron metal, regenerating ferrous chloride.

Initially it appeared that the low-grade ore problem was solved, but there were many practical difficulties, and the family fortunes could not be saved. In 1875 Douglas came to Phoenixville, Pa. as superintendent of the Chemical Copper Co. There he used a variant of the Hunt-Douglas process, but the company was not successful, and that plant was finally destroyed by fire. From contacts developed at Phoenixville, Douglas was sent by Phelps, Dodge and Co. to investigate the Copper Queen Mine in Bisbee, Ariz. He recommended purchase and was installed as manager.

At nearly 50 years of age Douglas found all of his varied talents put to use at last. He made technological improvements, found new ore deposits, and consolidated the Copper Queen with other holdings. He built railroads, first within the mine, ultimately to El Paso, Tex. He maintained highly progressive wage standards and labor policies, and expanded the company's holdings to include mines across the Rio Grande in Mexico. He broke down the mining industry's prevailing suspicion and hostility about trade secrets, welcoming all visitors and observers at the Copper Queen. His example gradually brought about so ready an exchange of information that the American mining industry became the most advanced in the world.

In 1908 Phelps, Dodge incorporated with Douglas as its first president. This brought him a sizeable personal fortune, and in his remaining years he took pleasure in the philanthropies it allowed him. Before and after his death he made substantial contributions to the American Institute of Mining Engineers, making it into a viable professional organization; to the American Museum of Natural History, the General Hospital at Kingston, and McGill University, among others.

Despite his full professional life Douglas found time to publish the historical studies "Canadian Independence, Annexation, and Imperial Federation," "Old France in the New World," and "New England and New France," and a volume of memoirs from his father's writings, "Journal and Reminiscences of James Douglas, M.D." Honors came late in Douglas's life: in 1906 he received the gold medal of the Institute of Mining and Metallurgy of Great Britain, and in 1916 the John Fritz Medal. Both Queens University (of which he was chancellor when he died) and McGill gave him honorary LL.D.s.

Douglas died June 25, 1918, aged 81 years. Two of his sons, Walter and James S., continued in the mining industry and in Phelps, Dodge.

A. R. Ledoux, *Bull. Amer. Inst. Mining Eng.* **109,** iv-ix (1916); R. W. Raymond, *ibid.* **141,** 1403-09 (1918); *Eng. Mining J.* **106,** 18-20 (1918); "Dictionary of American Biography," **5,** 396, Chas. Scribner's Sons (1930); "National Cyclopedia of American Biography," **23,** 22-23, James T. White & Co. (1933); "The John Fritz Medal, 1902-1922," 84-87, John Fritz Medal Board of Award (1922); *New York Times,* June 26, 1918.

ROBERT M. HAWTHORNE JR.

Silas Hamilton Douglas

1816-1890

Douglas (Douglass until 1872) was born in Fredonia, N.Y., Oct. 16, 1816. He attended an academy and New York University, went to Detroit in 1838 to study medicine with a physician, traveled to Baltimore to attend University of Maryland Medical School during the winter of 1841-42, and was licensed to practice medicine in Michigan in 1842.

He was appointed assistant to Douglas Houghton, professor of chemistry, mineralogy, and geology at University of Michigan Sept. 12, 1844. However, Houghton had not taught any chemistry classes since his appointment in 1839, and instruction in chemistry at University of Michigan dates from Douglas' appointment. In 1853 Douglas visited Ypsilanti to teach the first chemistry course at Michigan State Normal School, later renamed Eastern Michigan University.

The University of Michigan had no chemistry laboratory at the time of Douglas' appointment, and for several years he gave instruction in his own private laboratory. In May 1856 he was made superintendent of construction of a chemistry laboratory for experiments and instruction in analytical chemistry for which $2,500 was appropriated by the regents of University of Michigan. As in modern times, further costs were accrued and by the time the building was occupied in the autumn of 1856 the appropriations added up to $4,509.85. The university was one of the earliest schools in the Midwest to offer laboratory instruction in chemistry.

In 1870 Douglas was appointed first director of the Chemical Laboratory, his full title being Professor of Chemistry and Director of the Chemical Laboratory. He held this position until 1877, when his title was Professor of Metallurgy and Chemical Technology and Director of the Chemical Laboratory. He delivered lectures and gave laboratory instruction to collegiate and medical students. Victor C. Vaughan, who was attracted to the university upon reading Douglas' and Prescott's text and who became a chemistry teacher, described Douglas thus: "He was not a great teacher, either in the lecture room or in the laboratory, but he built the laboratory and saw that it functioned. In my opinion, he justly deserves the credit of introducing required laboratory instruction in chemistry into the curricula of medical schools."

Douglas severed relations with the university in 1877. Toward the end of his life he was infirm, and he died in Ann Arbor, Aug. 26, 1890.

Douglas published a few articles, two of them in *Peninsular Medical Journal* discussing the analysis of Michigan coal and the examination of waters. His chief work, in collaboration with Albert B. Prescott, was a text on qualitative chemical analysis, which passed through several editions.

E. D. Campbell, "History of the Chemical Laboratory of The University of Michigan" (1916); Victor C. Vaughan, "A Doctor's Memories," pp. 92, 232, Bobbs-Merrill Co. (1926); *Pharm. Rev.* **21,** 359-363 (1903); Albert B. Prescott, "Silas Hamilton Douglas, the Founder of the Chemical Laboratory," *Michigan Alumnus* **9,** 1-6 (Oct. 1902); information from Donald R. Hays.

LEIGH C. ANDERSON

Herbert Henry Dow

1866-1930

Dow was born Feb. 26, 1866, in Belleville, Ont., Canada. His father, Joseph Henry Dow, was an American mechanic-inventor employed temporarily in Canada; his mother was Sarah (Bunnett) Dow. Dow grew up in Derby, Conn., and Cleveland, Ohio. He attended Case Institute of Technology and after graduation taught chemistry at Huron Street Hospital College for a year.

During his studies at Case, Dow analyzed a sample of brine and sensed the value of the small quantity of bromine that was present. He developed a process in which bromine was freed by electrolysis and then blown from the solution by a current of air. At the age of 23 he organized Canton Chemical Co., near Canton, Ohio, to produce bromides. The plant turned out only a few carboys of ferric bromide and then went out of business. But Dow had gained experience, and with confidence in his ability to make the process work he organized in 1890 another firm, Midland Chemical Co. at Midland, Mich. The Midland firm was successful, but the directors' unreceptiveness to Dow's plans for expansion led him to organize in 1895 Dow Process Co. at Navarre, Ohio to produce caustic and bleach by electrolysis of salt solution and in 1897 Dow Chemical Co. (which absorbed the earlier two companies) to manufacture bleaching powder.

The availability of basic chemicals at his plants gave Dow incentive to prepare inorganic and organic compounds. He branched into the manufacture of pharmaceutical bromides, mining salts, sulfur chloride, and photographic chemicals, and he organized a second Midland Chemical Company (later taken into Dow Chemical) to synthesize chloroform and carbon tetrachloride.

Dow sought out topnotch chemists and gathered them into the firm. The team was responsible for the first American production of indigo in 1916, the commercial preparation of phenol from chlorobenzene in the 1920's, the manufacture of magnesium and magnesium alloys, and the extraction of bromine from sea water.

Besides the financial and engineering difficulties common to pioneers in chemical industry, Dow faced two additional hurdles. In the late 1890's the United Alkali Co., an English syndicate, tried to force him out of the bleach business by dropping their American prices to ruinously low levels. A few years later, while exporting bromides to Europe, he aroused the ire of the Bromkonvention, a cartel that controlled markets outside the United States, and for several years the Germans tried to beat him in a price war. Through all vicissitudes Dow hung on and built one of the largest chemical companies in the United States.

The odds against Dow's starting and developing a chemical firm were tremendous. His success was a result of his tenacity, hard work, ability, and vision. By 1930, the year of Dow's death, his company was producing 150 different chemicals at the rate of 800 carloads a month. Dow died Oct. 15, 1930, at Mayo Clinic, Rochester, Minn., following an operation.

One of Herbert Dow's children, Willard Henry Dow, was a chemical engineer. Born in Midland, Jan. 4, 1897, Willard graduated from University of Michigan in 1919, held various positions in the firm, and succeeded to the presidency in 1930. He and his wife were killed in an airplane crash Mar. 31, 1949.

Murray Campbell and Harrison Hatton, "Herbert H. Dow, Pioneer in Creative Chemistry," Appleton-Century-Crofts (1951); Williams Haynes, "Chemical Pioneers," D. Van Nostrand Co. (1939), pp. 259-278; sketch by Haynes with portrait in Eduard Farber, "Great Chemists," 1219-1232, Interscience (1961); "Dictionary of American Biography," **21,** 261-2, Chas. Scribner's Sons (1944); Don Whitehead, "The Dow Story: The History of the Dow Chemical Company," McGraw-Hill Co. (1968).

For Willard Henry Dow see *Chem. Eng. News* **26,** 1840 (June 21, 1948); **27,** 1073 (April 11, 1949), port.; *New York Times,* April 1, 1949; Don Whitehead, "The Dow Story: The History of the Dow Chemical Company" (1968).

WYNDHAM D. MILES

Henry Draper

1837-1882

Draper was born in Prince Edward County, Va., Mar. 7, 1837, son of John William Draper, professor of chemistry at Hampden Sidney College. When Henry was 2 years old his father accepted the professorship of chemistry at New York University, and the family moved to New York City. In his 'teens, Draper enrolled in the Academic Department at NYU but transferred to the Medical Department at the end of his sophomore year and completed his studies when he was 20. Too young to be granted his M.D. degree, he spent a year in Europe and returned in 1858 to graduate.

Draper was on the staff of Bellevue Hospital from 1858 to 1860, surgeon of the Twelfth Regiment, New York State Militia in 1862, professor of natural science in NYU Academic Department from 1860 to 1882, professor of physiology in the Medical Department from 1866 to 1873 and the Science Department from 1870 to 1882, and professor of analytical chemistry in the Science Department from 1862 to 1882. During the years he was at NYU, he and his brother, John Christopher Draper, assisted their father with his chemistry courses.

Draper married a wealthy heiress, Mary Anna Palmer, in 1864 and after the death of her father in 1874 took over the management of the immense Palmer estate. George Barker, a notable chemist who knew Draper well, said Henry had no superior in New York City as a business man.

In the 1870's, Draper began to take vacations in the form of long hunting trips on horseback in the northern Rockies. Exposure to cold and snow during a 2-month, 1500 mile ride through the Northwest in the autumn of 1882 undermined his health, and he died in New York City, Nov. 20, 1882.

Draper carried out his first published research, on the function of the spleen, when he was 20. During this investigation he made excellent photomicrographs. Shortly after this during his visit to Europe he saw the great (for that time) 6-foot reflecting telescope owned by the Earl of Rosse. This drew him to astronomy. Thereafter photography and astronomy were his lifelong interests. During the 1860's and 1870's he spent his spare time on astronomy. He constructed some of the best telescopes in the United States and mounted them in his father's observatory at Hastings-on-Hudson. He built a private astronomical laboratory and workshop in the upper story of his stable; and judging from the large, expensive electrical, optical, photographic, and spectroscopic equipment that he installed, it must have been the finest private laboratory of its kind in New York City, perhaps in the United States. Draper's eminence in celestial photography led the government to place him in charge of the Transit of Venus Commission in 1874. As a reward for his services Congress voted that a medal be struck in his honor.

Draper retired from teaching in 1882 at the age of 45, intending to spend the remainder of his life carrying on research. It was ironic that he died a few months later.

Draper contributed to the chemistry of photography, and he was a pioneer in the application of photography and spectroscopy to astronomy. As a memento of his service in chemical education he left behind a "Text-Book of Chemistry," published in 1866, which went through several editions. Among his score of articles are several relating to chemistry.

Biographies by George Barker in *Amer. J. Sci.* (3S) **25,** 89-96 (1883), with list of publications, *Proc. Amer. Philos. Soc.* **20,** 656-662 (1881-1883), and in *Biog. Mem. Nat. Acad. Sci.* **3,** 81-139 (1895), with list of publications; "General Alumni Catalogue of New York University, 1833-1907, Medical Alumni" (1908); "General Alumni Catalogue of New York University, 1833-1905, College, Applied Science, and Honorary Alumni" (1906).

WYNDHAM D. MILES

John Christopher Draper

1835-1885

Draper, son of chemist John William Draper, was born in Mecklenburg County, Va., Mar. 31, 1835. The family moved to New York City in 1839, when the father accepted the professorship of chemistry in the Medical Department of New York University. John Christopher attended NYU, graduating with an M.D. degree in 1857. After travelling in Europe for a year he became professor of analytical and practical chemistry in NYU Science Department and held this position until 1871. During the same period he taught chemistry at Cooper Union from 1860 to 1870 and was assistant surgeon in the field with the Twelfth New York Regiment in 1862. He was also professor of natural sciences at College of the City of New York, 1863 to 1885, and professor of chemistry in NYU's Medical Department, 1865 to 1885.

John Christopher published articles in medical and scientific journals and in magazines. He wrote "A text-book on anatomy, physiology, and hygiene" (1866), "A Practical Laboratory Course in Medical Chemistry" (1882), and "A text-book of medical physics" (1885). He was an industrious, competent teacher but not the equal of his father or brother, Henry, as a researcher. He died in New York, Dec. 20, 1885.

Joshua L. Chamberlain, ed., "Universities and their sons, New York University . . ." (1901), vol. 1, part 2, pp. 60-61, with portrait; "General Alumni Catalogue of New York University, 1833-1905, College, Applied Science, and Honorary Alumni" (1906); "General Alumni Catalogue of New York University, 1833-1907, Medical Alumni," (1908).

WYNDHAM D. MILES

John William Draper

1811-1882

Draper, born near Liverpool, England, May 5, 1811, received his early education and interest in science from his father, a Methodist minister. He entered University of London when it opened in 1829. Two of his favorite subjects were literature and chemistry. He published poetry in *Ladies' Magazine* and learned chemistry under the tutelage of Edward Turner, a noted British chemist, and by experimenting in his own little home laboratory.

He left the university before obtaining his degree, and with his mother and sister, his father having died, emigrated to the United States in 1832, settling at Christiansville, Va. Draper had hoped to teach in a Methodist institution but arrived too late to obtain the job. He continued to follow science and finally decided to study medicine. He attended University of Pennsylvania, where he extended his knowledge of chemistry under Robert Hare. While in Philadelphia he also studied with John K. Mitchell, professor of chemistry at Jefferson Medical College. At that time medical school terms covered only 4 months of each year. Draper spent the other 8 months of the year experimenting at home.

After receiving his M.D. degree in March 1836, Draper practiced medicine. In the autumn of 1836 he became professor of chemistry and natural philosophy at Hampden Sydney. In 1838 he was elected professor of chemistry and natural history at New York University and moved to New York City. He was professor of these subjects in the Collegiate Department until 1882, of chemistry in the Medical Department from 1841 to 1865, of physiology in the Medical Department from 1850 to 1867, and president of the Medical School from 1850 to 1873. In his later years he did not often officiate at the chemistry courses, but one of his sons was always present.

Draper contributed to physiology, history, physics, and chemistry. His interest in physiology, starting during his student days and stimulated by Justus Liebig's researches in biological chemistry, increased as time went by, resulting in his inaugurating a course on the subject in 1850 and publishing "Human Physiology, Statical and Dynamical," 1856, translated into several languages, including Russian. He also published "A Text-Book on Physiology" in 1866. Draper was one of the most authoritative professors of physiology in American medical schools at a time when the subject was taught generally by physicians without much knowledge of chemistry.

In middle age Draper became involved in history and published "History of the Intellectual Development of Europe," 1863, "Thoughts on the Future Civil Policy of America," 1865, and "History of the Conflict between Religion and Science," 1874. The latter subject was of such universal interest that "Science and Religion" went through many editions and was translated into French, Spanish, German, Dutch, Italian, Portuguese, Polish, Servian, and Russian. But Draper's most authoritative work was his three-volume "History of the American Civil War," 1867-1870, based on documents in the War Department, captured Confederate records, and oral accounts he received from political and military leaders.

In chemistry and physics Draper carried out investigations in capillary attraction, osmosis, electricity, the relationship between temperature and radiation of incandescent metals, spectroscopy, and the effect of light upon certain compounds. His interest in the latter subject led him to try Daguerrotyping as soon as news of the invention reached the United States, and he made the first Daguerrotype of a human face (his sister's), the first photo of the moon's surface, and the first of a diffraction spectrum. He is said to have taught some of America's earliest photographers.

Draper edited an American edition of "Elements of Chemistry" by the British chemist Robert Kane in 1842 and brought out his own book in 1846, "Text-Book on Chemistry," which went through several editions. He also published "Text-Book on Natural Philosophy," in 1847, and "A Treatise on the Forces which Produce the Organization of Plants," 1844.

Draper published his first article when he was 21, and the last of his at least 66 articles when he was 69. He wrote 11 books and published 15 pamphlets of his public addresses and lectures. A genius, Draper was highly regarded internationally. In 1869 youthful William H. Nichols, later founder of the giant Allied Chemical & Dye Corp., decided to study under Draper because he considered him "the most outstanding of all the chemists of that time." Draper was awarded the Rumford medals of the American Academy of Arts and Sciences in 1875 for his investigations on radiation and was elected first president of the American Chemical Society when the society organized in 1876.

As a teacher in a large medical school Draper earned considerable money, and in 1848 he purchased an estate at Hastings-on-Hudson, where he lived between terms and where he died Jan. 4, 1882. He had six children, two of whom, Henry and John Christopher, were his assistants in chemistry and later professors of chemistry in New York City institutions.

Much has been written about Draper, including biography by George F. Barker in *Biog. Mem. Nat. Acad. Sci.* **2,** 351-388 (1886), with list of publications; William H. Nichols, "Chemistry at New York University—a retrospect," *J. Chem. Educ.* **5,** 448-451 (1928); Donald Fleming, "John William Draper and the religion of science," Univ. Pennsylvania Press, Phila., (1950).

WYNDHAM D. MILES

Thomas Messinger Drown

1842-1904

Born in Philadelphia, Mar. 19, 1842, Drown was the youngest of three sons of a Philadelphia merchant, William Appleton Drown. He graduated from the local high school in 1859 and from the University of Pennsylvania Medical School in 1862. His chemistry professor was John Frazer and his thesis, "An

Essay on Urological Chemistry." He practiced the healing arts only once, on a ship during a round trip between Philadelphia and England.

In pursuit of chemical training Drown studied with George Brush and Samuel Johnson at Yale, and with Wolcott Gibbs at Harvard. With Gibbs he investigated the rare earths. He then went to Europe and studied under Plattner at Freiberg and with Bunsen at Heidelberg. While in England he met and married Helen Leighton in 1869.

Drown taught at Lawrence Scientific School, Harvard, as instructor in metallurgy for one year, returned to Philadelphia to work for Frederick A. Genth in a consulting laboratory in 1870, and went to Lafayette College in 1874. In Easton, Drown taught chemical analysis from 1874 until 1881. He also served as secretary of the American Institute of Mining Engineers and presided over the organization in 1897. He conducted industrial research and developed new methods of analysis of iron and steel. Perhaps his greatest contribution was training such students as Edward Hart and Porter Shimer.

Drown's most notable career as a chemist and teacher came during his association with Massachusetts Institute of Technology from 1885 to 1895. In the interim between Easton and Cambridge he returned to Philadelphia to help operate the family business of umbrella manufacturing. At MIT Drown became head of the Chemistry Department in 1888; he also assumed charge of the curriculum of chemical engineering in 1893. While there Drown built up the departments to include 21 instructors and 500 students. He also acted as chemist to the State Board of Health of Massachusetts. He began, with the aid of Ellen Richards and others, a chemical survey of water sources, and developed a "map of normal chlorine" for the State, which showed the normal composition of surface water in any given district.

In 1895 Drown was persuaded to return to Lehigh Valley and accept the presidency of Lehigh University in Bethlehem. He advocated "severe drill in mathematical and mechanical subjects, aided by laboratory practice" to teach the methods of original research and he urged a study of political economy to prepare students for the duties of citizenship. During his era the chief source of income, stock in the Lehigh Valley Railroad, was curtailed, and tuition, once free, had to be instituted. Drown's appeal to the state for funds was granted. Lehigh grew in stature as Drown brought strong faculty members, numbering 57 in 1904, to teach the 630 students.

Drown continued his association as consultant to the State of Massachusetts and also did major service in making Bethlehem water better. He gave valuable suggestions to the chemistry staff and frequently lectured on scientific subjects. He died Nov. 16, 1904 after an operation for an abdominal ailment not considered serious. A monument to his services was erected 2 years later in the form of a student center named Drown Memorial Hall.

Biographies by R. W. Raymond in *Trans. Amer. Inst. Mining Met. Eng.* **36,** 288-304 (1906) and R. D. Billinger in *J. Chem. Educ.* **7,** 2875-2886 (1930); Catherine D. Bowen, "History of Lehigh University," *The Lehigh Alumni Bulletin* (1924).

ROBERT D. BILLINGER

Charles Benjamin Dudley

1842-1909

Dudley, born July 14, 1842 at Oxford, N.Y., attended a country school, Oxford Academy and worked on farms and in shops until the Civil War. In 1862 he enlisted in the 114th Regiment, New York Volunteers, fought in several battles, and was shot and crippled for life at Winchester, Va.

He entered Yale in 1867 and worked his way through. After graduation he worked on a newspaper for a year to earn money, then returned to Yale as a graduate student, simultaneously being night editor on a newspaper. Receiving his Ph.D. in 1874 he taught physics at University of Pennsylvania, then science at Riverside Military Academy, Poughkeepsie. He planned to specialize in physiological chemistry but changed his mind in 1875 when the Pennsylvania Railroad asked him to organize a chemistry department.

Chemistry in those days was still on the

periphery of industry, and Dudley, apparently was the first head of a chemistry laboratory devoted exclusively to a railroad. He proceeded by ascertaining the shortcomings of railroad materials—rails, bearings, axles, paint, coal, lubricants, and so on—and then determining how they could be improved. This involved analysis, development of better materials, and drawing up of specifications. Standard methods of analysis of industrial products were just beginning to be promulgated, and Dudley pioneered in the field. He was a leader in the formation of the American Society for Testing Materials, its president from 1902 to 1909, and president of the International Association for Testing Materials in 1909. Specifications for purchasing materials were practically unknown when Dudley went to work, and it is said that he was "the first to embody a description of the methods of chemical analysis in specifications for materials."

Dudley wrote more than 90 articles, almost all of them concerning technical problems of railroads, he held 13 patents, and invented railroad devices that he did not patent. For 34 years he was chief chemist to the greatest railroad on earth. His laboratory grew into one of the largest in the country, with a staff of 34 at the time of Dudley's death.

Dudley was president of the American Chemical Society from 1896 to 1898, a delegate to the International Railroad Congresses of 1886 and 1900, and a collaborator of the U.S. Bureau of Forestry. Presidents Cleveland and McKinley appointed him to the U.S. Assay Commission. President Roosevelt placed him on the Advisory Board of Fuels and Structural Materials. At Roosevelt's request he delivered an address on natural resources to the Conference of Governors in May 1900. His testimony before a House committee in 1908 helped to bring about modern regulations for the safe transportation of explosives.

Dudley, despite the demands on his time by his profession and the Federal government, was a leader in civic affairs in Altoona, Pa., where he lived from 1875 onward, and where he died Dec. 21, 1909.

"Memorial Volume Commemorative of the Life and Life-Work of Charles Benjamin Dudley" (published by the American Society for Testing Materials, 1910), a 269-page book containing a poem by H. Wiley, addresses by E. F. Smith and other men, a portrait, a list of Dudley's articles and patents, and related biographical material; quotes are from this book; obituaries by H. P. Talbot in *Proc. Amer. Chem. Soc.*, 48-50 (1910); by H. Fay in *Amer. Chem. J.* **43,** 279-281 (1910); *J. Franklin Inst.* **169,** 70-71 (1910); "Dictionary of American Biography," **5,** 479-80, Chas. Scribner's Sons (1930).

WYNDHAM D. MILES

Samuel Pearce Duffield

1833-1916

Duffield, founder of Parke, Davis & Co., drug manufacturers, was born at Carlisle, Pa., Dec. 24, 1833. He was the son of Rev. George and Isabella (Bethune) Duffield. The family moved to Detroit when his father was called to become the minister of the First Presbyterian Church. He graduated from University of Michigan in 1854 and remained at the university to study chemistry and anatomy before entering the Medical Department of University of Pennsylvania. In 1856 he went to Berlin, Germany for treatment of failing eyesight. While in Germany, he attended Dr. Albrecht von Graefe's clinics and the lectures of Mitscherlich at the University for 3 months. He then went to Munich to study chemistry and physics at Maxmillian's University under Justus Von Liebig. Upon the advice of Liebig, he transferred to Ludwig III University at Giessen where he obtained his Ph.D. degree in 1858.

Returning to Detroit in 1858, Duffield started a drug store and operated an analytical laboratory. As early as 1863 he was manufacturing "photographic and medicinal chemicals." The partnership of Duffield & Parke, with Hervey C. Parke as manager, was formed in October 1866, to manufacture drugs. Duffield superintended the manufacturing and laboratory.

Long hours in the laboratory breathing noxious fumes affected Duffield's health. This coupled with changes in the tax laws, the lack of profit of the company, and the chronic illness of his wife, caused Duffield to sell his interest in the business in 1869. The firm assumed its present name, Parke, Davis and Co. in 1871.

Duffield established the chemical laboratory of Detroit Medical College in 1868 and was professor of chemistry and of medical jurisprudence and toxicology there from 1868 to 1881. He also studied medicine at the college, and received his M.D. degree in 1872. Thereafter he not only taught chemistry but practiced medicine in Dearborn.

In the winter of 1886 he went to Russia and studied the analysis and isolation of poisons from poisoned animals under George Dragendorff at the Imperial University in Dorpat. He served as Chief Health Officer of the City of Detroit from 1887 to 1893 and again from 1895 to 1898. He died in Highland Park, Mich., Mar. 25, 1916.

Robert B. Ross and George B. Catlin, "Landmarks of Wayne County and Detroit," Evening News Association, Detroit, 694 (1898); "Appleton's Cyclopedia of American Biography," D. Appleton & Co., **2,** 248 (1888); Paul Leake, "History of Detroit," Lewis Publishing Co., **2,** 269 (1912); Williams Haynes, "The American Chemical Industry: A History," **6,** 320 (1954), Van Nostrand; obituary, *Detroit Free Press,* Mar. 26, 1916.

Donald R. Hays

Eleuthère Irénée Du Pont

1771-1834

Eleuthère Irénée Du Pont de Nemours, founder of the gunpowder factory which was to become the largest chemical corporation in the United States, was born in Paris, France, June 24, 1771 and did not come to this country until he was nearly 30 years old. His father, Pierre Samuel Du Pont, was a noted economist and government official, assistant to Turgot, the Comptroller-General, who stood as Irénée's godfather and by whom his unusual Christian names were chosen. His mother was Nicole Le Dee Du Pont. (The "de Nemours" was added to the family name by the elder Du Pont in 1789—not as the direct result of his ennoblement in 1783 by Louis XVI for his part in drawing up trade and peace agreements with Britain and the new American nation, but simply because he wished to distinguish himself from three other Du Ponts in the national assembly. "de Nemours" was dropped by the American du Ponts in the nineteenth century.)

Irénée was educated privately at his father's estate, "Bois des Fossés," about 40 miles south of Paris, near Nemours. His tastes early turned to agriculture and science, and in 1788 Antoine Lavoisier, who was Pierre Samuel's close friend and director of the royal gunpowder works, took Irénée into the laboratory and works at Essonne and secured for him the right of reversion to the post of director.

The turbulent events surrounding the French revolution soon overturned the lives of both Irénée and his father. Lavoisier lost his directorship of the powder works in 1791, and Irénée went to Paris to run his father's recently founded conservative printing works. The next several years were a nightmare of repeated suppressions, imprisonments, and flights, with father and son escaping the guillotine in 1794 only because Robespierre died before getting around to them in the prison at La Force. In the last years of the century the elder Du Pont concluded that a better future lay in America and organized a company in Paris to exploit land in the valley of the James River in Virginia. In 1799 the Du Pont family set sail for America, arriving in Newport, R.I., on New Year's Day of 1800.

The land-development enterprise failed to prosper, and other ventures ate up half the capital which had been raised in Paris. In 1801 Irénée realized that the scarcity of high-grade English gunpowder and the price and inferiority of the native product left the market open to a man with his experience; but he had to go back to France for capital to put up a factory. Fortunately Napoleon's government was interested in any scheme which promised to weaken British trade, and Irénée was allowed to observe the latest techniques at Essonne while the government itself provided machinery at reasonable cost for the factory. The 95-acre property of Jacob Broom on the Brandywine Creek about 4 miles above Wilmington, Del. was chosen as the site for both the factory and a home for Irénée and his family. There was water power available, and cotton goods had formerly been produced at the site.

Powder production did not begin until 1804. Sales in this first year amounted to about $10,000, tripling the next year, and reaching $43,000 by 1807. The government contracts promised by Jefferson, a personal acquaintance of Irénée and his father, failed to materialize until the war of 1812, and even these did not ease Irénée's struggle appreciably. In the early years of the company—indeed, through most of the nineteenth century—both building capital and sheer cash were in exceedingly short supply in America, and business was conducted on long-term credit of 6 months or more. Perennially threatened with demands for redemption of the original shares of the company, Irénée threw his entire efforts into keeping his company solvent, freeing it of its original debt, and making the very best powder available in an increasingly competitive market. He died of a heart attack Oct. 31, 1834. His father had died in 1814, his brother Victor in 1827, and his much loved wife Sophie in 1828; all these sorrows added to the burdens of his struggling enterprise. He left behind him three sons, Alfred Victor, Henry, and Alexis Irénée, who took over the business in 1837 after an interim period in which it was managed by his son-in-law, Antoine Bidermann.

Under Alfred Victor du Pont, head of the firm until 1850, trade was extended through particular attention to sales and distribution methods. The next head, from 1850 to 1889, was Gen. Henry du Pont; under him was Lammot du Pont, who was responsible for an enormous number of technical innovations, both mechanical and chemical, which maintained the competitive position of the company during the post-Civil War boom years. It was Lammot who was instrumental in moving Du Pont into the high explosives market (nitroglycerine, dynamite, guncotton, smokeless powder) which began to grow in the 1870's. He was killed in an explosion in 1884.

E. I. du Pont de Nemours and Co. was incorporated for the first time in 1899, with Eugene du Pont as president. A du Pont would continue to head the firm until 1948. Du Pont's output concentrated on powders and blasting compounds almost up to the first World War. The first important manufacture of other materials was connected with cellulose derivatives as plastics, an outgrowth of du Pont's interest in guncotton, or nitrocellulose. From this small beginning in plastics has grown the present-day corporation with its incredible diversification in consumer goods, explosives, solvents, fine chemicals, and so forth.

"Dictionary of American Biography," **5,** 526-8, Chas. Scribner's Sons (1930); "National Cyclopedia of American Biography," **6,** 456, James T. White & Co. (1892); W. Haynes, *Chem. Industries* (now *Chem. Week*) **46,** 427-34 (1940); Bessie G. du Pont, "Life of Eleuthère Irénée du Pont," 12 vols., Univ. of Delaware Press, (1923-26) offers a picture of the man through his correspondence, as well as through the author's notes and introduction. The same author published "E. I. du Pont de Nemours and Company, A History, 1802-1902," Houghton Mifflin Co., (1920); a more compact history of the company for the same period and through the '20's, with broader emphasis on the explosives industry as a whole, can be found in Van Gelder and Schlatter, "History of the Explosives Industry in America," Columbia Univ. Press, (1927), Ch. 7. The du Pont family is treated popularly in John K. Winkler's "The Du Pont Dynasty," Reynal & Hitchcock, (1935). For a thorough and definitive biography of Pierre Samuel Du Pont, with much information about Irénée, see Ambrose Saricks, "Pierre Samuel Du Pont de Nemours," Univ. of Kansas Press, (1965).

ROBERT M. HAWTHORNE JR.

Saul Dushman

1883-1954

Dushman's research interests were mainly concerned with problems in the field of physics and relationships to chemistry, but he always considered himself a chemist and actively associated himself with colleagues in this profession. Much of his work was theoretical with considerable mathematical slant.

He was born in Rostoff, Russia, July 12, 1883. When he was 9 years old his family migrated to Toronto, Canada. After graduation from University of Toronto in 1904 he remained as a demonstrator in electrochemistry and later was a lecturer in physical chemistry. He received his Ph.D. degree from the university in 1911 and an honorary Sc.D. degree from Union College in 1940. He became a naturalized citizen of the United States in 1917.

About the time Dushman was in graduate school, Willis Whitney was in the process of expanding the development of the research laboratory of the General Electric Co. In 1912 Whitney invited Dushman to become a member of the research staff. Dushman accepted and moved to Schenectady where, on the suggestion of Irving Langmuir, he began experimentation on electron emission from hot filaments. This work was followed by studies of high vacuum devices which later developed into a series of vacuum tubes, the forerunners of the "tron" family and the basis of the cyclotron, betatron, and others.

He studied vacuum pumps and developed ideas that were important in the realization of the first high speed model. He devised the "Dushman Equation" for relations between the temperature of a filament and its emission of electrons. The work in electron emission and high vacuum devices stimulated his interest in the importance of theories of atomic structure in solving problems in physics and chemistry. This interest developed over the years and finally resulted in the publication of his book, "Fundamentals of Atomic Physics" (1951). Other published books include: "High Vacuum" (1923), "Elements of Quantum Mechanics" (1938), "Scientific Foundations of Vacuum Technology" (1949).

At the time of his death he was writing the second volume of this latter work. He also contributed numerous articles in scientific journals and chapters in scientific books. He was recognized as an international authority in high vacuum research.

From 1922 to 1925 he was director of research of the Edison Lamp Works in New Jersey, the first General Electric lamp factory. He divided his time between here and Schenectady. From 1928 to 1948 he was assistant director of the research laboratory in Schenectady and after his retirement in 1948 remained as a consultant.

As a member of the "core" of the research laboratory, which among others included William Coolidge, Albert Hull, Irving Langmuir, and Willis Whitney, he contributed his unique interest in human values and what was called the "spirit of the laboratory." As a modest, friendly person he expressed ideas in simple terms. He liked people and realized that cooperation and good will were essential in the operation of the laboratory. The weekly "Dushman luncheons" brought together the personnel of various departments whom he felt could contribute to each others progress. He was recognized under various titles such as: "a human catalyst," "dean of men," "morale builder," "a pillar of strength." He was a faithful attendant at monthly meetings of the local section of the American Chemical Society where he sparked discussions by well chosen questions and comments.

At his memorial service Albert Hull, of the laboratory staff said in part: "Dr. Dushman's most outstanding quality, in science as in life, was strict integrity, which was so much a part of him that his outspoken candidness never gave offense." He died at his home in Scotia, N.Y., July 7, 1954.

Irving Langmuir, "Saul Dushman—A Human Catalyst," *Vacuum* **3,** 112 (1954), also in the "Collected Works of Irving Langmuir," Guy Suits, gen. editor, Vol. 12, 409, Pergamon Press (1962); Lawrence Hawkins, "Adventures in the Unknown," Morrow (1950); obituary, *New York Times,* July 8, 1954.

EGBERT K. BACON

Hyppolite Etienne Dussauce

1829-1869

Dussauce was born in Paris, Christmas day, 1829. Interested in science from boyhood, he studied under the famous chemist, Michel Chevreul. In his twenties Dussauce became professor of industrial chemistry at Ecole Polytechnique and consultant to companies and government laboratories, among them the Botanical Garden, the Conservatoire Imperiale of Arts and Manufactures, and the Gobelins factory.

"Seized with a desire to travel," he resigned

his jobs and after spending a year or two in Great Britain, sailed for New York. He was hired by Tilden & Co., a manufacturer of pharmaceuticals and chemicals, and spent the remainder of his life in their employment at New Lebanon, N.Y. He died June 20, 1869.

In Tilden's laboratory Dussauce spent at least part of his time trying to isolate naturally occurring compounds from indigenous plants. His eulogist, whose judgment may have been swayed by the sad occasion, wrote that Dussauce "had perhaps no superior in the country" in his field.

In 1863 Dussauce became acquainted with Henry C. Baird, a noted publisher of industrial monographs. Dussauce compiled, translated, edited, and wrote for Baird several books on industrial chemistry, among them: "Treatise on the coloring matters derived from coal-tar; their practical application in dyeing cotton, wool, and silk. The principles of the art of dyeing and the distillation of coal-tar. With a description of the most important new dyes now in use" (1863); "A practical guide for the perfumer: being a new treatise on perfumery the most favorable to beauty without being injurious to the health, and the formulae of more than one thousand preparations, such as cosmetics, perfumed oils, tooth powders, waters, extracts, tinctures, infusions, spirits, vinegars, essential oils, pastils, creams, soaps, and many new hygienic products not hitherto described" (1868); "A general treatise on the manufacture of vinegar; theoretical and practical, comprising the chemical principles involved in the preparation of acetic acid and its derivatives, and the practical details of the various methods of preparing vinegar by the slow and the quick processes, with alcohol, wine, grain, malt, cider, molasses, beets, etc.; as well as the fabrication of pyroligneous acid, wood vinegar, etc., etc., together with their applications, and a treatise on acetometry" (1871); "A general treatise on the manufacture of soap, theoretical and practical; comprising the chemistry of the art, a description of all the raw materials and their uses, directions for the establishment of a soap factory, with the necessary apparatus, instruction in the manufacture of every variety of soap, the assay and determination of the value of alkalies, fatty substances, soaps, etc., etc." (1869); "A practical treatise on the fabrication of matches, gun-cotton, colored fires and fulminating powders" (1864); and "Blues and carmines of indigo; a practical treatise on the fabrication of every commercial product derived from indigo" (1863).

For a time Dussauce edited *Industrial Chemist,* a journal that existed from 1862 to 1864. He also edited *Journal of Applied Chemistry,* a periodical that ran from 1866 to 1875, and undoubtedly wrote some of the unsigned articles in the early volumes. He published articles in *"Druggists Circular"* and other journals. But his influence, whether it was large or small, was exerted chiefly through his industrial chemistry monographs which were used in factories for a generation or more.

Obituary in *J. Materia Medica* (n.s.) **8,** 219-222 (1869); Dussauce's books and articles.

WYNDHAM D. MILES

E

Amos Eaton

1776-1842

Amos Eaton was born May 17, 1776, the son of a farmer, Abel, of old New England stock, in Chatham, N. Y., just across the border from Massachusetts. His education was conventional and culminated in graduation from Williams College in 1799, with intervals of teaching in a country school. He then read law in New York City and was admitted to the bar. Eaton displayed an early interest in natural history, particularly botany, but until 1810 his major professional interest was law and land business in Catskill. Then tragedy struck in the shape of conviction for alleged fraud, and he was sentenced to prison. After 5 years he was pardoned, and a new career became necessary. In this unusual fashion Eaton was introduced to science in mid-life.

Eaton prepared for his new work by spending a year at Yale, studying with Benjamin Silliman. He then tested his skill as a scientific lecturer successfully at his alma mater, Williams College, and moved on to Albany and Troy to exercise his new craft. Here he spent the remainder of his life as a practitioner of science and won the support of many of the promoters of improvements in this area, among them Governor DeWitt Clinton and Stephen Van Rensselaer, the principal landlord of the region.

It is difficult to classify Amos Eaton in this early age of undifferentiated science. He was certainly no chemist in the modern sense, but chemistry was the major subject of his popular lectures. Eaton became an itinerant lecturer, whose travels reached up and down the Hudson Valley as well as east into New England and west into the Mohawk Valley. He brought science to the people and illustrated it with simple demonstrations and experiments.

In 1841, within a year of his death in Troy on May 10, 1842, Eaton summarized his scientific career. Since 1817 he had lectured almost daily, a total of some 6000 times, "loud and long." He had given 40 courses, averaging 300 experiments each, in chemistry. To this he added the arduous labor of compilation and authorship, preparing textbooks in various fields of science, botany, geology, chemistry, and zoology. One of his early productions was "Chemical Instructor," first published in 1822; its principal objective was "to bring down the sublime science of chemistry within the reach of the laboring agriculturalist, the industrious mechanic, and the frugal housekeeper." As John Torrey, his friend and a distinguished botanist, commended him: "Never mind what the cynics say, you are doing more for science than any ten philosophers in the country."

Eaton made his most important contribution to early American science when he persuaded his benefactor, Stephen Van Rensselaer, to finance a school for training in "the application of science to the common purposes of life." This was Rensselaer Polytechnic Institute, established in Troy in 1824. Wearied with his itinerant lecturing, Eaton was now able to transmit his skills and enthusiasm for science to a small but settled band of pupils. Chemistry figured prominently among the subjects taught, and one of the early chemical laboratories was established here for student use. Here too Eaton trained some of the best chemists in America, among them James Booth, Ezra Carr, and Eben Horsford. Their most distinctive trait as chemists was a pragmatic orientation toward useful ends. As James Hall, one of Eaton's disciples and a noted geologist put it: "Professor Eaton taught us the manipulations in science with the simplest materials so that a student could go into the forest and construct a pneumatic trough or balance, and perform there his experiments in chemistry or physics." What could be better suited to the conditions of the American frontier?

The principal source for Eaton is Ethel N. McAllister, "Amos Eaton, Scientist and Educator," University of Pennsylvania Press (1941); Samuel Rezneck, "Education for a Technological Society: A Sesquicentennial History of Rensselaer Polytechnic Institute." Both works contain ample bibliographies of writings both by and about Eaton.

SAMUEL REZNECK

Thomas Alva Edison

1847-1931

Edison, born in Milan, Ohio, Feb. 11, 1847, spent most of his boyhood in Port Huron, Mich. He had only a few months of formal education but was taught the three R's by his mother. Soon he took an interest in chemicals, telegraphy, and mechanical things generally. As a practicing telegraph operator he learned chemistry first hand from maintenance duties relating to the power supply, liquid primary batteries. By 1870 he was a manufacturer of telegraphic equipment in Newark, N. J. and carrying out independent research as time permitted. In 1876 he gave up manufacturing and established the first industrial research laboratory in the United States at Menlo Park, N. J.

Edison was the first inventor in the United States to employ a broad knowledge of chemistry in industrial research. This was particularly true of organic chemistry, which at that time was not even taught as a separate course in the nation's colleges. Of his small professional staff, Dr. Alfred Haid was a chemist. Haid functioned primarily as an analyst, and he purified certain chemicals obtained from suppliers.

While working on the electric light Edison discovered the strong adsorptive powers of platinum for gases; he developed carbon filaments; he developed improved electrical insulation so that electrical conduits for the first time could be placed underground; and he was the first to market a device containing a permanent vacuum.

A unique resinous material for phonographic records was developed consisting of a mixture of metallic soaps and fatty acids. Recordings on such compositions are still in usable conditions after 80 years. Edison developed molding and other fabrication techniques for records including the early use of phenolic resins.

Edison not only pioneered the concentration of magnetite iron ore by magnetic separation but also the use of rubber-fabric conveyors. In cement manufacture he introduced a new type of kiln. He developed improved cement compositions and manufactured them on a quantitative basis. He was the first to use foam concrete in construction.

Edison's alkaline storage battery, industry's first, is being manufactured today substantially as it was in 1908. For his battery he prepared highly pure iron powder by a direct hydrogenation process. He employed a unique electroplating process for making extremely thin nickel flakes.

During World War I Edison manufactured eight organic chemicals which had become unavailable because of the blockade and embargo by the Allies. He also constructed two recovery and purification plants to recover chemicals from the coking of coal. After the United States entered the war, Edison and a part of his staff worked on problems for the Navy, mostly on anti-submarine devices.

At the age of 80, Edison sought a domestic source of natural rubber which could be used in case of a national emergency. After analyzing thousands of plants for their rubber content, goldenrod was found best. Species of goldenrod were found which had up to 12.4% rubber in the dried leaves with yields of rubber per acre of over 200 pounds.

Late in 1887 Edison established new research facilities at West Orange, N. J. with floor space approximately 10 times that of the Menlo Park installation. These West Orange laboratories are now a National Historic Site. Other sites relating to Edison are his winter home at Fort Myers, Fla.; his birthplace at Milan, Ohio; and the restored Menlo Park laboratories at Greenfield Village, Dearborn, Mich. All are open to visitors.

Although Edison was America's first applied chemist of note, he had overall knowledge in science and engineering. The preserved library at West Orange contains approximately 10,000 books and bound periodicals, some of which have archival value. The restoration at Greenfield Village includes a library characteristic of that which Edison had at Menlo Park.

Edison founded the weekly magazine *Science* in 1880, hired a full-time editor, and gave it financial support. Edison was unable to provide financial support after one year. Beginning in 1883 *Science* was sponsored by Alexander Graham Bell.

The self-educated chemist received many awards and honors, including a Doctor of Science degree from Princeton University. Although Edison established many businesses based on his inventions, he had little interest in them once they were in operation. He was active in research up to a few months before his death, Oct. 18, 1931.

Matthew Josephson, "Edison," McGraw-Hill (1959); B. M. Vanderbilt, "Thomas Edison, Chemist," American Chemical Society (1971).

BYRON M. VANDERBILT

Gustav Egloff

1886-1955

Egloff deserved the appellation "Mr. Petroleum" which was often applied to him during his lifetime. His career began with the early development of petroleum technology to which he made many contributions. He deserves much credit for the great strides that the petroleum industry made between 1915 and 1955. It is probable that no other man was as familiar as he with the facts and figures of most of the phases of the industry.

He was born in New York City, Nov. 10, 1886 of Swiss parents. After obtaining his A.B. degree from Cornell University in 1912, he worked during the summer at Standard Oil Co., New York, and then became a graduate student at Columbia University from which he received his M.A. degree in 1913 and Ph.D. degree in 1916.

The first four of Egloff's publications appeared in *Journal of Industrial and Engineering Chemistry* during 1915. They dealt with the physical constants of petroleum fractions and with thermal reactions of aromatic hydrocarbons. Sixteen papers published in 1916, 11 in 1917, and 8 in 1918 serve as proof of the immense amount of work the energetic Egloff carried out as a graduate student.

His wide range of interest in hydrocarbon chemistry at a time when petroleum technology was largely an empirical art is apparent from an inspection of the subjects of these early papers. About half were concerned with cracking, particularly to produce aromatic hydrocarbons, olefins, or gasoline. Others dealt with catalysis, sulfuric acid treatment of gasoline, physical properties, analytical procedures, and economics of gasoline production.

On Feb. 15, 1917, reportedly largely because of his publications on cracking, Egloff was lured by Universal Oil Products Co. as director of research, a position he held until his death. When he began his work for UOP in Independence, Kans., the company was interested in commercializing Jesse Dubbs' cracking process which was based on the thermal treatment of petroleum fractions under pressure. He helped Carbon Petroleum Dubbs, son of Jesse, develop the clean circulation cracking process which permitted continuous conversion of heavy oils to gasoline; the principal alternate process of that era could not use heavy oils and had to be shut down after short periods of operation to permit the removal of coke deposits which prevented efficient cracking.

In the late 1920's Egloff invented the multiple coil process for cracking petroleum to produce high octane motor fuel. During World War II over two million barrels of oil were cracked daily in plants having two or more coils. His most valuable patents covered the thermal cracking of petroleum. Others were concerned with processing natural gas, coal, and shale oil, and with producing petrochemicals. The breadth of his interests and the boldness and originality of his ideas are shown by the 281 U.S. patents which bear his name.

Egloff foresaw the impact of catalysis on petroleum technology. While in Germany in 1930 at the World Power Congress, he met Vladimir Ipatieff, famous for his catalytic and high pressure research, and persuaded him to come to Chicago to organize research on catalysis and high pressure for UOP. The results were outstandingly important, and Egloff was understandingly proud of his accomplishment in bringing Ipatieff to the United States.

Well over 650 articles with Egloff's name as author or co-author appeared in chemical, engineering, technological, and business journals. His books include "Earth Oil" (1933), "Reactions of Pure Hydrocarbons" (1937), and "Physical Constants of Hydrocarbons" (five volumes, 1939-1953). He was a co-author of "Catalysis, Inorganic and Organic" (1940), "Emulsions and Foams" (1941), "Isomerization of Pure Hydrocarbons" (1942) and "Alkylation of Alkanes" (1948).

Many honors were bestowed upon Egloff, including the Octave Chanute Medal of the Western Society of Engineers (1939), the Gold Medal of the American Institute of Chemists (1940), the Medal of Merit of Columbia University (1943), the Washington Award (1953), and the Carl Engler Medal of the German Institute of Petroleum and Coal (1954). He held three honorary Doctor of Science degrees.

He was an efficient, tireless worker who rarely took a vacation. Although the official office hours began at 8:30 A.M., he was usually at his desk by 7:30 after a four mile walk from his apartment. His life was dedicated to making this a better world through utilization of petroleum. Thanks in no small part to him, there is immense truth in a statement he made in one of his many papers: "Nowhere in our world economy is there greater evidence of faith in research than in the petroleum industry."

In April 1955, Egloff entered a hospital in Chicago, presumably for a check-up. His illness was a secret he kept so closely that even his most intimate colleagues were unaware of its seriousness until less than a day before his death on April 29.

R. N. Hader, "Petroleum's Perapetic Prime Mover," *Chem. Eng. News,* **31,** 869 (1953) includes cover portrait; Egloff's chemical and technological achievements are discussed in talks given at a testimonial dinner in his honor in *Chemist,* **19,** 263-277 (1942); "Five Egloffs—All Gustave," *Science Illustrated,* Sept. 1947, the title being based on a statement made by R. J. Moore at the just-mentioned dinner because he felt that only by realizing that five Egloffs, all named Gustav, were born in November 1886, could one explain the ubiquitous nature of the man or the encompassing nature of his achievements; personal recollections; obituary and photographs, *Universal News* (bimonthly magazine published by Universal Oil Products Company), **9,** 2-7 (May-June, 1955).

LOUIS SCHMERLING

Otto Eisenschiml

1880-1963

Eisenschiml was born in Vienna, Austria, June 16, 1880 and died in Chicago, Dec. 7, 1963. He was educated in Vienna, graduating with his Ch.E. degree from Vienna Polytechnic Institute with highest honors in 1901. Promptly thereafter he embarked for the United States. His leaving Austria was facilitated by his being an American citizen because his father was American, naturalized by army service. His father had emigrated to the United States in 1848, intending to land in New York; but news reached the ship of the discovery of gold in California, and the captain changed his course and landed in San Francisco after sailing around South America. Seemingly his father was fortunate in his mining venture but then lost everything and engaged in a number of activities. He enlisted in the northern army in 1861. After the war he remained in the United States until 1872, returning to Vienna and marrying in that year.

Otto's first job was with the Carnegie Steel Co. mills in the Pittsburgh area. Within 3 years he advanced from bench chemist to chief chemist. He then resigned and moved to Chicago where, after several temporary jobs with firms that failed, he was hired by American Linseed Oil Co. at $60 a month, with the stipulation that he could devote one-half of his time to outside chemical activities that did not conflict with his job.

At that time little was known of the chemistry of linseed oil; commercial standards were inadequate, and adulteration was prevalent. Otto became a recognized authority on linseed oil, assisting legislatures of various states to enact laws to control the purity of commercial oil and acting as an expert in legal suits brought against "dopers."

He remained with American Linseed Oil Co. until 1912. During that time, by virtue of his contract, he devoted part of his time to consulting work, with marked success. Among his achievements was development of an oil for making transparent envelopes, an oil for waterproofing silk and other fabrics in place of rubber sheeting, a material for use on the inner soles of shoes, and an oil for use in painting freight cars. All of these products required special oils. Eisenschiml established Scientific Oil Compounding Co. to produce these special oils for customers. Ultimately he became a dealer in vegetable and plant oils; a broker's ticker was a feature of his office in later years.

In 1914 Eisenschiml became chairman of the Chicago section of the American Chemical Society. The writer recalls the first meeting at which he presided. He advocated a monthly publication and urged that $1500 be appropriated for the purpose. At that time the Chicago section had no funds. There were no dues. Expenses for the new publication were defrayed by a few of the more opulent members. Thus the *Chemical Bulletin* was born with Otto as its first editor. He signed his editorials with his initials, and after that he was commonly referred to as O.E.

The establishment of *Chemical Bulletin* was characteristic of O.E.'s efforts throughout his career to improve the public's appreciation of the chemist, to increase the chemist's remuneration, and to increase the chemist's influence in the economic life of the community. It was one of O.E.'s regrets that he was unable to have a Chicago street named after a chemist.

O.E. had many interests. In the days of the talking machine, before radio and television, he translated German songs into English, and vice versa, and sold the translations to Victor Talking Machine Co. This venture proved so successful that Victor wanted to hire him full time at a high salary.

O.E. was greatly interested in the Civil War. He visited many battlefields and read much of the literature. His extensive investigations of the political, social, and diplomatic aspects of that war culminated in his publishing "Why Was Lincoln Murdered?" in 1937. This was the first of a series of books he wrote about the War, including "In the Shadow of Lincoln's Death" (1940), "An American Illiad" (1947), "As Luck Would Have It" (1948), "The Case of A.L." (1948), "The Celebrated Case of Fitz John Porter" (1950), "The Story of Shiloh" (1952), "Eye Witness: the Civil War as We Lived It" (1956), and "The Hidden Face of the Civil War" (1961).

In 1942 he published his autobiography, "Without Fame," a book that is full of skillfully told little stories from a lifetime of varied experiences; it is one of the few and by far the most readable autobiography written by an American chemist.

Personal recollections; Otto Eisenschiml, "Without Fame," Alliance Book Corp. (1942); *Chemical Bulletin;* "Current Biography Yearbook," p. 128-30, H. W. Wilson Co. (1963); autobiography "O.E., Historian Without an Armchair," Bobbs-Merrill (1963).

DAVID KLEIN

William Elderhorst

1828-1861

Elderhorst was one of the first German-trained chemists to come to America in the mid-nineteenth century and to serve as an early link in transmitting German chemical skill and knowledge to the United States. Born Sept. 30, 1828 in Hannover, Germany, Elderhorst was the son of an army officer who later became postmaster of Hameln, where young William grew up. He began his career as a military cadet but was released because of weak eyesight. Elderhorst studied chemistry, mineralogy, and botany at the George Augustus University at Göttingen until 1850 and spent several months in the laboratory of the Horsley Fields Chemical Works in England. He was briefly assistant professor of chemistry at the Polytechnic School of Stuttgart but had to leave because of ill health.

In 1853 Elderhorst migrated to the United States and worked in drugstores in Charleston, S.C., and in New York City. He also offered private instruction in chemistry. In 1855 he was named professor of theoretical and practical chemistry at Rensselaer Polytechnic Institute, a position he held until his death. Originally founded for the "application of science to the common purpose of life" by Amos Eaton, Rensselaer was now undergoing a broadening of its goals and methods as a polytechnic institution by its director, Benjamin Franklin Greene. Elderhorst was to be an important instrument of this transformation, injecting an element of European scientific training and approach. While at Rensselaer Elderhorst also taught at the Vermont Medical Academy at nearby Castleton, Vt., which conferred an M.D. degree upon him.

In 1857 Elderhorst was appointed chemical assistant to David Dale Owen, son of the famous Robert Owen, who founded the unusual communal settlement of New Harmony, Ind., which was at once a gathering of idealistic and scientific individuals, both European and American. David Owen became principal geologist of Arkansas, and with the cooperation of Elderhorst and others made a geological survey of Arkansas, for which the latter did the chemical work in a laboratory at New Harmony. His "Chemical Report of the Ores, Rocks, and Mineral Waters of Arkansas" was published as part of Owen's "First Report of a Geological Reconnaissance" in 1858.

In 1856 Elderhorst prepared and published "A Manual of Blow Pipe Analysis and Determinative Geology" for the use of Rensselaer students. The method was recommended as faster, simpler, and more useful to the mineralogist, geologist, and mining engineer than to the chemist. The required equipment could be carried in a small, portable box and could be used to examine minerals on the spot. A second, enlarged edition appeared in 1860 and was used and commended highly by many teachers of chemistry. Third and fourth editions were prepared and brought out in 1874 and 1881 by Henry Nason, a colleague and successor of Elderhorst at Rensselaer, and Charles Chandler, of the Columbia College School of Mines.

One other contribution to the chemical literature of America was made by Elderhorst in 1857 in the form of a translation of a German work by F. J. Otto, "A Manual of the Detection of Poisons by Medico-Chemical Analysis."

Unhappily the promising career of the young chemist was brought to an early and untimely end at the age of 33. Elderhorst traveled to Maracaibo, Venezuela, in May 1861. There he was stricken with yellow fever and died, July 28, 1861.

A summary of Elderhorst's life and career is given by Henry B. Nason in "Biographical Record of the Officers and Graduates of the Rensselaer Polytechnic Institute" (1887); other incidental information is in Elderhorst's "A Manual of Blow Pipe Analysis" (1860), and in David Dale Owen, "First Report of a Geological Reconnaissance of Arkansas" (1858), especially pp. 143-90.

SAMUEL REZNECK

Charles Hodges Eldridge

1886-1952

Eldridge was born Dec. 4, 1886 in Fort Lyon, Colorado and graduated from Ohio State University in 1909 with the degree of mining engineer. He was then employed for about a year as assistant superintendent in a cyanide leaching plant at Guanojuato, Mexico.

Eldridge next worked from 1910 to 1914 as a research chemist at the General Electric Co. laboratories in Harrison, N. J. Here he had some association with Colin Fink since Fink was employed in the same laboratories for producing ductile tungsten lamp filaments from 1910 to 1917. From 1914 to 1917 Eldridge was an instructor in chemistry and electricity in high schools and technical schools.

From 1918 through 1921 Eldridge was a research chemist and metallurgist in the newly established laboratories of the Chile Exploration Company in New York City, of which Fink was the head. Eldridge was a member of a 10-man team which developed an insoluble copper-alloy anode for the electrowinning of copper, which has since been used in Chuquicamata, Chile. While Fink was the inventor of the anode (U.S. patents 1,441,567-8), Eldridge gave it its name, "Chilex." This anode was said to have resulted in savings of some millions of dollars.

In 1922 Eldridge was employed as a metallurgist by the U.S. Bureau of Mines in Pittsburgh, Pa. and his work resulted in a joint publication with R. J. Anderson on the effect of heat treatment on release of stress in bronze castings. At the same time Fink took up teaching electrochemistry at Columbia University.

Early in 1923 Eldridge became Fink's personal research assistant at Columbia University for a 2-year period. The immediate occasion was the development of a process for restoring antique bronze objects for the Metropolitan Museum of Art, published in 1925. This work proceeded favorably, but the treatment developed required long periods of time. Fink accordingly asked Eldridge to investigate chromium plating also in the second half of 1923, and by May 1924 the two men felt that they had discovered the basic principle of chromium deposition from chromic acid solutions, *i.e.* the presence of about 1% of the weight of chromic acid of sulfate radical or other similar radical to act as a catalyst.

Fink obtained financial backing from John T. Pratt, a trustee of the Metropolitan Museum of Art, and formed the Chemical Treatment Co. to exploit the discovery. Eldridge went with this company in 1925-1926, and continued with its successor companies the rest of his life in various capacities. He moved to Detroit, Mich. early in 1927 to go with the General Chromium Corp. and then with United Chromium Inc. which was formed shortly thereafter.

The Fink-Eldridge collaboration was a very close and productive one. No less than eight joint patents issued, and others failed to be granted, at a time when joint inventorship was not very common. The basic patents on chromium plating were in Fink's name alone. According to patent litigation testimony Eldridge had an agreement with Fink that they would share any royalties on an equal basis.

Eldridge devoted the rest of his life to helping numerous plants with the technique of chromium plating. Of his many contributions to the industry, the development of more corrosion resistant lead-tin alloy anodes might be mentioned.

Eldridge had a rather modest and unassuming personality and was well liked in the trade. He was quite active in The Electrochemical Society and the American Electroplater's Society, especially on the local level. He died Oct. 10, 1952 in Detroit, survived by one son, Colin C. Eldridge, and a daughter, Anne (Mrs. Henry R. Strickland).

Trans. Amer. Electrochem. Soc. **45,** 404-406 (1942); C. G. Fink and C. H. Eldridge, "The Restoration of Ancient Bronzes and Other Alloys," The Metropolitan Museum of Art, 53 pages (1925), and *Chem. Abs.* **19,** 3433 (1925); G. Dubpernell, *Plating* **47,** 35-53 (1960); U.S. Patent 1,975,227 (October 2, 1934); *Metal Finishing* **50,** 113-114 (March, 1952), and **50,** 115-116 (December, 1952); *Plating* **39,** 1304 (1952).

GEORGE DUBPERNELL

Charles William Eliot

1834-1926

Eliot, president of Harvard University for 40 years, started his career as a chemistry teacher and in that field is best known for the textbooks which he wrote with Francis H. Storer. Eliot was born in Boston, Mar. 20, 1834, where his father, Samuel, served at various times as mayor and a member of Congress. He attended Boston Latin School and graduated from Harvard in 1853, second in his class. He earned his M.A. degree from Harvard in 1856. During his undergraduate years, he was the first and only student admitted to the laboratories of Josiah Parsons Cooke, Jr. for individual study. He was also privileged to travel with Cooke in the summer months to various chemical plants in the Northeast. He was tutor at Harvard, 1854-58, assistant professor of mathematics, 1858-61, and assistant professor of chemistry, 1858-63. He was the first Harvard teacher to have the title of assistant professor.

His first laboratory courses for Harvard undergraduates were given in 1858-59. In 1861 his teaching duties were limited to chemistry, and he was placed in charge of the chemical laboratories of Lawrence Scientific School. His first innovation in teaching was to give written exams instead of oral quizzes.

Some time in this period, he decided definitely that teaching would be his life's work. Therefore, when his 5-year contract expired in 1863, he severed his connection with Harvard and for the next 2 years roamed through Europe studying the organization of technical schools and the prevailing methods of teaching chemistry and physics.

He was so determined to stick to the field of education that when in 1865 he was offered an annual salary of $5,000 as an industrial chemist, he turned it down to teach at the newly formed Massachusetts Institute of Technology for $2,000 per year. Here he found himself working with a former classmate, Frank Storer.

That fall, Eliot and Storer organized two courses in inorganic chemistry in which laboratory practice, including the construction and use of apparatus, was a vital part. Intending that each student "see, smell, and touch for himself," they wrote a new textbook, "A Manual Of Inorganic Chemistry," which laid out over 200 experiments in an explicit, cookbook style. Although books of experimental chemistry for students had been published before, for example by Amos Eaton, this was the first textbook to teach theory side by side with experiment. Even so, the authors' approach to theory was conservative, declaring "the existence of atoms is in itself a hypothesis and not a probable one . . . all dogmatic assertions upon such points are to be regarded with distrust." The way in which authorship of the book was assigned symbolizes the complete intertwining of contributions by both men. When time came to set type for the title page, they flipped a coin to see whose name should go first—Eliot won. However, on the spine of the first edition, the initials E and S are combined in a monogram in such a manner that it is impossible to know which name should be read first. In the revised edition, the names cross so that neither appears before the other; in still another edition, the names are superimposed in different type faces.

Eliot and Storer published a "Manual Of Organic Chemistry" in the fall of 1868 and also introduced their "Compendious Manual Of Qualitative Analysis." This book eventually went through 19 editions, the last being in 1899.

Eliot resigned from the Institute in July 1869 to become president of Harvard. He gave up chemistry. His election to the presidency of Harvard was, in a great measure, the result of papers he had published on educational problems as he saw them in the United States and in Europe. He was convinced that colleges paid too little attention to students' individual needs.

At Harvard, he reduced the emphasis on the study of the classics as entrance requirements and developed the elective system, whereby a student no longer pursued a rigidly prescribed curriculum but could choose courses for himself. Under Eliot's administration, Harvard enrollment moved from 1,000 to more than 4,000; the faculty increased from 60 to 600; the value of the physical plant increased by 60%; and endowments rose from $2 million to $20 million.

As a leader in the National Education Association, Eliot was highly influential in raising the standards of public schools and introducing manual training for those that would not go on to college.

After his appointment as president emeritus in 1909, he undertook preparation of the Harvard Classics. Popularly called "Dr. Eliot's five-foot shelf of books," the classics were a collection of 50 volumes embracing selections of the works of hundreds of writers throughout the history of civilization. The wide-spread circulation received by the series made it a milestone in adult education.

Eliot was the author of several books and was the recipient of nine honorary doctorates. He was married twice, first in 1858 to Ellen Peabody, of Boston; and after her death to Grace Hopkinson in 1877. He had four children by his first marriage. A liberal in religious and political matters, he lectured extensively on social problems. After World War I, he was a champion of the League of Nations.

He was vigorous to the final year of his life and died in Northeast Harbor, Maine, Aug. 22, 1926, in his ninety-second year.

Henry James, "Charles W. Eliot, President Of Harvard," Houghton Mifflin Co. (1930); Tenney L. Davis, "Eliot And Storer—Pioneers In Laboratory Teaching Of Chemistry," *J. Chem. Educ.* **6,** 868-879 (1929); "National Cyclopedia of American Biography," James T. White & Co. **6,** 421-423, (1929).

HERBERT T. PRATT

Carleton Ellis

1876-1941

Carleton Ellis was born Sept. 20, 1876 in Keene, N.H. His father, a florist, gave him a camera on his eleventh birthday; this aroused an interest in the chemical reactions involved in photography, which he studied in a home laboratory. He entered Massachusetts Institute of Technology in 1896, received his B.S. degree in 1900 and was an instructor there until 1902.

Ellis was a prolific inventor, obtaining 753 United States patents. During his senior year at MIT he developed his first invention, a paint and varnish remover. With borrowed money he began to manufacture the mixture. The venture was successful; Ellis' product was the most effective then available. In 1905 he formed the Chadeloid Chemical Co., a patent holding and licensing company which took over patents relating to paint and varnish removers.

In 1907 Ellis extended his research interests by founding with Nathaniel L. Foster the Ellis-Foster Co. in Montclair, N.J., an organization for research in industrial chemistry. With laboratories in Montclair and Key West, Fla. and a large staff of chemists and engineers, Ellis averaged over two inventions per month for over 25 years. His patented inventions included soaps, cosmetics, explosives, asphalt, floor tiles and waxes, food products, printing inks, dyes, fertilizers, and insecticides.

In 1913 Ellis received a patent for a method of making a cheap but good oleomargarine by hydrogenating vegetable oils. This invention was one of the foundations of the margarine industry. Probably his most important invention was the "tube and tank process" of cracking crude petroleum (1919). Standard Oil Co. of New Jersey took over the process and by 1940 over 40 billion gallons of gasoline had been produced by this method. During World War I Ellis synthesized commercially valuable isopropyl alcohol from waste gases produced by cracking oil and also invented a process for preparing acetone by catalytic oxidation of isopropyl alcohol. Other researches in petroleum by Ellis enabled automobile manufacturers to increase the power of engines and gasoline refiners to achieve higher octane gasoline.

The paint and varnish field continued to be a major concern to Ellis. In 1925 he produced the first durable lacquer for automobile paint. He developed several synthetic resins which led to many improvements in paints and lacquers. Studies on resins led him into the field of plastics, and he developed a urea-formaldehyde plastic from which he gained a large personal fortune. His patents were the basis for the Unyte Corp. (1932) which made urea-formaldehyde products.

Ellis served as consultant to American Cyanamide Co., Procter & Gamble, and Standard Oil. He was the author of nine books on industrial chemistry and recipient of many awards. During the last 2 years of his life he suffered from poor health. While travelling to his winter home in Nassau, he developed influenza and died Jan. 13, 1941 in Miami Beach, Fla.

"National Cyclopaedia of American Biography," **32,** 33, James T. White Co. (1945) with portrait; *Chem. Eng. News* **14,** 240 (1936); A. D. McFadyen in *Chem. Ind. (New York)* **46,** 488 (1940), with portrait.

ALBERT B. COSTA

Conrad Arnold Elvehjem

1901-1962

Elvehjem carried out significant work on the role of vitamins and minerals in nutrition. Particularly important was his work on the role of copper as a factor in nutritional anemia and his recognition of nicotinic acid as a curative factor for blacktongue in dogs. Born on a farm near McFarland, Wis., May 27, 1901 he was the son of Ole and Christine (Lewis) Elvehjem. His college work was done entirely at University of Wisconsin where he received his B.S. degree in agricultural chemistry in 1923 and his Ph.D. in 1927. Working with Edwin B. Hart and Harry Steenbock, he studied the influence of light on calcium and phosphorus metabolism in animals during lactation.

Upon receiving his doctorate he was made instructor in the department. He advanced rapidly, reaching a professorship in 1936. In 1944 he became chairman of the department which was now named biochemistry. Between 1946 and 1958 he doubled as dean of the Graduate School and in the latter year became president of the university.

He and Constance Waltz, whom he married in 1926, were the parents of Peggy Ann and Robert. Elvehjem died July 27, 1962 of a heart attack suffered in the president's office.

Elvehjem's research led to the publication of more than 800 papers. Eighty-eight students received their Ph.D. degrees under his guidance, many of them going on to make significant nutritional contributions of their own. He was well-read in the research literature and was a perceptive, hard-driving worker who was a natural leader in the laboratory.

Elvehjem and his students made studies on nearly every phase of vitamin and mineral nutrition, giving particular attention to the various members of the vitamin B complex. Early in his career, in association with Hart, Steenbock, and James Waddell, he revealed the role of copper in stimulating the uptake of iron in rats suffering from milk-induced anemia. His group later studied the role of zinc, manganese, cobalt, potassium, boron, molybdenum, fluorine, and arsenic in nutrition.

In 1937 when European students of metabolism showed the presence of nicotinic acid in Harden's coenzyme I and related coenzymes, Elvehjem and associates showed that nicotinic acid cured blacktongue in dogs. Blacktongue was known to be the equivalent of human pellagra. He later showed that the reason corn was even more prone to cause pellagra than other cereal grains was associated with the low lysine content of corn. Lysine proved to be a precursor of nicotinic acid.

Elvehjem was at the height of his biochemical career when the vitamin B complex was still a vast puzzle. His laboratory played a leadership role in unravelling the nature of the complex, working at one time or another with thiamin, riboflavin, biotin, pantothenic acid, folic acid, *p*-aminobenzoic acid, inositol, and cobalamine. He showed great perceptiveness in recognizing new approaches in vitamin studies and utilized a variety of new experimental animals. His laboratory quickly began studies on bacterial nutrition when it was recognized that such organisms offered unique values over animal experiments.

Elvehjem's unpublished papers are available in the Archives of the University of Wisconsin. They deal primarily with his departmental chairmanship, deanship, and presidency. His more than 800 scientific papers were published primarily in *Journal of Biological Chemistry, Journal of Nutrition, Proceedings of the Society for Experimental Biology* and *Medicine, and American Journal of Physiology*. For a short biography see A. J. Ihde in "Dictionary of Scientific Biography," **4**, 357-9, Scribner (1971), which includes reference to several autobiographical addresses dealing with his scientific work. The University of Wisconsin Biochemistry Department holds a bound set of his collected works with a full bibliography.

AARON J. IHDE

Charles Philip Engelhard

1867-1950

Engelhard was born in Hanau-am-Main, Germany, Mar. 8, 1867. His parents were Julius Engelhard and Susanne Holzmann Engelhard. He was educated at the Real-Pro-Gymnasium at Weinheim and began a career in banking in Frankfurt before emigrating to the United States in 1891 as an agent for W. C. Hereaus GmbH Platinum Works of Hanau. Although his formal training was in finance, his subsequent contributions to the precious metals industry probably depend in part upon the influences of his father, a noted jeweler, and his maternal grandfather, Philip Holzmann, founder of an important engineering firm.

Engelhard purchased Charles F. Croselmire Co. of Newark, N.J. in 1902 and with it founded American Platinum Works the following year. In 1904 he reorganized Baker and Co., a precious metals refiner founded by Cyrus C. Baker in 1875. Hanovia Chemical and Manufacturing Co., of Newark, was acquired in 1905, and Glorieux (later Irvington) Smelting and Refining Works of Irvington, N.J., in 1907. These enterprises engaged in separate but related aspects of the precious metals industry and were operated as independent businesses despite their close relationships. Engelhard's business interests, which by 1914 also included Charles Engelhard Inc., of Newark, D. E. Makepeace & Co., of Attleboro, Mass., East Newark Realty Corp., and subsidiaries in Canada, England and Japan were consolidated after his death by his son and successor, Charles William Engelhard (1917-71). The resulting enterprise is now known as Engelhard Minerals and Chemicals Corp.

Charles Philip Engelhard's decisive technical contribution was the creation of a strong industrial research effort concerning precious metals. This effort, conducted principally at Baker and Co. and at Hanovia Chemical and Manufacturing Co., resulted in effective catalysts for oxidizing ammonia to nitric acid, successful commercial electroplating of rhodium and rhodium alloys, development of precious metal dental alloys, and improved ultraviolet health lamps. Engelhard's 1924 monograph, "The Platinum-Palladium Controversy and its Relation to the Jewelry Industry," demonstrates his understanding and appreciation of some of the accomplishments of his technical staff without claiming any original technical contribution as his own.

Engelhard became a United States citizen in 1906. He was awarded an honorary D.Eng. degree by Stevens Institute of Technology in 1938, and an honorary D.Sc. degree by Newark College of Engineering in 1942. He died Dec. 1, 1950 at Bernardsville, N.J.

"The National Cyclopedia of American Biography," **41,** 402-3, James T. White Co. (1956); obituary, *New York Times,* Dec. 2, 1950.

MICHAEL B. DOWELL

Francis Ernest Engelhardt

1835-1927

Francis Ernest Engelhardt, son of Ernest Phillipp Engelhardt and Marie Antoinette (Schwachheim) Engelhardt, was born June 23, 1835 in Gieboldhausen, Hanover. He studied natural sciences at University of Göttingen from 1853 to 1857 and was an assistant to Friedrich Wöhler during his final year. Accepting William S. Clark's offer of an assistant professorship in the Chemistry Department at Amherst College, he arrived in Massachusetts in November 1857.

He left Amherst in 1858 to become chemist for Charles Ellis & Co., manufacturing and pharmaceutical chemists, in Philadelphia. From 1859 to 1860 he was chemist for the Eastwick Brothers Sugar Refinery, in the same city. He assisted Charles Joy at Columbia from 1861 to 1862, was professor of chemistry and natural science at St. Francis Xavier College, New York from 1862 to 1867, and professor of materia medica at New York College of Pharmacy from 1868 to 1869.

Engelhardt went to Syracuse, N.Y. in 1869 as chemist for the Onondaga Salt Co. and held this position for 40 years. During this time he was also a consultant, state chemist of the Onondaga Salt Spring Reservation from 1870 to 1890, chemist and milk inspector of the Syracuse Board of Health from 1877 to 1920, and Alcoholic Beverage Chemist of the New York State Board of Health in 1881, 1882, and 1885.

Engelhardt was a formidable local authority on public hygiene and zealous in pursuit of duty. He was a familiar figure on the streets of Syracuse, a small man with a full beard and principal actor in a drama that was reenacted daily. Early in the morning he would scurry after milkmen as they whipped up their horses to get away. When he caught them he would dip for samples into their milkcans, from which they dispensed in bulk. Said one Syracuse editor, "We will never know the whole safety which his indefatigable endeavors gave us, but they are a part of our history. . . . Health to the growing generation was his monument, even more necessary than piles of masonry."

Engelhardt was instrumental in the selection of Skaneateles Lake as a source of water for Syracuse. He is also credited with an important extension of the salt industry in New York State; his reports on geological formations and his identification of brine well sites are said to have convinced W. B. Cogswell to locate the Solvay soda plant at Syracuse, near the salt deposits in the Tully Valley and the limestone at Jamesville.

Engelhardt was married twice—in 1868 to Cornelia Titus, who died in 1869, and in 1870 to Anna Mary Miller, who died in 1918; five of his eight children lived to maturity. He died in Syracuse, Feb. 8, 1927, age 91.

W. M. Beauchamp, "Past and Present of Syracuse and Onondaga County, New York," S. J. Clarke Co. (1908), pp. 687-688; T. W. Herringshaw, ed., "National Library of American Biography," American Publishing Co. (1919), 2, p. 389; Syracuse city directories, newspaper clippings, and other information, in Onondaga Historical Association; personal recollections of Bessie D. Gould.

MICHAEL B. DOWELL
ROBERT F. GOULD

Henri Erni

1822-1885

Erni, whose name may be found spelled in a variety of ways, was born in Switzerland, Jan. 22, 1822. His father owned a silk and wool printing firm, and Henri, who was expected to become the manager, was educated in an industrial school and at University of Zurich, where he studied sciences and impressed the professor of chemistry. Typhus killed Henri's father, the family had to sell the business, and Henri emigrated to the United States late in 1848 in hope of finding a job in science.

Benjamin Silliman, Jr. hired him to assist in Yale's chemical laboratory in 1849. Seeking advancement he went to University of Tennessee as professor of chemistry, mineralogy, geology, botany, French, and German in 1850. Overworked, lowly paid, and with little equipment to teach and research, he resigned in mid-1852 and went to Massachusetts to become a traveling science lecturer for the Massachusetts Board of Education. He visited state normal schools, which at that time did not have much in the way of practical science teaching, and gave lectures and demonstrations, using a portable kit of chemicals and apparatus.

In 1854 Erni went to University of Vermont as professor of chemistry, toxicology, and pharmacy in the Medical Department and professor of chemistry and natural philosophy and instructor in Romance languages in the Academic Department. There was no laboratory course for undergraduates because Vermont, like the majority of American colleges, did not yet have a laboratory for students. There was, however, a private laboratory in the Medical Department for the professor of chemistry, and with this laboratory at his disposal Erni offered instruction in practical chemistry for students who desired it. The university awarded him an honorary M.D. degree in 1856.

The northern climate was too rigorous for Erni's wife, a southern girl whom he had married in Tennessee, and in 1857 he resigned from the University of Vermont and moved to Knoxville. Shelby Medical College in Nashville engaged him as professor of chemistry and medical jurisprudence in 1859. Chemistry laboratory instruction was not required for students, but Erni offered private laboratory instruction to students who desired it and also acted as a consulting and analytical chemist.

The turmoil of the Civil War soon caused Shelby to close. Erni had no job, and the conflict slowly impoverished his family. He sympathized with the North but had to remain in Tennessee for 3 years before he could get through the Confederate lines. He made his way to Washington and was hired by the Secretary of Agriculture as the department's chemist.

Erni was the only chemist in the recently created department. In a small, damp basement laboratory he studied fermentation and analyzed farm soils, guano, sugar beets, and other materials. At home he wrote a series of articles on petroleum for the Washington newspaper *Sunday Morning Chronicle*. These were published as a book in 1865, "Coal Oil and Petroleum: Their origin, history, geology, and chemistry, with a view of their importance in their bearing upon national industry."

Owing to ill health Erni resigned from the Department of Agriculture in 1866 and became an examiner in the U.S. Patent Office, handling chemistry and physics inventions. Simultaneously, from mid-1866 to early in 1868 he edited *Journal of Applied Chemistry.*

In 1869 President Grant appointed Erni consul at Basle, Switzerland. He took his family to Basle and lived there until 1878 carrying on his official duties. President Hayes did not reappoint Erni to the post, and in 1878 he returned to Washington. He opened a "Consulting Bureau for Practical Chemistry, Arts and Manufacture," and also offered his services as an expert in criminal cases and patent cases, and as a translator of European technical publications. Apparently he did not have enough clients to support his family, and he took a succession of part-time jobs, some as chemist and some doing other things, for several government agencies.

Erni published one book, several pamphlets, and at least 10 articles. He translated and enlarged F. von Kobell's "Mineralogy Simplified," which went through several American editions. He was one of the original members of the Chemical Society of Washington in 1884. He died in Washington, May 18, 1885.

Augustus C. Rogers, "Our Representatives Abroad," pp. 274-275, with portrait, Atlantic Pub. Co. (1874); Wyndham D. Miles and Louis Kuslan, "Washington's First Consulting Chemist, Henri Erni," *Records Columbia Hist. Soc. 1966-1968,* **66-68,** 154-166 (1969), with portrait.

WYNDHAM D. MILES

Gustavus John Esselen

1888-1952

Esselen was born in Roxbury, Mass., June 30, 1888, the son of Gustavus J. and Joanna Blyleven Esselen. All of his higher education was obtained at Harvard University where he was awarded an A.B. *(magna cum laude)* in chemistry in 1909 and a doctorate in 1912. In that same year he was married to Henrietta W. Locke who with three children survived him at the time of his death on Oct. 22, 1952.

Until 1921 he was a member of the research staff of General Electric Co. of Lynn and then of Arthur D. Little, Inc., of Cambridge, Mass. At the latter firm he was associated with Little and Wallace Murray in the fabrication of a "silk" purse from reconstituted collagen, in turn derived from a sow's ear. In 1921 he founded Gustavus J. Esselen, Inc., which subsequently became Esselen Research Corp. and then, following a merger, Esselen Research Division of United States Testing Co., Inc. During this period he was involved in solving a variety of problems submitted by industrial clients. Among these was the development of anhydride curing agents for epoxy resins and poly(vinylbutyral) as an improved material for safety glass, both of which enjoyed considerable commercial success.

Esselen was a member of the American Chemical Society for 43 years during which time his outstanding services to the society and to the profession of chemistry were recognized and honored not only in his native New England but throughout the United States.

Esselen was twice chairman of the Northeastern Section, ACS (1922-23), and served as councilor and director of the national organization, during which time he was a member of the ACS Council Policy Committee. His chairmanship of the national ACS meetings held in Boston in 1928 and 1939 was an outstanding service. He was on the advisory boards of I/EC and C&E News, 1946-48. In 1948 he received the James Flack Norris Honor Scroll as "the person who has done most to advance the interests of the Northeastern Section." In 1950 he was made an honorary member of the American Institute of Chemists for his services to the profession of chemistry and chemical engineering. From 1949 to 1951 he was chairman of the American Section of the Society of Chemical Industry. Prior to World War II he was a reserve officer in the U.S. Army's Chemical Warfare Service.

During the war he was a committee chairman with the Office of Scientific Research and Development.

Esselen's distinguished contributions to chemistry and chemical engineering were in accordance with the highest ethics of these professions; his recognition of the duties of a professional led to his exertion of a wise and beneficient influence on all the professional societies to which he gave so generously of his time and led to his active participation in church and civic activities in the Boston area.

Esselen was a very sensitive person, devoted throughout his life to the fine arts and music. His motto, contained on a tapestry in his office, was a quotation of Richard Willstätter, "It is our destiny, not to create, but to unveil."

Obituaries in *The Nucleus,* December 1952, p. 74; *Chem. Ind. (London),* Nov. 15, 1952, with portrait.

EDWARD R. ATKINSON

Ward Vinton Evans

1880-1957

Evans was born in Rawlinsville, Pa., June 8, 1880 to parents of modest means. He attended grammar school in Rawlinsville, working at odd jobs during the summers to earn money to add to the family's income. After graduation, he worked full-time for several years before deciding to continue his education. He did so well in two 14-week terms in Franklin and Marshall Academy that he was permitted to enroll at Franklin and Marshall College, without the usual high school education. He earned necessary money by serving as a conductor on the local trolley during the school years and summers.

After receiving his Ph.B. degree in 1907, Evans taught in a local high school for 2 years after which he accepted a position to teach mathematics and coach football at Mohegan Lake Military Academy near Poughkeepsie, N. Y. Because another man seemed to be more suitable as the mathematics teacher, the president of the school suggested that Evans teach chemistry and physics. He agreed and took a summer course in chemistry at Columbia University. He found the subject so captivating that while he continued to teach at the academy, he took morning courses at the university, returning to the academy by 1:00 P.M. after a 90 mile round-trip. Beginning in 1914 he spent full time at Columbia and received his Ph.D. degree in 1916 for work carried out under John L. R. Morgan. After a postdoctorate year at Columbia as Harriman Fellow, he left for Northwestern University to be an instructor in chemistry.

His teaching career was interrupted for a year because of World War I. He served as a lieutenant in the Army, testing high explosives, until his return to Northwestern in January 1919, as an assistant professor of physical chemistry. He became a professor in 1928 and served as chairman of the Chemistry Department from 1942 until 1945 when he retired. He spent the next year organizing the Chemistry Departments at the U.S. Army universities in Shrivenham, England, and in Biarritz, France. He then planned to retire completely, spending his time fishing and hunting at his country home near Lancaster, Pa. However, his love for teaching forced him to accept an invitation from Loyola University to teach and to serve as head of the Chemistry Department. He had an outstanding ability to teach chemistry; there was deep understanding and respect between him and his students. Even after his compulsory retirement at the age of 70, he continued to teach at Loyola as a "Special Lecturer" until his death on Aug. 2, 1957 while on vacation at his home in Pennsylvania.

Evans' research effort was minor and offbeat but had a surprising, although belated, industrial impact. It included study of detonation rates of high explosives, electromotive forces of non-aqueous solvents, critical solution constants of furfural, and some properties of Grignard reagents (chiefly their luminescence and their electrolytic behavior). He was attracted to the Grignard compound research by observing that *p*-bromophenylmagnesium bromide etherate leaking from a cracked reaction vessel glowed in the dark laboratory. He soon had a series of graduate students working on the electrical properties of the reagents. Two of the students, some

30 years later (and unfortunately after his death) developed a process for preparing the anti-knock agents, tetramethyl- and tetraethyllead, by substituting a lead anode for the platinum anode which was originally used when electrolyzing methyl or ethylmagnesium chloride; the process led to the world's largest Grignard reagent manufacturing plant.

His consulting work was largely in the field of explosives on which he was often a court witness. During World War II, he gave extracurricular lectures on explosives and poison gases.

He was quite active in the American Chemical Society, serving as chairman of its Chicago Section in 1929 and of its Division of Physical and Inorganic Chemistry, 1929-30. He was awarded the Honor Scroll of the American Institute of Chemists in 1946.

In April 1954, Evans began to serve as one of the three "judges" on the J. Robert Oppenheimer hearing board to decide whether the eminent physicist, the father of the A-bomb, was a security risk. After spending a week reading and studying the voluminous investigative report on Oppenheimer and more than 3 weeks listening to testimony of witnesses, the board members retired to their respective homes for 10 days of rest and thought. When they reconvened in Washington, Evans presented the minority report stating that Oppenheimer deserved restoration of complete security clearance. The majority report of the other two judges decided that while Oppenheimer was not only loyal but also unusually discrete with secrets, he was a security risk no longer entitled to his government's trust, a decision which has been termed "internally inconsistent." Evans' role in this investigation was immortalized in the Broadway play, "In the Matter of J. Robert Oppenheimer."

"Ward Vinton Evans," *Chem. Eng. News* **28,** 2423 (1950) contains biographical information and a cover portrait; personal recollections; "Minority Report of Ward Evans," *Chem. Eng. News* **32,** 2365 (June 14, 1954), a revealing first person minority report suggesting restoration of J. R. Oppenheimer's clearance; Philip M. Stern, "The Oppenheimer Case. Security on Trial," Harper and Row Publishers (1969) contains a biographical sketch based on interviews with Evans' family and friends as well as excerpts from Evans' questions and statements during his membership on the security board. The hearing is dramatized in the play, "In the Matter of J. Robert Oppenheimer" by Heinar Kipphardt, translated from German by Ruth Speirs, Hill and Wang, Inc. (1967), abridged in "The Best Plays of 1968-1969," Dodd, Mead and Co. (1969).

LOUIS SCHMERLING

William Lloyd Evans

1870-1954

Evans was born in Columbus, Ohio, Dec. 22, 1870 the son of Welsh immigrants, William Henry and Anne (Lloyd) Evans. He attended Ohio State University, from which he received his B. Sc. degree in 1892 and his M. Sc. in 1896. He was chemist with the American Encaustic Tile Co., Zanesville, Ohio, from 1892 to 1894. After receiving his M. Sc. degree, he was an assistant in the department of ceramics at Ohio State until 1898, and then a high school instructor in Colorado Springs, Colo. until 1902.

Evans received his Ph. D. degree in chemistry at University of Chicago in 1905 as a student of John U. Nef. He returned to Ohio State University as assistant professor of chemistry, was promoted to an associate professorship in 1908 and to a professorship in 1911. From 1911 to 1915 he concurrently served as lecturer in chemistry at Starling-Ohio Medical College.

Evans was widely known for his research contributions to the chemistry of carbohydrates. He began his work with studies on the mechanism of oxidation of such simple organic compounds as ethanol, acetaldehyde, 2-propanol, and acetone and extended it to the oxidation of the more complicated carbohydrates. As his work progressed he found it necessary to synthesize new sugars and other compounds not previously known. He was a member of the National Research Council Committee on Carbohydrate Research, 1926-27.

His teaching duties at Ohio State were principally in general chemistry. During his career he lectured to more than 45,000 students. He was co-author of several general chemistry laboratory manuals and qualitative analysis texts with William McPherson, William E. Henderson, Jesse E. Day, Alfred B. Garrett, Laurence L. Quill, and Harry H. Sisler.

Evans served with the Chemical Warfare Service at Edgewood Arsenal from 1917 to 1919, with the rank of major. In 1928 he succeeded William McPherson as chairman of the Department of Chemistry at Ohio State, and served in this capacity until his retirement from the university in 1941.

He was chairman of the Organic Division of the American Chemical Society in 1928, recipient of the William H. Nichols medal of the ACS in 1929, president of the American Chemical Society in 1941, and recipient of the gold medal of the American Institute of Chemists in 1942. He was awarded honorary degrees by Ohio State University and by Capital University.

Evans was highly musical and must have struggled with the decision of choosing chemistry over music for his profession. He had played the violin as the youngest member of the Columbus Symphony Orchestra and directed the Ohio State Glee Club while Mrs. Evans directed the Women's Glee Club. He organized choruses of men and women at Colorado Springs and edited *The Ohio State University Song Book.*

Evans died on Oct. 18, 1954 in Columbus.

Biography by Henry B. Haas, *The Chemist* (May 1942); obituary with portrait, *Chem. Eng. News,* **32,** 4408 (Nov. 1, 1954); biography, *Chem. Eng. News,* **18,** 1118-1119 (1940); personal recollections.

LAWRENCE P. EBLIN

F

Constantine* Fahlberg

1850-1910

The first of the synthetic sweeteners was saccharin, whose sweet taste was discovered by Fahlberg in 1879. Born in Tambow, Russia, Dec. 22, 1850, Fahlberg received his Ph.D. degree at Leipzig in 1873, worked as a chemist in Germany, New York, London, and British Guiana, and went to Johns Hopkins as a fellow in 1878. Ira Remsen, professor of chemistry, and Fahlberg collaborated in studying the oxidation of *o*-toluenesulfamide. They obtained a compound which Fahlberg found possessed "a very marked sweet taste, being much sweeter than cane-sugar."

A number of myths have sprung up among chemists concerning the discovery of the sweetness of saccharin. Fahlberg's account is as follows:

"One evening I was so interested in my laboratory that I forgot about supper until quite late and then rushed off for a meal without stopping to wash my hands. I sat down, broke a piece of bread and put it to my lips. It tasted unspeakably sweet. I did not ask why it was so, probably, because I thought it was some cake or sweetmeat. I rinsed my mouth with water and dried my moustache with my napkin. When, to my surprise, the napkin tasted sweeter than the bread. Then I was puzzled. I again raised my goblet and, as fortune would have it, applied my mouth where my fingers had touched it before. The water seemed syrup. It flashed upon me that I was the cause of the singular universal sweetness and I accordingly tasted the end of my thumb and found that it surpassed any confectionery I had ever eaten. I saw the whole thing at a glance. I had discovered or made some coal-tar substance which out-sugared sugar. I dropped my dinner and ran back to the laboratory. There, in my excitement, I tasted the contents of every beaker and evaporating dish on the table. Luckily for me none contained any corrosive or poisonous liquid.

"One of them contained an impure solution of saccharin. On this I worked for weeks and months until I had determined its chemical composition, its characteristics and reactions, and the best modes of making it scientifically and commercially.

. . . .

"I gave the name of saccharine to this body when, in 1879, I made the interesting observation that when mixed with starch sugar in suitable quantities the resulting product was very similar in taste to cane sugar."

Leaving Baltimore, Fahlberg worked at Gray's Ferry Chemical Works in Philadelphia and then as a consultant in New York. He was a first class chemist but more than that. Not many chemists have had the energy, business acumen, inclination, and luck to take a laboratory synthesis, in which compounds are prepared in fragile apparatus and quantities measured in grams, and develop it into a commercial process taking place in huge tanks, requiring hundreds of workers, with production measured in tons. Fahlberg hoped to build a saccharin factory in the United States, which he regarded as his adopted country, but the high cost of production and the high tariff on raw materials caused him to erect a plant in Germany, where he prospered. He died at Nassau, Germany, Aug. 5, 1910.

Fahlberg's account is from an interview by an anonymous journalist in *Amer. Analyst* **2,** 211, 249 (1886); the announcement of the sweet taste of saccharin is in articles by Fahlberg and Remsen, *Berichte* **12,** 469-473 (1879) and *Amer. Chem. J.* **1,** 170-175 (1879-1880); "J. C. Poggendorff's Biographisch-literarischer Handworterbuch."

WYNDHAM D. MILES

* sometimes spelled Constantin

Eduard Farber

1892-1969

Farber was born in Brody, Austria, Apr. 17, 1892, and grew to manhood in Leipzig. His father intended that he go into business, but bowing to Eduard's studious nature allowed him to attend University of Leipzig. An eye defect kept him from serving in the army during World War I. After receiving his doctor's degree in 1916, Farber assisted Carl Neuberg at Kaiser Wilhelm Institute in studying the formation of glycerine from carbohydrates by fermentation. On Neuberg's recommendation he was given the job of converting a portion of a fermentation plant into a glycerine plant.

After the war he went with Bergin A. G. and investigated the conversion of wood into carbohydrates. In the early 1930's foreseeing the tragedy that Hitler would bring upon Germany, he began to plan to emigrate to the United States, but 1938 arrived before he and his family were permitted to leave. With letters of introduction but no money, he arrived in the United States and sought a job. He opened a laboratory to develop new uses for waste paper for Polyxor Chemical Co. in New Haven, Conn., acted as a consultant, moved to Washington, D. C., in 1943 to take charge of chemical research for Timber Engineering Co., and retired in 1957. The results of his industrial research for various firms may be found in approximately 100 patents and 50 articles.

As a student Farber read Ernest Meyer's "Geschichte der Chemie." He was caught by the subject but could not agree with Meyer's way of presenting history. In spare time following the end of World War I, Farber wrote a history showing the evolution of chemistry. He showed the manuscript to Neuberg, who disapproved of a chemist's spending time on history but applauded Farber's initiative. Neuberg contacted a firm which published the work in 1921 under the title "Die Geschichliche Entwicklung der Chemie." Thereafter as Farber pursued his career as industrial chemist, he followed history as a hobby.

Farber was interested in all of history of science rather than a specific time period, specialty, country, or person. The path that he followed depended upon the subject that stirred him at the time, library facilities available to him, and promise or possibility of publication. Consequently the books and approximately 50 articles he published touched on a variety of topics.

In 1928 Farber and his brother translated into German and annotated Boyle's "Skeptical Chymist" for Ostwald's "Klassiker." After World War II, firmly established in the United States, he spent more time on history. He wrote "Evolution of Chemistry, a history of its ideas, methods and materials" (1952; 2d ed., 1969), "Nobel Prize Winners in Chemistry" (1953; new ed., 1963), "Oxygen and Oxidation Theories and Techniques in the 19th Century and the Beginning of the 20th" (1967); and edited "Great Chemists" (1961) and "Milestones of Modern Chemistry" (1966). For a number of years he taught a course in history of chemistry at American University.

Farber died of cancer in Washington, July 15, 1969. In accordance with his wishes his body was cremated.

This sketch is based chiefly on an interview of Eduard Farber tape-recorded in 1962 by Wyndham D. Miles; obituaries in *Chem. Eng. News* **47,** 54 (July 28, 1969), port.; in Washington *Evening Star,* July 17, 1969; by W. D. Miles in *Arch. Intern. Hist. Sci.* **22,** 63-65 (1969).

WYNDHAM D. MILES

Colin Garfield Fink

1881-1953

Fink was born in Hoboken, N.J., Dec. 31, 1881. His father had learned the drug and chemical trade in Germany and, coming to this country in 1867, was employed by Eimer and Amend until 1874. Then with Louis Lehn the father founded the wholesale drug house of Lehn and Fink, which among other activities was a supplier to Thomas Edison in the earlier days of the phonograph and the incandescent lamp. His father also collaborated with C. F. Chandler at Columbia University. According to Harold Hibbert, the father possessed one of the finest collections of the elements of the periodic system.

Fink was educated in private schools in New York City and graduated in 1903 with an A.B. degree at Columbia University with honors in chemistry, physics, and mathematics. He spent the next 4 years in the Ostwald laboratory of University of Leipzig, where he received the M.A. and Ph.D. degrees in 1907. His thesis with Max Bodenstein on the "Kinetics of Contact Sulfuric Acid" formed the basis of a new theory of surface reactions.

A brief description of the dynamic, productive, and influential career which followed is next to impossible. Over 250 of his publications are listed in *Chemical Abstracts*. Fink served as an abstractor for *Chemical Abstracts* beginning with its first year, 1907, and became an assistant editor in charge of the section on electrochemistry in 1909, continuing both of these activities until forced to drop them for reasons of health in 1950. Small wonder that he was fondly dubbed Mr. Electrochemistry by the *C.A.* staff.

Fink joined The Electrochemical Society in 1907 and immediately became active in its affairs. He published in its transactions and stimulated and participated in discussions which were a constructive characteristic of his period of activity in the society. Serving as president in 1917-1918, Fink became executive secretary, and editor of its publications in 1921. This comprised an unbelievable amount of activity including the editing of an 8- to 16-page bulletin to members every month and organizing and managing two national conventions every year. All of this was done with no salary and at no expense to the society. In fact Fink was said to have had out of pocket expenses for society business of several thousand dollars per year which were uncompensated.

Thus in casual conversation The Electrochemical Society was often referred to as "Fink's Society." It was a heart-breaking time when he decided to retire in 1947, under pressure for change in the organization. He was presented with a scroll and made Secretary Emeritus at a testimonial dinner in his honor Oct. 16, 1947.

It is particularly in connection with his research that we find that the kaleidoscopic nature of Fink's career defies adequate description in a brief account. Hardly more than developments of significant sociological or multi-million dollar economic importance can be mentioned. Starting as a research engineer with General Electric Co. in Schenectady in 1907, Fink devised procedures for making ductile tungsten for electric lamp filaments and other uses. From 1910 to 1917 he was research director at Edison Lamp Works of General Electric Co. in Harrison, N. J. Here he not only followed the practical utilization of tungsten lamp filaments but developed a copper-clad nickel-steel lead-in wire to take the place of platinum in vacuum bulbs, a very important contribution. Beginning in 1913, he wrote the annual chapter on tungsten for "The Mineral Industry" until it ceased publication in 1941.

In 1917-21, Fink became head of a new laboratory established in New York City by the Chile Exploration Co. Here he developed an insoluble copper alloy anode, the "Chilex" anode, which made possible an immense reduction in the cost of electrowinning copper from the ore at Chuquicamata, Chile.

From 1922 until his retirement in 1950, Fink taught at Columbia University, becoming head of the Division of Electrochemistry and professor of chemical engineering. Here he continued his research at an accelerating pace, both with his own hands and with the aid of graduate students and paid assistants in considerable number. One of the more important of the latter was Charles Eldridge in 1923-24. With Eldridge he worked out a process for restoring and authenticating ancient bronzes which has been used all over the world. This consists of low current density cathodic treatment in 2% caustic soda for a number of days until the patina is softened and cleaned away and some of it reduced back to metallic copper in the same place from which it corroded.

Also with Eldridge in 1923-24 he worked out the basic principles controlling the electrodeposition of chromium from chromic acid solutions and proceeded to found the chromium plating industry. Fink obtained financial backing for this purpose from John T. Pratt, a trustee of the Metropolitan Museum of Art, and formed the Chemical Treatment Co. The growing pains and patent suits of this infant industry demanded a good deal of Fink's time and attention for many years thereafter. The writer and others were employed in this way in ongoing research on chromium plating in Fink's laboratory at Columbia in 1925-27.

In the field of education, Fink was very popular as a lecturer and thesis advisor. He gave unstintingly of his time and attention to his students. They had first priority even in the face of other important engagements. His boundless enthusiasm for the possibilities of research was an inspiration to everyone. One hardly dared to mention that something could not be done, as this was an instant challenge to Fink and lead to an assignment to investigate further. Many of Fink's outstanding accomplishments came from attacking problems that were supposedly insoluble.

Fink's broad interest in all of the elements of the periodic system was often in evidence. He wrote articles on many unusual elements in the monthly *Bulletin of the Electrochemical Society,* and many of his students worked on the electrodeposition of rare or uncommon metals. Thus papers were published on the electrodeposition of manganese, tin, tungsten, molybdenum, chromium, selenium, lead-thallium and lead-bismuth alloys, rhenium, rhodium, antimony, indium, cuprous oxide, germanium, and numerous alloys. The range of his interests was boundless, from practical problems such as coating iron with aluminum or making abrasive-coated stair treads to art museum studies.

Fink had a pleasant personality and a dry sense of humor that was much in evidence. He possessed the ability to inspire confidence and a rare combination of other talents—executive ability, energy, initiative, and a flair for publicity or ability to popularize science and its accomplishments.

Fink received a number of honors including the Acheson Medal of The Electrochemical Society (1933), the Perkin Medal (1934), honorary D.Sc. degree from Oberlin College (1936), and the Modern Pioneer Award of National Association Manufacturers (1940).

He was married to Charlotte Katherine Muller June 6, 1910, and they had two sons, Frederick William and Harold Kenneth. Dr. Fink died in Red Bank, N. J., Sept. 16, 1953.

Electrical World **77,** 961 (April 23, 1921); Harold Hibbert, *Ind. Eng. Chem.* **26,** 232-234 (1934); *Metal Cleaning Finishing* **8,** 601-602 (1936); "National Cyclopaedia of American Biography," James T. White, current volume E, 290-291 (1939) and vol. 27; *Trans. Electrochem. Soc.* **92,** 1-9 (1947); *Chem. Eng. News* **28,** cover, 2958 (Aug. 28, 1950); *Chem. Eng. News* **31,** 925 (1953); obituaries: *New York Times,* Sept. 19, 1953, *N.Y. Herald Tribune,* Sept. 19, 1953; *Asbury Park Evening Press,* Sept. 18, 1953; *Chem. Eng. News* **31,** 4934, Sept. 28, 1953; *Chemical Abstracts* **47,** 9813 (1953); *Metal Finishing* **51,** 131-132 (1953); *J. Electrochem. Soc.* **100,** 317C (1953); *The Columbia Chemical Engineer,* No. 4 (March 1954); G. Dubpernell, *Plating* **47,** 35-53 (1960); H. B. Linford and H. K. Fink, *J. Electrochem. Soc.* **108,** 229C-231C (1961); C. A. Hampel, "Encyclopedia of Electrochemistry," pp. 600-601 (1964).

GEORGE DUBPERNELL

Richard Fischer

1869-1955

Fischer was born in New Ulm, Minn., Nov. 18, 1869, the sixth child of seven born to Richard Fischer, senior, and his wife, Anna G. Hollstein. His father like many others had fled (1848-49) the unsettled conditions in Germany and had eventually settled in New Ulm, a settlement which had been founded by such refugees. Fischer had an active boyhood, and was graduated with the first class of New Ulm high school, having received instruction in chemistry from an elder brother. He was apprenticed to a pharmacist, a friend of the family.

His interest in chemistry and its application to pharmacy led to his study at the alma mater of his benefactor, University of Michigan. There in 1892 he received his pharmaceutical chemist degree. Service as an assistant under Albert B. Prescott led to his B.S. degree in 1894. Edward Kremers, then head of the School of Pharmacy at Wisconsin, discovered him and appointed him as instructor in pharmacy.

During a 2-year leave of absence from

1898 to 1900 he studied under Emil Fischer at Berlin for one semester and then moved to University at Marburg. There under Ernst Schmidt, an authority in alkaloids, he was granted his Ph.D. degree with the dissertation: "Beitrage zur Kenntnis der Papaveracien Alkaloide."

Upon his return to Wisconsin he was promoted to assistant professor in pharmacy. In 1909 he transferred to the Department of Chemistry as professor. That year also marked his entrance upon other services to the state related to chemistry. In 1903 he had been appointed chemist to the Wisconsin Dairy and Food Commission but was permitted to continue teaching. In 1909 he was appointed State Chemist and Director of Laboratories. In 1913 he became consulting director, a position he retained until 1930 when the commission was consolidated with the State Department of Agriculture and Markets.

Pure Food Laws in Wisconsin dated from 1889 and had been strictly and vigorously enforced. James Wilson, U.S. Secretary of Agriculture, anticipating the enactment of a federal food and drug act, invited Fischer to serve with a committee of the Association of Agricultural Chemists "to establish standards of purity for foods and determine what are to be recognized as adulterations therein." Fischer was active with testimony and in other ways in prosecution of violators of state statutes which prohibited the sale of yellow oleomargarine in Wisconsin, whether colored by pigment or by oils and fats which would yield a yellow product in imitation of butter. He was also active in a case involving use of the term corn sugar in place of "glucose."

While in the School of Pharmacy Fischer was on the editorial board of *Pharmaceutical Review*. He was a charter member of the Wisconsin Section of the American Chemical Society and served as vice chairman and chairman in 1910 and 1911 respectively and councilor (1916-19).

Fischer's lectures were always very formal. Nevertheless his course in advanced organic chemistry, meeting once a week for a 100-minute session after the supper hour was, in jest, referred to as a personally conducted tour through the land of formulas.

As a youth he had become expert with the slingshot. This interest in shooting with slingshot and rifle remained with him throughout his life. Each fall he would return to hunt in Minnesota. His companions enjoyed his singing around the campfire the songs he had learned while a student in Germany. He could affect the high pitched voice of the Saxon and the broad accent of the Bavarian. He had no direct descendants but helped nephews and nieces in establishing themselves and bequeathed $10,000 to the Department of Chemistry to inaugurate a permanent scholarship fund for students majoring in chemistry.

He became professor emeritus in 1939. He died in Minneapolis, Feb. 17, 1955.

H. A. Schuette, "American Contemporaries: Richard Fischer," *Chem. Eng. News* **16**, 307-308 (1932); A. J. Ihde and H. A. Schuette, "The Early Days of Chemistry at the University of Wisconsin," *J. Chem. Educ.* **29**, 65-72 (1952); personal communications from Kurt R. Bell, Aaron J. Ihde, and M. Esser.

ERNEST R. SCHIERZ

Chester Garfield Fisher

1881-1965

Fisher, founder of a major chemical supply company, was born in Pittsburgh, Pa., Aug. 10, 1881. He attended University of Pittsburgh where he studied chemistry and received his degree in mechanical engineering in 1900.

When, by 1902, the introduction of chemical control in the iron and steel industry of the Pittsburgh area made it necessary for firms to install laboratories, Fisher founded Scientific Materials Co. to provide analytical instruments, apparatus, and chemicals. In 1914 he added a development laboratory to his firm to design instruments for the transition of chemistry from "wet" or test tube analysis to "dry" or instrumental analysis. This laboratory developed the first modular gas analysis equipment and the Fisher burner, the first burner to provide a high-heat flame from ordinary natural gas.

When the United States entered World War I in 1917, the Army had no chemical warfare organization. At the request of the War Department, Fisher hurriedly assembled sufficient apparatus and chemicals for a chemical warfare laboratory, seven full freight cars in all, provided a glass blower, an instrument maker, and clerks and sent it to the American Expeditionary Force which set up the laboratory in France.

In 1919 he was appointed a special adviser to the government tariff commission which was designed to protect the American industries built up during the war. He helped convince congressmen that the design, development, and manufacture of research tools was a specialized industry of importance to the nation, and hence he played a part in establishing the first protective tariff for the young American apparatus industry.

In 1926 Fisher changed his firm's name to Fisher Scientific Co. In 1940 he acquired Eimer & Amend, an old chemical supply house, and in 1957 the firm of F. Machlett & Son, both of which he merged into his company.

Fisher was a founding member of the Scientific Apparatus Makers Association in 1917 and president of the organization in 1931 and 1932. He received the Association's Award in 1962 "for highest achievement in developing that industry's capacity for serving the nation in the field of research, production, education, health, and defense." He also was awarded that year an honorary Doctor of Science degree by University of Pittsburgh. In 1947 he established the Fisher Award for outstanding work in analytical chemistry. He died in Pittsburgh, May 3, 1965.

Fisher's major avocation was the collection and preservation of art related to history of science, especially chemistry. He began collecting paintings and engravings of alchemical and chemical subjects in 1921. He amassed hundeds of items including the notable collection of paintings left by the British chemist, Sir William Pope. In 1953 he established his Pasteur Memorial Collection, a comprehensive assemblage of portraits, manuscripts, letters, books and glassware of Pasteur. He gave many talks on the history of chemistry. Fisher's collection, probably the finest specialized art collection of its kind in existence, is on view for the public in his Pittsburgh building.

Williams Haynes, "American Chemical Industry," D. Van Nostrand Co. (1949), vol. 6, pp. 165-166; Pittsburgh Award to Fisher, *Chem. Eng. News* **25,** 3849-3850 (1947), port. on cover, and *The Crucible* (Pittsburgh Section, ACS) **32,** 311-313 (December 1947), port.; obituaries, *Pittsburgh Press* (May 4, 1965), *The Pittsburgh Post-Gazette* (May 5, 1965), and Harry M. Schwalb in *The Crucible* **50,** 132-34 (June 1965); obituaries: *New York Times,* May 6, 1965, *Chem. Eng. News,* May 17, 1965, p. 91; personal recollections.

BENJAMIN R. FISHER

Harry Linn Fisher

1885-1961

Fisher was born Jan. 19, 1885 at Kingston, N. Y., the son of George Edwin and Emma (Bray) Fisher. He worked for 3 years after graduation from high school and this necessitated his taking a year of refresher work at Dwight Preparatory School to ready himself for Williams College. He majored in the classics, but in his junior year he became interested in chemistry. He received his A.B. degree in 1909 and obtained a scholarship to Columbia to study under Marston T. Bogert. He received his Ph.D. degree in 1912 from Columbia.

He remained at Columbia as an instructor in organic chemistry until 1919 when he resigned to do research for B. F. Goodrich Rubber Co. at Akron, Ohio. He remained there 7 years and for the next 10 years was research chemist with the U.S. Rubber Co. laboratories in New York and in Passaic, N. J. He became director of organic research for U.S. Industrial Chemicals in 1936 and remained there until he retired in 1950. After retiring, he was appointed administrative assistant for the National Research Council and in November 1951, he was appointed special assistant to the director of the Office of Synthetic Rubber. In 1953, he went back to teaching at the University of Southern California as head of the Department of Rubber Technology and director of the TLARGI Rubber Technology Foundation. He also served as vice-president, Ocean Minerals,

Inc., working on chemical methods to turn sea water into fresh water.

Fisher was a nationally known authority on the chemistry of vulcanization. He was the author of about 50 patents chiefly in the field of rubber technology. He was prolific writer of technical articles regarding the chemistry and technology of rubber. His best known books were: "Laboratory Manual of Organic Chemistry," "Rubber and Its Use," and "The Chemistry of Natural and Synthetic Rubbers." He predicted that automobile tires eventually would be made to last 100,000 miles and would come in colors to match the cars. He also foresaw highways of rubber. During World War I, he helped develop the first synthetic rubber not made from a hydrocarbon. During World War II, he was technical consultant to the Office of the Rubber Director.

Harry received many honors and awards. He was president of Phi Lambda Upsilon from 1916 to 1918; president of American Institute of Chemists from 1940 to 1942; and president of the American Chemical Society in 1954. He was chairman of the Rubber Division and secretary of the Organic Division of the American Chemical Society. In 1940, he received the Modern Pioneer Award from the National Association of Manufacturers for developing a method to attach rubber to metal. He received the Charles Goodyear Award from the American Chemical Society in 1949 and the Charles Frederick Chandler Medal from Columbia University in 1954 for his outstanding contributions to the chemistry of synthetic rubber. In 1941 he was the 16th Marburg Lecturer of ASTM.

Fisher was an enthusiastic hiker to whom a mountain was a challenge. He climbed most of the peaks in the Adirondaks. Only a few days before his death he inspected the Death Valley National Monument from the heights of Mahogany Flats.

Fisher married Nellie Edna Andrews in 1910. They had two daughters, Helen and Francis, and one son, Robert.

He died Mar. 19, 1961 at Claremont, Calif.

Chem. Eng. News **32,** 82 (1954); Williams College alumni files; ASTM *Bull.* May, 1941.

Lester Kieft

Francis Alexander James FitzGerald

1870-1929

Francis FitzGerald, the eldest son of Charles Edward FitzGerald, a noted surgeon, and Isabel (Clarke) FitzGerald, was born in Dublin, Ireland on June 1, 1870. He attended Trinity College, Dublin University, receiving his B.A. degree in 1892. It was his father's ambition to have him train for the medical profession, but his interest in chemistry and electricity led him into fields removed from medicine. A year after graduating from Trinity College, he made a trip to the United States. Liking the country, he remained and matriculated at Massachusetts Institute of Technology, earning his B.S. (electrical) degree in 1895.

On Sept. 16, 1895 he applied to Edward Goodrich Acheson for a position with the recently founded Carborundum Co., located in Monongahela City, Pa. In his letter of application FitzGerald listed the courses he had taken at the Dublin and Massachusetts schools and made known the fact that at the former institution he was moderator and medallist in experimental physics and chemistry. Acheson was impressed with FitzGerald's training and the subjects in which he had specialized, and without hesitation offered him the position of chemist at the Carborundum Co.'s plant.

Not long after FitzGerald's employment, Acheson's company moved to Niagara Falls, N. Y. to take advantage of the water power available there. FitzGerald elected to continue his work with Carborundum at the Niagara Falls site. In 1900 Acheson transferred FitzGerald to his graphite company. Here his assignment was to carry out extensive investigations on the conversion of various forms of amorphous carbon to graphite.

Now located in a city whose main industries were based on either electrothermal or electrochemical processes, FitzGerald sought to join the American Electrochemical Society just being organized. He was gratified to be named a charter member of that body.

At the Society's second meeting, in September 1902, FitzGerald read his first paper before a learned society, the subject of which was "On the Testing of Carbon Electrodes." Two months later his lengthy paper on "The Conversion of Amorphous Carbon to Graphite" appeared in the November issue of *Journal of the Franklin Institute.*

Anxious to broaden his activities in electrochemistry, FitzGerald tendered his resignation to the graphite company in July 1903, after what he termed "my eight years of apprenticeship with Acheson." He had in the course of his graphite work become friendly with P. McNiven Bennie, a fellow employee. The two men decided to pool their knowledge and set up a consulting firm. FitzGerald had been characterized by Acheson as "attentive, energetic, and very intelligent," and his departure was regarded as a distinct loss.

With his resignation it was important that FitzGerald surrender the directorships he held in Acheson's companies. "I feel," he wrote his former employer, "that the best interests of our clients compel me to abstain from executive connection with commercial enterprises."

His consulting work was centered about electric furnaces—many of them of his own design—and their role in formulating new processes and improving the efficiency of old. As a consultant FitzGerald ran the gamut of electrochemistry and electrothermics. His accomplishments were reported in the many papers he contributed to *Transactions of the American Electrochemical Society.*

In addition to membership in the American Electrochemical Society—of which he was president, 1916-17—FitzGerald served for many years on the Society's publications committee. He was a fellow of AAAS and member of American Institute of Electrical Engineers, Franklin Institute, and the Faraday Society.

Francis FitzGerald died of pneumonia at his home in Niagara Falls, Ont. on Oct. 26, 1929. He was survived by his wife, the former Frances Mary Winifred Knox, and three sons.

Obituary in *Trans. Am. Electrochem. Soc.*, **56,** 13-14 (1929); details of Fitzgerald's early work in electrothermics were obtained from the personal files of Edward G. Acheson; Raymond Szymanowitz, "Edward Goodrich Acheson," Vantage Press (1971).

RAYMOND SZYMANOWITZ

Edward Curtis Franklin

1862-1937

Franklin was born Mar. 1, 1862 in Geary City, Kans. His father operated a sawmill on the banks of the Missouri River. He and his brother William, later a well known physicist, grew up in what was then relatively a frontier countryside. According to his own account, his boyhood must have resembled that of Tom Sawyer's. Besides this, however, both he and his brother were interested in electricity and built themselves a number of pieces of electrical equipment. Between the ages of 15 and 22, Franklin worked in drugstores and job printing offices. In 1884, at the urging of his brother, he entered University of Kansas as a special student. His interest in chemistry developed quickly. When he graduated in 1888 he remained at the university to teach elementary chemistry and to carry on analytical work. During the year 1890-91 he attended lectures at University of Berlin. In 1894 he obtained his Ph.D. from Johns Hopkins University, where he studied with Ira Remsen.

He tried industrial work for short periods, working in a Louisiana sugar plantation in 1888 and at a gold mine in Costa Rica in 1896, but in each case he returned to Kansas, where he became professor of physical chemistry in 1898. In 1903 he moved to Stanford University as professor of organic chemistry. Except for the years from 1911 to 1913, which he spent as professor of chemistry at the Hygienic Laboratory of the Public Health Service, he remained at Stanford for the rest of his life, becoming professor emeritus in 1929. He received many honorary degrees and in 1923 he served as president of the American Chemical Society.

Franklin was always an enthusiastic outdoorsman. He was an ardent mountain climber and camper. In his later years he took extensive automobile trips, often alone. In the last year of his life at the age of 74, he drove over 13,000 miles by himself, giving lectures all over the country. The trip lasted 2 months.

Franklin was an outstanding lecturer and a dramatic showman when he wished to be. His lecture demonstrations with liquid air and glass blowing were famous. In particular his skill in glass blowing served him well in constructing the elaborate glass apparatus which he used in his laboratory investigations. He was primarily an experimentalist. He had an original and often unconventional mind which saw analogies and novel vistas wherever he looked. All this came to a focus when he began his experimental career. He did not really commence this phase of his activities until he reached the age of 35, for research facilities at University of Kansas were very limited when he first taught there. However, once he began his laboratory studies he soon became recognized as a leader among his contemporaries. His attitude toward research was expressed in the remark he once made, "It's being paid to play all your life."

One of his students, Homer Cady, had noticed that in certain salts ammonia of crystallization could replace water of crystallization. This suggested to Franklin that liquid ammonia might be a polar solvent like water. His experiments on conductivities in liquid ammonia solutions soon confirmed this. Beginning with these physical measurements, Franklin went on to erect his whole concept of an ammonia system of compounds, as extensive as the water system. Reasoning by analogy, and sometimes almost intuitively, he showed that both organic and inorganic compounds of nitrogen resembled their oxygen analogs. From the new insights suggested by this approach he was able to discover new types of reactions and new varieties of compounds. He published 91 articles on ammonochemistry, and in 1935 he summed up the results of his life work in a monograph, "The Nitrogen System of Compounds," published as ACS Monograph No. 68. This book is a classic in American chemistry. As a teacher, investigator, and friend, Franklin cannot be forgotten by those who knew him. He died at his home on the Stanford campus, Feb. 13, 1937.

Personal recollections; the best biography is the "Franklin Memorial Lecture" by Alexander Findlay in *J. Chem. Soc.* 1938, 583-595; Franklin's autobiographical notes were privately printed by his children; obituaries: *New York Times,* Feb. 14, 1937; H. M. Elsey, *J. Amer. Chem. Soc.* **71,** 1 (1949).

HENRY M. LEICESTER

Francis Cowles Frary

1884-1970

Frary was born in Minneapolis, Minn., July 9, 1884. His interest in chemistry began at an early age with experiments on electroplating a door key with copper and delving into various aspects of photography. He entered University of Minnesota and received his degree of analytical chemist in 1903 and his master's degree in 1906. He spent a year at University of Berlin, followed by a combination of teaching and research at Minnesota, and completed his Ph.D. degree in chemistry in 1912. His interest and competence in chemistry were greater than average, and he taught general chemistry, qualitative analysis, electrochemistry, glass blowing, and photography. By the time he received his Ph.D. at the age of 28, he had published 11 different technical papers. He continued his teaching as an assistant professor until 1915. He married Alice H. Wingate in 1908. Their two children were Faith (Mrs. Luke Saunders) and John.

His great interest in research and the recognition of his capabilities by industry led him to accept a position as research chemist with Oldbury Electrochemical Co. in 1915. He developed a special familiarity with phosgene production and was selected with Dana J. Demorest to build and operate a phosgene plant at Edgewood Arsenal for the Chemical Warfare Service in World War I. He became a major by the end of the war. Prior to the Armistice, he accepted a position as director of research for Aluminum Co. of America and assumed his duties in December 1918.

The successful development of a strong research program at ALCOA, which became recognized as one of the best in the metallurgical world, resulted from Frary's great leadership. In him were combined all the ingredients of human success—a warm personality, a great curiosity, an active imagination, a generous spirit with concern for the individual, an unusual physical stamina, a penchant for hard work and intellectual toil, coupled with a fine modesty and a keen sense of humor. He was called, affectionately and respectfully, "Mr. Aluminum Research."

For 34 years, Frary headed the ALCOA Laboratories and saw them grow from a staff of a few men on a second floor in the New Kensington Works to a group of independent research buildings (in 1928) with a total staff of over 200, a library, and a patent division. He led his staff to a broad spectrum of discoveries with over 30 patents to their credit. His leadership resulted in the electrolytic production of super-purity aluminum, an improved method for purifying alumina, successful investigations on alloying, casting, and working of aluminum into its various forms for commerce, and developing strong aluminum alloys. Whereas, in 1928, only 1.7% of aluminum went into aircraft manufacture, by 1945 over 70% of ALCOA's production went into this enterprise. Frary retired in 1952. He authored and co-authored many papers and several books. His chief book was "The Aluminum Industry" with Junius D. Edwards and Zay Jeffries, 2 volumes, 1930.

In recognition of his many accomplishments, Frary received the Pittsburgh Award of the American Chemical Society (1937), Acheson Medal of the Electrochemical Society (1939), Perkin Medal of the Society of Chemical Industry (1946), Gold Medal of the American Society of Metals (1948), and James Douglas Metallurgical Award of the American Institute of Mining and Metallurgical Engineers (1949). He served as president of the Electrochemical Society in 1929 and of the American Institute of Chemical Engineers in 1941.

Frary continued an active life after his retirement. He was a student of languages over the years (French, German, Italian, Norwegian, Swedish, Danish), and studied Russian at the age of 75. He translated many articles for Alcoa during his retirement. He was an avid gardener and fruit grower and was active in community and church affairs. He died Feb. 4, 1970 at the age of 85 at Oakmont, Pa.

Biographies in *Chem. Eng. News* **26,** 2493 (1948), with portrait; *Alcoa News* **36,** No. 1, 6-8 (1964), with pictures; *Lab Log* (Alcoa) **6,** Section 2, Jan. 29, 1952, with pictures; obituaries in *J. Electrochem. Soc.* **117,** 154C (1970); *Chem. Eng. News* **48** (8), 59 (1970), with portrait.

ROY G. BOSSERT

Herman Frasch

1851-1914

Born on Christmas Day 1851, at Gaildorf, Württemberg, Herman Frash completed the local elementary school, and his father, Johannes Frasch, a prosperous apothecary, sent him to the gymnasium at Halle, after completing which he was supposed to enter the university. Instead, he came to America, landing in Philadelphia in 1868, and immediately secured a position at Philadelphia College of Pharmacy as assistant to Professor John M. Maisch.

Sensing the growing importance of chemicals and the increasing opportunities for chemists, Frasch enthusiastically continued to study chemistry. Already excited about the possibilities of petroleum hydrocarbons, he determined to become one of the foremost authorities in the new field now beginning to boom in Pennsylvania.

Still continuing to teach, Frasch built up a consulting practice in Philadelphia until 1877, when he patented an improved process for refining paraffin wax. Selling his patent rights to Merriam and Morgan Paraffin Co., in which Standard Oil Co. had a substantial interest, he was assured of an annual retainer fee from Merriam and Morgan; he then moved to Cleveland and opened up an office and a small laboratory. He soon became known as the outstanding chemical consultant in the city, then a hotbed of petroleum activities.

The oil recently discovered in Ontario attracted his attention for its high sulfur content gave it an evil odor, and the kerosene refined from it burned badly, sooting up lamp chimneys and gumming up wicks. In 1882 Frasch sold the exclusive rights to three patents for improvements in refining this Canadian crude oil to Imperial Oil Co., Ltd., and in 1884 moved to London, Ont. The following year he bought Empire Oil Co., which had wells and a small refinery at Petrolia, Ont., believing that he could rescue the struggling company from bankruptcy by making its products marketable.

In August of 1886, however, Frasch was called back to Cleveland by Standard Oil. The famous Lima field of northwest Ohio and northeast Indiana had just opened up, and the sour petroleum produced there could be sweetened by processes covered in his patents. To maintain its dominant position in the industry Standard Oil was producing sour petroleum much faster than it could be marketed. Frasch was asked to make marketable this sour oil, which till now could be sold only as industrial fuel.

Standard Oil bought Frasch's Empire Oil Co. holdings and his patented process for refining sour oil, engaged him as a full-time consultant, and sent him to the brand-new Solar Refinery in Lima in charge of the first experimental research program ever undertaken in the American petroleum industry.

Frasch's plan adopted the simple concept of removing sulfur by treating petroleum with a metallic oxide, precipitating the sulfide formed, and recovering the oxide for reuse. His basic patent, issued Feb. 21, 1887, covered the use of oxides of copper, lead, iron, bismuth, cadmium, mercury, and silver, in all combinations.

Frasch's process was such an improvement over the old litharge treatment that the success of the Lima operation had far-reaching results. Frash was assigned to the new refinery at Bayonne as a special consultant to study crude oil and develop by-products. The policy of continuous research was adopted, and Frasch was recognized as the father of research in the petroleum industry. Now independently wealthy, Frasch refused an executive post with Standard Oil, continuing only as a lifetime consultant.

After perfecting a new method of recovering sulfuric acid used in refining heavy fractions of crude oil, Frasch obtained his first patents on his sulfur-mining process. As president and half-owner of Union Sulfur Co., which he had organized in 1892, he succeeded, after many failures, by using water heated to 230°F., in bringing tons of liquid sulfur to the surface, where in a matter of minutes it solidified to brimstone having an elemental sulfur content in excess of 99.5%.

As long as Frasch's basic patents were in force, Union Sulfur Co. had a virtual monopoly in the United States, enabling it drastically to lower the domestic price of sulfur and to invade the European market as well. Frasch initiated research to remove the fire and explosion hazards from brimstone and to improve the design of sulfur burners used to manufacture sulfuric acid. He was one of the first to enter into cooperative research with state agricultural experiment stations in studies of the use of sulfur as an insecticide and fungicide.

Frasch was awarded the Perkin Medal in 1912 and received many other honors before his death in Paris on May 1, 1914, but his greatest honor is the distinction of being a chemist whose name is indelibly associated with two great processes in two distinct fields —the Frasch process for making sour oil marketable and the Frasch process for producing sulfur.

Eduard Farber, "Great Chemists" (1961), pages 923-933; *Ind. Eng. Chem.* **4,** 138 (1912); *Chem. Met. Eng.* **10,** 78 (1912) (Perkin Medal award); W. W. Duecker, *Ind. Eng. Chem.* **42,** 286 (1950); A. E. Marshall, *Chem. Eng.* **57,** 293 (1950); and *J. Soc. Chem. Ind.,* May 30, 1914 (obituary).

JAMES C. COX, JR.

John Fries Frazer

1812-1872

Frazer was born to Robert and Elizabeth (Fries) Frazer in Philadelphia, July 8, 1812. He attended academies and University of Pennsylvania. At the university he assisted Alexander D. Bache, professor of chemistry and natural philosophy. After receiving his bachelor's degree in 1830, Frazer studied medicine for a while, then turned to law and was admitted to the bar. Law could not hold his interest and when the geological survey of Pennsylvania began, he persuaded Henry D. Rogers, one of the leaders, to employ him as a field assistant. He left the survey in 1836 to teach chemistry and physics at Central High School, Philadelphia, and in 1844 went to University of Pennsylvania as professor of chemistry and natural philosophy. From 1846 to 1850 he also taught chemistry at Philadelphia Medical Institute, a private, non-degree-granting school.

Frazer remained at Pennsylvania the rest

of his life. He was not a researcher but an influential planner, administrator, and teacher. One of the leaders in the rejuvenation of the university, he died of a heart attack Oct. 12, 1872, while running upstairs to his new laboratory the day after the building was dedicated.

Frazer was active in the Philadelphia scientific community, particularly in Franklin Institute and American Philosophical Society. He edited *Journal of the Franklin Institute* from 1850 to 1866. When National Academy of Sciences was created in 1863, he was one of the original 50 members.

J. L. LeConte, "Memoir of John Fries Frazer, 1812-1872," *Biog. Mem. Nat. Acad. Sci.* **1**, 245-256 (1877); records of University of Pennsylvania; "Dictionary of American Biography," **7**, 3, Chas. Scribner's Sons (1931).

WYNDHAM D. MILES

William Frear

1860-1922

Frear was born Mar. 24, 1860 in Reading, Pa., the first child of George Frear, a Baptist minister, and Malvina Rowland Frear. After attending public schools in Reading and Norristown, Pa. he enrolled at Bucknell University with the intention of becoming an engineer. He later decided to concentrate on chemistry and the natural sciences and was appointed an assistant in the natural science department of Bucknell upon receiving his B.A. degree from that institution in 1881. While working at Bucknell, Frear studied for his doctorate under the guidance of Illinois Wesleyan University. He also attended summer courses at Harvard University. Upon receiving his Ph.D. degree from Illinois Wesleyan in 1883, Frear was appointed assistant chemist under Harvey W. Wiley of the United States Department of Agriculture. While with the USDA, Frear conducted investigations of beet sugar and cereals.

In 1885 Frear was appointed assistant professor of agricultural chemistry at The Pennsylvania State College (now The Pennsylvania State University). A year later, he was promoted to professor and department head. In 1887 the State Experiment Station was established at State College and Frear became its vice-director and chief chemist, positions he held until his death. In 1908 the Department of Agricultural Chemistry was divided and Frear became professor of experimental agricultural chemistry.

Frear's research at the Experiment Station encompassed a variety of topics. He studied soil acidity, the production and use of lime, and the composition of foods, including commercial vinegars, and feeds, including forage crops. He conducted plow tests, feeding experiments, germination tests, and fertilizer analyses. In addition he studied the culture of tobacco and the composition of Pennsylvania limestone. Frear saw to it that his research was reported promptly through the issuance of Experiment Station publications.

With the creation of the Experiment Station in 1887, responsibility for weather observation was transferred from the Department of Physics to the station. The recording of meteorological data and the teaching of meteorology became added duties for Frear. He developed a keen interest in the subject and became an authority on climate in central Pennsylvania.

Frear made a significant contribution through his service on various commissions of the governments of Pennsylvania and the United States. He was chemist for the Pennsylvania State Board of Agriculture (1888-1919) and the State Department of Agriculture (from 1885). He was also chemist for the Cattle Feed Control Board (1902-1905) and served in the same capacity for the Dairy and Food Commission. In 1900 he was named chemist for the Food Standards Committee of the U.S. Department of Agriculture. When the Joint Committee on Food Definitions and Standards was created in 1914, Frear was named its chairman. His exhaustive studies and lucid writings on the subject of food quality and standards became important references for the writers of both state and federal food and drug legislation.

Frear was a member of many scientific societies. He was secretary of the Society for the Promotion of Agricultural Sciences from

1893 to 1895 and was its president in the year 1903. He edited *Agricultural Science* from 1892 to 1894 and was on the board of editors of *Journal of the Association of Official Agricultural Chemists* from 1920 until his death.

Frear was a community leader. He helped to organize the first utility companies (water, electricity, and ice) for the village of State College, Pa. He helped form the first transportation company, the first building and loan association, and was a director of the first bank in State College. In 1900 at age 40, Frear married Julia Reno. He is reported to have enjoyed his role as husband and father. It was at home, near his family and near the college he had served for almost 37 years, that Frear died Jan. 7, 1922.

Thomas I. Mairs, "Some Pennsylvania Pioneers in Agricultural Science," pp. 77-98, Pennsylvania State College (1928); I. K. Phelps, *J. Ind. Eng. Chem.* **14,** 247 (1922); H. Leffmann, *J. Frank. Inst.* **193,** 567 (1922); "Dictionary of American Biography" **7,** 5, Chas. Scribner's Sons (1931); information from Dixon Johnson, Depart. of Public Information, Penn. State.

GEORGE M. ATKINS, JR.

Paul Caspar Freer

1862-1912

Freer was born in Chicago, Ill. on Mar. 27, 1862. His father was a respected physician of that city and one of the founders and later president of Rush Medical College. Orphaned at an early age by a typhoid epidemic which claimed the life of his father, Paul received his early tutelage from his mother who was well versed in the classics as well as in French and German. He was later sent to Germany to complete his early studies.

On returning to Chicago, he graduated first in his high school class. Inspired perhaps by his late father, he went to Rush Medical College, graduating in 1883. During his medical studies, he had been interested in chemistry which he studied with Walter Haines; to pursue this interest, he returned to Germany.

At that time, Adolph von Baeyer, of University of Munich, was considered one of the leading chemists in Europe. Freer studied in his laboratory, receiving his Ph.D. degree *summa cum laude* in 1887.

While in Munich, he became acquainted with William H. Perkin, Jr., son of the founder of the coal–tar dye industry. His first paper was published with Perkin in 1886.

On completing his work in Munich, he went to England for a brief stay in Perkin's private laboratory. He left there soon for a short stay at Owens College, Manchester, where he worked in the laboratory of Harold B. Dixon. During this period he developed an interest in organic ring structures, which was to continue almost literally to the day of his death.

Having decided against a career in commercial chemistry, he returned to America. His first post was at Tufts College. However, the University of Michigan claimed him shortly. Starting in 1889 as a lecturer, the following year he was promoted to professor of inorganic chemistry, having a chair in both the Medical School and the School of Arts.

He was to hold these combined chairs for the next 12 years. His work was largely concerned with the sodium derivatives of various aldehydes and ketones, their formation and behavior. An important piece of work in 1890 concerned the constitution of aceto-acetic ether. Acetone, he found, while containing no methylene group, formed a sodium derivative, with reactions almost totally analogous to aceto-acetic ether.

In 1898 he completed a study on phenylhydrazones. These were prepared during a winter at Ann Arbor when the temperature reached 20°F below zero. Even at extremely cold temperatures, some of these compounds required very careful handling. In later years he expressed some amusement to a colleague at the difficulties which might be expected in trying to reproduce these results in the prevailing temperatures in the Philippines.

In 1901 he accepted his last and probably most important post as first superintendent of Government Laboratories, the Philippine Islands. He was chosen by Dean Worcester, then Secretary of the Interior for the Philippine Islands, who said, "never was a man more fortunate in his choice." Freer secured a year's leave of absence from his academic duties, which eventually grew to 3 years, until in 1904, he accepted the position permanently.

The new position hampered his scientific research because his duties were primarily administrative. His responsibilities included designing new laboratory facilities, overseeing their building, and equipping them. In addition, it was necessary to carry on routine public health work in connection with amoebic dysentery, Asiatic cholera, and bubonic plague.

The new Bureau of Government Laboratories successively absorbed the Bureau of Mines, projects in botany, entomology, fisheries and cement testing, and became shortly the Bureau of Science. In addition to heading the Bureau, Freer was the first editor of *Philippine Journal of Science,* to which he was also a contributor. Further, he was dean and professor of chemistry at the College of Medicine and Surgery, University of the Philippines.

According to his contemporaries, who respected and admired him, he was a man of keen scientific insight, administrative skill, and vision.

Although bothered by continuous ill health at the end, Freer worked almost until the day of his death, Apr. 17, 1912 at Baguio, Philippine Islands.

Chemical Abstracts; "National Cyclopedia of American Biography," **19,** 423, James T. White & Co. (1926); *Popular Sci. Monthly* **80,** 521-529 (1912); *Science,* **36,** 108-109 (1912); obituary, *New York Times,* Apr. 18, 1912; communication from M. P. Ramiro, Pasay City, Republic of the Philippines.

J. PAUL O'BRIEN

Benjamin Ball Freud

1884-1955

Freud was born Feb. 12, 1884 in Chicago, Ill. His father was of Czechoslovakian ancestry, his mother Dutch. He was educated in the Chicago public schools and was a member of the class of 1904 of the University of Chicago. Throughout his youth, he helped his family financially as much as possible by working at odd jobs.

In the university he came into contact with Julius Stieglitz, William Harkins, Alexander Smith, Albert Michelson, and Robert Millikan. His major subject was chemistry. After graduation he entered the Chemical Engineering Department of Armour Institute of Technology, where his assignment was training engineers in pure chemistry. In 1915 he took his Ch.E. degree from the institute. When the United States entered World War I, he volunteered and, after a short period of training, was sent to France as a captain in the First Chemical Regiment, where he had experience in the art of chemical warfare. After the war he returned to Armour Institute, advancing in time to the rank of professor. In 1924 he married Henrietta Zollman, also a chemist.

During his earlier career he was active as an engineer and consultant, and carried out investigations of sulfur chlorides. He became involved in the affairs of the Chicago Section of the American Chemical Society, acting as chairman of a number of committees and eventually as chairman of the section.

In 1927 he took his Ph.D. degree under William D. Harkins, working on the properties of monomolecular films on liquid surfaces and taking motion pictures of falling drops. After Armour and Lewis Institutes combined to form Illinois Institute of Technology he served for a time as dean of the Evening Division. In 1937 a separate Department of Chemistry was organized, and he became chairman. He remained in this position while the institute expanded rapidly, until he retired in 1947. From 1942 to 1945, as a colonel in the U.S. Army Reserves, he was mobilized to become liaison officer between the Fifth Corps Area, comprising Illinois, Michigan, and Wisconsin, and the Civil Defense of that region, managing the day-to-day affairs of that organization. In 1947 he became emeritus professor of University of Illinois at Navy Pier, which position he retained until 1952 when illness forced his final retirement. He died Dec. 12, 1955.

Throughout his career his primary interest was his students; however, he had secondary interests also. His military career is mentioned above. After World War I he resigned from the army but shortly after joined the chemical warfare reserves. He rose in rank to colonel and organized and trained a cadre of officers known as the 304th Chemical Regiment. When World War II broke out, the regiment was disbanded, and most of the personnel became chemical officers of other military units. He also assisted in training other groups, such as the Civilian Conservation Corps.

About 1908 his father had formed a connection with a neighborhood bank, and the son developed an interest in finance which continued throughout his life. For some time he served as director of a bank. While he was involved in the financial affairs of the Chicago Section, its Endowment Fund was established.

His chief intellectual interests were philosophy, music, and history, especially the history of chemistry. He gave a course in this subject sponsored jointly by the ACS Chicago Section and University of Chicago. He enjoyed motoring and spent a number of vacations exploring the United States and Canada.

Personal recollections.

HENRIETTA Z. FREUD

John Fryer

1839-1928

Fryer was born at Hythe, Kent, England, Aug. 6, 1839. He graduated from Highbury College, London. He was principal of St. Paul's College, Hong Kong, China, 1861-63; professor of English language and literature at Tung-Wen College, Peking, 1863-65; head master of the Anglo-Chinese school at Shanghai, 1865-67. His scholarly reputation and adeptness with the Chinese language brought him an offer by the government to organize a project for translating western scientific, technological, and military treatises, which he accepted in 1867. To qualify himself for this task, he purchased books on chemistry and studied them. He also acquired a set of chemicals for qualitative analysis and a collection of minerals. He thus became a self-taught chemist. He was engaged in this undertaking for 29 years and was the author or translator of nearly 100 scientific works, including chemistry and chemical technology, in the Chinese language.

For his services to China, the government awarded him the third degree of the civil brevet rank in 1873 and the first rank of the third degree of the Order of the Double Dragon in 1899. He received an honorary LL.D. degree from Alfred University, New York in 1889.

In 1896 Fryer left China to accept the position of the first Agassiz Professor of Oriental Languages and Literature at University of California, where he led a distinguished teaching career until he became emeritus in 1914. He died in Berkeley, Calif., July 2, 1928.

Archives, University of California, Berkeley; "National Cyclopaedia of American Biography," **21,** 246-7, with portrait, James T. White & Co. (1931); "Lamb's Biographical Dictionary of the United States," **3,** 202, James T. White & Co. (1892); Adrian A. Bennett, "John Fryer: The Introduction of Western Science and Technology into Nineteenth-Century China," Harvard East Asian Monograph No. 24, Harvard University Press (1967); obituary with portrait, *San Francisco Chronicle,* July 4, 1928.

MEL GORMAN

Casimir Funk

1884-1967

Casimir Funk, American biochemist of Polish birth, is best known for having given "vitamines" their unique name and for having provided the first insights into their structures. He was a prolific contributor to the early biochemical literature on hormones, and he was prominent in the commercial development of vitamin preparations, as well as in other biochemically-related fields.

Born Feb. 23, 1884 in Warsaw, Poland, the son of a physician, he received professional training as an organic chemist under Stanislaw von Kostaniecki at University of Bern, Switzerland. His dissertation on the chemistry of brasilin and haematoxylin led to his Ph.D. degree in 1904 and an opportunity to work with Gabriel Bertrand in the biochemistry department of Pasteur Institute in Paris from 1904 to 1906. He then worked 4 years in Emil Fischer's laboratories in Berlin, as an assistant to Emil Abderhalden, with heavy emphasis on amino acid and peptide chemistry. In 1910 disagreements between Funk and Abderhalden on data generated in experiments on the purported sufficiency of "synthetic" diets led Funk to travel to London, where he accepted an appointment at Lister Institute. His synthesis of *dl*-dihydroxyphenylalanine ("dopa") was well-regarded by his superiors. He was soon given the opportunity to investigate the nutritional chemistry of beriberi, a disease common in the East Indies among people whose diets consisted largely of polished rice. Beriberi was cured by eating rice hulls. Funk was able to isolate from rice hulls a curative crystalline material, which he characterized as having a pyrimidine structure. Later work by others on this vitamin (thiamine) showed the correctness of this view although Funk's isolate was not actually the desired pure substance. Funk also isolated in the same procedure, nicotinic acid, which does not cure beriberi and was not recognized as a vitamin until much later when Conrad Elvehjem showed it to be the pellagra-preventive factor.

Funk suggested that beriberi, scurvy, pellagra, and rickets were all due to deficiencies of specific nutritional factors which he assumed to be amines (hence "vitamines," a name first appearing in a 1912 publication). His work on the anti-beriberi vitamin led to a book on vitamins (published first in Wiesbaden, Germany in 1914) and to the D.Sc. degree at University of London in 1913. In 1913, the year of his marriage to Alix D. Schneidesch, a Belgian girl, he left Lister Institute for a biochemist position at Cancer Hospital. He remained there until 1915, when he emigrated to the United States. He acquired U.S. citizenship in 1920.

His principal work in New York during World War I was devoted to efforts to produce medicinal chemicals formerly available only from Germany. Working for Calco Co. (Bound Brook, N.J.) he developed industrial processes for making *beta*-naphthyl benzoate and cinchophen. Later, for H. A. Metz and Co., a former importer of German drugs, he developed successful procedures for making Salvarsan and epinephrine (adrenaline). After World War I he turned to making vitamin concentrates and carried out research at Columbia University dealing, in part, with the microbiological assay of vitamins. In 1923 he accepted a Rockefeller Foundation-funded appointment as biochemist at the National Health Institute in his native Warsaw. He again undertook to develop manufacturing methods for medicinal materials formerly available principally from Germany. This time it was insulin, and Funk was again medicinally successful although his attempts to produce the chemically pure hormone failed. By the end of 1927 the Rockefeller funding had expired, and Funk left Poland for Paris where in 1929 (in the suburb of Reuil-Malmaison) he founded his own company, Casa Biochimica, for research and production of the sex hormones, insulin, and liver extract used in treating pernicious anemia.

In 1939 with the outbreak of World War II Funk returned to New York. He continued to work on vitamin preparations, largely in connection with his position as consultant for the U. S. Vitamin Corp., and later, as head of the Funk Foundation Research Laboratories. His post-World War II work increasingly emphasized cancer-related problems. He died in New York, Nov. 20, 1967.

Benjamin Harrow, "Casimir Funk: Pioneer in Vitamins and Hormones," Dodd, Mead & Co. (1955); "Current Biography Yearbook," p. 210-12, H. W. Wilson Co. (1945); obituary, *New York Times,* Nov. 21, 1967.

CHARLES H. FUCHSMAN

Nathaniel Howell Furman

1892-1965

Furman was born in Lawrenceville, N.J., June 22, 1892. Upon graduation from Lawrenceville School at the top of his class, he entered Princeton University. He graduated in 1913 with his B.S.E. degree and Phi Beta

Kappa honors. He continued his studies at Princeton, obtaining his A.M. degree in 1915 and Ph.D. in 1917. After 2 years as an instructor in analytical chemistry at Stanford University, he returned to Princeton as assistant professor in 1919. Rising through the ranks, he was appointed to the Russell Wellman Moore Professorship in 1945 and served as chairman of the Department of Chemistry from 1951 to 1954.

Furman was an outstanding teacher and analytical chemist, specializing in electroanalytical methods. He contributed heavily to the development of this field with the publication of many papers on potentiometric and coulometric titrations and on Polarography. In addition, he did research on oxidation-reduction indicators, the use of mercury as a reducing agent, the use of concentration cells in quantitative analysis, ceric sulfate as an oxidant in volumetric analysis, electrodeposition methods, solvent extraction, and spectrophotometry.

During the first World War he served in the U.S. Chemical Warfare Service, conducting research on the detection of and protection against poisonous gases. In the second World War he worked for the Office of Scientific Research and Development and for the Manhattan Project. Here, he improved analytical methods for detecting impurities in uranium and its compounds, and later, for determining traces of uranium in anything with which it might have come in contact during its processing. He co-authored several textbooks of analytic chemistry which were widely used. He also co-authored "Potentiometric Titrations: A Theoretical and Practical Treatise" (with I. M. Kolthoff, 1926), and translated from the German, "Indicators. Their Use in Quantitative Analysis and in the Colorimetric Determination of Hydrogen-Ion Concentration" (1926) and "Volumetric Analysis" (1928/1929), both by Kolthoff. He was editor of "Scott's Standard Methods of Chemical Analysis" (5th ed., 1939) and contributed chapters to this work. He was associate editor of "Analytical Chemistry of the Manhattan District Project" (1958), and he contributed chapters to "Treatise on Physical Chemistry" (1924), "Sixth Annual Survey of American Chemistry" (1932), various editions of "Neue Massanalytische Methoden," "Trace Analysis" (1957), and Margosche's "Die Chemische Analyse." He served as associate editor of *Analytical Chemistry* and of *Journal of the American Chemical Society.*

In 1951 Furman served as president of the American Chemical Society, the first president to be nominated by petition and then to be elected by popular vote of the members. As president, he directed the Diamond Jubilee meeting of the Society in New York.

Among the many honors bestowed on Furman were the Fisher Award in Analytical Chemistry (1948), the Achievement Award of his Princeton Class of 1913 (1949), the Palladium Medal of the Electrochemical Society (1953), and an honorary degree of Doctor of Science from Boston University (1950).

Furman was an unselfish and modest individual who combined an urbane and witty nature with great drive and initiative. His hobby was book collecting, and he enjoyed playing golf, swimming, and travelling. He died on Aug. 2, 1965 in Burlington, Vt.

"National Cyclopedia of American Biography," James T. White & Co., **52,** 610 (1970); I. M. Kolthoff, *J. Chem. Education* **44,** 328-330 (1967); "Current Biography," H. W. Wilson & Co., pp. 219-221 (1951); *Chem. Eng. News* **43,** No. 32, pp. 7, 25 (Aug. 9, 1965); archives, Princeton University.

JOHANN SCHULZ

G

Frederick Augustus Genth

1820-1893

Friedrich August Ludwig Karl Wilhelm Genth, who later shortened his name to Frederick Augustus Genth, was born in Germany, May 17, 1820. He studied at University of Heidelberg from 1839 to 1841, University of Giessen from 1841 to 1843, and University of Marburg from 1844 to 1845. After receiving his Ph.D. at Marburg, Genth assisted Bunsen until 1848, then emigrated to the United States.

Genth opened a commercial laboratory in Philadelphia, and in 1849 moved to North Carolina to superintend Silver Hill mine; in 1850 he returned to his consulting practice in Philadelphia. He became well-known as an analyst and built a lucrative business. In his private laboratory he also instructed young men, several of whom are mentioned in his articles.

In the 1870's and 1880's before Pennsylvania established an agricultural experiment station, Genth was chemist for the Board of Agriculture, an important post because his analyses of fertilizers were official within the state. He was not a state employee but made analyses in his commercial laboratory. In the 1870's he was also mineralogist for the Pennsylvania Geological Survey, running analyses in his laboratory.

In 1872 University of Pennsylvania hired Genth as professor of chemistry and mineralogy but a disagreement with the administration led him to resign in 1888. Thereafter, he concentrated on his commercial work.

Genth was one of the corporate members of the American Chemical Society, a vice president in 1876, and president in 1880. He married twice and had 12 children. Toward the end of his life Genth was so fat he had difficulty moving around, but he carried on his analytical work almost up to the time of his death, Feb. 2, 1893 in Philadelphia.

Genth published more than 100 articles in journals, books, and official reports and carried out a myriad of unpublished investigations for clients. His major contribution to pure chemistry was his research on ammonia-cobalt compounds, begun in 1846, which laid the foundation for future work in this area. Later, he and Wolcott Gibbs expanded the knowledge of this class of compounds with a co-operative, classic investigation. His favorite study was mineralogical chemistry. For almost half a century his analyses appeared in publications. He discovered more than a score of new minerals. The eminent chemist George Barker said that Genth was "well-nigh without a peer" as an analyst. At the time of his death he was, to his colleagues in Philadelphia, "certainly the foremost mineral analyst this country has known."

Frederick Augustus Genth, Jr., born in Philadelphia on Feb. 12, 1855, followed his father in chemistry. He received his B.S. degree in 1876 and M.S. degree in 1878 from University of Pennsylvania, and an honorary doctor of pharmacy degree from Medico Chirurgical College, Philadelphia, in 1909. He was assistant in chemistry with Pennsylvania Geological Survey from 1877 to 1880, instructor and assistant professor at University of Pennsylvania from 1881 to 1888, chemist for the Dairy and Food Commission of Pennsylvania's Department of Agriculture from 1897 to 1903, lecturer on chemical jurisprudence at Medico-Chirurgical College from 1907 to 1908, and professor of mineralogy, assaying there from 1908 until his death. Concurrently he was associated with his father as a consultant from the 1870's until the latter's death and thereafter conducted his own consulting firm. He died in Landsdowne, Pa., Sept. 1, 1910.

Biographies of Genth by G. F. Barker in *Biog. Mem. Nat. Acad. Sci.,* **4,** 201-231 (1902), with portrait, list of publications; and in *Proc. Amer. Philos. Soc.*, **40,** x-xxii (1901), with portrait, list of publications; W. H. Wahl, H. F. Keller, T. R. Wolf, *J. Franklin Inst.*, **135,** 448-452 (1893), with portrait; obituary, *Philadelphia Ledger* (Feb. 3, 4, 1893).

Information on F. A. Genth, Jr., in "Who Was Who"; records of American Chemical Society; obituary in *J. Franklin Inst.*, **170,** 320 (1910).

WYNDHAM D. MILES

Lewis Reeve Gibbes

1810-1894

Gibbes was born in Charleston, S. C., Aug. 14, 1810, the eldest son of Lewis Ladson and Maria Henrietta (Drayton) Gibbes. He attended University of Pennsylvania Grammar School (1821-22), Pendleton Academy (1823-27), and South Carolina College (1827-29) from which he graduated with highest honors. He began his first course in medicine at Medical College of South Carolina in 1830 and tutored in mathematics at South Carolina College, 1831-35. He resumed medical studies in 1835 and received his M.D. degree in 1836. Subsequently, Gibbes journeyed abroad carrying botanical and conchological specimens which he exchanged with naturalists in France and Germany. He continued his medical studies in Paris and attended lectures by Jean Dumas, Pierre Dulong and others. Shortly after returning home in November 1837, Gibbes was appointed professor of mathematics and natural science at Charleston College, a post he held for 54 years.

He married a cousin, Ann Barnwell Gibbes on Sept. 21, 1848. She bore nine children and died in 1884 at age 61.

Although his primary teaching duties were in mathematics, Gibbes lectured in chemistry, physics, astronomy, mineralogy, natural history, philosophy, and political economy. He published about two dozen papers covering a range of scientific subjects. His achievements in astronomy include the invention of an "occulator" and a "portable heliotrope," calculations of asteroid and comet orbits, and observations of eclipses and a transit of Mercury. Gibbes discovered some post Pleiocene localities near Charleston and made the first extensive list of fossils in this formation. His contributions to biology are illustrated by his revision of lists of crustacea in American collections, a list of phenogamous plants of South Carolina and a report on crabs of the East Coast. He also worked with the U.S. Coast Survey on the determination of the difference in longitude between Washington, D.C. and Charleston.

His most significant work was the "Synoptical Table of the Chemical Elements." Gibbes' immediate aim was to construct a lecture room chart "to exhibit a synoptical view of the elements and their relations." He displayed the known families of elements in order of atomic weight on a series of horizontal lines numbered minus four to plus three to correspond to the principal valence of each group. Thus arrayed, the elements in adjoining families formed vertical columns in which Gibbes observed a "remarkable regularity in the succession of atomic weight values." With his scheme he was able to construct a table of elements which compared favorably with some of the early efforts of Mendeleev and Meyer. He placed 35 of the 63 known elements correctly and left gaps for several undiscovered elements. He distinguished between primary and secondary subgroups and observed that the elements carbon through fluorine were sufficiently different from other members of their respective families as to constitute a separate series. Although there is no explicit recognition of the Periodic Law in his paper he was aware of regularities in valence and electrochemical behavior as a function of atomic weight. The Synoptical Table was developed during the years 1870-74 at a time when Gibbes had little access to journals as a result of the economic difficulties of the Reconstruction. His paper was delivered to the Elliott Society in 1875 but was not published until 1886, long after it could have exercised a decisive influence on the development of the Periodic Table. Nevertheless, it was a remarkable document for its time and did satisfy his initial objectives.

Gibbes' outstanding characteristics were his wide range and depth of knowledge, his intellect, industry, habits of classification, and versatility. He was an admired and lucid teacher, a painstaking experimenter and keen observer. Gibbes was a founder of the Museum of Natural History in Charleston and of the Elliott Society of Natural History, which he served as president for 30 years. He retired in 1893 because of failing eyesight and died on Nov. 21, 1894 at the age of 84.

L. R. Gibbes, "Synoptical Table of the Chemical Elements" *Proc. Elliot Soc. of Charleston,* 77-90 (Oct. 1875); J. H. Easterby, "A History of the College of Charleston" (1935); J. H. Taylor, "Lewis Reeve Gibbes and the Classification of the Elements," *J. Chem. Educ.* **18,** 403-407 (1941); the library at College of Charleston has a Gibbes file containing a collection of his publications, personal notebooks, and newspaper and magazine articles concerning his work; communication from Edward Towell and Anne Moise.

JOHN E. FREY

Josiah Willard Gibbs

1839-1903

Josiah Willard Gibbs, the first great American mathematical physicist and ranked by many among the ten most influential physical scientists of the 18th and 19th centuries, was born in New Haven, Feb. 11, 1839. His father, Josiah Gibbs, was professor at Yale Divinity School. His mother was Mary Ann (Van Cleave) Gibbs, daughter of a trustee of Princeton University. Her fourth child and only son was named Josiah Willard, the middle name after Samuel Willard, who was acting president of Harvard College in 1701. After 4 years at Hopkins Grammar School in New Haven, Willard entered Yale and graduated in 1858. He won prizes not only in mathematics but also in Greek and Latin and continued his graduate studies at Yale under Benjamin Silliman, Jr. and Herbert A. Newton.

Two weeks before the outbreak of the Civil War Willard's father died, leaving an estate of $38,000. This enabled him to obtain his doctorate in 1863 and to accept a tutorship at Yale in Latin and later in natural philosophy. His eyes began to trouble him. He set to work to determine experimentally the formula for the glasses best suited for him. This was one of the few cases in which Gibbs resorted to actual physical experimentation.

In 1866 Gibbs went to Europe for further study and stimulation. He attended the lectures of the great figures in the world of physics and mathematics in Berlin, Paris, and Heidelberg. He returned to New Haven in 1869.

At about this time a committee at Yale recommended further stressing science education and formation of a chair in mathematical physics; in 1871 Gibbs was named to occupy this first chair of mathematical physics in America. That same year young Clerk Maxwell was appointed to a similar chair at Cambridge University, England. It marked the recognition of the need for more training in the sciences of heat and electricity by the new industrialists who were replacing the merchants and the ministers on the boards of trustees of our colleges.

For 32 years Gibbs held this professorship. He attended scientific meetings at which on rare occasions he lectured. He organized the Yale Mathematical Club and once, being an excellent horseman, read a paper before its members on "The Paces of a Horse."

He and his sister Anna lived with their sister Julia and her husband in their father's house almost across from Sloane Physics Laboratory where Gibbs had a small office. He lectured to a small group of perhaps a half dozen graduate students for about half an hour a day. He helped his sisters in their household duties, always insisting on mixing the salads on the ground that he was a greater authority on the equilibrium of heterogeneous substances. He attended church regularly on Sundays, had very simple tastes, was broad and tolerant, and showed none of the usual eccentricities expected of genius.

Gibbs worked out his own "Elements of Vector Analysis" between 1881 and 1884 in which he developed his "theory of dyadics."

For 14 years after the publication of his last paper on thermodynamics as applied to chemical equilibrium Gibbs did not concentrate in this field and did not even lecture on thermodynamics. Between 1882 and 1889 he published a number of papers in *American Journal of Science* on the electromagnetic nature of light.

In 1901 just about the time when the Royal Society of London was honoring Gibbs with the Copley Medal, Gibbs was busy on his last great work "The Elementary Principles of Statistical Mechanics" which he completed in 9 months.

Gibbs was never very strong, but a quiet, well-regulated life enabled him to carry on without serious illness up to within a very short time before his death, Apr. 28, 1903, at the age of 64.

Gibbs' most significant contribution was his 140-page paper "On the Equilibrium of Heterogeneous Substances," which was published in 1875, followed by a second paper of 181 pages published in 1877 and 1878 in *Transactions of the Connecticut Academy of Arts and Sciences.* Five pages of mathematical equations in his paper on Equilibrium of Mixtures and now known as the *phase rule* turned out to be of great practical value in dealing with certain industrial processes such as the commercial production of ammonia.

American science saw no genius in his highly abstract contributions, but James Clerk Maxwell, perhaps the greatest theoretical physicist of his day, was watching the work of Gibbs since his publication of his two papers in 1873: "Graphical Methods in the Thermodynamics of Fluids," and "A Method of Geometrical Representation of the Thermodynamic Properties of Substances by Means of Surfaces." But Maxwell died in 1879, and for 13 years the gem buried away in *Transactions of the Connecticut Academy* remained unknown. Then Wilhelm Ostwald in 1891 translated it into German and had it printed. It was not until 1906 that it was reprinted for the first time in English.

Henry A. Bunstead, Ralph G. Van Name, "The Collected Works of J. Willard Gibbs," Longmans, Green (1906; 2 ed., 1928); Lynde P. Wheeler, "Josiah Willard Gibbs," Yale Univ. Press (1951); Bernard Jaffe, "Men of Science in America," pp. 307-30, Simon & Schuster (revised ed., 1958).

BERNARD JAFFE

Oliver Wolcott Gibbs

1822-1908

Gibbs was born in New York City, Feb. 21, 1822. He dropped his first name early in life. His father, Colonel George Gibbs, for whom the mineral Gibbsite was named, was one of the earliest American mineralogists.

In 1837 Gibbs entered Columbia College. In his junior year at the age of 18, he published his first scientific paper, "Description of a New Form of Magneto-Electric Machine and an Account of a Carbon Battery of Considerable Energy." It was a description of a new form of galvanic battery in which carbon was used, probably for the first time, as the inactive plate. In those days American colleges were intensely classical in their aims, and science received a minimum of attention. Therefore, it was even more unusual for a student of 18 to make an original investigation than it would be today.

After receiving his bachelor's degree from Columbia in 1841, he went to Philadelphia where he served as an assistant in the laboratory of Robert Hare, the inventor of the compound blowpipe, who was then professor of chemistry in the Medical School of University of Pennsylvania.

Gibbs received his master of arts degree from Columbia in 1844 and a degree of doctor of medicine from College of Physicians and Surgeons in New York in 1845 but never practiced medicine. The study of chemistry was the main purpose of his life, and he studied medicine in order to qualify himself as a teacher of chemistry in medical schools.

He went to Europe in 1845 to study with Carl Ramelsburg and Heinrich Rose in Berlin and with Justus Liebig in Giessen. He attended the lectures of Auguste Laurent, Jean Dumas, and of Henri Regnault in Paris.

Returning to America in 1848, he became assistant professor at College of Physicians and Surgeons and also lectured at Delaware College in Newark, Del. In 1849 he was appointed professor of physics and chemistry at Free Academy of New York City, now City University of New York, where he remained for 14 years.

In 1850 Gibbs pointed out the interesting fact that compounds which change color when heated do so in the direction of the red end of the spectrum. In 1851 he became an associate editor and abstractor for *American Journal of Science*. Gibbs prepared a series of abstracts which brought the results of foreign investigations to the attention of American readers. He contributed 472 pages containing abstracts of 605 investigations in chemistry and physics. He held a similar position for American chemistry as correspondent to the German Chemical Society from 1869 to 1877.

The account of his thorough, systematic investigation of ammonia-cobalt compounds with Frederick A. Genth was published in December 1856. In this memoir they described 35 salts of the four bases roseocobalt, purpureocobalt, luteocobalt, and xanthocobalt, including adequate analyses. Later, Gibbs published papers alone in 1875 and 1876 in which he described many more ammonia-cobalt compounds. Among them were the salts of the new base, croceocobalt.

During the years 1861-64 he published researches on the platinum metals, relating mainly to analytical methods. In 1871 he published a brief note on the remarkably complex nitrites formed by iridium, and in 1881 he described a new base of osmium.

In 1863 Gibbs was elected to the Rumford Professorship of the Application of Science to the Useful Arts at Harvard University, a position that he held for 24 years. He inspired his students with a zeal for research and introduced them to laboratory methods of investigation. His students had the greatest admiration and affection for him.

Gibbs described his researches during the 1860's in a number of short papers, principally on chemical analysis. He developed new analytical methods and improved older ones. He introduced the electrical deposition of metals as a method of quantitative analysis. In 1867 in cooperation with Edward R. Taylor, he devised a glass and sand filter, which was the prototype of the devices invented by Charles Munroe and Frank Gooch, the former a student and the latter an assistant of Gibbs. In the same year he published a paper on atomicities or valences in which he developed the idea of residual affinities. In 1868 he discussed the constitution of uric acid and its derivatives, and in 1869 he described some products formed by the action of alkaline nitrites upon them.

He also produced several memoirs on optical subjects, including a map of the solar spectrum and another on the wavelengths of elementary spectral lines. He also studied interference phenomena.

In 1871 chemical instruction in Lawrence Scientific School was consolidated with that of Harvard, and Gibbs lost his teaching laboratory. He was relegated to the Physics Department and for the next 16 years he lectured on spectroscopy and thermodynamics. Fortunately he was wealthy enough to establish a small private laboratory in Cambridge and employ an assistant. In that laboratory he investigated complex inorganic acids formed by vanadium, tungsten, molybdenum, phosphorus, arsenic, and antimony. In his research it was necessary for him to work out difficult problems of analytical chemistry since the separation of many of the elements involved had never been attempted before. In all, he described complex salts belonging to more than 50 distinct series. The first paper on this subject appeared in 1877 and the last in 1896.

Gibbs invented the ring burner in 1873, his most important contribution to the apparatus of analytical chemistry. The same year he described his porous diaphragms for heating precipitates in gases.

In 1877 upon becoming professor emeritus, Gibbs retired to his home in Newport, R. I., where he continued his work on the complex acids in a laboratory which he built for that purpose. He was 71 years old when he published his work describing the separation of rare earths. In this paper he described a new method for determining atomic weights of rare earths, based on the analyses of their oxalates and oxides. He studied the physiological effect of isomeric organic compounds on animals with Hobart A. Hare in 1889 and later with Edward T. Reichert. He died in Newport on Dec. 9, 1908.

Gibbs was an American pioneer in research. Although primarily concerned with inorganic and analytical chemistry, he had a wide range of interests including theoretical, organic, and physiological chemistry, mineralogy, thermodynamics, and physics.

He was one of the founders of the National Academy of Sciences as well as its president from 1895 to 1900 and president of the American Association of the Advancement of Science in 1897. He was a member of many societies; most notably he was the only American to be elected an honorary member of the German Chemical Society. He was a member of the U.S. Sanitary Commission during the Civil War. The Union League Club of New York, devoted to the social organization of sentiments of loyalty to the union, was formed primarily because of his efforts. An extensive report on the instruments for physical research was prepared when he was commissioner to the Vienna Exposition in 1873.

Theodore W. Richards, "The Scientific Work of Morris Loeb," Harvard University Press, pp. 108-117 (1913); Edgar F. Smith, "Chemistry in America," D. Appleton & Co., pp. 264-271 (1914) with photograph; obituaries by Charles L. Jackson, *Amer. J. Sci.,* **27,** 253-259 (1909), T. W. Richards in *Science* n.s., **29,** 101-103 (1909), Edward W. Morley in *Proc. Amer. Phil. Soc.,* **49,** xix-xxxi (1910), and Frank W. Clarke in *Biog. Mem. Nat. Acad. Sci.,* **7,** 3-22 (1910) with bibliography and portrait.

ANN MARIE LADETTO

William John Gies

1872-1956

Gies, American biochemist, known particularly for his contributions to the chemistry of teeth and oral hygiene and to the reform of dental education, was born in Reisterstown, Md., Feb. 21, 1872. His father was a German immigrant and his mother the daughter of a Pennsylvania newspaper editor. After receiving his B.S. degree in 1893 from Gettysburg College, he earned his Ph.D. degree in biology in 1897 at Yale, where he also assisted Russell Chittenden in physiological chemistry. When, in 1898, Chittenden began his once-a-week professorial visits to the new Department of Physiological Chemistry at Columbia's College of Physicians and Surgeons, in New York, Gies was appointed instructor in that department.

His marriage May 24, 1899 to Mabel Loyetta Lark was followed by a summer in Bern, Switzerland, where he engaged in physiological research at the university, with Hugo Kronecker and Leon Asher. He spent the summers of 1901 and 1902 at Wood's Hole studying biochemistry with Jacques Loeb and Rodney True. Gies added to his teaching activities those of a consulting chemist to the New York Botanical Gardens, whose chemical laboratory he helped establish in 1902.

Chittenden resigned his professorship at Columbia in 1903. Gies, already an adjunct professor, was named acting head of the Department of Physiological Chemistry. He now seemed to expand in all directions. He was elected secretary of the Society for Experimental Biology and Medicine, a post he held until 1908. In 1904 he founded and became first editor of *Proceedings* of the society; he published his "Textbook of General Chemistry"; and he added to his teaching appointments a professorship in physiological chemistry at New York College of Pharmacy.

When, in 1904 Harvey W. McKnight resigned as president of Gettysburg College, Gies' candidacy was unsuccessfully promoted by the school's alumni. Although he retained his interest in the school's affairs, serving as trustee from 1908 to 1920, his career after 1904 was almost wholly identified with Columbia University. He was made professor in 1905, and appointed secretary of the faculty of College of Physicians and Surgeons. Gies' enormous capacity for work was not yet saturated. The first edition of his "Textbook of Organic Chemistry" appeared in 1905 while Gies was busily organizing a new section (K) of the American Association for the Advancement of Science. In 1906 he was elected secretary of the American Society of Biological Chemists, and in 1907 editor of the *Proceedings* of that society. The year 1906 also saw publication of his "Laboratory Work in Biological Chemistry."

Gies and his students investigated the biochemistry of collagen and mucoids, but the volume of their research diminished noticeably from 1908 to 1911.

In 1911 things had changed. Gies now was

editor of *Chemical Abstracts'* section on biological chemistry and had begun to edit the *Biological Bulletin.* But he had given up his other editorships. At the university he added a concurrent professorship in physiological chemistry at Teachers' College. His research publications increased markedly in number and diversity. He studied the role of proteolytic enzymes in edema, but he was also drawn increasingly to dental problems. He investigated saliva, the possible significance of the thiocyanate ion present in it, the composition of the organic components of teeth, and the role of bacteria in dental caries.

His interest in dental problems went beyond chemical research. Partly as a result of his efforts, Columbia University established its own Dental School in 1916. His publications between 1916 and 1919 included refutations of dentifrice advertising claims. He challenged the theory that pepsin in tooth powders could minimize tooth decay; he warned against the toxicity of potassium chlorate; he pointed out misstatements of fact on dentifrice labels.

In 1921 the Carnegie Foundation for the Advancement of Teaching asked him to head a study of dental education. This study, which resulted in fundamental reform in the training of dentists, virtually terminated Gies' activity in chemical research. While the study was in progress, his efforts led to the formation in 1923 of the American Association of Dental Schools. In 1926 Gies published his detailed and influential report, "Dental Education in the United States and Canada," calling for reconstruction of dental professional training along lines paralleling those of medical schools and asking for closer cooperation between the two professions.

"Biochemical Researches," an 8-volume series of his reprints which began to appear in 1903, terminated in 1927.

Gies retired from Columbia in 1937. Since 1926 his life had been marked by a long succession of awards and honorary degrees. Fellowships were founded in his honor by students and colleagues. Except for service on a New York State committee on fluoridation in 1944, he was virtually inactive professionally after 1940. He died May 20, 1956 of a cerebral hemorrhage, at his home in Lancaster, Pa. He was survived by his widow, two sons, and a daughter.

Private communication from Lillian H. Smoke, Gettysburg College; S. G. Hefelbower, "History of Gettysburg College, 1832-1932," by college (1932); *Alabama J. Med. Sci.* **2,** 337-41 (1965); J. Shrady, "The College of Physicians and Surgeons," **1,** 456-66, Lewis Pub. Co. (1903); obituary, *New York Times* May 21, 1956.

CHARLES H. FUCHSMAN

Harvey Nicholas Gilbert

1889-1971

Gilbert was born Aug. 22, 1889, in Chambersburg, Pa., where his father was a businessman, public official, and part-time inventor.

Educated in the public schools, Harvey entered Gettysburg College where he excelled in the physical sciences. A prize-winning research paper on a contemporary phenomenon, Haley's Comet, enabled him to purchase a microscope which was a life-long possession. He graduated in 1910 with a B.S. degree in chemistry, *summa cum laude.*

While he was an undergraduate, Gilbert worked summers as a mechanic's helper in the locomotive repair shops of Pennsylvania Railroad. After graduating from Gettysburg, he took his first job in chemical industry as an analyst trainee in the Du Pont Powder Co.'s Eastern Laboratory, 8 years after its founding at Gibbstown, N. J., by Charles Reese. He transferred later to the control laboratory of the adjacent Repauno Works, then the world's largest nitroglycerin and dynamite plant.

Convinced after nearly 1 year that he was professionally limited by lack of chemical training, Gilbert entered the Graduate School of Chemistry at Cornell University. Under an outstanding staff that included Wilder D. Bancroft, Arthur W. Browne, Emile M. Chamot, and Louis M. Dennis, he completed requirements for his Ph.D. degree in physical chemistry, particularly electrochemistry and colloid chemistry, in 1915.

He married Maude Fogle, a classmate at Gettysburg College, in Hazleton, Pa. in 1914.

He returned to chemical industry with the

A. D. Little Co., and acquired valuable experience in technical problem solving in field assignments around Boston.

Gilbert moved to Niagara Falls, N. Y., in 1916, accepting a job offer from Hector Carveth, then a recent Cornell Ph.D. and manager of the Niagara Electrochemical Co. Founded as a U.S. subsidiary of a German company, the firm, like many others in the region, based its manufacturing operations on the cheap, local hydroelectric power available for electrochemical and metallurgical processes.

For the next 36 years, Gilbert's professional career was continuously associated with processes for producing caustic soda, chlorine, sodium, sodium peroxide, hydrogen peroxide, sodium cyanide, and derived products. In the course of various assignments, including chief chemist in the plant and research manager in the laboratory, he was granted more than 165 domestic and foreign patents.

In 1930, he became reaffiliated with his original employer, the Du Pont Co., as a result of corporate acquisition of the Niagara Falls company, known by this time as the Roessler and Hasslacher Chemical Co.

In the early 1940's, Gilbert became interested in the chemical reaction of metallic sodium with hydrogen in molten caustic soda to produce sodium hydride, a powerful reducing agent. Searching for practical uses of the system, he developed a new chemical treating bath for removing scale from steel mill products such as wire, strip, and bar stock. After rapid reduction of the scale in the caustic bath, and quench, the clean metal stock was suitable for drawing and other fabrication. Demonstration of practical utility of the process led to mill installations and procedures that established a valuable technology in making specialty alloy steel products as well as a substantial new market for sodium.

Broadly interested in the physical sciences, Gilbert derived his greatest satisfaction from creative activity, seeking and producing workable technical solutions to practical problems. His formula for highest achievement combined an idea, sound testing in practice, and prompt adoption in use for improving an operation or a material. Typically, his frustration was greatest when scientific facts and simple economics alone were not enough to attain the acceptance of a new development.

In recognition of technical achievements in chemistry and metallurgy, Gilbert was awarded in 1946 the Schoelkopf medal of the American Chemical Society's Western New York Section. He received the Distinguished Alumni Certificate of Gettysburg College in 1965. He was a member of the American Chemical Society for 55 years and a member of the Electrochemical Society.

Gilbert retired from the Du Pont Co. in 1952, fully occupied with avocational interests including travel, old coins, Civil War history, and flower culture. He moved to Albany, N. Y., in 1969, where he died on Dec. 6, 1971. He was buried in Albany.

Personal recollections.

ARTHUR D. GILBERT

Charles Anthony Goessmann

1827-1910

Charles Anthony (Karl Anton) Goessmann exemplifies, along with many others, both German and American, the transit of chemistry from Germany to America during the nineteenth century. Trained in chemistry as a theoretical science under Friedrich Wöhler, he was able to make a practical application of it in the American context and thus bridged a wide geographical and professional gap.

Goessmann was born at Naumburg in Hesse Cassel, June 13, 1827, son of a physician, Heinrich, who had been a fellow-student and friend of Wöhler at Marburg, and his wife Helena Henslinger-Boediger Goessmann. He was apprenticed to a pharmacist and served several years as a pharmacist's assistant before entering the University of Göttingen in 1850. He could not decide whether to follow pharmacy or chemistry. Wöhler's influence prevailed, and Goessmann was awarded his Ph.D. in 1852. He remained at Göttingen until 1857, serving as an assistant in Wöhler's laboratory, where there were many American students. Organic chemistry was his principal interest, and some 20 papers were the fruit of his investigations during those years.

Although his prospects at Göttingen were good, Goessmann came to the United States in 1857 on the invitation of former American students, the Eastwick brothers, who operated a sugar refinery in Philadelphia. He became its chemist and general superintendent and dedicated himself to improving sugar production and studying its sources, not only cane but also sorghum and beets.

In 1861 Goessmann became chemist and superintendent of the Onondaga Salt Co., Syracuse, N.Y., and was especially interested in manufacturing pure dairy salt. Here too he combined a scholarly with a practical approach. In this salt phase of his career Goessmann was persuaded by a former fellow-student at Göttingen, Charles Chandler, to accept a professorship of chemistry and natural history at nearby Rensselaer Polytechnic Institute. He remained at Rensselaer from 1862 to 1864, and also retained his connection with the salt industry in Syracuse, where he married a local girl, Mary Anna Clara Kenny, and thus became fully Americanized.

In 1868 Goessmann entered the academic profession for good. Again it was because of the influence of a former Göttingen student, William S. Clark, now president of the newly established Massachusetts Agricultural College at Amherst. Goessmann became professor of chemistry, and he occupied this position for 40 years, until his retirement in 1907. He dedicated himself zealously to developing agricultural chemistry, not only as a teacher, but also as an investigator. Fertilizers were the first object of his study, and, when Massachusetts adopted the first legislation in 1873 for their regulation and commercial sale, Goessmann was named State Inspector of Fertilizers and a member of the State Board of Agriculture. He served in this capacity virtually to the end of his life in 1910.

When the new chemistry building was erected at Massachusetts Agricultural College (now University of Massachusetts) in 1921, it was named the Goessmann Laboratory.

As early as 1878 a private experiment station was established at Amherst, which obtained state and federal funds by 1882 and thus became a public institution. Goessmann was its first director and chemist, became honorary director in 1895, and retained his connection with it until retirement. His researches covered many topics, from various crops to use of fertilizers and feeding of animals. A stream of publications, in the form of reports, papers, and articles rather than books, poured from his laboratories, comprising a total of 362 items in the memorial volume published by the college in his honor in 1917.

Goessmann was elected president of the American Chemical Society in 1887. It was said of him that there was "not a better practical chemist in the United States." He was "a teacher in a wide sense," who taught and trained many agricultural chemists for other schools. But he was also an experimenter in the German tradition, who dedicated his skill and zeal to the "utility of science" in the American sense. He died Sept. 1, 1910.

"Charles Anthony Goessmann," Alumni of Massachusetts Agricultural College (1917) contains a full bibliography of 362 items; "Dictionary of American Biography," **7,** 354, Chas. Scribner's Sons (1931); obituary: *J. Amer. Chem. Soc. Proceedings,* December 1910.

SAMUEL REZNECK

Moses Gomberg

1866-1947

Moses Gomberg, one of the great organic chemists of his time, was born Feb. 8, 1866, in Elizabetgrad, Russia, son of George and Marie (Resnikoff) Gomberg, and died in Ann Arbor, Mich., Feb. 12, 1947, just 4 days after his 81st birthday. Almost nothing is known of his early life; he was so modest and self-effacing that even his closest associates in Ann Arbor knew nothing of his childhood. It is known only that in 1884 his father was accused of political conspiracy and, to escape imprisonment, had to flee Russia, sacrificing all of his property. The son, then 18 years of age, was also under suspicion and left Russia, but it is not known whether he came with his father, or followed soon after. The rest of the family probably came, also, although that is not certain. Moses' younger sister Sonia emigrated to America, and after her brother settled in Ann Arbor she kept house for him. Neither of them ever married.

In 1886, 2 years after his arrival in America, the young Gomberg enrolled as a freshman at University of Michigan, the institution with which he was to be connected for the rest of his life. He received his B.S. degree in 1890, M.S. in 1892, and Ph.D. in 1894. His thesis problem, which was concerned with the structure of caffeine, was done under the direction of Albert B. Prescott. He advanced from the rank of teaching assistant to that of instructor in 1893. Two years after he received the doctor's degree, he was granted a leave of absence, which enabled him to spend the 1896-97 school year in Germany—half of the time with Baeyer in Munich and half with Victor Meyer in Heidelberg. During his brief stay in Heidelberg, he succeeded in preparing tetraphenylmethane, which several well-known organic chemists had tried unsuccessfully to obtain. Victor Meyer, himself, was one of these, and he had become convinced that the material could not exist. He attempted, in vain, to dissuade Gomberg from spending his time on the project. Gomberg's yield of the tetraphenylmethane was very low (0.3g), so after he returned to Michigan, he repeated and extended the experiment to assure himself that his earlier results were correct. He then undertook the preparation of the next member of the series, hexaphenylethane. He attempted to effect this synthesis by treating triphenylchloromethane with fine silver powder (molecular silver)

$$2\emptyset_3CCl + 2Ag \rightarrow \emptyset_3CC\emptyset_3 + 2AgCl$$

Instead of the desired product, however, he obtained an oxygenated material which he identified as triphenylmethylperoxide, $\emptyset_3C$—O—O—$C\emptyset_3$. Upon repetition of the experiment in the complete absence of air, he obtained a highly reactive hydrocarbon which he felt must be hexaphenylethane. Its great reactivity convinced him that it was the free radical, triphenylmethyl, $\emptyset_3C$—. This postulate was quite contrary to the theories of the day, and although he stated in his first paper that he wished to reserve the field for himself, a number of chemists very quickly began experiments on the material, and advanced theories of its structure quite different from his. To refute their arguments, Gomberg was forced to perform a great variety of experiments, many of which were entirely novel and difficult to do with the equipment available at that time. Eventually, his ideas were universally accepted, and he became recognized as the discoverer of organic free radicals.

During the last few years of his career, Gomberg turned his attention from triphenylmethyl to the reducing system, $MgI_2 + Mg$. He believed this material was partially in the form of the free radical —MgI. This reduces ketones to benzoins through the formation of free radical intermediates, the organic ketyls.

$$\underset{\underset{O}{\|}}{\emptyset\text{—}C\text{—}\emptyset} + \text{—}MgI \rightarrow \underset{\underset{\text{ketyl}}{|}}{\emptyset_2C - OMgI} \rightarrow$$

$$\begin{matrix}\emptyset_2C - OMgI \\ | \\ \emptyset_2C - OMgI\end{matrix} \rightarrow \begin{matrix}\emptyset_2C - OH \\ | \\ \emptyset_2C - OH\end{matrix}$$

It is popular now to postulate "free radical mechanisms," but such ideas were revolutionary in 1925.

He also studied dyes, was first to synthesize tetraphenylethane, prepared benzyl ethers of carbohydrates, studied the $(ClO_4)_x$ radical, and synthesized biaryls by the diazo (Gomberg) reaction.

When Edward D. Campbell died in 1925, Gomberg reluctantly agreed to accept the chairmanship of the Chemistry Department at Michigan "until a new man could be found." As things developed, he was to retain that position until his retirement. It involved much of the kind of work which he did not enjoy, but he did it well.

Gomberg had hoped to spend his years of retirement in research and in travel, but his sister Sonia's health failed at about the time of his retirement, and he devoted the remaining years of his life to taking care of her.

The material in this sketch is abstracted from an article by this same author about Gomberg in "Biographical Memoirs of the National Academy of Sciences," Columbia University Press, Vol. XLI, pp. 141-165 (1970). There are biographies of Gomberg by C. S. Schoepfle and Werner Bachmann in *J. Am. Chem. Soc.* **69,** 2921 (1947), and an anonymous one in *Chem. Eng. News* **25,** 548 (1947).

JOHN C. BAILAR, JR.

Frank Austin Gooch

1852-1929

Gooch was born in Watertown, Mass., May 2, 1852 to Joshua G. and Sarah (Coolidge) Gooch. Both parents were descended from early colonial stock. His father, a lumber merchant and town assessor, took a considerable interest in science and encouraged his only son to set up his own laboratory.

He went to Harvard, intending to become a physician. However, he came under the tutelage of Josiah Parsons Cooke, professor of chemistry, whom he served as an assistant during his undergraduate years. He took his B.A. degree in 1872, and continued to assist Cooke for 3 years. In 1875 and 1876, he studied in Europe, spending most of his time at the Imperial Mineralogical Museum of the University of Vienna learning mineralogy and crystallography.

In 1877, he received his M.A. degree and his Ph.D. degree, the latter being the first doctoral degree in chemistry granted by Harvard. These degrees were granted in absentia since he had left to study thermochemistry with Julius Thomsen in Copenhagen. He soon lost interest in this topic and spent most of his time travelling through Europe, visiting the great chemical laboratories, visits which were to stand him in good stead when he was entrusted a few years later with designing a laboratory at Yale.

On his return in 1877, he was asked by Wolcott Gibbs, professor at the Lawrence Scientific School, to work with him. There was, however, no future for him at Harvard at that time, and he took a position as chemist for the Tenth United States Census in 1879. In this work, he made microscopic and chemical analyses of iron ores. In 1881, he became chief chemist of the Northern Transcontinental Survey for which he studied the uses and development of the natural resources of the Northern Pacific region. When Survey funds ran low in 1884, he went to work for the U.S. Geological Survey for which he analyzed water from Yellowstone Park springs.

In 1885, he was appointed professor of chemistry at Yale but remained with the Geological Survey until 1886. When he moved to New Haven, one of his first tasks was to help to plan Kent Chemical Laboratory. He was also instrumental in planning additions in 1902 and 1906. Gooch directed Kent Laboratory until he retired in 1918.

Besides directing the laboratory, he was chiefly responsible for organizing undergraduate and graduate chemistry courses, and providing opportunities and funds for research. In addition, he taught inorganic and analytical chemistry. He was a good lecturer, clear, well prepared, and his demonstrations usually worked. He gave much of his time to his graduate students because research was his first love. He was a natural director of research, intuitive, broadly knowledgeable, and kindly but demanding.

He devised the "Gooch Crucible," which he first described in *American Chemical Journal* in 1878, in a paper entitled "On a New Method for the Separation and Subsequent Treatment of Precipitates in Chemical Analysis." This filtering device saved vast amounts of time. It was not widely used at first, however, because most chemists were unable to find a proper grade of asbestos, and it did not become popular until the turn of the century.

In 1879, he published a paper in the *American Chemical Journal* "On the Estimation of Phosphoric Acid as Magnesic Pyrophosphate" in which he described how this method could be used to analyze phosphotungstates and phosphomolybdates. Other early work was with the quantitative separation of lithium from other alkali metals using amyl alcohol on the metallic chlorides and the estimation of boric acid with methyl alcohol distillation, followed by treatment with calcium oxide.

In his later research, he was particularly concerned with developing good methods for estimating molybdenum, vanadium, selenium, and tellurium; with the proper conditions for precipitating phosphates for weighing as magnesium or manganese pyrophosphate; and particularly in iodimetry, with accurately estimating elements and radicals through volumetric determination with iodine.

Gooch was one of the first to use the rapid

stirring method for the electrolytic determination of metals with relatively simple and inexpensive apparatus. His apparatus required less than a gram of expensive platinum. It was one of his delights to finish an elegant bit of research with more efficient and less expensive apparatus.

Throughout his life, he remained a devotee of the field of analytical chemistry which he believed could never be exhausted as long as chemists were intelligent, ingenious, and willing to work.

In addition to many papers which appeared in journals all over the world, he wrote three textbooks, one on qualitative analysis, one on quantitative analysis, and one on inorganic chemistry.

In 1880, Gooch married Sarah E. Wyman of West Cambridge, Mass. They had one child, a daughter. He retired in 1918 and died Aug. 12, 1929, in New Haven at the age of 78, after a lengthy illness.

Biography by Philip Browning, *Ind. Eng. Chem.* **15,** 1088-1089 (1923), port.; obituaries by Ralph G. Van Name in *Biog. Mem. Nat. Acad. Sci.* **25,** 105-135 (1931), port.; Philip Browning in *Amer. J. Sci.* (5S) **18,** 539-540 (1930); James F. Norris in *Proc. Amer. Acad. Arts Sci.* **70,** 541 (1935-1936).

LOUIS I. KUSLAN

Charles Goodyear

1800-1860

In 1839 Goodyear discovered the chemical reaction of rubber with sulfur at elevated temperatures known as vulcanization. At the time of Goodyear's discovery, the infant rubber industry was on the verge of collapse because products manufactured from crude rubber retained the faults of the rubber: they hardened in winter, softened and became sticky in summer, were attacked by solvents, and were smelly. Vulcanization, which eliminated these faults, became the basis for the entire rubber industry.

The son of Amasa Goodyear, inventor and hardware manufacturer, Goodyear was born in New Haven, Conn., Dec. 29, 1800. He was a member of the sixth generation of a family prominent in the founding of the Colony of New Haven and unique in history because it produced seven inventors over four generations. He attended school at Naugatuck until 1817 when he was apprenticed to a hardware manufacturer and importer in Philadelphia where he stayed until 1821. He then went to work for his father in New Haven and Naugatuck, and in 1826 they established in Philadelphia the first retail hardware store in the United States. Failure of this undertaking in 1830 because of overextended credit left Goodyear with debts that were to plague him and send him to debtors' prison several times in the coming years. Left with no resources to pursue the hardware field in which he was trained, he decided to make his living as an inventor. In the next 4 years, six patents on mechanical devices were issued to him.

Goodyear was early intrigued with rubber, or gum-elastic as he called it, which he had tried unsuccessfully to use in some of his mechanical inventions. In 1834 a rubber products salesman told him of the dire state of the rubber industry because of the instability of its finished products. Goodyear determined to find a way to improve rubber, for he looked on rubber as having certain properties, such as elasticity, plasticity, strength, durability, non-conduction of electricity, and resistance to water which could make it a very important industrial raw material. He had some apparent early success, such as rubber sheeting (containing magnesium and calcium oxides as additives to decrease stickiness). This sheeting, claimed by Goodyear to be the first ever made from rubber, won him a silver medal from the Mechanics Institute of New York in 1835. However, the aging properties of the rubber were not improved, and the sheeting became soft and sticky in summer.

In 1836 Goodyear found that sheets of rubber treated with nitric acid along with nitrates of bismuth and copper showed much improved properties. This process, known as the "acid gas process," was licensed and utilized for a number of years to make rubber sheeting for use in table covers, draperies, and other products. Goodyear himself, in 1837-38, made a quantity of sheet material by arrangement with the Roxbury (Mass.)

India Rubber Co. However, the "acid gas process" affected only the surface and was useless for thicker articles, a defect that led to the failure of the Roxbury company.

In 1838 Goodyear purchased the Eagle India Rubber Co. in Woburn, Mass. from Nathaniel Hayward and promptly hired him. It was Hayward who first introduced to Goodyear the use of sulfur as a useful additive to make rubber less tacky. At Goodyear's insistence, Hayward patented his additive and then sold his patent rights to Goodyear. In 1839, as one of his unceasing experiments, Goodyear put a piece of cloth coated with rubber compounded with lead oxide and sulfur on a hot stove. Although it was well known that crude rubber melts and becomes sticky when heated, to Goodyear's astonishment and delight the compounded rubber became leathery and was no longer tacky. Furthermore, it did not soften with heat nor become horny when chilled. He fully recognized the importance of his experiment although the several witnesses present were unimpressed; immediately he set out to apply his process, now known as vulcanization, to the manufacture of rubber articles. In this he persevered despite a period of extreme poverty, continuing ill health, and days spent in debtors' prison. By 1841 he was able to produce satisfactory sheet goods by his process, and by 1843 he was producing a variety of articles. He now had enough background to apply for a patent in 1843, issued June 14, 1844. He was once more sent to debtors' prison in 1843, and circumstances forced him to go into bankruptcy. However, profits from manufactured goods and license fees soon ended his poverty and enabled him to discharge all his debts.

Infringement of Goodyear's patent led to "The Great India Rubber Case," a suit brought in 1851 in Goodyear's name by some of his licensees against a manufacturer, Horace Day, one of the more flagrant infringers. Daniel Webster, then Secretary of State of the United States, served as chief attorney for the plaintiff, and his masterful closing argument to the jury brought complete victory to Goodyear in 1852.

In 1851 Goodyear displayed samples of his products at the "Great Exhibition" in London, England, and his exhibit was awarded the Council Medal, the highest honor. After the end of the Day suit in 1852, he returned to Europe where he spent 6 years promoting his process, mainly in England and France. He was denied a patent in England because of alleged prior art, and his French patent was invalidated by the French courts on a technicality. His large exhibit at the Exposition Universelle in Paris in 1855 was awarded a gold medal, and he himself was made an officer of the Legion of Honor by Napoleon III. Ironically, the symbolic cross was delivered to Goodyear in Clichy Prison where he was lodged because he was unable to produce money immediately to refund license fees after his French patent had been invalidated.

In 1858 Goodyear returned to New Haven, and in 1859 he moved to Washington. He died July 1, 1860 in New York, where his final illness had forced him to stop while on his way to New Haven to attend his daughter's funeral. Goodyear had married first Aug. 24, 1824 Clarissa Beecher, who died in 1853; they had nine children: Clarissa, Ellen, Cynthia, Sarah Beecher, Charles II, William, William Henry, Clara, and William Henry II. He married second May 30, 1854 Fanny Wardell, and they had three children: Alfred Wardell, Fanny, and Arthur. Of all these children Charles II had many patents on shoemaking machines and organized the Goodyear Welt Shoe Machinery Co.; William Henry, a noted archeologist, was a curator at The Metropolitan Museum of Art for a few years and then for 35 years was curator of fine arts for The Brooklyn Museum of Arts and Sciences.

The depth of Goodyear's knowledge about rubber and his great ingenuity regarding the use of rubber can be grasped from his two-volume treatise: Volume I (published in 1855): "Gum-Elastic and Its Varieties, with a Detailed Account of its Applications and Uses, and of The Discovery of Vulcanization"; Volume II (published 1853): "The Applications and Uses of Vulcanized Gum-Elastic; with Descriptions for Manufacturing Purposes." Here will be found what was known about rubber: occurrence, recovery, properties, and chemical constitution, and Goodyear's account of the discovery of vulcanization and his statement: ". . . this process is not simply the improvement of a substance; but it amounts, in fact, to the production of a new material . . ." which shows his full awareness of the nature of his discovery. Here will also be found descriptions of over 600 suggested products and processes using rubber vulcanization. Nearly every use of rubber today is to be found in "Gum-Elastic."

"A Centennial Volume of the Writings of Charles Goodyear and Thomas Hancock," The American Chemical Society (1939), containing reproductions of Goodyear's two volume "Gum-Elastic" and Hancock's "Personal Narrative of the Origin and Progress of Caoutchouc or India-Rubber Manufacture in England"; P. W. Barker, "Charles Goodyear," privately printed by Godfrey L. Cabot, Inc., Boston, Mass. (1940); Mary Batten, "Discovery by Chance," Funk & Wagnalls, pp. 137-156 (1968); A. P. Miller, Jr., "Those Inventive Americans," National Geographic Society, pp. 80-99 (1971); Ralph F. Wolf, "India Rubber Man," B. F. Goodrich Co. (1939).

GUIDO H. STEMPEL

Neil Elbridge Gordon

1886-1949

Gordon, founder of *Journal of Chemical Education* and the Gordon Research Conferences, was born in Spafford, N.Y., Nov. 7, 1886, the son of William J. and Ella C. (Mason) Gordon. He attended Homer Academy, Homer, N.Y. He received his Ph.B. degree in 1911, A.M. in 1912 and B.Ped. in 1922 from Syracuse University. He began his teaching career in Baltimore, attending Johns Hopkins University in his spare time. After receiving his Ph.D. from Johns Hopkins in 1917, he was assistant professor of inorganic chemistry at Goucher College until 1919. He then joined the faculty of University of Maryland as professor of physical chemistry, becoming director of the department in 1921. This position automatically made him chief chemist for the state of Maryland, carrying with it responsibility for control of the quality of feeds, fertilizers, and lime sold in the state.

In 1921 the Division of Chemical Education of the American Chemical Society was organized largely through the efforts of Gordon to provide a means for teachers of chemistry to meet and discuss their pedagogical problems. Shortly after this was accomplished, Gordon led in founding *Journal of Chemical Education* (1923) and edited the periodical until 1933. Gordon's preeminence as a leader in chemical education led Johns Hopkins to call him there as professor of the subject in 1928. This was the first and perhaps the only professorship of chemical education.

Gordon's belief that scientists would profit by knowing each other better and by having an opportunity to discuss their interests in a leisurely and informal manner amid pleasant surroundings led him to organize the first Gibson Island Conference in 1931 on Gibson Island in the Chesapeake Bay. The conferences were established on a permanent basis in 1938 under the auspices of the American Association for the Advancement of Science with Gordon, then secretary of the AAAS Section on Chemistry, as director. To honor Gordon, they were renamed Gordon Research Conferences in 1948.

In 1936 Gordon left Johns Hopkins and went to Central College, Mo. as chairman of the Chemistry Department. Through Gordon's efforts, and with the aid of benefactors, Central was enabled to purchase the great private library left by noted sugar chemist Samuel Hooker. The Chemical Foundation paid the expense of moving the library from Brooklyn, N.Y. to Central College, and Gordon organized and directed the Friends of the Hooker Library to help support it.

In 1942, Gordon became chairman of the Chemistry Department at Wayne University. He secured financial backing largely from the Kresge Foundation to purchase and move the Hooker library to Detroit where it is now the Kresge-Hooker Scientific Library of the Wayne State University Libraries. He instituted the doctoral program in chemistry at Wayne in 1945. Ill health forced him to resign the chairmanship of the Chemistry Department in 1947. He did, however, continue as professor of chemistry, director of the Friends of the Kresge-Hooker Scientific

Library, and editor of *Record of Chemical Progress*. His publications included three college chemistry textbooks, 26 research papers between 1921 and 1936, and editorship of *Journal of Chemical Education* and *Record of Chemical Progress*. His patents included a thermoregulator for constant temperature baths, and methods for preparing cadmium sulfide, and recovering potassium permanganate.

Gordon married Hazel A. Mothersell, June 29, 1915, and had two children, Fortuna Lucille and Neil Jr. He committed suicide May 30, 1949 after two years of ill health by jumping from the twelfth story of the hotel in which he was living.

Obituaries in *Record Chem. Progress* **10,** (4) ii (1949), and *Chem. Eng. News* **27,** 1742 (June 13, 1949); remarks by Gordon on founding *J. Chem. Educ.* in that journal **20,** 369 (1943); *Detroit Free Press,* May 31, 1949, p. 19.

DONALD R. HAYS

John Gorham

1783-1829

Gorham was born in Boston, Feb. 27, 1783. He graduated from Harvard College in 1801, then took up medicine and received his bachelor of medicine degree from Harvard Medical School in 1804. He went to Europe for 2 years and studied in Paris, London, and Edinburgh. While his main purpose was to learn more about medicine, he liked chemistry, and he attended lectures on the subject.

Returning to Boston, Gorham began to practice medicine. In 1808 he married the daughter of John Warren, M.D. Through Warren he became friendly with Aaron Dexter, M.D., Erving Professor of Chemistry at Harvard, and whose pupil Gorham had been. Dexter, busy with his medical practice and not having time (and perhaps the inclination) to keep abreast of rapidly advancing chemistry, offered to use his influence to have Gorham engaged as adjunct professor of chemistry if Gorham would concentrate on the science. Gorham agreed, and was appointed adjunct professor of chemistry and materia medica in 1809.

In 1816 Dexter resigned and Gorham succeeded him as Erving Professor of Chemistry. He taught chemistry in the college and Medical School until 1824 and in the Medical School from 1824 to 1827. He was succeeded by his student, John Wight Webster.

At the time of his appointment to the Harvard faculty and for many years later, Gorham spent a considerable portion of his time with chemistry. In 1810 and a few succeeding years he offered courses of public lectures on chemistry to which citizens of Boston were admitted for a fee. In 1819-20 he published a 2-volume text, "The Elements of Chemical Science." But like the majority of teachers of chemistry in medical schools, he earned the major portion of his income from his medical practice; and as his medical reputation grew, and as more and more patients sought his service, his remuneration from this source became larger and larger. Concurrently he spent a decreasing amount of time with chemistry.

He was one of the founders in 1812 of *New England Journal of Medicine and Science,* one of the editors during the 14 years of the journal's existence, author of medical articles, and undoubtedly a writer of many of the journal's unsigned notes and articles. During the same period he published only five articles on chemistry, none of them requiring great skill, chemical knowledge, or insight.

Finally Gorham had to choose between chemistry and medicine and he picked the latter. He resigned in 1827 because the Harvard authorities decided that Erving professors should reside in Cambridge; and Gorham, who had built a lucrative medical practice in Boston, was loath to move. A popular and successful physician, he contracted pneumonia and died in Boston, Mar. 27, 1829.

"Quinquennial Catalogue of . . . Harvard University" (1900); Harold C. Ernst, "The Harvard Medical School 1782-1906," (1906); Thomas F. Harrington, "The Harvard Medical School," Lewis Publishing Co. (1905), 3 vols., with portrait; I. Bernard Cohen, "Some Early Tools of American Science," Harvard University Press (1950); obituary in *Boston Med. Surg. J.* **2,** 107-109 (1829-30).

WYNDHAM D. MILES

Ross Aiken Gortner

1885-1942

Gortner was an early investigator in the emerging field of biochemistry, but perhaps his greatest influence was as an inspiring teacher in the association of organic and physical chemistry in biological systems. His greatest satisfactions were in his students and in passing on to them his contagious enthusiasm for the many facets of biochemistry.

Gortner was born on a homestead farm in Nebraska in 1885. His father was a circuit rider Methodist minister. Two years later, his brother and he were taken by their parents to Garraway Station, a missionary post in Liberia, Africa. When Ross Gortner was three, his father died there of "African fever." This event placed severe hardships on the family throughout the years ahead but also reinforced the desire expressed by his father to his mother to "give the boys an education."

Gortner was strongly influenced by several professors as he trained for his life's work in chemistry. At Nebraska Wesleyan University where he graduated in 1906, Frederick J. Alway crystallized his resolve to be a chemist.

At the University of Toronto, he received his M.A. degree in 1908 under W. Lash Miller, who brought the field of physical chemistry into Gortner's training and into his future career.

In 1909, a Ph.D. degree was awarded to Gortner for his work under Marston T. Bogert at Columbia University. Thus, an outstanding leader in organic chemistry made Gortner at home in this field.

At the Carnegie Station for Experimental Evolution at Cold Spring Harbor, Gortner became exposed to biological problems, which helped synthesize his earlier experiences into the field of biochemistry. During his years there from 1909 to 1914, he became associated with the eminent botanist, J. Arthur Harris. This close professional and personal friendship continued throughout their lives.

The University of Minnesota attracted Gortner to teaching in the Division of Soils in 1914 and 2 years later to the Division of Agricultural Biochemistry. A year later Gortner became professor and chief of the division, a position which he held until his death.

These 25 years of developing and leading an internationally known department was Gortner's greatest satisfaction. He loved his close association with the students, and he proved to be an inspiring teacher. He was not a narrow specialist. He had the enthusiasm of a true scientist in inquiring into the phenomena underlying the colloidal and chemical mechanisms controlling biological processes. His scientific contributions of more than 300 papers were often classics in the rapidly developing field of biochemistry. They encompassed the properties of proteins; the chemistry of wheat flour, of soils, and of wood; and particularly the colloidal properties and the role of water in biological processes.

Biochemistry developed rapidly during these years. Gortner, through his research, his writings, and his teachings, had an important leadership role in this growth. An outstanding contribution to this expanding field of science was Gortner's "Outlines of Biochemistry," a book which appeared in 1929 and which he revised in 1938. The book was widely known and used as a text for advanced students and as a reference work for teachers and researchers concerned with biological chemistry, with colloidal phenomena, and with agriculture.

Recognition and honors came to Gortner from many directions. He was elected to the National Academy of Sciences in 1935. Lawrence College conferred on him an honorary Sc.D. degree. Gortner served as national president of Phi Lambda Upsilon, 1921-26, the American Society of Naturalists in 1932, and the Society of Sigma Xi, 1941-42. He held key posts on scientific journals. He was invited to be Wisconsin Alumni Foundation Lecturer in 1930, Priestley Lecturer at Pennsylvania State College in 1934, and George Fisher Baker Lecturer at Cornell University, 1935-36. He received the Thomas Burr Osborne medal from the American Association of Cereal Chemists in 1942.

On Sept. 30, 1942, Gortner died of a heart attack in Minneapolis. He had just passed the twenty-fifth anniversary of his appointment as chief of the Division of Agricultural Biochemistry at Minnesota.

Biographies by L. S. Palmer in *Science* **96,** 395-97 (1942); C. A. Browne in *Scientific Monthly* **55,** 570-73 (1942), port.; S. C. Lind in *Biog. Men. Nat. Acad. Sci.* **23,** 149-180 (1944), port.; personal recollections.

WILLIS A. GORTNER

Eugene Ramiro Grasselli

1810-1882

Caesar Augustin Grasselli

1850-1927

Eugene Ramiro Grasselli emigrated to the U.S. in 1837. He was born in Strassbourg, France, Jan. 31, 1810, grew up and attended school there. His all-consuming interest, chemistry, he pursued, not in the laboratory among beakers and flasks, but in the plant in manufacture and production. It was only natural that chemistry should arouse such a keen interest in young Eugene; it had been the family trade for generations. Since the middle ages, the Grassellis had been druggists and chemists, makers of medicinals and perfumeries, gunpowder, and various chemicals. It was his father, Giovanni Angelo, who left the ancestral homestead in northern Italy, at Torno, to settle in Strassbourg, to set up his own chemist's shop.

Eugene R. studied chemistry and instrument making at University of Strassbourg, completing his education at University of Heidelberg. After serving an apprenticeship in his father's shop, he journeyed to America. Arriving in the United States, he decided to stay in Philadelphia to become acquainted with the new land, its people, their customs and their manner of doing business. He took a job for 2 years, another sort of apprenticeship. During this time his plans matured. Of the many communities he chose Cincinnati, Ohio, to establish himself and his business. Cincinnati, the "Queen of the Prairie," was already quite a manufacturing center and a ready-made market for his chemicals. Sulfuric acid, using the lead chamber process, was his first product, followed by soda ash, artificial alkali. Other chemicals followed. His venture was marked with success, not only because of a growing market, but mainly because of his knowledge of chemistry and instrument making. Improving his process, equipment, and control, he was able to achieve a better, more uniform product with fewer impurities. He met competition by increasing productivity and reducing cost of manufacture.

Eugene's son, Caesar Augustin, was a man cut from the same cloth as his father. He had ambition and determination, boundless energy, and was a tireless worker. Caesar was born in Cincinnati, Nov. 7, 1850, fifth in a family of nine children. A child still, he was interested in everything his father was doing; chemistry fascinated him. By the time he was 15 he quit school to work for his father full time. To complete his education he studied chemistry with a private tutor, a professor from the University of Karlsruhe. Reminiscing in his later years, Caesar recalled that he could not remember the time that he was not interested in chemistry and did not expect to follow his father in business.

The elder Grasselli put his son through a thorough apprenticeship, technical and commercial, nothing was left out, no matter how menial. It served him in good stead. It was the time when Civil War raged throughout the country. The business got into serious difficulties, particularly at the end of the war when an economic crisis developed. There were some very lean years, but the son stuck close to the father, never losing faith for a moment. The war brought about a difficult situation for Cincinnati. Shipments of goods in and out of that city became erratic. In 1865 Eugene decided to build a new plant in Cleveland for several reasons. The developing oil and steel industry had opened up a new market. Petroleum refining and steel treating had created a demand for sulfuric acid. Shipments of raw materials were received with more assurance and finished chemicals could be delivered with more certainty.

The building of the new plant, to plans by Eugene Grasselli, was put into the hands of Daniel Bailey, a mechanical engineer who had married Grasselli's daughter Lucretia. Caesar Augustin was sent along as an assistant. Shortly thereafter he took over the leadership in building the plant and beginning production. By 1867 everything was running smoothly; the first shipments rolled out of the door. Company headquarters were established at the new location. Eugene Grasselli moved his family to Cleveland. Within the next few years his son, Caesar Augustin, showed such adeptness and business acumen that by 1873 his father admitted him into the business as a full partner. The firm's name became E. Grasselli and Son. Not long before on his 21st birthday he married Johanna Ireland, a school mate of his sisters, the daughter of a Cincinnati merchant. In addition to Lucretia, mentioned above, Caesar had four other sisters, and two of them, Frances and Ida, married Irelands; Mamie married a Glidden, Victorine a Sprankle. Caesar's two brothers were Albert and Eugene. On June 31, 1882, when the father died suddenly, the full responsibility of the business fell on the son. Caesar was barely 30 years old but already a seasoned businessman and executive.

In 1885 the Grasselli Chemical Co. was incorporated with a capital of $600,000. Caesar Augustin was elected president. During the next 30 years of his administration and leadership the company grew; its assets expanded to over $30 million. He subsequently became chairman of the board; his son Thomas succeeded him in the presidency.

These 30 years were the most intense period in the American industrial revolution. As new opportunities arose and new markets developed, he turned to new fields of production. Caesar Augustin was an aggressive, resourceful executive at a time when as a chief, he personally had to be ready to meet the complex problems of production, selling, transportation, and finance. He was a remarkable man, perceptive, with an insight to what the situation demanded and an ability to deal with people.

The Grasselli Chemical Co. grew by leaps and bounds, to some extent by acquisitions, but mainly because it set up production centers near major markets to save on cost of transportation, to gain a competitive edge. The Grassellis pioneered in the use of large tank cars for shipments of acid.

Another innovation derived from Caesar Augustin's leadership was the use of technical literature in support of the products the company manufactured. Previously it had been the practice to quote testimonials in support of sales. An early example of Grasselli technical literature was: "Curing Concrete Roads, Streets and Bases with Grasselli Silicate of Soda." Of somewhat later date was: "$ZnCl_2$, The Standard Salt Wood Preservative." These pamphlets gave technical and test data, methods of preparation, and handling. There are indications that there existed some form of technical service in support of sales.

Caesar Augustin's interests and activities ranged beyond the chemical industry and business. He participated actively in civic and community affairs. He initiated the reorganization of the Chamber of Commerce in Cleveland. He played a prominent role in the William McKinley campaign for president and the enactment of a tariff to enable the chemical industry to manufacture an increasing number of products to reduce dependence on imports.

A devout Catholic and active in a number of lay organizations, he gave much of his time and money in support of many projects. The crowning achievement was the establishment of a center for physically handicapped children. For the latter he donated the old Grasselli family home, a stately mansion. In later years, for his activities among the Catholic laity he received a decoration from Pope Pius XI, who made him a commander of the Order of St. Gregory the Great. King Victor Emanuel III, who previously had made him a Knight of the Order of the Golden Crown of Italy, raised him to be a commander of the same Order.

He helped found two banking and trust institutions in Cleveland. He became a patron and trustee of the Western Reserve Historical Society and an ardent supporter of the Museum of Natural History, Museum of Fine Arts, and others. He also received an honorary degree of Doctor of Science from Mount St. Mary's College at Emmitsburg, Md., for his contributions to the development of the chemical industry and for his keen support of science. He attended Mount St. Mary's College in his early youth.

Typical of a man of his stature and verve, Caesar Augustin Grasselli was active to the very end of his life. He went to his office at the Guardian Trust Building in Cleveland, every day. He died at the age of 76 on July 23, 1927, as a result of heart disease. He was a friendly, outgoing person; a man sensitive to others, with integrity, motivated by a keen sense of honor in all his relations.

Williams Haynes, "Chemical Pioneers," D. Van Nostrand Co. (1939); "Dictionary of American Biography," Charles Scribners Sons, vol. 2 (1931); obituary in *Cleveland Plain Dealer,* July 27, 1927; *Trans. Western Reserve Historical Soc.,* Annual Report No. 110 (1929).

MILLE STAND

Jacob Green

1790-1841

Green, born in Philadelphia, July 26, 1790, attended University of Pennsylvania, started to study medicine with a physician of Philadelphia, found he was too tenderhearted for surgery, opened a bookstore in Albany, N. Y., was not particularly successful because he read books when he should have been selling them, took up law, and was admitted to the New York bar.

Around 1816 Jacob left Albany for Princeton, N. J., where his father was president of the university. Following his father and grandfather, prominent Presbyterian clergymen, he began to study theology.

Green's hobby from childhood had been science. He and a friend had published "An Epitome of Electricity & Galvanism" when he was only 19. At Princeton the teacher of natural philosophy hired him as an assistant. This swung Green away from religion and back to science. The trustees established a professorship of experimental philosophy, natural history, and chemistry in 1818 and elected Green to the post. He held this job until the university abolished the professorship in 1822.

Green moved to Philadelphia and gave popular lectures on chemistry during the winter of 1822-23. He joined several physicians in founding Jefferson Medical College in 1825 and was professor of chemistry there the remainder of his life.

During several summers Green traveled to western Pennsylvania to lecture at Washington and Jefferson College, and one summer to Easton to teach at Lafayette. A student at Jefferson Medical College remembered him as "a simple-minded man, not deeply versed in the science which he professed but an agreeable and instructive lecturer with a good deal of sophomoric flourish and a mild gentlemanly address."

In 1828 Green visited Europe and met Faraday, Gay-Lussac, and other scientists. Letters he wrote home were published as a book, "Notes of a Traveller, during a tour through England, France, and Switzerland." He was disappointed in not seeing Sir Humphry Davy, whom he admired. In 1830 he brought out an edition of Davy's last book, "Consolations in Travel."

Green was a conscientious teacher. For his students he wrote "A Text-Book of Chemical Philosophy," "Syllabus of a Course of Chemistry," and "Chemical Diagrams." Three of his articles, "On Some Chemical Arts known to the Aborigines of North America," "On the Metals Known to the Aborigines of N. America," and "Some Remarks on the Pottery used by the Aborigines of North America," (*American Journal of Pharmacy,* 1834), point to him as a pioneer investigator of archaeological chemistry in the United States. However, Green's chief interest lay in natural history rather than in chemistry. One has the feeling he would have preferred to concentrate on botany or biology if he could have earned a living in those fields. He wrote a monograph on trilobites, a small volume on plants of New York State, and articles on amphibians, salamanders, snails, and shells. His investigations added, in a small way, to the early knowledge of natural history of the United States.

Jacob Green died of heart disease in Philadelphia, Feb. 1, 1841.

The primary source of information on Jacob

Green is a sketch by his father Ashbel Green in James F. Gayley, "History of the Jefferson Medical College of Philadelphia" (1858), pp. 31-34, port.; Edgar F. Smith, "Jacob Green, Chemist, 1790-1841" (1923), a pamphlet, abridged in *J. Chem. Educ.* **20,** 418-427 (1943); "Dictionary of American Biography"; G. W. Bennett, "Old Jakey Green at Canonsburg," *Proc. Pa. Acad. Sci.* **23,** 218-221 (1949); the quote is from "Autobiography of Samuel D. Gross" (1887), vol. 1, p. 36; Gross spoke of Green as "an old bachelor." This statement has been copied by other writers. Actually Green married late in life and had two children.

WYNDHAM D. MILES

Traill Green

1813-1897

Green was one of the many people with a medical education, who became prominent in science and in science education in the United States in the 19th century. He was born in Easton, Pa. May 25, 1813. His father, Benjamin Green, had moved to Easton in 1793. His maternal grandfather, Robert Traill, had taken an active part in the Revolution and held many prominent positions, including that of associate judge.

Green received his secondary education in Union Academy and in Minerva Academy, both in Easton. He became interested in science in his youth, mainly through reading Buffon's "Natural History." He entered University of Pennsylvania where he studied under John K. Mitchell, a physician and chemist, and received his M.D. degree in 1835. After graduation he served a year at Fifth Street Dispensary in Philadelphia, where he acquired the habit of keeping complete records of all the cases he treated, a practice which he continued throughout his medical career.

He returned to Easton in 1836 and opened his medical office. During his studies at the university he had developed an interest in chemistry and he decided to teach it. For this purpose he gathered at his office a group of interested young people. Very soon he was invited to teach at Lafayette College in the same city and in 1837 was appointed professor of chemistry. The same year he became a trustee of the college.

In a conflict between the board of trustees and the faculty, although he was a member of the board, Green sided with the faculty. As result of this disagreement he left Lafayette in 1841. From 1841 to 1848 he was professor of natural science at Marshall College, Mercersburg, Pa. While teaching at Marshall he did not practice medicine.

Green returned to Easton in 1848. In 1849 he was reappointed professor of chemistry at Lafayette. This position with some changes in title he occupied till he became professor emeritus in 1891. For 2 years from 1875 to 1877 he was also professor of medical jurisprudence. From 1881 to the year of his death he was dean of the Pardee Scientific Department, and for 1 year (1890-1891) he was acting president of the college. It can be safely said that Green devoted the later part of his life primarily to Lafayette College. During his stay at Lafayette the college went through two crises. The first was during the Civil War, when the student body dwindled, and the faculty and administration decreased to nine. At this time Green worked without salary. The second crisis was the smallpox epidemic in 1871-72, when the students panicked and left the college, despite Green's efforts to reassure them that the danger in Easton was no greater than in their home towns.

Green was not only a teacher and an administrator, he was also a benefactor of the college. His largest gift was $15,000 for the observatory, which was nicknamed "Star Barn."

Parallel to his teaching at Lafayette Green went on with his medical practice and became rather prominent in his profession. Believing that the medical profession is a benevolent one, he was very willing to give assistance without monetary return. The list of his medical publications is extensive. He died Apr. 29, 1897.

Green was a very religious man. His interests were broad. Among other things he was involved in the founding of Easton Cemetery, at the gate of which his statue still stands. He was one of the pioneers in medical education for American women. He was instrumental in introducing gas lights on the streets of Easton. He belonged to many societies, including the American Bible Society. He was one of the founders of the American Association for the Advancement of Science, the American Academy of Medicine, and the Northampton County Medical Society. He was the first president of the two latter organizations. In 1841 he received an honorary M.A. degree from Rutgers, and in 1866 an LL.D. degree from Washington and Jefferson College. A school in Easton bears Green's name.

David B. Skillman, "The Biography of a College," Lafayette College, 1932; biography by George C. Laub in *Lafayette Alumnus* **15,** 12-13 (May 1945); obituary in *Proc. Med. Soc. Northampton County,* June 18, 1897, reprinted from *Lehigh Valley Med. Mag.*

GEORGE SIEMIENCOW

William Houston Greene

1853-1918

Greene was born Dec. 30, 1853, in Columbia, Pa. He graduated from Central High School, Philadelphia, in June 1870 and from Jefferson Medical College (now Thomas Jefferson University) as doctor of medicine in 1873. At Jefferson he was assistant to B. Howard Rand, professor of chemistry and, 2 years later, demonstrator of chemistry.

In 1877 he left for Paris, where he was a pupil under Charles Adolphe Wurtz, devoting 2 years there to research in organic chemistry in the company of Friedel and other chemists of distinction.

Upon his return to Philadelphia he served as demonstrator of chemistry in the School of Medicine at the University of Pennsylvania for 1 year, then became professor of chemistry at Central High School, succeeding his close friend Elihu Thomson in 1880. From that year until 1892 his work as a teacher was marked by distinction. He was an original thinker in developing successful methods of instruction in lecture demonstration and laboratory practice. Based upon the ground work of his predecessor, Thomson, who states that he had "made special laboratory developments, introducing and fitting out special work tables, one table for each student, equipped with reagents, etc. . . . rather a novel proceeding in those early days . . .," Greene is said to have been the first to equip a high school laboratory with desks having water and gas lines and reagents for each student, and he drew up a complete course of chemical experiments. Central High School thus pioneered as a secondary school in which laboratory work was carried on by students as a part of regular instruction in chemistry.

In addition to being a successful teacher, beloved by students and esteemed by colleagues, he carried on chemical researches during his 12 years on the faculty, publishing about 45 papers, mostly on organic chemistry, some being joint publications with colleagues. During that period he translated Wurtz's "Elementary Lessons in Modern Chemistry" and wrote "Lessons in Chemistry," both of which passed through many editions. He also wrote "A Practical Handbook of Medical Chemistry Applied to Clinical Research and the Detection of Poisons" and was American editor of "First Steps in Scientific Knowledge" by Paul Bert. He found time to serve as a consultant in medical and industrial chemistry.

Greene and Thomson went to great lengths to expose psychic frauds prevalent during the late 19th century. They attended all of the spiritual séances of John W. Keely, the leading psychic of that time, who displayed a mysterious motor supposed to be driven by spirit forces, and they denounced Keely as a fraud. Together the two friends listened patiently to many spiritualist mediums and were ejected from every séance in the city.

In 1892 Greene abandoned chemistry and entered his father's printing firm. Thomson once said of him that had he not left the field of science he would have achieved very high distinction in chemistry for he had "all of the qualities of mind for a great scientific career."

He died in Wenonah, N. J., Aug. 8, 1918.

Obituaries in *J. Franklin Inst.* **186,** 387-392 (1918), port., and by P. Maas and A. Henwood in *Ind. Eng. Chem.* **16,** 529 (1924); Harold J. Abrahams and Marion B. Savin, "Selections from the Scientific Correspondence of Elihu Thomson" (1971).

HAROLD J. ABRAHAMS

Roger Castle Griffin

1883-1956

Griffin was born at Montclair, N. J., on Sept. 11, 1883. His father, Roger B. Griffin, was superintendent of the first sulfite paper mill in the United States. In 1886 the elder Griffin and his partner, Arthur D. Little, founded the firm of Griffin and Little in Boston, Mass., to provide technical services to the pulp and paper industry. The concern later became Arthur D. Little, Inc., following the death of Griffin as the result of a laboratory accident in 1893.

The younger Griffin was educated at Harvard, B.S. 1904, A.M. 1905, M.S. 1906. After working briefly at Mallinckrodt, 1908, he joined Arthur D. Little, Inc., in 1909 and was chief analyst, 1911-17, laboratory director from 1917, assistant treasurer, 1920-35, treasurer and assistant secretary from 1935; he was vice president when he retired in 1949. He died at Needham, Mass., May 27, 1956.

Griffin is best known for his book, "Technical Methods of Analysis as Employed in the Laboratories of Arthur D. Little, Inc.," first published in 1921. This work included details of 340 procedures and was the first such treatise to include specifications of industrial materials as a guide to interpreting analytical results. Over 100 pages were devoted to wood and wood products alone. Griffin also was the author of 20 articles in technical journals.

Throughout his professional career Griffin was active in the affairs of the Technical Association of the Pulp and Paper Industry (TAPPI) and the American Society for Testing Materials (ASTM). He was the first chairman of the ASTM committee on Paper and Paper Products and served on many committees of both organizations. In 1953 he was awarded the Gold Medal by TAPPI, of which he had been a charter member.

Griffin's associates recall him as a very careful worker, who was always most appreciative of help given by his coworkers. After business hours he was a valued friend and social companion.

Tappi **36** (1), 71A (1953); *ibid.* **39** (8), 136A (1956) (with portrait); "American Men of Science," 6th Edition, Science Press (1938).

EDWARD R. ATKINSON

John Griscom

1774-1852

Griscom, born at Hancock's Bridge, N.J., Sept. 27, 1774, educated in country schools, began to teach in a log cabin near Salem when he was 17. Teaching in Burlington at the age of 20 he learned physics from books but found chemistry difficult until he borrowed a copy of Lavoisier's "Elements." He then "understood everything clearly" and discovered that chemistry was "the most interesting pursuit [he] had ever engaged in." He sent to Great Britain for apparatus to experiment and journeyed to Philadelphia to hear James Woodhouse lecture. He converted a room in his house into a laboratory and here "the more advanced pupils were taught chemistry, perhaps for the first time in any of the common schools in that part of the United States."

Shortly thereafter Griscom moved to New York City. He presented a course of public lectures on chemistry and erected a school building equipped with imported apparatus where chemistry was one of the subjects. Griscom's school was the only institution in the town, other than Columbia University and College of Physicians and Surgeons, where chemistry was taught in the early 19th century.

He operated his private school from 1808 to 1818, founded the Monitorial High School, taught there from 1825 to 1831, and was professor of chemistry and natural philosophy at Rutgers Medical School in New York City from 1811 to 1816 and again from 1826 to 1830. During this time he delivered courses of public lectures on chemistry and other sciences to large audiences, using "probably, the most extensive and costly apparatus . . . then owned in this country." A colleague, paying tribute to Griscom's ability as a teacher of chemistry, said that "for 30 years Dr. Griscom was the acknowledged head of all other teachers among us."

In 1818, overworked, he sailed across the Atlantic for a vacation. He described his travels in a popular book, "A Year in Europe," in which he gave accounts of prominent scientists whom he met.

In 1832 Griscom moved to Providence,

R.I., and superintended a boarding school where he taught chemistry and other sciences for 2 years. Thereafter he moved a number of times, occasionally giving courses of public lectures. As late as 1843, when he was 69 years of age, he lectured on chemistry in Salem, N.J. He died at Burlington, N.J., Feb. 27, 1852.

Griscom prepared abstracts and translations for *American Journal of Science* and *Journal of The Franklin Institute.* He edited an American edition of C. Irving's text, "A Catechism of Practical Chemistry," in 1829. I have found only one article by him, an analysis of mineral waters of New York State in the *American Mineralogical Journal,* 1814. He was not a practicing chemist; he was the most influential of our early teachers of elementary chemistry.

One of John's children, John Haskins Griscom, M.D. (1809-1874), followed in his father's footsteps for a time. He delivered public lectures on chemistry in New York City, taught chemistry at New York College of Pharmacy from 1835 to 1842, but found medicine more congenial and became a leader in public health.

"Memoir of John Griscom, LL.D., Late Professor of Chemistry and Natural Philosophy," R. Charter & Bros. (1859), with portrait, compiled by John H. Griscom from his father's autobiography; quotes are from this book; "John Griscom," Barnard's *Am. J. Educ.* **8,** 325-347 (1860), based mainly on Griscom's "Memoir," has the quote "for 30 years"; Edgar F. Smith, "John Griscom" (1925), a pamphlet, with portrait, abridged in *J. Chem. Educ.* **20,** 211-218 (1943); F. B. Dains, "John Griscom and his Impression of Foreign Chemists in 1818-1819," *J. Chem. Educ.* **8,** 1288-1310 (1931).
For John Haskins Griscom see sketch by J. W. Francis, *Med. Surg. Reporter* (New York) **15,** 118-122 (1866); Curt P. Wimmer, "College of Pharmacy of the City of New York" (1929), pp. 41-45; "Appleton's Cyclopedia of American Biography," D. Appleton & Co., **3,** 2 (1888).

WYNDHAM D. MILES

Samuel Guthrie, Jr.

1782-1848

Among early nineteenth-century American chemists Guthrie was well known as a manufacturer of various forms of gunpowder and of fermentation products, but today he is best remembered as the first to prepare chloroform. The significance of the latter substance was recognized before his death, but it was not until 100 years after the discovery that historians gave him appropriate recognition.

Guthrie was born in Brimfield, Mass. in 1782, the son of a physician who fought in the Revolutionary War. He began to study medicine under his father at the age of 18 and to practice at Smyrna, N. Y. 2 years later. He acquired formal training at College of Physicians and Surgeons in New York City and at University of Pennsylvania. By 1817 he had settled in the small town of Sacket's Harbor located at the eastern end of Lake Ontario and a few miles north of Watertown, N. Y. During the War of 1812 the town had been the scene of a sea-land engagement at which Guthrie had served as a surgeon for the United States forces.

Guthrie soon developed a flourishing manufacturing business. By 1835 he was producing each year 100,000 gallons of alcohol and 120,000 gallons of vinegar. During this period he developed procedures for preparing sugar and molasses from potato starch and for commercially producing potassium chlorate and mercury fulminate, both of which were related to his development of a very popular percussion primer that displaced the then common flintlock device for firing rifles. His manufacturing operations were accompanied by many accidental explosions, which so disfigured him that he forbade the preparation of a portrait by the new Daguerrotype process. Probably as a result of his many injuries he became an invalid during his later years and died at Sacket's Harbor Oct. 19, 1848.

The record of his discovery of chloroform is contained in Benjamin Silliman's *American Journal of Science and Arts.* In 1830-31 Guthrie treated whiskey with chlorinated lime and obtained a distillate containing alcohol and "chloric ether" (chloroform). A formal description of the work appeared in Silliman's journal in July 1831. Independent discoveries of chloroform were published by Liebig in Germany (1832) and Soubeiran in

France (1831). Although Guthrie soon became aware of the anesthetic properties of chloroform, he did not anticipate its first use in surgical anesthesia by James Simpson in Edinburgh (1847). Following its use by Queen Victoria during the birth of her seventh child (1853) chloroform became a popular anesthetic, and 1.5 million pounds of it were used during battlefield surgery in the Civil War. Guthrie was a prolific correspondent, and extensive excerpts from his writings are contained in the following sources and references cited therein: Tenney L. Davis, *Archeion,* **13,** 11 (1931); F. H. Getman, *J. Chem. Educ.* **17,** 253-259 (1940); J. R. Pawling, "Samuel Guthrie: Discoverer of Chloroform," Brewster Press, Watertown, N. Y. (1947).

EDWARD R. ATKINSON

H

William Jay Hale

1876-1955

Chemist, teacher, innovator, chemurgist, man of the world, raconteur—and always a student—such was Billy Hale. He was born Jan. 5, 1876, in Ada, Ohio to Rev. James Thomas Hale, the clergyman of a local church, and, as a youth, he worked as a janitor in his father's church.

In 1897, he left Miami University in Oxford, Ohio with his A.B. and A.M. degrees and entered Harvard University where in 1902 he was awarded his Ph.D. degree in chemistry and a traveling fellowhip in Europe. He pursued advanced studies in organic chemistry and in languages at Göttingen and at Heidelberg.

Hale became a chemistry instructor at University of Michigan in 1904, an assistant professor in 1908, and an associate professor in 1915. In January 1917, he left the university to join The Dow Chemical Co. in Midland, Mich., where he became director of organic chemistry research. He founded the Dow Co.'s extensive science library.

Hale's research at Dow contributed to improving processes for indigo, chloracetic acid, and phenylethyl alcohol, and to developing Dow's unique process for producing phenol from chlorobenzene. Some 45 patents were granted him for work in Dow's research.

In 1917, he married his former student, Helen Dow, the eldest daughter of Herbert H. Dow, founder and president of Dow Chemical Co. The Hales had one daughter, Ruth. Mrs. Hale died in 1918, shortly after Ruth was born. Hale never married again.

Finding industrial research a bit too confining for his tremendous curiosity about what was going on in world science, Hale assumed the role of research consultant in 1934 and devoted more time to writing, traveling, lecturing, and in associations with scientists in a variety of endeavors. His earlier writings included a number of college textbooks. Later he became deeply interested in agriculture and its problems and in conservation. He proposed that man learn to live chiefly on what the earth could produce currently rather than using up what nature had stored over the centuries.

Hale was deeply interested and involved in many of the early attempts to convert farm products into ethyl alcohol at costs which would permit its use in gasoline to produce a clean and high octane gasoline.

Hale was the father of "chemurgy" (a word coined from the Greek for chemistry and work and meaning chemistry at work) and in 1935 he founded the National Farm Chemurgic Council. Along with Hale in this endeavor were Henry Ford, Francis P. Garvan, Charles H. Herty, Wheeler McMillen, and other men of vision. The Farm Chemurgic Council and the interest it developed throughout the country were largely responsible for the establishment by the U. S. Department of Agriculture of the four Regional Agricultural Research Laboratories to develop industrial uses for farm crops. These laboratories in New Orleans, San Francisco, Peoria, and Philadelphia are testimonials to the devoted work of many earnest scientists of which Hale was an articulate leader.

At the fourth conference of the Farm Chemurgic Council at Omaha, Hale was presented the "Pioneer's Cup" for his "distinguished service to the American people through the farm chemurgic movement." He wrote several books on chemurgy. In "Farmward March" published in 1939 he made one of his strongest cases for chemurgy. He stated: "National prosperity is dependent upon a full agriculture. National prosperity, therefore, is dependent upon ever-increasing chemurgic activities."

After World War II, Hale became particularly interested in the physiological properties of chlorophyll. He patented a number of applications of chlorophyll's healing properties, and set up the Verdurin Co. to handle his interests. A chlorophyll containing chewing gum and a chlorophyll containing cigarette were marketed based upon Hale's patents.

Hale had been interested also in the National Agrol Co. In the early 50's he dissolved Agrol—turning over the alcohol production facilities to the Gateway Chemurgic Co. of Lincoln, Nebr. Thereafter, he was president of Verdurin Co. and on the board of directors of Chlorophyll Chemical Corp. of McAllen, Texas, and of Trenton Chemical Co. of Trenton, Mich.

Shortly after Hale's death on Aug. 8, 1955, the new chemical laboratory at Connecticut College for Women was dedicated to Hale. He had once been a visiting professor of chemurgy at the college, which his daughter Ruth had attended, and had contributed effort and substance to the new laboratory. At the dedication ceremony, Hale's son-in-law, Wiley T. Buchanan, U.S. Ambassador to Luxembourg, gave a moving eulogy to Hale in which he revealed Hale's motto—*Principia, Non Homines,* or "Principles, not Men." Hale's former secretary, reflecting on this speech, said: "I am afraid I cannot agree with Wiley completely on *Principia, Non Homines.* When Dr. Hale had a goal, he was relentless, selfish, and sometimes determinedly unreasonable and stubborn. But he loved mankind deeply. He was a born teacher, and he was interested tremendously in people—particularly young people—his whole object—to make a better world. We desperately need today the kind of inspiration he contributed. There was no middle ground with Hale—you either loved him or hated him and some days you did both."

Obituary in *Cosmos Club Bulletin,* May 1956; "Address of the Hon. Wiley T. Buchanan dedicating the Hale Chemical Laboratory at the Connecticut College for Women, Dec. 7, 1955"; biographical data from the files of The Dow Chemical Co.; W. J. Hale, The Farm Chemurgic Movement, in Williams Haynes, editor, "American Chemical Industry," Vol. V. (1954); *Progress Guide,* May 1944; communication from Gertrude Winfield.

LEONARD C. CHAMBERLAIN

Lloyd Augustus Hall

1894-1971

Hall was born in Elgin, Ill., June 20, 1894 and graduated with honors from the local high school. His father was a Baptist minister who occupied a position of importance in his community. His paternal grandfather was pastor of the first Black church in Chicago. Hall was married in 1919. The family later adopted two children.

While Hall was often made aware of discrimination against Blacks, that did not prevent him from participating actively in the many professional, civic, and social organizations to which he belonged. He graduated in 1916 from Northwestern University with a B.S. degree in chemistry and took frequent graduate courses at University of Chicago.

During the First World War, from 1917 to 1919, Hall served as a lieutenant in the Ordnance Department inspecting explosives at a plant in a small Wisconsin city. His color subjected him to such indignity that he asked to be transferred. Besides his war service, he held four positions with industrial food laboratories and the Chicago Department of Health between 1916 and 1924. By that time he was convinced that his best professional opportunity lay in private practice as a consultant.

One of Hall's clients, Griffith Laboratories, Inc., in 1924 offered to provide a laboratory for him to do their work and also to continue his consulting practice. After 1929 he devoted himself entirely to Griffith, remaining with them until he retired in 1959. He was, at the time, technical director and head of the laboratory. His work at Griffith permitted him to serve on many state and government advisory boards such as the Science Advisory Board on Food Research, Quartermaster Corps, U.S. Army from 1943 to 1948 and the Illinois State Food Commission from 1944 to 1949. He was also consultant to the Carver Research Foundation and was on the advisory board of *Chemical and Engineering News.* After his retirement he spent 6 months in Indonesia as a consultant to the United Nations food and agricultural organization.

President Kennedy appointed him to the American Food for Peace Council.

Among the civic, social, and religious organizations in which Hall participated were Hull House, of which he was a trustee; Chicago Conference of Chrisians and Jews, the YMCA, the NAACP, and the Urban League.

He was a founder and honorary member of the Institute of Food Technologists. He received three honorary degrees as well as the Honor Scroll of the Chicago Chapter of the American Institute of Chemists in 1956.

Hall's chief interest as a chemist was in food technology. He developed methods of curing meat, processes using antioxidants to prevent fats and oils from becoming rancid, and a process for sterilizing spices, often contaminated by bacteria, by exposure to ethylene oxide. He was granted more than 100 patents and wrote more than 50 papers on food technology.

Shortly after retirement from Griffith, Hall moved to Altadena, Calif. where he continued to be active in chemistry and community affairs. He was a director of the Pasadena Chamber of Commerce, of Pasadena Beautiful, of the Red Cross and was a consultant to the United California Bank. He died in Altadena on Jan. 2, 1971.

Personal recollections; biography by Samuel P. Massie, *Chemistry* **44**, No. 3, 21 (March, 1971); obituaries in *Chem. Eng. News,* Jan. 25, 1971, p. 56, *Chemical Bulletin,* February 1971, p. 5, and *The Chemist,* February 1971, p. 47.

BERNARD E. SCHAAR

Lyman Beecher Hall

1852-1935

In 1880, as the first step in a program to strengthen the science course at Haverford College by hiring experts in each of the basic areas, Hall was called from an instructorship in chemistry at Johns Hopkins to the John Farnum Chair of Chemistry and Physics at the college. The timing was propitious for 2 years later Theodore William Richards, who was to be awarded the Nobel Prize for chemistry in 1914, enrolled at Haverford. Hall was well prepared for his new position and for the young man who was to become his most distinguished protegé.

Hall was a New Englander, born in New Bedford, Mass., Jan. 16, 1852 to Isaac D. and Hannah (Norris) Hall, and when the time came for him to go to college he went to a New England institution, Amherst. After earning his A.B. degree there in 1873, he chose Göttingen for advanced work in chemistry. At Göttingen the professorship was held by Hans Hübner who had succeeded Friedrich Wöhler in 1857.

Hübner's chief interest was in the rapidly developing area of aromatic chemistry, and so Hall did research on the nitration of salicylic acid and oxidation of various aromatic sulfamides. Upon completing this in 1875, he was awarded his Ph.D. degree, returned to the United States, and was invited to join the select group working under Ira Remsen at Johns Hopkins. A series of publications describing research on various mesitylene derivatives followed. In 1879 Hall was appointed instructor of chemistry at Johns Hopkins, and the next year he moved to Haverford College.

Hall immediately impressed his new colleagues with the vigor and efficiency which he brought to chemistry. He gradually expanded the curriculum and began to provide a solid technical background for his students. Richards was among the earliest of these. It was clear even then that he was destined for distinction in some area, but to Hall he always gave the credit for arousing his interest in chemistry, channeling his energies into it, and providing him with a sound grounding in the discipline.

In 1888 Hall was relieved of his responsibilities for physics, which up until then had been an adjunct of chemistry at Haverford, and for the next 29 years, he devoted himself to teaching chemistry, pouring his intellectual energies entirely into his students. He was known as a demanding teacher and a stern disciplinarian, with a reputation for little patience toward those students who lacked seriousness about chemistry and for notable kindness and fairness toward those who shared his love of his subject and his respect for truth. Over the years he influenced many

Haverford undergraduates. His most notable student after Richards was his own son, Norris Flagler Hall, a distinguished physical chemist and a member of the faculty at University of Wisconsin from 1929 to 1955, when he retired.

Lyman Beecher Hall retired in 1917. As marks of the esteem in which he was held at Haverford, the old chemistry building was named in his honor, and in 1923 the Haverford class of 1898 endowed a prize in chemistry in his name. He died at Madison, Wis., Jan. 20, 1935.

Alumni Association of Haverford College, "History of Haverford College 1830-1890" (1892); Haverford College Class of 1916, "The Record," (a yearbook); *Haverford News,* Mar. 13, 1917 and Feb. 4, 1935; Rufus M. Jones, "Haverford College—A History and an Interpretation," MacMillan (1933); Isaac M. Sharpless, "The Story of a Small College," John C. Winston Co. (1918).

COLIN MACKAY

William Thomas Hall

1874-1957

Hall was born Aug. 4, 1874 in New Bedford, Mass. He received his bachelor's degree in chemistry from Massachusetts Institute of Technology in 1895 and then went to Germany for further study. From 1895 to 1897 he was at University of Göttingen where he received his Ph.D. degree. The following year Hall returned to MIT as an assistant in analytical chemistry. He was promoted to instructor (1900), assistant professor (1911), and associate professor (1918). He retired, still at the rank of associate professor, in 1940. From 1942 to 1943 he was head of the Science Department at Thayer Academy in South Braintree, Mass. At the time of his death, Jan. 4, 1957, he was living in Rochester, Mass.

Over an academic career which spanned more than 40 years, Hall published little in the way of original research—only seven papers. He was much more active as a translator particularly of German and an abstractor. His translation of F. P. Treadwell's two-volume "Analytical Chemistry" (1903, later editions in 1932 and 1937) made "Treadwell and Hall" a standard reference work in this country. By 1915 he had translated nine texts and monographs, including works by Ostwald, Abderhalden, and Biltz in addition to that by Treadwell. He also wrote "Textbook of Quantitative Analysis," which went through three editions (1930, 1935, 1941), and co-authored "The Chemical and Metallographical Examination of Iron, Steel, and Brass" (1921). He revised the 3rd edition (1939) of Forris J. Moore's "History of Chemistry," which appeared after Moore's death.

From its inception in 1907, Hall was an abstractor for *Chemical Abstracts,* an activity he maintained until a few months before his death. He prepared a large share of the abstracts in the field of analytical chemistry during most of his 50 year period. In addition he served as a section editor of *C.A.* in charge of metallography from 1907 until 1921, when he became section editor for analytical chemistry. He continued in this capacity until the end of 1956, at which time he was made an honorary section editor. He was, indeed, one of *C.A.'s* "iron men" and an example of a chemist who worked primarily as a "transmitter," facilitating the flow of information from the research worker to the rest of the profession.

Obituaries in *Technology Review* **59,** 196 (1957), and *Chemical Abstracts* **51,** 2449 (1957); photograph in Brown and Weeks, "A History of the American Chemical Society," ACS (1952), p. 363.

RUSSELL F. TRIMBLE

Robert Hare

1781-1858

Hare was born in Philadelphia, Jan. 17, 1781. He learned chemistry by self-study and attending lectures of James Woodhouse at University of Pennsylvania. While seeking a means of attaining higher temperatures in his laboratory in 1801, he invented the oxy-hydrogen blowtorch, the ancestor of the acetylene and other torches. It made it possible to melt platinum and refractory materials and formed the basis of the limelight and Drummond light. The Chemical Society of Philadelphia, of which Hare was a secretary, published an account of the device in "Memoir of the Supply and Application of the Blow-Pipe" in 1802.

In 1810 Hare became professor of natural philosophy in the University of Pennsylvania Medical School. This was a new, trial professorship. The subject was an elective, medical students refused to take it, and the trial was unsuccessful. Hare resigned in 1812.

During his youth and while he was nominally professor at the university, Hare managed the family brewery. He worked with science only in his spare time. The business upheaval accompanying the War of 1812 ruined the firm around 1815. Hare tried to start a plant for manufacturing illuminating gas in New York City but was not successful.

College of William and Mary hired him as professor of natural philosophy and chemistry in early 1818. Later that year he went to University of Pennsylvania Medical School as professor of chemistry. Hare's lectures at Pennsylvania were characterized by a large quantity and variety of demonstrations. An expert craftsman, he designed and constructed much of his apparatus. He preferred large pieces of equipment, easily observable by everyone in his large lecture room. Illustrations of the quaint-looking but effective devices may be seen in his articles, in his text, and in a two-volume book he issued in 1826, "Engravings and Descriptions of a Great Part of the Apparatus Used in the Chemical Course of the University of Pennsylvania." The text appeared in 1822 under the title "Minutes of the Course of Chemical Instruction in the Medical Department of the University of Pennsylvania," and later as "Compendium of the Course," (1828; 2nd ed., 1834; 3rd ed., 1838; 4th ed., 1840). Hare occasionally made instruments for other teachers.

Hare devised novel lecture demonstrations. He developed the deflagrator, a battery having large plates for producing powerful and rapid combustion; and the calorimotor, a plunge battery producing powerful heating effects. He constructed an electric furnace in which he produced phosphorus, calcium carbide, and other compounds. He experimented with electrolytic methods of isolating elements, using a mercury cathode. Occasionally he was called upon as a toxicologist in poison cases.

Hare resigned in 1847. As a teacher for 29 years in the country's best-known medical school, he taught chemistry to more medical students than any other teacher of the period. One cannot help but have the feeling he raised the standard of chemical education in medical schools by his influence on students who went forth from his classes. Several of his pupils became teachers of chemistry. He gave advanced instruction to a number of Americans who did not want or could not afford to go to Europe.

Among Hare's other interests were banking and currency reform, the accidental explosion of nitre warehouses, and the cause of storms and tornadoes. He wrote poetry and in 1850 a historical novel, "Standish the Puritan," under the pen name Eldred Grayson. In later life he came to believe in spiritualism and wrote pamphlets and a book on the subject. He died in Philadelphia, May 15, 1858.

One of Robert Hare's six children, John Innes Clark Hare (1816-1905), was a chemist for a time and published four articles, then switched to law and became a notable judge in courts of Philadelphia, professor of law at University of Pennsylvania, and writer of legal works.

Edgar F. Smith, "The Life of Robert Hare," J. B. Lippincott Co. (1917), illustrated, has much information but no references to sources; Edgar F. Smith, "Chemistry in America," D. Appleton & Co. (1914), pp. 152-205, reprints Hare's "Memoir . . . of the Blow-Pipe" (1802); biography with portrait by W. D. Miles in: Eduard Farber, editor, "Great Chemists," Interscience Publishers (1961), pp. 420-433; "Dictionary of American Biography," **8,** 263, Chas. Scribner's Sons (1931).

For J. I. Clark Hare *see* "Dictionary of American Biography," *ibid.*, p. 263.

WYNDHAM D. MILES

William Draper Harkins

1873-1951

Harkins was born in Titusville, Pa., Dec. 28, 1873, the son of Nelson Goodrich and

Sarah Eliza (Draper) Harkins. He began his college work later than usual and so received his A.B. degree in 1900 from Stanford University. For the next 7 years he carried on graduate work both at University of Chicago and Stanford University and received his Ph.D. degree from the latter institution in 1907. However, during this period he was an instructor in chemistry at Stanford, 1898–1900, and professor and head of the Department of Chemistry at University of Montana, 1900–12. Also, Harkins spent a year (1909) in Fritz Haber's Laboratory of Institut für Physikalische Chemie, Karlsruhe, Germany and a year (1909–10) in the Research Laboratory of Physical Chemistry at Massachusetts Institute of Technology under Arthur A. Noyes, who strongly influenced the development of physical chemistry in the U.S.

While at University of Montana Harkins was located in the midst of large copper smelters and subsequently became interested in the polluting effects of smelter smoke on surrounding lands, which became a serious problem for farmers. He was appointed chemist in charge of the smelter smoke investigation for the Anaconda Farmer's Association, 1902–10, which led to publication of a number of papers jointly with Robert E. Swain on analyzing smelter smoke.

On several occasions later, while at University of Chicago, Harkins remarked that when he began his research career in physical chemistry there were two areas he vowed to avoid, namely investigations on solubility and also on catalysis and surface phenomena. However, while at MIT his research was devoted primarily to the former, and throughout his long and productive career he became a pioneer and developer of surface chemistry.

Harkins went to University of Chicago as instructor in 1912 and remained there as an active research investigator until his death on Mar. 7, 1951. In the intervening years he was promoted to associate professor (1914), professor of physical chemistry (1917), and Andrew MacLeish Distinguished Service Professor (1935). During this period he published several books and 270 research papers, mostly in collaboration with his doctoral and postdoctoral students. He remained active in his laboratory until the day of his death.

His interest in atomic structure began early in his career. In 1915 he published papers on energy relations, changes in mass and weight, and structure of complex atoms. Other contributions in this area included the evolution of the elements, stability of complex atoms, the building of atoms and the new periodic system, the hydrogen, helium, H_3, H_2 theory of atomic structure, and stability of atomic nuclei and the whole number rules, the separation of the isotopes of chlorine and of mercury, and the synthesis and disintegration of atoms as shown by the photography of Wilson Cloud Chamber tracks. He predicted the existence of the heavy isotopes of hydrogen and of the neutron but never carired out experimental programs to identify them. This was left to later discoveries by Harold Urey and James Chadwick.

Shortly after Ernest O. Lawrence developed the cyclotron at the University of California at Berkeley in the early 1930s Harkins proceeded to build one at Chicago with a 20″ pole piece for the magnet. Very limited resources for the program were available to him, but with great enthusiasm and determination, together with a dozen or so graduate and postdoctoral students, he accomplished the task by the mid-1930s with considerable success. Using the cyclotron as a tool, he published many papers in collaboration with his students until he reached the age of retirement about 1940. In 1939 the cyclotron was turned over to the Department of Physics under the direction of S. K. Allison. When World War II became imminent and the S-1 Committee of the OSRD was formed and later the Manhattan Project was initiated as a follow-up of this Committee, the cyclotron became an unusually important accelerator. It was used during the war for producing radioisotopes for the Project, in particular for the Metallurgical Laboratory at University of Chicago.

Harkins' research in the area of surface chemistry began in 1915. Much of his early work was devoted to methods for measuring surface tension. He made many contributions to surface physics and chemistry, including studies of the orientation of molecules in surfaces, surface energy, adsorption, surface catalysis, molecular attraction, adhesion, liquid spreading, monomolecular films, stability of emulsions, wetting of pigments, surface pressures and potentials, thermodynamics of films, the film balance, emulsion polymerization, etc. Without question his major and most numerous contributions were devoted to this field for which he was internationally known and respected.

Harkins received many honors throughout his career. In 1928 he was awarded the Willard Gibbs Medal of the Chicago Section of the American Chemical Society. He was a consultant to several chemical industries and government agencies. During the War he was associated with NDRC and the OSRD and contributed substantially to the synthetic rubber program. In 1936–37 he was the George Fisher Baker Lecturer at Cornell University. He was also a member of the National Academy of Sciences.

Chem. Eng. News, **27,** 1146 (1949); W. D. Harkins, "The Physical Chemistry of Surface Films," Reinhold Publishing Corp. (1952); personal recollections.

WARREN C. JOHNSON

John Harrison

1773-1833

Harrison was born in Philadelphia, Dec. 17, 1773, sixth of nine children of Thomas and Sarah (Richards) Harrison. Thomas, a Quaker from Carlyle, then London, England, had arrived in Philadelphia in 1763 and had married there. His father, Thomas, was listed variously in early American directories as a pharmacist, "cobbler and currier," and as a successful merchant.

John received his early education in Philadelphia, and at about the age of 16 he was apprenticed to a druggist, Thomas Speakman. Some historians say John spent about 2 years in Europe (1791-93 ?) studying chemistry with Joseph Priestley and also learning the business of manufacturing chemicals.

Upon his return in 1793, John formed a partnership with Samuel Betten. The firm of Betten and Harrison advertised itself as "wholesale and retail chemists and druggists." After surmounting many difficulties, they established a laboratory to manufacture aqua fortis and most of the chemicals which were formerly imported.

With developing commercial interests in Philadelphia, Harrison recognized that there would be an increasing market for chemicals. He believed that it was possible to supply these materials at somewhat lower costs because of rising ocean freights and the uncertainty of delivery of products from abroad. John began a series of experiments which resulted in practical scale manufacture of nitric, hydrochloric, and sulfuric acids. Little is known of the details of these operations, but we do know that his was the first successful commercial sulfuric acid plant in this country.

In November of 1802 John married Lydia Leib, daughter of a prominent local doctor and politician; they had at least four children—two sons and two daughters.

In 1801 the partnership with Betten was dissolved, but Harrison continued the business until 1804. This tiny plant, at first, seemed a success technically but not financially. The high price of certain materials and the initial small demands made the venture difficult. During the 1800–04 period sales increased. Customers were chiefly physicians, but a recorded item states that in May 1803 the U.S. Mint bought 20 pounds of oil of vitriol. Harrison gradually built a reputation for producing and delivering chemicals, and he began to do well financially. About this time John had his portrait painted by the fashionable artist, Rembrandt Peale. In 1896, Mary J. Peale produced another oil of John, modeled after that of Rembrandt Peale. This 22" x 27" painting is now in the possession of the Chemistry Department of the University of Pennsylvania.

In 1806 Harrison discontinued the apothecary and drug business. Some accounts state that he opened an enlarged plant on Green Street and that he added the manufacture of white lead, colors, and successively various other chemicals. The exact locations, products produced, and scale of operations are uncertain, perhaps because fires at several times destroyed records and some plants.

By 1807 Harrison had built a sizeable lead chamber plant. It was a continuous process and increased the production of sulfuric acid to over 3500 carboys per year. The acid was concentrated in glass retorts which broke frequently and increased costs of manufacture.

In early 1808 Harrison sponsored a petition to Congress, asking them to support higher tariffs on chemical imports, especially on sulfuric acid. Affidavits supporting the quality of his products were included with the petition. He mentioned that 20 products were manufactured in his plants. In order to expedite matters, John Harrison wrote a letter to Thomas Jefferson, mentioning his plans, his troubles, and his desire to have Jefferson's help in bringing the matter to a favorable vote.

As early as 1807, Harrison began laying plans to expand and to move "out of the city" to the Kensington-Northern Liberties area. A plot of 14 acres was purchased by John, his father, and his father-in-law. John moved into a house on Frankford Road, and it is believed that shortly thereafter construction began on the new plant.

In 1809, the Green Street plant was destroyed by fire. Within a short period of time the Kensington plant was described as a "veritable giant chemical industry." John designed much of his apparatus, especially modifying the lead chambers. He also was quick to take advantage of other technical discoveries. Thus Harrison became the first sulfuric acid manufacturer to replace glass stills with platinum.

Eric Bollman, a German emigrant to Philadelphia, had familiarized himself with the working of large amounts of platinum by Wollaston in London. He therefore built to Harrison's specifications a still from platinum available in Philadelphia. This expensive apparatus fully justified itself not only because of its continuous operation without breakage but also because it was able to yield a more concentrated acid and a greater production.

Again fire struck, this time in 1827 at the Kensington plant, destroying most of the remaining records and much of the plant. Soon thereafter the entire manufacturing process was moved from Kensington to Gray's Ferry Road, where new, modified, and much larger plants were constructed. In 1831 Harrison admitted his two sons into the partnership and changed the name to John Harrison and Sons. After Harrison's death in 1833, the name was changed to Harrison Brothers and Co. Capacities and the line of products were increased over the years to include alums, pigments, paints, oils, and varnishes.

The sons and grandsons succeeded to the business of John Harrison, and finally, in 1917, it became a part of the duPont de Nemours and Co.

John Harrison, aside from his industrial activities, found time for other interests. For various periods of time he was captain of the Philadelphia Militia, a member of the Schuylkill Fishing Co., a recorder of the City and County of Philadelphia, an active politician, a member of the first Board of Managers of The Franklin Institute, and a secretary of a commission to attend to the vaccination of citizens in his area.

Through the efforts of Charles Custis Harrison, provost of University of Pennsylvania (1894–1910), and his two brothers, Alfred C. and William W., funds were provided to the university to build a laboratory as a memorial to their grandfather. This building, begun in 1892 and occupied in 1894, was named John Harrison Laboratory of Chemistry. It was demolished in the Fall of 1969.

Another descendant, Thomas Shelton Harrison, in 1919 left funds and details for casting a 10-foot statue of John Harrison to be placed on a suitable pedestal in Fairmount Park. It was finally cast in 1931 by the Gorham Co. Somehow it was never erected in the park but was given to the University in 1934 and is on the plot of ground adjacent to the chemistry building.

John Harrison Papers in the E. F. Smith Memorial Library, Univ. of Penna.; J. Thomas Scharf and Thompson Westcott, "History of Philadelphia—1609-1884," III, p. 2273; "The Chemistry of Paints," Harrison Bros. and Co., (1902); Evolution of the Chemical Equipment Industry in "Equipment's Contribution to Chemical Progress," pp. 5-10, (1936), published by *Chemical Industries; The Penn Chemist, No. 1,* p. 4-5, July 1968; *DuPont Magazine,* **53,** No. 1, p. 23-24, 1959.

CLAUDE K. DEISCHER

Benjamin Harrow

1888-1970

Harrow, American biochemist, teacher and author, was born in London, England, Aug. 25, 1888. His name, originally Benjamin Horowitz, was Anglicized to Harrow shortly after he completed his doctoral studies at Columbia University in 1913.

He had studied at Finsbury College in London from 1904 to 1906. Having come to the United States in 1907, he attended Columbia University where he obtained his B.S. degree in 1911, A.M. degree in 1912, and Ph.D. degree in 1913. His doctoral thesis dealt with the reaction of ammonia with thymol. After assisting in organic chemistry at Clark University (1911) and in biochemistry at Columbia (1912-13), he served as assistant professor at Fordham University Medical School (1913-14). In 1914 he was appointed instructor (and later associate) in physiological chemistry at Columbia's College of Physicians and Surgeons, where he remained until 1928. In 1917 he married Caroline Solis of Philadelphia; they had one daughter, Mrs. Margaret Lebenbaum.

He did little research while at Columbia. He was, however, a prolific author, principally of popularly written books on the history of chemistry, and on his favored biochemical subjects—vitamins, enzymes, and hormones. Harrow's most durable book of this period was his "Eminent Chemists of Our Time." Appearing in 1920 as a collection of biographical sketches of nine well-known chemists, the book was expanded in 1927 to include detailed accounts of the scientific accomplishments of the biographees. Other books of Harrow's Columbia period are exemplified by "Vitamins, Essential Food Factors" (1921), "Glands in Health and Disease" (1922), and "Romance of the Atom" (1927).

Appointed assistant professor at the College of the City of New York in 1928, Harrow devoted himself largely to teaching. The two popular books, "The Making of Chemistry" and "Romance of Chemistry" which he published in 1930 and 1931 respectively, were the last of the type.

By 1933 when he was promoted to associate professor, a new more professional Harrow had emerged. He now participated extensively in collaborative research work. He had met Casimir Funk, whose studies of vitamins and hormones further stimulated Harrow's interests in these fields. Working with Funk, Harrow conducted studies on the male hormone, demonstrating that injection of the hormone could induce comb growth in castrated male chickens. Harrow published, jointly with younger members of the City College staff, studies on hyperglycemia and on biochemical detoxication mechanisms, particularly those involving glucuronic acid. In place of earlier popular books, Harrow now wrote textbooks which reflected his own teaching experience. Ironically, his largest and most ambitious work, "A Textbook of Biochemistry" (1935), co-edited with Carl P. Sherwin, was not a textbook at all but a collection of monographs which quickly became obsolete for reference purposes, and which was unsuitable for classroom teaching, even in Harrow's hands. (Harrow later used the same title for the second and subsequent editions of his more successful book, "Biochemistry for Medical, Dental and College students.")

In 1939 Harrow became professor and, in 1944 chairman of the Chemistry Department. He continued small-scale research while revising earlier textbooks and laboratory manuals. He retired in 1954, remembered by students and colleagues as a soft-spoken and sensitive person and as a careful and effective teacher. In 1955 he published a small biography, "Casimir Funk, Pioneer in Vitamins and Hormones." It was a work of gratitude to a friend and half-forgotten trailblazer from the heroic period of biochemistry. Harrow's early popular and readable style was once again evident in this his last publication. Harrow died in New York at the age of 82, Dec. 9, 1970.

Obituary in *New York Times* Dec. 11, 1970; personal recollections; *Chem. Abstr.* 1913-55; Benjamin Harrow, "Eminent Chemists of Our Time," D. Van Nostrand (1920); "Text Book of Biochemistry," W. B. Saunders (1935); "Casimir Funk," Dodd, Mead (1955).

CHARLES H. FUCHSMAN

Edward Hart

1854-1931

Edward Hart—teacher, inventor, editor and author—was a dominant figure in American chemical education for 50 years. On Oct. 16, 17, and 18, 1924, an Edward Hart Celebration was held at Lafayette College to honor this man who was brought there in 1874 to assist Thomas Drown.

Hart was born to George and Mary Longstreet (Watson) Hart in Bucks County, Pa., Nov. 18, 1854. He graduated from the Doylestown English and Classical Seminary. Before pursuing chemistry he worked as a surveyor and also studied law in his uncle's office for 2 years. After learning elementary chemistry at home, he entered the consulting laboratory of Thomas Drown in Philadelphia in 1872 as a student and helper. When Drown went to Lafayette College as professor of analytical chemistry in 1874, Hart accompanied him and for the next 2 years was concurrently a student, an assistant, and a tutor in chemistry. He never took a chemistry course at Lafayette but was awarded an honorary bachelor's degree by the college some years later.

In 1876 he received a fellowship from Johns Hopkins, acquired his Ph.D. degree in 1878, and returned to Lafayette where he succeeded Drown as head of the Chemistry Department in 1881. His status at Lafayette was assistant and tutor from 1874 to 1876, adjunct professor from 1878 to 1882, professor from 1882 to 1924, and dean of Pardee Science Department from 1909 to 1924.

"Buddy" Hart, as he became known among the students, was literally a bursting bud of versatility. To him chemical knowledge, unless applied, was like a cow that was worshipped instead of milked. In 1881 he and John Baker, one of his students, began to make pure chemicals. With capital of only $500 they produced pure mineral acids and ammonia. George Adamson, another of Hart's students, joined the firm in 1884, bringing added capital of $2,000. The firm was incorporated in 1890 as Baker and Adamson Chemical Company; Hart's name was omitted because Lafayette College preferred that faculty members not be publicized in business ventures. Hart invented a nitric acid boiler and condenser. Eventually he held 10 chemical patents.

Hart undertook the printing and publishing of *Journal of Analytical Chemistry* when it was founded, using a foot-powered printing press. He published the journal during its existence, organizing the Chemical Publishing Co. to handle the printing in 1887. His firm also published *American Chemical Journal* and *Journal of the American Chemical Society*. Hart edited the latter journal from 1893 to 1902 when he withdrew from publishing activities following the death of his only son in an accident. Harvey F. Mack and Charles A. Hilburn assumed management of the firm which later became Mack Printing Co., which has for years printed ACS journals and magazines.

Meanwhile, the Chemistry Department at Lafayette grew through Hart's inspiration, teaching, and contacts. A new laboratory was provided by James Gayley, an alumnus noted for advances in blast furnace technology. Hart obtained additional funds for a chemical library and a research fellowship.

Hart wrote and published several texts: "Volumetric Analysis" (1876), "Chemical Engineering" (1920), and "The Silica Gel Pseudomorph" (1924). He invented a mineral wax bottle for handling hydrofluoric acid. He received the John Scott Medal from the Franklin Institute. In his later years he devoted time and study to agriculture and wrote a book entitled "Our Farm in Cedar Valley." He showed an active interest in civic affairs and local and county historical societies. He had some ability as a poet and frequently expressed his moods and reflections in verse. He died June 6, 1931 in his 77th year.

D. B. Skillman, "The Biography of A College," Lafayette College (1932); R. D. Billinger, "Chemical Pioneers of Lehigh Valley," *The Octagon* **30,** 52 (March 1947); William M. and Mary H. Perry, "Dr. Edward Hart, A Dynamo in Chemistry," *ibid.* **43,** 31-32, 42 (March 1960); A. J. Barnard, Jr. and W. C. Broad, "Lehigh Valley Pioneers," *ibid.* **52,** 51-53 (March 1969); "Dictionary of American Biography," **21** (Supplement 1), 376, Chas. Scribner's Sons (1944); E. C. Bingham, *Ind. Eng. Chem.* **15,** 974-975 (1923), with portrait; recollections of Paul Mack.

ROBERT D. BILLINGER

Edwin Bret Hart

1874-1953

Hart was born on a farm in Erie County near the town of Sandusky, Ohio, on Christmas day, 1874. During his high school days, he became interested in scientific studies; thus he pursued a major in chemistry and medicine upon enrolling in the University of Michigan in 1892. After graduation in 1897, he accepted the position of assistant chemist at the New York Agricultural Experiment Station in Geneva where he was closely associated with Lucius L. Van Slyke. His research was in the field of dairy chemistry, with emphasis on studies of proteins. His interest in proteins, coupled with a desire for additional education, led him to Germany in 1900 to study under the eminent physiological chemist, Albrecht Kossel, at University of Marburg. Hart's transfer to University of Heidelberg, when Kossel was appointed head of the Physiological Institute at that University, resulted in loss of academic credits. This coupled with personal problems resulted in Hart's stopping his studies before he received the Ph.D. degree. The end of his studies in Germany terminated his formal education. In 1902 he resumed work at the Geneva station as an associate chemist.

In 1906 Hart accepted the position of professor of agricultural chemistry at University of Wisconsin and chemist in the Wisconsin Agricultural Experiment Station. Shortly thereafter, he succeeded Stephen Babcock as head of the Agricultural Chemistry Department and initiated what was frequently referred to by many as the "Hart Era." His responsibilities encompassed three areas: research, teaching, and administration.

As an investigator, Hart was involved in some phase of investigative work dealing with nearly all the vitamins and nutritive minerals studied during the first half of this century. His most notable research was in the field of trace nutrients. Together with Harry Steenbock, Conrad A. Elvehjem, and James Waddell he showed that a trace amount of copper was needed to promote the utilization of iron by animals to produce hemoglobin (1928). Earlier, he had been the chief organizer of the famous "single-grain experiment," initiated in 1907 and carried out in association with Elmer V. McCollum, Harry Steenbock and George C. Humphry. This experiment underscored the importance of other entities besides proteins, fats, and carbohydrates in animal nutrition and dramatized the need for new experimental approaches to nutritional studies. During his career Hart authored or co-authored more than 300 scientific articles in nutritional chemistry.

Hart's role as a teacher tends to be minimized because of other outstanding facets of his professional career. E. B., as he was affectionally known to his students, was considered one of the outstanding teachers at University of Wisconsin. He was famed for his ability to ask challenging questions that inquired with amazing acuity into a student's knowledge of the subject matter. His greatest asset as a teacher was his ability to integrate current scientific discoveries into everyday teaching. He was credited with propelling the scientific seminar series in his department to a place of prominence in university teaching circles.

As an administrator, Hart was successful in advancing the Agricultural Chemistry Department (name changed to Biochemistry in 1938) to a place of renown in nutritional research. Under his leadership the department grew in size and importance. Hart had an uncanny ability to pick competent men for his staff. This, coupled with his philosophy of fair play and stern leadership, made him an effective department chairman and research director. In the course of his 38 years as chairman, members of the department published more than 1200 scientific articles.

Hart was recipient of many honors and awards during his career. The most cherished, by his own admission, was the honorary doctorate bestowed upon him in 1949 by University of Wisconsin.

After his retirement from the faculty in 1945, he continued his association with the department and university as emeritus professor. He was in his office nearly every day until his sudden death on Mar. 12, 1953.

In July 1972, The Edwin Bret Hart Professorship of Biochemistry was established to honor the man who was instrumental in establishing the prestigious Biochemistry Department at University of Wisconsin. The professorship underscores Hart's academic philosophy which placed equal emphasis on research and teaching.

Charles H. Trottman, "Edwin Bret Hart: Agricultural Chemist" (Ph.D. Dissertation, University of Wisconsin, 1972); Conrad A. Elvehjem, "Edwin Bret Hart, 1874-1953," *Biog. Mem. Nat. Acad. Sci.,* **28,** 117-161 (1954); "The Life and Accomplishments of Edwin B. Hart," a symposium with papers by H. T. Scott, C. A. Elvehjem, S. Lepkowsky, E. M. Nelson, and K. G. Weckel dealing with various aspects of his life and career, *Food Technology,* **9,** 1-13 (1955); Ellery H. Harvey, "Edwin Bret Hart," *Chem. Eng. News,* **22,** 435 (1944); University of Wisconsin, Madison, has a collection of Hart papers and memoirs; there are also personal papers in possession of Hart's daughter, Mrs. Russell H. Larson, Madison, Wis.

CHARLES H. TROTTMAN

Ernst Alfred Hauser

1896-1956

Hauser was born in Vienna, July 20, 1896. After graduating from the Academisches Gymnasium in 1914 he joined the Austrian army, from which he was discharged as captain in 1918. He then continued his education at the University of Vienna where he received his Ph.D. degree in 1922.

From 1921 to 1923 Hauser was an assistant chemist at Göttingen, first to Max Born and then to Richard Zsigmondy. In 1923 he became a research chemist at Rohstoff Trocknungs Gesellschaft in Frankfurt. In 1925 he was appointed chief chemist of the colloid chemical laboratory of the Metallgesellschaft in Frankfurt, a position he held until 1933 when he became head of the research department of the Semperit Austro-American Rubber Co. In the meantime, in 1928-29 Hauser was non-resident associate professor of chemical engineering at Massachusetts Institute of Technology. In 1935 he came to MIT full time and in 1948 was promoted to professor and appointed director of the Division of Colloid Chemistry. He also served as visiting professor of colloid chemistry at Worcester Polytechnic Institute, 1948-52, where he was awarded an honorary D.Sc. in 1952.

Hauser's first entry in *Chemical Abstracts* refers to an article in "India Rubber World" volume 68 (1924) concerning the use of an aniline dye, opal red, to make rubber particles more visible under the microscope. His interest in and contributions to the colloid chemistry of rubber latex were to continue as long as he lived. In 1925 Herbert Freundlich and Hauser proposed that the colloidal stability of latex is the result of adsorption on the particles of a polar layer made up largely of proteins. This general concept, still valid, followed from their finding that the protein separated from latex coagulates in the same pH range as the latex itself.

In 1926 following J. R. Katz's demonstration in 1925 that stretched rubber gives an x-ray interference pattern characteristic of crystalline materials, Hauser and Herman F. Mark reported the first extensive quantitative x-ray study of the crystal structure of stretched rubber. They showed that the x-ray interference spots have the required independence of aspect from degree of elongation necessary to characterize crystallinity. They also showed that the intensity of interference spots increases with elongation on stretching. They assigned the unit cell of stretched rubber to the rhombic system and were the first to assign values to the unit cell parameters: a = 8.0 A.U.; b = 8.6 A.U.; and c = 7.68 A.U. (direction of stretch). Although they are of the right order of magnitude, these values have since been refined.

In addition to many publications on the properties and structure of rubber and on the colloid chemistry of latex, clays, and antibiotics, Hauser was author of: "Latex: Its Occurrence, Collection, Properties, and Technical Applications" (1927), (English translation by W. J. Kelly (1930)); "Colloid Chemistry of the Rubber Industry" (1928); "Colloidal Phenomena" (1939); "Experiments in Colloid Chemistry" (1940); and "Silicic Science" (1955). He also edited the important handbook "Handbuch der Gesamten Kautschuktechnologie" (1935) to which he contributed sections on latex and the use of rubber in automotive manufacture.

During World War II Hauser served as a consultant to the Quartermaster General and as technical adviser to the Baruch synthetic rubber committee.

Hauser was married to Vera M. Fischer, Apr. 8, 1922. They had three sons: Ernst F., Wolf Dieter, and George W. Hauser died Feb. 10, 1956 in Cambridge, Mass. and was buried at Cathedral of Pines in New Hampshire.

Rubber Age **78** (6), 398 (1956); *Rubber World* **6,** 849 (1956).

GUIDO H. STEMPEL

Elwood Haynes

1857-1925

Haynes, American scientist, chemist, designer of the earliest commercial gasoline-power automobile, and inventor of valuable alloys, was born in Portland, Ind., Oct. 14, 1857. He had a keen interest in books and read, when about 12 years of age, David Wells' "Principles of Natural Philosophy and Chemistry." It was in the latter subject that he became most interested, and it gave him a preliminary insight and curiosity to know more about the fundamental properties of metals.

At the age of 20, having taken only 2 years in high school, he went to Worcester, Mass. and entered the Worcester Polytechnic Institute from which he graduated 3 years later, in 1881. Returning from that institution to Portland, he taught in the district school for 1 year and served as principal in the high school for 2 years. In 1884-85 he took a year of post-graduate work in chemistry at Johns Hopkins University. After returning from that institution, he taught 1 year in the Eastern Indiana Normal School.

In 1886 natural gas was struck in Portland and Haynes was involved in organizing a company for supplying it to the town. At this time he devised a method for determining the amount of gas flowing through apertures of various sizes under various pressures. He also invented a small thermostat for regulating the temperature of a room heated by natural gas. This apparatus worked perfectly in service, and he afterward used it for about 14 years in his own home.

In 1890, Haynes took charge of the gas field of the Indiana Natural Gas and Oil Co., which was piping gas from the Indiana field into Chicago. In the summer of 1893, he decided to begin building a small road vehicle and completed the drawing for the machine in the early fall of that year. A machine was made at the Kokomo Riverside Machine Works, owned by Elmer Apperson. After it was ready for trial on July 4, 1894, it was taken out into the street. The machine was pushed forward a short distance when the motor started under its own power and the little buggy, America's first "mechanically successful" gasoline-powered automobile, was under way for its first trip. It carried three passengers a distance of about 1.5 miles before it was stopped. It was then turned about and driven all the way back into Kokomo without making a stop. This little machine was run, all told, a distance of perhaps 1,000 miles. It is now in the Smithsonian Institution, Washington, D.C.

In 1898 the Haynes-Apperson Co. was formed. In 1902, the name of the firm was changed to Haynes Automobile Co. Elwood Haynes and Elmer Apperson worked together to solve many of the basic problems inherent in the construction of automobiles. Together, they were granted many of the basic patents in the automobile industry.

While Elwood Haynes is credited as an inventor and developer of America's automobile industry, his major contribution was in the creation and advancement of the so-called "superalloy" metallurgy. Many of his alloys (exactly as he invented them during 1899-1920) are in use today throughout the world known as the Stellite cobalt-base alloys.

In 1899, Haynes developed alloys of nickel and chromium and shortly afterward an alloy of cobalt and chromium. A paper was read in 1910 before the American Chemical Society at San Francisco describing these alloys and their properties. Shortly afterward Haynes found that by adding tungsten or molybdenum to the cobalt-chromium alloy a still harder composition could be produced. In 1913 patents were issued for these compositions. The alloys quickly proved to be a practical success for lathe tools, and the business of their manufacture as commercial products grew rapidly.

Haynes was married to Bertha B. Lanterman, of Portland, Ind.; they had two children, a son and daughter. Haynes died at his home in Kokomo, Ind., Apr. 13, 1925.

Information from Stellite Division, Cabot Corp., Kokomo, Ind.; and Howard County Historical Society, Kokomo, Ind.

J. J. PHILLIPS

Williams Haynes

1886-1970

(Nathan Gallup) Williams Haynes, chemical economist, author and historian, editor and publisher, was born in Detroit, Mich., July 29, 1886, the son of David Oliphant and Helen Williams Haynes. As a special student at Johns Hopkins, 1908-11, he studied chemistry, biology, and economics. Following the bent of his father he then began a journalistic career, first as a foreign correspondent and briefly as editor of the *Northampton (Mass.) Herald.*

In 1916 he joined the D. O. Haynes (his father's) Publishing Co. as secretary and editorial director of its drug trade magazines and a few years later struck out on his own to establish the weekly, *Drug and Chemical Markets.* This proved to be the launching pad for a brilliant career in recording and interpreting the progress and achievements of the American chemical industry. From this nucleus he developed a half dozen chemically related magazines, of which perhaps, *Chemical Week, Modern Plastics,* and *Drug and Cosmetic Industry* are best known today. All were designed, he often said, "to serve the chemically minded business man and the business minded technical man." In a letter he wrote to the writer a few months before he died on Nov. 16, 1970, in a nursing home in Westerley, R. I., he said "Throughout my life my theme song has always been the impact of chemistry on our whole way of life, especially that of man-made synthetics, plastics, fibers, elastomers, drugs and medicines, vitamins and hormones."

Growing out of his life-time interest in chemistry's role in modern civilization was literally a 6-ft shelf of books interpreting the contributions of science and technology to laymen throughout the world. Many were to become best sellers, such as "This Chemical Age" (with three printings in English and 5 foreign translations), "The Chemical Front" (in two printings and special Spanish and Portuguese editions for Latin America), "Men, Money and Molecules," "Sulphur, the Stone That Burns" (2 editions) and "Cellulose, the Chemical That Grows."

Williams Haynes' *magnum opus,* his greatest contribution to the lasting literature of our times, was the six-volume history of "The American Chemical Industry, 1609 to 1948." This was published between 1945 and 1954 by D. Van Nostrand, Inc. with the loyal support of a volunteer committee headed by William J. Hale of Dow Chemical Co. At that time it was hailed as "the most comprehensive and authoritative chronicle of any American industry, a scholarly source of information for all concerned with the origin and development of the major chemical companies." Copies may be found on the reference shelves of all major libraries and are widely used by high-school and college students and the general public, as well as by historians and commentators.

In later years the Haynes pre-Revolutionary home and great library in Stonington, Conn. became an important center for New England historians. Its owner authored, edited, and published many widely quoted papers and books on the early history of the American colonies, especially his ancestral state, Connecticut.

But the works of Williams Haynes were not confined to the printed word. He gave literally hundreds of stimulating and informative talks to service clubs and to social and civic organizations throughout the country. He lectured on chemical economics at several universities and business schools. Probably the most prestigious of these was his famous Brackett lecture at Princeton in 1935.

Haynes was a likeable, gregarious man. He was the founder and first secretary of the Chemical Salesmen's Association. He edited and published the first "Chemical Who's Who" in 1928 and supervised the 1937 and 1951 editions. To the writer of this brief biography "Billy" Haynes will always be remembered as a dear friend and challenging contemporary, a brilliant and sometimes argumentative conversationalist, and a resourceful editor who has left his bench-mark on the lasting literature of the chemical industry.

Personal recollections; "Chemical Who's Who," 3rd ed., Lewis Historical Publishing Co. (1951); *The Percolator* (published by Chemists Club, N.Y.), June-July 1961, pp. 15-18, port.; obituary in *Chem. Week* **107**, No. 2, 12 (Nov. 25, 1970).

SIDNEY D. KIRKPATRICK

Rowland Hazard

1829-1898

Hazard, the "father of the American alkali industry," was not a chemist but a textile manufacturer with sound technical understanding and a rare foresightedness in business. He was born in Newport, R.I., Aug. 16, 1829 son of Rowland Gibson and Caroline (Newbold) Hazard, in the eighth generation of the Hazard family which founded Newport and shaped the business and political affairs of Rhode Island from the mid-17th century onward. In 1833 the family moved to Peace Dale, R.I., on the west side of Narragansett Bay, named after a great-grandfather, Isaac Peace.

Hazard was educated locally, then at Friends' School at Westtown, Pa., at Haverford College, and at Brown University. He graduated from Brown in 1849, having won the first university premium in mathematics, second in mechanical philosophy, the Jackson premium in intellectual philosophy, and the first university premium in astronomy. After a year of travel abroad, he entered the family woolen mills at Peace Dale, succeeding to the business in 1866 along with his brother John Newbold. Hazard was master of all aspects of textile production: he knew the machinery and could operate a loom; as an artist he designed the shawls for which the mill was famous; and as craftsman he laid out the patterns for the machines. In 1872 he planned and constructed a large worsted mill and introduced a system of profit-sharing with the employees.

In 1870 the Hazard interests bought Mine La Motte, the oldest and largest lead mine in the Ozark mountains in southern Missouri. With the combined energy, business acumen, and humanitarianism so characteristic of Quaker businessmen, Hazard plunged into modernization of the mine operations and improvement of the lot of the workers by shortening working hours, instituting a bonus system, and building schools and living quarters. His policies, always tolerant and nonintrusive, paid off, and in 1875 he was able to leave the mines in the hands of William B. Cogswell, a young mining engineer who shared Hazard's views of management and worked with equal energy. At the end of 4 years, Cogswell left Missouri for a well-earned vacation, hoping to find a position with the Hazards in the East.

As Cogswell knew nothing of textile manufacture, he decided to try to interest Rowland Hazard in a new enterprise, the manufacture of sodium carbonate, or soda ash, by the Solvay process, which he had just heard of in a paper presented before the 1879 meeting of the American Institute of Mining Engineers. The process, still in use, consists of adding carbon dioxide to a saturated aqueous sodium chloride-ammonia solution until sodium bicarbonate precipitates; this is recovered by filtration and heated to form sodium carbonate. The process produces unusually pure soda ash, and the raw materials—salt, coke, and limestone—were readily available near the Erie Canal system in central New York State. Thus, he proposed setting up a plant near Syracuse, at what eventually became Solvay, N.Y. Cogswell and Hazard succeeded in convincing the Belgian brothers, Eugene and Alfred Solvay, of the potential of the American market, and by 1884 the first Solvay process soda ash was produced in this country. In the years that followed, the company extended its operations to include its own coking facilities; the ammonia for the reaction was a by-product of the coking process. Ultimately, the by-product business led them into production of phenol and other aromatics. His son, Frederick Rowland, graduated from Brown University, A,B. 1881, A.M. 1884, studied chemistry in Europe, came into the business in 1884, and succeeded to the presidency on his father's death. Another son, Rowland Gibson, joined him in the operation of the Solvay plant as well as heading the mills at Peace Dale.

Hazard led an extraordinarily active life outside his business activities. He was president of the Washington County Agricultural Society, Rhode Island, from its founding in 1876 until his death. Among his many other activities, he was for many years moderator of the South Kingston town meeting, state representative and state senator, unsuccessful candidate for governor, trustee of Brown University, and founder of Peace Dale public library. He married Margaret Anna Rood in 1854; they had five children, of whom beside the sons mentioned, the daughter Caroline, fifth president of Wellesley College, is noteworthy.

Hazard died on his 69th birthday, Aug. 16, 1898 at Watkins, N.Y.

W. Haynes, *Chem. Ind. (New York)* **42,** 248-53 (1940); "National Cyclopedia of American Biography," **12,** 221, James T. White & Co. (1904); excellent accounts of Rowland and other Hazards are gathered in one place in "Lamb's Biographical Dictionary of the U. S.," **3,** 620-621, James T. White & Co. (1900); and Caroline E. Robinson, "The Hazard Family of Rhode Island, 1635-1894," private printing (1896); biography of Rowland Gibson Hazard can be found in "Dictionary of American Biography," **8,** 471, Chas. Scribner's Sons (1931); an obituary notice of Frederick Rowland Hazard appeared in *Ind. & Eng. Chem.* **9,** 413-14 (1917).

ROBERT M. HAWTHORNE JR.

Fred Harvey Heath

1883-1952

Heath was born in Warner, N.H., Feb. 25, 1883 the only child of Benjamin Franklin and Julia Augusta (Wadleigh) Heath. Following his elementary and secondary education in Warner, he attended New Hampshire College, now University of New Hampshire, where he was an excellent student and where his later interest in the rarer elements was stimulated by Charles L. Parsons, and by Charles James, a world-famed authority on the rare earths. After graduating in 1905, Heath began graduate work at Yale University under Frank Austin Gooch and received his doctorate in analytical chemistry there in 1909. During the summer semester of 1908 he studied at Universität Marburg in Germany.

Heath taught general, analytical, physical, and theoretical chemistry at Massachusetts Institute of Technology (1909-10), Case School of Applied Science (1910-11), Wesleyan University of Connecticut (1911-12), the University of North Dakota (1912-17), the University of Washington (1917-23), and University of Florida (1923-51). He was in charge of the University of Florida's general chemistry program from 1926 until his death, Jan. 26, 1952. During the summers of 1914 to 1916, he conducted analytical research on natural waters at North Dakota Biological Station at Devil's Lake.

Heath carried out research on the iodometric determination of copper (the subject of his doctoral dissertation, 1907), the iodometric determination of arsenic and antimony (1908), and the iodometric determination of hydrogen sulfide in natural waters (1923). Long a prominent authority in the study of war gases, he was a major in the Chemical Warfare Service Reserves and was chemical warfare gas consultant for the Florida State Defense Council during World War II. He insisted that war gases were more humane than other weapons, and he emphasized the impossibility of prohibiting chemical warfare by legislation. Together with Waldo L. Semon, he first synthesized symmetrical tetrachlorodiethyl selenide, the selenium analog of mustard gas (1920).

Heath was an authority in several fields including geology, photography, astronomy, sponges, phosphorescent materials, fluoroborates, and the rare earths. He undertook research for the United States Submarine Base at Key West and was a consultant to many industrial firms. He had a long record of service with the American Chemical Society, having been chairman of the Puget Sound Section; secretary, alternate councilor, and chairman of the Florida Section; and a member of the Division of Chemical Education's Committee on Tests and Examinations. Despite his varied research activities, his major interest lay in chemical education; he developed several analytical procedures for use in the undergraduate laboratory.

For many years on the University of Florida campus, Heath's name has been linked with the discovery of element 43. While at University of Washington, sometime between 1917 and 1923, Heath and J. D. Ross, an electrical engineer, were said to have succeeded in preparing compounds of a hitherto unknown element from ores which they received from British Columbia. They believed this to be element number 43, the long sought ekamanganese whose existence had been predicted by Mendeleev as early as 1872. They submitted their results to *Journal of the American Chemical Society,* whose editor, Arthur B. Lamb, requested x-ray spectra as final proof of their discovery. At that time, the equipment needed for this work was not available to them, and consequently their report was not published.

When Heath went to University of Florida, he began work on an emission spectrograph, but in June of 1925, before he could produce the required information, Walther Noddack and Ida Tacke of the Physico-Technical Testing Office in Berlin, together with Otto Berg of the Werner-Siemens Laboratory, announced their discovery of elements 43 and 75, which they named masurium and rhenium, respectively. Despite extensive investigation, their discovery of element 43 was never confirmed and was not accepted. However, their discovery of element 75, rhenium, was recognized.

Since only about one-billionth gram of element 43 was isolated from 5300 g. of Congo pitchblende by Kenna and Kuroda (1961) and since Heath and Ross had isolated salts in weighable quantities, it seems extremely improbable that the element that Heath and Ross reported was technetium. However, it might possibly have been rhenium, in which case they would still have antedated Noddack, Tacke, and Berg's discovery. On the other hand, since the chemical literature contains many unsubstantiated claims to the discovery of the elusive element 43, their substance might have been a mixture of elements.

G. B. Kauffman, "Fred H. Heath (1883-1952)," 124th National Meeting, ACS, Sept. 1953; G. B. Kauffman, "Fred H. Heath and the Discovery of Element 43," *Quarterly J. Florida Acad. Sci.* **26,** 1-3 (1963).

GEORGE B. KAUFFMAN

Lawrence Joseph Henderson

1878-1942

Henderson was born June 3, 1878 in Lynn, Mass., the son of Joseph and Mary Reed (Bosworth) Henderson. He graduated from Harvard University in 1898 with a major in chemistry and received his M.D. degree from Harvard Medical School in 1902. After working for 2 years in the laboratory of Franz Hofmeister at Strassburg, he returned to the United States in 1904 and accepted a position in T. W. Richards' laboratory at Harvard. He was soon appointed lecturer in biochemistry and he remained on the Harvard faculty until his death on Feb. 10, 1942.

Among his contributions to the university may be noted the instrumental role he played in the development of the Fatigue Laboratory (1927) and the Society of Fellows (1932).

He was a man of broad interests which included physiology, medicine, philosophy, sociology, and history of science as well as chemistry. In spite of his diverse interests, his work exhibits in retrospect a fundamental unity in its outlook and methodology. During the course of his research, he became impressed with the need to examine whole systems and to study the mutual interactions between the variables and also with the apparent orderliness of certain systems. Henderson reflected as well as contributed to an organismic, holistic trend which played an important part in the thought of the early twentieth century.

His early work involved the application of physical chemistry to the problem of acid-base equilibria in the body and led to development of his famous buffer equation, first published in 1908,

$$(H^+) = k\frac{(acid)}{(salt)},$$

where (H^+) represents the hydrogen ion concentration and k represents the dissociation constant of a weak acid. The equation describes, in an approximate manner, the action of buffer solutions and makes it clear that a weak acid and its salt will act most effectively as a buffer at a hydrogen ion concentration equal to the dissociation constant of the acid. It was converted into the logarithmic form by K. A. Hasselbalch in 1916, and the Henderson-Hasselbalch equation is still a useful mathematical device for treating problems dealing with buffer solutions.

Henderson's equation explained why carbonic acid and monosodium phosphate, which have dissociation constants of about 10^{-7} moles per liter, acted so effectively in preserving the approximate neutrality of body fluids. This work greatly impressed him with the "fitness" of substances like carbonic acid for various physiological processes and led to publication of two philosophical books, "The Fitness of the Environment" (1913) and "The Order of Nature" (1917). He concluded that the properties of carbon, hydrogen, and oxygen and their compounds uniquely favor the evolution of complex physicochemical systems such as living organisms, and that a kind of "order," whose origin could not be explained in mechanistic terms, exists in nature. The most lasting significance of these books was not in their philosophical statements but in their clear expression of the view that the inorganic world has placed certain restrictions on the direction that organic evolution can take.

His studies on the buffer systems of the body and on acidosis, which contributed significantly to understanding these subjects, also impressed him with the organization or pattern of living beings. This interest in biological organization strengthened his belief in the importance of studying the regulatory processes of the organism, such as the regulation of neutrality. It also served to focus his attention on the wholeness of the organism and on the interdependence of its parts and processes. When the physicochemical relations of the various components of blood began to be studied in his laboratory in 1919, he searched for a graphical device to describe the interrelations between a number of variables. He accidentally stumbled upon the Cartesian nomogram, which was later simplified by transforming it into an alignment chart of the type invented by D'Ocagne. With the nomogram he was able to plot several different variables on a single two-dimensional graph. Much of the work of Henderson and his collaborators on blood as a physicochemical system was summarized in 1928 in his classic book, "Blood: A Study in General Physiology."

In the last decade of his life, he devoted most of his time to sociology. He was impressed by the attempt of Vilfredo Pareto to apply the methods of the physical sciences to the social sciences, and he propagated and elaborated upon Pareto's conception of society as a system in equilibrium.

John Parascandola, *J. Hist. Biol.* **4,** 63-113, 115-118 (1971); John Parascandola, *Medizinhist. J.* **6,** 297-309 (1971); W. B. Cannon, *Biog. Mem. Nat. Acad. Sci.* **23,** 31-58 (1943), with portrait and bibliography; John H. Talbott, *J. Amer. Med. Assoc.* **198,** 1304-1306 (1966), with portrait; Jean Mayer, *J. Nutr.* **94,** 1-5 (1968), with portrait; John Edsall, "Dictionary of American Biography," **23,** 349-52, Chas. Scribner's Sons (1973).

JOHN PARASCANDOLA

William Edwards Henderson

1870-1962

Henderson was born at Wilkinsburg, Pa., Jan. 29, 1870. He attended The College of Wooster, from which he received his Bachelor of Arts degree in 1891. He served as professor of natural science at College of Emporia for the next 2 years. He then began graduate studies at Johns Hopkins University, where he received his Ph. D. degree in chemistry in 1897. He taught chemistry at Ohio University for 2 years, and in 1899 joined the faculty of Ohio State University as assistant professor of chemistry. He was advanced to an associate professorship in 1901. He was made professor of inorganic and physical chemistry in 1908, and in 1912 he became professor of inorganic chemistry.

Henderson's researches included pioneering work in physical and inorganic chemistry, such as investigations of thionic acids and their salts, the action of anhydrous aluminum chloride on unsaturated compounds, the hydrolysis of ethyl acetate by neutral salt solutions, and the decomposition of diazo compounds with sulfur dioxide.

Henderson taught the first science survey course at Ohio State. There he developed the courses in physical chemistry, advanced inorganic chemistry, inorganic preparations, history of chemistry, chemical biography, and chemical bibliography.

He was in a position to realize that the first years of this century were momentous ones for the development of chemistry—and the need for greater understanding of the fundamental physical aspects of chemistry. With his colleague, William McPherson, he developed first-year chemistry textbooks written in the light of the newer physical chemical concepts. The first McPherson-Henderson textbook was "Elementary Study of Chemistry," published in 1905. A laboratory manual, "Exercises in Chemistry," followed in 1906. Altogether, 21 different texts and manuals, in many editions, were published for college and high school use during the succeeding 40 years. The McPherson-Henderson textbooks are among the most widely used chemistry textbooks ever published. They were adopted by hundreds of colleges, and the number of students who used one or more of them is estimated to approach two million.

Henderson was secretary of the council of the American Association for the Advancement of Science in 1915, and the following year he was general secretary of AAAS. In 1917 he was secretary of the Division of Inorganic and Physical Chemistry of the American Chemical Society and the following year was chairman of the division. At Ohio State he was dean of the College of Arts, Philosophy, and Science from 1921 until 1926. He retired from the university in 1941.

Henderson was awarded an honorary D. Sc. degree by The College of Wooster in 1925, and an honorary LL. D. degree by Ohio State University in 1950. He was Ohio State's oldest emeritus professor at the time of his death at his home in Columbus, Sept. 30, 1962, at the age of 92, following several years of poor health.

Obituary with portrait, *Columbus Citizen-Journal,* Oct. 1, 1962; obituary with portrait, *Columbus Dispatch,* Oct. 1, 1962.

LAWRENCE P. EBLIN

William Franklin Henderson

1892-1962

Henderson was born in Decatur, Ill., July 10, 1892. He died in Florida, Nov. 16, 1962. He received his A.B. degree from James Milliken University in 1914 and did graduate work there from 1917 to 1920. He received his M.S. degree in 1921 and Ph.D. degree at University of Pittsburgh in 1922.

In 1920, Henderson became a fellow at Mellon Institute, Pittsburgh. He worked on a fellowship established by Visking Corp. (E. O. Freund), Chicago, Ill., to perfect a process to manufacture a commercial cellulose viscose casing for sausages. At the completion of the fellowship in 1926 he was employed by his sponsor at their plant in Chicago, to supervise making the plastic casing. He became chief chemist until his retirement because of illness in 1952, after which he was a consultant for the company until his death. The last 10 years of his life were spent at his Wistful Village cottage, Fort Meyers Beach, Fla.

Henderson was a trustee of Milliken University. In 1944 during the war he worked for the government as a civilian. Between 1921 and 1928, he wrote several papers pertaining to the chemistry and manufacture of cellulose sausage casing. The first, with John C. Hessler, was published in *Journal of the American Chemical Society,* March 1921, "Butyl- and Isobutyl-Cyano-Acetic Acids." Subsequent papers were published in *Industrial and Engineering Chemistry:* "The Chemical Properties of Cotton Linters," August 1923; "Cellulose Sausage Casings," with Harold E. Dietrich, November 1926; "The Manufacture of Cellulose Sausage Casings," June 1928. Thus he covered the entire process from raw material to final product, including the chemistry involved. In his papers he spoke of the difficulties he encountered in going from laboratory method to plant production of a type of viscose made from cotton linters and suitable for casings which could be peeled off frankfurters to yield the skinless sausage.

Henderson's avocation was pursued with the same singleness of purpose and scientific precision which he gave to his profession. It resulted in assembling one of the most beautiful collections of butterfly specimens in private hands. Some time before his death, he presented the collection to a natural history museum.

Henderson's interest in collecting butterflies naturally brought him in contact with the flora visited by butterflies and insects, and with other aspects of nature. On one of his trips through the Smoky Mountains of Tennessee and North Carolina, he visited Alum Cave. The park naturalist told him there was a deposit in the cave which was thought to have been used during the Civil War, from which gunpowder was made. Henderson scraped up some which he analyzed when he returned home and found it to be largely potassium nitrate.

The writer warmly remembers Henderson as a fellow observer of the natural scene. He was well liked by his colleagues.

Personal recollections; information from Mellon Institute.

BERNARD E. SCHAAR

Waldersee Brazier Hendrey

1900-1962

Hendrey was born in Oklahoma City, Aug. 28, 1900. His grandparents had been early homesteaders in what was then the Territory of Oklahoma. His father, William B. Hendrey, was a deputy marshal and later became Chief of Police of Oklahoma City in the years 1898-1904. His mother was Rosebelle Narron. He had two brothers, Verne (who predeceased him) and William, and a sister, Mrs. C. I. Cooper. While the family was in Oklahoma City, a congenital tumor of Waldersee's right kidney was diagnosed, and the kidney was removed in 1901. This was a pioneering operation in those days.

In 1904 the family moved to Trinidad, Colo. where his father again joined the police force. In 1916 they moved to Bigelow, Ark. where Waldersee graduated from high school. From there he attended University of Arkansas, 1918-22, obtaining his A.B. degree with a major in chemistry and a minor in English. He was on the varsity baseball team and was editor of the school paper. He did not decide on chemistry as a major until well into his third year. He always gave credit to Professor Harrison Hale for his interest in the science.

Waldersee of necessity had to finance his own education. Consequently, on graduation from University of Arkansas, he had a series of jobs, including instructor in freshman chemistry at Texas A&M, and professor of science and math and coach at Morrisville College. He entered Yale University in 1924 as a candidate for the Ph.D. degree in chemistry, which he received in 1928. While at Yale he married Vere Hazel Bradley. They had no children and were divorced in 1933.

On receiving his degree he was hired by Texas Co. as a research chemist—reputedly their first Ph.D.— in the Research Department in Beacon, N. Y. He became assistant director of the Grease Specialties Department and was later director of the Organic Department. During Hendrey's employment with Texas Co., his research in lubricants, additives, and solvent refining led to a series of 18 patents and a number of papers. Hendrey left Texas Co. to go with Spencer Kellogg & Sons as assistant director of research at their laboratories in Buffalo and was responsible for much work on catalytic polymerization, hydroxylation, and utilization of edible oils. His work at Spencer Kellogg resulted not only in some patents but in articles on determination of soybean flour, which appeared in *Industrial & Engineering Chemistry,* as well as other articles on analytical procedures in *Journal of the Association of Official Agricultural Chemists.* In the fall of 1940 he left Spencer Kellogg and joined Edwal Laboratories of Chicago as chief chemist, later vice president.

He helped guide the company through the difficult war years, and it was at this point that his interests in marketing and sales of fine chemicals developed. This led him in 1947, to resign from Edwal and help found Delamar-Hendrey Chemical Co. In 1949, Delamar-Hendrey was dissolved and Waldersee set up Hendrey Chemical Co., representing a number of small, noncompetitive companies including Fine Organics, Inc., Metal Salts, and Pfahnstiel Chemical Co. As a result of the latter association, he joined Pfahnstiel in 1952 and was there until he died on Aug. 25, 1962, while visiting friends in Madison, Wisc.

Waldersee's personal life revolved about his family and his many friends. While at Beacon he visited New York City frequently, and there met Irene Krause, whom he married Sept. 12, 1935. They had four sons, Robert, William, David, and Thomas. Only Thomas followed in his father's footsteps in chemical industry; he receiving his B.S. degree in chemical engineering from University of New Mexico.

Shortly after the Hendreys arrived in the Chicago area, they bought a home in Wilmette and lived in that suburb until his death. In addition to a devoted family life, he was active in many organizations; among them he served as chairman of the Chicago Chapter of the American Institute of Chemists and was active in AAAS, Chicago Drug & Chemical Association, Chicago Chemists Club, and Chemists' Club (New York). His friends remember his warmth and sincerity, his fine sense of humor, and his ability as a raconteur.

Personal recollections; information from Irene Hendrey.

WALTER S. GUTHMANN

Ellwood Hendrick

1861-1930

Hendrick was "an essentially social type; he was a sort of collector of friends." He was born Dec. 19, 1861 in Albany, N.Y. to James and Anna Hendrick. His father came from England to New York in 1839 and settled in Albany, where he practiced law and became president of several corporations. One of these was Albany Aniline & Chemical Works—one of three known dye plants in United States before 1882.

Ellwood, the eldest son, attended University of Zurich. He studied chemistry under Victor Meyer, Victor Merz, and Wilhelm Weith but left school before receiving a degree. He numbered among his close friends Albert Einstein and Lafcadio Hearn. Upon returning home, he began his career as assistant superintendent of Albany Aniline & Chemical Works.

In 1884 he left the dye plant to become agent for a London insurance company; and from 1900 spent 15 years in the stock brokerage business in New York City. Meanwhile, he was spending his spare time with chemistry. From 1917 onward he edited the house paper for Arthur D. Little, Inc., and at the same time served as consulting editor of *Chemical & Metallurgical Engineering*. In 1924 he resigned both positions.

Hendrick was appointed curator of Chandler Museum of Columbia University in 1924. When Charles Frederick Chandler joined Columbia School of Mines in 1864, he began to collect materials which would show growth and progress of applied chemistry. On his retirement in 1910, the museum fell into neglect. It was not until Hendrick took over that conditions were improved, and he retained this position for the rest of his life.

It was Hendrick's ability to popularize chemistry that helped to make him popular and admired. He continued to promote the potentiality and value of chemists and scientists, before, during, and after World War I. He wrote articles and books promoting research and championing the "unsung" who spent their lives doing research work. Among his books were "Everyman's Chemistry" (1917), "Opportunities in Chemistry" (1919), "Percolator Papers" (1919), "Life of Louis Miller" (1925), and "Modern Views of Physical Science" (1925). Hendrick frequently wrote on scientific and philosophical subjects for magazines and newspapers, such as *The Atlantic Monthly, Harpers Magazine, The North American Review, The New York Times,* and such publications as *Chemical &*

Metallurgical Engineering and *Industrial & Engineering Chemistry.*

Hendrick received many honors: Franklin and Marshall College conferred on him the honorary Doctor of Science degree; he was president of the Chemists Club, chairman of the American Section of the Society of Chemical Industry, fellow of the American Association for the Advancement of Science, member of the American Chemical Society, Franklin Institute, and la Société de Chemie Industrielle. He was a trustee and member of executive committee of the Research Corp. As president of the Eastern Association on Indian Affairs, he advocated justice for the Indians. During 1918 he was chairman of the Chemical Division of National War Savings Committee.

Hendrick had recently returned from Europe when he contracted pneumonia and died at his home, 139 E. Fortieth St., New York City, Oct. 29, 1930. He was survived by his wife and two children.

I. F. Stone, compiler, "The Aniline Color, Dyestuff, and Chemical Conditions from August 1, 1914 to April 1, 1917," privately pub.; "Ellwood Hendrick," *J. Ind. Eng. Chem.* **20,** 978-979 (1928); *New York Times* (Oct. 30, 1930); *Nature* **127,** 28-29 (1931).

DAVID H. WILCOX, JR.

Walter Scott Hendrixson

1859-1925

From a small tobacco farm in the backwoods of southern Ohio came a man of character and decision who, for 35 years, exerted forceful leadership in chemistry at Grinnell College and in the state of Iowa, a man who left his mark on the chemical profession that persists to this day.

Hendrixson was born near Felicity, Ohio, Jan. 11, 1859, of English-German stock, the son of Eber Adkins and Sarah Hoover Hendrixson. He was orphaned in his late teens, and Walter used his modest inheritance for his education. He attended Union Christian College, graduating in 1881. From 1882 to 1890 he was instructor and later professor of chemistry, physics, and botany at Antioch College. In 1888 he obtained leave from Antioch to become assistant at Harvard where he received his A.M. degree in 1889 but continued his studies another year.

In 1890 he became professor and head of the Chemistry Department at Grinnell College but continued work toward his Ph.D. degree which he received from Harvard in 1893. A few years later he obtained leave and studied at Berlin and Göttingen. As a student of Walther Nernst at Göttingen during the summer of 1895 he extended Nernst's famous distribution law to systems undergoing association and dissociation in different solvents. His classic paper on the distribution of benzoic acid and of salicylic acid between water and benzene, published in *Zeitschrift für anorganische und allegemeine Chemie* in 1897 was referred to in physical chemistry texts for decades.

Hendrixson was always a colorful person, both in and out of the classroom; his marriage, Apr. 18, 1906, resulted from the gallant rescue of Bessie Bradley from a runaway carriage while he was on vacation in the Colorado mountains. To this union were born two children: Ellen in 1911 and Philip in 1914.

During the period following his marriage, Hendrixson continued his research on the halogen acids and published several papers, including one in 1912 on perchloric acid in electrochemical analysis. In 1915 he began a rigorous investigation of potassium acid phthalate as a primary standard for alkalimetry. A series of papers extending to 1920 established potassium bithalate as the best standard for solutions of sodium hydroxide. Its endorsement by the National Bureau of Standards indicated the far-sighted quality of Hendrixson's research. His work during the 1920's on potentiometric titrations did much to establish the validity of these methods. During his lifetime he wrote 22 papers and three brief, locally published, textbooks. An excellent and rigorous teacher and lecturer who enjoyed demonstrations of the popbang variety, he stimulated all with whom he came in contact. Many of his students reached eminence in the chemical profession, among them Albert Noyes, Jr. a president of the American Chemical Society.

Hendrixson served terms as president of the Iowa Academy of Science and of the Iowa Association of Colleges. In addition to his teaching at Grinnell, he lectured at University of Iowa during the summer of 1902, at University of Illinois, while on leave in 1917, and was research fellow at Johns Hopkins University 1920–21. He died alone of cerebral hemorrhage while opening his cottage at Portage Point resort near Manistee, Mich. His body was discovered June 30, 1925.

Obituaries in *Grinnell and You,* June-July 1925, *Grinnell Scarlet & Black,* Sept. 9, 1925, and *Grinnell Register,* July 2, 1925; information from Philip Hendrixson, from Antioch College, Harvard University, and Grinnell College; personal recollections.

WILLIAM C. OELKE

John Brown Francis Herreshoff

1850-1932

Herreshoff was born Feb. 7, 1850 in Bristol, R.I., son of Charles Frederick Herreshoff 2nd and Julia Ann (Lewis) Herreshoff. He entered Brown University in the class of 1870 but instead of completing his course he became instructor of analytical chemistry, 1869-72. Brown later awarded him three honorary degrees: M.A. (1890), Ph.B. (1905), and Sc.D. (1909).

Herreshoff moved to New York in 1872 and spent his entire professional life near that city. He worked for about 1 year with Charles A. Seeley, joined Silver Spring Bleaching and Dyeing Co. as a chemist in 1873, worked with William M. Habershaw during 1874 and 1875, and in 1876 became superintendent of Laurel Hill Chemical Works of G. H. Nichols and Co. at Newton Creek, L. I. He remained an associate of William Nichols throughout his career, becoming vice-president for operations of Nichols Chemical Co. in 1890, vice-president of Nichols Copper Co. in 1899, and honorary vice-president and consulting engineer of General Chemical Co., a successor, in 1924. His major achievements include a method for manufacturing sulfuric acid on a large scale and improved methods for extracting and analyzing ores.

Herreshoff patented three major advances in sulfuric acid technology. U.S. Patent 335,699 (Feb. 9, 1886) describes replacement of the brick linings used in the chamber process absorption tower with linings of crushed glass or quartz. Brick linings are attacked by acid but glass and quartz linings are not. Substitution of glass for brick thus results in a longer tower life and a purer product. Properly laid crushed glass liners also prevent contact of aqueous sulfuric acid with the tower's lead sheath. U.S. Patents 357,528 (Feb. 8, 1887) and 369,790 (Sept. 13, 1887) describe respectively a process and an apparatus for concentrating sulfuric acid made in Herreshoff's modified chamber process. A continuous stream of sulfuric acid is distilled from a pan which has an insulated cover. The vapor is condensed at an appropriate temperature as very pure, 66° Be′ (93.5 wt. %) H_2SO_4 while the residue, consisting of 98 wt. % impure H_2SO_4, is recycled. This process made very pure, concentrated sulfuric acid available economically in large quantities for the first time. Moreover, shipment of sulfuric acid in large quantities became practical because sulfuric acid of this high concentration attacks steel containers much less rapidly than does dilute acid. For this achievement, Herreshoff was awarded the Perkin Medal of the Society of Chemical Industry in 1908.

Herreshoff made a number of important contributions to extractive metallurgy. He was awarded a U.S. patent in 1883 for a copper smelting furnace having a moveable well, which facilitated the separation of matte and coke from molten copper. With William C. Ferguson, he developed a process for removing minute quantities of impurities from copper as well as several analytical methods for determining copper. In 1896 he devised a smelter having a rabble end which could be agitated to prevent concretion of the ore and which could be replaced without shutting down the furnace. This furnace was first used to smelt iron pyrites, producing iron ore and fertilizer-grade sulfuric acid. It is used today to make activated carbon, and is still known as a Herreshoff furnace.

Herreshoff was a founding member of the American Chemical Society in 1876 and was one of five founders honored at the Society's Golden Jubilee meeting in Philadelphia in 1926. He was a director of American Institute of Chemical Engineers, president of the American Society of Mechanical Engineers, and chairman of the American section of the Society of Chemical Industry.

He was married four times: to Grace E. Dyer, to Emily Duval, to Carrie Ridley Enslow, and to Irma Ridley Grey. He died in Atlanta, Ga., Jan. 30, 1932.

Samuel Carter III, "The Boatbuilders of Bristol; the story of the Amazing Herreshoff Family of Rhode Island," Doubleday (1970); biography, H. Wigglesworth, *Ind. Eng. Chem.* **19,** 1205-1206 (1927); "National Cyclopedia of American Biography" **24,** 96, James T. White & Co. (1935); obituaries: *New York Times,* Jan. 31, 1932; "Who Was Who," vol. 1, p. 555 (1943).

MICHAEL B. DOWELL

Karl Marx Herstein

1896-1961

Herstein was born Nov. 29, 1896 in Elizabethport, N.Y. He was the son of Bernard Herstein, a chemist who received his Ph.D. degree in 1890 from University of Berlin. The senior Herstein worked for the U.S. Bureau of Chemistry as technical expert and as chief of appraisers of chemicals for New York City under the U.S. Treasury Department. He was at the time of his death chief technologist of U.S. Industrial Alcohol Co.

With a father who specialized in economic aspects of industrial problems, it was not surprising that his son followed a similar course of action in his lifetime. Graduating from Columbia in 1917, he was a chemist for the U.S. Navy, the U.S. Army, Sackett and Wilhelms Lithographing Co., American Tobacco Co., and Hochstadter Laboratories, Inc. In 1935 he formed a partnership, Kenney-Herstein, Inc., and in 1937 his own firm, Herstein Laboratories, Inc. He was coauthor of "The Chemistry and Technology of Wines and Liquors" (1935) and was chairman of the New York Chapter of the American Institute of Chemists, as well as a national councilor of the American Chemical Society. In 1958 he received the honor scroll of the New York Chapter of the Institute.

Most of his research was in the fields of textiles, paving materials, tobacco and natural colloids. During the last 5 years of his life his laboratory concentrated on sucrochemistry. This research resulted in many sucrose ester combinations.

He regarded his work for Acrolein Research Corp. as his most outstanding success. He developed a process for producing acrolein from propylene, a special process for regenerating the reagents, and a process for converting acrolein to the salts of acrylic acid. The success of this work was measured by the fact that in the late 1940's these processes were licensed to a major chemical company.

His development of a process for gluing salt to pistachio nuts was his most humorous assignment. His client sued a company that began using the same process. The hearing was held in a private chamber of a judge, and a large bowl of pistachio nuts, as the evidence, was on the desk. "We sat around and discussed the legality of the question," said Herstein. "The nuts were within everyone's reach and the group found themselves nibbling once and again. By midafternoon we realized, much to our horror, that we had eaten up almost all the evidence. At any rate the infringers testing the validity of the patent lost. Except for the amount of free nuts distributed that afternoon our client was able to sell most of his output amounting to an annual volume of over a half million dollars."

The book, "The Chemistry of Wines and Liquors," was started by Thomas C. Gregory. For some reason or other, Gregory never finished it and the publisher, knowing Herstein and his technical competence, asked him to take the existing notes and complete the work. As far as is known, Herstein at the time had no particular expertise on the subject other than from the standpoint of a consumer, although he subsequently became knowledgeable from the chemist's viewpoint. The publication of the volume had an amusing aftermath. His father was a rabid White Ribbon "Prohibitionist." Many people mistakenly congratulated him on his son's new book, and the reaction in chemical terms was quite violent.

Karl Herstein was also a raconteur and lover of limericks and was able to recite literally hundreds without benefit of notes. As he lost all of his hair at an early age in a bout with a chemical reaction, his Yul Brynner appearance caused him to be known among his close associates as "Curly."

Herstein died June 1, 1961 in New York City after a long illness.

Personal recollections; obituaries in *New York Times,* June 3, 1961, *The Chemist,* July 1961, and *The Percolator,* July and Sept. 1961.

RICHARD L. MOORE

Christian Archibald Herter

1865-1910

Herter was born Sept. 3, 1865, in Glenville, Conn., into a family of some note, son of Christian and Mary (Miles) Herter. His father operated Herter Brothers, architects and interior designers to post-Civil War industrialists such as Morgan and Vanderbilt; his nephew would later be Secretary of State under President Dwight Eisenhower.

Herter's early education was largely private and strongly influenced by his father's breadth of culture. His grounding in literature and the arts colored and lent pleasure to the rest of his life; in particular, Herter was an accomplished amateur musician. But he turned to medicine for a career, receiving his M.D. degree from Columbia University's College of Physicians and Surgeons in 1886 at the age of 21. This was followed by postgraduate study with William Welch at Johns Hopkins and with Auguste Forel in Zürich.

In 1886 Herter married Susan Dows, who was to bear him two sons who died in infancy and three daughters. Returning from Europe in 1888, Herter settled with his new family in New York in a large house at 819 Madison Avenue where he would spend the rest of his too-short life. He practiced medicine privately in these early years, specializing in neurological diseases. From this experience he produced "The Diagnosis of Diseases of the Nervous System" in 1892, when he was 27, and a chapter on "Diseases of the Cranial Nerves" in Francis Dercum's "Textbook of Nervous Diseases by American Authors" (1895).

During this time Herter developed the conviction which was to dominate his professional life and have far-reaching effects in American medicine and biochemistry. Medical practice and scientific research cannot be divorced, he decided, but must contribute equally to the science of medicine—an idea common enough today but not widely held in the '80's when there was little real research carried out in medical schools and none in industry. Herter established his own laboratory on the fourth floor of his house, and over the next 20 years he and various assistants produced more than 70 publications in bacteriology, chemistry, pharmacology, and pathology.

Herter's interest in neurology waned as he took up the biochemical problems which seemed capable of solution in the laboratory. His positions reflect this change of interest: visiting physician, New York City Hospital, 1894 to 1904; professor of pathological chemistry, Bellevue Hospital Medical College, 1898 to 1903; professor of pharmacology and therapeutics, College of Physicians and Surgeons, 1903 until his death. The Bellevue chair resulted in his second book, "Lectures on Chemical Pathology in Its Relation to Practical Medicine" (1902), which emphasized the relation between research and practice. Herter also worked briefly with Paul Ehrlich in Frankfurt in 1903-04.

His most important medical-chemical investigation dealt with the bacteria of the gastro-intestinal tract, the chemical agents they produce, and the chronic diseases caused by chemical toxins arising from abnormal intestinal flora. Two books resulted from this work: "The Common Bacterial Infections of the Digestive Tract and the Intoxication Arising from Them" (1907), and "On Infantilism from Chronic Intestinal Infection" (1908), a classic which described what came to be known as "Herter's infantilism." Herter's final work was published posthumously: "Biological Aspects of Human Problems" (1911).

Herter's insistence on medicine as science

had three other consequences felt to the present day. In 1905 he and John Abel, of Johns Hopkins, founded *Journal of Biological Chemistry,* the first such publication in English; Herter served as editor until his death. Second, he helped found Rockefeller Institute for Medical Research, served as director and treasurer and insisted on the establishment of its teaching and research hospital, where he was a visiting physician. Finally, Herter and his wife brought major European scientists to Bellevue and Johns Hopkins by establishing the Herter Lectures at those schools.

In addition Herter helped to found the Harvey Society in New York, helped organize the American Society of Biological Chemists in 1908, and was one of President Theodore Roosevelt's five appointed referees to advise the Department of Agriculture in enforcing the pioneering Food and Drug Act of 1906.

Despite his heavy burden of professional activity Herter always made time for hospital medical practice, and more important, for encouraging younger men in the field. Herter's health failed in his last year, and he died of pneumonia in New York City, Dec. 5, 1910, only 45 years old.

Biochem. J. **5,** xxi-xxxi (1911); "Dictionary of American Biography," **8,** 597, Chas. Scribner's Sons (1931); *J. Biol. Chem.* **8,** 437-9 (1910); *Johns Hopkins Hosp. Bull.* **22,** 161 (1911); *Science* **33,** 846-7 (1911); *J. Amer. Med. Assn.* **55,** 2077 (1910); *N. Y. Times,* and *N. Y. Herald,* Dec. 6, 1910.

ROBERT M. HAWTHORNE, JR.

Charles Holmes Herty

1867-1938

Charles Holmes Herty, Jr.

1896-1953

Charles Holmes Herty, like chemists of old, could do and did everything well that he tried. His son preferred to be an excellent specialist.

Charles Holmes Herty was born Dec. 4, 1867, at Milledgeville, Ga. His father, Bernard Herty, enlisted in the Confederate Army at age 17 and rose to rank of captain by the end of the Civil War. Thereafter, he became a druggist. His son retained an interest in pharmacy until he became a graduate student. When Charles was 11 years of age his parents died, leaving him and his younger sister under the care of an aunt. He graduated from Georgia Military Academy and then from University of Georgia with his Ph.B. degree in 1886. That fall he entered Johns Hopkins University. His extracurricular activities included the glee club and baseball team. In 1890 he received his Ph.D. degree for a dissertation entitled, "The Double Halides of Lead and the Alkali Metals." Herty became assistant chemist at Georgia State Experiment Station that year. In 1891 he became instructor at his alma mater, advanced to the rank of adjunct professor in 1894, and held the appointment until 1902.

In 1899 he took his wife and two young children with him to Berlin. Here he listened to the lectures of the great synthetic dye chemist, Otto N. Witt. He spent the following year at Zurich studying with Alfred Werner. His investigations of complex inorganic compounds were related to Herty's own dissertation. This culminated in a joint authorship of a paper in *Zeitschrift für Physikalische Chemie* in 1901: "Beiträge zur Konstitution anorganischer Vergindungen." It was the first publication of an American with Werner.

While Herty was in Berlin, Witt had made some caustic remarks about Southern methods of obtaining turpentine. Upon returning to Georgia in 1901 Herty contacted the U.S. Bureau of Forestry, visited the pine belts, and found that our naval stores industry was declining. He thereupon invented the "cup" which displaced the old "boxing" method and thus saved money for those collecting turpentine. From 1902 to 1904 he was a regular member of the Forestry Bureau's staff.

During 1904 and 1905 he was with Chattanooga Pottery Co. developing and manufacturing his turpentine cup. This invention made him financially secure.

Herty accepted the position as head of the Chemistry Department at University of North

Carolina in 1905. Here he continued his researches. He published at least 83 papers, of which 23 were on pine and its products. He served as dean of the School of Applied Science at Carolina from 1908 to 1911.

Herty was elected president of the American Chemical Society for the years 1915 and 1916. In 1916 he became editor of *Industrial and Engineering Chemistry* and held this post into 1921. He continually campaigned for a strong chemical industry and in the post-war years for a strong protective tariff to save the chemical industries that had arisen during the war. It was a hard struggle. In September 1919, President Wilson sent Herty abroad to arrange for the purchase of impounded dyes. The Textile Foundation grew out of this trip.

In 1921 Herty relinquished his editorship to become president of the Synthetic Organic Chemical Manufacturers Association and retained the office until 1926.

He helped promote the Ransdell bill of 1930 which established the National Institutes of Health. This was an outgrowth of the mutual interest of Herty and Francis P. Garvan on the relation between chemistry and medicine.

The American Chemical Society's prize essay contests for high school and college freshman were financed by the Garvans as a memorial for their daughter, and Herty helped organize the contests. In 1926 he became advisor to the Chemical Foundation, presided over by Garvan, and held the post until 1935.

During this period Herty's mind turned to the use of Southern pines for newsprint. The pines grew much faster than Canadian spruce. With help from the Chemical Foundation and the State of Georgia he organized the Savannah Pulp and Paper Laboratory. His hopes met with success and on Mar. 31, 1933, *The Soperton News,* a weekly, was printed on paper made by Herty. Now, many American newspapers use paper made from Southern pine. The laboratory that he established is now known as the Herty Foundation Laboratory.

Many honors came to Herty, including seven honorary degrees and many medals. After 3 weeks in a Savannah hospital he died July 28, 1938, survived by two sons and a daughter and by the aunt who had taken care of him as a boy. A Charles Herty Gold Medal Award presented annually to a chemist working in the South is sponsored by the Georgia Section of the American Chemical Society.

Charles Holmes Herty, Jr. was born Oct. 6, 1896 in Athens, Ga. After attending Asheville School, he studied chemistry, played shortstop, and strummed the guitar at University of North Carolina, from which he graduated with his B.S. degree in 1918.

He served a year in the Chemical Warfare Service. At the end of the war he entered Massachusetts Institute of Technology, and obtained his M.Sc. degree in chemical engineering in 1921. Soon he was learning steel-making metallurgy at the Lackawanna Plant of Bethlehem Steel Co. and, thereby, earned his D.Sc. degree in 1924. He became engrossed with the metallurgy of steel and was author or co-author of some 80 papers on the physical chemistry of metallurgy. He received many prizes and awards relating to metallurgy. In 1946 he was president of the American Society for Metals. Lehigh University gave him an honorary D.Sc. degree. Herty was vice president of the steel division of Bethlehem Steel Co. from 1942 until his death on Jan. 17, 1953.

Sketches of Charles Holmes Herty in: "Golden Jubilee Number," *J. Amer. Chem. Soc.,* vol. 48 (August 20, 1926); and by Frank K. Cameron, *ibid,* **61,** 1619-1624 (1939); David H. Wilcox, Jr., "Werner and Dyes," *Advan. Chem. Ser.,* **62,** 86-102 (1967); personal communication from Williams Haynes; Mark O. Lamar, letter, *Chem. Eng. News,* Jan. 22, 1968, p. 7; biography of Charles Holmes Herty, Jr. by Paul D. Merica in *Biog. Mem. Nat. Acad. Sci.,* **31,** 114-126 (1958), with portrait.

DAVID H. WILCOX, JR.

William Francis Hillebrand

1853-1925

Hillebrand was born Dec. 12, 1853 in Honolulu, son of William Hillebrand, a physician and botanist. Young William attended Oahu College near Honolulu and the College School in Oakland, Calif. for his elementary and secondary education. He entered Cornell University uncertain of his career plans in 1870 and remained there until November 1872, when at his father's suggestion he transferred to Heidelberg to study chemistry. He studied for five semesters under such prominent scientists as Robert Bunsen and Gustav R. Kirchhoff and received his Ph.D. degree *summa cum laude* in March 1875. Fifty years later, shortly before his death on Feb. 2, 1925, Heidelberg awarded him an honorary Doctor of Natural Philosophy degree for his work in chemical geology.

Hillebrand remained at Heidelberg for two additional semesters before going to Strassburg for three semesters' work in organic chemistry with Rudolph Fittig and in petrography. He then took a course in metallurgy and assaying at the Mining Academy at Freiberg. He returned to the United States in 1873 and became a partner in a Leadville, Colo., assaying firm. His partners retired shortly after he joined the firm, and Hillebrand ran the business alone until July 1880. At that time he became a chemist for the Geological Survey and was stationed at Denver until he was transferred to Washington in November 1885.

In 1892 he became professor of general chemistry and physics at National College of Pharmacy in Washington (now part of George Washington University), concurrently holding his job with the survey. He remained at that position until 1910. In 1908 he left the Survey to become Chief Chemist at the National Bureau of Standards, succeeding William A. Noyes. He remained in this position until his death.

Hillebrand's first chemical research was carried out at Heidelberg on the metals cerium, lanthanum, and didymium. His second paper pointed out that these metals were trivalent rare earths and not divalent calcium-like metals as had previously been believed. Three of Hillebrand's studies dating from his Colorado period are significant. In 1890 he isolated nitrogen from the gases released when uraninite is dissolved in sulfuric acid, and he narrowly missed discovering helium. Hillebrand also observed that the igneous rocks of the Rocky Mountains contain a much greater percentage of strontium and barium than do the rocks to the east and the west. He also devised a practical method for analyzing silicate rocks. The U.S. Geological Survey issued several bulletins, authored by Hillebrand and very well received, on "Some Principles and Methods of Analysis Applied to Silicate Rocks," *Bulletin* **148** (1897), and later "The Analysis of Silicate and Carbonate Rocks," *Bulletin* **305** (1907), 200 pp.; and *Bulletin* **700** (1919), 285 pp. At the time of his death he was preparing a comprehensive treatise on analytical methods in collaboration with Gustav E. F. Lundell.

Hillebrand is probably best known to many members of the American Chemical Society for his service to the Society. In 1896 he became an organizing member of the Committee on Coal Analysis. Until the year of his death he served as chairman of the Supervisory Committee on Standard Methods of Analysis, which must formally approve all analytical procedures before they can be published as standard. Hillebrand was associate editor of *Journal of the American Chemical Society* from 1900 until 1922, assistant editor of *Chemical Abstracts* from 1907 until 1910, and assistant editor of *Journal of Industrial and Engineering Chemistry* from 1909 until 1916. He also served as president of the Society in 1906.

Biographies by C. E. Waters in *Proc. Amer. Chem. Soc.*, **47,** 53-60 (1925), and E. T. Allen in *J. Chem. Educ.* **9,** 72-83 (1932).

SHELDON J. KOPPERL

Ethan Allen Hitchcock

1798-1870

Hitchcock, born at Vergennes, Vt., May 18, 1798, son of Samuel and Lucy Caroline (Allen) and grandson of Revolutionary War hero Ethan Allen, entered the United States Military Academy when he was 16 and embarked on a military career that lasted 40 years. He taught at West Point, fought against the Seminoles in Florida, and was breveted for bravery in the Mexican War.

He retired from the Army in 1855 and turned scholar—he had been one of the most studious officers in the Army, lugging trunks of books from post to post.

In 1854 while browsing through a book shop in New York, he came upon an alchemical tome and purchased it out of curiosity. He thought he recognized hidden meanings in the text, and before long conceived a theory that alchemical works were really treatises upon religious education. "The alchemists were [religious] *Reformers* in their time," he wrote, "obliged indeed to work in secret. . . . They lived, for the most part, in an age when an open expression of their opinions would have brought them into conflict with the superstition of the time, by which they would have been exposed to the stake. . . . The works of the *genuine* Alchemists—excluding those of ignorant imitators and mischievous imposters—are all essentially religious; and . . . the best external assistance for their interpretation may be found in a study of the Holy Scriptures, and chiefly in the New Testament."

In 1855 less than a year after he had started his alchemical theorizing, he published his ideas in a pamphlet, "Remarks upon Alchymists, and the Supposed Object of Their Pursuit; Showing that the Philosopher's Stone is a Mere Symbol, Signifying Something Which Could Not Be Expressed Openly without Incurring the Danger of an Auto de Fe."

Thereafter he galloped ahead on his hobby horse, reading the works of Van Helmont, Paracelsus, Raymond Lully, Hermes Trismegistus, Jacob Bohme, and other adepts. After the bookstores of America had disgorged their old alchemical tomes, he sent to Europe for books. When the ship *Mary Green* from Liverpool went down in 1856, it carried to the bottom 20 volumes ordered by Hitchcock. During the winter of 1856-57 he expanded his pamphlet into a book, "Remarks upon Alchemy and the Alchemists, Indicating a Method of Discovering the True Nature of Hermetic Philosophy; and Showing That the Search after the Philosopher's Stone Had Not for its Object the Discovery of an Agent for the Transmutation of Metals."

During the Civil War Hitchcock was offered the command of armies, but his health was so poor he declined active duty and remained in Washington as military adviser to Stanton and Lincoln. Even during the conflict he pursued his illusions about alchemy. His obsession led him to believe that he had discovered hidden alchemical meanings in Shakespeare's sonnets, Dante's poems, and Spenser's poems, and he published several books on the subject. A passage from his 1866 diary reads: "I saw, a moment since, what the Philosopher's Stone signifies . . . a great number of passages in books of alchemy seem perfectly clear now. . . . It is a kind of revelation, but, when seen, has an effect something like looking at the sun."

Hitchcock died at Sparta, Ga., Aug. 5, 1870. His theory was ridiculed here and abroad, but in the 20th century it caught the attention of a few European psychologists, and they accepted some of his beliefs. Hitchcock's old book on alchemy thus stimulated the writing of works such as Carl Jung's "Psychologie und Alchemie."

W. A. Croffut, "Fifty Years in Camp and Field: Diary of Major-General Ethan Allen Hitchcock," G. P. Putnam's Sons (1909) contains selections from Hitchcock's diaries and letters; Hitchcock's diaries, letters, manuscripts, books, clippings and other material are in the W. A. Croffut papers, manuscript division, Library of Congress; I. Bernard Cohen, "Ethan Allen Hitchcock, Soldier-Humanitarian-Scholar, Discoverer of the 'True Subject' of the Hermetic Art," *Proc. Amer. Antiquarian Soc.* (April, 1951), pp. 29-136; "Dictionary of American Biography," **9,** 73, Chas. Scribner's Sons (1932).

WYNDHAM D. MILES

Lauren Blakely Hitchcock

1900-1972

Hitchcock was born in Paris, France on Mar. 18, 1900 of American parents. His father, Frank L. Hitchcock, a professor of mathematics at Massachusetts Institute of Technology, whose Ph.D. was in chemistry, provided a milieu well-suited to channeling Lauren's interests. Lauren's bachelors degree (1920), masters degree (1927), and doctor of science degree (1933), all from Massachusetts Institute of Technology, were in chemical engineering.

His interest in teaching (University of Virginia, 1928-35; University of Buffalo, 1963-72) bracketed several positions in industrial research (Hooker Electrochemical Co., Quaker Oats, National Dairy Products Corp.). In 1954 he was made president of the Los Angeles Air Pollution Foundation, a private organization dedicated to identifying and eliminating the sources of air pollution. Hitchcock's vision in recognizing and bringing order to the multiplicity of factors contributing to a complex problem found its greatest challenge and opportunity in the attempt to remove the blight of urban air pollution.

He brought together experts in the fields of meteorology, chemistry, physics, and engineering to integrate their contributions in the study of the behavior of parts per million gases, culminating in the discovery of the role of the incomplete combustion of petroleum products (e.g., internal combustion engine exhaust) as acted upon by the sun's energy under particular meteorologic conditions such as temperature inversions, to produce noxious chemical products. His group further showed that these chemicals were related not only to adverse reactions upon species of the animal and plant kingdoms but hastened the deterioration of materials.

Hitchcock advocated that the role of the chemical engineer be sufficiently broadly conceived as to encompass not only the areas of special technical knowledge contributing to the understanding of a problem but also the relevant social and political forces without knowledge of which any recommendations for remedy might well be unrealizable.

Upon completing his work in Los Angeles (1954-1957), Hitchcock returned to New York City where he established a chemical engineering consulting firm, Lauren B. Hitchcock Associates. During this time, he elaborated and articulated the principles of management, which he believed would best serve the chemical industry and society.

In 1963 he joined the engineering faculty of University of Buffalo where he remained active in the field of environmental pollution and in continuing education for industrial research and management.

Hitchcock joined the Buffalo firm of Ecology and Environment, Inc. as director of development three months before his death, which occurred on Oct. 16, 1972.

In 1947 he was a founder and first president of the Commercial Chemical Development Association, later the Commercial Development Association, Inc. He was chairman of the American Section of the Society of Chemical Industry in 1953 and in 1971 received the Professional Achievement Award of the Western New York Section of the American Institute of Chemical Engineers.

Hitchcock published many articles in science and management journals, and edited "The Fresh Water of New York City: its Conservation and Use" (1967).

Personal recollections; obituary in *New York Times* (Oct. 17, 1972).

JOHN HITCHCOCK

Calm Morrison Hoke

1887-1952

C. M. Hoke, as she always signed herself, was born in Chicago, the daughter of Samuel W. and Utopia (Wright) Hoke, July 25, 1887. She attended the local schools, started college at Wittenberg, Ohio and finally obtained her B.A. (chemistry) degree at Hunter College in 1908. Much interested in botany, she returned to Chicago, took a B.S. degree in biology in 1910, and did other graduate work at Columbia and New York University.

One summer when she came home, she found that her father had bought a business that made blowtorches, and that he expected her to run it. She did a complete mental somersault, turned to the Columbia School of Mines for all the courses in engineering, mining, and metallurgy into which she could pry admission and emerged in 1913 with her master's degree—the first in engineering that Columbia had ever granted to a woman. Concurrently, she had been learning her new business, and except for 2 years during World War I when she taught at Columbia, her entire professional career was devoted to it.

This was Hoke, Inc., and the associated Jewelers Technical Advice, which her father had started in 1912. The company soon had several patents for her inventions in fine small torches, valves, and other devices useful in the delicate work of jewelers. Their advertisements, with the OK of the name encircled and raised as a superscript, soon became familiar to many thousands who never knew the originator.

Calm became an authority on precious metals, especially the reclamation of wastes. Methods of treating gold and silver were old and well-tried, but platinum was new and expensive, so her contributions were of great economic importance. She was chief chemist of Hoke, Inc., until 1926, when she became vice president and chief consultant. The business was sold in 1934.

Through Jewelers Technical Advice, C. M. Hoke instructed jewelers, dentists, refiners, and others in the melting, refining, salvaging, and finishing of all the precious metals. She also gave generously of her time and knowledge for talks to schools, technical groups, and clubs, on what she called "gemology" and "metal book-keeping." She introduced the use of platinum for jewelry and designed many beautiful pieces, with and without precious stones, to show its possibilities. She had many articles published in professional and trade journals; and her two books: "Testing Precious Metals" (1932), and "Refining Precious Metal Wastes" (1940) are classics in their field.

As a professionally minded person, Calm Hoke was an inspiration to anyone. She was one of the prime movers in the founding of the American Institute of Chemistry, as it was first called. Her signature is on the charter. Many of the early meetings were held in the office of Hoke, Inc.; and there Calm ground out the mimeographed sheets that served as the *Bulletin* of the AIC for the first 5 years.

Calm enjoyed her crowded, well-rounded life. She had a pixie-ish sense of humor and a comforting, resilient disposition. On her birthday in 1927 ("It took the curse off becoming forty") she married T. Robert McDearman, a civil engineer from North Carolina, for many years with the New Jersey State Highway Department. They made their home in Palisade, N.J. where they could indulge their common love of nature in gardening, camping, touring, and rearing cats. As a secret hobby, not learned until someone found a scrap book of clippings after she died, Calm wrote bits of beautiful poetry.

She was a rabid "Lucy Stone" in her insistence on being known and addressed as "Miss Hoke." She was a member of many groups, including ACS, AAAS, AIC, and AIMME.

Calm died July 13, 1952, survived only by her husband. As a devoted wife and companion, a good neighbor, a loyal friend, an intellectually honest scientist, a skilled technician, and a trustworthy consultant, she will long be remembered by those that were in any way associated with her.

Personal recollections; *Chemist,* Apr. 1948, Oct. 1952; *Mining Engineering,* Nov. 1952.

FLORENCE E. WALL

Harry Nicholls Holmes

1879-1958

Holmes was born at Fay in Lawrence County, Pa., July 10, 1879, the son of John Pattison and Eliza (Nicholls) Holmes. He attended Westminster College, Pa. from which he obtained his B.S. degree in 1899 and his M.S. in 1904. He received his Ph.D. degree in chemistry at Johns Hopkins University in 1907. Westminster College awarded him an honorary LL.D. degree in 1941.

Holmes was head of the Chemistry Department at Earlham College from 1907 to 1914 and professor of chemistry and head of the Chemistry Department at Oberlin College from 1914 until his retirement in 1945. He served with the United States Army in 1918.

Holmes was well known for his work in colloid and physical chemistry and for his researches on emulsions and vitamins. He conducted the first laboratory class in the U.S. on the chemistry of colloids (1910), wrote the first lab manual of colloid chemistry used in this country (1921), and with J. H. Mathews organized the first Colloid Symposium (1923). In his research he dealt with dialysis, gels, catalysts, adsorption, emulsions, and finally vitamins and biochemistry. With Ruth E. Corbet he was the first to isolate crystalline vitamin A in 1937. He and co-workers at Oberlin isolated butyl alcohol from bone marrow. He also introduced the laboratory technique of chromatography into the United States.

Holmes was a member of the National Research Council from 1923 to 1929 and was chairman of the NRC subcommittee on colloid chemistry, 1919 to 1925. He was secretary of the Division of Physical and Inorganic Chemistry of the American Chemical Society in 1919 and chairman of the division in 1921. He served as chairman of the Cleveland Section of ACS in 1924.

Holmes wrote two series of general chemistry textbooks with accompanying laboratory manuals which went through several editions. He was also the author of a colloid chemistry text. His laboratory manual of colloid chemistry went through several revisions. He wrote several popular books on science. One, entitled "Out of the Test Tube," went through five editions. Holmes took a vital interest in athletics. His versatility was further demonstrated by his skill in painting.

Holmes was president of the American Chemical Society in 1942. He served as consultant to the National Defense Research Council, was a member of the War Production Board in 1942, and served as a civilian with the Office of Scientific Research and Development in 1944.

Holmes received the Oberlin alumni medal in 1945, the gold medal of the American Institute of Chemists in 1951, the Kendall Co. award in colloid chemistry (an ACS award) in 1954, the James F. Norris award of the ACS Northeastern Section for outstanding achievement in the teaching of chemistry in 1955, and the Westminster College alumni award in 1957. He died July 1, 1958 in Oberlin.

Harry N. Holmes, "The Growth of Colloid Chemistry in the United States," *J. Chem. Educ.* **31,** 600-602 (1954); *Chem. Eng. News,* **20,** 1623-1624 (1942); obituary with portrait, *Chem. Eng. News,* **36,** 122 (July 21, 1958).

LAWRENCE P. EBLIN

Elon Huntington Hooker

1869-1938

When Hooker, founder of the Hooker Electrochemical Co., one of America's leading chemical companies, was born, electrochemistry was little more than a laboratory exercise. By the time of his death, it was a vast industry in its own right.

Hooker was born in Rochester, N.Y., Nov. 23, 1869 the third of eight children born to Susan Huntington and Horace B. Hooker, a direct descendant of Thomas Hooker, well-known as the founder of Hartford, Conn., in 1638. Although the Hooker family always maintained an excellent social position, there were times, according to Elon's mother, when the family was "dreadfully pinched for money."

A confident and charming young man, Hooker was educated in the Rochester schools and was awarded his B.A. degree by University of Rochester in 1891. At first he was not a very good student and was inclined to be more interested in sports and social affairs than serious study, but during the summer vacations he worked for Emil Kuiching, Rochester city engineer, whose motto "Bring things to pass" so inspired Hooker with driving force that it became his motto, too, and served him well in later years.

After he graduated from University of Rochester, Hooker worked under Kuiching for a year and then in the fall of 1892 entered Cornell University, graduating in 1894 with his B.S. degree in civil engineering. The following fall he was back at Cornell on a graduate fellowship in hydraulic engineering; the first 6 months of the fellowship he spent at Cornell, the last six months in Europe at the Polytechnicum in Zurich and the Ecole des Ponts at Chausees in Paris.

In 1895 Hooker returned to Cornell to finish his dissertation and was awarded his Ph.D. degree in 1896. After engaging in private engineering work, he was named in 1898 a member of a commission charged with the responsibility of inspecting the proposed canal routes across Panama and Nicaragua.

From 1899 to 1900 he served under Governor Theodore Roosevelt as deputy superintendent of public works for the state of New York, and from 1900 to 1901 he served under Governor Odell in the same capacity. This position opened up to Hooker political and business contacts with many people of great influence, with whom he was able to establish a reputation which was to prove valuable when he later decided to establish himself in a business of his own.

Jan. 25, 1901, Hooker married Blanche Ferry, daughter of Dexter Ferry, a Detroit industrialist, and in July of that year, a friend of the Ferry family, impressed by Hooker's ability, made him vice-president of The Development Co. of America, a company organized to locate potentially profitable but undercapitalized enterprises and then supply needed backing and management to put them on a profitable basis. This was the type of training that prepared him for organizing Hooker Electrochemical Co.

Hooker remained with The Development Co. of America less than 2 years and then established his own company, Developing and Funding Co., in January 1903. Having completed the organization and raised the necessary capital, he began the search for an enterprise suitable for investment. After examining more than 250 projects, he decided on the manufacture of bleaching powder and caustic soda, using the recently discovered Townsend electrolytic process.

In November 1904 Hooker brought in as consultants his brother, Albert Huntington Hooker, then chief chemist for a Chicago paint company, and Leo H. Baekeland, already famous as the inventor of the phenolic known as Bakelite and of Velox photographic film.

Experimental plants in Brooklyn and Cleveland proved the process, and in 1905 he selected Niagara Falls as the site for the company's plant, Jan. 9, 1909 the plant went into operation. By June of that year the operation was up to the scheduled production of 5 tons of caustic soda and 11 tons of bleaching powder a day. Convinced that the plant was too small to be profitable, Hooker quadrupled the capacity of the plant Nov. 6, 1909, and renamed it Hooker Electrochemical Co.

When the supply of German chemicals was shut–off during World War I, Hooker decided to branch into the manufacture of hydrochloric acid, chlorobenzene, picric acid, and 15 other leading chemical products needed in the war effort. Again in 1929 Hooker expanded its operations and opened up its new and modern plant in Tacoma, Wash. By 1973 Hooker Chemical Co. produced hundreds of chemical products and intermediates used by every important chemical industry in America.

Hooker served as first president of Hooker Electrochemical Co. and thereafter, until his death, was on its board of directors. Although this required the principal effort of his life, he found time to preside over the Manufacturing Chemists Association from 1923 to 1925.

In 1912 Hooker served as the chairman of the American Defense Society and later as the president and a director of the Research Corp., as a member of the National Industrial Conference Board, and as a trustee of University of Rochester. He died May 10, 1938.

Robert E. Thomas, "Salt and Water, Power and People, A Short History of Hooker Electrochemical Company," Hooker Electrochemical Co. (1955).

JAMES C. COX, JR.

Samuel Cox Hooker

1864-1935

Hooker, born in England, Apr. 19, 1864, was certainly the most versatile of the chemists who chose careers in the sugar industry of the United States in the latter part of the 19th Century.

Hooker graduated with highest honors at Royal School of Science, London, at the age of 20. He then went to Munich, where he studied under Eugen Bamberger and obtained his Ph.D. degree in the remarkably short period of 1 year. Four articles by Bamberger and Hooker in 1885 on the composition of retene showed the proficiency of the junior author in organic chemical research, a field to which he planned to devote his entire career.

Hooker went directly from Munich to the United States where he hoped to obtain a position in a university with a view to continuing his research. This type of work proved unavailable so he became chief chemist of Franklin Sugar Refinery in Philadelphia, late in 1885. This started Hooker's 30-year career in sugar.

His work at Franklin Refinery permitted him to carry out investigations unrelated to sugar, one of the most important of which was his proof by chemical analysis that the Philadelphia water supply was polluted by water from the Delaware River. He also published many papers on his results in organic chemical research in various German, British, and U.S. journals between the years 1887 and 1896. His 11 articles on lapachol and the derivatives and structure of this compound, extracted from bethabara wood from Guiana, became the basis for his research in later years.

Meanwhile American Sugar Refining Co. purchased the Franklin Refinery. The president of American was Henry O. Havemeyer, the dominating figure in the sugar refining industry during the last decade of the 19th century. He soon recognized Hooker's ability and technical knowledge of refining and made him one of the important consultants of the expanding corporation. Havemeyer invited Hooker to meetings of the board of directors as a technical adviser. This was an unusual development because chemists had not then attained the prestige in the sugar industry that came later.

The beet sugar industry, just developing in the West, attracted Havemeyer's attention, and he acquired controlling interests for his company in many of the new beet sugar factories. Although Hooker was then only 32 years old and had never seen a beet operation, he was assigned in 1896 the colossal task of organizing the technical operations of all beet sugar plants under Havemeyer's control. Hooker is credited by his biographers with having saved the beet sugar industry from its disorganized state and placing it on a high level of efficiency. He also manifested great business ability in the organization of the Great Western Sugar Co. of which he was a director from 1909 to 1913. His towering personality both mentally and physically (his height is variously given as 6 feet 6 inches to 6 feet 10 inches) caused many to be overawed in his presence. He was a perfectionist who expected a job to be performed faultlessly, a standard he also demanded of himself. This strict training was obviously beneficial to his assistants, many of whom later obtained positions of great responsibility, both in cane and beet sugar work.

In spite of Hooker's unquestioned success in both branches of the industry he never published anything on sugar. This may have been because of the ill-advised policy of "secrecy" in the larger refining companies which dictated that not even minor details of refinery operations could be published or discussed by employees. Whatever Hooker's reasons for his silence, his name is practically unknown to men now in the industry to which he contributed so commendably.

In 1915 when he was 51 years old, Hooker severed all connection with sugar work to continue his research on lapachol and its derivatives about which he had published so much 20 years before. Working as a private investigator in a well-equipped laboratory behind his residence in Brooklyn, he resumed his earlier research, but for some reason, variously explained by his biographers, he did not publish any of his findings on lapachol and related compounds before his death Oct. 18, 1935. These appeared as 11 articles under Hooker's name in *Journal of the American Chemical Society* during 1936. A memorial volume, including all the publications by Hooker relating to lapachol, (22 in all) was described by Louis F. Fieser, the editor, as "a remarkably complete and unified exposition of a brilliant chapter in organic chemistry." Fieser also said, "The appearance [of this series of papers] in the year following his death after a silence of 40 years, came as a complete surprise to most followers of chemical literature."

Hooker's only publicity came from his unusual interest in stage "magic" and illusions. As a schoolboy he became deeply involved in this subject and gave exhibitions at age 16, but he set this activity aside to pursue his chosen career in chemistry. After his retirement Hooker resumed the development of his old hobby with his usual zeal and application. He soon acquired a high degree of proficiency and became a member of the American Society of Stage Magicians. Occasionally he gave private exhibitions to fellow-craftsmen and many acts and illusions of his own invention completely baffled professionals in the art. Hooker's accomplishments in this field attracted wide attention. His feats and technical devices in magic were described in newspapers, magazines, and professional journals. Sidney J. Osborn, in a biographical sketch in *Industrial and Engineering Chemistry* devoted half a column to Hooker's ability in this unusual hobby. One writer termed him "the greatest living magician." Hooker had an extensive library on magic, including many early German, Dutch, and French books on the subject. His collection also included apparatus used by famous magicians. This enigmatic man, who did not publish a line about his high-caliber accomplishments in chemistry and technology for 40 years, became a noted figure in the realm of stage entertainment!

Of greater importance was Hooker's energetic accumulation of scientific books which one biographer described in 1936 as "one of the most complete chemical libraries in the world." After Hooker's death Neil Gordon, professor of chemistry at Central College, Mo., with the aid of benefactors, raised money for Central to buy the library. When Central found maintenance too costly, Gordon managed to locate other benefactors, including the Kresge Foundation, to enable Wayne University, where Gordon had gone as professor, to purchase the library from Central. It is now called the Kresge-Hooker Scientific Library, and seems the only permanent monument bearing Hooker's name.

Hooker married Mary E. Owens in 1887. Mary had learned chemistry from Frank W. Clarke at University of Cincinnati and then continued her studies at Royal School of Science, where she and Hooker were fellow students. Mary published at least three articles before she left chemistry to marry Hooker and raise a family. The couple had four children. She died June 21, 1936, less than a year after her husband.

Biographies by C. A. Browne in *J. Chem. Soc.* pp. 550-555 (1936), and by S. J. Osborn in *Ind. Eng. Chem.* **23,** 828-829 (1933); L. F. Fieser, ed., preface and table of contents, Memorial Volume, Collected Works of S. C. Hooker on Lapachol Chemistry (1936); *Chem. Eng. News* **22,** 946 (1944); personal recollections.

GEORGE P. MEADE

Worthington Hooker

1806-1867

Hooker, writer of a successful chemistry text for children, was born in Springfield, Mass., Mar. 3 (or 2), 1806. He graduated from Yale in 1825, received his M.D. degree from Harvard in 1829, then settled in Norwich, Conn. and practiced medicine. Yale engaged him as professor of theory and practice of medicine in 1852, and thereafter he resided in New Haven. He was active on committees of the American Medical Association and served as vice president of that organization in 1864.

In 1849 Hooker published a book on medical ethics. He proved to be a competent writer with ability to express himself clearly and understandably, and thereafter he turned out a medical work or a science text almost every year. A mother who read Hooker's "Child's Book of Nature," to her daughter was so pleased with the book that she suggested that he write a similar book on chemistry. He tested her proposal in the following manner:

"I selected a few of those schoolrooms in the public schools of New Haven in which the scholars were from 11 to 13 years of age. I visited these rooms from time to time, talking to the pupils for half an hour on chemistry, without trying any experiments but illustrating the subject largely from common every-day phenomena. At each visit I questioned them upon what I had told them at the previous visit and allowed them to ask me questions. In this way I found out what they could understand and what they wanted to know about chemistry. I was surprised to see how much of this science was within the reach of their capacity and, at the same time, could be made very interesting to them. During all this time I jotted down my results and at length put them into the shape in which they now appear, so that the book was almost literally made in the school-room."

The little volume that resulted, "First Book of Chemistry. For the Use of Schools and Families," appeared in 1862. It was reprinted through the 1860's and early 1870's, and a second edition was issued in 1877. In the book Hooker gave his youthful readers directions for performing simple experiments.

Hooker soon followed his "First Book of Chemistry" with a set of three volumes under the title, "Science for the School and Family," published between 1863 and 1865. Volume 2 of this set was "Chemistry," intended for slightly older readers. This text was used in many schools and remained in print for 20 years.

With only a knowledge of chemistry learned as an undergraduate and as a medical student, reinforced by what he assimilated from books during his busy life, Hooker was an interesting figure in American chemical education because of his skill in composing readable, stimulating, elementary texts for youngsters. He died in New Haven, Nov. 6, 1867.

Obituaries in *Trans. Amer. Med. Ass.* **19,** 442 (1868); *Med. Rec.* (New York) **2,** 453 (1867); and by Henry Bronson in *Proc. Conn. Med. Soc.* (2 S) **3,** 397-402 (1868-1871). Quote is from preface to "First Book of Chemistry," Harper & Bros. (1877).

WYNDHAM D. MILES

B Smith Hopkins

1873-1952

B Smith Hopkins was born in Owosso, Mich., Sept. 1, 1873 the son of Loren and Clara Norgate Hopkins. He received his A.B. degree in 1896 and A.M. degree in 1897, both from Albion College. He studied at Columbia University during the 1900-1901 school year and received his Ph.D. degree from Johns Hopkins University in 1906. Albion College awarded him an honorary Sc.D. degree in 1926, and Carroll College (Waukesha, Wisconsin) granted him an honorary LL.D. degree in 1940. He won many other honors for his chemical work.

Hopkins began his teaching career in the public schools of Menominee, Mich. in 1897. He became principal of the high school the following year and superintendent of schools in 1901. During his high school teaching career he taught science, classics, and coached the football team. After receiving his doctor's degree in 1906, he entered the college teaching field at Nebraska Wesleyan University. Three years later he went to Carroll College. He joined the faculty of University of Illinois in 1912 and remained there until his retirement in 1941.

Upon joining the staff at Illinois, Hopkins joined forces with Clarence W. Balke, who already had underway a series of researches on beryllium, yttrium, columbium (now called niobium), tantalum, and the rare earths. When Balke left the University in 1916, Hopkins carried on with this research, specializing more and more in the chemistry of the rare earths. It was in this field that he made his great contributions to chemistry. At that time separation of the rare earths from each other was a long and tedious task, depending upon repeated recrystallizations of the double magnesium nitrates, the bromates, and other salts. For some separations, there had to be several thousand recrystallizations. Each of the graduate students enrolled in Hopkins' course spent a portion of his time on the separation and purification of materials which later students would use. He devoted the rest of his time to his thesis research on purified materials which earlier students had prepared and purified. These studies included spectral and magnetic measurements, atomic weight determinations, synthesis of new compounds, and numerous other studies in rare earth chemistry. In 1926 Hopkins, with Leonard Yntema and J. Allen Harris, announced the discovery of the long sought element 61, which they named "illinium." Additional work did not concentrate this element any further, and after the fission reaction was developed during World War II, it was found that element 61 is highly radioactive, and most chemists came to the conclusion that it did not exist in nature. Professor Hopkins was bitterly disappointed that his discovery of illinium, which he considered to be the climax of his work, was not accepted. In any event, his contributions to rare earth chemistry were enormous and laid the groundwork for much of the research that followed.

Hopkins had expected to retire on his sixty-eighth birthday in 1941 but was urged to stay on, first to direct the organization of a new general science curriculum and then to teach in the Army Student Training Program. Even after he was released from these duties, he served on university committees and was active in faculty organizations.

In addition to the many scientific articles which he wrote, Hopkins was the co-author of a high school textbook, three general chemistry texts for colleges, and two chemical reference books, both concerned with the less familiar elements.

Throughout his life, Hopkins took an active part in his church and in civic affairs. He was president of the Urbana Board of Education from 1932 to 1944.

Hopkins suffered a heart attack during the night Aug. 27, 1952 and died a few hours later.

The material in this sketch was drawn from personal recollections, based on long and close acquaintance with Professor Hopkins. Dates were taken from official records.

JOHN C. BAILAR, JR.

Cyril George Hopkins

1866-1919

Hopkins, agricultural chemist, agronomist, and almost-missionary of permanent soil fertility, was born on a farm near Chatfield, Minn., 22 July 1866, son of George Edwin and Caroline (Cudney) Hopkins; he was one of nine children. He was educated, and later taught, in the district schools but left for South Dakota Agricultural College to earn a B.S. degree in chemistry in 1890. He served as an assistant in chemistry at the Agricultural College and Experiment Station at Brookings, S.D., 1890-92, then spent 1892-93 at Cornell University, which awarded him an M.S. degree in 1894. Returning to Brookings in 1893, he married Emma Matilda Stelter and was acting professor of pharmacy at the college in the ensuing year.

In 1894 Hopkins was appointed chemist of the Agricultural Experiment Station of University of Illinois, Urbana. Here he began the chemical investigation of corn which ultimately became "The Chemistry of the Corn Kernel," the thesis for which Cornell granted him his Ph.D. degree in 1898. During the years up to 1900 Hopkins published a number of investigations, mostly as bulletins of the experiment station, in which his chemical background was applied to various agricultural systems. He started a series of notes on equipment and techniques which would later describe the Hopkins condenser and distilling apparatus and the Hopkins limestone tester. He began a program of genetic selection of strains of corn with high and low oil content and protein content, which was continued for many decades with desired properties increasing over the generations.

In 1899-1900 Hopkins worked with Bernhard Tollens at University of Göttingen to increase his knowledge of carbohydrate chemistry. While there he was appointed professor of soil fertility and head of the newly organized Department of Agronomy at Illinois, a position offered and accepted by cable. He held it for the rest of his life, adding the vice-directorship of the experiment station in 1903.

It was in this position, between 1900 and his death in 1919, that Hopkins developed the notions of soil fertility and permanent agriculture which became his dominant concern. Illinois was at that time filled with worn-out farmland, and livestock farming was on the decline. Hopkins focussed on grain production and insisted that farmers must supply soil nutrients such that their soil should be, over the years, a little better than they found it. He identified calcium, magnesium, potassium, phosphorus, and nitrogen as elements which must be replaced for continuing soil productivity and recommended ground limestone and phosphate rock, potash and, for nitrogen enrichment, crop alternation between grains and legumes. He had a contempt for commercial fertilizers which stimulate larger crop yields but do not fully replace the elements that are removed by the plants.

Hopkins' formidable industry, experimental ability, and dedication to his vision of a steady-state system of soil fertility carried all before him. He organized a series of soil surveys by counties in Illinois and appeared as senior investigator and author in the first 18 of them. He spoke and wrote indefatigably, producing the books "Soil Fertility and Permanent Agriculture" (1910), "The Story of the Soil, from the Basis of Absolute Science and Real Life" (1911), and "The Farm That Won't Wear Out" (1913), as well as more than 170 technical publications over his lifetime. His name became a household word among grain-belt farmers. He demonstrated his principles on his own "Poorlands Farm," an abandoned tract in southern Illinois which he made to yield as well as the best.

Hopkins was a member of nearly a dozen scientific societies and president of the Association of Official Agricultural Chemists in 1905-06. In 1918 he was asked by the Red Cross to direct the restoration of agriculture in war-torn Greece. He embarked on the job with characteristic energy, completing soil, climate, and other surveys of the entire country in just 10 months and producing a final report titled "How Greece Can Produce More Food," 200,000 copies of which were printed and distributed by the Greek government. For his services the King of Greece awarded him the seldom-granted Order of Our Savior. Hopkins set sail to return to America in early October 1919 but was stricken with malaria, contracted in Greece, 4 days later. He died in the British Military Hospital at Gibraltar, Oct. 6, 1919, only 53 years old.

"Dictionary of American Biography," **9**, 207, Chas. Scribner's Sons (1932); Lester S. Ivins and A. E. Winship, "Fifty Famous Farmers," pp. 268-73, MacMillan Co. (1924); *Ill. Agriculturist* **31**, 167 ff. (1927); obituaries in *New York Times*, Oct. 9, 1919; *Orange Judd Farmer*, Oct. 18, 1919; *Breeder's Gazette*, Oct. 23, 1919; *Science* **50**, 387-88 (Oct. 24, 1919); and *33d Annual Report, Ill. Ag. Expt. Station*, p. 11 (1921). The best single source is the volume "In Memoriam, Cyril George Hopkins," Univ. of Ill. Press (1921); the sketch by DeWitt C. Wing, editor of *Breeder's Gazette*, *Breeder's Gazette* **76**, 971-72 (Nov. 6, 1919) presents some sound criticisms which give an extra dimension lacking in the other accounts.

ROBERT M. HAWTHORNE JR.

William Dodge Horne

1865-1960

Many chemists of high standing have devoted their entire careers to sugar chemistry and technology. Horne is a notable example. Born in Brooklyn, Jan. 25, 1865, young Horne lived for several years in the island of Barbados where his father had some connection with cane culture or sugar manufacture. The family returned to the United States when Horne was 10 years old, and in 1886 he graduated from Columbia University where he studied under such early specialists in sugar chemistry as Charles Chandler, Samuel Tucker, and Paul Casamajor. Eight years later he received his doctorate from Columbia, presenting a thesis on "Advances in Sugar Manufacture and Machinery."

Immediately following his graduation in 1886 he was employed at Oxnard-Fulton Sugar Refinery in Brooklyn. Horne expressed the opinion in later years that he was the first full-time chemist privately employed in refinery control work in the United States and possibly in any cane sugar work in the Western Hemisphere. Most sugar chemists of that time were college professors in consulting capacities, Department of Agriculture specialists, or the like.

Other refineries of the Oxnard-Sprague interests employed Horne during those formative years, and in 1893 he became chief of the Chemical Department of National Sugar Refining Co. with headquarters at Yonkers, N.Y. He remained with that corporation for 27 years.

In 1903 Horne achieved prominence in world sugar circles when he presented a paper before the International Congress of Applied Chemistry at Rome giving details for his method for clarifying sugar solutions with anhydrous lead subacetate. The use of lead subacetate in solution to clarify sugar solutions for polarimetric analysis had been universal for many years, but the formation of the lead precipitate before the standard solution was made to 100 cc volume had disadvantages. The effect of the "volume error" or "lead error" in sugar polarizations had long been a matter of debate, and Horne's suggested method merely increased the controversy.

Meanwhile, Horne and others found two other outlets for the powdered lead subacetate, resulting in the use of the dry reagent in amounts reaching several tons a year. Routine "water purities" using Casamajor's table of factors, the basis of sugar refinery control, were greatly simplified by the dry lead clarification. Beet sugar factories found the dry lead method simpler than the dilution required by subacetate of lead in solution. Of greater use still, the dry reagent proved the ideal preservative for juices awaiting analysis in cane sugar factories. At the writing of this book all factories in the tropics still based their control on juice samples preserved with Horne's dry lead. He became "Horne, the dry lead man" to thousands of workers who had never met him. For this outstanding contribution to sugar laboratory technique he received the silver medal of the Association des Chemistes de Sucrerie de France in 1905.

He was active in the American Chemical Society from its early days and a frequent contributor to its journals. He was a leader in the opposition to the formation of the American Institute of Chemical Engineers, a controversy which was ludicrous in the light of after-events but which loomed large at a time when the American Chemical Society had a membership of about 1800. Vol. I, No. 1 of *Journal of Industrial and Engineering Chemistry* listed Horne as a member of its editorial board and carried a review article by him on the cane sugar industry. The first public call for a nation-wide organization of sugar chemists was issued by Horne, and in 1919 this resulted in the formation of the Division of Sugar Chemistry and Technology, later known as the Carbohydrate Division, of the American Chemical Society. In recognition of this and for his many innovations in sugar analysis, Horne received the C. S. Hudson Award in 1949 at the 30th anniversary of the division of which he was in effect the founder.

While at National Refinery he did outside consulting work and became health commissioner and chemist-bacteriologist for the city of Yonkers, N.Y. He devised a method for removing algal odors from water with potassium permanganate and treated the city water supply with chlorinated lime as his own idea although he later learned that chlorination of water was suggested by others several years before.

In 1920 Horne joined the Hersey sugar interests in Cuba where he developed many innovations in refining and manufacturing. He patented a soluble phosphate called "Amigon" to aid cane juice clarification; he was among the first to recognize the role of P_2O_5 in this operation. He continued publishing his results until he retired in 1935. He died following an attack of pneumonia at Princeton, N.J., Dec. 6, 1960.

Charles A. Browne, Mary E. Weeks, "History of the American Chemical Society" (1952); G. P. Meade, "American Contemporaries," *Ind. Eng. Chem.* **22,** 80 (1930); personal recollections; obituaries in *Int. Sugar J.* **63,** 32 (1961), in *Sugar J.* **24,** 16 (1961), and in *New York Times,* Dec. 7, 1960.

GEORGE P. MEADE

Eben Norton Horsford

1818-1893

Horsford exemplifies two important trends and traditions in the development of chemistry in America during the 19th century. One was the dependence on European, and especially German, education as a factor in the training of American chemists. The other was the pragmatic orientation of American chemistry and its application to agriculture and industry. Horsford was, indeed, an important figure in the establishment of industrial chemistry in the United States.

Horsford was born July 27, 1818 at Moscow (now Livonia), N.Y. He was the son of Jedediah and Charity Maria (Norton) Horsford, who had migrated from Vermont and combined farming with missionary activity among the Seneca Indians. To cap a conventional education, Horsford enrolled in the Rensselaer Institute in 1837. Here, under the tutelage of Amos Eaton, an inspired advocate of a new type of education in the "application of science to the common purposes of life," Horsford was imbued with a zeal for practical science generally. After graduation, he spent several years in several capacities, serving on the New York geological survey and teaching science at the Girls' Academy in Albany.

In Albany Horsford became part of a scientific circle which grew up around James Hall, a state geologist, and Luther Tucker, editor of *Cultivator,* an agricultural journal. They befriended young Horsford and helped to finance further education in Germany. In 1844 Horsford traveled to Giessen where he was admitted as a student of Justus von Liebig, already world famous in the application of chemistry to agriculture. Here Horsford spent 2 years carrying on research in the chemistry of plants and beginning a promising career of publication. Without completing work for a German university degree, Horsford was named Rumford professor of chemistry and applied science at Harvard University in 1847, thanks to the influence and help of friends, among them James Hall and John W. Webster, professor of chemistry at Harvard.

Horsford's primary interest was nutrition, in which he made his most promising efforts. Probably motivated by a desire for material gain as well as practical results, Horsford succeeded in developing a phosphate baking powder to be used in place of yeast. In association with George Wilson, a manufacturer, he turned his energy to the creation of an industrial plant, the Rumford Chemical Co., at Rumford, near Providence, R.I. This venture in industrial chemistry prospered, and Rumford baking powder was used widely in America. Horsford resigned his professorship at Harvard in 1863 although he retained his residence in that academic environment.

The Civil War offered Horsford an opportunity to combine his chemistry with the spirit of patriotism and entrepreneurial enterprise. He developed and offered to the government of the North a compact ration, composed of portions of bread and meat, to be used by an army on the march. It was, however, not successful. In 1873 he performed an official service for American science when he was United States Commissioner to the Vienna Exposition, out of which came a work on Vienna bread.

In his later years, Horsford dedicated his energy and wealth to broader intellectual and philanthropic purposes, as a patron of newly founded Wellesley College for women and as a zealous searcher for the routes and sites of Viking migration and settlement in New England. Thus did a farm boy of central New York display and apply versatile talents both to the pursuit of chemical science and the creation of a successful new chemical industry as well as a family fortune by the time of his death, Jan. 1, 1893.

S. Rezneck, "The European Education of an American Chemist and its Influence in Nineteenth Century America," *Technology and Culture* **11,** 366-388 (1970), port.; S. Rezneck, "Horsford's Marching Ration for the Civil War Army," *Military Affairs* **33,** 249-255 (1969); H. S. Van Klooster, "Liebig and His American Pupils," *J. Chem. Educ.* **33,** 493-497 (1956); C. L. Jackson, *Proc. Amer. Acad. Arts Sci.* **28,** 340-346 (1892-93); "Dictionary of American Biography," **9,** 236, Chas. Scribner's Sons (1932); many of Horsford's business and professional papers are in the archives of Rensselaer Polytechnic Institute.

SAMUEL REZNECK

William Hoskins

1862-1934

Hoskins was born July 15, 1862, in Covington, Ky. He had no formal education in chemistry. He attended high school only 2 out of the 3 years then offered, yet he could not be called a drop-out, as the term is used today. He became interested in the microscope when he was a boy. At the age of 13 he became a member of the Illinois State Microscopical Society and at 17 was elected secretary of the organization. Throughout his life the microscope was a constant ally in his vocation and avocation. He prepared thousands of slides of hair from animals from all parts of the world, and also took beautiful photographs of ice crystals.

In 1880 at the age of 17, William began his association with Guy Mariner at a salary of 5 dollars a week. His job was to prepare samples of materials to be analyzed. Mariner had graduated from Lawrence Scientific School of Harvard University in 1854 and had established himself in Chicago as a consulting and analytical chemist in 1856. In 1880, there were only two chemists in Chicago, besides Mariner, who sought to make a living as commercial chemists. Young Hoskins remained with Mariner. Ultimately he became partner in the firm, which changed its name to Mariner and Hoskins, and he married Mariner's daughter.

Hoskins learned his first chemistry from Mariner. The rest he acquired through intensive reading and experimenting. He became a recognized expert witness in lawsuits, an inventor with at least 37 United States patents, and an international figure through the part he played in the development of Nichrome, carried out in his laboratory by Albert L. Marsh.

Illustrative of the wide range of his accomplishments were his formulation of a superior billiard cue chalk and the materials used on the track of the Washington Park Race Track in Chicago, which the owners wanted to equal, if not surpass, the fastest track in the United States, that at Churchill Downs. He developed a safety paper for bank checks, a method for destroying weeds, a process for making Keane's cement, and a laboratory assay furnace using the Hoskins gasoline blow torch. With the advent of Nichrome, the gasoline torch furnace became obsolete, but the writer vividly recalls using it in his college days.

Hoskins was active in the affairs of the American Chemical Society, being a charter member of the Chicago Section and second president of the section, a director of the ACS, and a member of the board of editors of the chemical monograph series. He was a member of the Division of Chemistry and Chemical Technology of the National Research Council and during World War I was an associate member of the Naval Consulting Board.

Hoskins gave freely of his time and knowledge to those who sought his help. Many a chemist owed his success to Hoskins' counsel. On one occasion when a fire destroyed the laboratory of a competitor, Hoskins made his facilities available to the unfortunate chemist during the reconstruction of a new laboratory.

Hoskins was an active public servant in La Grange, Ill.; he was on the school board for many years. The auditorium of the high school was named Hoskins Hall in his honor. His home in La Grange was a show place, with a magnificent lawn, profuse flowers, and trees. His colleagues in Chicago held him in high esteem and issued The William Hoskins Number of the *Chemical Bulletin,* February 1930, in his honor.

In later life Hoskins was a striking figure with white hair, white beard, and moustache. He and several other chemists met monthly for lunch, calling themselves the Roundtable and talking about subjects of interest. During a discussion of birth control Hoskins said, "I believe in birth control, but it should be made retroactive," a remark characteristic of his sapient, dry humor. Hoskins died in La Grange, May 18, 1934. His name is continued in the title of Hoskins Manufacturing Co.

Personal recollections; *Chem. Bull.* Feb. 1930; *Ind. Eng. Chem.* **19,** 181-182 (1927); eulogy by Carl Marvel in *Chem. Bull.* **21,** 146 (1934).

DAVID KLEIN

Douglass Houghton

1809-1845

Houghton was born Sept. 21, 1809, in Troy, N.Y. the son of Jacob and Mary (Douglass) Houghton. He attended Fredonia Academy, Fredonia, N.Y. soon after it was founded in 1824. Among his classmates was Harriet Stevens, whom he later married, and his cousin Silas H. Douglass (later shortened to Douglas), who followed him as professor of chemistry at University of Michigan. As a boy, Houghton experimented in chemistry and other sciences at home and was badly injured when a percussion cap that he made exploded. At the age of 16, while still a student at Fredonia Academy, he began to study medicine with a local physician.

He entered Rensselaer Polytechnic Institute in the spring of 1829 and studied chemistry under Amos Eaton, participating in one of the earliest laboratory courses taught in America. There he also became proficient in geology and botany. Houghton graduated with his B.A. degree in October 1829, six months after he entered Rensselaer. Upon graduation he was appointed assistant instructor in chemistry and natural history, and in February 1830 was promoted to assistant professor. He spent the summer of 1830 on a geological field trip through New Jersey and New York, carrying books, equipment, and supplies on steamboats and canal boats for laboratory experiments along the way.

Houghton delivered a series of public lectures on chemistry in Detroit, Mich., during the winter of 1830-31. The success of this series led to a second series, "Mechanical Philosophy and Natural History," covering geology, botany, chemistry, electricity and magnetism. In 1831 he received his license to practice medicine from the Medical Society of Chautauqua County, N.Y. and practiced medicine in Detroit for the next 6 years. He also served as surgeon and botanist for the 1831 Schoolcraft expedition to the headwaters of the Mississippi River. In 1841 he served as mayor of Detroit, Mich.

From 1837 to 1845 Houghton served as Michigan state geologist. He analyzed water and geological samples collected on state geological surveys. He was instrumental in developing the salt springs and wells which later became the foundation of Michigan's chemical industry. The Michigan Geological Survey located the copper and iron ores which were the sources of Michigan's mining industry.

In May 1842 Houghton was appointed the first professor of geology, mineralogy and chemistry at University of Michigan. His duties with the Geological Survey prevented his teaching regularly on the campus, and Silas Douglass replaced him as assistant professor of chemistry.

While Houghton was working on the Michigan survey, Oct. 13, 1845 his small boat capsized in a squall on Lake Superior and he drowned. He had located many of the sources of Michigan's mineral wealth and helped lay a foundation for science education at The University of Michigan.

E. Rintala, "Douglass Houghton," Wayne University Press (1954); A. Bradish, "Memoir of Douglass Houghton," Raynor & Taylor (1889).

DONALD R. HAYS

Harrison Estell Howe

1881-1942

Howe was born in Georgetown, Ky., Dec. 15, 1881, and during the 60 years of his life became one of the most noted and articulate spokesmen of the chemical profession. Though he wrote numerous books and papers, both popular and learned, and though he was in constant demand as a speaker, his voice was heard most strongly through his 21 years' editorship of *Industrial and Engineering Chemistry,* which he served to the time of his death.

In 1901, Howe received his B.S. degree from Earlham College. Later he did graduate work at University of Michigan and earned his M.S. degree in chemistry at University of Rochester in 1913.

In 1902 he began his working career as a chemist for Sanilac Sugar Refining Co. in Croswell, Mich. Two years later he joined Bausch and Lomb Optical Co. in Rochester, N. Y. as a chemist, later serving the company as office manager and finally as editor. In 1905 he married May McCaren.

During the following years, Howe became affiliated with Arthur D. Little, Inc., in Boston and Arthur D. Little, Ltd., in Montreal. From 1916 to 1917 he worked as chemical engineer and assistant to the president of the Little firm in Canada. He also served as consultant to the Nitrate Division of Army Ordnance during World War I.

In 1919 he was named chairman of the Division of Research Extension of the National Research Council and in this capacity raised a large part of the money that went to construct and equip the Marine Biological Laboratories at Woods Hole, Mass. He remained with the NRC until 1921; in that year he began his long and noteworthy service as editor of *Industrial and Engineering Chemistry.*

The duties of an everyday occupation

never absorbed all of Howe's energies, however. In 1922, as chairman of the committee on work periods of the American Engineering Council of the Federated Engineering Societies, he played a leading part in shortening the 12-hour work day to the 8-hour standard. He also served as a trustee of Science Service and as a member of Purdue Research Foundation. He was a member of the advisory board of Lalor Foundation and served the Institute of Politics in Williamstown, Mass., as a general conference leader from 1926 through 1929.

Howe held the rank of colonel in the Chemical Warfare Reserve, U.S.A. and was a member of the Army's ACS Advisory Committee. He also chaired the Chemicals Group of the Chemical Priorities Committee, Office of Production Management, later becoming chairman of the advisory committee of the Chemical Section of the World War II War Production Board.

Beyond merely holding membership in numerous organizations, Howe's broad interests and energies led him to become actively involved in all. In his affiliation with American Institute of Chemical Engineers, he served as director for two terms and as the Institute's representative on the American Engineering Council for 10 years, during 8 of which he was also treasurer. For the American Chemical Society, he acted in numerous capacities in addition to his function as editor. He took a genuine interest in Rotary International which he served at all levels from the presidency of the Washington, D.C. unit to membership on the international board of directors in 1936-37.

Many honors came to Howe during his lifetime. He received honorary doctorates from University of Rochester, 1927; Southern College, 1934; Rose Polytechnic Institute, 1936; and South Dakota State School of Mines in 1939. Italy honored him as an Officer of the Crown in 1926, and in 1942 he received the Chemical Industry Medal from the American Section of the Society of Chemical Industry.

Howe's lively interest in the applications of science, especially chemistry, to the problems of daily life led him to write and edit several popular books. Among them were "The New Stone Age," "Profitable Science in Industry," "Chemistry in the World's Work," "Chemistry in the Home" and, with E. M. Patch, a series of six Nature and Science Readers for School Children. Articles also appeared under his by-line in scientific journals and lay publications. In 1924-25 he edited two volumes of "Chemistry in Industry" in connection with the ACS Prize Essay Contest. Howe added to his printed works a high reputation as a public speaker on chemistry throughout the United States.

He died Dec. 10, 1942, at his home in Washington, D.C.

Personal recollections; F. J. Van Antwerpen, *Science* **97,** 82-84 (1943); Marston T. Bogert, *Chem. Eng. News* **21,** 678 (1943); Walter J. Murphy, *Chem. Eng. News* **24,** 464-467 (1946); obituary in *Chem. Eng. News* **20,** 1617-1618 (1942).

FRANKLIN J. VAN ANTWERPEN

James Lewis Howe

1859-1955

James Lewis Howe was born in Newburyport, Mass., Aug. 14, 1859, the son of Francis Augustine and Mary Frances (Lewis) Howe. He attended Brown High School in Newburyport and Amherst College. As was the custom in those days, after graduation Howe went to Germany to complete his studies. From 1880 to 1882 he studied at Universität Göttingen under Hans Hübner, J. Post, and Friedrich Wöhler. After receiving his doctorate in the spring of 1882 he left Göttingen for Universität Berlin where he studied under Liebermann, Liebreich, Sell, Baumann, Tiemann, Paulsen, Curtius, Gabriel, and Döbner.

On returning from Germany, Howe became instructor of science at Brooks Military Academy in Cleveland, Ohio, where he remained for a year. In 1883, he became professor of chemistry (later of physics and geology as well) at Central University, Richmond, Ky. where he remained for 11 years. On Dec. 27, 1883 he married Henrietta L. Marvine of Scranton, Pa. The Howes became the parents of two daughters, Guendolen and Frances, and a son, James Lewis, Jr.

The greater part of Howe's career was spent at Washington and Lee University in Lexington, Va. He accepted the chair of chemistry in 1894, a position which he occupied until his retirement in 1938. For the first 15 years of his tenure Howe was a one-man Chemistry Department.

Aside from some miscellaneous research on organic, analytical, and inorganic chemistry, his experimental work dealt with the least known of the platinum metals—ruthenium—particularly its cyanide and halide complexes on whose chemistry he was an undisputed world authority. Howe's *magnum opus,* his monumental "Bibliography of the Metals of the Platinum Group," together with his unchallenged position as the leading American authority on the chemistry of the platinum metals, led directly to his appointment in 1917 as chairman of a special sub-committee on platinum of the National Research Council. He subsequently received presidential appointments on three occasions to commissions for assaying the coinage of the United States. During World War II, Washington and Lee recalled Howe from retirement to teach German and chemistry.

Howe was a member and officer of a number of professional and honorary organizations. He joined the American Chemical Society in 1893 and for a number of years contributed his well-known annual reviews of "Recent Work in Inorganic Chemistry" for the society's journal. In 1937 he received the ACS Georgia Section's Charles H. Herty Medal for his bibliography. At the time of his death on Dec. 20, 1955 at the age of 96, he was one of the ACS's oldest members.

Among his many extra-scientific activities, Howe served as deacon and elder of the Lexington Presbyterian Church, and his son, himself a chemist who spent a number of years in China involved in missionary work, always felt that his father placed his religion above his secular life. Howe was an active Mason and also held positions in the Lexington Town Council and the People's National Bank of Lexington.

George B. Kauffman, *J. Chem. Educ.* **45,** 804-811 (1968); S. C. Lind, *Ind. and Eng. Chem.,* **18,** 434 (1926); *Alumni News,* Amherst College, 1955; *Lexington Gazette,* Dec. 21, 1955; *Rockbridge County News,* Dec. 22, 1955; *Chem. Eng. News,* **34,** 482 (Jan. 30, 1956); *The Alumni Magazine,* Washington and Lee University, Jan. 1956, p. 16.

GEORGE B. KAUFFMAN

Claude Silbert Hudson

1881-1952

Hudson, the second child of William James and Maude Celestia (Wilson) Hudson, was born in Atlanta, Ga., Jan. 26, 1881. At the age of 16 he entered Princeton University, intending to become a Presbyterian minister. His interests, however, shifted to chemistry, and it was in that subject that he graduated in 1901. He continued with graduate studies at Princeton, from which he received his M.S. degree in 1902. The following year found him in Göttingen and Berlin in the laboratories of Nernst, van't Hoff, and Tammann. He spent a third year with Arthur Noyes at Massachusetts Institute of Technology. He then served as instructor in physics at Princeton (1904-05) and at University of Illinois (1905-07). He received his Ph.D. degree from Princeton in 1907 for his physicochemical studies on the forms of milk sugar (lactose).

Hudson's first important assignment with the United States Government was with the Bureau of Chemistry of the Department of Agriculture, from 1908 to 1911 and later from 1912 to 1918 (he was on leave of absence from 1911 to 1912 as acting professor of physical chemistry at Princeton). It was there that he began his studies on enzymes, particularly invertase; on the preparation and purification of sugars from natural sources; and on "The significance of certain numerical relations in the sugar group" (1909). For the many subsequent papers on numerical relationships it became necessary for Hudson and his collaborators to prepare not only the pure sugars but also their acetates, glycosides, phenylhydrazides, lactones, and amides, and Hudson's transformation from a physical chemist into an organic and carbohydrate chemist had begun.

After a brief period of research on preparing activated carbon for gas masks in 1917 for the War Department and 4 years as a commercial consulting chemist from 1919 to 1923, Hudson joined the National Bureau of Standards. There his work with only a few collaborators resulted in the publication of the first 23 papers in the now classical series entitled "Relations between rotatory power and structure in the sugar group." In them he evolved the rules of isorotation (optical superposition) which, although not as rigorous as he had originally hoped, nevertheless became extremely valuable in permitting the assignment of configuration (α-D—, β-L—, etc.) to the anomeric carbon atoms of carbohydrates and their derivatives.

The final phase of Hudson's career in the service of the United States Government began with his appointment in 1929 as professor of chemistry (the fifth and last of that title, which was granted by an Act of Congress) in the Hygienic Laboratory (later to be called the National Institutes of Health). He was allowed to continue fundamental research in carbohydrate chemistry and was fortunate in having about 35 professional collaborators (though not all at the same time) between 1929 and his retirement early in 1951. More than 200 publications appeared during this period while his complete bibliography runs to 314 papers and patents.

In addition to the subjects mentioned previously, Hudson and his collaborators at the National Institutes of Health published additional papers on enzymes (invertase, amylase, emulsin) and on relations between rotatory power and structure; also on the lactone, phenylhydrazide, amide, and benzimidazole rules; on ketoses (including heptuloses) and their preparation by the biochemical oxidation of polyhydric alcohols; on the synthesis of higher-carbon sugars and their derivatives; on periodate oxidations, especially to determine ring structure of sugar derivatives; on neolactose, celtrobiose, and D-altrose; on anhydro sugars and sugar alcohols; on acetals of polyhydric alcohols; on phenylosotriazoles derived from sugars; and on many other topics.

In honor of Hudson's 65th birthday, the Division of Sugar Chemistry and Technology, ACS, held a 4-day "Hudson Celebration" at the Chicago meeting of the Society in September 1946. An award, given annually thereafter to an outstanding carbohydrate chemist, was subsequently designated the "Claude S. Hudson Award."

Hudson received many honors and awards including the Nichols Medal, the Willard Gibbs Medal, the Hillebrand Prize, the Richards Medal, the Borden Medal and Award, the Cresson Medal, the first prize of $10,000 of the Sugar Research Foundation Grand Prize (1950), and the first Federal Security Agency Award for Distinguished Service (1950).

Hudson served as an associate editor of *Journal of the American Chemical Society*, as co-editor of "Advances in Carbohydrate Chemistry," and as chairman of the Post Office Department Advisory Committee.

In his "Autobiography," Hudson wrote "I have been married four times and must regretfully record that the first three of these marriages ended in divorce." His first was to Miss Alice Abbott, whom he married in April 1906; they had three children—William, Alice, and Sally. Later, he married Mrs. Olive G. Gale, Mrs. Mabel Felix Hazard, and finally, in April 1942, Mrs. Erin Gilmer Jones.

Hudson died of a heart attack Dec. 27, 1952 in Washington, D.C. He was buried in the Presbyterian Cemetery at Princeton, N.J.

C. S. Hudson, "Autobiography [to June 1945]," in R. M. Hann and N. K. Richtmyer (Eds.), *The Collected Papers of C. S. Hudson*, Vol. 1, Academic Press Inc. (1946), with portrait; M. L. Wolfrom, "Claude Silbert Hudson, 1881-1952," *Advan. Carbohyd. Chem.* **9,** xiii-xviii (1954), with portrait; E. L. Hirst, "The Hudson Memorial Lecture," *J. Chem. Soc.* 4042-4058 (1954), with portrait; L. F. Small and M. L. Wolfrom, "Claude Silbert Hudson, 1881-1952," *Biog. Mem. Natl. Acad. Sci.* **32,** 181-220 (1958), with portrait, complete bibliography.

NELSON K. RICHTMYER

Mary Hannah Hanchett Hunt

1831-1906

Mary Hunt was born in South Canaan, Litchfield County, Conn., June 4, 1831. Her father, a grandson of the first discoverer of iron ore in the United States, was of Welsh descent and an iron manufacturer. He was active in the reforms of his times and was vice-president of the first temperance society formed in the United States.

Mary completed her education at Patapsco Institute, near Baltimore, where she learned chemistry and other sciences from Almira H. L. Phelps. After graduation she remained at the institute as a teacher of chemistry and physiology and collaborated with Mrs. Phelps in writing science texts.

In 1852 she married Leander B. Hunt, a manufacturer of East Douglas, Mass. In 1870 she observed some chemical experiments made by her teen-age son on alcohol and became attracted to the scientific study of the nature and effects of alcoholic drinks. She was convinced that the only remedy for alcoholic intemperance is abstinence through early education as to the bad effects of such beverages; as part of the study of general hygiene, she proposed compulsory instruction on that subject for all pupils in public schools. During the 3 years from 1879 to 1882 she visited almost every state in the Union delivering lectures in support of her views and plans.

She drafted many of the laws for compulsory temperance education on the statute-books of the United States, and these were taken as models for similar legislation in other countries. The Woman's Christian Temperance Union helped to pass and enforce these statutes, and she was the world and national superintendent of scientific instruction of this organization. Sixteen years after the first publication of Mrs. Hunt's plans, temperance education laws had been enacted in 41 of the 45 states of the Union and by the United States Congress for the benefit of the military, naval, territorial, and other schools under Federal control. Thus a teacher of chemistry became a leader in the temperance movement and was highly influential in having the subject of hygiene introduced into and taught in American schools.

Mary Hunt's son, Alfred Ephraim Hunt, 1855-99, became a chemist and metallurgist of note.

Mary died at her home in Boston, Mass. Apr. 24, 1906.

"National Cyclopedia of American Biography," **9,** 156-7, James T. White & Co. (1899); "Dictionary of American Biography," **9,** 388, Chas. Scribner's Sons (1932); obituary, *New York Times,* April 25, 1906; Edward T. James, Janet W. James, Paul S. Boyer, eds., "Notable American Women 1607-1970," **2,** 237-39, Harvard Univ. Press (1971).

DOROTHY B. DEAL

Thomas Sterry Hunt

1826-1892

Hunt was among the American chemists who assembled at Northumberland, Pa. to celebrate the Priestley Centennial on Aug. 1, 1874. He presented one of the two papers. It was entitled: "A century's progress in chemical theory." Chemistry was Hunt's first love, and he always looked at geology and mineralogy from a chemist's view.

He was born in Norwich, Conn., Sept. 5, 1826. During his early childhood the family moved to Poughkeepsie, N. Y. His father died there when Sterry (he generally used his middle name) was 12 years old, and his mother took her family of six children back to their old home in Norwich. For a time Sterry attended the public school but soon dropped out to help support the family. He worked in a printing office, an apothecary shop, and a book store, remaining about 6 months with each employer. Next he obtained a clerkship in a country store where the owner allowed him to do pretty much as he wished. Sterry kept a skeleton and some home-made chemical apparatus under the counter at the store, and studied at every opportunity.

In 1845 Hunt attended the meeting of the Association of American Naturalists and Geologists in New Haven and reported the meeting for a New York newspaper. While there he visited Benjamin Silliman, Sr. Impressed by the youth's knowledge of chemistry and mineralogy, Silliman helped him to be admitted to Yale. Hunt spent about one and a half years at Yale and served as assistant to the younger Silliman. During this time he contributed 18 papers to Silliman's *American Journal of Science* and wrote the organic chemistry section for Silliman Junior's "First Principles of Chemistry." In the 1852 edition of this text he was the first American to define organic chemistry as the chemistry of carbon compounds.

In 1846 Hunt became chemist to the geological survey of Vermont and a year later resigned to work for the geological survey of Canada. On moving to Montreal he became closely associated with the chief geologist, Sir William Logan. The team of Logan and Hunt lasted for 25 years. Hunt not only did all of the routine analytical work at the laboratory but carried out field work in geology. During this period he wrote many chemical and geological reports and published several articles on speculative theory. In 1849 when he was 23 he published an article, "On the decomposition of aniline by nitrous acid." During his experiments he did not notice anything happen in the cold. He heated the reaction mixture and isolated phenol but thus missed the presence of diazobenzene in the cold solution. He could have anticipated Peter Griess by 9 years in finding a reaction basic to azo dye chemistry.

During part of his stay in Canada he lectured on chemistry at Laval and McGill. In 1872 Hunt became professor of geology at Massachusetts Institute of Technology but resigned in 1878 to devote full time to consulting and writing. The previous year he had married, but upon finding that it interfered with his scientific interests he and his wife decided to live apart.

Like most men of genius he had his eccentricities, one of which was to be obstinate in his beliefs. This caused him to engage in controversies both in his writings and at scientific meetings. He was elected a fellow of the Royal Society of London in 1859 and served as president of many scientific societies, including the American Chemical Society in 1879 and 1888. He was also a delegate to European and American expositions of the time.

Early in his career Harvard conferred on him an M.A., and later Laval and Cambridge the LL.D. degrees. Hunt published about 160 scientific articles, and several books, including "Chemical and Geological Essays" (1875, 1878), "Special Report on the Trap Dykes and Azoic Rocks of Southeastern Pennsylvania" (1878), "Mineral Physiology and Physiography" (1886), "A New Basis for Chemistry: A Chemical Philosophy" (1887), and "Systematic Mineralogy" (1891). At the time of his death he was writing "The History of an Earth."

Hunt died Feb. 12, 1892 in Park Avenue Hotel, New York City, of heart trouble.

B. Silliman, "Thomas Sterry Hunt," *Amer. Chemist* 105-6 (Aug., Sept. 1874); James Douglas, "Biographical Notice of Thomas Sterry Hunt," *Trans. Amer. Inst. Mining Eng.* **20,** 400-410 (1891); Anon, "Thomas Sterry Hunt," *Proc. Amer. Acad. Arts Sci.* **27,** 367-372 (1891-2); Raphael Pumpelly, "Memorial of Thomas Sterry Hunt," *Bull. Geol. Soc. Amer.* **4,** 379-385 (1892); Persifor Frazer, "Thomas Sterry Hunt," *Amer. Geol.* **11,** 1-13, (Jan. 1893); Edward R. Atkinson, "The Chemical Philosophy of Thomas Sterry Hunt," *J. Chem. Educ.* **20,** 244-5 (1943).

DAVID H. WILCOX, JR.

Matthew Albert Hunter

1878-1961

The life and labor of Matthew Hunter exemplify the great capacity of America to accept, absorb, and utilize talent from the whole wide world. Hunter was born Nov. 9, 1878, in Auckland, New Zealand. He attended school in Auckland and graduated from Auckland University College with an honors degree in chemistry. After receiving his master's degree he served a brief apprenticeship as an instructor of science and metallurgy at Thames School of Mines. In 1901 he went to Britain on a scholarship. At University of London he pursued chemistry under the guidance of Sir William Ramsay. He received his doctorate of science in 1904 and spent another year in study on the European continent. His travels took him to Paris, Göttingen, and the Technische Hochschule at Karlsruhe; he came under the aegis of such scientists as Moissan, Nernst, and Haber.

Thus equipped in the best European fashion, Hunter came to the United States in 1905, partly for sentimental reasons, to renew a friendship with and to marry an

American girl, Mary Pond. He also came to begin a scientific career in the American environment, in his own self-deprecating words, "as an impecunious alien [who] had to get a job." His first regular employment was at the General Electric Laboratory in Schenectady, N. Y. Here the search had just begun for a metal suitable for use as a filament in the electric light. Hunter went to work on titanium, and by mid-1907 he succeeded in developing a workable method for reducing this metal, which, however, long continued without any adequate use in industry. It was, nevertheless, a great, and probably his most dramatic scientific triumph. Now known as the Kroll-Hunter process, it is basically still in use in titanium reduction.

In 1908, Hunter's association with General Electric was severed, but good fortune brought him to Rensselaer Polytechnic Institute in near-by Troy as an assistant professor of electrochemistry and physics. The Institute was then embarking on a period of growth and curricular expansion, and Hunter's association with it was to continue for more than half a century. Here he developed his talents in four directions, in each of which he excelled. He was, in the first place, an effective teacher, particularly in non-ferrous metallurgy, who commanded the respect and devotion of many generations of students. He was secondly, an administrator, for whose services Rensselaer found many uses, first as head of the Department of Electrical Engineering, then of Metallurgy when that was established separately. His crowning achievements were as director of graduate studies and finally as the first dean of a reorganized Rensselaer faculty during World War II.

Hunter also found an opportunity to direct his efforts and energy into research and consultant activity. His first consulting connection was with Titanium Alloy Manufacturing Co. as early as 1912. His longest and most fruitful association was with Driver-Harris Co. of New Jersey, which lasted from 1919 to 1961. He had other consulting employment, including the Chemical Warfare Service during World War I and the War Metallurgy Committee in the second world war. His contributions to metallurgy were especially noteworthy in the alloy field, as indicated by such products as "Invar," "Manganin," "Constantan," and "Nichrome," in whose development he participated.

Hunter's professional society affiliations and honors were numerous. The culminating reward of his long career was undoubtedly the gold medal awarded to him by the American Society for Metals in 1959, in his 81st year.

Hunter continued his work as consultant and researcher for many years after his retirement from Rensselaer in 1949. He died in Troy, N. Y., Mar. 24, 1961.

Floyd Tifft's appreciative profile of Hunter on his retirement in *Rensselaer Alumni News,* September 1949; Samuel Rezneck, "Man of Metals" in *Rensselaer Alumni News,* December 1961; *New York Times,* March 25, 1961; personal recollections.

SAMUEL REZNECK

Ernest Hamlin Huntress

1898-1970

Huntress was born in Laconia, N.H. May 9, 1898. His entire career was spent at Massachusetts Institute of Technology. He received his B.S. degree in 1920 and became an assistant in chemistry the same year, rising to instructor in 1922. While serving as an instructor he earned his Ph.D. degree in 1927 for work with Forris J. Moore on unsymmetrical phenanthriodones. He was promoted to assistant professor in 1929, to associate professor in 1935 and professor in 1941. He retired in 1963. In addition to his teaching and research, Huntress also performed administrative tasks for the department and the institute. He served for 7 years as head of the department's undergraduate work in organic chemistry and for 11 years as chairman of its graduate committee. From 1952 to 1956 he was director of the Summer Session and for 2 years beginning in 1950 he was deputy dean of the Graduate School. He was secretary of the Graduate School from 1956 to 1963 and also a member of the board of the Technology Press. From 1941 to 1945 he was a technical advisor for the developmental laboratory of the U.S. Army Chemical Warfare Service.

His research work was directed primarily towards problems of synthesis and identification; he was not a theoretical or physical organic chemist. Huntress was best known for his work on the identification of organic compounds, a continuation of Samuel P. Mulliken's work. This finally culminated in his book "Organic Chlorine Compounds" (1949). This work entailed a great deal of library research, and Huntress became very interested in problems of chemical literature and nomenclature. His "Brief Guide to the Use of Beilstein's Handbuch der organischen Chemie," published in 1938, was a useful introduction and remained in print for several decades. He also developed the use of fluorescent compounds in inks for invisible laundry markings.

Huntress's undergraduate organic chemistry lectures were, as might be expected from his research emphasis on classification, extremely well-ordered. The same attention to order, classification, and nomenclature was to be found in his textbook "Problems in Organic Chemistry," (1938). Perhaps related to his extensive literature work was his interest in the history of chemistry which resulted in the publication of several lists of chemical anniversaries and centennials, which are of pedagogical value. The largest of these, in *J. Chem. Educ.* **14**, 328-44 (1937), gave a chemical event for each day of the year.

He was a familiar figure to many chemistry graduate students who had to have their programs approved by him, a large man with ruddy complexion and a cigarette dangling in the corner of his mouth. He was also a familiar figure in the library where he worked long hours collecting the literature data that went into the tables in "Organic Chlorine Compounds." He died, at the age of 71, Feb. 1, 1970.

Obituary in *Technology Review,* **72,** No. 5, 101 (1970); pictures in *Technology Review,* **65,** No. 5, 4 (1963); personal recollections.

RUSSELL F. TRIMBLE

I

Vladimir Nikolaevich Ipatieff

1867-1952

Ipatieff was born in Moscow, Russia, Nov. 21, 1867. His early training was in military schools. Science was his favorite subject and much of his early knowledge of chemistry was self-taught. While he was a student at Mikhail Artillery Academy in St. Petersburg, he carried out his first research, which was concerned with the structure of steel used in artillery. After graduation in 1892 with the military rank of captain he taught inorganic chemistry and qualitative analysis at the academy until 1896.

Ipatieff found time during this period to take his first organic chemistry course at University of St. Petersburg and to carry out his first organic research, which culminated in his dissertation on the isomerization of allenes and acetylene. This work resulted in his being awarded a Minor Butlerov Prize of the Russian Physical-Chemical Society and in his appointment as assistant professor at the academy. He was sent to Munich for a year's study under Adolf von Baeyer; two of his classmates were Richard Willstätter and Moses Gomberg. Work on caronic acid was followed by independent research on the proof of structure and synthesis of isoprene.

From 1899 to 1930, Ipatieff was professor of chemistry and explosives at the academy. In 1908 he received the degree of doctor of chemistry from University of St. Petersburg to which he presented his thesis, "Catalytic Reactions Under High Pressures and Temperatures." This was based on the observation he made in 1900 that, contrary to general belief, the course of the reaction of organic compounds at high temeprature can be controlled and, what is more important, can be caused to go in specific directions with catalysts. In 1903 he made the technologically important discovery that ethyl alcohol decomposes in the presence of aluminum powder at 600° to yield butadiene.

In 1903–04 Ipatieff developed his well-known high pressure rotating autoclave, which has a simple disk gasket closure and which was, and still is, used with safety as a test tube for carrying out reactions at temperatures up to 500° and pressures up to 450 atm. In 1912 he discovered the joint (or promoter) action of catalysts.

Ipatieff's scientific work was interrupted by World War I. As a lieutenant-general in the Russian Army he was chairman of the Chemical Committee of the Chief Artillery Administration and was responsible for developing chemical industry and war materials. After the war, he held a number of offices under the Soviet regime and continued his scientific investigations but felt that he could not carry out his research so freely and effectively as he did before the revolution. While in Berlin in June 1930, as a delegate to the Second World Power Congress, he was therefore quite pleased when Gustav Egloff offered him the opportunity of living and working in the United States. He and his wife arrived in Chicago a year later, and he began a new research career in the chemical laboratories of the Universal Oil Products Co. He also lectured on catalytic chemistry at Northwestern University. He became a U.S. citizen in 1936.

By applying catalysis to petroleum technology, Ipatieff made discoveries of marked theoretical and practical importance. He and his co-workers discovered and developed a new catalyst, a composite of phosphoric acid and kieselguhr, for polymerizing gaseous olefins to liquid olefins, hydrogenation of which produced high octane gasoline.

A most significant contribution to hydrocarbon chemistry was the discovery by Ipatieff and Herman Pines in 1932 that isoparaffins are not chemically inert but can be catalytically alkylated with olefins, and, a few years later, that *n*-paraffins can be isomerized to branched-chain paraffins. Many other fundamental reactions of saturated and of aromatic hydrocarbons were also described by the Ipatieff group.

At Northwestern University Ipatieff helped finance a laboratory, which the university named "The Ipatieff High Pressure and Catalytic Laboratory," for instruction in the chemistry and techniques of catalytic and pressure reactions.

Ipatieff published about 350 chemical papers, almost 150 of which were published in the United States, and a number of books. More than 200 U.S. patents were issued to him as sole or joint inventor.

Among the many awards and honors which were bestowed on Ipatieff in Europe and in the United States were election to membership in the Russian Academy of Sciences (1916) and in the National Academy of Sciences, U.S.A. (1939), award of the Lenin Prize (1927), the Berthelot Medal (1928), the Lavoisier Medal (1940), the Willard Gibbs Medal (1940), and at least two dozen other awards.

Ipatieff set up a trust fund to establish the Ipatieff Prize to promote work in high pressure and catalytic chemistry. This is administered by the American Chemical Society and is awarded triennially to a chemist less than 40 years old.

Still actively engaged in research, Ipatieff died in Chicago, Nov. 29, 1952. His wife, *nee* Varvara Ermakova, whom he had married in Moscow in 1892, died only 10 days later. They were survived by two of their four children; a son, Vladimir, a professor of chemistry in Leningrad, and a daughter, Anna.

V. N. Ipatieff, "The Life of a Chemist," Stanford University Press (1946), a 658 page book covering his life and research in Russia (contains many photographs, editor's notes, and a detailed index); Aristed V. Grosse, "Vladimir Ipatieff," *J. Chem. Educ.,* **14,** 553-4, 597 (1937); *Chemist,* **19,** 315-341 (1942) contains photographs and talks which were presented at a testimonial banquet sponsored by the American Institute of Chemists Nov. 20, 1942 in honor of three milestones in Ipatieff's career—his 75th birthday and the 50th anniversary of his wedding and of his first chemical publication; the talks were on Ipatieff's influence on industry (Gustav Egloff) and on world chemistry (Frank C. Whitmore), on some biographical items (Ward V. Evans), and on his 12 years in the U.S.A. (V. N. Ipatieff); Herman Pines, "Ipatieff: Man and Scientist," *Science,* **157,** 166-170 (1967).

LOUIS SCHMERLING

Samuel Isermann

1878-1949

Isermann, a pioneer manufacturer of synthetic organic chemicals and aromatic materials, was born near Riga, Latvia, Mar. 23, 1878, the son of Isadore and Fanny (Wolfson) Isermann.

He emigrated to the United States in 1893, took a job in a drug store, and avidly studied all the pharmacy and chemistry he could find, both by himself and at New York College of Pharmacy. Like most chemists and pharmacists of the time, he soon became "Doctor" to his friends and associates.

In 1902 Isermann and Louis A. Van Dyk established the Van Dyk Co. to make synthetic organic chemicals, first in New York, eventually in Belleville, N.J. Van Dyk left the company in 1911, but Samuel Isermann's brother, Max, had already joined the company and remained with it until he died in 1947.

In those days there was practically no market for coal tar chemicals, especially for synthetic aromatic perfumery materials, but on a small scale Van Dyk made many potentially useful compounds. When the war cut off European (especially German) supplies from manufacturers in all industries, Van Dyk and Synfleur (Monticello, N.Y.) were the only independent American manufacturers of synthetic aromatic materials. A vigorous educational program was started.

Many of the chemical companies which had been importing practically all their supplies —even glassware—were faced with the alternatives of making their own raw materials or going out of business. In 1915 Isermann organized Chemical Company of America to make dyes and intermediates and soon made available aniline and several derivatives, benzaldehyde, nitrobenzene, nitrotoluene, toluidine, and other compounds. They introduced many rare organic compounds, pure chemical reagents, and several synthetic flavorings. The educational program went on, but the resistance to synthetic compounds continued unabated.

Isermann was in the forefront of the struggle by the industry to meet the demand for American dyes, intermediates, and other chemicals during a turmoil concerning tariffs on imported aromatic materials. He served on several important committees assembled by the President and government agencies, notably for controls on industrial alcohol (especially trying after the introduction of prohibition). He was active in the American Dyes Institute and in the Dyestuffs Manufacturers Association until these were absorbed into the Synthetic Organic Chemical Manufacturers Association in 1921. He worked closely with Francis P. Garvan and the Chemical Foundation in its efforts to maintain a sound position for the growing American chemical industry.

At one time Van Dyk was the only available source of benzyl chloride and benzyl cyanide, which the government needed for "dopes" for the cloth wings of airplanes and for training in chemical warfare. During World War II also, Van Dyk put much of its production at the service of the government, notably plasticizers, flame retardants, insect repellants, and sunscreens.

In 1925 Isermann organized Patent Chemicals, Inc., to produce oil-soluble dyes, suitable for coloring gasoline; Synthetic Chemicals, Inc., to make wetting agents and detergents; and Summit Chemical Co., to make ammonium thioglycolate for cold permanent waving. During the depression, he served on President Hoover's committee on the status of the industry.

"Dr. Sam" was a dynamic personality—frank, resourceful, and generous with his vast and varied technical knowledge. He was the author of the section "Perfumes and Flavors" in "Chemistry and Industry" (H. E. Howe, Ed.); and of a house organ called "Progressive Perfumery and Cosmetics." He loved to travel and divided his time between homes in New Jersey and in Tucson, Ariz. He died at San Antonio, en route to Tucson, Feb. 2, 1949. He was married twice and was survived by two sons of his first marriage, his second wife (Centa Cossman), and their daughter, Joanne C. Isermann.

Personal recollections and information from journal and press clippings in possession of the author.

FLORENCE E. WALL

Martin Hill Ittner

1870-1945

Ittner, an indefatigable researcher on soaps, perfumes, and cosmetics, was born May 2, 1870 in Berlin Heights, Ohio, the son of Conrad S. and Sarah (Hill) Ittner. He attended local schools and then went to Washington University, where he received his Ph.B. degree in 1892 and B.S. in 1894. Continuing at Harvard, he obtained his M.A. in 1895 and his Ph.D. in 1896.

Directly from Harvard Ittner started a lifelong career with Colgate and Co. (later Colgate-Palmolive-Peet) as chief chemist and director of research. He inherited a well-established body of business in soaps, tooth paste, and glycerol, but by the 100th anniversary of the company in 1906, he had in charge nearly 200 kinds of soap, over 600 perfumes, and many other items. He held numerous foreign and domestic patents on hydrogenation of oils, glycerol, counter-current hydrolysis of fats, and soaps from petroleum hydrocarbons. Some of his special studies were on rancidity, catalysts for hydrogenation, and the conversion of acid compounds from the oxidation of petroleum into useful products.

Ittner was much interested in the development of synthetic aromatic materials. In 1913 when the eruption of Mt. Etna in Sicily destroyed large quantities of stored bergamot oil and acres of trees, he produced an acceptable synthetic substitute within months. During World War I when supplies of essential oils from Europe were cut off, many American manufacturers slighted American-made aromatic compounds, but he was a strong champion of American synthetic raw materials for perfumery. Colgate did all possible to further developments in the new, struggling chemical industry in general. To challenge the stubborn preference of American women for imported perfumes, the company ran a much publicized "blindfold test" in which a panel of over 100 women sniffed and compared three imported French perfumes with three of their own products (all unlabeled). One of the Colgate perfumes came out first.

Despite his intensive program, Ittner was also most active professionally. During World War I, he was practically "Mr. Industrial Alcohol." He was on committees in all corners of the industry. In the American Chemical Society, he was chairman for years of its Industrial Alcohol Committee, which handled all problems of industrial alcohol, denaturants, etc., and he often was its delegate at important conferences. He was chairman of the New York Section in 1923 and councilor in 1930.

In the American Institute of Chemical Engineers, also, he served on many committees, notably that for unemployment during the depression, and as director, treasurer, and president (1936-37). In 1936 he was its delegate to the first International Chemical Engineering Congress in London. He was president of The Chemists Club in 1936 and 1937.

Honors came to him: the Sc.D. degree from Colgate University in 1930; LL.D. from Washington University in 1938; the Pioneer Award of the National Association of Manufacturers in 1940; and the Perkin Medal of the Society of Chemical Industry in 1942.

Ittner was married twice: in 1900 to Emilie Younglof, who died in 1933; in 1934 he married Hildegarde Hirsche. They made their home in Spring Lake, N.J., where he died Apr. 22, 1945. He was survived by a son, Irving H., and a daughter, Mrs. E. B. Sullivan; and by his second wife, and their son, Robert A. Ittner.

Personal recollections; *Chem. Eng. News,* April 20, 1937, p. 177 and May 10, 1945, p. 834; *Chemist,* Nov. 1943, May 1945; *Soap San. Chem.,* Dec. 1941, May 1945, July 1945; obituary, *New York Times,* Apr. 24, 1945.

FLORENCE E. WALL

J

Charles Thomas Jackson

1805-1880

Charles Jackson was born in Plymouth, Mass., June 21, 1805. He graduated from Harvard Medical School in 1829. Three years of study and travel in Europe added to his foundation as a physician, chemist, mineralogist, and geologist. He opened in Boston one of the earliest analytical and consulting laboratories in the United States. Mining and textile firms were among his clients. He also taught practical chemistry to private pupils (very few colleges offered laboratory instruction at that time) and several of his students became notable chemists. He was one of the pioneer American geological surveyors. From 1836 to 1848 he was successively state geologist of Maine, Rhode Island, and New Hampshire, and United States geologist of the Lake Superior region.

He published at least 100 articles, mostly reporting analyses, carried out a score of investigations for the Department of Agriculture and published several book length reports of geological surveys.

While in Europe Jackson had purchased instruments for experimenting on electricity. On the voyage home in 1832 he talked with a fellow passenger, Samuel F. B. Morse, about electrical and magnetic phenomena. His conversation so stimulated Morse that the latter spent the rest of the voyage working out his idea of the electromagnetic telegraph. Years later, after Morse's telegraph proved successful, Jackson tried to claim some of the credit for the invention.

From 1846 onward Jackson was involved in a controversy with two dentists and a physician over who should get credit for the use of ether as an anesthetic. In 1844 a dentist, Horace Wells, observed that a severe wound in the leg of a young man had caused him no pain because he had inhaled some nitrous oxide at a public demonstration of its effects just before he was injured. Wells immediately saw the significance of this event. The next day he asked another dentist to extract an infected tooth after he had inhaled enough nitrous oxide to become unconscious. Following this demonstration of the effectiveness of nitrous oxide in alleviating pain, Wells extracted teeth from several patients, all without pain. He then obtained permission to make a demonstration before an audience at Massachusetts General Hospital in Boston. Wells was nervous and excited and pulled the tooth before the gas had taken effect. The patient howled with pain and the audience jeered at Wells.

Two years after Wells' unfortunate demonstration, Sep. 30, 1846, William T. G. Morton, a former partner and student of Wells, asked Jackson, who had been his medical preceptor, how to make nitrous oxide. Jackson said it was too complicated but that ether would act just as well as an anesthetic. He gave him some and showed him how to apply it. Morton used it the same day on a patient, extracting a firmly rooted tooth without pain.

The next day Morton asked a patent attorney to apply for a patent on his use of ether. The attorney advised him that inasmuch as Jackson had suggested its use, it would be wise, in order to avoid possible litigation, to share the patent with him. Jackson refused the offer, saying it was against the ethics of the medical profession to patent such an idea. However, Jackson signed the application for his friend, but transferred his rights to Morton. When Morton became famous after a successful operation had been performed, Jackson felt that he had been left out and started a controversy with Morton. Wells entered the controversy, saying that he was the originator of anesthesia since it made no difference whether gas or a liquid was used to induce unconsciousness. Crawford W. Long, a physician of Georgia, then entered his claim that he was the originator since he had performed several operations with ether, the first Mar. 30, 1842. Until then Jackson apparently did not know that Long or anybody else had used it before he suggested it to Morton.

Jackson's many claims and controversies regarding priority and his repeated changes of interest and location would seem to indicate some instability in later life. He was committed to a mental hospital in Somerville, Mass., and died 5 years later Aug. 28, 1880.

Thomas E. Keys, "The History of Surgical Anesthesia," pp. 25, 26, 107, Schuman's (1963); Bernard E. Schaar, "A Strange Chapter in Anesthetics," *Chemistry* **39,** 18-19 (Oct. 1966); M. D. K. Bremner, "The Story of Dentistry," Dental Items of Interest Pub. Co. (1946); "Dictionary of American Biography," **9,** 536, Chas. Scribner's Sons (1932); obituary, *Proc. Amer. Acad. Arts Sci.* (n.s.) **16,** 430-432 (1880-81).

BERNARD E. SCHAAR

Walter Abraham Jacobs

1883-1967

Jacobs, an early investigator of nucleic acids and a prolific contributor to the chemistry of medicinally important glycosides, alkaloids, and steroids, was born Dec. 24, 1883 in New York City to Charles and Elizabeth (Friedlander) Jacobs. Walter's father, a tailor, actively encouraged his son's interest in science. The young man, after receiving his A.B. degree in 1904, remained at Columbia University for another year, studying under the distinguished organic chemist and teacher, Marston T. Bogert, and received his master's degree in 1905. Accepted by Emil Fischer for doctoral work at Friedrich-Wilhelms-University in Berlin, Jacobs proved to be a brilliant investigator. His resolution of *dl*-serine, *dl*-isoserine, and *dl*-diaminopropionic acid into their optically active components, provided the basis for his Ph.D. degree in 1907. On his return to the United States, he was accepted as a fellow at Rockefeller Institute for Medical Research, mainly at the urging of Phoebus A. Levene. In 1908 Jacobs married Laura F. Dreyfoos, whose encouragement and intellectual stimulation were to be a source of strength throughout her husband's long life. They had two children, Walter C. and Elizabeth R.

Working with Levene on nucleic acids, Jacobs identified ribose as the characteristic sugar in what is now known as RNA while another group under Levene isolated and identified desoxyribose (from DNA). Jacobs' work on the analysis of sugar moieties was extended to inosinic and guanylic acids.

By 1912 Jacobs was promoted to associate member of the Institute with his own chemotherapy laboratory, independent of Levene's, where he worked on hexamethylenetetraminium salts for treating poliomyelitis. With the outbreak of war in Europe in 1914, Jacobs turned to the synthesis of Salvarsan, cut off from German sources by the British blockade, and other arsenicals. Jacobs' principal new discovery was Tryparsamide, effective in treating African sleeping sickness. The discovery was belatedly honored in 1953 by the Belgian Government when Jacobs and his colleagues were made Officers of the Order of Leopold II.

Although Jacobs was very active in sulfanilamide chemistry, he considered sulfanilamide more as a convenient building block for chemotherapeutic agents and failed to discover its importance as a bactericide. In his synthetic work on arsenicals and sulfanilamides, his principal collaborator was Michael Heidelberger.

In the post-war period Jacobs changed his approach to chemotherapy, from one based on systematic synthesis to one based on structural elucidation of natural products of known physiological activity (*e.g.* digitalis, strophanthus, squill, and convallatoxin). The active substances proved to be steroids, the basic ring structure of which was not established until 1933. In the later stages of these structural studies, Jacobs worked closely with Robert C. Elderfield, who was at Rockefeller Institute from 1930 to 1936.

Jacobs' great work on ergot began about 1932. By 1934 he had isolated lysergic acid and subsequently confirmed its structure by synthesizing dihydrolysergic acid. In his later years at the Rockefeller Institute he studied the veratrine alkaloids (sterols) and the aconite alkaloids (diterpenes), whose structure was finally established after the revolution in analytical instrumentation that occurred shortly after his retirement in 1949. He remained active in an emeritus status from 1949 to 1959, continuing to publish (e.g. on the structures of atisine and cevine). In 1961 he moved to Los Angeles, where after 6 years of relative inactivity he died July 12, 1967.

George W. Corner, "A History of the Rockefeller Institute, 1901-1953 Origin & Growth," Rockefeller Inst. Press (1964); Lyman C. Craig, *The Rockefeller University Review,* Nov.-Dec., 1967, pp. 23-5; Alfred E. Mirsky, a personal communication.

CHARLES H. FUCHSMAN

William Stephen Jacobs

1772-1843

Jacobs, author of one of the early American chemistry texts, was born in Belgium, Mar. 4, 1772. In his youth he left home and went to Austria, where he began to study medicine and served in the army. In 1792 he traveled to Paris to continue his medical studies and was drafted into the French army as an assistant surgean. He soon deserted, made his way to Amsterdam, and worked in military hospitals. He emigrated to the United States in 1794, reaching Philadelphia in late 1794 or early 1795.

The University of Pennsylvania Medical School engaged Jacobs as dissector. He continued to study medicine and obtained his M.D. degree in 1801. Interested in chemistry, he wrote his thesis on the analysis of six intestinal stones, three from human beings and three from animals. To assist himself in studying, he took notes from the texts of Chaptal, Fourcroy, and other chemists. His fellow students persuaded him to publish these notes for the use of medical classes. He did so in a small book, "The Student's Chemical Pocket Companion," which appeared in 1802, with a second edition in 1807. During his residence in the United States he was a member and librarian of the Chemical Society of Philadelphia.

Jacobs preferred a milder climate, and in 1803 he moved to St. Croix, Virgin Islands, where he settled down, married, raised a family, practiced medicine, and died Dec. 30, 1843.

Biography by Wyndham D. Miles in *J. Chem. Educ.* **24,** 249-250 (1947).

WYNDHAM D. MILES

Charles James

1880-1928

James was born Apr. 27, 1880 at Earls Barton near Northampton, England. His formal training was at University College, London, where he studied under Sir William Ramsay and J. N. Collie. He graduated from the Institute of Chemistry, 1904, and came to the United States in 1906 to work for National Refining Co. That same year he was appointed to the chemistry faculty of University of New Hampshire. In 1912 he became professor of chemistry and succeeded Charles L. Parsons as head of the department, a position he held until his death on Dec. 10, 1928. An honorary Sc.D. degree was conferred on James by the university in 1927, and new chemistry laboratories, which he had looked forward to for several years, were dedicated as Charles James Hall in November 1929. As early as 1911 he had received the Nichols Medal for his research in rare earth chemistry.

At Durham, James conducted extensive investigations in the chemistry of the rare earth and related elements, the results of which were published in about 60 papers. He displayed a remarkable ingenuity for devising new methods for separating these elements, whose physical and chemical properties are so similar. Among these methods, that involving fractional crystallization of the double magnesium rare earth nitrates won him international acclaim as the James Method. His development of fractional crystallization techniques was of necessity accompanied by a parallel development of procedures for quantitative analysis of rare earth elements. He also determined the atomic weights of thulium, samarium, and yttrium. In this work it was often necessary to process kilogram quantities of various salts of the rare earths and other rare elements, and he was conspicuously generous in supplying other workers in the field with material unobtainable elsewhere.

In the first edition of Mary E. Weeks, "The Discovery of the Elements," published 5 years after James' death, he was credited with the independent discovery of elements 71 and 61. At the time that the news of

Georges Urbain's discovery of element 71 (lutetium) reached America, James had on hand a large supply of the pure oxide. He suppressed his disappointment and accepted Urbain's priority without question. James' preparation (with Hemon Fogg) of salts containing small amounts of element 61 admixed with large amounts of neodymium was carried out during the years 1923–26. The x-ray spectra, recorded by James Cork at University of Michigan, were published in 1926, shortly after B Smith Hopkins and Joseph Harris at University of Illinois announced the discovery of the element named illinium. At about the same time an independent discovery was claimed by L. Rolla and L. Fernandes at University of Florence. In all of this work what they all thought to be the element was concentrated from minerals that occurred in nature.

In 1938–41 Laurence E. Quill and Marion Pool at Michigan State University claimed the synthesis of element 61 by the cyclotron bombardment of neodymium and praseodymium. Element 61 was isolated as a fission product of uranium and was prepared by the neutron bombardment of neodymium by J. A. Marinsky, L. E. Glendenin and Charles D. Coryell at the Oak Ridge atomic pile laboratories in 1945. The latter workers were able to use ion-exchange techniques to separate rare earth elements, a procedure that has made obsolete the tedious fractionation technique of James and other early workers. The name promethium was accepted for element 61 by the International Union of Pure and Applied Chemistry in 1949.

After arriving at University of New Hampshire, James soon acquired the nickname "King James" and wore his crown throughout his years as a teacher and researcher, during which period his enthusiastic attack on the most tedious tasks was an inspiration to all. In 1915 he married Marion Templeton, who shared with him a common interest in floriculture. At their home in Durham was a collection of rare orchids and an entire acre of delphinium. James was recognized as an expert apiarist; his wide-ranging interests extended to all aspects of nature.

B S. Hopkins, L. A. Pratt, and M. M. Smith, *The Nucleus,* **9**, 11-16, 39-42, 63-66 (1931); M. E. Weeks and H. M. Leicester, "Discovery of the Elements," Journal of Chemical Education (1968) 7th ed., pp. 693-696, with portrait; biography by H. A. Iddles in *J. Chem. Educ.* **7**, 812-813 (1930), with portrait.

EDWARD R. ATKINSON

Edward Hopkins Jenkins

1850-1931

Jenkins, born May 31, 1850, at Falmouth, Mass., to John and Chloe (Thompson) Jenkins was involved for his entire professional life with the agriculture of the state of Connecticut, beginning as chemist in 1876 and formally retiring in 1923 as director of the experiment stations at New Haven and Storrs. Jenkins was educated at Phillips Andover and at Yale, from which he received his A.B. in 1872. He stayed on to do graduate work at Yale, moved to Leipzig University in 1875, and spent some time at the Forest School, Tharandt, in what is now Sachsen-Anhalt, East Germany. He finally received his Ph.D. degree from Yale in 1879.

During this time Jenkins took a position as chemist at the Connecticut Agricultural Experiment Station, newly established in 1875 in Middletown. In 1877 the station became a state institution and was transferred to New Haven with Samuel W. Johnson as director; Jenkins went along as chemist, working together with Henry Armsby. This team proved a remarkable success, with Armsby serving as the laboratory partner and Jenkins acting as outside man who maintained contact with farmers and legislators. "In those days applied agricultural science was on trial; every 'Farmer's Institute' was a court of inquiry, and the burden of proof was always on the station. (Jenkins) . . . proved to be a successful advocate in such a cause."

In recognition of his particular talents Jenkins was made vice-director of the station in 1884; he succeeded Johnson as director in 1900 and was made director emeritus on his retirement in 1923. Under his extraordinarily able administration the station added departments of entomology, forestry, and genetics and established the tobacco substation at Windsor, at the southern end of the fertile Connecticut River valley tobacco country which extends through Massachusetts. Jenkins was instrumental in promoting tobacco farming in Connecticut, publishing many Experiment Station Reports on growing and curing from 1893 to 1904. He introduced shade-growing of tobacco (under canopies of cheesecloth suspended 10-15 feet above the plants); acre upon acre of this peculiar mode of cultivation, which imparts properties desirable in cigar wrapper leaf, can be seen today in the Connecticut valley.

Tobacco was only one of Jenkins' concerns, however. His Experiment Station Bulletins covered a variety of subjects, including chemical composition of fertilizers, feedstuffs, food, and drugs; in addition his papers in the chemical literature touched on general analytical methods and on analysis of dairy products. In 1912 Jenkins was appointed director of the Storrs Agricultural Experiment Station. He effectively consolidated the management of the two stations under him, and his administration was a model of economy and enlightened cooperation with other agencies. He supported work on proteins and on animal diseases, working in collaboration with the medical faculty at Yale. He served as chairman of the State Sewage Commission and as State Food Administrator during World War I. He was a charter member of the Association of Official Agricultural Chemists, one of its early presidents, and a member of its first Food Standards Committee. He belonged to a number of agricultural and professional societies and was president of some.

Despite his demanding professional life Jenkins maintained many outside interests. He was a delightful companion, with a rare wit and a gift for the well-turned phrase; for his skill in this direction he was much in demand as a toastmaster, as well as a speaker and proselytizer on agricultural matters. His tastes were highly literary; the title page of his Station Bulletin, "Studies of the Tobacco Crop of Connecticut," bears a quotation from Ben Jonson in the humorous Elizabethan vein of praise for tobacco. He contributed anonymous sketches to literary and humorous magazines and wrote the section on Connecticut Agriculture in Norris G. Osborne's "History of Connecticut" (1926). He was an editor of the first edition of the "Century Dictionary," responsible for the chemical terms, and he contributed more than a dozen articles to the "Dictionary of American Biography."

Jenkins was married in 1885 to Elizabeth Elliot Foote; they had no children. He remained active after his retirement in 1923, right up to his death, Nov. 6, 1931, at age 81.

W. E. Britton, *Science* **74**, 537-38 (1931); E. M. Bailey, *Ind. Eng. Chem.* **17**, 874 (1925); "Dictionary of American Biography," **10**, 44, Chas. Scribner's Sons (1933); "National Cyclopedia of American Biography," **24**, 424, James T. White & Co. (1935); H. S. Graves, *Conn. Expt. Station Bull. 280* (Report No. 49, 1924-25), 651-52, with portrait, p. 650, and further sketch, p. 626.

ROBERT M. HAWTHORNE JR.

Samuel William Johnson

1830-1909

Johnson, the son of Abner Johnson, a prosperous farmer, and Hannah G. Johnson, was born July 3, 1830, in Kingsboro, N.Y. At the age of 14 one of his teachers interested him in chemistry and gave him a copy of the English translation of Fresenius' analytical chemistry which spurred him to further studies. His father encouraged him and helped him to equip a laboratory in an old building. To earn money, he taught in local schools and then at an academy in Brooklyn, N.Y.

His goal focused on agricultural chemistry, at that time a profession which could scarcely be said to exist in the United States but with which he had become familiar through reading Liebig's agricultural works. He entered the Yale School of Applied Chemistry in January 1850 to study with John Pitkin Norton, then at the height of his influence in agricultural chemistry. In 1851-52 he taught science at State Normal School in Albany, N.Y., and then returned to Yale for more study.

In 1853 he entered Erdmann's laboratory at Leipzig. In 1854 he travelled to Munich to work with Liebig, von Kobell, and von Pettenkofer and spent much of his free time visiting farms and agricultural institutes in order to bring back to the United States all of the latest information. In 1855 he studied with Frankland at Owen's College in London. In that year he was appointed first assistant in the analytical laboratory of the Yale Scientific School and soon rose to the rank of assistant professor of chemistry where he "superintended the laboratory, gave instruction, collected fees, paid the expenses, and when needful, tended stoves, washed apparatus, and swept up the floor."

In 1856 Johnson was named professor of analytical chemistry at the Scientific School and began his career as advocate of scientific farming. In 1857 he read a paper entitled "Frauds in Commercial Manures" at a meeting of the Connecticut Agricultural Society. This paper attracted wide attention because of his vigorous denunciation of the fertilizer frauds then bedeviling farmers. He advocated establishing a state chemist "to analyze . . . all the various manures that come into the Connecticut market" in order to protect farmers.

Johnson, who had been appointed chemist for the State Agricultural Society just prior to this meeting, undertook to make these systematic analyses in order to convince the agricultural community that it could save money by applying scientific research. This understanding had to be reached before farmers would support general agricultural research. To his analyses he appended a dollar value for each fertilizer so that farmers could compare them.

In 1864 Sheffield Scientific School was officially appointed the land-grant school in Connecticut in fulfillment of the Morrill Land Bill of 1862. In 1874 after long agitation by Johnson and many members of the State Agricultural Society, the Connecticut legislature approved the establishment of the first agricultural experiment station in the nation. This station was first located at Wesleyan University, whose laboratories and facilities were open without charge to the station, and from 1875-77 the station, without Johnson's active participation, attended mainly to fertilizer analyses. In 1877 it was reorganized to meet its goals, Johnson was appointed director, and it then became a true agricultural research station. For 5 years, the agricultural station was housed at the Sheffield Scientific School but moved into its own building in 1882. Johnson retired from the directorship of the station in 1900.

He was an excellent administrator, and it was as a result of his direction that the station was widely emulated. He soon ceased to carry on research, but he was an excellent guide for research performed by others. The roster of great chemists and agriculturalists who worked under him is exemplified by such names as E. H. Jenkins, W. O. Atwater, H. P. Armsby, A. L. Winton, R. H. Chittenden, and T. B. Osborne. From the point of view of his own research, he is best known for improving the Kjeldahl method for determining nitrogen in proteins so as to permit rapid determinations of large numbers of samples.

Johnson wrote more than 200 papers and 7 books. He produced three excellent translations of Fresenius' analytical works from 1864 to 1883. He was best known as a writer of an extraordinary book, "How Crops Grow," which passed through many editions and translations and made him world famous. It was written to be used and understood by an intelligent farmer, and it served for many years as a textbook on agricultural science.

Johnson was elected to the National Academy of Sciences at age 36. In 1881 he became the first president of the Association of Official Agricultural Chemists. Despite his many honors, the only college degree he ever received was an honorary master of arts degree from Yale when he was appointed professor of analytical chemistry in 1856.

With justice, Johnson was called "The father of American agricultural research." He died in New Haven, July 29, 1909.

Elizabeth A. Osborne (ed.), "From the Letter-Files of S. W. Johnson," Yale Univ. Press (1913); Edgar F. Smith, "Chemistry in America," D. Appleton & Co., 275-276 (1914); Hubert B. Vickery, "Liebig and Proteins," *J. Chem. Educ.* **19,** 75-76 (1942); Russell H. Chittenden, "History of the Sheffield Scientific School of Yale University," Oxford Univ. Press (1928).

LOUIS I. KUSLAN

Treat Baldwin Johnson

1875-1947

Johnson spent most of his life in and near Bethany, Conn., the town in which he was born Mar. 29, 1875. His father, Dwight Lauren Johnson and his mother, Harriet Adeline Baldwin, were descended from early colonial stock. Mr. Johnson was a farmer, and Treat, the oldest of three boys, was brought up on the usual farm chores.

At Yale University Johnson came to know Henry L. Wheeler, who was then well along on his lifetime career of synthesizing organic nitrogen compounds. He studied physical chemistry under Bertram B. Boltwood and analytical chemistry under Harold L. Wells. In 1898, when Johnson was appointed as a laboratory assistant, he published his first paper with Wheeler, "On the Non Existence of Four Methenylphenylparatolyl Amidines" in *American Chemical Journal.* In 1901, when he received his Ph.D. degree, he co-authored five papers with Wheeler and one by himself. In 1902, he was appointed instructor of chemistry at Yale and continued his research there. At retirement, he had published 358 papers, either alone or as co-author.

His early papers reported investigations of substituted ureas and thioureas and their uses in synthesizing hydantoins and thiohydantoins. From 1903 to 1906 he collaborated with Wheeler in establishing the structure of naturally occurring pyrimidines. They confirmed the chemical structures assigned by Kossel some years earlier. In addition, working with samples of wheat nucleic acid, they isolated cytosine and thereby proved that cytosine is present in animal and plant nucleic acid.

Beginning in 1903 Johnson assigned his students various pyrimidine syntheses. By 1915 46 papers had appeared on this subject. The work in 1908 when he treated 2,6-dioxypyrimidines with fuming nitric acid was particularly noteworthy. The thymine derivative was found to be soluble in alcohol and thus easily separated from the relatively insoluble uracil derivative. In 1907 he developed a color test to differentiate uracil and cytosine from thymine. This work on the pyrimidines was thorough and sound. Robert R. Williams, who demonstrated the structure of the vitamin thiamin in 1935 and who synthesized it in 1936, profited substantially from the pioneering work of Johnson and his collaborators who had synthesized a number of chemical relatives of thiamin and who had devised procedures which were readily adaptable for Williams' purpose.

In 1911 Johnson extended his interests to proteins and amino acids, with particular emphasis on hydantoins, and in 1912 he, assisted by one of his students, synthesized phenylalanine through a much simplified and less expensive approach than previously used.

His work on the chemistry of silk, which derived from his interest in proteins, led to considerable research sponsored by dyers and textile groups. In 1922 he embarked on still another line of research. Supported by funds from the National Tuberculosis Association, he investigated nucleic acids of tuberculosis bacilli. He and co-workers separated the bacillus nucleic acids and analyzed the remaining cell proteins.

Another line of his research was the preparation of synthetic drugs, such as ephedrine in 1929 and hydantoins with hypnotic properties, as well as the germicidal action of substituted phenols. His preparation of alkyl resorcinols with antiseptic properties led to patents and to the commercial production of hexylresorcinol.

Pyrimidine research continued along with the other research topics he found so interesting, and his life's work with pyrimidines ended with a total of 182 papers in 1944. Often, 5 to 10 publications, mainly on the pyrimidines, emerged yearly from his laboratory.

In addition to his work at Yale, Johnson maintained a private laboratory in Bethany. "Bethwood," his small but well equipped laboratory, was mainly used for commercially sponsored work carried on by paid assistants. One of his clients was A. C. Gilbert Co. of New Haven, to which he was chief consultant for over 30 years in the design of their toy chemistry sets.

Johnson was honored with the Nichols

medal of the New York Section of the American Chemical Society in 1918 for his work on pyrimidines. His acceptance speech constituted an excellent review of his work on the subject. He also published a more extensive review of certain derivatives of pyrimidines (together with Dorothy Hahn) in *Chemical Reviews* in 1933 as well as the chapter on pyrimidines, purines, and nucleic acids in the 1938 edition of Henry Gilman's "Organic Chemistry" and the chapter on the development of organic chemistry in America in the Fiftieth Anniversary Commemorative volume of the ACS in 1926. He and Hahn translated and revised Ferdinand Henrich's "Theories of Organic Chemistry," in 1922.

Johnson, although not an inspiring lecturer, was thorough, clear, and scholarly. He was at his best, however, as a director of research, and 94 graduate students received their degrees under him. He died July 28, 1947, in Bethany, Conn.

Biography by Hubert B. Vickery, *Biog. Mem. Nat. Acad. Sci.* **27,** 83-119 (1952), with portrait; Russell H. Chittenden, "The Development of Physiological Chemistry in the United States," Chemical Catalog Co. (1930); "Chemistry and the Toymaker," *Chem. Eng. News* **30,** 5238 (1952).

LOUIS I. KUSLAN

Walter Rogers Johnson

1794-1852

The federal government in the early years did not have scientists and laboratories as it has today. When a problem arose that required a scientist, a department hired an expert to carry out an investigation. Before the Civil War, when living was simpler, this method of handling scientific problems was quite satisfactory.

Johnson was one such scientist called on by the Treasury, Navy, and Interior departments in the 1830's and '40's. Born in Leominster, Mass., June 21, 1794, educated at Harvard, Johnson taught school in the Philadelphia area from 1821 to 1837 and was professor of chemistry and physics in Pennsylvania Medical College, 1839 to 1843. Thereafter he was a free-lance scientist, making surveys and analyses for mining companies, examining city water supplies, and carrying out assignments for the government.

His initial work for the government came about in the 1830's. He was on Franklin Institute's Committee on Explosions of Steam Boilers. Steamboats were a major transporter of goods and persons at that time, but boilers were poorly engineered and often unsafe. Tragic boiler explosions occurred, sinking ships and drowning passengers. The Treasury Department, concerned with navigation, financed investigations which lasted several years. A 254-page "Report of the Committee . . . on the explosions of steam boilers . . . made at the request of the Treasury Department" signed by Johnson appeared in 1837.

In the 1830's Johnson examined Pennsylvania coal areas for mining firms. He became an expert on coal and in 1841 suggested to the Secretary of the Navy that various coals be tested to determine their fuel value so that the Navy would have a basis for purchasing coal for warships. The Secretary engaged Johnson, who turned out a 600-page "Report to the Navy . . . on American coals applicable to steam navigation, and other purposes" (1844). Three-quarters of a century later Samuel Parr, noted coal expert, said of Johnson's work: "One cannot go through the pages of this report without marveling at the indefatigable energy that must have characterized the author, the skill with which he developed a technic along lines having no established precedents . . . and the wisdom with which he arranged and interpreted his results."

Subsequently Johnson carried out a number of government projects, the nature of which is indicated in titles of his reports: "Report of the Commissioners appointed by the Secretary of the Navy to examine the several plans of floating docks" (1842); "Report of the Secretary of the Navy . . . in relation to the invention of Thomas S. Easton, for preventing explosions of steam boilers" (1842); "Report to . . . Secretary of the Navy . . . of researches on the character and tests of sheathing copper" (1842); "Letter from . . . Johnson on . . . further experiments on iron, copper, and coal. In report of the

Secretary of the Navy" (1844); "Report of the Secretary of the Navy communicating the report of a board of examiners to make experimental trials of inventions and plans for preventing explosions of steam boilers" (1844); "Report to the Secretary of the Navy . . . respecting Forest Improvement Coal" (1845).

Finding most of his work centered in Washington, Johnson moved there in 1848. He edited a 2-volume "Chemical Technology" by Friedrich Knapp and compiled a book, "The Coal Trade of British America." Occasionally he traveled in Maryland, Virginia, West Virginia, and North Carolina, examining mine properties for clients. He published more than 100 articles on scientific and other subjects, wrote school books on geometry, Greek, physics, and chemistry ("An Elementary Treatise on Chemistry, together with Treatises on Metallurgy, Mineralogy, Chrystallography, Geology, Oryctology and Meteorology" (1846)), and he held at least two patents.

Johnson, from a contemporary description, was "a friend and helper to the poor and the oppressed . . . an ardent lover of liberty, both civil and religious. . . . Although he had never labored with the primary purpose of accumulating wealth, he had usually been liberally paid for his professional exertions; his rule of moderate expenditure enabled him to live in respectable independence and generous hospitality and to gather a comfortable provision against future contingencies."

Johnson's last government assignment was published as the "Report of . . . Johnson on the building stone used in constructing the foundations of the extension of the United States Capitol, made at the request of a select committee of the House of Representatives." A few weeks after completing the report and while still carrying out some experiments, he died Apr. 26, 1852.

Biography by George E. Pettingill, *J. Franklin Inst.* **250,** 93-113 (1950), with portrait, list of Johnson's publications; Samuel W. Parr, *Ind. Eng. Chem.* **18,** 94-98 (1926); quote is from obituary in Washington newspaper, *Daily National Intelligencer,* Apr. 27, 1852.

WYNDHAM D. MILES

Joseph Jones

1833-1896

Jones was born in Liberty County, Ga., Sep. 6, 1833, the second son of a well known Presbyterian minister and author, Charles Colcock Jones, Sr. After attending South Carolina College at Columbia for a year he went to Princeton University, receiving his A.B. degree in 1853. He then attended University of Pennsylvania Medical School where he studied under Joseph Leidy and Samuel Jackson. His first two research papers, on the effect of poisons on animal secretions and organs, were published in 1855 in *American Journal of Medical Sciences.* Jones was awarded his M.D. degree in 1856.

He began his teaching career in 1856 at Savannah Medical College. Simultaneously he started investigating malaria in the Marine Hospital in Savannah.

In January 1858 he became professor of natural philosophy at Georgia State University, Athens, but within a year he accepted the chair of chemistry at Medical College of Georgia in Augusta. When the cotton planters' convention met in 1860, he was appointed chemist and was the author of the first report submitted to that group.

With the outbreak of the Civil War he enlisted in the Liberty Independent Troop of Cavalry as a private but was shortly (April 1, 1862) commissioned full surgeon.

During the war years Jones studied malaria, yellow fever, gangrene, typhoid and typhus. He correctly identified "camp fever" as typhoid and noted the typhoid bacillus but, bound by the contemporary miasmatic theory, failed to see its relationship to the disease.

Because of his activities as investigator of the military prison at Andersonville, Ga., where nearly 11,000 Federal prisoners died, Jones was called as a witness in the trial of Henry Wirz, commander of the prison. Jones denied the charge that Wirz had used poisonous vaccines to kill Federal prisoners and pointed out that much of the disease prevalent in the camp was due to overcrowded conditions, brought on because the federal government had stopped the exchange of prisoners.

In 1866 the Medical College at Augusta, disbanded in 1861, was reorganized. Jones returned there, but in 1867 he accepted the professorship of pathology at University of Tennessee. In 1868 he became professor of chemistry and clinical medicine at University of Louisiana (later Tulane University), and concurrently physician at Charity Hospital, New Orleans. He taught chemistry and medicine until 1894 and thereafter chemistry and medical jurisprudence until he died.

During the three decades Jones lived in New Orleans, he carried out scientific research, especially on fevers current in the South. He claimed priority in the identification of the malarial organism, but he did not publish an account of his work until 1887, six years after Laveran.

From 1880 to 1884 Jones was president of the Louisiana Board of Health. In this capacity he waged a jurisdictional dispute with the National Board of Health over quarantine to keep Louisiana free of yellow fever while it was epidemic elsewhere. His efforts were supported when the Supreme Court of Louisiana (in 1884) and later the U. S. Supreme Court (in 1886) upheld the decision that quarantine was a legitimate exercise of police powers by individual states.

Jones was elected president of the Louisiana State Medical Society in 1887 and was prominent in the Ninth International Medical Congress in Washington in the same year. He was appointed Surgeon General of the United Confederate Veterans in 1890.

Jones' ideas on the chemical education of a physician and some of his observations relating chemistry and pathology may be found in his four volume "Medical and Surgical Memoirs," which he published himself between 1876 and 1890. He died in New Orleans, Feb. 17, 1896.

Biographies by Stanhope Bayne-Jones in *Bull. Tulane Univ. Med. Faculty* **17,** 223-230 (1958), by Joseph Krafka, Jr., in *J. Med. Assn. Ga.* **31,** 353-363 (1942), and by H. D. Riley, Jr., in *J. Tenn. State Med. Assn.* **53,** 493-504 (1960); James O. Breeden, "Joseph Jones, Confederate Surgeon," Ph.D. dissertation, Tulane Univ., 1967, excerpted in *Bull. Tulane Univ. Med. Faculty* **16,** 41-48 (1967); correspondence and manuscripts of Joseph Jones, Howard-Tilton Library, Tulane Univ.; Charles E. Jones, "Biographical and Historical Memoirs of Louisiana," **1,** 498-503, Goodspeed (1892); R. Manson Myers, ed., "The Children of Pride, a True Story of Georgia and the Civil War," Yale Univ. Press (1972).

VIRGINIA F. MCCONNELL

Thomas P. Jones

1773-1848

Practically nothing is known of the early life of Thomas P. Jones, a chemist, physician, and superintendent of the U.S. Patent Office. Born in England in 1773 or '74, he turned up in North Carolina many years later. Somewhere along the way he acquired an M.D. degree and a knowledge of chemistry.

Jones' first appearance in the role of scientist was on the lecture platform. From the mid-1700's to the late 1800's entertainment and education were provided by itinerant science lecturers who went from town to town delivering series of lectures to which men, women, and youngsters were admitted for a fee. Jones was on the lecture trail for a number of years. In the winter of 1809–10 he presented lectures on chemistry in Albany, N.Y. The following year he lectured in Philadelphia and was so successful he remained in that city for several years. One of his ads in the Philadelphia newspaper *Aurora,* January 1813, read thus:

> "On Wed., the 20th inst. at 7 o'clock in the evening, at Dr. Jones' Chemical Lecture Room, S.W. cor. of 4th in Chestnut St., a Lecture will be delivered on the properties of Nitrous Oxide, or the Exhilarating Gas, accompanied with a number of experiments—a large quantity will be prepared to exhibit its effects when inhaled. Tickets at 50 cents each may be had at A. Finley's Bookstore, S.E. corner of Chestnut in 4th St., or at the Lecture Room, on the stated evening."

Laughing gas demonstrations were a favorite with audiences and one cannot help but marvel at how closely Jones and other chemistry lecturers came to anesthesia without discovering it.

Jones became a noted lecturer and in 1814 College of William and Mary hired him as professor of chemistry and physics. He remained at the institution 3 years then resigned, presumably to go lecturing again.

Jones eventually returned to Philadelphia. He became professor of physics and mechanics at Franklin Institute in 1824, and in 1825 he conceived and founded at his own expense the Institute's *Journal*. He made the *Journal* the country's leading news magazine for inventors and engineers and edited it until 1847. His prominence in this endeavor led President John Quincy Adams to appoint him superintendent of the U.S. Patent Office in 1828. He immediately moved to Washington and took up the job. According to the old Washington newspaper *National Intelligencer*, Jones "was so well informed on [American inventions] that, with his strength of memory, when a supposed improvement was laid before him he could refer to the work and to the very page where the same thing had been described and claimed long before. In this way he could often save the applicants all further trouble and expense."

Jones was in office only a year when Andrew Jackson was elected President. In the job turnovers that took place Jones found himself transferred to the State Department's Bureau of Archives, Laws, and Commissions. He remained there in an administrative position until 1837 when he switched back to the Patent Office as an examiner. He resigned in 1838. Jones then became a patent agent and soon was one of the best known agents in the country. One of his prominent clients was Charles Goodyear, noted for his inventions in rubber.

Jones had yet another career in Washington. He was professor of chemistry and pharmacy in Columbian College Medical School (now The George Washington University Medical School) from 1828 to 1834 and 1839 to 1840 and a trustee of the college from 1840 to 1847. He brought out two texts, "Conversations on Natural Philosophy" and "New Conversations on Chemistry," which a Washington reviewer called "two of the best treatises on chemistry and mechanical philosophy which have appeared." Jones' "Chemistry," judging from the number of editions (at least nine between 1831 and 1848), was one of the most popular texts in the United States at the time.

A picture of Jones gives the impression of a slight, wiry person. The *National Intelligencer* said of him that "he passed through the world always possessing a happy, cheerful disposition, generous, humane, simple and unassuming, unreserved and communicative even when overwhelmed with business, and never uncourteous or impetuous." He died Mar. 11, 1848, in Washington.

Information from Elmer L. Kayser, archivist of The George Washington University; catalogs of Columbian College medical school, 1830's; records in the National Archives; Philadelphia newspapers, including *Aurora*, January 1813; Washington, D. C., newspapers, including obituaries in *National Intelligencer* Mar. 13, and Apr. 3, 1848; Washington city directories, 1840's; biography by Francis Fowler, *J. Franklin Inst.* **130,** 1-7 (1890), portrait.

WYNDHAM D. MILES

Webster Newton Jones

1887-1962

Web Jones, son of a Welsh coal miner, was born July 29, 1887 in Rich Hill, Mo. In this environment he learned achievement by hard work, respect for fellow workers, acceptance of discipline, and the value of higher education. At his mother's insistence he attended University of Missouri, leaving in 1909 with a master's degree.

This launched Jones into teaching chemistry at Harvard, Maine, Missouri, Radcliffe, and Montana, earning for him by 1918 the rank of assistant professor. More importantly, he acquired, as his chemistry skills increased, sympathetic understanding of the student and his problems. He liked people, had strict standards of his own, and believed strongly in sound research.

In 1918 during World War I he became a first lieutenant in the Chemical Warfare Service and in 1919 served as an expert for the War Trade Board. On Aug. 31, 1918, he married Nettie Donald Haire who was thenceforth his companion, adviser, and intellectual stimulus. They had two sons, Webster Newton, Jr. and William Cary.

In 1919 Jones joined B. F. Goodrich Co. Temporary facilities were secured for him at University of Chicago until the Goodrich research laboratories were completed. In discussing plans with Goodrich vice president William C. Geer, Jones elected to study the fundamentals of rubber oxidation. This work was interrupted only by a leave of absence at Harvard, where he received his Ph.D. degree in 1920. Returning to Goodrich, he made excellent progress until July 1921, when the depression halted the work. Five months later, he resumed his research with much more limited objectives.

After 1924, departing from basic research, Jones chose to influence the flow of men and materials. He was successively manager of general chemical laboratories, technical superintendent, and general manager of the processing division of B. F. Goodrich. He directed the technical training of young men and distributed many of them throughout the production departments, a move beneficial to the men and the company. He sought to alleviate the serious handicap of inadequate processing controls by introducing such controls as were feasible at that time.

In 1932 Jones returned to the academic world as director of the College of Engineering and Science at Carnegie Institute of Technology, becoming vice president of industrial and government relations July 1, 1953 and retiring from this job in June 1956.

Although he was almost 69 years old, he became an adviser to Pittsburgh Plate Glass Co. on the selection and training of professional personnel. There, with typical energy and enthusiasm, he began developing a program, but within a month he suffered a stroke which incapacitated him until his death on June 8, 1962 in Pittsburgh.

Everywhere Web Jones went he left his mark. His reorganization of the College of Engineering and Science at Carnegie was eminently satisfactory to students, faculty, and administration. He found ways to inspire students toward research. He served on the governor's staff for civil defense in Pennsylvania. He was advisor to the Baruch Committee on Synthetic Rubber, a member of the National Inventors' Council, and he organized the Office of Production Research and Development in World War II. He contributed to the engineering and rubber chemistry literature.

He was president in 1939-40 of the American Institute of Chemical Engineers and in 1944 of the Engineers' Society of Western Pennsylvania. His participation was always active and important, whatever the organization to which he belonged. He received many honors, including honorary degrees from Stevens Institute of Technology and Norwich University and the Award of the Pittsburgh Section of the American Chemical Society.

He was beloved by his friends, his students, and associates, who delighted to pay him honor on every suitable occasion. One of his enduring accomplishments was the inspiration of close associates—students or other—who achieved importance and even fame, as in the case of Harold C. Urey. They all, in some degree, bore the stamp of Web Jones.

Book of speeches and letters for presentation of 1941 Award of ACS Pittsburgh Section; "Who's Who in Engineering," Lewis Historical Publishing Co. (1954-62); *The Crucible* (publication of Pittsburgh Section, American Chemical Society) February 1942; *Pittsburgh Press,* April 18, 1956; *The Carnegie Tartan,* April 17, 1956; Mrs. Webster N. Jones; Dr. Harlan L. Trumbull; personal recollections.

RAY P. DINSMORE

Charles Arad Joy

1823-1891

Joy was born in Ludlowville, N.Y., Oct. 8, 1823. He attended academies at Ovid, N.Y. and Lenox, Mass. In 1844 he graduated from Union College and then attended Harvard Law School where he received his LL.B. degree in 1847. Influenced by Louis Agassiz and Charles Jackson, he turned from law to science and in the summer of 1848 was a member of the U.S. geological team that surveyed the Lake Superior region under Josiah Whitney.

Joy's interest in the minerals he collected led him to study chemistry in Göttingen, Berlin, and Paris from 1849 to 1853. He worked on selenium compounds in Wöhler's laboratory at Göttingen and received his Ph.D. degree in chemistry there in 1852.

At Union College President Eliphalet Nott had visions of developing practical education at higher levels. He planned a special Department of Chemistry. Joy went to Union in 1854 to develop such a department. He was sent to Europe to purchase equipment and chemicals and to consult with architects on plans for the new laboratory. When he returned in the fall of 1855 he found that delays in constructing the laboratory prevented occupancy until the early part of 1857. Charles Chandler came in the winter of 1857 to be Joy's assistant. There had been much bickering and agitation over the attention being given to chemistry and the new curriculum. The costs of the laboratory had greatly exceeded the budget. Partly because of this Joy resigned in the spring of 1857 to become professor of chemistry at Columbia leaving Chandler in charge of the Union laboratory.

At Columbia he began the trend toward modern laboratory instruction. He was invited to develop the School of Mines but declined in favor of Chandler who came to Columbia in 1864. Until ill health forced him to retire in 1877, he remained head of the Department of Chemistry and professor of general chemistry in the School of Mines.

Joy had fine social qualities. His services as a writer and lecturer were much in demand. He wrote popular articles on science for numerous publications, including *Scientific American,* Frank Leslie's periodicals, and "Appletons' American Cyclopedia." While he contributed a number of analyses of minerals for Dana's "System of Mineralogy," his publications of original research were limited.

His professional activities included: editor of *Scientific American* and *Journal of Applied Chemistry* for many years, president of the New York Academy of Sciences in 1866, president of the American Photographic Society, chairman of the Polytechnic Association of the American Institute, and foreign secretary of the American Geographical Society. He was a member of the juries at World's Fairs in London, Paris, and Vienna and at the Centennial Exposition in Philadelphia in 1876.

After his retirement from Columbia he spent considerable time in Europe in the localities of his student days. In 1890 he returned to his country home in Stockbridge, Mass., where he died May 26, 1891.

"Appletons' Cyclopedia of American Biography," **3,** 477, D. Appleton & Co. (1887); M. Benjamin, *Pop. Sci. Mon.* **43,** 405-409 (1893), with portrait; obituary, *Amer. J. Sci.* (3S) **42,** 78 (1891).

EGBERT K. BACON

Arthur Edgar Juve

1901-1965

Juve was born June 17, 1901 in Merrill, Wis. where his family lived until he was 12 years old. The family moved frequently for the next 3 years, finally settling on a farm near Ravenna, Ohio in 1916, where Arthur finished high school. He graduated from Ohio State in 1925 with excellent grades, the degree of bachelor of chemical engineering, and a commission of second lieutenant in the Army Reserve Corps. He married Ernestine Grove July 9, 1927.

Juve began his career in the Akron plant of B. F. Goodrich Co. Aug. 3, 1925 as a technical compounder in the Mechanical Goods Division. He became familiar by first-hand experience with basic rubber technology, raw materials, and processes. He moved from factory operation to development work as his knowledge and skill increased. He helped develop rubber covered rolls, offset blankets, typewriter platens, and engraver's gum for the printing industry. He developed rubber mixtures for special uses, including jar rings, gaskets, tank linings, and hose. An especially important job required strong adhesion of rubber to metal for a lining of a pump used in the petroleum industry. During this 17-year period, he had several illnesses and long convalescences. He used the time well to form lifelong habits of extensive technical reading and thoughtful planning concerning his work.

In 1933 he began to develop methods of applying the new specialty rubbers, Thiokol and neoprene, to hoses and other products requiring hydrocarbon resistance. During the thirties he worked on compounding and use of German Buna rubbers and a proprietary nitrile butadiene rubber developed by Goodrich for oil resistance. In 1942 he began research on synthetic rubber, emphasizing methods of test and evaluation. In 1948 he was appointed director of Compounding Research at the new Research Center at Brecksville, Ohio. In 1954, he became director of Technical Services, heading an organization of 65 persons working on adhesives, rubber, plastics, physical testing, and chemical analysis. His last appointment came in 1961 as director of corporate rubber technology. He was responsible for research in advanced rubber technology and for liaison with all branches and divisions on rubber processing and compounding.

Juve's technical interests were wide, and his insight was frequently sought both inside and outside of his company. He published prolifically on poorly understood phases of rubber technology and creation of new testing methods. He ranked 45th in a world list compiled by John McGavack concerning contributions to the literature of rubber technology. Juve liked people and showed remarkable talents in local and national affairs of the American Chemical Society, the ACS Rubber Division and rubber groups. He served as chairman of the ACS Rubber Division in 1956, presided at symposia, and was instrumental in the creation of the annual *Rubber Reviews* as the fifth issue of the quarterly *Rubber Chemistry & Technology*.

Perhaps Juve's greatest professional interest was testing of rubber. He was a member of the American Society for Testing Materials' Committee D-11 on Rubber and was especially active in standardization work on natural and synthetic rubbers, abrasion, aging, processability, and immersion tests. As chairman of Subcommittee XXIX on Compounding Ingredients, he worked with the National Bureau of Standards to make available certified lots of important compounding ingredients. He helped organize ASTM Committee D-24 on Carbon Black and presided at numerous ASTM symposia on various aspects of testing. In 1950 he became interested in International Standards Organization Technical Committee 45 on rubber and participated actively. In 1964 he was awarded the Charles Goodyear Gold Medal by the Rubber Division and the ASTM Award of Merit, both highly deserved in recognition of 40 years of outstanding service.

Arthur Juve was a brilliant scientist and gifted technologist. In spite of laurels heaped upon him by colleagues and organizations, he never lost his charming humility and tireless zeal to undertake the next essential step. He had a rare combination of a guileless spirit, unobtrusive determination and quiet enthusiasm which together were unconquerable. He died Mar. 10, 1965.

Biography by Arthur W. Carpenter in *Rubber Chem. Technol.* **39,** xxv-xxxii (1966) with portrait; *Rubber World* **152,** No. 1, 34 (1965) with portrait; *Rubber Age* **97,** No. 1, 164 (1965); *Rubber Journal* **147,** No. 5, 40 (May 1965); personal recollections.

W. C. WARNER

K

Louis Albrecht Kahlenberg

1870-1941

Kahlenberg was a pioneer in American physical chemistry and a severe critic of the Arrhenius theory of ionization. His parents, Albert and Bertha (Albrecht) Kahlenberg left Germany as youths and settled in Two Rivers, Wisc. where Louis was born Jan. 28, 1870. He was christened Louis Albrecht Kahlenberg but in later years did not use his middle name. His paternal grandfather was a sea captain, and his father had been a sailor before he settled in America as a butcher.

Louis attended the German Lutheran School in Two Rivers and graduated from high school as valedictorian at age 15. His father wished to have him enter the U.S. Naval Academy, but his mother, who disliked the sea, persuaded him to become a teacher. After studying for a year at Oshkosh Normal School he taught county school for 2 years. In 1889 he entered Milwaukee Normal School; a year later he transferred to University of Wisconsin where he majored in chemistry and received the B.S. degree in 1892. He was then granted the department's first fellowship and studied the solubility of metallic oxides in organic acids with attention to the influence on optical rotation. This work led to his M.S. degree in 1893.

Kahlenberg gave up an instructorship offer at Wisconsin to pursue graduate studies in Wilhelm Ostwald's laboratory at Leipzig. He continued his research on solubility of metallic oxides in organic acids while attending courses given by Ostwald in physical chemistry, by J. Wislicenus in organic chemistry, by Gustav Wiedemann in physics, and by F. Pfeffer in botany (osmotic pressure). His Ph.D. degree was conferred *summa cum laude* in 1895. Upon returning to University of Wisconsin he was told there were no vacancies in the Chemistry Department. Edward Kremers, director of the School of Pharmacy, thereupon appointed him instructor in physical chemistry and pharmaceutical technique. A year later he was taken into the Chemistry Department as instructor. He was promoted to assistant professor in 1897, professor in 1900.

In 1907 Kahlenberg became chairman of the department and director of the chemistry course. These positions were taken from him in 1919 as a reaction toward views held by Kahlenberg which were considered critical of the United State's position in the first World War. In subsequent years he was no longer associated with physical chemistry courses but taught introductory chemistry courses to the engineering students, solution chemistry, and a course in the history of chemistry up to the time of his retirement in 1940. He died of a heart attack in Sarasota, Fla., Mar. 18, 1941.

Kahlenberg married Lillian Belle Heald in 1896. They had three children: Hester, Herman, and Eilhard. Herman, studied chemistry at Wisconsin, taking his B.S. degree in 1922 and his Ph.D. degree under his father in 1925. At that time he became vice-president of Kahlenberg Laboratories, founded at Two Rivers to exploit certain pharmaceuticals developed out of his father's research. The company, of which Louis Kahlenberg was president, later moved to Sarasota, Fla.

Boats were Kahlenberg's principal source of relaxation. His father made him familiar with sailing vessels of various forms. Louis taught himself navigation and throughout his life sailed a variety of boats. In his later years he purchased a yacht in Boston and sailed it up the St. Lawrence, through the Great Lakes to Two Rivers.

Kahlenberg returned from Germany in 1895 with great enthusiasm for the new physical chemistry he had studied there. He began a program of research on solutions and by 1899 had graduated the first of 21 Ph.D.

students. His research soon led him to doubt the theory of ionization and he became a vigorous opponent of the theory, a position he maintained to the end of his life. In his earlier years he debated the subject vigorously in the literature and at chemical meetings. While new research led to modifications in the theory, Kahlenberg was unconvinced by the new interpretations and gradually found himself drifting from the mainstream of chemistry.

Besides his work on solutions, his research activities included studies on cellulose, passage of boric acid and other compounds through skin and other membranes, electrodeposition of metals, activity of metals, and use of colloidal gold in malignancies. He invented Equisetene, a suture material.

It was as a teacher that Kahlenberg was most famous. Although he spoke with a German accent, he was clear, forceful, and dramatic in the lecture room. The richness of his experience, coupled with his enthusiasm and his flair for anecdotes, made the subject an exciting one for his students.

He was president of Section C of the American Association for the Advancement of Science in 1908, served as president of the Wisconsin Academy of Sciences, Arts, and Letters from 1906 to 1909 and was president of The Electrochemical Society in 1930.

Biography by Norris F. Hall, *Trans. Wis. Acad. Sciences, Arts, Letters* **39,** 83-96 (1949) and **40,** 173-183 (1950), contains a full bibliography of his publications; other sketches are: A. J. Ihde in "Dictionary of Scientific Biography," Charles Gillispie, ed., Charles Scribner's Sons, Vol. 7 (1973), pp. 208-09; and A. T. Lincoln, *Chem. Eng. News* **16,** 336-337 (1938); Kahlenberg papers are held by University of Wisconsin Archives and State Historical Society of Wisconsin.

AARON J. IHDE

Martin Kalbfleisch

1804-1873

Kalbfleisch, chemical manufacturer, sometime mayor of Brooklyn, and United States congressman from New York State, was born in Flushing, The Netherlands, Feb. 8, 1804. The son of John and Petronella (Van Pollja) Kalbfleisch, a solid burgher family, he was educated in private schools until he was 18 when he took passage for Sumatra on an American vessel skippered by a New England Yankee who was to alter the entire direction of his life. Finding the Dutch East Indies racked by a cholera epidemic, the captain returned to Antwerp, sold the ship, and with young Kalbfleisch established a trading partnership in Paris.

It was here that Kalbfleisch took up chemistry seriously, studying at the Sorbonne. Here also he married Elizabeth Harvey of Southampton, England, the first of two wives who would ultimately bear him 11 children. By 1826 at 22 and with one child already, he was ready to move to the United States. Once in New York, he found employment in the very-much-infant chemical industry and became superintendent for the New York Chemical Manufacturing Co., makers of sulfuric acid, blue vitriol, lead acetate, and other chemicals. Here he stayed for 3 years, through plant expansion and a move to a new factory, adding practical experience to his theoretical knowledge of chemistry until in 1829 he struck out on his own.

His first venture was the production of paint pigments in a tiny plant far uptown in Harlem. After 6 years of successful operation he was finally tempted by rising Manhattan land prices to sell out in 1835 and to transfer to Bridgeport, Conn. This proved to be a financial error, as the new plant was too far from his metropolitan markets; but it had the advantage of freeing his conscience from any qualms about direct competition with his former employers, which he had consciously avoided in Harlem. He began to manufacture sulfuric acid and soon branched out into sulfates and other acids and their salts.

In 1840 Kalbfleisch moved back into the city, establishing a plant in Greenpoint, Brooklyn. The works would be moved yet once more in 1850 to Bushwick, a village which was later incorporated into Brooklyn. His firm adopted the motto "Quality First," written into its trademark; and this came to be a statement of manufacturing principle. Kalbfleisch determined to make the purest, strongest sulfuric acid possible, using pure brimstone as starting material, and he succeeded to the point that his product became a standard in the industry.

Kalbfleisch's sons were introduced into the family business at an early age, being given responsibility for plant inspection rounds at midnight and two o'clock and reporting certain important gauge readings to their father at breakfast. Three of the sons, Albert M., Charles H., and Franklin H., became managers of the firm in 1869, when the elder Kalbfleisch retired and the company name was changed to Martin Kalbfleisch and Sons. Franklin, the youngest, came ultimately to manage the company by himself. A peculiar mixture of hard-headed trade-war businessman, researcher with patents in his own name, and paternalistic supporter of his employees' families, Franklin finally retired in 1920 after engaging the company in great services to the government during World War I. In 1932 the Kalbfleisch firm was absorbed by American Cyanamid Co.

In the years after 1850, when the family finally settled in Bushwick, Martin Kalbfleisch was drawn into substantial civic and political activity. Finding no suitable school in Bushwick for his large family, he set about to organize a school district, to rent a building, and to hire a schoolmaster; in the end he secured a new, specially built school for the district. From this he went on to become supervisor of Bushwick, and when the town was incorporated into Brooklyn in 1853 he was one of the commission appointed to prepare a charter for the consolidated city. From 1855 to 1861 he served as alderman from Brooklyn's 18th Ward, and in 1861 he was elected mayor. He served simultaneously as representative from the 2d District of New York, being elected in 1862. Except for the 1865-67 term he was Brooklyn's mayor until 1871, and in that tumultuous and graft-ridden era he earned the name of "the Honest Dutchman" for his rigorously upright political dealings. Kalbfleisch died in Brooklyn, Feb. 12, 1873.

W. Haynes, "Chemical Pioneers," D. Van Nostrand (1939; reprinted 1970), pp. 42-56; *New York Times*, February 13, 1873; "Lamb's Biographical Dictionary of the United States," **4**, 470, James H. Lamb Co. (1901); "National Cyclopaedia of American Biography," **15**, 323, James T. White & Co. (1914).

ROBERT M. HAWTHORNE JR.

Oliver Kamm

1888-1965

Kamm, American pharmaceutical chemist, was born Dec. 6, 1888 in Highland, Ill. into a family which had emigrated there from Switzerland in the 1830's. After receiving both his undergraduate and graduate training at University of Illinois, where he received his Ph.D. degree in 1915, he taught for a year at University of Michigan. While a graduate student he married Minnie Watson, a graduate of Olivet College, who was to become a parasitologist, a connoisseur of "Pattern glass pitchers," and an author of books in both her vocational and avocational specialties.

Kamm was back at Illinois in 1917, first as instructor and then as associate professor of organic chemistry. He left academic life in 1920 to join Parke, Davis and Co. in Detroit, as director of chemical research.

His principal publications prior to 1928 were his textbook "Qualitative Organic Analysis," which appeared in 1923; and Volume IV of "Organic Synthesis" (1925), of which he was the editor. His text-book, based on methods developed at Illinois, was an academic standard for many years and served as a model for subsequent works on the subject.

Kamm's spectacular work on the hormones of the posterior pituitary gland brought him great and well-merited prominence in 1928. There had been, at the time, considerable controversy over the nature of the secretion of the posterior pituitary gland, whose activity was generally measured by its oxytocic property (the induction of contraction of uterine muscle); and by its pressor activity (the elevation of blood pressure). Prior to Kamm's work, the prevailing biochemical opinion was that a single hormone was responsible for both effects. By careful and tedious separation, Kamm was able to prepare two fractions of high potency, one of which (Pitocin) had only oxytocic properties, and the other (Pitressin) only pressor properties. Both fractions combined accounted quantitatively for virtually the entire activity of the initial pituitary extract. He was further able to provide initial characterization of the small differences in the amino acid residues in each hormone. His work on the pituitary gland was recognized by an award from the American Association for the Advancement of Science.

Kamm served as chairman of Division of Medicinal Chemistry of the American Chemical Society in 1931-32, and as chairman of the Detroit Section of the society in 1936.

As early as 1929 he cooperated with Edward Doisy, at St. Louis University, on the isolation of estrone (Theelin). But Kamm's main achievement in steroid chemistry centered on his collaboration with Russell Marker, then at Pennsylvania State University. In a series of over 20 papers by Marker, Kamm and other collaborators, published from 1935 to 1938, the chemical pathways linking the principal male and female steroid hormones were clarified and procedures were developed which permitted controlled inversion of particular asymmetric centers in their molecular structures.

Kamm was elevated to the post of scientific director of Parke, Davis in 1938, after which his name appeared less frequently as an author. But his research activity did not slow down. He provided the collaboration of Parke, Davis with Doisy in the latter's work on the chemistry of Vitamins K_1 and K_2. Similar cooperative programs included work with Albert Hogan of University of Missouri on anti-anemia factors and folic acid. Other cooperative programs extended into antisyphilitic organoarsenicals and chloramphenicol.

Kamm exemplified the leadership of unstructured research in medicinal chemistry. He sought and found talented and highly motivated chemists. Within broad limits he allowed them to choose their own problems and the approaches to solving those problems. His professional interests showed a preference for isolating and identifying natural products. In his view each new natural product meant one less to be discovered. His non-professional interests ranged from paintings and Oriental rugs, both of which he collected, to English literature, Dickens and Chaucer being among his favorites.

The lack of visible organizational discipline and Kamm's tendency to absent-mindedness led to his characterization as a relatively poor administrator.

By 1949 his administrative role at Parke, Davis came to an end. He remained as a consultant for the company until his formal retirement in 1953. In the years that followed he traveled widely and provided consultant services for many chemical and pharmaceutical companies. He ceased all professional activity in 1964. He died at Grosse Pointe Farms, Mich., Dec. 5, 1965, one day before his 77th birthday.

Kamm's articles, including an article on pituitary hormones in *Science* **67,** 199 (1928); Parke, Davis & Co., "Scientific Contributions from the Laboratories, 1866-1966" (1966); biography in C. M. Burton and M. A. Burton (eds.), "History of Wayne County and the City of Detroit," Vol. IV, 590-2, S. J. Clarke Pub. Co. (1930); obituaries in *The Detroit Chemist,* 13, Feb. 1966, and *Chem. Eng. News,* 84, Jan. 17, 1966; private communication from Leon A. Sweet.

CHARLES H. FUCHSMAN

Joseph Hoeing Kastle

1864-1916

Kastle was born Jan. 25, 1864 in Lexington, Ky. He attended University of Kentucky where Robert Peter turned his interest toward chemistry and where he received his B.S. degree in 1884 and his M.S. degree in 1886. He obtained his Ph.D. in organic chemistry at Johns Hopkins in 1888. From 1888 to 1905 he taught chemistry at University of Kentucky. From 1905 to 1909 he was "professor of chemistry," as the chief chemist was then called, of the U.S. Public Health Service's Hygienic Laboratory (predecessor of the National Institutes of Health) in Washington. He resigned from the Hygienic Laboratory to teach at University of Virginia, returned to University of Kentucky as research professor in the Agricultural Experiment Station in 1911 and in 1912 was appointed director of the station and dean of the College of Agriculture, a post he held until he died in 1916.

At University of Kentucky from 1888 to 1905, Kastle, without any help except that of a student assistant, taught chemistry to approximately 200 students a year, delivering three or four lectures a day and giving laboratory instruction in elementary, analytical, organic, and physical chemistry. He also directed the research of advanced students. Laboring under these conditions for 17 years in a small, poorly equipped laboratory he, nevertheless, contributed 51 articles to the literature and wrote a text, "The Chemistry of Metals." In the Hygienic Laboratory he carried on investigations in physiological and organic chemistry and cooperated with scientists of other disciplines in studying the epidemiology of typhoid fever in the District of Columbia. During his last years he reorganized and enlarged the Kentucky Agricultural Experiment Station, established soil experiment fields in various sections of the state, and stimulated investigations in animal nutrition, breeding, and disease.

Kastle died in Lexington, Sept. 24, 1916 of Bright's disease. A building at the university was named in his memory.

Archives University of Kentucky; records of the U.S. Public Health Service; *The Lexington Herald,* Sept. 25, 1916.

WYNDHAM D. MILES

Alexander E. Katz

1887-1957

Katz was a pioneer in the development of American flavor chemistry and industry. He was born May 6, 1887 in Odessa, Russia. His father, M. Jacob Katz, was German by birth and his mother, Fanny, was Russian. At the time of his birth, his father was already soundly established in the flavor and aromatic chemical industry as partner in F. Ritter & Co., Leipzig, Germany, founded in 1876.

Alex, as he was known throughout the industry, attended preparatory school in Odessa and the Technological Institute in Kharkov. In 1904 he was sent to Switzerland to study at University of Lausanne where he obtained his B.S. degree in chemistry in 1907. He was married in Lausanne and upon graduation moved to New York where his father had already transplanted family and business because of unsettled conditions in the old homeland.

He worked in the business while attending Columbia University. His Ph.D. thesis under Morris Loeb dealt with saccharin and various by-products. He assumed full charge of F. Ritter & Co. in 1913, and in the pre-World War I era he and his associates were the first company in the United States to manufacture saccharin on a commercial scale. By what appears to be an accident they also may have invented the first mechanical washing machine since on Mar. 6, 1922 he was granted a patent for a "Chemical Filtration and Washing Apparatus" which turned out to be the first inverted suction washing machine. This patent was in fact sold to the Mayflower Washing Machine Co.

A gifted musician in his youth, he had questions periodically as to whether science or music would ultimately win his attention. A man of both the arts and sciences, Alexander Katz numbered many famous musicians among his friends.

During World War I he served with the Alcohol Tax Unit as consultant on denaturing alcohol. Later, he served as consultant for the newly established Federal Food and Drug Administration, helping develop standards and test methods for important essential oils. From 1945 through 1947 he worked with the California State Department of Agriculture and the California Polytechnic School toward developing economic agriculture in California and developing the essential oil industry in the United States. He published many articles urging the cultivation here of botanicals and other source materials useful for producing essential oils.

Working with Monroe Kidder and Alvah Hall, Katz produced lignin and vanillin from yucca grown in California. He served as the first chairman of the Scientific Research Committee of the Flavoring Extract Manufacturers Association of the United States. He directed a study of the dermal irritating properties of essential oils and of synthetic aromatic chemicals used in flavoring and in perfumery. He was constantly traveling, lecturing, running a business, and working in the laboratory; with his associates he isolated many flavor-giving compounds from raw materials and then set out to synthesize these to make them available as new and less expensive tools for producing flavors and fragrances. In *American Perfumer and Essential Oil Review,* October 1949, Abraham Seldner, one of Katz's associates, listed approximately 100 new essential aromatics that had been synthesized in their laboratories. In the same journal, May 1951, some of the later results of work resulting from the economic agriculture that Katz had urged were described. An extract made from the crop of transplanted foenugreek seed was described as "as good an imitation maple base as that obtained from Moroccan seed." They also described work done with rose geranium oil, with oil of sweet basil, with summer savory oil, and with labdanum. Their work in developing synthetic aromatic chemicals, for example the intense grape character which cinnamyl anthranilate imparts to food material, was also described. They indicated that they had synthesized methyl beta methylthiolpropionate which they had first identified at the rate of one gram per ton present in Hawaiian pineapple.

In 1947 Katz discovered vanifolia growing in the Hawaiian Islands and worked out an ultra-violet, controlled laboratory cure for these vanilla beans. This led to speeding the cure of these beans to days instead of weeks. In 1945 he set up the first production facilities for flavor manufacture on the West Coast. During World War II he served as consultant on war gases with the National Office of Civilian Defense, particularly with regard to identifying them. He was a cofounder of Florasynth Laboratories and started a number of other companies in producing flavor raw materials. He died in Los Angeles, Mar. 18, 1957, survived by two sons, Leonard and Allan and his widow, Mrs. Isabelle Katz.

M. H. Baker, "A Pioneer of American Flavors," *American Perfumer and Aromatics,* pp. 43-45, July 1958, with portrait; personal recollections; files of F. Ritter & Co., in possession of Allan Katz.

MICHAEL H. BAKER

Charles C. Kawin

1877-1957

Without any formal education, a child forced to work at an early age became a pioneer chemist in an industry that welcomed his technology. This was Charles Kawin, born in Peoria, Ill., Aug. 5, 1877, who began work at the age of 12 as a sample runner at Illinois Steel Co. in Chicago. (Illinois Steel Co. became the South Works of United States Steel Corp.) This first job of carrying samples from the open hearth furnace to the laboratory introduced young Kawin to chemistry. Showing an intense interest in analytical procedures, it wasn't long before he learned some of the routine chemical analyses and was given an opportunity to work in the laboratory.

At the age of 19 in 1896, he left the steel company to become chief chemist of Griffin Wheel Co. in Chicago.

Although there are few records concerning Kawin's early work, there is preserved a carefully hand-written notebook dated Dec. 15, 1898, signed C. C. Kawin, Griffin Wheel Co. In an unusually legible writing, it contains procedures in complete detail that Kawin followed for analyzing silicon, manganese, sulfur, graphitic carbon, slag, coke and coal, water, ferro-manganese, sand, Babbitt alloys, brass, and a number of other materials.

After 7 years at Griffin Wheel Co. Kawin recognized the need to supply technical service to the burgeoning foundry industry which operated on hit and miss methods, except in the larger companies that could afford chemical laboratories. In 1903 he established Charles C. Kawin Co. in Chicago.

Within a relatively few years Kawin had hundreds of clients, mostly from the midwestern states and Canada. Among them were some from today's leading industrial names or their predecessor companies. There can be little question that Kawin's initiative and skill contributed to the scientific production of metals during an important period of our nation's industrial growth.

Originally Kawin carried out wet analyses, but as technology developed he added instruments including the metallograph, spectrograph, and x-ray. Because of the large number of the same type of analyses, Kawin was among the first to develop mass production line procedures in the laboratory. As a result, clients could receive same day or next day results on their products. As the demands for Kawin's services increased, he opened branch laboratories in Cleveland, Cincinnati, Buffalo, and San Francisco.

Charles Kawin was a warm, friendly, well-liked individual, a raconteur whose stories delighted those who had the good fortune of knowing him. He remained active in his business until his death in Chicago Jan. 26, 1957, at the age of 79. The Kawin tradition was carried on by two grandsons, Charles Silver and James Schwarz.

Obituary, *The Chemical Bulletin* (publication ACS Chicago Section), March 1957; personal recollections; information from the Kawin family.

LEO A. RAUCH

William Hypolitus Keating

1799-1840

Keating is best known for his account of the Long Expedition into the Minnesota territory. In April 1823, when Keating was "professor of chemistry and mineralogy as applied to agriculture and the arts" in the Faculty of Natural Science at University of Pennsylvania, Major Stephen Long was sent by the War Department to explore the region west of Lake Superior. Long assembled a party of several persons, Keating among them, and set forth from Philadelphia through the Midwest to Fort Snelling. From there the men followed the Minnesota and the Red rivers, took to bark canoes and paddled to Lake Winnepeg, then turned southeast and passed through Lake of the Woods and Rainy Lake to Lake Superior. Keating returned to Philadelphia in October after a journey of 4,500 miles. His report, published in 1824 under the title "Narrative of an Expedition to the Source of St. Peter's River, Lake Winnepeek, Lake of the Woods, Etc.," reprinted in London in 1825 and 1828, translated into German at Jena in 1826, and reprinted in Minneapolis in 1959, is regarded as an authoritative source of information on the region, particularly about the Indians before their culture was changed by white settlers.

Born in Wilmington, Delaware, Aug. 11, 1799, Keating received his bachelor's degree at University of Pennsylvania in 1816, then spent 2 years in Europe studying at the School of Mines in Paris and inspecting mines in France, Germany, Holland, Switzerland, and Great Britain. When he returned he was one of the few professionally trained mining engineers and mineralogical chemists in the United States. He wrote a monograph, "Considerations upon the Art of Mining, to which are added, Reflections on its actual state in Europe, and the advantages which would result from an Introduction of this Art into the United States" in 1821, went on short journeys into New York, New Jersey, and Pennsylvania to study geological formations and collect minerals, and published a few articles.

In 1822 he began to teach at University of Pennsylvania. His course was elective, 30 lectures long, and cost students a fee of $12.00, which was Keating's only remuneration. He published a "Syllabus of a Course of Mineralogy and Chemistry, as applied to Chemistry and the Arts" in 1822 for his pupils. The university permitted him to use a room in the basement as a laboratory, and there he offered instruction "in the various branches of chemistry applied to the arts, and to analysis, on the plan of the most approved European institutions." Keating was one of the earliest teachers in the United States to offer laboratory instruction, probably preceded only by William MacNeven at College of Physicians and Surgeons in New York.

Keating was absent for 6 months with the Long Expedition in 1823. Upon his return he resumed his course at the university, helped organize Franklin Institute, where he taught elementary chemistry to mechanics, and published an edition of Jane Marcet's "Conversations on Chemistry." In 1826 a firm hired him to superintend silver mines near Temascalapa, Mexico. In his spare time he studied the natural history of the region and collected Indian relics, some of which he sent back to the American Philosophical Society.

Around 1830 Keating returned to Philadelphia, entered politics, and was elected to the state House of Representatives for a 2-year term in 1832. His experience in politics and government led him to study law. He became a member of the bar and built a lucrative practice.

The 1830's was the decade in which railroad firms were organized in all settled areas of the United States. Keating, with experience in engineering, science, politics, and law, was well-suited for a managerial position in the new industry. The Little Schuylkill Navigation, Railroad, and Coal Co. elected him president. A few years later Reading Railroad named him one of its managers. He was sent to London in 1839 to raise capital for the Reading and died there suddenly May 17, 1840.

W. D. Miles, "A Versatile Explorer: A Sketch of William H. Keating," *Minnesota Hist.* **36,** 294-299 (1959), with portrait; W. D. Miles, "William H. Keating and the Beginning of Laboratory Instruction in America," *Library Chronicle* (Univ. Penna.) **19,** No. 1, 1-34 (1952-53); "Dictionary of American Biography," **10,** 276, Chas. Scribner's Sons (1933); London *Times,* May 20, 1840, gave the date of Keating's death as May 17, whereas the Philadelphia *Public Ledger,* June 22, 1840, reported it as May 15; the quote regarding laboratory instruction is from Keating's *Syllabus.*

WYNDHAM D. MILES

Lyman Frederic Kebler

1863-1955

Kebler, American analytical chemist, physician and energetic campaigner against adulteration of drugs, was born at Lodi, Mich., June 8, 1863. After teaching chemistry at Iowa State University in 1888-89, he enrolled at University of Michigan where he received his B.S. degree in 1891 and his M.S. degree in 1892. He went to work for Smith, Kline and French Co. in Philadelphia, where he remained as chief chemist until 1903.

Kebler's early publications were concerned with analytical procedures, particularly as they related to assay of drugs. By 1897, his writings voiced increasing concern for the detection of drug adulterants and the hazard of toxic impurities. The basic themes of drug purity and drug safety were to become the hallmarks of his long career.

The Philadelphia years were years of great professional growth. He published over 60 papers. He expanded his education by attending Jefferson Medical College in 1898 and Temple University from 1899 to 1903. By 1901 he had been elected chairman of the scientific section of the American Pharmaceutical Association, and in 1902 he served as president of the chemical section of the Franklin Institute. Kebler thus had a considerable reputation when, in 1902, Harvey Wiley, Chief of the Bureau of Chemistry of the U.S. Department of Agriculture, was looking for a man to head the Bureau's newly formed Drug Laboratory. John Uri Lloyd, himself one of the most distinguished pharmaceutical chemists of his day, recommended Kebler to Wiley as being "worth to you twice any other man." Kebler took the Civil Service examination, received the highest score of 27 candidates for the position, and began work in Washington March 2, 1903. His first discoveries were that: 1) he had no laboratory space; and 2) that most of the chemical reagents then in use in the government analysis of drugs were themselves mislabeled, adulterated, contaminated, or diluted. Wiley was shocked and impressed. Using the minimal space Wiley managed to provide, Kebler embarked on his program of analytical rigor and prosecution of fraud.

Kebler's duties quickly grew from those of a chemist to those of an administrator of a large chemical organization, from those of a detector of adulteration to those of a public campaigner for drug safety.

In 1906 he had obtained his M.D. degree at George Washington University. He now coupled his activities in pharmaceutical organizations with corresponding activity in the medical field. He contributed several articles to *Journal of the American Medical Association* and served from 1905 to 1914 on the Association's Council on Pharmacy and Chemistry. From 1910 until 1920 he was a member of the U.S. Pharmacopeia Revision Committee, and for the next 10 years he was secretary of the U.S. Pharmacopeial Convention. In 1917 he was vice-president of the American Chemical Society.

A departmental report on Kebler's efficiency, dated May 1, 1912, rated him at 100% for scientific and professional ability, adaptability, accuracy, personality, punctuality, deportment, and quantity of work and 95% on capacity for "administrative, executive, or supervisory duties."

Despite his many responsibilities Kebler now added teaching to his activities. He accepted an assistant professorship at Georgetown University, where from 1912 to 1932 he lectured on pharmacology and materia medica.

From the beginning of his government employment Kebler was called upon as an expert witness in Post Office Department prosecution of drug fraud cases involving use of the mails. The Drug Laboratory, which had been enlarged and renamed the Drug Division in 1907, following the Food and Drug Act of 1906, was further renamed "Special Collaborative Investigations" in 1923 in recognition of its special police and regulatory role.

In a major reorganization July 1, 1927 Kebler and his "Special Collaborative Investigations" were removed from the Bureau of Chemistry and made part of the new Food, Drug, and Insecticide Administration.

Kebler resigned from the U.S. Department of Agriculture June 24, 1929 to engage in private work. He became medical director of Tennessee Products Corp., acted as a chemical-medical consultant to other organizations, and from 1930 to 1932 served as a radio broadcaster on food, dietetics, and health. His attempt to gain recognition as a food authority was promoted by his book, "Eat and Keep Fit," which he wrote and published in 1930.

Kebler was active as a lobbyist. In 1931 he vigorously defended the wood alcohol industry against what he considered to be attempts by the grain alcohol manufacturers to legislate industrial methanol out of existence.

As long as Harvey Wiley was alive, Kebler's attitude toward his long-time supervisor was formally correct and friendly. But in private correspondence after Wiley's death in 1930, Kebler revealed his bitterness toward the man who had long dominated the Pure Food and Drug scene and who was often inconsiderate and unappreciative of the effort of Kebler and his colleagues.

An interest in the history of pharmaceutical chemistry, evident in Kebler's writings as early as 1926, found increased expression in the 1930's when as an old man "living on borrowed time," as he expressed it, he sought to save from oblivion the names and works of men who had been important in their own time. Kebler's papers and addresses dealt with Andrew Craigie, James Henry Shepard, Francisco Redi, Samuel Latham Mitchill, and others. His later publications and reminiscences lacked none of his delicate humor, characterized by understatement, as when he alluded to problems of mail fraud prosecution against a purveyor of a purported aphrodisiac.

Kebler married Ida E. Shaw, by whom he had three children, Mabel Alice, Victor L., and Ruth W. He died in Washington, D.C., Mar. 4, 1955, at the age of 91.

Private communication from S. B. Detwiler, Jr.; obituary in *New York Times*, Mar. 5, 1955; "Who Was Who in America" (1951-1960), p. 464; unpublished correspondence in the Kremers Reference Files, Pharmacy Library, University of Wisconsin; G. Griffenhagen, "Bibliography of Papers . . . (from) the American Pharmaceutical Association's . . . Section on Historical Pharmacy, 1904-1957"; biography in *J. Amer. Pharm. Assoc.* **29**, 379-383 (1940).

CHARLES H. FUCHSMAN

Robert Clark Kedzie

1823-1902

Kedzie was born in Delhi, N.Y., Jan. 28, 1823. He resided in Michigan from 1826 until he died, except when he was attending Oberlin College in Ohio. He worked at various jobs and taught in district schools to pay his way through college since his father had died when he was 3 years old, leaving seven children for his mother to raise. He graduated from Oberlin in 1847.

Kedzie taught at Rochester Academy, Mich. for 2 years before entering The University of Michigan Medical School and graduating with the first medical class in 1851. When the medical school opened in 1850, Kedzie was one of a group of students who assisted professor Moses Gunn in teaching anatomy.

After practicing medicine for 11 years, he joined the 12th Michigan Infantry as a surgeon after the Civil War started. Captured at the battle of Shiloh, he was released and invalided home. He resumed the practice of medicine at Lansing and in 1863 began to teach chemistry at Agricultural College of the State of Michigan (now Michigan State University). He was made professor in 1867.

From his viewpoint as a chemist and physician, Kedzie saw the value of public health to his state. He entered this field and became one of the leading sanitarians of the Midwest. He was a member of the Michigan Board of Health from 1873 to 1881 and president from 1877 to 1881. He worked hard to educate the public in the ways of good sanitation; he exposed the composition and methods of manufacture of useless patent medicines; he analyzed and publicized adulterated foodstuffs; and he used his influence to have laws passed compelling oil companies to supply pure kerosene to consumers instead of dangerous kerosene-gasoline mixtures.

He was a leader in bringing chemistry to the aid of agriculture. He founded the local farmers' institute which became popular in rural areas and thus fostered the spread of chemistry among farmers. He was involved in the introduction of beet sugar into the United States. He was influential in securing legislation compelling accurate analyses of fertilizers, thus protecting farmers from dishonest manufacturers.

Kedzie did not publish any major works, but he was a prolific writer of popular and technical articles on chemistry, agriculture, and public health for the periodical literature.

He was elected to membership in the American Chemical Society in 1876, resigned in 1879, and rejoined in 1893. He was a member of the Michigan Legislature in 1867, president of Michigan State Medical Society in 1874, vice president of American Medical Association, president of Association of Agricultural Colleges, and president of Association of Official Agricultural Chemists in 1899. He was a man of genial disposition and possessed unflinching determination and indomitable will power. He was a great favorite with students. He retired in 1902 as professor emeritus and died at Lansing, Nov. 7, 1902.

Biography by L. S. Munson, U.S. Dept. Agriculture Bulletin No. 73 (1903); W. J. Beal, "History of the Michigan Agricultural College and Biographical Sketches of Trustees and Professors," E. Lansing Agricultural College (1915); William B. Atkinson, "Biographical Dictionary of Contemporary American Physicians and Surgeons," 2nd ed., D. G. Brinton (1880) gives birthdate as Jan. 23; Charles A. Browne, M. E. Weeks, "A History of the American Chemical Society," ACS pub. (1952).

Leigh C. Anderson

Harry Frederick Keller

1861-1924

Keller was born in Philadelphia, Dec. 15, 1861. He attended public school in Philadelphia, the gymnasium in Darmstadt, Germany, and University of Pennsylvania. After receiving his bachelor's degree in chemistry in 1881, he worked for Edgemoor Iron Works, Chester, Pa., until 1883, then became an assistant in chemistry at University of Pennsylvania for 3 years. To continue his education he went to University of Weisbaden to study under the famous analyst, Karl Remigius Fresenius, a friend of his father, and later to University of Strassburg where he received his Ph.D. in organic chemistry under Rudolph Fittig in 1888.

Returning to Philadelphia he became instructor at University of Pennsylvania. He collaborated with Edgar F. Smith in revising the atomic weight of palladium and in writing "Laboratory Experiments in General Chemistry."

In 1890 he went to Michigan School of

Mines as professor of chemistry. He became an expert in analysis of copper, investigated copper minerals, and served as an umpire in disputes between Anaconda Copper Co. and its customers.

He returned to Philadelphia in 1892 to teach chemistry at Central High School. He and William H. Greene brought out several revisions of Greene's "Lessons in Chemistry," a high school chemistry text, and of Greene's translation of Carl Adolph Wurtz's "Elements of Modern Chemistry."

Busy as he was developing secondary school science courses, Keller found time for research and published articles on organic chemistry, mineralogy, and analysis of metallic artifacts from Indian mounds.

In 1915 Keller became principal of the newly established Germantown High School, a position he held until his death in Philadelphia, Feb. 5, 1924.

Obituary by Philip Maas and A. Henwood, *Ind. Eng. Chem.* **16,** 529 (1924), with portrait; Minutes of the Board of Managers relative to the death of Dr. Harry F. Keller, *J. Franklin Inst.* **198,** 115-117 (1924), with portrait.

WYNDHAM D. MILES

Louise Kelley

1894-1961

Louise Kelley was born Oct. 10, 1894 in Franklin, N.H., daughter of Elmer and Etta Kelley. Her elementary and high school education was received in the public schools of Franklin after which she entered Mt. Holyoke College in 1912, receiving her A.B. degree in 1916 and A.M. in 1918. From 1917 to 1918 she was instructor in chemistry at Wheaton College. At Cornell University she was awarded her Ph.D. degree in organic chemistry in 1920. Her dissertation "*p*-Hydroxybenzoyl-*o*-benzoic Acid and Some of Its Derivatives," was carried out under William Orendorff. While at Cornell she held the Sage, DuPont, and Woolley Fellowships.

With her appointment as assistant professor of chemistry at Goucher College in 1920, Miss Kelley began an association which lasted 39 years. She was promoted to associate professor in 1923 and to professor in 1930. In 1929 she studied with Fritz Pregl at the Medicinisch-Chemisches Institut of University of Graz. With G. Albert Hill of Wesleyan College she published the first edition of their much used text "Organic Chemistry" in 1932. She alone authored the second edition of this work in 1943. In 1931 she married Edward S. Hopkins; they were divorced in 1937.

In 1935 Prof. Kelley was invited to deliver the lecture on the place of women in chemistry at the Fifteenth Exposition of Chemical Industry in New York. In March 1941 she delivered the Marie Curie Lecture at Pennsylvania State University and was made honorary member of the Palladium Chapter of Iota Sigma Pi. From 1941 to 1944 she served as a member of Garvan Award Committee of the American Chemical Society and was its chairman in 1946 and 1947. While continuing to teach, she served as acting dean of Goucher College from 1947 to 1949 and as chairman of the Chemistry Department from 1946 until her retirement. From June 1942 through March 1947 Prof. Kelley served with particular distinction as senior technical aid, Divisions 8-11, National Defense Research Committee of the Office of Scientific Research and Development in Washington, D.C. For this work she was awarded the President's Certificate of Merit in November 1949, one of nine women so honored.

Early in her academic career she began editorial work on ACS journals and served as assistant editor of *Journal of Physical and Colloid Chemistry* from 1937 to 1959 and of *Chemical Reviews* from 1931 until her death.

At its centennial celebration in 1952 Mt. Holyoke presented Louise Kelley with a citation recognizing her distinguished record as a teacher and scientist; in 1959 she received one of the six awards made by the Manufacturing Chemists Association honoring men and women for excellence in college science teaching. On her retirement Goucher College conferred upon her the honorary degree of Doctor of Science. The large lecture hall added to Hoffberger Science Building at Goucher College in 1968 is named in her honor.

After retirement from Goucher College in 1959, she returned to Franklin, N.H., built a home, continued her prodigious personal correspondence, and served as director of the New Hampshire Council for Better Schools and as a trustee of Franklin Hospital Association until her death, Nov. 12, 1961.

The following paragraph is a partial quotation from the minutes of the Goucher College faculty meeting of Feb. 10, 1962:

"Dr. Kelley came to Goucher College in 1920 and students of the college immediately recognized and praised her skill as a teacher. They commended her personal interest in each as an individual, her inexhaustable patience, her challenging presentation of material. Miss Kelley's life at Goucher was a life of service and of outstanding devotion to the needs of the student and of the college. She served her community with equal vigor and dedication. For these sterling qualities she will long be remembered."

Personal recollections; various issues of *Goucher Weekly,* and *Goucher Alumnae Quarterly;* obituaries in *Boston Globe,* Nov. 13, *Baltimore Sun,* Nov. 14, and *Baltimore News Post,* Nov. 14, 1961.

JAMES L. A. WEBB

Alfred L. Kennedy

1818-1896

Kennedy was born Oct. 25, 1818 in Philadelphia. After receiving his elementary education he became a student and assistant of John Millington, a chemist-civil engineer-mining engineer, for 3 years. In 1839 he was appointed assistant professor of chemistry at Pennsylvania Medical College and in 1842 also taught botany and toxicology there. In 1843 he lectured on medical chemistry at Philadelphia School of Medicine. Teaching at these schools may have aroused his interest in medicine for he studied with a physician, attended the University of Pennsylvania's Medical School, and in 1848 received his medical degree. The following year he lectured on industrial botany at Franklin Institute and taught medical chemistry at Philadelphia College of Medicine. He then went to Europe where he studied medicine, physiology, geology, and botany in Paris, and physiological chemistry and pathological chemistry with Karl Lehmann at University of Leipzig. Returning to Philadelphia in 1852 he taught a course in agricultural chemistry at Franklin Institute. He practiced medicine from 1853 to 1865.

Kennedy was a very ambitious person. While carrying on all the activities mentioned above, he also founded and developed an educational institution. In 1842 he established "Philadelphia School of Chemistry," where students could learn "chemistry, practically in its connection with medicine, agriculture, or the arts." His school was located in the second story of a building 60 foot long and contained "a full assortment of carefully selected, pure chemical tests and re-agents; also apparatus of approved European model and construction, whereby the most delicate analytical processes may be performed, and the student be made acquainted with the best methods." Students were accepted by the year, quarter, or month, daily or triweekly, at a fee of $20.00 for daily students. A special course in chemical analysis was held in the evenings for students attending the medical and pharmaceutical colleges of Philadelphia.

He conducted Philadelphia School of Chemistry concurrently with his other activities until 1853 when he and a group of associates transformed it into "Polytechnic College of the State of Pennsylvania," with him as president and professor of industrial, agricultural, and analytical chemistry, and metallurgy. The polytechnic offered degrees in several fields and was the first institution in the United States to have a curriculum leading to a B.S. degree in mining engineering. Although the college, for reasons not known to us, did not gain support and become a major educational institution as, for example, did Massachusetts Institute of Technology, established shortly thereafter, it was said to have influenced the development of similar schools.

Toward the end of his life Kennedy lived alone in rooms in an office building in Philadelphia, where he was surrounded by his papers and manuscripts. He accidentally started a fire among his papers and was burned to death, Jan. 31, 1896.

James McGivern, "Polytechnic College of Pennsylvania," *J. Eng. Educ.* **52,** 106-112 (1961); Thomas T. Read, "The Development of Mineral Industry Education in the United States," Amer. Inst. of Mining and Metallurgical Engineering (1941), 36-39; Howard A. Kelly and Walter Burrage, "American Medical Biographies," Norman, Remington Co. (1920), 691-92; William B. Atkinson, "Biographical Dictionary of Contemporary American Physicians and Surgeons," 2nd ed., D. G. Brinton (1880); announcements of Polytechnic College of Philadelphia in Library Company of Philadelphia; pamphlets by Kennedy in College of Physicians of Philadelphia, including an announcement of Philadelphia School of Chemistry from which the above quotes are taken.

HAROLD J. ABRAHAMS

Morris Selig Kharasch

1895-1957

The recognition of the homolytic cleavage of covalent bonds by Gomberg in 1900 might have opened a whole new field of organic chemical reactions. But, in fact, it was 30 years before that potential began to be realized. At that time Morris Selig Kharasch selected the addition of hydrogen bromide to allyl bromide for detailed study. The literature concerning the reaction was chaotic in experimental fact and naive in theoretical explanation. Generally such a situation might be regarded as a most unpromising point of departure for a young man's career.

The basic problem was variability in the composition of the product mixture. Both 1,2- and 1,3-dibromopropane occurred in varying amounts, but the source of the shifting product ratio was not known. Kharasch, along with his student Frank R. Mayo, established a relationship between the age of the

$$CH_2{=}CHCH_2Br + HBr \rightarrow$$
$$\underbrace{CH_3CH(Br)CH_2Br + BrCH_2CH_2CH_2Br}_{\text{in varying amounts}}$$

allyl bromide sample used and the product composition. This finding led to the appreciation of the effect of traces of peroxides on the course of the reaction. In time, radical intermediates were recognized as the key to formation of the anti-Markownikoff product —1,3-dibromopropane.

Typically, Kharasch published the experimental observations at once since they could be put to good use by other chemists. Just as typical of the man was the fact that publication of the theory was delayed for 4 years while many very carefully devised experiments were conducted to test its validity. With the peroxide effect firmly established, Kharasch and a host of brilliant students developed many of the synthetic applications of radical chemistry. Important chemists of this century who studied with Kharasch include: Herbert C. Brown, Wilbert H. Urry, Andrew Foro, Frank H. Westheimer, Cheves Walling, Philip S. Skell, George Büchi, and Nien-Chu Yang. So impressive were his accomplishments as a teacher and researcher that the University of Chicago, where he was educated and spent his life, appointed him director of the Institute of Organic Chemistry created in his honor.

Kharasch had a varied and productive scientific career in addition to developing radical chemistry as a synthetic method. For example, he made pioneering studies of organomercurials important in agriculture (seed disinfectants) and medicine (Merthiolate). He and a collaborator isolated ergonovine from ergot and showed it to be the active principle in an important medicinal application. One important result of his radical work was a greater understanding of polymerization chemistry which led to significant improvements in both products and processes. Finally, Kharasch wrote an authoritative treatise on the Grignard reaction.

Kharasch not only used and contributed extensively to the chemical literature; he helped to create sources of it. He was one of the founders of the *Journal of Organic Chemistry*. Late in his life, he became the American editor for the new international journal of organic chemistry, *Tetrahedron.* This concern with the needs of his science as a whole was typical of the man.

Kharasch was born in Kremenetz in the Ukraine, Aug. 24, 1895. His parents sent him to the care of an older brother in Chicago when he was only 13 for the educational opportunities of the United States. All of his university education was obtained at the University of Chicago. After teaching at the University of Maryland for 2 years, he returned to Chicago in 1924 and remained there until his death in Copenhagen, Oct. 9, 1957. Kharasch married Ethel May Nelson, June 24, 1923, and they had a son and a daughter. Kharasch was a careful researcher, a prolific writer, and a master teacher. He seemed to be somewhat antisocial and rarely gave public talks; yet his students and colleagues found him a congenial and accomplished conversationalist. The insight required to identify the significant problems no matter how well hidden in the underbrush of triviality never seemed to desert him. He served his students well with his apparently unlimited patience.

Biographies by F. H. Westheimer, *Biog. Mem. Nat. Acad. Sci.* **34,** 123-152 (1960), and R. D. Billinger and K. T. Finley, *Chemistry* **38,** 19-20 (June 1965); *Chem. Eng. News* **27,** 343 (1949).

K. THOMAS FINLEY

David Herbert Killeffer

1895-1970

Killeffer was a chemist, a public relations consultant to chemical industry, and a skilled science writer.

Killeffer was born Dec. 22, 1895, in Columbia, Tenn., son of Alexander C. and Louise (Ayres) Killeffer. He obtained his B.S. degree in chemical engineering from University of North Carolina in 1915, then returned to Tennessee to work a year as chemist with Brown Laboratories and 2 years with Nashville, Chattanooga, & St. Louis Railway. In 1918 he went to New Jersey as chemist for Calco Chemical Co.

Killeffer once said, "I spent most of my life dodging around, over, and under handicaps." The first big handicap changed his career at age 24. He contracted dinitrobenzene poisoning, which impaired his usefulness as a chemist. His former chemistry professor, Charles H. Herty, then editor of *Industrial & Engineering Chemistry,* redirected Killeffer, who had an aptitude for writing, into technical journalism.

In 1920 Killeffer became marketing editor for *Drug & Chemical Markets,* in New York. Two years later he became associate editor of *Industrial & Engineering Chemistry,* and his book, "Eminent American Chemists," was published. He began writing the "Chemistry in Industry" section of *Scientific American,* a column he continued for 24 years. He helped build the News Service of the American Chemical Society, and in that capacity he covered ACS general meetings for years, writing news releases, interviewing chemists, and writing anecdotal reports of the meetings.

He changed his career temporarily in 1928 by joining Dry Ice Corp. of America as technical director; there he obtained 14 patents in the field of solid carbon dioxide refrigeration.

He returned to writing full-time in 1931. He was by then highly skilled in interpreting the esoteric language of science into accurate and understandable English. His contacts with the growing chemical industry were extensive. His writings became prolific. "Engineering—a Career, a Culture" was published in 1932. He began a 20-year association as editor of *Chemical Digest* for Foster D. Snell. He was a "pen for hire"—the ghostwriter of over 30 books in addition to speeches, sales literature, adventure stories, articles on subjects from atomic weapons and rubber to biography. Unsigned texts included those on carbon black, synthetic resins, and Mexican economics.

During World War II he was consultant to the Office of Production, Research & Development of the War Production Board. Afterward, he resumed private consulting, this time with emphasis on public relations. His clients included American Cyanamid, Columbian Carbon, General Electric, General Foods, Mallinckrodt, Monsanto, and Union Carbide.

Another handicap came—illness requiring major surgery. While he lay discouraged in the hospital, a friend speeded his recovery by putting him to work on another book, "The Genius of Industrial Research," published in 1948.

Killeffer believed that "If one brings together a scientist with something to say and a writer with the trained ability to say it, the result can be a true synergy." This he demonstrated in "Molybdenum Compounds" (1952) written in collaboration with that expert on molybdenum, Arthur Linz.

"Two Ears of Corn, Two Blades of Grass" was his next book (1955), a report of scientific discoveries that benefit humanity. It was reprinted in 10 languages.

A long-time member of The Chemists' Club (his wedding reception was held there in 1922), Killeffer wrote, "Six Decades of The Chemists' Club" in 1957. Previously, he wrote the "First Fifty Years," a history of St. John's Church in Tuckahoe, N. Y.

His retirement, in 1958 to Clearwater, Fla., was as active as his previous life. He taught sailing in the Clearwater Power Squadron and organized the Windjammers of Clearwater. He was a member of the Library Board, aiding the Wickman Memorial Collection of books on the sea. He was on the Science Advisory Committee of the Pinellas County School Board; science advisor to the Chamber of Commerce, and he wrote the biography of Fernley H. Banbury, "Banbury, the Master Mixer," in 1962. He also wrote two books in the Chemistry in Action Series that is aimed at the high school level—"Chemical Engineering" in 1967 and "How Did You Think of That?" in 1969. Altogether, he wrote about 50 books and 100 articles.

Among his honors were: the first impression of the James T. Grady Medal of the American Chemical Society (1956) for "promoting public understanding of chemistry and chemical engineering," the Honor Scroll of the Florida Chapter of the American Institute of Chemists for his "excellence as a scientist," and two Civic Service Awards from the Florida Section of the American Chemical Society.

He died in Clearwater, Fla., Jan. 15, 1970, survived by his wife, Dorothy (Savage) Killeffer, and his son, Robert Ayres Killeffer.

The Percolator (Dec. 1962), (Dec. 1965), (Aug. 1967); *The Chemist* (Feb., July, Aug. 1952), (July 1955), (Mar. 1961), (Mar. 1962), (Sept., Dec. 1965), (Sept. 1967); "Chemical Who's Who," 4th Ed. (1956); personal acquaintance.

VERA KIMBALL CASTLES

Leonard Parker Kinnicutt

1854-1911

Kinnicutt, director of the Chemistry Department of Worcester Polytechnic Institute, was at his death in 1911, one of the country's leading authorities on sanitary chemistry. He was born in Worcester, May 22, 1854 to Francis Harrison and Elizabeth (Parker) Kinnicutt in the eighth generation of a family associated with eastern Massachusetts since 1650. He was educated in the Worcester schools, then at Massachusetts Institute of Technology, from which he received his B.S. degree in 1875. Thereafter, he went to Germany to study first at Heidelberg under Bunsen, from whom he learned gas analysis and an appreciation for sound analytical data; later at Bonn Kekulé introduced him to organic chemistry. His stay in Germany began a series of international acquaintances which he would maintain throughout his life —most particularly with Richard Anschütz, who became a close friend.

Kinnicutt's earliest publications, in 1878, were investigations with Anschütz of phenylglyceric and phenyltribromopropionic acids. A family emergency compelled him to return to America in 1879 before he could finish his doctoral work at Bonn. He studied for a year under Ira Remsen at Johns Hopkins, then served as instructor and personal assistant to Wolcott Gibbs at Harvard for 3 years. Here he continued his organic research, explored the analytical chemistry of nitrous acid derivatives, and received his Sc.D. degree from Harvard in 1882.

Kinnicutt then returned to Worcester, accepting an instructorship in chemistry at Worcester Polytechnic. For the rest of his life Worcester would be his base and center; the roots of family and profession in that city supported a flowering of international dimensions. In 1883 he was promoted to assistant professor and 3 years later to professor; in 1892 he became head of the department. During this time the city of Worcester established a sewage disposal works, and Kinnicutt turned to the chemistry of waste waters. He was quickly led to consider not just municipal wastes but manufacturing effluents in that most industrialized part of the United States; this ultimately led him into matters of watersheds and water supply, of atmospheric purity, and analytical methods of determining it. His publications in the chemical literature reflect this shift, turning from pure organic chemistry to sanitary chemistry, carbon monoxide determination, milk contamination analysis, and occasional examination of mineral products.

During the last 20 years of his life Kinnicutt attained international authority on sewage and waste waters. His presentations before various conferences became, when published, definitive statements on the problems they addressed: "Sewage and sewage disposal, with a description of the sewage works at Worcester, Mass." (1890); "The present status of the sewage problem in England" (1902); "The action of the septic tank on acid iron sewage" (1902); and the volume which became the standard work in the field, "Sewage Disposal" (1910), written with Charles E. A. Winslow of MIT and R. Winthrop Pratt of the Ohio State Board of Health; this book went into a second edition in 1919, after Kinnicutt's death. In addition, Kinnicutt was much in demand as a water and waste treatment expert and gave testimony in such cases as *The State of Missouri* v. *the State of Illinois and the Sanitary District of Chicago,* dealing with alleged pollution of the Mississippi River by Chicago sewage; and *Jersey City* v. *Jersey City Water Supply Co.*

He spent every other summer in Europe to keep abreast of experts there and their work.

Kinnicutt's strong sense of duty and personal commitment was reflected in all his activities. He helped establish a safe summer supply of milk for Worcester's babies. During the Worcester "water famine" of 1910-11 he directed testing and treatment from his sickbed. An excellent classroom teacher, he also established close and candid relations with his students at Worcester Polytech, helping with personal problems, legal and financial, and closely following their subsequent careers. Kinnicutt married twice; first in 1885, to Louisa Hoar Clarke, who died in 1892, and again in 1898, to a cousin of his first wife, Frances Ayres Clarke. He had no children. In 1910 tuberculosis, dormant since his student days in Germany, interrupted his busy life. He took to his bed and died Feb. 6, 1911, only 56 years old.

Technol. Rev., **13,** 235-39 (1911); R. Anschütz, *Chem. Ber.,* **44,** 3567-70 (1911); W. L. Jennings, *Proc. Acad. Arts Sci.,* **53,** 821-24 (1918); C. Richardson, *Ind. Eng. Chem.,* **3,** 142-43 (1911). For much information about the Kinnicutt family, *see* C. Nutt, "History of Worcester and Its People," **4,** 724, Lewis Historical Pub. Co. (1919); in addition, "Dictionary of American Biography," **10,** 418-419, Chas. Scribner's Sons (1933); "National Cyclopedia of American Biography," **25,** 160, James T. White & Co. (1936); *J. Worcester. Poly. Inst.,* **14,** 155-61, 353-55 (1911); *J. Assn. of Engrg. Soc.,* **46,** 387-89 (1911); *Proc. of New England Water Works Assn.,* **25,** 187-89 (1911); *Am. Chem. J.,* **45,** 411-12 (1911).

ROBERT M. HAWTHORNE JR.

Raymond Eller Kirk

1890-1957

Kirk was born on a farm in Hamilton County in southeastern Nebraska, June 24, 1890, the son of Joseph Alexander and Virginia Eads (Eller) Kirk. He showed an early interest in mathematics and science in Nebraska public schools and continued to show interest in science, especially chemistry, during his years as a student at Nebraska State Normal School in Kearney (now Kearney State University) from 1910 until 1913.

In 1915 Kirk graduated from University of Nebraska with his bachelor of science degree in chemistry, and 2 years later he earned his master of science degree in chemistry from Iowa State University. He then became an instructor in chemistry at Iowa State and remained there until 1920 when he accepted a position as assistant professor of chemistry at University of Minnesota. During the years from 1917 to 1919 he also served as a civilian inspector with the Ordnance Department of the United States Army and from 1923 until 1930 held a reserve commission as a captain in the Ordnance Corps and from 1930 until 1942 as a major.

On June 30, 1930, Kirk married Beth Sibley. Their two children were named Virginia and Josephine Alvira.

Having completed his Ph.D. degree in

chemistry at Cornell University, where he was a Grasselli Research Fellow in 1927, Kirk was promoted to the rank of associate professor of chemistry at University of Minnesota. He carried on investigations in collaboration with the Minnesota Engineering Experiment Station. In 1927 he collaborated with Maynard C. Sneed, also of University of Minnesota, in writing a manual for students of general chemistry, published as "Laboratory Manual in Inorganic Chemistry."

Two years later Kirk accepted the position as professor of chemistry and head of the department at Montana State University, serving at the same time as state chemist in the oil and gas laboratory.

In 1931 Kirk went to Polytechnic Institute of Brooklyn as professor of chemistry and head of the department, a position he held until his retirement in 1955. From 1936 to 1942 he was also director of Shellac Research Bureau, and during the 11 year period from 1944 until 1955 was dean of the Graduate School at Polytechnic Institute of Brooklyn. In 1954 he received the Scientific Apparatus Makers' Award for outstanding achievement in chemical education.

A frequent contributor to scientific journals, Kirk collaborated with Donald Othmer, head of the Department of Chemical Engineering at Polytechnic Institute of Brooklyn, during the years from 1947 to 1956 in publishing the first 15 volumes of "Encyclopedia of Chemical Technology." His own areas of research specialization include the investigation of complex ions, the oxidation of hydrazine, and the manufacture of Portland cement.

Raymond Eller Kirk died Feb. 6, 1957.

Obituary with portrait, *Chem. Eng. News,* **57,** Feb. 18, 1957; and *New York Times,* Feb. 7, 1957.

JAMES C. COX, JR.

Sidney Dale Kirkpatrick

1894-1973

Kirkpatrick, one of the best known editors in the field of chemical engineering, was born in Urbana, Ill., April 2, 1894 where his father operated a restaurant on the University of Illinois campus. The career of Kirkpatrick spanned an era that saw U.S. chemical process industries rise from an immature post-World War I level to full-blown world leadership in the application of chemical technology for the benefit of mankind.

As editor-in-chief (1928-50) and editorial director (1950-59) of *Chemical Engineering,* Kirkpatrick exerted influence on the unfolding of these industries and their chemical engineering technological base. He had joined the magazine in 1921 as an assistant editor, following graduation from University of Illinois with his B.S. degree in chemical engineering in 1916, then service in World War I, and a post-war stint as chemical advisor to the U.S. Tariff Commission. At that time, the 19-year-old weekly known as *Chemical and Metallurgical Engineering* was making a determined effort to cover the economic and business as well as the technical aspects of its expanding field. By 1925, a decision had been made to concentrate on serving the technical side of the field and to convert to monthly publishing frequency. In November 1928, Kirkpatrick was appointed editor-in-chief.

During the difficult depression years that followed, Sid used his publication as an instrument to help revitalize the chemical process industries. Many of the present editorial features and established practices of the publication were born at that time, including the biennial award to a company for chemical engineering achievement. During the World War II years, Kirkpatrick and his magazine were in the thick of transferring technological information to chemical-engineer readers who were building a $3-billion synthetic-rubber industry, developing new explosives, drugs, metals, and the greatest technological project of the era—the war-ending production of the atomic bomb. Shortly after the war, *Chemical and Metallurgical Engineering* became *Chemical Engineering* and in the early nineteen fifties he became editorial director of *Chemical Engineering* and *Chemical Week.* Under his guidance, these two publications attracted reader support that enabled them to establish and maintain a preeminent position in their field.

All through the years that Sid was leading his publications to their strong positions, he was also building the chemical engineering profession's textbook literature, first as consulting editor and later as a director and vice president of McGraw-Hill Book Co. Along with his primary responsibilities at McGraw-Hill, Sid engaged in a heavy schedule of outside professional activities that led to his election in 1942 as president of the American Institute of Chemical Engineers and in 1944 as president of the Electrochemical Society. Further, he was one of the founders of the Chemical Marketing Research Association.

He was a director of General Aniline & Film Corp., Michigan Chemical Corp., Carus Chemical Co. and Roger Williams Technical & Economic Services. During World War II he was a consultant to the War Manpower Commission, the War Production Board, the U.S. Army Chemical Corps, and the Quartermaster Corps. Several months before the end of the war, Sid went to Europe as a member of the Technical Industrial Intelligence Commission to obtain information about German industry. In 1946, he was chosen as a consultant to the Secretary of War for Project Crossroads, the Bikini Atoll atomic bomb test. In 1950-55, he was chairman of AEC's advisory committee on information for industry.

Among Kirkpatrick's many honors were doctorates from Clarkson Institute of Technology and Polytechnic Institute of Brooklyn; the Silver Anniversary Award of AIChE (1932); the Chemical Industry Medal of the American Section of the Society of Chemical Industry (1945); honorary membership in the American Institute of Chemists (1952); the Founders Award of AIChE (1958); the Memorial Award of the Chemical Marketing Research Assn. (1959); the Crain Award for a Distinguished Editorial Career of the American Business Press (1969); and the John R. Kuebler Award of Alpha Chi Sigma, national chemical fraternity (1972). At the time of the Alpha Chi Sigma award dinner, ill health prevented Kirkpatrick from being present to accept the award. However, a telephone hookup enabled him to address the audience from his bedside in Short Hills, N.J.

When Jerome Alexander, who was noted for his poetry as well as his chemical knowledge of colloids, wrote of Ellwood Hendrick, a former president of the Chemists' Club, he called him in a eulogy in verse, a "great lion of a man." Such a phrase also characterizes Kirkpatrick. Sid collected people and his friends were legion. He had a warmth and love of people that made everyone his friend. He was called upon time and again to preside at dinners because he was truly the toastmaster's toastmaster and gave many organizations unstintingly of his time. His spontaneous wit and geniality endeared him to all generations. His vast collection of scrapbooks was full of pictures and memorabilia of Sid with his family and friends from all corners of the globe. Perhaps no man in the entire industry was more widely known in a period of chemical progress spanning the last half century.

Kirkpatrick died of heart failure Feb. 24, 1973.

Personal recollections; obituaries in *Chem. Eng.*, March 19, 1973, and *New York Times*, Feb. 26, 1973.

RICHARD L. MOORE

John Gamble Kirkwood

1907-1959

Kirkwood was born in Gotebo, Okla. on May 30, 1907 son of John Millard and Lillian Gamble Kirkwood. He received his elementary and secondary education in Wichita, Kan., entered California Institute of Technology in 1923 before he finished high school, and later transferred to University of Chicago, where he received his B.S. degree in 1926. He was awarded his Ph.D. degree in chemistry in 1929 from Massachusetts Institute of Technology for his experimental work with Frederick G. Keyes on the dielectric constants of two polar gases, carbon dioxide and ammonia, as functions of density and temperature.

As a National Research Council fellow at Harvard University in 1929-30, Kirkwood began his studies in theoretical chemistry, for which he was to become world-famous. In 1931-32 he was an International Research Fellow with Peter Debye in Leipzig, with Arnold Sommerfeld in Münich, and with Fritz London in Berlin. As a research associate at MIT 1930-31 and 1932-34, he began theoretical work on the physical chemistry of solutions with applications to proteins and other macromolecules. During this period, with the aid of quantum mechanics, he also studied the nature of intermolecular forces and the relationship between quantum and classical statistics. As with all the research problems on which he ever worked, Kirkwood returned to these studies again and again, continually developing a more advanced or a more general theoretical understanding.

Kirkwood started his teaching career as assistant professor of chemistry at Cornell University in 1934. He was appointed associate professor of chemistry at University of Chicago in 1937 and returned in 1938 to Cornell as Todd Professor of Chemistry. While at Cornell he developed a molecular-distribution-function approach to the equilibrium theory of simple fluids, in which he correlated the thermodynamic properties of matter with the potential of intermolecular force by utilizing statistical mechanics. He also began theoretical studies on dielectric behavior, irreversible processes in solutions of macromolecules, electrolytic and nonelectrolytic solutions, cooperative phenomena, and optical rotation, on all of which he continued to work throughout his lifetime. During World War II he worked on explosives and shock waves for the U.S. Navy at Woods Hole, Mass. After the war he continued his work on shock waves in addition to his many other interests.

In 1947 Kirkwood was appointed Arthur Amos Noyes Professor of Chemistry at California Institute of Technology where he developed and applied a new method of protein fractionation by differential electrophoresis across a narrow cell combined with convection. He also began his fundamental studies on the statistical-mechanical theory of transport processes, which he continued at Yale University, where in 1951 he became Sterling Professor of Chemistry and chairman of the Department of Chemistry. In 1956 he also assumed the position of Director of Sciences at Yale.

For his research accomplishments Kirkwood received three American Chemical Society awards: the Award in Pure Chemistry in 1936, the Theodore Williams Richards Medal of the Northeastern Section in 1950, and the Gilbert Newton Lewis Medal of the California Section in 1953. He received an honorary Doctor of Science degree from University of Chicago in 1954 and from Free University in Brussels in 1959. He was an active member of the American Chemical Society, serving as vice president and chairman of the Division of Physical and Inorganic Chemistry, 1940-41.

In 1958 Kirkwood was struck with a fatal illness but nevertheless continued with his active research interests and even held the Lorenz Visiting Professorship at University of Leiden in the spring of 1959. He died in New Haven, Conn. on Aug. 9, 1959, and was buried in the Grove Street cemetery in New Haven near the grave of J. Willard Gibbs, who founded statistical mechanics on which so much of Kirkwood's work was based.

Personal recollections; *Chem. Eng. News*, **31**, 4900 (Nov. 23, 1953); G. Scatchard, "John Gamble Kirkwood 1907-1959," *J. Chem. Phys.*, **33**, 1279 (1960); obituary, *New York Times* (Aug. 11, 1959); obituary, *New Haven Register* (Aug. 11, 1959).

DONALD D. FITTS

Henry Granger Knight

1878-1942

Knight was born on a farm near Bennington, Kans., July 21, 1878, the son of Edwin R. and Elva (Edwards) Knight, and a capsule review of his life reads like the abstract of an up-dated story by Horatio Alger. The 10-year-old boy who promised his dying mother that he would "be somebody"; his struggle for a formal education, complete with Ph.D.; progress in teaching from a one-room rural school to the deanship of important agricultural colleges; and finally the growth of the new chief of a bureau in the U.S. Department of Agriculture into the director of the world's largest research organization.

He worked hard and studied hard, and as soon as he acquired some knowledge he started to teach it to those who had less. He decided early that he wished to be a chemist and enrolled in the University of Washington where he began as an assistant in the Department of Chemistry, became an instructor, and earned his B.A. degree in 1902. After a year of study and teaching at University of Chicago, he returned to Washington as assistant professor and took his M.A. in 1904.

From 1904 to 1910 Knight was professor of chemistry at University of Wyoming and also state chemist. From 1910 until 1918 he was director of the Wyoming Experiment Station and Farmers' Institutes, and from 1912 he doubled as dean of the College of Agriculture. He took a year off to obtain his Ph.D. at University of Illinois in 1917, and when he returned to Wyoming he found himself director of the State Council for Defense.

From 1918 to 1921 he was dean and director of Oklahoma Agricultural College; after that he had a year as honorary fellow at Cornell. From there he went to West Virginia, where he was director and research chemist at the University Experiment Station and dean of the College of Agriculture for 1926-27.

When the Bureaus of Chemistry and of Soils and the Fixed Nitrogen Laboratory of the U.S. Department of Agriculture were combined in 1927, Knight became chief of the new Bureau of Chemistry and Soils. One of his first programs was on salvaging wastes to ease the sad plight of the farmers. Everything possible was to be saved and converted into cellulose, for which there was an increasing demand.

Knight was full of bright ideas. For instance, in 1904, during the agitation for a Pure Food and Drug Law, he set up at the St. Louis World's Fair an exhibit of cotton swatches, dyed in various tints extracted from colorings in the cans and jars of goods to which they were attached. He was most appreciative of new ideas of others, and many interesting "firsts" appeared on his desk. An unostentatiously efficient executive, with a rare gift for organization and administration, and friendly yet dignified, he soon won the cooperation and devotion of all who worked with him. He loved the outdoors, with all it could give him of hunting, fishing, motoring, and golf. He especially enjoyed the "Tung Chautauqua" tours he and Charles Concannon of the U.S. Department of Commerce had to make periodically. He was fond of good music and dancing, was an excellent amateur photographer, and a delightful *raconteur,* with a keen but kindly sense of humor.

Knight was active in the American Chemical Society and a fellow of the American Institute of Chemists, in which he was a councilor and president, 1932-34. He was awarded the AIC medal in 1941.

In 1939, Knight and some well chosen assistants had the privilege of spending $4 million on planning and building the four great regional laboratories of the U.S. Department of Agriculture. They stand as monuments to him, for his great contributions to the applications of chemistry to agriculture and the linking of agriculture to industry. In the same year his Bureau became Agricultural and Industrial Chemistry.

Knight died in Washington, July 13, 1942; he was survived by his wife, the former Nelly Dryden, and one son, Richard D. Knight.

Personal recollections; newspaper and journal clippings in possession of Florence E. Wall; *Chemist,* June 1932, March 1941, Summer 1942; obituaries in *New York Herald Tribune* and *New York Times,* July 14, 1942.

FLORENCE E. WALL

Fred Conrad Koch

1876-1948

Koch was born in Chicago, Ill., May 16, 1876. When he was five, the family moved to Elmhurst, where he and seven sisters spent most of their childhood. After graduation from Oak Park High School, he attended University of Illinois where he received his B.S. degree in 1899 and remained as instructor in chemistry for 2 years. From 1902 to 1909 he was research chemist at Armour &

Co. in Chicago. Feeling the need for more fundamental training he accepted a graduate fellowship at University of Chicago in the Department of Physiological Chemistry, where he received his Ph.D. degree in 1912 and remained in the department, first as instructor, in 1926 as professor and chairman, finally as Frank P. Hixon Distinguished Service Professor of Biochemistry.

Following retirement from the university in 1941, he rejoined the staff at Armour & Co. and continued his researches with unprecedented vigor and enthusiasm. When, after a heart attack in 1946, he was advised by his physicians to relax, he tried to work fewer hours each day and managed to take longer and more frequent excursions to his summer home in Wisconsin; but it was difficult for a man of his vitality and enthusiasm to slacken his pace. While recuperating at home during his last week, he had conferences with his research associates.

Koch's research interests were chiefly in the fields of hormones, vitamins, enzymes, and quantitative analytical methods. His work on the male sex hormone is probably the best known and most spectacular of his achievements. In 1926, with Lemuel C. McGee, he first demonstrated male hormone activity in extracts of bulls' testicles; later he developed methods for extracting the hormone from human urine. With Thomas F. Gallagher he developed a quantitative method for assay of the male hormone, based on the increased size of the capon's comb, which was widely used and which finally led to the isolation and synthesis of androsterone by Butenandt in 1931 and later also of testosterone and other androgens. In collaboration with zoologists and physicians he guided many studies on the physiology and clinical use of the male hormones. Other research included secretin and other gastrointestinal hormones, thyroid and pituitary hormones, pepsin, rennin, trypsinogen activation, and vitamin D's. He and his wife, Elizabeth (Miller) Koch, first observed the conversion of heat-treated cholesterol into provitamin D in 1925. In the field of blood and urine analysis his micro-Kjeldahl method for total nitrogen estimation and his enzymatic uric acid method are best known. He was author of a valuable teaching manual, "Practical Methods in Biochemistry," which appeared in five editions.

Koch had great influence as a teacher and trainer of biochemists. Many of the 40 doctors of philosophy and 20 masters of science who earned their degrees under his direction occupied places of responsibility and distinction in educational and research institutions of this country. The annual departmental teas, held each autumn in the Koch's apartment at 1534 East 59th Street, were, for returning and current students and staff, occasions of friendliness and charm, to be cherished and long remembered. In 1939 students and friends contributed toward a bust of Koch, which was presented to the university and which stands in the biochemistry lecture room in Abbott Hall, a symbol of respect and affection.

His long record of distinguished service was recognized by the award of many honors as officer of several local and national scientific societies and as distinguished lecturer. In 1935 he was chosen this country's delegate to the League of Nations Conference on Standardization of Sex Hormones and in 1941 he was this country's representative at the Pan-American Congress on Endocrinology in Uruguay. In 1942 he received the Squibb Award of the American Association for the Study of Internal Secretions; in 1943 the annual award of the Chicago Chapter of the American Institute of Chemists.

A devoted companion in his scientific as well as personal life was his charming wife, also a chemist, who collaborated with him in several scientific studies. He died Jan. 26, 1948.

Obituary by Martin E. Hanke, *Science* **107,** 671-672 (June 25, 1948); obituary in *New York Times,* Jan. 27, 1948; personal recollections.

MARTIN E. HANKE

Vaman Ramachandra Kokatnur

1886-1950

Kokatnur was born at Athani in the Bombay Presidency in India, Dec. 16, 1886, the son of Ramachandra A. and Krishnabai

Kokatnur. He received his B.S. degree from Fergusson College (Bombay University) in 1911 and spent 1 year as chemist in the Ranada Industrial Institute at Poona. He came to the United States in 1912, studied 1 year at University of California, and then went to University of Minnesota where, as the first foreign Shevlin Fellow, he earned his M.S. in 1914 and his Ph.D. in 1916. From 1915 to 1917 he was research assistant in chemistry at Minnesota.

From 1917 to 1921 Kokatnur did research for the Mathieson Alkali and the Niagara Alkali companies and was chief of the vat dye group for National Aniline and Chemical Co. After a year of special investigations for the By-Product Steel Corp. and for Du Pont, he settled down in New York as a consultant in 1922. In 1928 he was adviser to the Russian government on chlorine and caustic soda. He returned to India in 1930, and remained until 1933, as general manager of Sri Shakti Alkali Works, in Dhrangadhra; technical adviser to the American Trade Commission and Consulate, and technical director of Kopran Chemical Co. Finally, in 1934 he established Autoxygen Co. in New York to develop his own inventions.

For these Kokatnur held over 30 patents. He was active (captain) in the Chemical Warfare Service and made many valuable contributions—in airplane dopes, war gases, the M-59 bomb, jellied gasoline used in flame-throwers, gasoline-water mixtures for motor fuel, etc. Some of his other skills were in organic synthesis, esterification and sulfonation, indanthrene dyes, indigo, free-flowing sodium hydroxide, benzoic acid and derivatives, bleaching and preserving of food, batik printing, and organic peroxides. In 1938 and again in 1946 he was a consultant to the Bureau of Aeronautics of the U.S. Navy. In 1940 he was honored during the 150th anniversary of the Patent Office and was also inscribed in the Wall of Fame at the New York World's Fair.

For all his unassuming, placid appearance, "Koke" was an intensely human being, and his greatest contribution was sociological, in the way he merged his two cultures—Indian and American. Always interested in the activities of the Indian colony and in the welfare of visiting Indian students, he served on the Advisory Board of the Watamull Foundation for educational fellowships. His fine library was given to his Alma Mater, Fergusson College, and his Indian friends established a scholarship as a memorial.

He was one that always "kept a growing edge." He was particularly devoted to his hobby, the history of science—especially Indian science; to delve more deeply into this lore he could draw on an exceptional knowledge of ancient languages and dialects. It is unfortunate that his translations of many early manuscripts were never published.

Kokatnur died Apr. 14, 1950. He was survived by his wife, the former Helen Graber of Minneapolis; a daughter, Urmila, and a son, Arvind Kokatnur.

Personal recollections; *Chemist,* May 1951; obituaries in *New York Times* and *New York Herald Tribune,* Apr. 16, 1950.

FLORENCE E. WALL

Charles August Kraus

1875-1967

Born Aug. 15, 1875, in Knightsville, Ind., Kraus received his B.S. degree in engineering from University of Kansas in 1898. His undergraduate background contained but a single course in chemistry, and he developed his experimental skill working with Edward Franklin on the physical properties of electrolytes in liquid ammonia. After holding fellowships at Kansas and Johns Hopkins he became instructor in physics at University of California. He accepted an assistantship in chemistry in 1904 at Massachusetts Institute of Technology where he continued his work, with Arthur Noyes, on the properties of electrolytes in solution. As a result of this work he received his Ph.D. degree in 1908. He remained at the institute until 1914, first as research associate, then as assistant professor of physical chemistry research.

Kraus then went to Clark University as professor of chemistry and director of the chemical laboratory and remained there until 1924. His skill in research and his interest in graduate students resulted in his appointment in 1924 as professor of chemistry and director of chemical research at Brown University. He became professor emeritus in 1951.

In his early days at Brown, Kraus and his students worked on the second floor of the Newport Rogers laboratory while undergraduate instruction was given in the new Metcalf laboratory. He, alone, directed practically all the graduate work in both classroom and laboratory for many years. In the early 1930's he received assistance from W. Albert Noyes Jr., Raymond Fuoss, and Lars Onsager who, for a while, were members of the Brown faculty.

Greatly admired and respected by his students, Kraus was a hard but benevolent task master. To him the laboratory was a home. Holidays and weekends were opportunities for work. On Saturday evening he held research conferences at which times students presented reports, partook of refreshments and participated in a German card game called skat. Kraus and a few of his students occasionally spent weekends working his farm in Princeton, Mass.

At times between 1918 and 1924 he was a consultant for the U.S. Bureau of Mines, the Chemical Warfare Service, and the Fixed Nitrogen Laboratory. He served in a number of capacities for the National Research Council from 1931 to 1935.

Kraus was president of the American Chemical Society in 1939. He received the Nichols Medal in 1923, the Gibbs Medal in 1935, the Richards Medal in 1936, the Franklin Medal in 1938, the Priestley Medal in 1950, and honorary doctor's degrees from Kalamazoo College, Colgate, Brown, and Clark.

Much of his lifetime research interests are contained in his book, "Properties of Electrical Conducting Systems." He was well known for his research on chemical reactions in vacuum systems and metallo-organic compounds of gallium and germanium. He paved the way for high-compression automobile engines by developing a process for making tetraethyllead commercially.

He died in Providence, R.I., June 27, 1967, at the age of 91.

J. Chem. Educ. **6,** 4-7 (1929), with portrait; biographies: F. G. Keyes in *Chem. Eng. News* **18,** 763-64 (1940); *Chem. Eng. News* **28,** 2704 (1950), portrait on cover; R. M. Fuoss in *Biog. Mem. Nat. Acad. Sci.* **42,** 119-159 (1971); obituaries: *Chem. Eng. News* **45,** 59 (July 17, 1967), portrait on cover; *New York Times,* p. 45, June 28, 1967; personal recollections.

EGBERT K. BACON

Paul John Kruesi

1878-1965

Kruesi was an industrialist who formed and developed manufacturing concerns largely in the fields of electrical ceramics and electrometallurgy. He was born in Menlo Park, N.J., Feb. 3, 1878.

His family moved to Schenectady, N.Y. in 1886 where his father, John Kruesi, in the employ of Thomas A. Edison, was appointed first general manager of Edison Machine Works, which by several mergers became General Electric Co. in 1900.

Paul graduated from public high school in Schenectady in 1896 and then entered Union College where he was an active student for 2 years. During the summer of 1896 his first job was that of office boy in the payroll department of the Edison Co. The two following summers he worked in the treasury department of the Edison-General Electric Co. and in the Edison laboratory in Orange, N.J. In the fall of 1898 he did not return to college but went to Chicago where he was employed in the statistical division of the accounting department of the Chicago Edison Co. The following year he returned to New York to become assistant to the general manager of the General Incandescent Arc Light Co. In 1902 he left for Tennessee to start his career as an industrialist.

Although he did not graduate, the college considered him a member of the class of 1900 and later in 1935 conferred on him an honorary Sc.D. degree. The early association with Edison resulted, in 1950-51, in his election as vice-president of "Edison Pioneers"—those who were associated with Edison prior to 1900. Edison said, "Paul Kruesi was the first of a second generation of his principal associates to enter his employ."

Kruesi's life in Chattanooga, Tenn. was a

productive one. First, at the age of 24 he founded American Lava Corp., a company largely concerned with technical ceramics used as insulators in the electrical industry. His next important industrial venture was the founding in 1917 of Southern Ferro Alloys Co., the earliest southern maker of ferrosilicon by electrofurnaces. In addition, he was founder and an executive of numerous Tennessee industries, as well as a director of a railroad, an insurance company, a bank and several manufacturing associations. He was an excellent organizer and had the ability to select the technical and scientific personnel that made his companies successful.

Because of his broad background and experience in the fields of management and industrial techniques he was called upon to serve in diverse capacities. In national affairs he was assistant to the Secretary of Commerce in 1922 and was one of the original members of the board of directors of the U.S. Chamber of Commerce. During both wars he served on a number of industrial committees of the War Production Board, on the executive committee of war loans, as well as many other services at the national or local level.

Both the community and the state regarded him as a great leader and a devoted and loyal citizen. In politics he participated in numerous activities of the Republican Party. In civic affairs he was active in various capacities in areas of education, religion, and local government. He was honored by election as president of The American Electrochemical Society in 1928.

He died Nov. 29, 1965 in Chattanooga, Tenn.

"Who's Who in America," **28** (1954-55), his preferred biographical reference; "Who's Who in Engineering" (1941); "National Cyclopedia of American Biography" **B,** 465, James T. White & Co. (1927); *Trans. Amer. Electrochem. Soc.,* 53 (1928), portrait; obituaries, *New York Times,* Dec. 2, 1965; also extensive write-up in *Chattanooga News—Free Press,* Dec. 1, 1965; archives of Union College.

EGBERT K. BACON

L

Frederick Burr La Forge

1882-1958

La Forge, a chemist who dealt with naturally occurring organic compounds, was born in Bridgeport, Conn., Mar. 12, 1882. He received his degree of bachelor of science from Princeton University in 1905. He continued his graduate study at Berlin University, received his Ph.D. degree in 1908, and then served as an assistant to Robert Pschorr in the Ferit Chemical Institute of the university until 1909. Upon his return to the United States he joined the staff of Rockefeller Institute for Medical Research where from 1910 to 1915 he conducted research on nucleic acids, sugars, and other biological products.

La Forge began his 37 years service with the United States Department of Agriculture in 1915. His research in the Carbohydrate Laboratory of the Bureau of Chemistry resulted in his discovery in 1916 of a new 7-carbon natural sugar, mannoketoheptose, an important step in the synthesis of carbohydrates. He followed this up the next year with the discovery of a second heptose, sedoheptose, in the sedum plant.

During World War I he began his research on developing useful materials from substances generally considered waste products. He perfected methods for extracting adhesives, cellulose, and furfural from corncobs and other crop wastes, utilizing steam digestion, that reduced the cost of production to the point where the method could be adopted commercially.

Upon completing these experiments, he transferred to the Insecticide Laboratory of the Bureau of Chemistry. Although he made significant progress in his investigation in extracting and identifying the active principle of pyrethrum with a view to making it or some closely related product of equal insecticidal value synthetically, a foreign paper was published that covered the subject. Shifting his viewpoint somewhat, he continued his search for new insecticides. In 1932, he and his associates won the race with the English, German, and Japanese chemists who were seeking to discover the chemical structure of rotenone, a material poisonous to insects but not to warm–blooded animals. This achievement opened the way to making it synthetically on a commercial scale. For this, La Forge and Herbert L. J. Haller received the Hillebrand Prize of the Chemical Society of Washington. They continued their research to discover rotenoids, such as deguelin, tephrosin, and toxicarol, related compounds that contribute to rotenone toxicity.

Meanwhile, on reexamining pyrethrum, La Forge and his associates found two additional esters, cinerin I and cinerin II, and proved its chemical structure to be more complicated than previously considered. Foreign supplies of pyrethrum were reduced during World War II, with the imports restricted to use by the Army and Navy. The importance of pyrethrum and of the work of La Forge and his associates was recognized after the war by a commendation from Ross McIntire, Surgeon General of the Navy. Meanwhile, in 1944 they had found that "pyretholone" was not a pure compound but a mixture of two alcohols. "Pyretholone" was retained as the name for the major component and "cinerolone" was given to the other. Four years later La Forge, Milton S. Schechter, and Nathan Green developed a method to synthesize pyrethrum-like esters, for which they received a Superior Service Award from the Department of Agriculture in 1951 and an achievement award from the Chemical Specialties Manufacturers. One of these synthetic esters, allethrin, is used extensively in insecticidal aerosols and sprays.

La Forge officially retired from the Department of Agriculture in March 1952, when he reached mandatory retirement age. However, he continued his laboratory work with those who discovered that one of the eight optical isomers of allethrin was about ten times as powerful an insecticide as the pyrethrins.

On Sept. 4, 1958 La Forge died in Washington, D. C. His research on furfural, rotenone, pyrethrin, and allethrin had given him a world-wide reputation. He had served in an advisory and consultant capacity to manufacturers and government agencies. He had 10 patents issued to him in the field of organic chemistry and had published about 125 articles in professional journals dealing with the chemistry of complex materials.

Personnel File, Agricultural Research Service, USDA and Federal Records Center, St. Louis; obituary in *J. Economic Entomology,* **52,** 180 (1959).

VIVIAN WISER

Arthur Becket Lamb

1880-1952

Lamb was born in Attleboro, Mass., Feb. 25, 1880, to Louis and Elizabeth (Becket) Lamb. The family owned a jewelry manufacturing business in the area. While in high school Arthur spent much of his spare time studying and experimenting in astronomy, chemistry, and physics. As an undergraduate at Tufts College he majored in biology although he was greatly influenced by the organic chemist Arthur Michael. After receiving both his B.A. and M.A. degrees in 1900, Lamb continued his studies at Tufts with Michael. Although he did not receive his Ph.D. degree from Tufts until 1904, he completed his thesis research by 1902. At this time Lamb became interested in physical chemistry and at Michael's recommendation he studied thermochemistry with Theodore William Richards at Harvard where he received a second doctorate in 1904.

A year of study in Germany reinforced his interest in physical chemistry. Upon his return in 1905 he became instructor of electrochemistry at Harvard. In 1906 he was appointed assistant professor and director of the laboratory at New York University. He returned to Harvard in 1912 as assistant professor, became professor in 1920, Erving Professor in 1929, and dean of the Graduate School of Arts and Sciences in 1940. He retired from undergraduate teaching in 1948 although he retained his office and laboratory. He died unexpectedly in Cambridge, May 15, 1952. He married Blanche A. Driscoll in 1923 and was the father of one son and one daughter.

Lamb's early chemical investigations included several organic projects with Michael and calorimetric studies with Richards. He translated Fritz Haber's classic book on gas thermodynamics into English as "Thermodynamics of Technical Gas Reactions" (1908); and he wrote "Laboratory Manual of General Chemistry" (1914-16). His research at Harvard dealt mainly with the physical chemistry of transition metal coordination compounds. During World War I he studied the absorption of carbon monoxide by various substances in an attempt to minimize the danger from gas below the decks of ships. From 1921 onward he had charge of elementary chemistry at Harvard and for 20 years did nearly all of the lecturing in the course.

Lamb is probably best known for his long and devoted term as editor of *Journal of the American Chemical Society.* In 1917 when he accepted the editorship, Lamb had a staff of 10 associate editors and a publication rate of some 200 papers per year. He made wide use of the referee system, which had been initiated by William A. Noyes. Within 10 years the journal was publishing over 700 papers annually. At the age of 69 Lamb resigned as editor in 1949 but remained as consulting editor to smooth over the transition period. The January 1950 issue of the Journal was dedicated to him and came as a pleasant surprise. He received numerous awards from the ACS both national and sectional, in recognition of his devoted service.

Biographies in *Chem. Eng. News* **27,** 2840, 2876 (1949); by G. S. Forbes in *J. Chem. Soc.* 1322-1324 (1953); by Allen D. Bliss in *J. Amer. Chem. Soc.* **77,** 5773-5780 (1955), with a bibliography and portrait; by F. G. Keyes in *Biog. Mem. Nat. Acad. Sci.* **29,** 201-234 (1956), with portrait.

SHELDON J. KOPPERL

Victor Kuhn La Mer

1895-1966

An outstanding educator and long-time professor of chemistry at Columbia University, La Mer was born at Leavenworth, Kans., June 15, 1895, the son of Joseph Secondule and Anna Pauline (Kuhn) La Mer. He earned his bachelor of arts degree from University of Kansas in 1915 and his doctor of philosophy degree in chemistry from Columbia University in 1921.

He married Ethel Agatha McGreevy July 31, 1918. Their three daughters were Luella Belle, Anna Pauline, and Eugenia Angelique.

After graduation from University of Kansas, La Mer spent a year as an instructor of high school chemistry and followed this by a summer of graduate study at University of Chicago. The year 1916-17 La Mer spent as a research assistant at Carnegie Institution of Washington, D.C. and then he served as a lieutenant in the Army Sanitation Corps from 1917 till 1919.

Continuing his graduate studies at Columbia University in 1919, La Mer secured an assistantship in the Department of Chemistry and soon thereafter became an instructor in food chemistry as well. In 1920 he became instructor in general chemistry at Columbia, where he was awarded his doctor of philosophy degree in chemistry in 1921.

The year 1922 brought La Mer a Cutting Traveling Fellowship. He spent the first year of it at Cambridge University and the second at University of Copenhagen. While working with Brønsted at Copenhagen in 1924, he provided the first experimental verification of the Hückel theory of electrolytes. His other major research contribution was the discovery of higher order Tyndall spectra in 1941. His other research efforts were in the fields of nutrition, physical chemistry, oxidation-reduction potentials, thermodynamics, theory of electrolytes, reaction kinetics, deuterium isotopes, spectroscopy, aerosols, fluorescent quenching, colloids, and acid-base catalysis.

Returning to Columbia as assistant professor of chemistry in 1924, La Mer became associate professor in 1928 and professor of chemistry in 1935. In 1961 he became professor emeritus at Columbia. He served as a visiting professor of chemistry at Northwestern University in 1928 and at Stanford University in 1932 and in 1950 was made honorary professor of chemistry by the University of San Marcos in Lima, Peru.

La Mer served as the Priestley lecturer at Pennsylvania State University in 1932, as a Fulbright lecturer in Australia in 1959 and as distinguished lecturer at the Shell Development Co. in 1963. He also served as a National Science Foundation lecturer in the Colloid School at the University of Southern California in 1962-63, and in the National Science Foundation Colloid School at Lehigh University in 1965.

He was the translator and editor of "Fundamentals of Physical Chemistry" by Arnold Eucken (with Eric Jette) in 1925, associate editor of *Journal of Chemical Physics* from 1933 to 1936, and editor-in-chief of *Journal of Colloid Science* from 1946.

During World War II La Mer served as a civilian with the Atomic Energy Commission, the Office of Scientific Research and Development, and the National Defense Research Committee. He received a Presidential Certificate of Merit in 1948 and the Kendall Award of the American Chemical Society in 1956.

La Mer was a member of the National Academy of Sciences, Royal Danish Academy of Science, Faraday Society, Sigma Xi, Phi Lambda Upsilon, Phi Chi, and Epsilon Chi. He was a fellow of the New York Academy of Sciences and served in its offices, culminating in the presidency in 1949. He also served as president of the Leonia Civic Conference, 1940-41.

Death came to La Mer Sept. 26, 1966, and he is buried in St. Joseph Cemetery, Hackensack, N.J.

Chem. Eng. News **34,** 2117, April 30, 1956, with portrait; *J. Colloid Interface Sci.* **21,** 263-265 (1966); information from News Office, Columbia Univ.; obituary, *New York Times,* Sept. 28, 1966.

JAMES C. COX, JR.

Walter Savage Landis

1891-1944

Landis was born in Pottstown, Pa., July 5, 1891, the son of Daniel and Clara (Savage) Landis. He attended the public schools of Pottstown and of Orlando, Fla. and Bethlehem Preparatory School. He graduated from Lehigh University with the degree of metallurgical engineer in 1902. He at once took a position as assistant in the Department of Metallurgy at Lehigh. Along with his teaching duties he was able to study for the degree of master of science, which he received in 1906. In 1905 and 1906 he went to Heidelberg, Germany, where he studied crystallography and mineralogy, also working for a time at Krupp Institute in Aachen.

While on the metallurgical teaching staff at Lehigh, Landis rose through the various grades, becoming associate professor in 1910. But he was soon to leave Lehigh and the teaching profession. In 1912 he became chief technologist of American Cyanamid Co. This was the beginning of a distinguished career in chemical industry, which caused him later to be described as a "man of extraordinary achievements."

Landis now turned to research, invention, and management. In 1923 he was elected vice-president of American Cyanamid, an office which he held during the remainder of his life. He established the company's first research laboratory in 1913. He was also a director in five other companies. He received more than 50 patents and led in development of several important processes and manufacturing plants. Among these was the production of sodium cyanide from cyanamid. First marketed in 1916, this product soon was used by most gold and silver mills in America, as it is today. Other processes he developed were those for producing ferrocyanide, dicyandiamid and urea, and hydrocyanic acid. He took part in the first commercial production of argon and aided the ammonium phosphate and explosives industries. During the first World War he was engaged in building the first American plant to produce ammonia from cyanamid and to oxidize ammonia to nitric acid. He also designed a portable hydrogen generator for the United States Army. In the early nineteen thirties he worked on improvements of the DeLaval electrothermic zinc process at Trolhätten, Sweden and Sarpsborg, Norway. The increasing demand for magnesium led to work by Landis on the electrothermic process for magnesium at Radentheim, Austria.

While Landis was at Lehigh, Joseph W. Richards was head of the Metallurgical Department, and it was in 1902, the year of Landis' graduation, that Richards joined with a small group to organize The American Electrochemical Society. Richards was made its first president and in 1920 Landis became the seventeenth president, having previously served as assistant secretary of the society and as chairman of the New York section. His term of office as president ended in the year of Richards' death (1921).

The Electrochemical Society in 1932 established the Joseph W. Richards Memorial Lectureship. Fittingly Landis was chosen in 1936 to give the third of these lectures; it was entitled "Joseph W. Richards, The Teacher—The Industry."

Landis was also active in the American Chemical Society. He was chairman of its New York section in 1932. He was also president of the Chemists' Club of New York.

Landis received a number of awards of merit. These included the Chemical Industry Medal of the American section of the Society of Chemical Industry (1936), the Perkin Gold Medal of the Society of Chemical Industry (1939), and the Medal of the American Institute of Chemists (1943). In 1922 he was awarded an honorary degree of doctor of science by Lehigh University.

No biography of Landis would be complete without mentioning his admirable personal characteristics. Of these we may emphasize his modesty, his generosity, and his willingness to serve. He made many trips abroad in technological work but took advantage of the opportunities afforded to study the economies of the countries visited, becoming an authority on this subject, evidenced by a large number of addresses and published articles. He deplored the trend toward socialism both abroad and at home as well as the attempts of political leaders to sidestep economic laws.

Landis also maintained an active interest in education, which he believed should be broad based rather than narrowly technical. He was a trustee of Lehigh University at the time of his death. His last public address was given Feb. 12, 1944, at the fiftieth anniversary celebration of the Lehigh Valley section of the ACS. He died at his home in Old Greenwich, Conn., Sept. 15, 1944, at the age of 53.

Brief biography, *Trans. Electrochem. Soc.* **66,** 6 (1934); obituary, *idem.,* **86,** 47-9 (1944); personal recollections.

ALLISON BUTTS

Norbert Adolph Lange

1892-1970

Lange was born Aug. 4, 1892 in Sandusky, Ohio. He received his A.B., M.S., and Ph.D. degrees from University of Michigan, where he published papers on organic chemistry with William Hale and Moses Gomberg. His doctoral dissertation with Hale was on cyclic ureas.

He was an instructor in chemistry at Michigan from 1917 to 1919; an assistant professor and later an associate professor of organic chemistry at Case Institute of Technology from 1919 to 1934; a lecturer in chemistry at Cleveland College of Western Reserve University from 1925 to 1952. From 1920 to 1932 he was co-editor of "Handbook of Chemistry and Physics." In 1934 Lange was one of the founders of Handbook Publishers, Inc. and served as the company's secretary and editor until the sale of the company to the McGraw-Hill Book Co. in 1958. From 1959 to 1967 he acted as a consultant to McGraw-Hill.

He is probably best known as the author of 11 editions of the internationally known Lange's "Handbook of Chemistry," the first edition of which appeared in 1934. In his work of compilation he always emphasized accuracy over volume, and as errors were pointed out by users of the handbook they were carefully checked and corrected in the next printing. Dozens of his friends and acquaintances, some of them world-famous scientists, were happy to be associated with Lange's Handbook by preparing tables or articles dealing with their particular specialties.

In his researches Lange always emphasized experimental results, and he performed his own analyses. After he left Michigan his collaborators were either undergraduates or former students. In his laboratory at Case he investigated the use of malic, maleic, and fumaric acids as primary standards in volumetric analysis and also devised a colorimetric method for determining small amounts of iodine as iodide and iodate. He then returned to his cyclic ureas, this time expanding the ring to six members and finally opening it up altogether. He had become interested in the taste of some of his compounds and prepared a series of ureas and thioureas using vanillylamine and *p*-phenoxyaniline to examine the effect of various groups on taste. This work did not turn up any new sweetening agents, all the compounds being tasteless, bitter, or pungent.

In 1929, following correspondence with Marston Bogert of Columbia, Lange began working with water-soluble ethers of quinazoline. He found several new reactions with these compounds, such as the reaction of 2-chloro-4-ethoxyquinazoline and sodium methoxide in methanol to give 2,4-dimethoxyquinazoline. The long sought 2-methoxy-4-ethoxy, and 2-ethoxy-4-methoxyquinazolines were finally obtained by treating the dimethoxy and diethoxy derivatives with sodium ethoxide and sodium methoxide respectively. This alkoxy interchange at the 4-position in quinazoline has been confirmed by several subsequent investigators. From 1930 to 1939 Lange published eight papers on quinazolines, five of them with Fred E. Sheibley, his close friend and associate.

Lange was active in the American Chemical Society, serving the Cleveland Section in various offices including that of chairman and as editor of *Isotopics,* regional bulletin of ACS local sections.

A fellow of the American Institute of Chemists, he was given its Achievement

Award in 1956. In 1958 the Cleveland Chemical Profession honored him with the Certificate of Merit in recognition of his enhancement and interpretation of the chemical profession in the Cleveland area.

The Western Reserve Historical Society made him an Honorary Member in 1959 when the Society's Publication No. 115 appeared. This was "Sandusky—Then and Now," a translation from German to English by Lange and his wife of von Schulenburg's "Sandusky Einst and Jetzt" of 1889.

An excellent musician, Lange played the French horn in University of Michigan band and helped pay his tuition by playing the violin in dance orchestras at Ann Arbor. In his later years he often would play chamber music with his wife and another musician forming a trio. For a time he played with the Sandusky Symphony.

A man of great integrity and erudition, he was admired by his many friends for these qualities and for his quiet sense of humor. He died Aug. 2, 1970, survived by his wife, Marion (Cleaveland) Lange, a former professor of chemistry at Case Western Reserve University.

Obituary by F. E. Sheibley in *Isotopics,* November 1970; N. A. Lange, autobiographical notes; communications from Marion C. Lange.

GORDON M. FORKER

Irving Langmuir

1881-1957

Irving Langmuir, Nobel laureate and scientist extraordinary, was born in Brooklyn, N.Y., Jan. 31, 1881. As a boy he traveled extensively in Europe with his family but also attended public school in Brooklyn, a boarding school in Paris, and an academy in Philadelphia. In 1898 he graduated from Manual Training High School of Pratt Institute and then enrolled in School of Mines, Columbia University, where he received a degree in metallurgical engineering in 1903.

His youthful interests in outdoor activities such as bicycling, boating, hiking, mountain climbing, skating, and skiing remained as important activities throughout his life. Later he became interested in flying and owned an airplane. He did not like competitive hobbies but was fond of those that involved an understanding of the mechanism of simple and familiar natural phenomena. He kept detailed records of what he did, had a high regard for accuracy, and sharp powers of observation. To him science was fun, and he considered it the highest and noblest calling. His whole career was one of action that reflected these interests and attributes.

It seemed only natural that Langmuir should pursue graduate work, which he did at University of Göttingen in 1904. He became interested in the dissociation of various gases by a hot platinum wire, worked with Walther Nernst, and received his Ph.D. degree in 1906. On his return to the United States his early passion for pedagogy led him to accept a position on the faculty of Stevens Institute of Technology where he taught analytical chemistry until 1909 when he resigned. He had no patience with unwilling learners, had little time for research, and hence became disenchanted with a teaching career. However, his teaching interests still remained, and he always welcomed this sort of contact, particularly with young people.

He spent the summer of 1909 as a guest researcher at the General Electric Co. laboratory in Schenectady. Here he was delighted with the atmosphere and the freedom given for his personal investigations in science. Willis Whitney, the director of the laboratory, was also delighted with the research potential of the young scientist and persuaded him to accept a position on the staff. Langmuir accepted, became an associate director in 1929, retired in 1950, but remained as a consultant until his death in 1957.

His first work at General Electric was concerned with a detailed study of conditions that existed in a vacuum system with a high temperature tungsten filament. After about 3 years of work he had a guilty feeling that the company was getting no return from the cost of these investigations. Whitney, assured him that this should be of no concern of his but rather that of the director.

These studies eventually gave handsome

returns. They resulted in improved vacuum techniques, the effect of gases on tungsten filaments and a practical gas filled lamp, the atomic hydrogen torch, vacuum tubes in general, improved x-ray tubes, new electronic devices, and many others. On the theoretical side they were the foundation for an understanding of plasmas, thermionic phenomena, catalytic function, a new theory of adsorption, new concepts of valence and the structure of atoms and molecules, and the discovery of monomolecular films. The latter resulted in studies in surface chemistry which were the basis of the Nobel prize.

Langmuir served the government in both World Wars. In the first he worked with William D. Coolidge, retired vice president and director of research for General Electric, to develop a series of highly effective listening devices for detecting submarines; in the second he and his assistant, Vincent J. Schaefer, collaborated to produce a smoke-making machine that was used by Army and Navy for screening troops and ships from enemy observation. The government support given in the second war was continued in studies on the problem of icing conditions on airplanes and was then extended to the general area of meteorology, which included cloud nucleation and the broader aspects of weather control. These latter studies were Langmuir's last investigations and were subject to some criticism from the professional weather men. These excursions into rather complex areas such as atmospheric phenomena and protein and enzymes structures were related to his successful investigations in surface chemistry. To him the results were not completely satisfying and no doubt ended in a certain degree of personal frustration.

Langmuir was in much demand as a semi-popular lecturer. He enjoyed expressing his views on the philosophy of science and the interrelationship of science and social and political problems. Much of this appears in printed articles.

He received 15 honorary doctors degrees and 22 medals and awards, one of which was the Nobel prize in chemistry for 1932. These decorations are on display in General Electric Research and Development Center in Schenectady. He was president of the American Chemical Society in 1929 and of the American Association for the Advancement of Science in 1943.

Many considered Langmuir as a non-social individual of personal conceit. This was only an outer impression of his unique personality. His active mind and physical energy left little opportunity for small talk. However, he had close social contacts with his direct family, relatives, and selected friends. He died in Woods Hole, Mass., Aug. 16, 1957.

Biographical references are so numerous that a selection is difficult. An outstanding one is; "The Collected Works of Irving Langmuir," 12 volumes, C. Guy Suits, Gen. Ed., Pergamon Press (1962); Vol. 12, "Langmuir the Man and the Scientist," by Albert Rosenfield, contains a remarkably complete, interesting, biography largely based on Langmuir's personal diaries, letters, and reminiscences by his family and associates; and a tabulation of 134 references of biographical source material as well as numerous complete popular and semi-popular articles and addresses by him.

EGBERT K. BACON

Samuel Allan Lattimore

1828-1913

The later years of the nineteenth century were very important ones for the development of modern chemical thought, and this progress had a marked effect on education in science. For over four decades Samuel Allan Lattimore dominated the chemistry program at University of Rochester. When he was first appointed professor of chemistry in 1867, his responsibilities included geology, zoology, and physics as well as chemistry. This extra teaching load continued until 1883.

Lattimore was very much a chemist-educator as we now view that term. He was most insistent upon the need for student laboratory work. His most important predecessor, Chester Dewey, had made some progress in collecting apparatus for lecture demonstrations, but it was only through the most remarkable improvisation that Lattimore was able to begin at once to offer a practical course in analytical chemistry. His skill as a chemist and as a teacher allowed him to compensate for wholly inadequate facilities. It took 20 years for his dream of a chemistry building to be realized.

In the meantime, the makeshift arrangements proved to be immensely popular. The University's Nineteenth Annual Catalog (1868-69) carried the following new note:

"Analytical Laboratory"

"In addition to the instruction given to undergraduates in General Chemistry, an additional room has been provided and furnished with the requisite apparatus, for the use of those who may wish to pursue a more extended study under the direction of the Professor."

Not only regular and special college students, but Rochester physicians, pharmacists, mechanics, and farmers came looking for chemical training and understanding.

The practical aspect of Lattimore's approach to chemical education is evident in his frequent free public scientific lectures. He spoke to the workingmen of Rochester and the socially significant at the Chautauqua Institute. His concern for people and science, his teaching skill, and his use of illustrative experiments all contributed to his ability to communicate the understanding and significance of charcoal, explosives, iron, etc., to his diverse audience.

This great concern for the practical side of science and its application for the benefit of the community is also characteristic of Lattimore's own scientific work. The best known example is his report to the Executive Board of the City of Rochester dated Sept. 29, 1877. The Board had asked Lattimore to study the water supply of the city; a task we today might regard as very desirable. After a thorough program of sampling and testing, Lattimore submitted ". . . to the consideration of all thoughtful citizens of Rochester . . ." that their water supply was:

1) ". . . *extremely liable to pollution from surface drainage* . . ." (emphasis Lattimore's)
2) served by ". . . so-called sewers, which, . . . are simply covered ditches . . . filling to their own level the wells, often for great distances on each side of their course. . ."
3) found to contain common salt at ". . . *the surprising figure of 16.78 grains per gallon!* . . . more than three times the maximum quantity . . . pronounced compatible with safety. . ."

Lattimore was realistic enough to understand that ". . . The arguments and opinions expressed in this Report may not find universal acceptance. . ." He took an unequivocal stand,

"Long ago, I became so thoroughly convinced of the danger of using well water, that I discarded it totally from my own household. . ."

and offered neighborly advice,

". . . and I must earnestly commend, as a prudential measure, the same course to my fellow-citizens."

Lattimore was born in Union County, Ind., May 31, 1828. He graduated from Indiana Asbury (De Pauw) University in 1850. After teaching Greek at his alma mater for 10 years, he was elected professor of chemistry at Genesee College in Lima, N.Y. where he taught for 7 years before going to Rochester. Lattimore married Ellen Frances Larrabee July 28, 1852 and raised five daughters. Iowa Wesleyan University, Hamilton College, and De Pauw honored him with honorary degrees.

He died in Rochester, Feb. 17, 1913. Lattimore was devoted to his profession, students, and community. This high esteem was amply returned; for example, one of his first students said of him, "He was a chemist of exceptional ability, and state-wide reputation, and a fine teacher with a charming personality, enhanced by old school courtesy. He had broad, cultural interests; I doubt whether any member of the teaching staff exercised a finer or stronger influence upon the community life of Rochester."

Documents in University of Rochester archives, including A. J. May's unpublished history of the university; "National Cyclopedia of American Biography" **12,** 244, James T. White & Co. (1904).

K. THOMAS FINLEY

Louis Hyacinth Laudy

1842-1905

That the Civil War stands out more vividly in the memory of the United States than earlier wars is owing in a large measure to the dedicated photographers who roamed the camp and battlefield and snapped imperishable pictures. One of those men was Louis Laudy, who later taught chemistry at Columbia and who entertained the students in his classes with his reminiscences of the war.

Laudy was born about 1842, but it is not known where. Nothing seems to be known about his youth other than that he followed pharmacy and photography. After the Civil War he studied chemistry and other subjects at Cooper Institute night school in New York, graduating in 1868. He continued his education at New York Veterinary College, receiving his doctor of philosophy and doctor of veterinary science degrees in 1879.

Cooper Institute hired him as instructor in physics in 1869 and assistant in chemistry in 1870. In 1872 he became instructor in chemistry and photography at Columbia and remained at the university as instructor and later tutor in chemistry. He assisted Charles Chandler for so many years that Columbia students said jokingly that Chandler must have graduated "summa cum Laudy." According to Chandler, Laudy was "remarkably ingenious and successful in devising apparatus for lecture illustration." In 1886 he published the pamphlet, "The Magic Lantern and its Applications."

In the summer of 1874 Laudy attended the Priestley Centennial in Northumberland, Pa. He photographed the Priestley House, the chemists who attended the meeting, and Priestley's apparatus, which he located in the attics of Priestley's grandchildren. Copies of these photos, which Laudy advertised for sale in *American Chemist,* still exist in collections.

Laudy died in New York City, Aug. 17, 1905, at the age of 63, still connected with Columbia University.

Records of Columbia University; obituary by Charles Chandler, *Columbia Univ. Quart.* **8,** 42-43 (1905), with portrait; Robert Taft, "Photography and The American Scene," Macmillan Co. (1938) (reprint 1968); *American Chemist* **5,** 1874-75; the quote, "summa cum Laudy," is from M. Bogert, "Chandler Centenary," *Ind. Eng. Chem.* **30,** 118 (1938).

WYNDHAM D. MILES

Matthew Carey Lea

1823-1897

Lea, a pioneer in the chemistry of photography, was born in Philadelphia, Pa. on Aug. 28, 1823. He was the second son of Isaac Lea, a Quaker publisher, and grandson of Matthew Carey, the first publisher of "Encyclopedia Americana." Isaac Lea was also a distinguished naturalist who served as president of the American Association for the Advancement of Science and the Philadelphia Academy of Natural Sciences.

Although Lea never went to school or college, his education was most liberal. He had the best tutors, books, and science equipment that money could buy. He traveled in Europe with his parents, read classical literature, studied art, learned four languages, and got to know many of his father's scientist friends. In his teens, he joined the Philadelphia firm of Booth, Garrett, and Blair, the nation's first consulting chemical laboratory. After a few years, however, he gavc this up to study law and was admitted to the Bar in 1847 at the age of 24. Evidently he did not find law to his liking because, after touring Europe for a year or more, he went back to Booth's laboratory. His next move was to a private laboratory in his home at Chestnut Hill.

Lea was a wealthy, sickly, anti-social, one-eyed, recluse. Few chemists knew him personally but those who did found him a brilliant conversationalist and an intense lover of truth who never expressed an opinion not based on fact. He was married twice; first in 1852, to his cousin, Elizabeth Jaudon, and after her death in 1881, to Eva Lovering. He had one son. The only organization he belonged to was the Franklin Institute, which he joined in 1848. As a result of his scientific work with photography, he was elected to the National Academy of Sciences two years before his death on Mar. 15, 1897. At his request, his laboratory notebooks were destroyed after his death and his laboratory equipment and library given to the Franklin Institute. He also left an endowment to the Institute library.

Lea's first research on the "Southern Coal Fields of Pennsylvania," which was published when he was only 18, inaugurated a lifetime of scientific activity that resulted in 400 papers published in American and European journals. Three hundred of these were on the chemistry of photography, over 100 of which appeared in *American Journal of Science* between the years 1860 and 1897. His one book, "Manual of Photography," was published in 1868, with a second edition in 1871.

Lea was much concerned with the relationship between atoms and undertook to show that the number "44.45 plays an important part in the science of stoichiometry." This relationship was found to extend to no less than 48 of the elements. Although the germ of the periodic law was in his thoughts, he never clearly enunciated it.

In analytical chemistry, Lea developed tests for detecting gelatin, prussic acid, traces of iodine, ruthenium in the presence of iridium, traces of sulfuric acid in the presence of sulfates, and sodium hyposulfite.

In organic chemistry he studied methods for preparing ethylamines, urea, napthylamine, and the reaction of gelatin with mercury salts. He found a new method for preparing picric acid and at the beginning of the Civil War recommended to the United States Government its use as an explosive. However, this use was not appreciated for another 50 years.

In the chemistry of photography, Lea was a true pioneer. He studied colloid/silver halide emulsions, developing agents, and the influence of colored light on the reduction of silver salts. Eight papers were published on what he regarded as "allotropic" forms of silver of various colors. His work with the colored compounds formed by silver salts and aniline dyes foreshadowed color photography.

Biographies by George F. Barker in *Biog. Mem. Nat. Acad. Sci.*, **5**, 155-203 (1905), with a list of 101 of Lea's most important papers, and by Edgar F. Smith in *J. Chem. Educ.*, **20**, 577-579 (1944); obituary in *J. Franklin Inst.*, **145**, 143-147 (1898).

HERBERT T. PRATT

Paul Nicholas Leech

1889-1941

Leech was born in Oxford, Ohio, Aug. 12, 1889. He received his A.B. degree from Miami University in 1910, his master of science degree from University of Chicago in 1911, and his Ph.D. in 1913. He also held the honorary degree of master of pharmacy from Philadelphia College of Pharmacy and Science. From 1911 to 1913 he was assistant in the Department of Chemistry at University of Chicago, working with Julius Stieglitz. Even following his work at the university he maintained a close personal and working relationship with Stieglitz. During his university career, Leech contributed to research in the fields of formic acid, molecular rearrangements, medicinal chemistry, and drug development.

In 1913 he joined the staff of the American Medical Association as a chemist and in 1923 became director of the chemical laboratory, working under William A. Puckner, who was the secretary of the AMA Council on Pharmacy and Chemistry. Following the death of Puckner in 1932, Leech became secretary of the Council on Pharmacy and Chemistry; and when the AMA Board of Trustees created the Division of Foods, Drugs, and Physical Therapy, he was appointed director.

In his position as secretary of the Council on Pharmacy and Chemistry, he edited many publications dealing with drugs, vitamins, glandular products, and pharmacy. Under his direction the AMA chemical laboratory became one of the early centers making use of microanalytical examination and spectrographic determination of drugs. Reports of analyses carried out in the laboratory appeared in *Journal of the American Medical Association*, other journals, and in "Nostrums and Quackery," and "The Propaganda for Reform in Proprietary Medicines," published by the AMA. The effective campaigns of the American Medical Association against worthless patent medicines and against every type of fraudulent medical claims as well as the constructive work of its Council on Pharmacy and Chemistry depended in large measure on the accuracy and reliability of the research work carried out in its chemical laboratory.

Leech was a councilor of the American Chemical Society from 1922 to 1936 and was chairman of the Chicago Section in 1926. Much of the established mode of operation of the section was due to Leech, and many of his ideas were broadly adopted. Throughout his society membership, he was an active proponent of giving more responsibility to local sections.

Leech was a charter member of the Chicago Chemists Club and trustee from 1925 to 1932. His contributions to medicine were recognized by associate fellowship in the American Medical Association and membership in the Institute of Medicine of Chicago. On several occasions, Leech represented the American Medical Association in the U.S. Pharmacopeial Convention. For more than a quarter of a century he served the physicians of this country and earned for himself a nationwide reputation for integrity, scientific judgment, and high ideals in the advancement of pharmacy and chemistry.

One of Leech's prime areas of interest was in the establishment of the Food and Drug Administration under the U.S. Department of Agriculture. This was exemplified by his logical, technical defense of the FDA in 1930 before the U.S. Senate Committee on Agriculture and Forestry.

Leech died in Chicago, Jan. 14, 1941.

Chemical Bulletin (publication of ACS Chicago Section), March 1941; *J. Amer. Med. Ass.* **116**, 327 (1941); information from American Medical Association.

ERNEST H. VOLWILER

Albert Ripley Leeds

1843-1902

Leeds was born in Philadelphia, June 27, 1843. After graduating from Central High School, he attended Haverford College and then Harvard College. Upon graduation from the latter in 1865, he became lecturer on chemistry in Franklin Institute and professor of chemistry in Philadelphia Dental College. To these duties he added in 1868 a similar position at Haverford College, where he was involved in establishing a laboratory for chemical analysis and research. The heavy work-load imposed by these positions impaired his health, forcing him in the fall of 1869 to take a 2-year rest. Leeds spent this period travelling in Europe and studying at the School of Mines and University of Berlin. On his return, he was asked to organize the Department of Chemistry at the newly established Stevens Institute of Technology. He became professor of chemistry and held this position for the rest of his life. He was an outstanding teacher who influenced generations of students by his zeal for research and his personal character. He was a man with wide-ranging interests, which included foreign languages, literature, and art.

His early research interests were in mineralogy and lithology. Subsequently, he investigated the properties of ozone and hydrogen peroxide and the effects of these substances as well as of chlorine and oxides of nitrogen on aromatic compounds. He also published several papers on photochemistry. Much of his later work was in the field of water analysis and sanitary science, and he became widely known in these fields. He introduced artificial, mechanical aeration of water supplies, and promoted mechanical filtration of water with the aid of alum as a means of purification. Still later, he concerned himself with the analysis of milk and various foods.

For many years, Leeds was actively engaged in helping various municipal water works solve their problems in water purification. He was chemist to the water boards of Newark and Jersey City and to the Hackensack Water Co. He investigated the water supplies of Albany, New London, Jamestown, Philadelphia, Reading, Plymouth, Wilmington, and Ottawa. In 1896 he made a detailed investigation of the water supply of the city of Brooklyn. He gave a start to the development of sanitary science in the state of New Jersey and served as a member of the State Board of Health and as chairman of its Council of Analysts. In recognition of these services, Princeton University conferred upon him an honorary doctor of philosophy degree.

Leeds was present at the meeting at which the American Chemical Society was organized in 1876. For many years he served as one of the vice-presidents and member of the ACS Board of Directors, in addition to being the chairman of the Publication Committee.

In later years, increasing deafness forced him to give up frequent attendance at meetings and class-room recitations, but he continued to lecture until a few weeks before his death, which occurred at his home in Philadelphia, Mar. 13, 1902.

G. C. Whipple, *J. Amer. Chem. Soc.* **24,** 53-57 (1902); *Stevens Institute Indicator* **19,** 105-109, 173-184, 304, 414 (1902); "The Morton Memorial: A History of the Stevens Institute of Technology," Stevens Institute of Technology (1905), pp. 223-228.

JOHANN SCHULZ

Henry Leffmann

1847-1930

"Henry Leffman was born Sept. 9, 1847 in Philadelphia. His ancestry on his father's side was partly Russian Jewish, on his mother's side partly Welsh Quaker. He was educated in the public schools, completing the course of B.A. at Central High School, Philadelphia, and subsequently receiving the degree of M.A. *honoris causa.*"

It is to be noted that the above paragraph lies within quotation marks because, when Henry Leffmann was once asked about his earlier days, he said he would like his biography to begin as quoted above. His reason was "in these days of genetics and eugenics it is worth while to show that a mongrel may have some merit."

To amplify Leffmann's own beginning statements, he was the fourth son of Henry Leffmann, a German Jew, and Sarah Ann Paul of Doylestown, Pa., a Hicksite Friend. His early education was completed within the Philadelphia public school system, and he should have qualified for graduation from the Central High School in 1861, but illness prevented him from being a member of that graduating class and he received his diploma in 1865. The Central High School in Philadelphia was unique in granting its graduates Bachelor of Arts degrees.

Following receipt of his high school certificate of graduation he became associated with the Jefferson Medical College in Philadelphia as assistant to the professor of chemistry there and continued in that position until 1870. In the meantime he undertook courses of instruction at Jefferson and was awarded his M.D. degree in 1869. Other degrees conferred upon him were an honorary Ph.D. from Wagner Free Institute of Science in 1874 and a Doctor of Dental Surgery from Philadelphia College of Dental Surgery in 1884.

Leffmann spent much of his professional life teaching, becoming a quiz master at Jefferson Medical College in 1870 and a lecturer on toxicology in 1875, a post that he held until 1880. He was a demonstrator of chemistry there during the 1884-85 session and was pathological chemist at Jefferson Medical College Hospital from 1887 until 1905. He had an extended teaching affiliation with Wagner Free Institute of Science of Philadelphia, being lecturer on botany during the 1874-75 session, lecturer on chemistry from 1875 to 1885, and professor of chemistry from 1885 to 1903. He became a member of the Board of Trustees at Wagner in 1903.

His other teaching commitments included the post of assistant professor of chemistry at Central High School from 1876 until 1880, professor of chemistry at the Artisans' Night School from 1877 until 1879, demonstrator of chemistry at Philadelphia College of Dental Surgery from 1882 until 1884, professor of chemistry at Philadelphia Polyclinic from 1883 to 1898, professor of chemistry at Philadelphia College of Dental Surgery from 1884 until 1899, professor of chemistry at Woman's Medical College from 1888 to 1916, and lecturer on research at Philadelphia College of

Pharmacy and Science from 1920 until his death.

Leffmann's professional duties within the community included those of chemist to the coroner of the city of Philadelphia from 1875 to 1880 and again from 1885 until 1897, microscopist of the Pennsylvania State Board of Agriculture from 1877 until 1905, port physician of Philadelphia from 1884 to 1887 and again in 1891-92, and member of the Assay Commission in 1889 and of the Philadelphia Quarantine Commission in 1892.

Leffmann's editorial and organizational expertise was utilized in his work as editor of *Medical Bulletin* in 1880-81, editor of *Philadelphia Polyclinic* from 1883 until 1888, president of the Engineers' Club in 1901, vice president of the Society of Public Analysts of Great Britain in 1901 and 1902, president of the Medical Jurisprudence Society in 1890, and membership on the Board of Directors of the Mercantile Library Company of Philadelphia starting in 1902.

He wrote articles on the history of the Revolutionary War period and a book, "About Dickens: Being a Few Essays on Themes Suggested by the Novels" (1908).

Leffmann had wide influence in public health through his more than 400 articles and several books. He was concerned with the examination of water, safeguarding of water supplies, disposal of waste, adulteration of food and liquor, inspection of milk, and other topics concerning the health problems of the citizen and community. His "Examination of water for sanitary and technical purposes" (1889), "Analysis of milk and milk products" (1893), and "Select methods in food analysis" (1901), all written with William Beam, went through several editions and were used widely.

Leffmann died in Philadelphia, Christmas 1930.

Joseph W. England, ed., "The First Century of the Philadelphia College of Pharmacy," 429-30, published by the college (1922); *Amer. J. Phar.* **103,** No. 3 (March 1931); *Ind. Eng. Chem.* **18,** 648-649 (1926); *J. Franklin Institute* **211,** 257-260 (1931); "Dictionary of American Biography" **11,** 142-3, Chas. Scribner's Sons (1933); "Outline Autobiography of Henry Leffman, A.M., M.D., Ph.D., D.D.S., of Philadelphia" (1905).

JOHN E. KRAMER

Phoebus Aaron Theodor Levene

1869-1940

Levene was born in Sagor, Russia, Feb. 25, 1869 and died in New York, Sept. 6, 1940. When he was about 2 years old, the family moved to St. Petersburg. He attended the Classical Gymnasium and later specialized in medicine at the Imperial Military Medical Academy in that city. He was one of the few Jewish students allowed to enter his class. He studied organic chemistry under Alexander Borodin's son-in-law, Alexander Dianin. About this time, in 1891, his family, because of growing anti-Semitism in Russia, decided to emigrate to the United States. The young Levene accompanied them but returned to St. Petersburg to complete his studies for the medical degree, then rejoined his family in March 1892.

Although he practiced medicine on the lower East Side in New York until 1896, Levene simultaneously carried out investigations in physiological chemistry in the laboratory of John G. Curtis at the College of Physicians and Surgeons of Columbia University. In 1896 he was appointed associate in physiological chemistry at the Pathological Institute of the New York State Hospitals but in November of that year he developed tuberculosis and went first to Saranac Lake, N.Y., and then to Davos, Switzerland. After a year of rest, he returned to the institute but his work was interrupted by the closing of the laboratory.

He then trained in Berne under Drechsel, in Marburg under Kossel, the authority on nucleins, in the electrochemical laboratory of H. Hofer in Munich, and at University of Berlin under Emil Fischer, who had established the fundamentals of carbohydrate and purine chemistry and who was at that time studying proteins and amino acids.

Following a study with Fischer on the hydrolysis of gelatin, Levene returned to New York to work in the chemical laboratory of the Pathological Institute, which had reopened under the direction of Adolf Meyer.

The Rockefeller Institute for Medical Research was founded about this time, and in January 1905 came the great opportunity for

Levene and the fortunate circumstance for American biochemistry, his appointment as head of the biochemical laboratory. The appreciation of his activities and genius was shown in 1907 when Levene was made a member of the institute in charge of the Division of Chemistry. In July 1939 he retired with emeritus status but continued his research into life processes.

The span of Levene's scientific activity stretched from his first paper of 1894 to a group of posthumous papers of 1941; over 700 in all, for the most part on the chemical structure of tissue constituents. His fields might be listed as: autolysis and enzymes, proteins and amino acids, especially nucleoproteins and glycoproteins.

Levene began his career with an investigation on the conversion of proteins into carbohydrates by the animal system. Two of his early studies of significance to protein chemistry were improved analytical methods developed with Donald Van Slyke and the isolation of the first crystalline intermediate products of protein hydrolysis with George B. Wallace. His isolation of prolylglycine anhydride among the products of the tryptic digestion of gelatin was a challenge to Fischer's peptide chain theory. Levene's systematic study of the racemization of a series of diketopiperazine rings and peptide chains revealed the difference in behavior of these two groups of compounds and demonstrated the validity of the peptide chain theory.

The major emphasis of Levene's protein studies, however, was on the non-protein constituents of the nucleoproteins and glycoproteins. The problem of the stereochemistry of the hexosamines caused him to evolve methods to determine their points of linkage and the elucidation of the ring structures of sugars and their derivatives. His technics were essential in unraveling the complicated formulas of nucleic acids. The identification of the sugar in thymonucleic acid was finally achieved in 1929. The solution of configurational relationships involved extensive studies on optical rotation, a field in which Levene won eminence, for he was outstandingly successful in applying the methods of physical chemistry to biochemistry.

Interspersed in these investigations were the studies on phospho sugars and phospho hydroxyamino acids. Systematic studies in the field of lipoids revealed that three main groups of these substances exist, two of which are phosphorus-containing. During the last years of his life, the chemistry of gums and pectins attracted his interest.

Levene's work was recognized by the award to him of the Willard Gibbs and the Nichols medals of the American Chemical Society. Many workers sought his advice on various problems, not only because of his brilliance but also because of his genial demeanor and generosity.

Biographies by Lawrence W. Bass, *Science* **92,** 392-395 (1940); Donald D. Van Slyke and Walter A. Jacobs, *Biog. Memoirs, Nat. Acad. Sci.* **23,** 75-126 (1944), with portrait; Donald D. Van Slyke and Walter A. Jacobs, *J. Biol. Chem.* **141,** 1-2 (1941); R. Stuart Tipson, *Advan. Carbohyd. Chem.* **12,** 1-12 (1957).

VIRGINIA F. MCCONNELL

Gilbert Newton Lewis

1875-1946

Lewis was born at Weymouth, Mass., Oct. 23, 1875. When he was 9 years old his family moved to Lincoln, Nebr., and he received his elementary education there. He attended University of Nebraska for 2 years but then transferred to Harvard University, from which he graduated in 1896. He taught for a year at Philips Andover Academy and then returned to Harvard for his M.A. in 1898 and his Ph.D. in 1899. After a year as instructor in chemistry at Harvard he went to Europe, studying at Leipzig and Göttingen in 1900-01. He came back to Harvard again as instructor from 1901 to 1906. However, for the year 1904-05 he was on leave to serve as superintendent of weights and measures in the Philippine Islands and chemist to the Bureau of Science in Manila. In 1907 he was appointed assistant professor of chemistry at Massachusetts Institute of Technology. There he rose to the rank of professor. In 1912 he was called to University of California as professor of physical chemistry and retained this position for the rest of his life. During World War I he held the rank of colonel in the Chemical Warfare Service and worked on defenses against war gases.

While at MIT Lewis began the work on thermodynamics and the determination of the free energies of many compounds to which he devoted most of his life. The results of his efforts were summed up in the well known text book "Thermodynamics and the Free Energy of Chemical Substances" which he published in 1923 with his colleague, Merle Randall. He was also responsible for notable advances in the theory of the chemical bond. As early as 1902 he had developed his theory of the cubical atom in a series of notes which he did not publish. In 1916 he expanded these ideas into one of the first satisfactory explanations of the nature of the non-polar chemical bond. His concept of the shared electron pair was highly original and led to many further advances in the electron theory of chemical structure. As in the case of his work on thermodynamics, he summed up his theories in a book, "Valence and the Structure of Atoms and Molecules," published in 1923. His versatility was also shown by his concept of acids and bases as electron pair acceptors and donors. Toward the end of his life he had begun to develop a theory of color.

Lewis was essentially a research chemist. He was called to Berkeley to stimulate a department which had devoted very little attention to graduate work. He succeeded brilliantly in carrying out his new responsibilities. Only four Ph.D. degrees had been awarded during the chairmanship of his predecessor, Willard Rising. Under Lewis 290 were granted. As part of his planning for his department he brought in a staff of very capable young instructors and arranged for the construction of a new chemistry building. He organized a weekly graduate seminar in which students and staff participated actively. He himself did not teach in the classroom, but he guided the course of the research seminar in such a way that it became one of the chief factors in the reorganization of his department. As one of the leaders in chemical investigations, Lewis was always very active in the laboratory, and it was while working in the laboratory that he died suddenly Mar. 23, 1946.

The best account of Lewis' life is by Joel H. Hildebrand in *Biog. Mem. Nat. Acad. Sci.* **31,** 210-235 (1958), reprinted from *Obituary Notices of the Fellows of Royal Society*. A shorter account of his life by G. Ross Robertson is in *Chem. Eng. News* **25,** 3290-3291 (1947).

HENRY M. LEICESTER

Winford Lee Lewis

1878-1943

Lewis was born in Gridley, Calif., May 29, 1878. When a grown man he wrote a paper entitled, "Why I Became A Chemist, And If So, To What Extent." In this he said of his early surroundings; "I was born and reared near a small California town of about 1,200 inhabitants. In this town there were eighteen saloons, one postoffice, a grocery store, a black-smith's shop, and several anemic churches, putting up a feeble resistance to the more firmly entrenched godless elements. The men patronized the saloons, the women the churches and the horses the blacksmith's shop."

He attended Stanford University and graduated in 1902. At Stanford he mainly studied law because his father practically drove him toward this profession. Lewis then studied chemistry at University of Washington, receiving his master's degree in 1904. He wrote; "My father was much displeased with my decision to study chemistry. In fact the word never passed his lips, and was a sad experience between us, for due to the fact that I had lost my mother very young, he was both father and mother to me. I never blamed him for I knew that his vision of a chemist was limited to the local purveyor of milk shakes and Radam's Microbe Killer at The Sign of the Red and Blue Bottles. I was never able to explain it all to him for his disappointment was so great that we could not discuss it."

Lewis moved from Washington to University of Chicago where he did research in the field of carbohydrates with John Nef. He obtained his Ph.D. degree in 1909.

For a short time he was an assistant chemist in the United States Department of Agriculture. In 1909 he went to Northwestern University as instructor in chemistry. He quickly discovered that Evanston had no city chemist. He volunteered to serve and the offer was immediately accepted. He was appalled at the contamination of Evanston's water supply. There was massive pollution and, very significantly, a high incidence rate of typhoid fever and related diseases. He presented his conclusions so forcefully to the city fathers that they built an up-to-date filtration plant.

He was promoted to assistant professor in 1913 and associate professor in 1917. For the duration of World War I he worked as a captain in the Chemical Warfare Service in Washington. It was here that he gained international fame as the inventor of Lewisite. Its discovery stemmed from finding a thesis of Nieuwland describing an experiment with acetylene and arsenic chloride which produced a toxic arsenic-containing mess. Lewis isolated the compound that his associates named Lewisite. Preparations to manufacture the compound were under way when the Armistice intervened. Lewisite was never used on the battlefield, and later developments in chemical warfare rendered it obsolete. The war over, Lewis proposed to leave the academic world and to accept a very attractive industrial position. But Northwestern called him back as professor and head of the Chemistry Department. When he accepted the chairmanship he stated that he had two principal aims: to stiffen the course in general chemistry and to increase graduate study. He succeeded in both aims. The building-up of Northwestern's Chemistry Department to what it is now, one of the country's great departments, was begun by Lewis.

In 1924 he accepted the invitation of The Institute of American Packers, (later the American Meat Institute) to serve as director of their newly organized Department of Scientific Research. He said he took the position not because of the increased remuneration but because he felt it was an excellent opportunity to teach a great industry the value of research.

While Lewis was head of chemistry at Northwestern he very successfully continued research on carbohydrates. A contemporary wrote; "The contributions of Lewis and his students have been widely accepted by carbohydrate chemists as classic."

His war work with arsenic compounds led him to continue research in this field. He worked out the mechanism of Lewisite formation and did other synthetic work with arsenic compounds, hoping to prepare substances of medical value. Even while he was scientific director of the Meat Packers' Institute, he spent evenings directing research by graduate students at Northwestern.

At the Meat Institute Lewis organized both a research and a service laboratory. The research he directed dealt with problems fundamental to meat packing. One was nitrite curing of meat which revolutionized the curing process. He also studied in a very original way the souring of hams, finding that offending bacteria even existed in the blood of live hogs. During his Meat Institute days Lewis cooperated with other active and socially inclined chemists to organize the Chicago Chemists's Club, which has flourished ever since. He was chairman of the Chicago Section of the American Chemical Society, 1919-20, and often wrote for its *Bulletin.*

Lewis' brilliance in science was accompanied by a mastery of English and a keen sense of humor. A close friend, the author of several successful books, said of him; "Lee Lewis would have made an equally enviable name for himself as a writer or public speaker. His choice of words was something at which I always marvelled."

Lewis was handicapped during his last years by a stroke. He died Jan. 20, 1943, as a result of injuries sustained when he fell from his porch roof.

W. Lee Lewis, "How I Became A Chemist," *Chem. Bull.* **11,** 100 (Chicago Section, ACS, publication) (1924); Lewis' essays and editorials in *Chem. Bull.;* scientific contributions in *Chem. Bull.* **30,** 221 (1945); obituary, *Chem. Bull.,* 47, Feb. 1943; personal recollections.

CHARLES D. LOWRY

Samuel Colville Lind

1879-1965

One of Tennessee's most famous scientists, Lind was born in McMinnville, Tenn., June 15, 1879. After graduation from the public schools in 1895 he entered Washington and Lee University. Partly to satisfy graduation requirements he enrolled in the beginning course in chemistry in his senior year. He found the subject so fascinating that after graduating in 1899 he remained another year, taking as much chemistry as was available. He continued his studies at Massachusetts Institute of Technology, receiving his bachelor's degree in 1902, and then studied physical chemistry at University of Leipzig in Wilhelm Ostwald's laboratory.

After receiving his Ph.D. degree in 1905 he returned to the United States and accepted an instructorship at University of Michigan where he remained until 1913. Here he became interested in radioactivity. He spent part of the year 1910-11 on leave in Madame Curie's laboratory in Paris and at the Radium Institute in Vienna. This experience sparked a lifetime interest in radiation chemistry.

To expand his interest he accepted in 1913 a position in the new U.S. Bureau of Mines laboratory in Denver, Colo., where he worked on the extraction of radium from carnotite ore, a potassium, uranium vanadate mineral contained in sandstone. At first most of the work was of an analytical nature which was followed by production procedures to separate out radium in the form of radium chloride from by-products vanadium and uranium. After proof that the carnotite was a valuable source of radium the work was continued under contract for an eastern corporation known as Radium Institute that desired the radium for therapeutic purposes. The by-product vanadium was sold to a steel company. The 30 tons of uranium oxide eventually was shipped east and dropped out of sight.

The total amount of radium produced in the form of radium chloride was 8½ grams which at that time had a market value of $120,000. The Radium Institute gave the bureau ½ gram which was entrusted to Lind for his radiation studies throughout his active life.

At the end of 3 years he moved as head of the bureau's station to Golden, Colo. where free from production problems he carried out radiation studies, particularly the synthesis of water by radon action on hydrogen and oxygen.

In 1920 the bureau station was moved to Reno, Nev. Here in addition to further radiation research he became interested in investigations on the coloring of diamonds by exposure to radon. He found it possible to convert yellow tinted diamonds to those of a desirable green color which greatly enhanced their value. The artificially colored, radioactive stones resulted in considerable turmoil in the legitimate gem market, and the work was dropped.

In 1923 Lind moved to Washington, D.C. where he became chief chemist in the bureau. In 1925 he accepted the position of assistant director of the Fixed Nitrogen Laboratory but resigned the next year to become director of the School of Chemistry at University of Minnesota. He was appointed dean of the Institute of Technology at the University of Minnesota in 1935, and he remained in this position until he retired in 1947.

Lind had too much energy to become inactive and in 1948 he moved to Oak Ridge, Tenn., as consultant to the Union Carbide Co. which was under contract with the Atomic Energy Commission to operate the plants and laboratories. In 1954 he assumed for awhile the duties of the director of the Chemistry Division of the Oak Ridge National Laboratory but continued with his research in radiation chemistry as senior consultant until the time of his death.

During his employment at the bureau stations he was continuously exposed to radiation and at one time to an excessive amount. In those days radiation shielding was not properly developed. It was believed that small amounts of radiation were beneficial to the human system. It was estimated that he permanently carried about ¼ microgram of radium in his body. However, there was no evidence that these exposures resulted in any serious after effects during his long life.

In addition to numerous scientific papers published from 1903 to 1964 he authored or co-authored three books concerned with radiation and electrochemistry. For many years (1933-50) he was editor of *Journal of Physical Chemistry,* and was on the editorial staffs of a number of other scientific publications. He received many honors, including doctor's degrees from four universities, the Nichols Medal (1926), and the Priestley Medal (1952). He was president of the American Electrochemical Society (1927), and president of the American Chemical Society (1940).

Lind was a friendly person of considerable charm. He enjoyed people with common interests regardless of their social background. His favorite hobby was trout fishing, and while pursuing this hobby he met his death in the Clinch River, Tenn., Feb. 12, 1965. He was 86 years old.

Charles A. Browne and M. E. Weeks, "A History of the American Chemical Society," ACS pub. (1952); American contemporaries sketch by A. D. McFadyen, *Chem. Eng. News* **25,** 1664 (1947); Memoirs of Samuel Colville Lind, *Jour. Tenn. Acad. Sci.* **67,** 1-40 (1972), with portrait; obituary, *New York Times,* Feb. 14, 1965.

EGBERT K. BACON

Arthur Dehon Little

1863-1935

Little was an innovator in chemistry in the era when laboratory research was first mixed with economic forces to quicken the rate of industrial change. He believed that scientific research could provide new opportunities for industry and a fuller life for mankind. He applied this conviction in the founding of the research, engineering, and management consulting firm that bears his name; in developing a new philosophy of chemical engigeering education; and in developing the idea of "unit operations" which sought for the first time to differentiate chemical engineering from industrial chemistry.

Little was born in Boston, Mass., Dec. 15, 1863 to Thomas Jones and Amelia (Hixon) Little. In 1881, he entered Massachusetts Institute of Technology where he studied chemistry and helped establish *Technology Review,* an alumni journal.

After graduation, he went to work for the Richmond Paper Co., at Rumford, R.I., the first company in the United States to use the sulfite process. Little soon became a leading authority on this means of making paper. While there, he secured his first patent, on a vessel known as a digester, used to soften pulp with chemicals.

In 1886, along with Roger B. Griffin, he opened Griffin & Little, an analytical chemistry firm in Boston. Competition in analyzing the substances that came through the busy port of Boston was keen. To diversify their services, Messrs. Griffin and Little served as paper mill consultants. At one time, they were retained by more than 65 mills to advise on basic chemistry and engineering and plant construction and to undertake the technical appraisal of processes and products. Together, the two men wrote "The Chemistry of Paper Making" (1894), the first American textbook on the subject.

Among Little's early projects were the production of casein by precipitation from skimmed milk and research on industrial applications of cellulose. The latter resulted in commercial processes for nonflammable wire insulation and nonflammable motion picture film. Fibers of cellulose acetate were first woven into a fabric by Little's associates.

After Griffin was killed in a laboratory explosion in 1893, Little controlled the firm until 1900 when he went into partnership for 5 years with William H. Walker as Little & Walker. In 1909 he reorganized the firm as Arthur D. Little, Inc., which became the largest chemical research firm in the United States.

As the scope of Little's organization broadened in the early 1900's, a policy of adding specialists to the staff was inagurated so that the company could respond better to the increasing interests of industry in applied science. Investigations conducted in this period included the production of paper from bagasse, the use of bagasse mulch paper to keep down weeds in Hawaiian sugar and pineapple fields, the production of alcohol from wood waste, and the use of Southern long-leaf yellow pine for making kraft paper.

In the early 1920's, Little began focusing his and his company's attention on petroleum though industry was deriving most of its organic chemicals from coal. He envisioned a great industry based on ethylenes, butylenes, and propylenes and foresaw possibilities of tailor-making fuels and developing gasoline's antiknock qualities. Meanwhile he began to recognize the commercial value of food research, and he added kitchen laboratories to his company to pursue scientific study of odor and flavor.

Little was honored for his contributions to chemistry on many occasions by the award of honorary degrees, the Perkin Medal in 1931, and other prizes. He was elected president of the American Chemical Society in 1912 and 1913, of the American Institute of Chemical Engineers in 1919, and of the Society of Chemical Industry in Great Britain in 1928.

He retired as chairman of the board of Arthur D. Little early in 1935. He died at Northeast Harbor, Maine of a heart attack, Aug. 1, 1935.

Obituary by F. G. Keys in *Proc. Am. Acad. Arts Sci.* **71**, 513-519 (1936-37); "Dictionary of American Biography," **21** (Supplement 1), 500, Chas. Scribner's Sons (1944); sketch by W. Haynes in Eduard Farber, "Great Chemists," *Interscience* (1961), pp. 1192-1201, with portrait; records of Arthur D. Little, Inc.

GEORGE W. BAKER

John Uri Lloyd

1849-1936

Lloyd was born Apr. 19, 1849 in West Bloomfield, N.Y., the son of Nelson and Sophia (Webster) Lloyd. When the boy was four the family moved to Kentucky because his father, a civil engineer, contracted to make a survey in connection with a proposed railroad. Financial considerations forced a halt to the project, and Mr. and Mrs. Lloyd turned to operating a school in their home as a source of income. Subsequently, they taught in the Kentucky public schools.

John's early education was largely obtained at home, and he never received the benefit of a high school or college education. At the age of 14, he was apprenticed to a pharmacist in Cincinnati and spent the next few years of his life working as an apprentice and later as a drug clerk. In 1871, at the urging of an eclectic physician John King, he joined the pharmaceutical firm of H. M. Merrell of Cincinnati as a chemist. This company specialized in the botanical drugs favored by practitioners of the eclectic sect. Eventually Lloyd and his two brothers gained control of the company, which became known as Lloyd Brothers.

Lloyd was a versatile and energetic man who involved himself in many other activities besides his business. He served as professor of chemistry at Cincinnati College of Pharmacy from 1883-1887 and at Eclectic Medical Institute from 1878-1907. He and his brothers created the Lloyd Library in Cincinnati, noted for its collections in pharmacy, botany, medicine, natural history, and related fields. A prolific writer, his thousands of publications include scientific papers and books, historical studies, and works of fiction. His novels, such as "Stringtown on the Pike" (1901), dealt with Kentucky life and folklore, except for the fantasy "Etidorpha" (1897), which exhibited a curious mixture of science and metaphysics. Among the important pharmaceutical textbooks and reference works he authored were "Drugs and Medicines of North America" (two volumes, 1884-87; in collaboration with his brother, C. G. Lloyd), "Origin and History of All the Pharmacopoeial Vegetable Drugs, Chemicals, and Preparations," Vol. I, "Vegetable Drugs" (1921) and "The Chemistry of Medicines" (1881).

He also managed to find time to pursue scientific research. His commitment to improve upon and expand the vegetable materia medica led him to investigate the medicinal and chemical properties of numerous indigenous plants. He was especially interested in separating the fundamental constituents of plant substances without destroying their integrity and searched for relatively gentle manipulative processes which would enable him to accomplish his goal. In this connection, he developed and patented his "cold still," which extracted the soluble constituents of plants with a minimum of heat. He also discovered the ability of hydrous aluminum silicate (Lloyd's Reagent) to adsorb alkaloids from solution. His research often found practical application in the production of the "specific medicines" marketed by Lloyd Brothers.

Physical chemistry was still in its infancy in the late nineteenth century, and Lloyd was one of the first to apply this science to pharmaceutical techniques. In a series of papers published in *Proceedings of the American Pharmaceutical Association* from 1879 through 1885, he discussed adsorption, capillarity, and other physical phenomena in relation to the preparation of fluid extracts. These papers were republished by Wolfgang Ostwald, one of the founders of the science of colloid chemistry, in his journal, *Kolloidchemische Beihefte,* in 1916. Ostwald felt that they contained material which was significant, original, and pertinent to the study of colloidal phenomena.

Lloyd was married twice and had three children by his second wife. He died April 9, 1936 in Van Nuys, Calif.

"Dictionary of American Biography," **22** (supplement 2), 389-390, Chas. Scribner's Sons (1958); Roy Bird Cook, *J. Amer. Assoc., Practical Pharmacy Edition,* **10,** 538-544 (1949), with portrait; George Beal, *Amer. J. Pharm. Educ.,* **23,** 202-206 (1959); several biographical articles were reprinted in the May, 1936 memorial issue (vol. 96, no. 5) of *Eclectic Med. J.,* with portraits; Corinne Miller Simons, "John Uri Lloyd, His Life and Works, 1849-1936, with a History of the Lloyd Library" (1972).

JOHN PARASCANDOLA

Morris Loeb

1863-1912

Loeb was born May 23, 1863 in Cincinnati, Ohio, but the family moved to New York while Morris was still quite young. His father, Solomon Loeb, was one of the most prominent financiers in the country, being founder of Kuhn, Loeb & Co. Morris received his primary education under Julius Sachs and entered Harvard University when he was 16. He graduated *magna cum laude* in 1883 with "Honorable Mention" in chemistry and English. The same year he went to University of Berlin to study under August Wilhelm von Hofmann and received his Ph.D. degree in 1887. The ensuing year was spent at the Universities of Heidelberg and of Leipzig, studying physical chemistry under Wilhelm Ostwald and Walter Nernst. He was Ostwald's first American pupil.

Upon returning to America in the fall of 1888, Loeb easily could have joined his father's firm but chose instead to become a volunteer assistant to Wolcott Gibbs who had retired from the Rumford professorship at Harvard to establish a private chemical laboratory.

In 1889 Loeb was appointed lecturer in physical chemistry at Clark University and, as such, became one of the pioneers in America in this new field of science. In August of the following year he was one of those presenting papers at the first general meeting of the American Chemical Society which was held at Newport, R.I.

In 1891 he was elected to a professorship at New York University and by 1894 had developed a Department of Chemistry from what had been a single year's course. The new Havemeyer Chemical Laboratory was completed under his direction in 1894. A complete curriculum in chemical engineering was adopted in 1898. Loeb's influence extended beyond his own department and the group system of teaching, as opposed to free electives, was adopted largely through his initiative. Loeb resigned from the university in 1906 at age 43.

Loeb's resignation from the university was not because of a lack of interest in chemistry but rather because of pressure of other demands on his time. Although he was a member of a half dozen professional organizations, his favorite enterprise was the Chemists Club of New York of which he was president twice—1909 and 1912. During his first presidency, plans were laid to house the club in its own building. Loeb envisioned not only a social center but a center of chemical knowledge complete with library, small laboratories for individual use, a museum of new chemical substances, and living quarters for transient chemists. He donated land on 41st Street valued at $175,000 and was one of the chief stockholders in the 10-story building. He eventually gave all of his shares to the club.

Loeb was author of 30 varied papers, reviews, and essays as well as a "Laboratory Manual of Inorganic Chemistry." His chemical publications covered the reactions of phosgene, dye chemistry, gas analysis, analysis of bronzes, the conductivity of silver salts, the effect of magnetism on chemical action, electrolytic dissociation, and the use of osmotic pressure in the determination of molecular weights. He showed that the molecular weight of iodine increased continuously with concentration so that there was no point in the narrow limits between extreme dilution and saturation at which molecular weight was constant. He perfected a more accurate method for measuring weak electrical currents by depositing silver on an asbestos bottomed Gooch crucible instead of the platinum crucible normally used. In studies of the crystallization of sodium iodide from aliphatic alcohols, he showed that the molecular proportion of alcohol assimilated by the iodide decreased as the series ascended.

Loeb was a staunch supporter of his alma mater. In 1909 he served on the committee appointed by the Board of Overseers of Harvard College to report on the needs of the chemical laboratory. It was upon his initiative through a gift of $50,000 that the Wolcott Gibbs Memorial Laboratory for research in physical and inorganic chemistry was built. He also designated $500,000 in his will to be used by Harvard for the advancement of chemistry and physics.

Loeb was the first secretary of the New York Section of ACS and was chairman in 1909. He also served as an abstractor in industrial chemistry for *Chemical Abstracts.* In 1912 he was a delegate to the Eighth International Congress of Applied Chemistry meeting in New York, and some believed that his death from typhoid and pneumonia Oct. 8, 1912 was the result of overexertion on behalf of the Congress.

Loeb was married to Edna Kuhn of Cincinnati on April 3, 1895; they had no children. Much of his life was given to charitable work, and he was especially devoted to associations which sought the improvement of the great mass of Hebrew poor in New York. At sundry times he was either member, office holder, trustee, or creator of five schools and organizations devoted to furthering the musical, theological, agricultural, or technical education of Jewish people. His only political activity was one term on the New York School Board. He was awarded a D.Sc. degree by Union College in 1911.

T. W. Richards, "Scientific Work of Morris Loeb," Harvard University Press (1913); Arthur E. Hill, Morris Loeb as a Teacher, in "Morris Loeb, 1863-1912, Memorial Volume," The Chemists Club (1913); "The National Cyclopedia of American Biography," **26,** 10, James T. White & Co. (1937); obituary, *J. Ind. Eng. Chem.* **3,** 846-849 (1912).

HERBERT T. PRATT

John Harper Long

1856-1918

Long was born near Steubenville, Ohio, Dec. 26, 1856 and was orphaned in boyhood. In 1873 he enrolled in the newly organized University of Kansas and by the beginning of his sophomore year an inspired teacher had aroused in him an absorbing interest in chemistry. In 1879 he graduated and in the same year published in the *Transactions of the Kansas Academy of Science* his first scientific paper entitled, "A Method for Determining the Velocity of the Wind."

Like many other American students preparing for careers in chemistry, Long went to Germany for further study. He spent 3 years in Tubingen working with the "generous, helpful and wise" Lothar Meyer. In 1880 he returned to the United States and for a year was assistant to Wilbur Atwater, a pioneer in agricultural and food chemistry at Wesleyan University. In the next year he began his long association with Northwestern University. At first he was a teacher of chemistry in the College of Liberal Arts in Evanston and also gave some instruction in chemistry to students in the Medical School. He was soon made a full time member of the Medical School faculty, an association which lasted until the end of his life.

Although the dependence of medical science on chemistry was beginning to be recognized, the teaching of chemistry in medical schools at this time was "lamentable." Long, himself, said in an address before the American Association for the Advancement of Science, "It is hard to tell who has been most to blame for the warped and stunted conception of chemistry held even at the present time by the great majority of medical men in this country. It is likely that much of the fault lies in the weak and wholly unsatisfactory manner in which chemistry has been presented in most of our medical schools. A professor of chemistry was usually a physician who, as a rule, was not considered sufficiently strong to fill a chair of practice, obstetrics, or surgery, but who might teach acceptably the less important branch of chemistry." The first systematic laboratory course in the Medical School was Long's course in chemistry, and it was principally through his untiring efforts and influence that the departments of anatomy, physiology, bacteriology, pathology, and pharmacology were reorganized and gradually enabled to develop laboratory courses with men in charge who devoted their whole time to teaching and research.

Long devoted his life to developing the chemical phases of medicine. He published a number of books on general chemistry, qualitative analysis, volumetric analysis, and physiological chemistry. In addition he published over 100 papers, on subjects ranging from pure organic, inorganic, physical, and biochemistry to the analysis of sewage.

He served as president of Chicago Institute of Medicine. He was a member of the Council of Medicine and Pharmacy of the American Medical Association. He was consulting chemist to the State Board of Health of Illinois and the Board of Health in Chicago. He was active in the organization of the Congress of Chemists at the World's Columbian Exhibition in 1893, a congress which led to the formation of a Chicago section of the American Chemical Society in 1895 with Long as treasurer. He was president of the ACS in 1903.

For 20 years Long was concerned with the problem of sewage disposal for the city of Chicago. The city reversed the flow of the Chicago River and built a drainage canal which swept the whole mass of sewage down the Illinois and Mississippi rivers. St. Louis and other downstate cities feared for the safety of their water supplies. They brought suit against Chicago to prevent this method of sewage disposal. Long was engaged as a consultant and made a painstaking and lengthy study of the condition of the flowing water downstream from Chicago. He sampled the water at a number of stations all the way from Chicago to St. Louis, personally collecting the first samples from each location. In 1905 the Supreme Court ruled in favor of Chicago—strongly influenced by the years of study, by Long, of the destruction of sewage in running streams.

Long was a member of the Referee Board of the Department of Agriculture of which Ira Remsen was chairman. Remsen wrote of Long, "If the services of the Referee Board were valuable to the Country a large share of the credit is due to Dr. Long. His industry, conscientiousness and wisdom were constantly in evidence and we who worked with him gratefully acknowledge the value of his cooperation."

Long died June 14, 1918 after an illness of several months.

Chemical Bulletin **5,** 126 (1918); "John Harper Long," a brochure of tribute by his associates, published by Northwestern University (1918); *J. Amer. Chem. Soc.* **41,** 69-82 (1919); *Bull. Univ. Kansas* **26,** 11-12 (1925); J. H. Long, "Some points on the early history and present condition of chemistry in the medical schools of the United States," *Science* (n.s.) **14,** 360 (1901).

CHARLES D. LOWRY

Howard Johnson Lucas

1885-1963

Lucas was born in Marietta, Ohio, Mar. 7, 1885. His early years were spent in Ohio, and he received his B.A. (1907) and M.A. (1908) degrees in chemistry from Ohio State University. In 1953 Ohio State honored him with the doctor of science degree. Lucas spent 2 years as a graduate instructor and a year as a fellow at University of Chicago before joining the Chicago Bureau of Chemistry, U.S. Department of Agriculture, as a chemist in 1910. His interest in the chemistry of food continued after he left the Chicago Bureau, and he later served as a consultant to Kelco Co., to whom he assigned a patent concerned with the processing of dairy products (1942).

Lucas joined the faculty of Throop College of Technology (now California Institute of Technology) as an instructor in 1913 and was promoted to associate professor in 1915 and to professor in 1940. For more than 40 years, Lucas was in charge of the undergraduate organic chemistry course at Cal Tech. He enjoyed working with undergraduate students and spent many hours outside of class with them. He was the sponsor of the chemistry club which later became a student affiliate chapter of the American Chemical Society. In recognition of his outstanding contributions as a teacher, Lucas received the Scientific Apparatus Makers Award in Chemical Education in 1953.

Although Lucas never sought a Ph.D. degree himself, he directed the Ph.D. dissertation research of several outstanding chemists. He was a brilliant pioneer in the field of physical organic chemistry, and his painstaking care in experimentation as well as his logical interpretation of data in terms of modern electronic theory served as models for later work.

Beginning in 1924, Lucas published a series of papers in which he explained the course of many fundamental reactions, such as addition of hydrogen halides to olefins, dehydrohalogenation, and aliphatic and aromatic substitution, in terms of "electron displacement" which had been proposed by Gilbert N. Lewis to explain the variation in strength of organic acids. Lucas studied simple model systems, for which he used scrupulously pure substrates and precise analytical techniques, which was unusual in his time. In analysis of product mixtures, Lucas used not only fractional distillation and refractive index measurements but differences in chemical reactivity and dielectric constant. The practical problems associated with analyses of mixtures of isomeric alcohols led him to the discovery of the Lucas reagent (hydrochloric acid–zinc chloride) in 1930. In 1926, Lucas proposed that the electron attraction of radicals was best measured by the dissociation constants of the para-substituted benzoic acids.

In 1934 Lucas reported the first kinetic study on the acid-catalyzed hydration of an olefin, a topic which generated much interest in subsequent years.

The most important of Lucas' research was carried out in the decade 1935-45. In a series of elegantly conceived and carefully worked experiments, Lucas and his students, notably Saul Winstein, presented evidence for the existence of cyclic halonium ions which were formed in the course of substitution reactions of 3-halo-2-substituted butanes and similar substrates. Such intermediates had been postulated by Roberts and Kimball to explain the course of halogenation of olefins, and Lucas provided many examples of the importance of these species in substitution reactions. The cyclic halonium ions were required to explain the stereochemical consequences of the substitutions, and the studies demonstrated the importance of stereochemical information in elucidating reaction mechanisms. Lucas described the cyclic ions in terms of the resonance theory which had recently been developed by his colleague Linus Pauling.

Lucas' interest in understanding organic chemistry in a rational manner based on electronic theory was shown in his widely-used undergraduate textbook published in 1935. He published a laboratory manual coauthored by David Pressman in 1949.

After World War II Lucas devoted less time to research. He served as dean of the faculty at Cal Tech and was appointed professor emeritus in 1955. He maintained a deep interest in chemistry after his retirement. He died June 25, 1963.

Anon., *Chem. and Eng. News* **31,** 778 (1953); information from William G. Young.

MARTIN FELDMAN

Hiram Stanhope Lukens

1885-1959

Lukens was born in Philadelphia, Pa. Sept. 30, 1885 the only son of Joseph C. Lukens, a merchant, and his wife, M. Louisa (Stanhope) Lukens. He attended a private school and later the public schools in Philadelphia. He received his B.S. and Ph.D. degrees in chemistry from University of Pennsylvania in 1907 and 1913. In 1905 and 1906 he was a member of the university lacrosse team. His free time was occupied with other outdoor sports—cricket, golf, soccer, ice skating, and sailing. The latter sport remained as a hobby until the time of his death.

Upon graduation he was appointed instructor in chemistry. In 1913 he advanced to the rank of assistant professor, professor in 1921, and became Blanchard Professor of Chemistry in 1937. In 1932 Lukens was named director of the Department of Chemistry and Chemical Engineering, a position he held until 1952. During the 1937-38 period he also served as acting dean of Towne Scientific School. In 1953 he retired after serving the university for a total of 45 years.

During his long service to the university, he was a counselor to thousands in his profession, confidant of hundreds, and a real friend to any who sought his advice or help in any matter. His greatest pride was to head a department which produced capable and potentially successful undergraduate and graduate students with a sound academic training. Lukens was a modest and quiet-spoken individual, who devoted much of his extra time and energy to professional societies, fraternities, and clubs for many years.

He was a member of the American Chemical Society, serving on many Philadelphia Section and national committees over the years. He was chairman of the Philadephia Section in 1924-25, one of its trustees from 1933 to 1948, and a national councilor from 1926 to 1947. He was a member of the Division of Chemistry and Chemical Technology, National Research Council, from 1934 to 1937.

Lukens was also an active member of the American Institute of Chemical Engineers. He served on the executive committee of the Philadelphia-Wilmington Section from 1933 to 1939, was its vice-chairman in 1938, chairman in 1939.

The Electrochemical Society also claimed his support and action. He served the Philadelphia Section as secretary from 1930 to 1931, was chairman from 1931 to 1932, a member of its executive committee from 1930 to about 1950, and chairman of the electrochemical division from 1931 to 1932. Lukens became the president of the national society in 1934-35.

Lukens taught a limited number of undergraduate and graduate courses in areas of inorganic, analytical, and electrochemistry. He directed the research of 10 graduate students, all of whom received their Ph.D.'s. In addition he had 23 patents (U.S., Canadian, British and German) to his credit, nearly all of them in the electrochemical field, from 1925 to 1938, and 20 publications, plus at least eight others (Ph.D. dissertations), on which he would not allow his name to be listed as co-author.

Nine of his patents dealt with magnesite cements, the remainder with pen points and methods for gold plating chromium alloy steels.

His publications included electrolysis studies of the halides of the alkaline earths, electrodeposition of chromium and cadmium, deposition of metals on aluminum from fluoride solutions, the use of mercury and tantalum as cathodes, the electrochemical oxidation of benzene and the reduction of nitrobenzene, the effect of oxygen on the electrodeposition of copper, and atomic weight determinations of scandium, mercury, and phosphorus.

Outside of the university he served as technical director of Royal Electrotype Co. from 1918 to 1932 and vice-president and technical director of Solidon Products, Inc. from 1923 to 1932.

Lukens received the ACS Philadelphia Section service award in 1950 and the Philadelphia Professional Chapter, Citation of Merit Award from Alpha Chi Sigma in 1959.

As a memorial tribute to Lukens' academic, administrative and professional record, and his devoted life to chemistry and chemical education, many loyal and liberal contributors made possible the establishment of a fund known as the Hiram S. Lukens Memorial Scholarship Fund, the income from which, in the form of a tuition scholarship, is awarded to a senior student with a major in chemistry at University of Pennsylvania.

Lukens died on Nov. 25, 1959. He was survived by a wife, Marguerite (Perrine), and a son, William Perrine Lukens.

Records in department of chemistry, Univ. of Pennsylvania; personal recollections; obituaries in *J. Electrochem. Soc.* **107,** No. 3, 81C (Mar. 1960); *The Catalyst* **45,** 5 (1960); *The Philadelphia Inquirer,* Nov. 26, 1959.

CLAUDE K. DEISCHER

M

Charles Frederic Mabery

1850-1927

To Mabery belongs the distinction of first putting petroleum chemistry on a sound quantitative footing in the nineteenth century; but this is only one aspect of the life of a man who built the Chemistry Department at Case Institute of Technology and had a hand in founding the electrolytic aluminum and bromine industries as well.

Mabery was born Jan. 13, 1850 in New Gloucester, Maine, son of Henry and Elizabeth A. Mabery. He was educated in the public schools of the area and thereafter taught in the local common schools and academies until 1873. By that time his teaching of chemistry, natural philosophy, and mathematics at Gorham Seminary had awakened in him a desire for further education. He went to Harvard University's summer school to study chemistry and in the fall of 1873 enrolled in Lawrence Scientific School to work under Josiah P. Cooke. Mabery received his B.S. degree in 1876 and his degree Sc.D., also at Harvard, in 1881. While a graduate student he was appointed assistant teacher in the Harvard chemical laboratory and developed summer courses in chemistry, which were attended by teachers from all over the country.

In 1883 Mabery accepted a professorship at Case Institute of Technology, where he built the Chemistry Department and chaired it until his retirement. The school was brand-new at the time; it had been founded in 1880 and commenced instruction in 1881. Classes were then taught in the old Case homestead, with chemistry housed in a brick barn. Conditions were primitive and difficult; the permanent chemistry laboratory, designed by Mabery, was not completed until 1892. Nonetheless Mabery, his assistants, and in many cases his students, carried out research and produced a series of papers striking in their breadth of interest. Long after his retirement from active teaching in 1911 Mabery continued his schedule for full-time laboratory work, pausing only in the summers, which he and his wife spent in Maine. Mabery's final papers were still appearing in 1927, the year of his death.

Mabery's earliest investigations, dating from his time at Harvard, dealt mainly with unsaturated and halogenated organic acids. During the 1880's his output was rather slender, probably because of the Spartan conditions in the early days at the Case School. Still, he became interested in electrochemistry, acted as consultant to the Cowles brothers in their electrolytic production of aluminum, was associated with them in patents, and published his own works as well, analyzing products from their electric furnace operations and working out a method for preparing anhydrous aluminum chloride.

In the 1880's Mabery became interested in petroleum chemistry, both sources and products. The oil industry, begun in the 1850's with the discovery of oil in northwestern Pennsylvania, gradually shifted its center to Cleveland, where John D. Rockefeller incorporated the original Standard Oil Co. in 1870. When Mabery started his investigations in the 1880's, the industry's approach was still largely empirical and non-chemical. Mabery carefully analyzed crude oils from all over the world, with particular attention to sulfur and nitrogen content, hydrocarbon distribution, and lubricant properties and over a period of three decades became an authority in this field. His activities in oil chemistry led him into some interesting byways: examination of the salt brines often associated with petroleum deposits produced a publication with one of his students, Herbert H. Dow, which the latter subsequently developed into the Dow process for recovering bromine. Water chemistry of this kind further led Mabery into analysis of minerals in ground waters, which was of use in the problem of public water supplies. Mabery's lifetime output ran to over a 100 papers, of which nearly 60 were devoted to petroleum chemistry.

Mabery was married in 1872 to Frances A. Lewis, of Gorham, Maine; their only child died in infancy. The Maberys were active patrons of the arts in Cleveland, taking part in the development of the Cleveland School of Art and the Cleveland Art Museum. In the winter of 1926 Mrs. Mabery died following an illness; Mabery survived her by less than a year, succumbing on June 26, 1927.

"Dictionary of American Biography," **11,** 540, Chas. Scribner's Sons (1933); "National Cyclopedia of American Biography," **10,** 411-12, James T. White & Co. (1900); W. R. Veasey, *J. Chem. Educ.* **5,** 1117-20 (1928); A. W. Smith, *Ind. Eng. Chem.* **15,** 314 (1923); *Ind. Eng. Chem. News Ed.* **5,** 5 (1927); *Proc. Am. Acad. Arts and Sci.* **78,** 5-6 (1949); "J. C. Poggendorff Biographisch-literarischer Handworterbuch für Mathematik, Astronomie, Physick, Chemie und verwandte Wissenschaftsgebiete," **5,** 781, Verlag Chemie (1926).

Robert M. Hawthorne Jr.

Duncan Arthur MacInnes

1885-1965

MacInnes was one of the leading electrochemists that the Western Hemisphere has produced. In an active scientific life extending over a period of 47 years, he published over 100 papers on electrochemistry and a book, "Principles of Electrochemistry," that was used widely.

MacInnes was born in Salt Lake City, Utah, Mar. 31, 1885 to Duncan and Frances Charlotte (Sayers) MacInnes. At the age of 13 he had a nearly fatal streetcar accident in which he lost two fingers on his left hand and injured one leg very badly. At 16 he entered the preparatory school of University of Utah and had the good fortune to board at the same house with Solomon Acree, later to become head of the pH Standards Division of the National Bureau of Standards. This early association with an active research scientist developed in him an interest in chemistry, and in 1903 he enrolled at Utah in a chemistry class taught by William Ebaugh, who continued to stimulate his growing desire for a career in chemistry. He lived in Utah until he graduated from University of Utah in 1907 with a B.S. degree in chemical engineering.

He completed his research for his Ph.D. degree in chemistry at Illinois under Edward Washburn in 1911 with a thesis on aqueous salt solutions. He accepted an appointment as instructor in physical chemistry at Illinois and remained there until 1917. During this time he published nine papers, five of which were in the field of electrochemistry. In 1917 he moved to Massachusetts Institute of Technology in the Research Laboratory of Physical Chemistry directed by Arthur A. Noyes where he remained for 10 years. During his M.I.T. period he published 14 papers concerned with electrochemistry, several of them in collaboration with men like Noyes, James A. Beattie, Edgar R. Smith, and Theodore Shedlovsky. During this period he digressed briefly from his concentration on electrochemistry and worked on x-rays part of the time.

In 1925 MacInnes took a sabbatical leave from M.I.T. for study and travel in Europe. He made his headquarters in Paris and visited several laboratories from there. For 2 months he visited workers in physical chemistry in England, particularly F. G. Donnan.

In 1926 he was invited to join Rockefeller Institute for Medical Research in New York City, where he remained for the remainder of his life, becoming a member in 1940 and emeritus member in 1950. He was freed by this change from formal teaching duties and provided with a generous budget and technical assistance for research.

Among MacInnes' important contributions to electrochemistry were his work on the moving boundary method for determining transference numbers, potentiometric methods of analysis, the glass electrode and measurement of acidity (pH), ionization constants, and activity coefficients. MacInnes himself considered perhaps his most important work to be his experimental confirmation of the Debye-Hückel theory of interionic attraction in solutions of electrolytes. He was an early believer in the idea that strong electrolytes are completely ionized in aqueous solutions and that the Debye-Hückel theory explained best the departure of their behavior from the ideal.

MacInnes was very active in the New York Academy of Sciences, being elected president of that society in 1944. He worked with Detlov Bronk and others in setting up a series of small conferences in the period from 1945 to 1953 at Rams Head Inn, Shelter Island, Long Island. These Academy Conferences were limited to about 25 participants and allowed sufficient time for discussion of the papers presented. MacInnes considered the role he played in initiating these conferences to be one of his major contributions to science in the United States.

The Electrochemical Society elected him president 1935-37 and awarded him its Acheson Medal in 1948. He also received the Nichols Medal of the American Chemical Society's New York Section in 1942. During World War II he was director of a group which did work for the Chemical Warfare Service, for which he was awarded the Presidential Certificate of Merit in 1948. He was elected to the National Academy of Sciences in 1937.

MacInnes was an active mountain climber and a member of the American Alpine Club and the Appalachian Mountain Club. He died Sept. 23, 1965.

Biography by Lewis G. Longsworth and Theodore Shedlovsky, *Biog. Mem. Nat. Acad. Sci.* **41**, 295-317 (1970).

JOSEPH A. SCHUFLE

Edward Mack, Jr.

1893-1956

Mack was born May 10, 1893 at Goldboro, N.C. He attended University School in Cincinnati and entered Centre College in 1909. In 1911 he transferred to Princeton University, from which he graduated B.A. *magna cum laude* in 1913 with highest honors in chemistry. He continued at Princeton and specialized in physical chemistry under the guidance of George A. Hulett. He received his M.A. degree in 1914 and his Ph.D. in 1916. From 1915 to 1917 he was an instructor in chemistry at West Virginia University. At the outbreak of World War I he entered the army as a first lieutenant and was promoted to major by the end of the war.

In 1919 Mack was appointed assistant professor at Ohio State University and placed in charge of the courses in physical chemistry. He was promoted to a professorship in 1925. He was elected head of the Department of Chemistry at University of North Carolina in 1935 and remained there until 1939 when he left to assume charge of research education at Battelle Memorial Institute in Columbus. In 1941 he was elected chairman of the Department of Chemistry at Ohio State University. He was granted a leave of absence in 1943 to direct research in certain phases of the Manhattan Project. In 1946 Mack resumed his duties as chairman at Ohio State and so continued until 1955 when he resigned the chairmanship with the intention of returning to teaching and to a program of research on biochemical problems sponsored by the Kettering Foundation. Ill health intervened, and he died June 4, 1956.

Mack was married to Louise Matson of Columbus, Ohio, Aug. 23, 1921. To them were born a son and two daughters.

Prior to becoming involved in administrative work, Mack was very actively engaged in research in pure physical chemistry and in directing the researches of students working toward graduate degrees in that field. His own researches and those of his students were characterized by their originality and diversity. Among the topics investigated were the determination of the cross-sectional areas of molecules by gaseous diffusion, the calculation of the flash points of organic compounds, the influence of centrifugal force on rate of evaporation, the vapor pressures of copper and copper oxide, solid solutions of alkali and ammonium halides, and the spacing of molecules in crystal lattices. Although he continued to direct the researches of graduate students to some extent after he became chairman at Ohio State, both this activity and his participation in class teaching gradually dwindled with the increase in the burdens of this office.

Mack was very active in the affairs of the American Chemical Society. He was at various times chairman and counselor for the Columbus Section and the North Carolina Section and also served as councilor-at-large. For various periods he was an assistant editor of *Chemical Abstracts* and an associate editor of *Journal of the American Chemical Society* and *Journal of Physical Chemistry*.

He was the co-author of various textbooks, all of which were successful. His first book, with Wesley G. France, was "A Laboratory Manual of Elementary Physical Chemistry," the first edition of which was published in 1928. The others were textbooks in general chemistry that appeared in various editions beginning in 1940.

As a chemistry department chairman, Mack worked with an unusual degree of energy and unselfish devotion. He was an ardent champion for his colleagues and for the advancement of chemistry. He was especially sensitive to the needs of students, both undergraduate and graduate. To many of these who found themselves in financial difficulty he loaned, and even gave, money out of his own pocket. It was especially appropriate that after his death colleagues and friends established at The Ohio State University the Edward Mack Memorial Loan Fund, which over the years has been of great help to needy students. His work in chemistry and for the advancement of chemistry was recognized by various honors such as honorary doctorates from University of North Carolina (1944) and Centre College (1949).

Earle R. Caley, "History of the Department of Chemistry," The Ohio State University (1969); personal recollections.

EARLE R. CALEY

John William Mallet

1832-1912

Mallet was born in Dublin, Ireland, Oct. 10, 1832. His father, a fellow of the Royal Society, encouraged him toward science and set up a chemistry laboratory for him at home. He attended Trinity College, University of Dublin, and University of Göttingen, in the meantime publishing a number of articles.

In 1853 Mallet came to the United States on business for his father, was persuaded by William Clark to teach French and German at Amherst, and remained here. He switched to teaching analytical chemistry at Amherst, went south to teach chemistry at University of Alabama, and then moved to Medical College of Alabama.

At the outbreak of the Civil War he joined the Confederate Army, became aide to General Rodes, and saw action at the battles of Williamsburg and Seven Pines. In 1862 General Gorgas appointed him superintendent of Confederate States Ordnance Laboratory. He supervised the testing of raw materials and finished products, sought substitutes for scarce materials, developed war devices, reorganized southern arsenals and depots, and planned the central ordnance laboratory at Macon, Ga. While inspecting a Charleston fort during the siege of 1863 he received a minor wound.

When peace returned Mallet had $1,400 in Confederate currency, worth roughly 14 pounds of tobacco. Financiers hired him for $300 a month in gold to prospect for oil, and he set out on horseback along the coast of Louisiana into eastern Texas. Contracts arranged for salt and petroleum properties came to naught when his employers ran afoul of the government.

In 1865 he became professor of chemistry in University of Louisiana Medical School. During the summers of 1866, 1867, and 1868 he and General Gorgas attempted to revive the Briarfield iron works in Alabama, wrecked by Wilson's cavalry in 1865, but abnormal business conditions of the postwar period ruined the venture.

In 1868 Mallet moved to University of Virginia where he inaugurated a course in industrial chemistry, one of the earliest in the United States. He went to University of Texas in 1883 to start the Chemistry Department, hoping the climate would improve his son's health. His son died, and Mallet resumed his former position at Virginia.

Scarcely had Mallet reached Charlottesville when he was offered a professorship at Jefferson Medical College. He moved to Philadelphia, did not like the position, and returned to Virginia where he died Nov. 7, 1912.

Mallet served as a toxicologist in criminal cases, an expert witness in commercial cases, and chemist for firms dealing in various products. He testified before a Senate committee on pure food legislation and before the War Department about the purity of rations provided during the Spanish-American War. His most significant research was the determination of the atomic weights of aluminum, gold, and lithium. He served as a judge of chemical products at the Centennial Exhibition of 1876, a member of the Mint's Assay Commission in 1886, 1888, and 1896, as chairman of the Inorganic Section of the International Congress of Science and Art, St. Louis, 1904, and president of the American Chemical Society in 1882.

Desmond Reilly, "John William Mallet, 1832-1912," *Endeavour* **12,** No. 5 (January 1953); Desmond Reilly, "John William Mallet (1832-1912), His Earlier Work in Ireland," *J. Chem. Education* **25,** 634-636 (1948); *Alumni Bulletin of The University of Virginia* **6,** 2-47 (1913) has a notice of Mallet's death, resolutions of university groups, two portraits, and an extremely informative sketch by William H. Echols; Otto Eisenschiml, "John W. Mallet, ACS President in 1882," *Chem. Eng. News* **29,** 110-111 (Jan. 8, 1951); "Dictionary of American Biography," **12,** 223, Chas. Scribner's Sons (1933). In the National Archives are many documents relating to Mallet's service as a Confederate officer, including drawings of shells, of an electromagnetic device for determining the rate of burning of fuses, and of other devices Mallet developed or modified.

WYNDHAM D. MILES

Edward Mallinckrodt

1845-1928

Edward Mallinckrodt was born in Saint Louis, Jan. 21, 1845. His interest in chemistry began when he read some of Justus Liebig's works on applying chemistry to agriculture which were in his father's bookcase. His father, Emil, encouraged him by setting up a laboratory for him in an old brick outbuilding. Here Edward worked during his infrequent spare time while he was managing the family farm during the absence of his father who had taken part of the family to visit in Europe. Edward's expectation of being a farmer changed to a desire to become a chemist.

In 1864, Edward and his brother, Otto, entered Wiesbaden Agricultural Institute to study chemistry with Karl Fresenius. A plan for establishing a chemical works in Saint Louis had originated early in the minds of Edward and his brothers, Otto and Gustav. Gustav had gone to work for Richardson Drug Co. when he was 17. After 2 years at Wiesbaden, Otto went to work in a chemical plant at Oeynhausen, and Edward was apprenticed to the de Haen Chemical Works at List near Hanover where he spent a year and a half.

They returned to America in 1867 to begin the work of erecting a factory for manufacturing chemicals with a capital of $10,000, which their father raised by mortgaging his land. The firm was established under the name of G. Mallinckrodt and Co., Manufacturing Chemists. Gustav was the general administrator and sales manager. Edward was in charge of the factory, and Otto was the analyst, purchasing agent, and shipping clerk. Their products were in the fine chemical field. They worked very hard trying to break into a market dominated by established firms in the East and to find new markets in the expanding West. They established the highest possible standards for their products and packaging. They were well trained, courageous, intelligent, and frugal. They worked hard and long. Often success seemed doubtful, but it came slowly after about 10 years. Then within a 6 month period in 1876-77, Edward's two brothers died, and he was left to carry on the business alone. The company was incorporated in 1882 and the name changed to Mallinckrodt Chemical Works.

Edward's keenest interest was in manufacturing fine chemicals, and he had great ability in business management and merchandising. A great factor in his success was his ability to look ahead and grasp opportunities in new chemical fields. From the beginning the company produced bromides and as early as 1873 he visited the salt works along the Ohio River and contracted directly for "bitter liquor," then the source of crude bromides. The manufacture of photographic chemicals was taken up as soon as the first portable camera was developed. The use of anhydrous ammonia as a refrigerant increased, and soon Mallinckrodt Chemical Works was the largest producer in the United States. Finally National Ammonia Co. was incorporated in 1889 by consolidation of various independent organizations, and Edward Mallinckrodt was elected president and general manager. He was active in Phosphorus Compounds Co. at Niagara Falls.

He was a member of the board of directors of the Union Trust Co. in 1891. He invested in real estate. The chemical works prospered, his business activities diversified, and his wealth increased.

He gave many paintings to the Saint Louis Art Museum of which he was vice-president of the board. He was a member of Washington University board, a trustee of the Missouri Botanical Garden, and a president of the Mercantile Library.

Edward Mallinckrodt was generous in giving his time, personal interest, and money. He gave large sums of money to reorganize Washington University Medical School. The gifts to this school established the Mallinckrodt Departments of Pharmacology, Pathology and Bacteriology, and Pediatrics. His largest contribution made possible the Edward Mallinckrodt Institute for Radiology. In memory of his wife, he made an endowment to Saint Louis Children's Hospital. He was president of the board of Saint Luke's Hospital in Saint Louis. He devoted a great deal of time and interest to the hospital, and he assisted financially to make up its deficits. His benefactions were numerous but often anonymous.

In 1923 Edward donated $500,000 toward a new laboratory at Harvard, the alma mater of his only son, Edward Jr. He hoped that other donations would be forthcoming, and they were. He died Feb. 1, 1928. The laboratory opened in September 1928 and is named for Edward Mallinckrodt.

George D. Stout, "Edward Mallinckrodt, A Memoir" (privately printed in Saint Louis, Mo., 1933, with photographs); Williams Haynes, "Chemical Pioneers," D. Van Nostrand Co. (1939); *Ind. Eng. Chem.* **16,** 1194 (1924), with photograph; "Dictionary of American Biography," **12,** 224, Chas. Scribner's Sons (1933).

ANN MARIE LADETTO

Edward Mallinckrodt, Jr.

1878-1967

Edward Mallinckrodt, Jr. was born Nov. 17, 1878. He studied chemistry at Harvard, receiving his bachelor of arts degree *cum laude* in 1900 and his master of arts degree in 1901. He then joined Mallinckrodt Chemical Works established by his father and uncles. He succeeded his father as chairman of the board in 1928. Under him the company expanded to a major producer of fine chemicals for industry and medicine. He was responsible for important developments in the fields of anesthesia, atomic research, and the purification and conversion of ores.

Soon after joining his father's company he became interested in processes for manufacturing ether for anesthesia. He recognized that ether made and packaged at that time deteriorated because of peroxides that formed, and he studied this for most of his active life. He held 16 patents relating to purification, packaging, and preservation of the anesthetic. He supported research at Harvard and Washington University Medical Schools, where he established chairs in anesthesiology. In conjunction with the celebration of the 100th anniversary of the first use of anesthesia in surgery, he had a film made of a reenactment of the event in Massachusetts General Hospital.

After 1908 he became interested in the use of x-rays for diagnostic purposes and had prepared at the Mallinckrodt Works barium sulfate of purity necessary for use in the digestive tract. Later he supported a group at Washington University, headed by Evarts Graham, who found that iodinated phthaleins could be used to visualize the gall bladder by x-rays. His firm prepared the compounds. A number of other contrast media were found through research which he encouraged.

The early interest that Mallinckrodt took in atomic energy was mentioned by Henry Smyth in "Atomic Energy," and by Arthur Compton in "Atomic Quest." The Mallinckrodt ether extraction process for producing uranium metal and its oxides made it possible to produce the materials on a commercial scale with a degree of purity seldom achieved previously.

In 1952 he received the Midwest Award of the American Chemical Society, given annually as public recognition for outstanding achievement in chemistry in the midwest area. The award cited Mallinckrodt for his research, particularly in ether purification, and for his sense of responsibility as a scientist and a leader of industry.

Mallinckrodt was a mountain climber, and scaled major peaks in Europe and North America. He hunted in Alaska and Africa. On one of his hunting trips he carried with him a movie camera invented by one of his friends in St. Louis and became much more interested in shooting animals with a camera than with a gun. He was an amateur astronomer and built an observatory at his farm in Pike County, Mo., to house a 10-inch telescope.

Conservation was one of Mallinckrodt's particular interests. Over the years he worked hard to preserve primitive areas such as the Panther Lake area in New York State and the region now known as Dinosaur National Monument. He fought the building of dams which would inundate beautiful sections of canyons and valleys. In 1962 he was awarded the Horace Albright Medal for leadership in the field of scenic preservation.

Edward Mallinckrodt, Jr., whose philanthropies in the fields of medicine and education were carried out quietly and often anonymously, was honored as the 1962 recipient of the Saint Louis Humanities Award. He gave many millions of dollars to Harvard, St. Louis University, and Washington University and their medical schools and to the Mallinckrodt Institute of Radiology in St. Louis. He endowed the Mallinckrodt Ward at Massachusetts General Hospital in Boston involving a project devoted to the study of baffling diseases. He died in St. Louis, Jan. 19, 1967.

George Dumas Stout, "Edward Mallinckrodt, A Memoir," privately printed in Saint Louis, Missouri, 1933, with photographs; Williams Haynes, "Chemical Pioneers," D. Van Nostrand Co. (1939) pages 143-164, with photograph; *Ind. Eng. Chem.* **16,** 1194 (1924) with photograph; "Dictionary of American Biography," **12,** 224, Chas. Scribner's Sons (1933); information from A. H. Homeyer, Mallinckrodt Chemical Works.

ANN MARIE LADETTO

James Jay Mapes

1806-1866

Mapes, agricultural chemist and manufacturer, chemical consultant, expert court witness, academic lecturer in chemistry, *bon vivant,* and skilled amateur painter, managed to pack half-a-dozen lifetimes into his brief 59 years. Born in Maspeth, Long Island, May 29, 1806, the son of Maj. Gen. Jonas and Elizabeth (Tylee) Mapes, he attended a classical school operated by Timothy Clowes but was in the main self-educated. Mapes early showed his interest in science when, on attending a lecture at the age of eight, he went home and devised a retort from a clay pipe to produce illuminating gas.

While still in his teens Mapes entered his father's mercantile house in New York City. General Mapes, who had been military commander in New York City during the War of 1812, died in 1827, and James, who married Sophia Furman in that year, opened his own business as a cloth merchant. During the 7 years of this enterprise he studied and experimented unceasingly; and in 1831-32 he perfected the first of many inventions—a chemical process for refining sugar from cane, machinery for this process, and a process for making sugar from West India molasses. For a time he engaged in sugar refining in the West Indies, with a notable lack of success.

By 1834 Mapes was ready to hang out his shingle as a chemical consultant in New York, one of the earliest consultants in the country. This proved an occupation perfectly suited to his eclectic tastes, for over the next 12 years he gave his attention to an incredible variety of chemical and manufacturing processes. With the New York State Senate as client he conducted analyses of beer and wines and established methods which were standards for many years. He manufactured Epsom salts (magnesium sulfate) from natural magnesium silicate and invented a synthetic leather tanning agent to replace that obtained from hemlock. He introduced methods of chemical control into the daguerrotype and electroplating industries and worked with Seth Boyden in the experiments which led to producing malleable iron. He produced new artists' pigments of unusual brilliancy and permanence. Above all, precisely because of the generality of his interests, he became a much sought-after witness in cases of patent litigation, from which much of his income was derived. In this function he was composed and unshakable, extroardinarly lucid, and could—as many trial lawyers found to their dismay—exercise a good-natured but devastating wit.

During these years also, Mapes served as professor of chemistry and natural philosophy at National Academy of Design and later at American Institute. As professor and as president of Mechanics Institute he vigorously promoted night schools for adult education.

But this is not yet to deal with the career for which Mapes is best remembered. Gradually he became interested in scientific agriculture; learning of the pioneering work on agricultural chemistry in Germany, he concluded that "our farming methods are primitive even to the verge of barbarism," and he set out with characteristic energy to change this situation. To combat the entrenched resistance of dirt farmers to changes suggested by "book farmers" in 1847, he purchased a worn-out farm in New Jersey, and by sub-soil drainage, crop rotation, and judicious fertilization he brought it to a peak of production. Here he conducted intensive 1-month courses in scientific agriculture and founded the magazine *The Working Farmer,* which he edited until his death and which continued publication until 1875. He advocated the establishment of a federal Department of Agriculture, with a secretary of cabinet status.

During this time he began manufacturing the chemical fertilizers which would bear his name for nearly a century—superphosphates from bones treated with sulfuric acid; the first complete mixed chemical fertilizer, which he patented in 1859; and later, when the Mapes Formula and Peruvian Guano Co. was in the hands of his son Charles Victor, the earliest special crop formulas. James Mapes was not without his detractors, some of whom, on failing to obtain Mapes' striking crop improvements, suggested that his motives were more commercial than scientific. Time, and more complete knowledge, appears to have proved him more right than wrong. The company continued operations until 1935, when it simply closed its doors without any successor.

Mapes died Jan. 10, 1866, survived by his son and three of his five daughters. Of these one is noteworthy: Mary Mapes Dodge, for many years editor of *St. Nicholas* magazine and author of the perennial children's classic "Hans Brinker."

W. Haynes, "Chemical Pioneers," 74-87, D. Van Nostrand Co. (1939, reprinted 1970); "Dictionary of American Biography," **12,** 264, Chas. Scribner's Sons (1933); "National Cyclopedia of American Biography," **3,** 178, James T. White & Co. (1891); "Lamb's Biographical Dictionary of the United States," **5,** 351-52, Federal Book Co. of Boston (1903); these sources also contain biographies of Charles Victor Mapes. *New York Times,* Jan. 12, 1866.

ROBERT M. HAWTHORNE JR.

Arthur Hudson Marks

1874-1939

Marks was born in Lynn, Mass. in 1874. He was educated at Lynn Classical School and attended Harvard University from 1893 to 1895. He then became assistant chemist at Boston Woven Hose & Rubber Co., Cambridge, Mass. Two years later he resigned to become chief chemist for Revere Rubber Co., Chelsea, Mass. In 1898 he went to Akron, Ohio, as superintendent of Diamond Rubber Co. In 1902 he was elected vice president and general manager of Diamond.

In 1899 Marks was granted a patent for the first successful process for reclaiming scrap rubber. This was accomplished by heating scrap rubber containing fabric in an autoclave with dilute sodium hydroxide at 360°F. for 20 hours. Fabric, uncombined sulfur, and some of the lead compounds were removed by the sodium hydroxide, and the rubber was plasticized by depolymerization from the heat. The product was useful as a replacement for crude rubber in many applications. The process was widely used until the advent of butadiene-styrene synthetic rubber which cannot be reclaimed by the action of sodium hydroxide.

With supplies of natural rubber scarce, this high-quality low-cost reclaimed rubber was quickly accepted by the trade. To exploit the patent, several new companies were formed including Alkali Rubber Co., Northwestern Rubber Co. of England, Pan American Rubber Co., and International Process Co. Marks was president of all of these companies while retaining his position at Diamond. Licenses were also granted to companies in Germany, France, Italy, and Canada. Marks collected a royalty on every pound of reclaimed rubber sold.

Marks's first great contribution to the industry was reclaimed rubber. His second was the introduction of the scientific approach to practical rubber chemistry in the United States. To this end he employed chemists, a rarity in the rubber industry, to carry out research and development at Diamond. He himself, chemically trained, found some time for experimentation. He first observed that high-quality para rubber, which vulcanizes relatively rapidly to give a product of excellent quality, is converted by acetone extraction to a low-quality rubber difficult to vulcanize. He further observed that addition of the material recovered from the acetone extract to a low-quality rubber markedly increases the rate of vulcanization and improves the quality of the vulcanizate. This was one of the clues which led to the discovery of organic accelerators.

In 1905 Marks hired George Oenslager, a Harvard trained chemist, to study ways to use low-cost rubbers. Oenslager quickly found the answer in organic accelerators which revolutionized the industry. Marks also hired David Spence from Liverpool, England in 1909. Spence did some of the earliest truly scientific work in rubber chemistry in the United States. The Diamond success story—reclaimed rubber and organic accelerators—did a great deal to convince the industry that chemists could be profitable investments.

After Diamond merged with B. F. Goodrich Co. in 1912, Marks served as vice president and general manager of the consolidated company until 1917 when he offered his services to the government. He served as lieutenant commander in the U. S. Naval Reserves in the Bureau of Construction and Repair, then as director of the Army Chemical Warfare Service. Illness forced him to resign in 1918. Recovering, he became vice president of Curtiss Airplane and Engine Co. until after the armistice.

In 1920 Marks became a director and vice president of Van der Linde Rubber Co., Ltd., Toronto, Ontario, Canada. At the same time he was elected president of Aeolian Skinner Organ Co., a post he held until his death in 1939. Retaining his other responsibilities, Marks returned to Goodrich as a director in 1930 and was named vice chairman of the board in 1937.

Marks collapsed while playing tennis and died in Palm Beach, Fla., May 1, 1939. He was buried in New York City.

Rubber World **100** (3), 57 (1939); *Rubber Age* **45,** 110 (1939); obituary in *New York Times,* May 2, 1939.

GUIDO H. STEMPEL

Albert Edward Marshall

1884-1951

Marshall was born in Liverpool, England, May 18, 1884. He attended Liverpool Institute and University College and upon graduating received honors in chemistry. He was also presented the Guild Medal in chemical technology at South Kensington, London.

Beginning his career in England, Marshall became a research chemist with United Alkali Co. and assistant manager of the Fleetwood Works. In 1910 he became assistant manager of the Thermal Syndicate, Newcastle-on-Tyne; following a visit to the United States in 1911 he returned the next year to become the manager of the Syndicate's New York branch. In 1916 he moved to Baltimore to become works manager of Davison Chemical Co., at that time a leading producer of sulfuric acid in the United States.

From 1921 through 1938, Marshall practiced as a consulting engineer, maintaining offices first in Baltimore and, after 1929, in New York. During that time he contributed greatly to the development of industrial uses of Pyrex glass and the manufacture of synthetic resins. In that period, too, he served as 1934-35 president of the American Institute of Chemical Engineers.

In 1938 Marshall became president of Rumford Chemical Works in Providence, R. I., remaining in that capacity until 1950, when Rumford sold its assets to Hulman and Co. He resigned from the firm then and was named vice president of Heyden Chemical Corp. in New York.

During World War II, Marshall managed the Agfa Ansco Division of the General Aniline and Film Corp., aiding the Office of the Alien Property Custodian in restructuring the German-controlled firm. He performed similar duties in connection with E. Leitz, Inc., another German-owned company in New York.

He served as director of several business and industrial firms, including American Potash and Chemical Corp., Trona, Calif.; Niro Corp., Copenhagen, Denmark; Technifinishing Laboratories, Inc., Rochester, N. Y.; Investors Trust Co. and Rhode Island Hospital Trust Co., both of Providence; and Thiokol Corp., Trenton, N. J.

Among the offices he filled were those of chairman of the New England Interstate Water Pollution Control Commission; member of the advisory council of the Engineering Experiment Station, University of Rhode Island; councilor and chairman of the American Section of the Society of Chemical Industry; chairman of the advisory council of the Chemical Engineering Department, Princeton University (1941-46); director of Providence Engineering Society (1941); chairman of Rhode Island Section of the American Chemical Society (1946-47); and president of the New England Industrial Research Foundation.

He also served as president of the Rhode Island Section of the New England Council (1945-46); member of the corporation of Northwestern University; Rhode Island state chairman of the Committee for Economic Development (1944-47); vice president of the Coffin School, Nantucket, Mass.; and director of the Rhode Island Philharmonic Orchestra.

Marshall wrote many technical papers for professional publications as well as articles on the history of the manufacturing arts, which appeared in technical and educational publications. His hobby was collecting old chemistry books; his collection was dispersed at auction after his death.

He died in Providence, Sept. 15, 1951, leaving his widow, the former Ruth Hildebrandt of Baltimore, and two sons.

Obituaries in *The Evening Bulletin* (Providence, R.I.), Sept. 15, 1951; *The New York Times,* Sept. 16, 1951; personal recollections.

FRANKLIN J. VAN ANTWERPEN

William Pitt Mason

1853-1937

Mason was born Oct. 12, 1853 in New York City into a family with a great tradition. He bore the name of the great English prime minister of the eighteenth century, William Pitt, who was a cousin of his great-grandmother. Mason's paternal grandfather was a founder and president of the Chemical Bank, one of the early banks in New York City. William graduated as a civil engineer from Rensselaer Polytechnic Institute in 1874 and spent a year thereafter studying chemistry at Harvard University. Then he became an assistant in chemistry and natural science at Rensselaer, where he stayed half a century until his retirement in 1925. He interrupted his teaching frequently, however, to continue his studies and for travel. In 1881 he obtained his M.D. degree at nearby Albany Medical College, which provided in a sense an orientation for his life-work in the field of water analysis and supply, and he prepared himself for it by further study.

Thus in 1887 he investigated the water supply systems of Europe and was in Messina, Sicily during a cholera epidemic. In 1889 and again in 1893 he took courses at the Pasteur Institute in Paris. In the summer of 1908 he was a student of Otto Zacharias at the German Hydrobiological Institute at Plön, in Holstein, and he became interested in plankton and other growths that produce taste and odor in water. In 1893 he was named professor of chemistry at Rensselaer and subsequently headed its Department of Chemistry and Chemical Engineering for many years; personally, he engaged in the pursuit of sanitary engineering. It became, in fact, an important branch of study at Rensselaer, thanks to Mason's effort and engineering background. Mason's dedication to the quest for good water coincided with the growth of American cities and their need of adequate sanitary services. Mason won international recognition as one of the world's leading sanitary engineers, serving as a consultant to many cities, designing systems, and championing in print the cause of a safe and adequate water supply.

His contributions to the field of sanitary engineering were manifold. He trained many engineers and chemists, and he wrote many text-books, articles, and reports. In 1912 he was president of the Hygiene Division of the Eighth International Congress of Applied Chemistry held in Washington. He was a member of many scientific societies in America and abroad and in 1909 was the first person from the academic world to preside over the American Water Works Association.

Among Mason's many writings, two books deserve special mention. Between 1896 and 1916 his work on "Water Supply" reached four editions and was acclaimed as one of the important texts in the field. Another book, "Examination of Water," first published in 1899, appeared in a sixth edition by 1931 and was long a practical and valuable guide to government chemists and sanitary engineers. Mason accumulated many honors, among them honorary degrees from Lafayette and Union Colleges, and he was a member of the commission for the revision of the United States Pharmacoepia in 1890 and of the United States Assay Commission in 1896.

Mason was a cultivated man, witty and kindly, a brilliant teacher, known affectionately to his students as "Billy." He combined professional gifts with personal charm in a degree perhaps unusual in a technical institution like Rensselaer, priding itself on a tradition of rough and ready engineering. His death, Jan. 25, 1937, ended a long life and a distinguished career in a field related to chemistry.

"National Cyclopedia of American Biography," **27,** 469, James T. White & Co. (1939); *Rensselaer Alumni News* **27,** 16 (1960); biog. by F. W. Schwartz, *Ind. Eng. Chem.* **16,** 93 (1924).

SAMUEL REZNECK

William Williams Mather

1804-1859

Mather was born in Brooklyn, Conn., May 24, 1804. He descended from Reverend Richard Mather who escaped from English persecution and arrived in America in 1635. Both his father's and mother's families were illustrious and notable families of this period. Very little is known of his younger years. He went to Providence, R.I. to study medicine and returned interested in chemistry.

In 1823, Mather entered the United States Military Academy at West Point. He proved to be very proficient in carrying out chemical analysis of ores and minerals. In 1826 he assisted in preparing John White Webster's "Manual on Chemistry." He graduated in 1828. Mather served as acting assistant professor of chemistry, mineralogy, and geology at West Point from 1829 until 1835. During this period he taught also at Wesleyan University, published a geological survey of Windham County, Conn., and published several papers and a successful geology textbook. He married his cousin, Emily Maria Baker during this period. They produced six children. In the early 1830's he determined the atomic weight of aluminum as 27, deduced from the ratio of AgCl to $AlCl_3$.

This is thought to have been the first time an atomic weight was determined in the United States.

In 1836 Mather was appointed geologist in charge of surveying one of four districts in New York State. During the 7 years required to complete this work, he also was chief geologist for the State geological survey of Ohio and the survey of Kentucky. His mineral and geological collection contained 26,000 specimens.

He held the post of professor of natural science in Ohio University at Athens from 1842 to 1845. He was acting president in 1845 and from 1847 to 1850 vice president of that university. During the period from 1845 to 1847, he examined mineral lands in many states for mining companies. In 1845 he carried out a series of experiments that indicated the feasibility of extracting bromine from the waters near the salt works in Athens, Ohio. Bromine was selling for 16 dollars an ounce at this time. A plant at Pomeroy, Ohio was established to produce bromine.

From 1850 to 1854 Mather was agricultural chemist for the state of Ohio and editor of *Western Agriculturist*. His wife died in 1850, and he married again in 1851 to Mrs. Mary Curtis. This marriage produced one son. Mather was large, robust, gentle, and considerate, and he had a great capacity for physical and mental labor. He died Feb. 26, 1859, at the age of 54 while president of University of Ohio.

Throughout his professional life, Mather contributed to *American Journal of Science* and other scientific periodicals, and many of his reports were published. He received an honorary LL.D. degree from Brown University in 1855. Mather was a member of many scientific organizations and for 15 years a trustee of Granville College (now Denison University).

W. J. Youmans, "Pioneers of Science in America," 402-409, D. Appleton & Co. (1896); obituary in *Amer. J. Sci.*, Series 2, vol. 27 (1859); "National Cyclopaedia of American Biography," **8,** 146, James T. White & Co. (1898); G. W. Cullum, "Biographical Register of the Officers and Graduates of the U. S. Military Academy at West Point," D. Van Nostrand (1868); Edgar F. Smith, "Chemistry in America," 225, D. Appleton & Co. (1914). Cullum gives Feb. 27 as date of death; other sources give Feb. 26.

GEORGE J. VIRTES, JR.

Joseph Howard Mathews

1881-1970

Mathews was born Oct. 15, 1881, son of Joseph and Lydia (Cate) Mathews, operators of a dairy farm near Auroraville, Wisc. He lacked enthusiasm for farm life and for the cheese factory operated by his older brother. After graduation from Omro High School he earned his B.S. degree in chemistry at University of Wisconsin in 1903. While still an undergraduate he and eight classmates (Edward Matke, Bart McCormick, James Silverthorn, Joseph Holte, Harold Eggers, Frank Petura, Alfred Kundert, and Raymond Conger) founded Alpha Chi Sigma in 1902. Mathews took an active part in establishing chapters of the professional chemistry fraternity in other schools and was active in the organization throughout his life.

Following graduation, he spent a year as an analytical chemist in the consulting laboratory of H. S. Mitchell in Milwaukee. He then returned to Wisconsin where he obtained his M.S. degree in 1905 and entered Harvard to study under Theodore Richards. His M.A. degree was conferred in 1906 for studies on compressibilities. Since Richards was then to spend a year in Europe, Mathews took a teaching position at Case Institute. He returned to Harvard where his Ph.D. degree was granted in 1908 for studies on heats of vaporization of a large number of organic compounds. During the course of these studies he and Richards developed an electrically heated adiabatic calorimeter.

He then began his long faculty tenure at University of Wisconsin. In 1909 he married Ella Gilfillan. With their two daughters, Joan and Marian, they led a warm and close family life. The remainder of his activities were devoted to his professional life.

When the United States entered World War I, Mathews took leave from the university to become a captain, later major, in the Ordnance Department. He served in

England and France in connection with chemical warfare. After the war he returned to the university where he became chairman of the Chemistry Department in 1919.

During the next three decades he focused his attention on building one of the strongest chemistry departments in the nation. He was a good judge of men and showed astuteness in attracting promising young chemists (such as Homer Adkins, Farrington Daniels, Samuel McElvain, John W. Williams, John Willard, William S. Johnson, Alfred Wilds, John Ferry). He was a strong administrator and was able to provide professors with facilities to develop their graduate programs. He secured funding for and planned two substantial additions to the chemistry building and was talented in the design of facilities, whether it was a piece of scientific or photographic apparatus, a laboratory, or a building.

He directed particular attention toward the development of three areas of chemical instruction: the undergraduate physical laboratory, instrumental analysis, and colloid chemistry. His interest in the physical chemistry laboratory resulted in the devolopment of a demanding course which gave students an opportunity to work with research-type equipment. With several colleagues, Mathews became author of a laboratory manual which went through numerous editions, each reflecting new developments in the field.

As chemical instrumentation advanced, Mathews obtained such equipment for the department. It was available not only for campuswide research but for undergraduate instruction. He encouraged Professor Villiers W. Meloche to develop one of the first courses in instrumental analysis.

In 1923 Mathews brought The Svedberg to Wisconsin as a visiting professor, and it was there that Svedberg developed the prototype of the ultracentrifuge. A decade later Mathews was able to obtain funding for the purchase of a Svedberg ultracentrifuge which was used extensively by John W. Williams' research group in the study of proteins. In 1923 Mathews was also responsible for establishing the Colloid Symposium, sponsored by the American Chemical Society. The symposium became an annual event, meeting at Wisconsin every tenth year.

In 1922, Mathews was called upon to assist in solving a crime in a nearby locality, a murder caused by a bomb. His investigation and testimony led to the conviction of the suspect and led to lifelong activities in developing scientific approaches to the study of criminal activities. He lectured widely on the subject. His interest in rifling marks and percussion pin impressions resulted in development of new comparison equipment and led to the publication of a multivolume treatise on the identification of hand guns. For more than a decade he taught a sociology course dealing with scientific methods of criminal investigation. He was active in making law enforcement officers aware of science as an aid in crime detection and helped create the Wisconsin Crime Laboratory. For many years he was a member of Madison's Fire and Police Commission. During his retirement years he carried on an extensive study of hand guns. He was not only a clever instrumentalist but a superb photographer whose photographs of firearms were surpassed only by his nature photographs made at the cottage at Trout Lake where the family spent many summers. His firearms studies were continued up to within two months of his death which occurred April 15, 1970.

He was a man of firm opinions and strong loyalties. Opposition lawyers found him a clear headed and tough expert in the witness chair. His faculty found him to be a demanding chairman who, once the importance of a given project was established, was indefatigable in gaining necessary support.

Papers dealing with Mathews' chairmanship are preserved in University of Wisconsin Archives; A. J. Ihde, "J. Howard Mathews," *Badger Chemist* (Newsletter of University of Wisconsin Chemistry Department), No. 18, 1971, pp. 3-6, with cover portrait; "Joseph Howard Mathews 1881-1970, Founder, GMA 1908-1914," *Hexagon of Alpha Chi Sigma,* **60,** 110-114 (1970); E. M. Larsen, "Joseph Howard Mathews, Alpha, A founder of Alpha Chi Sigma," *ibid.,* **41,** 282, 330-331 (1951); "Remarks of J. H. Mathews at the Opening of the Fortieth National Colloid Symposium," *J. Colloid Interface Sci.,* **22,** 409-411 (1966).

AARON J. IHDE

George William Maynard

1839-1913

George W. Maynard studied chemistry and mineralogy in America and Europe and became a mining engineer whose work carried him into many parts of the world and into a variety of enterprises. He was born June 12, 1839 in Brooklyn, N.Y. and was graduated from Columbia College in 1859. After serving for a year as an assistant to the Columbia professor of chemistry, Maynard went to Europe and studied at Göttingen University under Wöhler and at the Royal Mining Academy at Clausthal in Germany. His interests were in chemistry, mineralogy, and mining. In 1863 he spent a year at Wicklow, Ireland devising successfully a plan for treating pyrite minerals.

In 1864 Maynard returned to the United States and, with another chemist named Tiemann, opened an engineering office and consulting chemical laboratory in New York City, but Maynard spent much of the next 3 years operating one of the first engineering and assay offices in Colorado and working in gold mines during the prevailing gold boom. He then returned east and briefly headed a plant for manufacturing sulfuric acid on Staten Island. Between 1868 and 1872 Maynard further expanded his already broad technical experience by serving as professor of metallurgy and practical mining at Rensselaer Polytechnic Institute. It was a new experimental program, and Maynard all by himself trained a small number of mining engineers. The course was, however, discontinued abruptly, and Maynard was left without employment.

With characteristic resilience, Maynard went to England, where he opened an office in London as a consulting engineer to steel works. Here he became associated with Sidney Gilbert Thomas and helped him develop the practical application of his famous basic process for manufacturing steel. It was particularly suited to the phosphoric iron ores of Alsace-Lorraine. His work took him to Russia where he supervised the construction of a copper smelting plant for a British company. In 1879 Maynard returned to the United States, where he remained the rest of his life. His base was an office in New York City, but his consulting services carried him widely to mining operations in Canada, Cuba, and Mexico. He was particularly involved in developing metallurgy, principally in introducing the Thomas Process into the United States; it was eventually sold to the American Bessemer Association.

Somehow, in spite of a busily diversified practical career in metallurgy, Maynard found time and opportunity for professional and literary activity. He was vice-president of the American Institute of Mining Engineers in 1904 and 1905. He wrote frequently for technical journals, principally contributions to the *Transactions* of the American Institute of Mining Engineers on a wide range of topics. Among these were the "Introduction of the Thomas Basic Steel Process in the United States" (1910), "Chemical Laboratories in Iron and Steel Works" (1910), and "Late Developments in the Siemens Direct Process" (1897).

Maynard was taken sick in Denver while on a professional journey in 1913 at the advanced age of 74 years and died in Boston, Feb. 12, 1913. He was acclaimed as the "dean of mining engineers" in America. Ironically, Maynard had never earned a degree in mining engineering.

H. B. Nason, "Biographical Record . . . of the Rensselaer Polytechnic Institute," W. H. Young (1887), 154-55; *Bulletin of the American Institute of Mining Engineers* (No. 76, April 1913), pp. XXXIV-XXXVI; "Dictionary of American Biography," **12,** 458-59, Chas. Scribner's Sons (1933); T. T. Read, "The Development of Mineral Industry Education in the United States" (1941), 69-70, Amer. Inst. of Mining and Metallurgical Engineering.

SAMUEL REZNECK

Leonard Amby Maynard

1887-1972

Maynard was born Nov. 8, 1887 on a farm in the town of Hartford, Washington County, N.Y. After eighth grade in the Hartford village two-room school, he received a classical education in language, literature, 1911, *cum laude*.

The rural environment of Maynard's and mathematics at Troy Conference Academy, Poultney, Vt. Then in 1907 he enrolled in Wesleyan University and graduated in youth provided the foundation of his interest in plants and animals and formed the basis for his life-long work in biology and agriculture. Walter P. Bradley's course in chemistry at Wesleyan inspired the direction of his future. In Bradley's course he learned of the pioneer work of Wilbur Olin Atwater, who established and directed the first agricultural experiment station in the United States at Middletown in 1875. Fascinated by the accounts of Atwater's varied research activities in applying chemical knowledge and techniques to the problems of agriculture and human and animal nutrition, Maynard determined to specialize in chemistry.

Following a period of 2 years' service as an assistant in chemistry in the agricultural experiment stations of Iowa and of Rhode Island, Maynard entered Cornell University in the fall of 1913 as a graduate major in chemistry. During graduate study, he was greatly stimulated by Wilder D. Bancroft whom he described as a teacher whose "facile mind, familiarity with both classic and current literature of chemistry," and whose "wealth of ideas for research and enthusiasm made contacts with him, both in lectures and conferences, of outstanding interest and value."

After completing studies for his Ph.D. degree in chemistry in 1915, Maynard was appointed as an assistant professor of animal nutrition and began a career of teaching and research which lasted for 40 years until his retirement and appointment as emeritus professor of nutrition and biochemistry in 1955.

Maynard initiated his research career in the laboratory of animal nutrition, which he designed and equipped for chemical and biological studies in the Department of Animal Husbandry. It had just been established that adequate nutrition required the presence in the diet of specific amino acids, vitamins, and minerals, as well as proteins and the energy-yielding fats and carbohydrates. Since this "newer knowledge of nutrition" was the result of discoveries made with small laboratory animals receiving diets put together from purified ingredients, Maynard's first efforts were directed to the establishment of a rat colony. He was aware that the purified diet technique with laboratory animals had solved previously baffling nutritional problems. He recognized, however, that such a procedure was not feasible with large animals. Consequently, he conceived an integrated plan whereby pilot experiments with rats and the purified diet method were used to guide studies with farm animals to solve practical problems of animal husbandry.

Beginning in 1916 Maynard initiated a program to develop a "milk substitute" for weaning calves, which if nutritionally satisfactory and reasonably priced, would enable the farmer to profit from the milk thus replaced. By 1923 the comparative rat and calf experiments had led to the development of a "calf meal" that has found extensive use in the dairy farms of the northeastern United States.

In 1922 Maynard began a study of an ailment of growing pigs of unknown cause which resulted in posterior paralysis and lameness. It was shown that the physical symptoms of the trouble were reflected in an abnormal chemical and histological structure of the leg bones. The cause of the condition was found to be a dietary deficiency of calcium. Furthermore, accompanying experiments showed why pigs were more susceptible to the ailment when housed indoors and demonstrated the beneficial effect of sunlight in producing normal bones.

At the time Maynard undertook his swine studies, he was aware of the discovery in 1919 that ultraviolet rays improved calcium deposition in rachitic children. Thus, Maynard's investigations extended this finding to pigs and contributed to the subsequent explanation that sunlight produced the antirachitic vitamin (vitamin D) in the body of the animal.

Following post doctoral studies at Yale University with Lafayette B. Mendel, at University of Strassbourg with Emile Terroine, and at Ecole Veterinaire, Lyon, France with Charles Porcher, Maynard initiated a program at the end of the 1920's on the biochemistry of lactation. These studies covered a period of nearly 20 years and combined experiments with rats on purified diets with parallel investigations on goats and cows. It was found that the triglycerides of the blood were precursors of milk fat and that a minimum level of fat in the diet was essential for maximum milk secretion.

In addition to the above, the more than 100 original research papers which Maynard published report that fluorine as a contaminant of mineral supplements caused bone troubles in swine, that purified rations can be developed and used successfully to study fat and protein nutrition of ruminants, and that caloric restriction prolongs the length of life in the rat.

Besides his busy career as teacher and researcher, Maynard served not only Cornell but numerous other organizations as an effective administrator. From 1939 to 1945 he administered the U.S. Plant, Soil, and Nutrition Laboratory as the first director. In 1941 he helped to establish the School of Nutrition and did the same for the Department of Biochemistry in 1944. He directed both of these Cornell units until his retirement.

Maynard was elected to the National Academy of Sciences in 1944. He served as chairman of the academy's Food and Nutrition Board from 1951 to 1955 and of its Division of Biology and Agriculture from 1955-58. Just prior to his death he was actively engaged in revising the classic and widely used textbook of "Animal Nutrition" for a seventh edition.

He died June 22, 1972.

Personal recollections; biography, Cornell University Symposium on Biochemistry and Nutrition, 1955, published by the N.Y. State College of Agriculture (1956); "Modern Men of Science," pp. 351-352, McGraw-Hill (1968); in Memorial Tribute, *Graduate School of Nutrition News*, Vol. VIII, No. 2, August 1972; "Necrology of the Faculty of Cornell University, 1971-72," in press.

HAROLD H. WILLIAMS

Elmer Verner McCollum

1879-1967

McCollum was born Mar. 3, 1879 on a farm near Fort Scott, Kans. where he spent his first 17 years. Exposure to hard work, perseverance, and creativity which the successful management of a farm requires, formed the character and disposition of one who was to become the "Father of Nutrition" in America.

Despite the irregularity of his early schooling, which depended on the ability of neighboring farmers to pay the teacher, he was admitted in 1896 to high school in Lawrence, Kans. In 1900 McCollum entered University of Kansas. There, influenced by Edward Bartow, Homer Cady, and Edward Franklin, he determined to seek his career in organic chemistry. Having distinguished himself as an undergraduate by election to Sigma Xi as a junior, he was urged to continue his studies. He received his A.B. degree in 1903 and his M.A. degree in 1904. Knowledge of the work of Treat Johnson and Henry Wheeler led McCollum to apply for and receive a scholarship for study at Yale. Two years later he received his Ph.D. degree, having won the Loomis award along the way. During his study and 1 year postdoctoral experience at Yale, McCollum, under the aegis of Russell Chittenden and Lafayette Mendel, became proficient in biochemistry.

In 1907 McCollum accepted the post of instructor in agricultural chemistry at Wisconsin Agricultural Experiment Station. There he undertook the task of formulating the ideal feed for cows and soon noted the supplementary relationships between oats, corn, and wheat in such a diet. Encouraged by Stephen Babcock, then retired, he persevered in his belief that nutritional discoveries would come from studies in the rat. Despite administrative obstacles, he succeeded in establishing the first rat colony in America for the study of nutrition. In 1912 he announced the discovery of "fat-soluble A" and in 1915 postulated the existence of "water-soluble B." He thus established the value of his biological methods of food analysis.

In 1917 McCollum left his post at Wisconsin to become the first professor appointed at the newly formed School of Hygiene and Public Health at Johns Hopkins University in Baltimore. He served as head of the Department of Chemical Hygiene (later Biochemistry) until his retirement in 1946. In 1922 with Edwards Park he announced the discovery of Vitamin D. With his students he described the nutritional roles of manganese, magnesium, zinc, and other trace minerals as well as of B-complex vitamins and vitamin E. His work with trace minerals led in 1949 to the establishment in his honor of the McCollum Pratt Institute at Johns Hopkins University. Believing that nutritional discoveries were of little value unless implemented by the public, he disseminated correct nutritional knowledge through popular lectures and articles, and as a member of League of Nations committees, Food and Nutrition Board of the National Research Council, consultant to National Dairies, and other means. The several editions of his textbooks, "The Newer Knowledge of Nutrition" and "Food, Nutrition, and Health" influenced nutritionists for many years. He spent the last 2 decades of his life writing his comprehensive "History of Nutrition" and an autobiography as well as in research on the separation of amino acids and the metabolites of Vitamin E. He died in Baltimore, Nov. 15, 1967 at age 88.

E. V. McCollum, "From Kansas Farmboy to Scientist," Univ. of Kansas Press (1964); biographies by H. G. Day, *Biog. Mem. Nat. Acad. Sci.*, Volume 45, 1972; and Agatha A. Rider in *J. Nutr.* **100**:1 (1970).

AGATHA A. RIDER

Herbert Newby McCoy

1870-1945

McCoy was born in Richmond, Ind., June 29, 1870. He received his bachelor's and master's degrees from Purdue, worked for Swift and Co. as a chemist for a year, was instructor at Fargo College for 2 years, then was a graduate student under Julius Stieglitz at Chicago from 1895 to 1898. After receiving his Ph.D. degree he remained as research assistant for a year (during which time he developed his molecular weight apparatus), taught at University of Utah for 2 years, then returned to Chicago where he taught from 1901 to 1917. During those 16 years he published more than 30 papers on fundamental research, initiated many new courses, prepared the way for texts (published later with Ethel M. Terry, his wife), and engaged in extensive commercial activities.

McCoy was the first to prove that the alpha ray activity of a uranium compound is directly proportional to its uranium content, this being the first quantitative proof that radioactivity is an atomic property. His researches on alpha ray activity revealed the quantitative relationship between range and activity, from which he developed a standard of measurement—the McCoy Number.

McCoy's work on uranium, followed by studies with William H. Ross on thorium and radiothorium, established the existence of elements different in identity but of identical chemical properties. Such elements were later called isotopes by Frederick Soddy. McCoy also proved experimentally that uranium is the parent of radium. Bertram Boltwood of Yale and R. J. Strutt of England, reached this same conclusion at about the same time in 1907.

With various collaborators McCoy made the first quantitative study of the equilibrium between carbonates, bicarbonates, and carbonic acid in water solution. This fundamental work was important to the study of geologic formations, acid-base balance of blood, and the manufacture of trona.

McCoy was the first to use buffered solutions, by means of which he worked out the values of various low ionization constants, such as those of carbonic acid and phenolphthalein and the secondary ionization constants of various weak acids. One of his most spectacular researches was the preparation (with William C. Moore) of the first known organic amalgams, the first organic substances possessing metallic properties.

It was in the field of radioactivity that McCoy found contact with industrial chemistry. In 1909 he obtained from Lindsay Light and Chemical Co. of Chicago 2 pounds of ash of the scraps of gauze left from gas mantles. From this he extracted the trace of mesothorium for his research, and then as a courtesy to the company, transformed the ash into valuable thorium nitrate. From this exchange of favors he was retained as consultant and eventually became vice-president of the company.

In 1912 he suggested producing thorium nitrate from monazite sand. This work was completed July 4, 1914, about 4 weeks before the British blockade became effective. The mantle companies depended on Germany for thorium nitrate. Under McCoy's supervision a plant was put in operation Nov. 1 which supplied about half of the thorium nitrate used by the Allies during the war. Later he developed methods in his own laboratory for extracting mesothorium from the waste of thorium nitrate manufacture. This product was sold to the luminous watch-dial industry.

In 1914 McCoy patented a method for preparing radium salts from carnotite. About a year later Carnotite Reduction Co. was organized to extract radium from Colorado ores. McCoy served this company as consultant and was president during the years 1917 to 1920.

In 1937 the Willard Gibbs Medal was bestowed on McCoy. His address was on "The Separation of Europium from Other Rare Earths and the Properties of the Compounds of This Element." At that time Madame Curie called him the foremost American authority on radioactivity.

Having observed that too many men continued professional careers until unfitted to enjoy life, McCoy planned to do double duty during his prime with the intention of retiring at 47. This plan he carried out, retirement meaning to him freedom to carry on research, along with various other interests. Above his two-car garage behind his Spanish style home in Los Angeles he built a laboratory jiggered with a profusion of pet mechanical devices of his own design. Here, between bird trips to the tropics or the California wilds for the Los Angeles Museum of Natural History, he finished an exhaustive study of the extraction and separation of europium from the other rare earths.

On May 21, 1943 McCoy received a letter from Arthur H. Compton. "We should like especially," the letter said, "to discuss with you problems associated with the chemistry and metallurgy of radioactive materials and of the rare earths." On a Thursday Compton outlined a problem to McCoy saying that if he could find the answer in 6 months it would be of great value to what is now known as the atom-bomb project. On the following Monday McCoy went to Compton and said: "Here's the answer to your problem."

McCoy had prior warnings of heart trouble, and on May 7, 1945 he became one of the casualties of World War II. He was a kindly and viable man, an accomplished and successful business man though personally very generous, a great creative chemist, a "pioneer in a greater number of fundamental discoveries than any but three or four living American chemists" (Willard Gibbs citation), . . . a giant in his field who became one of the chief founding fathers of the atomic age.

G. Ross Robertson, "Herbert Newby McCoy, 1870-1945" (privately printed, Los Angeles, 1964), with a complete record of the scientific writings of McCoy; personal recollections.

LILLIAN EICHELBERGER

Richard Sears McCulloh

1818-1894

McCulloh, born in Baltimore, Mar. 18, 1818, attended University of Maryland for one year, and graduated from Princeton where he was one of the "best pupils" of John Torrey, professor of chemistry. He examined the water supply of the mountains of Western Maryland for engineers of the Chesapeake and Ohio Canal, studied analytical chemistry in James Curtis Booth's laboratory during the winter of 1838-39, was observer in the magnetic observatory of Girard College in Philadelphia, taught natural philosophy, mathematics, and chemistry at Washington and Jefferson College from 1841 to 1843, and delivered a course of public lectures on chemistry in Pittsburgh in 1843.

Resigning from Washington and Jefferson,

McCulloh went to Washington where he drew up a plan of organization for the Naval Observatory for the Secretary of the Navy, then undertook an investigation of sugar, molasses, and hydrometers for the Treasury Department. President Polk appointed him Melter and Refiner of the Philadelphia Mint in 1846. He superintended chemical operations there for 3 years, modifying the method of refining gold bullion by introducing the use of zinc. Feeling he had not received sufficient credit or remuneration for his innovation, he stirred up a controversy and wrote at least five pamphlets to present his side of the argument.

He went to Princeton as professor of natural philosophy in 1849 and to Columbia as professor of natural and experimental philosophy and chemistry in 1854. Sympathizing with the South he abruptly left Columbia in the fall of 1863, made his way to Virginia and offered his services to the Confederate government. There is some doubt as to the actual title he was given (one southern newspaper referred to him as a brigadier general; in later days he was called colonel), and there is some doubt as to his position, but he was connected with the Confederate Nitre and Mining Bureau.

He is said to have suggested a new gunpowder, which Confederate artillery may or may not have tried. He developed incendiary devices for sabotage and was authorized by Confederate Secretary of War Seldon to organize a company of men to destroy Union supplies. Word of the incendiary spread and made McCulloh notorious in the North. As a result Secretary of War Stanton ordered him captured after the war, and kept him imprisoned for almost a year before releasing him on parole.

Robert E. Lee then hired him as professor of natural philosophy at Washington and Lee. There he remained until 1877, then went to Louisiana State University as professor of general and agricultural chemistry, and also taught physics, zoology, botany, and anatomy at times before retiring in 1888. He died at Glencoe, Md., Sept. 15, 1894.

Biography by Milton H. Thomas in *Princeton University Library Chronicle* **9,** 17-29 (1947); "Testimonials of the Qualifications of Richard S. McCulloh" (1844 ?), and other pamphlets which McCulloh wrote, give bits of biographical information, as do official government reports of his investigations; W. D. Miles, "Chemists and Chemistry in the War Between the States," *Chem. Eng. News* **39,** 115 (April 3, 1961), with portrait.

WYNDHAM D. MILES

Charles Francis McKenna

1861-1930

McKenna was born June 4, 1861 in New York City. He received his A.B. degree in 1879 and his A.M. degree in 1880 from St. Francis Xavier College in New York City, his Ph.D. degree in 1883 from Columbia University School of Mines, and his Ph.D. degree in 1894 from Columbia.

From 1886 to 1889 he was chemist for Johnstown Steel Co., in Pennsylvania. After this he engaged in the production of lime from oyster shells. He had the shells taken by barge up the Hudson River to Edgewater, thence by mule team to kilns, of which he had six. For the output of these kilns he found ready sale, principally to gas companies in the East.

McKenna directed the Laboratory of Physical Testing of the City of New York from 1893 to 1895. He was chemist of the Passaic Zinc Co. from 1895 to 1897. In 1897 he purchased the building, laboratory equipment, and consulting practice that had been established by Gideon E. Moore. For many years he occupied three floors of the building; one being used as his office, library, and chemical laboratory; another entirely as a chemical research laboratory; and the third floor entirely for the physical testing of materials of construction and for experimental furnace work.

For a quarter of a century McKenna was engaged as a consulting chemist and chemical engineer. He became pre-eminent as an expert in materials of construction and explosives. During the time when he was actively in practice, there was a tremendous mass of steel and concrete construction throughout the length and breadth of New York City.

In that period the first subway was built, the first tunnels under the rivers that bounded Manhattan were cut, bridges were thrown over the East River, and a very large number of important private and public buildings were erected. McKenna acted as consultant on the testing of cement in the construction of the first subway and passed on the fitness of an enormous amount of the cement that was used in it. He inspected cement that was used in the anchorages of the Manhattan Bridge. He worked as an expert on cement and steel for many of the large foundation and construction companies. He inspected some of the cement that was destined for use on the Panama Canal. He was called on by practically all of the eastern cement mills at one time or another for his opinion and advice on cement making practice. He sent his inspectors as far west as Missouri to sample, test, accept, or reject, and seal carloads of cement shipped to many important undertakings in sections of the United States.

McKenna engaged extensively in the testing and analysis of steel, brass, nickel, vanadium, titanium, and alloys of various metals. He advised Anaconda Copper Co., and New Jersey Zinc Co., on troubles that had been observed but not solved at their plants. Other materials of construction that he was asked to consider, analyze or test and report upon, or to suggest methods for avoiding defects of manufacture in, were building brick, fire brick, sewer pipe, paints, oils, and varnishes. Such other diverse matters, as the making of artificial pearls, utilization of tin scrap and tinfoil, wool scouring, and bleachery practice claimed his attention for various clients.

During World War I the French Government employed McKenna to inspect and accept or reject, on its behalf, explosives it purchased in this country.

McKenna was kind, thoughtful, and courteous to those employed in his laboratories. He was honorable in his dealings with clients. No report went out from his office that had not been thoroughly considered and carefully stated because, realizing that his reports were to be used by persons who would depend upon them, he sought neither to understate nor overstate the facts and the conclusions to be drawn from them. He was always ready to help anyone who needed help. If business interfered with help that he thought he should give, he sacrificed his business to render the help.

He put many a chemist to work in his laboratories and paid him a living wage even though he had really no need to employ the particular man at the particular time. At times he continued a chemist in his employment who he had ceased to need and who was allowed all facilities for study of some problem for his own advancement in his particular field of chemistry while he sought permanent employment elsewhere. Many an investigator in some field of chemical engineering would call on McKenna and discuss his problems at length and be given the freedom of his laboratory for the purpose of making some series of experiments for which he did not have the equipment.

McKenna was a founder and a president in 1914 of The Chemists' Club. He served as chairman of the committee entrusted with the erection of the club's building. He was a founder and in 1910 president of the American Institute of Chemical Engineers.

He was active in the Society of St. Vincent de Paul, the great Catholic charity organization. He helped organize the Catholic Home Bureau for the placement and adoption of orphaned and destitute Catholic children in Catholic homes. Governor Hughes of New York appointed him to the State Probation Commission.

In 1885, McKenna had married Laura O'Neill, and they had seven children, Mary, Morris, Charles, Helen, Laura, Horace and William. His wife died in 1900. In 1903 McKenna married Julia Harlin and they had one daughter Elizabeth.

After his retirement from practice McKenna suffered for several years from the slow progress of a disease that eventually rendered him a complete invalid. He died Apr. 25, 1930 survived by his widow and all his children.

Personal recollections and information from Charles F. McKenna, to whom the writer served as secretary; biography by R. K. Meade, *Ind. Eng. Chem.* **21,** 987-988 (1929), with portrait.

JOHN P. WALSH

William McMurtrie

1851-1913

McMurtrie was born Mar. 10, 1851, on a farm near Belvidere, N.J. to Abram and Almira Smith McMurtrie. In 1867 he entered Lafayette College, enrolling in the mining engineering course, the one nearest to chemistry offered by the college. Graduating in 1871 he became assistant chemist in the United States Department of Agriculture in 1872 working under Ryland T. Brown, chief chemist of the department. He succeeded to Brown's position in 1873. In 1875 he received his Ph.D. degree from Lafayette College. Three years later, he was named superintendent of the agricultural section of the Exposition Universelle in Paris and from 1879 to 1882 served as special agent of the Department of Agriculture.

During his service in the Department, he wrote reports on grape vines, sumac, silk, wool, sugar beets, and other subjects. His report on sugar beets helped establish the industry in the United States.

In 1882 McMurtrie became professor of chemistry at University of Illinois, in 1884 chemist of the Illinois State Board of Agriculture, and in 1886 chemist of the Agricultural Experiment Station.

In 1888 McMurtrie went to New York as chemist of the New York Tartar Co., manufacturers of Royal baking powder. His task was to improve and reduce costs of the product. At that time cream of tartar was produced in a manner which led to its contamination with copper. McMurtrie solved the problem and greatly reduced the costs of producing a pure product. He then undertook the reorganization of the company, establishing successful manufacturing procedures.

McMurtrie became interested in the reorganization of the American Chemical Society, which was undertaken in 1893 when Harvey W. Wiley became president. During the next few years, membership increased substantially and the society's publication program became current. McMurtrie was elected president of the society in 1900. McMurtrie's last publication, "Disposal of Sewage with Recovery of Elements of Plant Food for Use in Agriculture," which appeared in 1913, is indicative of his concern with long-range problems.

He died May 24, 1913, survived by Helen M. Douglas, whom he had married in 1876, and by a son, Douglas C. McMurtrie.

"Dictionary of American Biography," **12,** 146-147, Chas. Scribner's Sons (1933); Edward Hart, "William McMurtrie," *Science,* Aug. 8, 1913; U.S. Department of Agriculture, "Annual Report of the Commissioner," 1874-1882; Douglas C. McMurtrie, "Bibliography of Scientific Writings by the Late William McMurtrie," private printing (1913).

WAYNE D. RASMUSSEN

William McPherson

1864-1951

McPherson was born on a farm 4 miles southwest of Xenia, Ohio, July 2, 1864, the tenth child of William and Mary Ann (Rader) McPherson. After completing grade school in a little red schoolhouse on the back corner of the farm, he attended high school at Xenia, sometimes driving the family buggy but usually riding the plow horse. He entered Ohio State University as a freshman in 1883, just 10 years after the university opened its doors.

There were not yet any organized athletics at Ohio State, but with his superb physique McPherson became a leader in field sports, notably in running and the broad jump. His scholastic work was of the highest grade and was recognized later by his election to Phi Beta Kappa and Sigma Xi. He received his B.S. degree in 1887, M.S. in 1890, and Sc.D. in 1895. He obtained his Ph.D. degree in chemistry at University of Chicago in 1899. He was awarded an honorary degree LL.D. by Wittenberg in 1927 and by Ohio State in 1940.

After receiving his B.S. degree, McPherson taught physics in Toledo High School for 2 years, then chemistry and Latin for the next 3 years. He returned to Ohio State University in 1892 as an assistant to Sidney A. Norton, then chairman of the Department of Chemistry. He was advanced to an assistant professorship in 1893 and to an associate professorship in 1895. In 1897 he was promoted to a professorship and was named to succeed Norton as chairman. During his chairmanship the department grew from a very small department to a very large one. He remained chairman until 1928 when he was succeeded by William Lloyd Evans.

The first years of the present century were momentous ones in the development of chemistry, and there was a pressing need for better texts on the fundamental physical aspects of the science than had previously been available to beginners. McPherson and William E. Henderson addressed themselves to writing first year textbooks in the light of the newer physical chemistry. The famous McPherson-Henderson series of textbooks and laboratory manuals began with publication of "Elementary Study of Chemistry" in 1905.

McPherson's own researches were on the formation of carbohydrates in the vegetable kingdom, reactions between substituted hydrazines and quinones, and the bearing of asymmetric syntheses on the doctrine of vitalism.

When William A. Noyes founded *Chemical Abstracts* at University of Illinois, he invited Austin M. Patterson of Xenia to join him as associate editor. When Patterson succeeded to the editorship, he asked McPherson if the *Chemical Abstracts* office could be housed at Ohio State. McPherson immediately offered the facilities and hospitality of the university to the new journal. Consequently, Patterson moved *Chemical Abstracts* to Columbus where it was housed for many years in the same building as Ohio State's Department of Chemistry.

The Graduate School of The Ohio State University was established in 1911, and McPherson was appointed the first dean. He remained in this position until his retirement in 1937. In this period the Graduate School grew from 300 students in 15 departments to 3300 in 56 departments.

In 1917 McPherson took leave of his academic duties to serve in a civilian advisory capacity with the United States Army. In 1918 he was commissioned captain in the Chemical Warfare Service. He left the army with the rank of lieutenant colonel in 1919 and returned to Ohio State. In 1923 death claimed his first wife, Lucretia Heston, whom he had married in 1893 and who bore him two children, William Heston McPherson and Gertrude May (Mouat). In 1924 he served as acting president of the university, and in 1925 he married Henderson's sister, Mary Browning Henderson, as his second wife.

McPherson served as president of the American Chemical Society in 1930. He was called out of retirement to serve for a second time as acting president of Ohio State University in 1938. After a long illness, he died at Harding Sanitorium, Worthington, Ohio, Oct. 2, 1951.

Obituary, *Chem. Eng. News* **29**, 4314 (Oct. 15, 1951); biography by William Lloyd Evans, *J. Amer. Chem. Soc.* **74**, 859-862 (1952), with portrait and list of publications; *Chem. Eng. News* **22**, 100 (1930).

LAWRENCE P. EBLIN

Richard Kidder Meade

1874-1930

Meade was born in Charlottesville, Va., Nov. 28, 1874. His father was Rev. Francis Alexander Meade. Richard studied at University of Virginia from 1892 to 1894, received his B.S. degree in chemistry from Lafayette College in 1899, and an honorary M.S. degree from the same college in 1908. He was city editor of *Independent-Herald*, Hinton, W. Va. in 1894. From 1896 to 1897 he worked as a chemist for Longdale Iron Co., Allegheny Co., Va., from 1897 to 1902 was an assistant in chemistry at Lafayette College, and after 1902 devoted most of his life to the cement industry, either working for various companies or as a consultant. He started as chief chemist for Edison Portland Cement Co. in New Jersey, established by Thomas A. Edison, and then worked with Northampton Portland Cement Co., Dexter Portland Cement Co., and Tidewater Portland Cement Co., being general manager of the latter from 1911 to 1912. From 1907 to 1911 he ran his own firm, Meade Testing Laboratories in Allentown, Pa., and from

1912 until he died Oct. 13, 1930 he was a consulting chemical engineer in Baltimore.

Meade carried out research on the volumetric analysis, technical analysis, and chemistry of Portland cement, and on the chemical composition of, and causes of the setting of, hydraulic cements. During World War I he developed methods of recovering alkali from cement plants. He was also involved in building plants for separating potash from green sand marls found in southern New Jersey and for the production of barium and strontium compounds. He invented a multi-tubular dryer for the cement industry.

Among Meade's publications should be mentioned: "The Chemists' Pocket Manual" (1900); "The Chemical and Physical Examination of Portland Cement" (1901); "Portland Cement" (1906); "The Design and Equipment of Small Chemical Laboratories" (1907); "Tables for Determination of Economic Minerals" (1907); and "The Technical Analysis of Brass" (1911). "Portland Cement" became an authority in the field and revised editions of it appeared, the last in 1926.

From 1904 to 1910 Meade was editor of "The Chemical Engineer," which he founded. He was one of the organizers, an auditor, and a vice president of the American Institute of Chemical Engineers. In the American Chemical Society he was a secretary of the Lehigh Valley section, and in the American Society of Mechanical Engineers a vice president of the Baltimore section. He also belonged to The Electrochemical Society and to ASME.

R. D. Billinger, "Richard Kidder Meade," *The Octagon* **30,** 72 (1947); R. D. Billinger, "Early History of Cement in Pennsylvania," Commonwealth of Pennsylvania, Department of Internal Affairs, *Monthly Bulletin* **19,** 7-13 (1951); D. Arthur Hatch (editor), "Biographical Record of the Men of Lafayette," Lafayette College, 1948.

GEORGE SIEMIENCOW

Grace Medes

1886-1967

Medes was born Nov. 9, 1886 in Keokuk, Iowa. She earned her B.A. degree at University of Kansas in 1904 and her M.A. degree in 1913. She was awarded her Ph.D. degree in zoology by Bryn Mawr College in 1916. That year she was appointed instructor in zoology at Vassar College and became assistant professor of physiology there in 1919. In 1922 she was appointed associate professor at Wellesley College. She joined the Department of Physiological Chemistry at University of Minnesota Medical School in 1924 as a fellow, becoming assistant professor the following year. Here she engaged actively in clinical physiological chemical research until 1932. During this period she discovered the human metabolic disorder which she named "tyrosinosis." Detailed experiments had established that a human, otherwise normal individual, excreted large amounts of tyrosine and *p*-hydroxyphenylpyruvic acid. For many years this was the only case on record. Later work, particularly in Norway, revealed the abnormality to be not as rare as was originally thought. A symposium on tyrosinosis was held in her honor June 2 and 3, 1965 in Oslo, Norway and proceedings were published as a Norwegian monograph on medical sciences in 1966.

In 1932 Medes became a member of the cancer research staff of Lankenau Hospital Research Institute, Philadelphia (merged in 1950 with the Institute for Cancer Research, Fox Chase, Philadelphia). Stanley P. Reimann, pathologist for the hospital and director of the Institute, had assembled a group of chemists, biochemists, and zoologists to investigate aspects of normal and malignant growth, and in particular, the relationships of different sulfur groups in compounds to stimulation and retardation of cell proliferation. In this atmosphere Medes began a biochemical study which contributed to the clarification of pathways of cysteine metabolism. In those days when "physiological chemistry" was becoming "biochemistry" and before paper chromatography, analytical chemistry problems in amino acid work presented grave stumbling blocks. These things she handled with patience.

With the availability of carbon isotopes in the mid-forties her work on fatty acid metabolism, undertaken with associates of the Institute, led to a series of papers which did much to clarify the intermediate steps in metabolism. Among these studies was a series of papers that demonstrated the participation of acetyl groups in acetoacetate synthesis, which helped to pave the way for the later discovery of acetyl coenzyme A. After 1950 isotope tracer studies were extended to the metabolism of fatty acids in cancer cells.

There were times through the years that Medes, at risk to her health, used herself as a "guinea pig" for her research. But despite these risks she continued active in chemical research even following her retirement in 1956, when she was given a post as visiting scientist in the Fels Research Institute, Temple University. She resumed her work on tyrosinosis and collaborated with Gerald Litwack on a paper which appeared in *Analytical Biochemistry* in 1967, "An Isotope Tracer Method for Determination of Tyrosine Transaminase Activity in Biochemical Material."

In 1955 the American Chemical Society awarded Medes the Garvan Medal which honors annually a woman chemist of distinction.

A modest woman of simple tastes with a ready smile, whose work and hobby were fused in an interest in biochemistry, she did, none-the-less, take a bit of time for cabinet-making, gardening, and camping. The niece of her friend, Albina Howell Read of Rosemont in suburban Philadelphia, with whose family Grace spent many holidays, recalled that Grace once said that she "liked to get to bed early and couldn't wait for the next day to come so that she could get on with her work." For a time she cared for a brother in Philadelphia until his death.

Aging and in failing health following a serious auto accident, she struggled to continue her work at Fels Institute laboratory. She was fortunate that the institute director, Sidney Weinhouse, took a friendly and compassionate interest in her during her final illnesses. She died in Philadelphia on Dec. 31, 1967.

Obituary by Sidney Weinhouse, Director, Fels Research Institute, Temple University, for release to the press; communications from Jean Read, and Clifford J. D. Read; personal recollections.

ETHEL ECHTERNACH BISHOP

George Merck

1867-1926

George Wilhelm Merck

1894-1957

In the year 1889 George Merck was dispatched by his father Wilhelm, head of the old and prosperous firm of E. Merck, Darmstadt, Germany, to open an American marketing outlet for Merck chemicals and pharmaceuticals. George was only 22 at the time, son of Wilhelm and Caroline (Moller) Merck, born in Darmstadt, Aug. 15, 1967 and educated there, privately and at the Latin School, with visits to England for language study. E. Merck, which had a company history going back as far as 1668 when Friedrich Jacob Merck took over the Engelapotheke pharmacy in Darmstadt, was in 1889 an extraordinarily sound organization which had based its great expansion in the early part of the nineteenth century on the manufacture of morphine, codeine, cocaine, and other alkaloids. This production was based on thorough research which resulted in dozens of papers and patents in the German chemical literature, mostly with some member of the family as principal investigator.

George Merck set up Merck and Co. in New York, with Theodore Weicker, a Darmstadt native who had acted as E. Merck's representative since 1887, as senior partner. Merck and Co. was conceived by the parent corporation as a marketing firm only, but George Merck had other ideas. By 1899 he acquired the land in Rahway, N.J., which is still the site of the company headquarters and principal research facilities. Here the manufacture of chemicals was begun in a single building. By 1904 Weicker (who would later be head of E. R. Squibb and Sons) withdrew; and in 1908 George Merck incorporated the firm in New York State with a capital of $250,000 and an issue of $750,000 debentures and became the first president. In the decade before World War I Merck and Co. became one of the leading producers of bismuth salts, iodine preparations, chloral hydrate, acetanilide, salicylates, and a variety of narcotics, alkaloids, disinfectants, and prescription, photographic, and reagent chemicals. During this time the company expanded by acquisition and construction: in 1903 the St. Louis firm of Herf and Frerichs was purchased; in 1910 a plant was built in Midland, Mich.; and in 1911 a branch office was opened in Montreal.

At the start of World War I George Merck deposited with the United States Alien Property Custodian a quantity of common stock sufficient to cover that portion of the company held by German interests. When the Custodian sold this stock the corporation was recapitalized by sale of preferred stock, with George Merck retaining controlling interest and all German holdings eliminated. Since that time the company has grown steadily to the major producer of pharmaceuticals and fine chemicals which it is today.

George Merck was married in Darmstadt in 1893 to Friedrike Schenck, of Antwerp, Belgium. In 1895 he became a citizen of the United States. He retained his position as president of the corporation until 1925, when he became chairman of the board and handed over the presidency to his only son and the eldest of his five children, George Wilhelm. George Merck died in West Orange, N.J., Oct. 21, 1926.

Under George Wilhelm, Merck and Co. changed character gradually, deepening the research orientation and becoming the almost scholarly organization which publishes the well-known "Merck Index" of drugs and chemicals, the "Merck Manual of Therapeutics and Materia Medica," and the "Merck Report" (now *Merck, Sharp, and Dohme Seminar Report*), a quarterly on pharmacy and medicine.

Born Mar. 29, 1894 and graduated A.B. from Harvard in 1915, George W. assumed the presidency of the company in its present form in 1927 when it merged with Powers-Weightman-Rosengarten of Philadelphia to become Merck and Co., Inc. In 1933 new and greatly enlarged research laboratories were dedicated, and Merck Institute for Therapeutic Research, an independent nonprofit organization, was established. Merck and Co., Inc. was a leader in synthesis and production of vitamins in the years before World War II and was one of the early developers of sulfa drugs, penicillin, and streptomycin, as well as cortisone and other steroids. George W. served during World War II as special consultant to the Secretary of War on biological warfare and was director of the American Forestry Association. He stepped down from the presidency of Merck and Co., Inc. in 1950, becoming chairman of the board, and died Nov. 9, 1957.

George Merck: "National Cyclopedia of American Biography," **37,** 17, James T. White & Co. (1951). George W. Merck: *ibid.*, "Current Biography" (1946) 387-90, H. W. Wilson Co.; *Am. Forests* **63,** 46 (1957); Merck & Co.: W. Haynes and E. L. Gordy, "Chemical Industry's Contribution to the Nation: 1635-1935," *Chem. Ind.*, May 1935 (Supplement); W. Haynes, "American Chemical Industry," Vol. VI, pp. 271-75, D. Van Nostrand (1949); *Fortune* **35,** 104 ff. (June 1947); *Time* **60,** 38 ff. (August 18, 1952).

ROBERT M. HAWTHORNE JR.

Otto Meyerhof

1884-1951

Meyerhof was born Apr. 12, 1884, in Hanover, Germany, the son of Felix and Bettina May Meyerhof. He received his M.D. degree from University of Heidelberg in 1909 with a thesis in the field of psychiatry. After graduating, he worked in the clinic of Ludwig Krehl, where he met Otto Warburg, who played an influential role in Meyerhof's shift of interest from psychology and philosophy to physiology (especially cellular physiology). He joined the staff of the physiological laboratory at University of Kiel in 1912, becoming professor extraordinarius in 1918. In 1924 he moved to the Kaiser-Wilhelm Institute for Biology in Berlin-Dahlem. When the Institute for Medical Research was established in 1929 in Heidelberg, he became head of its Department of Physiology. But Meyerhof was a Jew, and in 1938 he and his family were forced to leave Germany. They went to France, where he was made director of research at the Institut de Biologie Physico-Chimique in Paris. When the Nazis invaded France in 1940, the Meyerhofs escaped over the Spanish frontier and came to America. Meyerhof was made professor of physiological chemistry at University of Pennsylvania, a post which he held until his death, Oct. 6, 1951.

Meyerhof's research interests centered about the chemical and energy transformations which take place in the organism. As early as 1913 he published a theoretical paper on the energetics of cellular processes, which expressed his interest in the problem of how the potential energy of foodstuffs is made available to the cell.

In the 19th century the formation of lactic acid had become associated with muscular activity and fatigue, and certain investigators had suggested that glycogen might be the source of lactic acid. It was not until 1907, however, that really reproducible values were obtained for the lactic acid content in the different states of muscle by Walter M. Fletcher and Frederick G. Hopkins. They clearly demonstrated a marked increase in the concentration of lactic acid in fatigued muscle over resting muscle and showed that if the muscle were allowed to recover in oxygen the lactic acid disappeared. There was also some indication that the lactic acid might not all be removed by oxidation. In 1912-13 Archibald V. Hill showed that oxygen was needed for the recovery stage but not for the actual contraction. His data also indicated that during recovery not all of the lactic acid was oxidized. The fate of the lactic acid was not definitely settled until the work of Meyerhof in 1919-20. Meyerhof showed that the lactic acid consumed during recovery was only equivalent to one-fourth to one-third of the lactic acid disappearing. He confirmed the observation of Jacob Parnas and Richard Wagner (1914) that the glycogen that disappeared during muscle activity was equivalent to the lactic acid formed and further demonstrated that the increase in glycogen during recovery is equivalent to the amount of lactic acid which is not oxidized. These facts established that during recovery part of the lactic acid is oxidized to provide energy to convert the rest of it back to glycogen. For this work Hill and Meyerhof shared the Nobel Prize in Medicine and Physiology for 1922.

In 1926 Meyerhof reported the preparation of a potassium chloride extract from muscle which was capable of carrying out all of the reactions of glycolysis with added glycogen or hexosephosphates. Meyerhof's laboratory contributed greatly to the elucidation of the steps involved in the metabolism of hexoses, and this sequence of reactions is often referred to as the Emden-Meyerhof or Emden-Meyerhof-Parnas scheme.

Meyerhof originally maintained that lactic acid was directly involved in the mechanism of muscular contraction. In 1927 a labile phosphorus compound was discovered in muscle by Cyprus Fiske and Yellapragada SubbaRow and by Phillip Eggleton and Grace P. Eggleton. Later identified as phosphocreatine, it was shown to break down during contraction and also to be resynthesized during recovery. Einar Lundsgaard observed in 1930 that muscle poisoned with iodoacetate did not form lactic acid but was still capable of contraction. This forced Meyerhof to abandon his lactic acid theory of contraction and focused the attention of investigators on phosphocreatine breakdown as the source of energy for muscular contraction.

In 1929 another phosphorus compound, adenosinetriphosphate, had been isolated in Meyerhof's laboratory by Karl Lohmann. Meyerhof established the high-energy content of such phosphorus compounds and recognized their significance in energy transfer. He and his coworkers later demonstrated that the breakdown of adenosinetriphosphate actually provides the energy for muscular contraction, with lactic acid production and phosphocreatine breakdown participating only indirectly by providing energy for the resynthesis of adenosinetriphosphate.

Dorothy Needham, "Machina Carnis: The Biochemistry of Muscular Contraction in its Historical Development," Cambridge Univ. Press (1971); David Nachmanson, Severo Ochoa and Fritz Lippmann, *Science* **115,** 365-368 (1952), reprinted with an appended bibliography in *Biog. Mem. Nat. Acad. Sci.* **34,** 153-182 (1960), with portrait; R. A. Peters, *Obit. Not. Fellows Roy. Soc.* **9,** 175-200 (1954), with portrait and bibliography; Dorothy Needham, *Nature* **168,** 895-896 (1951).

JOHN PARASCANDOLA

Arthur Michael

1853-1942

There is little disagreement that as the twentieth century opened Arthur Michael was an outstanding American chemist. Michael is a true study in contrast. While his scientific accomplishments are widely recognized, his personal characteristics are virtually unknown. His synthetic contributions are almost incidental to his all consuming interest in organic theory. Students knew Michael as a stern, demanding scientific mentor and a charming, delightful host. Such was the make up of this exceedingly complex personality.

Michael had most of the advantages available to man. The most important of these was a brilliant, inquiring intellect coupled with an extraordinarily retentive memory. However, there were others, including the affluence needed to travel over Europe to work in the laboratories of the foremost chemists of his day—August Wilhelm von Hofmann at Berlin, Robert Wilhelm Bunsen at Heidelberg, and Charles Adolphe Wurtz at Paris. Not only the professors but his fellow students resemble a "Who's Who" of modern chemistry—Ira Remsen, Charles L. Jackson, Siegmund Gabriel, and Thomas F. Norton. Undoubtedly these men influenced Michael; challenging and sharpening his keen intellect. All of these advantages culminated in a research career of the highest order. So intense was Michael's research effort that much of the work was done in private laboratories fitted out at his own expense. The first of these was in his boyhood home in Buffalo, N. Y. This laboratory and a concerned teacher played a key role in his development since no formal instruction was given in chemistry at that time. Later private laboratories were located on the Isle of Wight and at Newton Center, Mass. Even after his students began to carry out their studies at Harvard's Converse Laboratory, they came to Professor Michael at Newton Center to report. This physical aloofness, along with Michael's reluctance to attend meetings, read papers, and in any way popularize his views, surely contributed to his being known well by only a limited number of physical-organic chemists. These same traits may have made his significant contributions possible.

The nature of Michael's contribution was very much related to his times. His career began just as organic structural theory and thermodynamics had achieved spectacular successes. The application of sound physical chemical reasoning to organic chemical reactions, now placed on an understandable basis, proved to be a masterstroke. Perhaps the chief deficiency of the theories of organic reactions current in Michael's day was their mechanical application of structural theory. His application of the concepts of free energy and entropy did much to bring the earlier theories into accord with new experimental evidence—much of it from his own laboratory.

The study of unsaturated compounds, active methylene synthesis, tautomeric equilibria, and addition reactions occupied a great deal of Michael's interest; for example:

$$\mathrm{PhCH{=}CH{-}\overset{\overset{\displaystyle O}{\|}}{C}OEt} \rightleftharpoons \mathrm{\underset{+}{Ph\overset{}{C}H}{-}CH{=}\overset{\overset{\displaystyle O^-}{|}}{C}OEt}$$

$$(EtO_2C)_2CH_2 + EtO^- \rightleftharpoons$$
$$(EtO_2C)_2CH^- + EtOH$$

$$\mathrm{Me\underset{+-}{\overset{\overset{\displaystyle ONa}{|}}{C}}CHCO_2Et + \underset{+-}{MeI}} \rightarrow$$

$$\mathrm{Me\overset{\overset{\displaystyle O}{\|}}{C}\underset{\underset{\displaystyle Me}{|}}{C}HCO_2Et + NaI}$$

His understanding of "positive-negative properties" (related to electronegativity) and "chemical neutralization" (an increase in entropy) are quite clear in the third example.

It is essential to realize both the importance of and the reason for Michael's contribution to synthetic organic chemistry. A useful insight on the former question is the extensive (377 pages) review of the Michael condensation published in "Organic Reactions," **10** (1959). The reason for this impressive accomplishment lies in Michael's insistence that unless a synthetic study had in it the possibility of establishing a theoretical principle, it was to be abandoned for a more promising approach. The Michael condensation proved, in his hand, to be both synthetically and theoretically rewarding:

$$PhCH{=}CH\overset{\overset{O}{\|}}{C}OEt + (EtO_2C)_2CH_2 \xrightarrow{\text{base}} \underset{\underset{CH(CO_2Et)_2}{|}}{PhCH}CH_2CO_2Et$$

The proper place of synthesis in organic chemistry is important in understanding Michael and his time. Few of his contemporaries acquired facility to apply his line of reasoning or to understand organic reactions as he did. It appears that we have still another example of contradiction in Michael's career. There is solid evidence that his methods were readily available and their importance appreciated; yet apparently the task of mastering the intellectual technique was too demanding.

This curious state of affairs led to still another of Michael's major contributions to chemistry; *e.g.*, his role as an important critic. His investigation of trans-addition, the benzilic acid rearrangement, steric hindrance, and ring strain theory all helped to clear up muddy thinking and refine or replace theories that suffered from the limitation of a purely mechanical interpretation.

Arthur, the son of John and Clara Michael, was born in Buffalo, N.Y., Aug. 7, 1853. His university education was obtained in Germany and France. He became professor of chemistry at Tufts College in 1880 and except for the period 1891-1894, taught there until 1907. The three missing years were spent on the Isle of Wight after an abortive effort as department chairman at Clark University. In 1912 he was elected professor of organic chemistry at Harvard where he became emeritus in 1936. He married one of his brilliant young students, Helen Abbott, in 1889; they had no children. He died Feb. 8, 1942 in Orlando, Fla.

Michael was an extraordinary synthetic organic chemist who at the same time can reasonably be called the grandfather of physical-organic chemistry. He was a scientific hermit who, through his writing and students, brought organic chemistry into the age of the electron. Michael was a teacher who brought students to the ". . . verge of despair by his exacting standards . . ." and then delighted them in his home with the breadth of his knowledge of art, music, and world affairs. Such was the complex nature of this American chemist who won international fame.

Biographies by W. T. Read, *Ind. Eng. Chem.* **22,** 1137-1138 (1930); E. W. Forbes, L. F. Fieser, and A. B. Lamb, *Harvard University Gazette,* 246-248, May 22, 1943, and A. B. Costa, *J. Chem. Educ.,* **48,** 243-246 (1971).

K. THOMAS FINLEY

Leonor Michaelis

1875-1949

Michaelis was one of the most influential scientists of the first half of the twentieth century when the methods of physical chemistry were being introduced into biology and medicine.

He was born Jan. 16, 1875, in Berlin. In 1893 he graduated from Köllnisches Gymnasium, which, although humanistic, had a chemical and physical laboratory. Hence, when he entered University of Berlin in the summer of the same year he was able to start studying organic chemistry under Emil Fischer. In 1896 he obtained his state license in medicine and in the next year passed the examinations in medical specialties.

While still a student in Berlin, Michaelis published a paper on histology with Oskar Hertwig and a paper on cytology. Upon his return to Berlin in 1897 he worked under Hertwig, then for a year was private assistant to Paul Ehrlich, at which time he discovered that Janus green stains mitochondria. He then became assistant in one of the municipal hospitals in Berlin.

In 1904 a paper by Michaelis attracted the attention of Ernst von Leyden who engaged him as research assistant at the newly created Institute for Cancer Research at the university clinic in the "Charite" hospital in Berlin. In the course of his employment there, he was sent to Jena to secure the first ultramicroscope built in the Zeiss optical works, an instrument which gave him an opportunity to study colloid chemistry so that in 1903 he acquired the position of *Privatdocent* at University of Berlin.

In 1905 Michaelis was promoted to professor extraordinarius, but he felt that any advancement in an academic career was hopeless because of his Jewish lineage so he accepted the position of bacteriologist in a municipal hospital in Berlin. From 1905 to 1914 Michaelis accomplished a remarkable series of researches in biochemistry and biophysics. He discovered the dependence of enzyme activity on pH simultaneously with Sørensen. The Michaelis-Menten constant is a factor representing the affinity of an enzyme for its substrate. He worked on the quantitative purification of proteins by their isoelectric points and the acid agglutination of certain bacteria, each at a characteristic pH. Together with Peter Rona he showed that the glucose of blood exists in the free state, not combined with a protein. They distinguished two classes of enzyme inhibitors and studied the absorption of electrolytes.

During World War I Michaelis engaged in hospital work. In 1921 the German government offered him a position at University of Berlin but without salary or laboratory, hence, he accepted a temporary professorship in biochemistry at the medical school in Nagoya, Japan. Here he did important research on the permeability of membranes, which attracted the attention of Jacques Loeb.

A lecture tour in the United States led to a resident lectureship for 3 years (1926-29) and then in 1929 to membership in Rockefeller Institute where he became head of a physical chemistry laboratory. His research there dealt with oxidation-reduction processes and especially the role of semiquinone radicals in reversible oxidation systems. The idea of the existence of such free radicals was at first rejected by the editors of American journals. However, Michaelis demonstrated the paramagnetism of the substances in question, which left no doubt as to the validity of his views. A later discovery was the radical of vitamin E.

Among other topics taken up in Michaelis' laboratory were studies on iron metabolism and the discovery that keratin is converted into soluble proteins when thioglycolic acid reduces its disulfide bonds. He learned later that the latter publication was the cornerstone for the "cold" permanent wave.

Michaelis became an emeritus member of the Rockefeller Institute in 1941, but he was elected a member of the National Academy of Sciences in 1943 and kept up active research until his death, Oct. 8, 1949.

Obituary in *Chem. Eng. News* **27,** 3432 (1949); biographies by D. A. MacInnes and S. Granick in *Biog. Mem. Nat. Acad. Sci.* **31,** 282-321 (1958), by W. M. Clark in *Science* **111,** 55 (1950), and by S. Granick in *Nature* **165,** 299 (1950); George W. Corner, "History of the Rockefeller Institute" (1964), pp. 178-180, Rockefeller Institute Press.

VIRGINIA MCCONNELL

Peter Middleton

Died 1781

Middleton, born in Scotland, practiced medicine in New York from about 1750 until his death, Jan. 9, 1781. He was one of the founders of Columbia University Medical School where he was professor of pathology and physiology from 1767 to 1776 and professor of chemistry and materia medica from 1770 to 1776.

He was one of the earliest American professors of chemistry, being preceded only by John Morgan, James Smith, and Benjamin Rush. During his day he was a leading New York physician, but nothing whatsoever is known about him as a teacher of chemistry.

"Dictionary of American Biography," **12,** 602-03, Chas. Scribner's Sons (1933); James Thacher, "American Medical Biography," DeCapo Press (1828, reprint 1967) **I,** 384-385; Milton H. Thomas, "Columbia University Officers and Alumni" (1936).

WYNDHAM D. MILES

Thomas Midgley, Jr.

1889-1944

Thomas Midgley, Jr., the man who invented Ethyl gasoline, was born in Beaver Falls, Pa., May 18, 1889. Just as Sir Humphrey Davy called Michael Faraday the greatest of his discoveries, so Charles F. Kettering, the founder of General Motors Research Laboratories, called Midgley one of his greatest discoveries. Although Midgley had a degree in mechanical engineering from Cornell University (1911), he won his greatest fame as a chemist. Yet, like Michael Faraday, he got his knowledge of chemistry mostly by studying it himself and working at it. For years Midgley said he "ate and slept chemistry."

His father, Thomas Midgley, Sr. was a prolific inventor, notably in automobile tires. His mother, Hattie Lena (Emerson) Midgley, was a daughter of James Emerson, inventor of the inserted-tooth saw. When Midgley was six, his family moved to Columbus, Ohio. There he attended public schools and became a boyhood friend of Evan J. Crane, later editor of *Chemical Abstracts.*

Upon being graduated from Cornell, Midgley took a job with National Cash Register Co. in Dayton, Ohio, as a draftsman and designer. Charles F. Kettering had been employed by the same company a short time before. Kettering had left to start his own firm, Dayton Engineering Laboratories Co. At the age of 27 Midgley went to work with Kettering, and the next 15 years were the time in which his most creative work in chemistry was done.

First he investigated the knock in the gasoline engine. In 1922 he discovered tetraethyllead as a gasoline antiknock preventative compound after others had tried an estimated 33,000 different compounds without success. Tetraethyllead became an important ingredient in the Ethyl gasoline of ordinary commerce. Midgley later was vice-president of Ethyl Gasoline Corp.

With Albert Henne of Ohio State University he began a search for compounds to replace the noxious ammonia gas used as the common refrigerant in refrigerators and air-conditioning units. They finally prepared dichlorodifluoromethane, now called Freon, which proved to be stable, non-inflammable, non-toxic, and yet to be eminently suitable as a refrigerant fluid.

His lifelong interest in rubber led him to undertake research on the composition of natural rubber, the chemistry of vulcanization, and the synthesis of artificial rubber. He considered this work his best scientific effort and made this the subject of his address, "Concepts in Rubber Chemistry" which he gave when he received the Gibbs Medal of the American Chemical Society in 1942.

Midgley gave great service to chemistry through the American Chemical Society. He was a member of the Board of Directors from 1930 until his death in 1944 and was elected president of the society for 1944.

Midgley's work was widely recognized during his lifetime. He received four of the principal medals awarded for chemical achievement: the Nichols Medal (1922), the Perkin Medal (1937), the Priestley Medal (1944), and the Gibbs Medal (1942). In addition he received the Longstreth Medal of the Franklin Institute (1925).

Among Midgley's other accomplishments, he played a dominant part in the production of high-octane gasoline, used in military aircraft in World War II. He discovered one of the first known catalysts for cracking hydrocarbons to yield aromatic compounds, iron selenide. He also developed the Midgley optical gas engine indicator and the widely used bouncing-pin indicator. He held 117 patents and took an active role in the commercial development of ideas which he had conceived.

Aug. 3, 1911 Midgley was married to Carrie M. Reynolds of Delaware, Ohio. They had two children, Jane (Mrs. Edward E. Lewis) and Thomas Midgley, 3rd.

He contracted infantile paralysis in 1940 and spent the last 4 years of his life as an invalid. He devised a harness with pulleys to assist himself in arising from bed. Nov. 2, 1944 at his home in Worthington, Ohio, he became entangled in the device and was strangled.

Midgley was a great showman. When he presented his paper on antiknock agents before the American Chemical Society, he had the stage of Carnegie Music Hall in Pittsburgh full of apparatus. With it he made impressive demonstrations of engine knocking, both in a glass tube and in an engine, and of how it could be stopped altogether by chemical means. In presenting his first paper on Freon, he demonstrated in one operation both its non-toxic and its nonflammable nature. Taking his lungs full of the gas he then slowly exhaled and extinguished a candle with his breath.

Biographies by Williams Haynes in Eduard Farber, "Great Chemists," Interscience (1961), pp. 1588-1597, with portrait; and by T. A. Boyd, *J. Amer. Chem. Soc.* **75,** 2791-2794 (1953); obituary, *New York Times,* Nov. 3, 1944.

JOSEPH A. SCHUFLE

John Millington

1779-1868

Millington was born in London, May 11, 1779. He attended Oxford but had to leave before graduating because of financial difficulties. He took up law and was admitted to practice in 1803 although he seems to have used his knowledge of legal matters only in acting as a patent agent for inventors. For a time he studied medicine, and then he turned to engineering. He found the latter to his liking. He worked with McAdam, the road builder, helped install the first gas lights on London streets and was associated in other engineering enterprises.

In 1815 the Royal Institution engaged him to present a series of lectures on natural philosophy. Two years later he was appointed professor of mechanics. From 1815 to 1829 he gave courses of lectures at the institution on natural philosophy, mechanics, and astronomy. There Humphry Davy, Michael Faraday, and other famous scientists were among his friends. During this period Millington also taught chemistry and natural philosophy at Guy's Hospital Medical School and mechanics at London Mechanics' Institution. For his students he wrote a text, "An Epitome of the Elementary Principles of Mechanical Philosophy" (1823). He was appointed professor of mechanics on the first faculty of the University of London in 1827, but he resigned before the university had opened.

Largely self-taught in engineering and science, Millington was a highly competent engineer and an excellent teacher. John Griscom, an American chemist who heard Millington lecture at Guy's in 1819, said this of him: "Professor M. appears to be an able teacher. He has devoted himself to science, *con amore,* and by the efforts of a strong native genius has acquired great facility in delivering instruction in the various branches of physical science."

In 1829 or 1830 Millington sailed to Mexico to superintend a group of silver mines and a mint owned by an English company. Silver Mexican coins from this mint are marked with his initials, J.M. While living in Guanaxuato he edited the second edition, 1930, of his "Epitome of . . . Philosophy." In 1832 or 1833 he left Mexico and settled for a time in Philadelphia. He opened "an extensive depot and manufactory for all kinds of philosophical, chemical, optical, and mathematical instruments, apparatus, machines, and implements, including chemical tests and reagents, minerals, and all articles connected with scientific instruction, investigation, and amusement." He promised that his products would be "of the greatest purity and perfect for chemical and philosophical experiments and . . . fully equal to any . . . imported from Europe." He also practiced engineering on occasion and taught the profession to private students. One of his students was Alfred Kennedy, who became an early advocate of teaching chemistry by laboratory instruction.

In 1835 Millington was appointed professor of chemistry and natural philosophy at College of William and Mary, and he moved to Williamsburg. Apparently he was a popular teacher, for the enrollment in his subjects increased greatly during his tenure. He made a collection of mineral and geological specimens for the students to study. He also taught classes in medicine, pharmacy, and surveying, and wrote a large text, "Elements of Civil Engineering" (1839). A quarrel among the faculty, which forced the college to close during the session of 1848-49, caused him to leave Williamsburg and accept the professorship of natural science at the newly organized University of Mississippi. He was also head of the geological survey of the state of Mississippi although the field work was done by his associates.

In 1853 Millington became professor of chemistry and toxicology at Memphis Medical College. His versatility was exemplified by his teaching a course in natural history upon the sudden death of the regular professor and also giving students an unscheduled course in geology because they requested it. The appreciative students presented him with a cane and a book.

Millington retired to live at La Grange, Tenn. in 1859 when he reached the age of 80. The turmoil of the Civil War caused him to move to Philadelphia, and after peace returned he lived with his daughter in Richmond, Va. where he died June 10, 1868.

Galen Ewing, "Early teaching of science at the College of William and Mary," *J. Chem. Educ.* **15,** 3-13 (1938); records of Franklin Institute, Philadelphia; *Memphis Medical Recorder,* vols. 2-6, 1853-1857; "Dictionary of National Biography," **13,** 441-2, Oxford Univ. Press (1921-2); "Dictionary of American Biography," **12,** 647, Chas. Scribner's Sons (1933); Wyndham Blanton, "Medicine in Virginia in the 19th Century," Garrett & Massie, Inc. (1933), pp. 15-16; John Griscom, "A year in Europe," Collins & Hannay, N. Y., H. C. Carey & J. Lea, Philadelphia, Richardson & Lord, Boston (1824), pp. 151-152; *Register and Library of Medical and Chirurgical Science* **1,** 12 (1833).

WYNDHAM D. MILES

Carl Shelley Miner

1878-1967

Carl Miner was born in State Center, Iowa, Aug. 5, 1878. After a year at Coe College, he transferred to University of Chicago, studied under Julius Stieglitz, and received his bachelor's degree in 1903. After graduation Miner was employed as research chemist by a firm which later became Corn Products Co. In 1906, fortified by his own venturesomeness, the encouragement of Stieglitz, and a loan of $10,000 from his father (repaid in 2 years), Miner established Miner Laboratories, which was to become the pre-eminent chemical consulting firm in Middle America.

At the start of this century, little research was done in industrial laboratories. Chemists were mainly employed to analyze raw materials. The newly enacted Food and Drug Law of 1906 required chemical control which the new Miner Laboratories were equipped to supply. Much of Miner's early work was concerned with determination of protein, fat, and fiber in agricultural feedstuffs.

In time, the larger food-handling companies turned to food processing and encountered the problems of by-products of their operations. To solve these problems, chemical research, rather than analytical control alone, was required, and this led to a major break-through by Miner. As an example, Quaker Oats Co. had great quantities of oat hulls which were a disposal nuisance. Miner first attempted to convert the hulls to sugars by acid hydrolysis; eventually, instead, he developed a practical process to manufacture furfural, a substance then unknown commercially but having today extensive industrial uses. Another early project of the Laboratories was corrosion inhibitors for glycerine antifreezes [ethylene glycol came later in this application].

Miner was a very accurate and meticulous man with a bent for legal activities. He became involved in patent infringement cases and was often an expert witness in trials involving not only patents but civil and criminal matters. The Laboratories also served as training school for many scientists who went on to important technical and administrative positions. By his training of and advice to chemists who came to him for guidance, Miner played a significant role in the burgeoning technology of the twentieth century. Ahead of his time, he took an active interest in promoting the professional improvement of chemists.

Miner had facility with, and respect for, the language. He wrote and spoke precisely, and he had a feeling for the right word in its proper place. He insisted on clarity in research reports, and his dictum was that reports should be written "not only so that they can be understood, but so that they cannot be misunderstood." He was a pioneer in the establishment of one of the earliest American Chemical Society local section publications, the *Chemical Bulletin* of the Chicago Section. He served as editor from 1917 to 1920 and as advisor for many years thereafter. He was chairman of the Chicago Section, 1922-23, and continued to serve the Society as a leader and elder statesman.

He dabbled skillfully in poetry, and his annual New Year's verse was eagerly awaited by his colleagues. Miner was a good athlete and an accomplished swimmer and tennis player, almost to the time of his death. Later in life, he and his wife became interested in shallow sea life and regularly spent winter vacations in Florida pursuing this hobby, studying the chemistry and biology of sea animals. Their collections and photographs of these organisms were outstanding in quality.

Miner received numerous honors and recognitions. Coe College awarded an honorary D.Sc. degree. He received the Modern Pioneer Award, the Perkin Medal, the Honor Scroll of the American Institute of Chemists, and a tribute for 25 years of service on the Board of Presbyterian St. Lukes Hospital in Chicago. He was a lieutenant colonel in the Chemical Warfare Reserve, a member of the Baruch Rubber Committee of World War II, and a dollar-a-year member of the OPRD Chemical Referee Board.

In 1956, Miner sold his firm to Arthur D. Little, Inc. When Universal Oil Products Co. was reorganized and became a publicly owned corporation in 1959, Miner became a member of the board and played an important role in guiding the firm in new directions. He also carried on a consulting practice, practically until his sudden death at age 89 on Oct. 22, 1967.

One of Miner's great gifts was his ability to see the essence of any problem, personal or business. He could quickly cut through extraneous material and get straight to the essentials. He never lost his enthusiasm for chemists and chemistry.

Chemical Bulletin (publication of Chicago Section, A.C.S.), Nov. 1955, June 1956, Oct. 1963, Dec. 1967; personal recollections.

ERNEST H. VOLWILER

John Kearsley Mitchell

1793-1858

Mitchell, a physician and chemist, was born in Shepherdstown, Va. (now W. Va.), May 12, 1793. His father sent him to Scotland, the family's homeland, for his elementary and collegiate education, and John received his bachelor's degree from University of Edinburgh. Returning to the United States, he studied medicine and obtained his M.D. degree from University of Pennsylvania in 1819.

To improve his health, Mitchell went to sea as a ship's surgeon for 3 years, making several voyages to the Orient. His health restored, he settled in Philadelphia and began to practice medicine. Soon he was asked to teach chemistry in Philadelphia Medical Institute, a school which had been organized to give medical lectures during the summer when the university was on vacation. In 1833 Franklin Institute asked him to deliver lectures on chemistry applied to the arts.

Mitchell was said to have been an eloquent, imaginative speaker, and apparently he illustrated his lectures with many demonstrations. One of his demonstrations was to solidify carbon dioxide, the first time this was done in the United States.

He published many medical writings and a book of poetry but little on chemistry. His significant contribution to chemical literature was an edition of Michael Faraday's "Chemical Manipulations, being instructions to students in Chemistry on the Methods of Performing Experiments of Demonstrations or of Research with Accuracy and Success" in Philadelphia in 1831; this was to serve as a manual for students who wanted to practice chemistry on their own because, in Mitchell's words, "In the United States there is not a single laboratory devoted to the instruction of students in analytical chemistry."

Mitchell taught chemistry at Philadelphia Medical Institute and Franklin Institute until 1840, concurrently practicing medicine. In 1841 he became a professor of medicine in Jefferson Medical College and thereafter concentrated on medicine. He died in Philadelphia, April 4, 1858.

Biography in *Boston Med. Surg. J.* **41,** 38-41 (1849); obituaries in *Trans. Med. Soc. State Penna.,* 11th annual session, part IV, 93-98 (1859), *N. Amer. Medico-Chirurgical Rev.* **2,** 588-590 (1858), *Charleston Med. J. Rev.* **13,** 122-127 (1858), *Med. Surg. Reporter* **11,** 307-311 (1858), and by R. Dunglison in *Proc. Amer. Philos. Soc.* **6,** 340-342 (1854-1858); S. H. Dickson, "The Late Prof. J. K. Mitchell," inaugural lecture, Jefferson Medical College, 1858, 40 p. pamphlet; quote is from preface to Mitchell's edition of "Chemical Manipulations."

WYNDHAM D. MILES

Thomas Duché Mitchell

1791-1865

Mitchell was born in Philadelphia in 1791. He attended academies, then studied medicine under a physician, Joseph Parrish. Parrish desired that his students learn pharmacy thoroughly. Mitchell, therefore, worked part-time for nearly a year in an apothecary shop and laboratory owned by Adam Seybert. Parrish and Seybert were competent chemists. Parrish delivered courses of public chemistry lectures in Philadelphia from 1807 to 1810, and Seybert published articles on mineralogy and chemistry. Either of these men may have kindled Mitchell's interest in chemistry. Mitchell studied under Parrish for 3 years, concurrently attending University of Pennsylvania Medical School and receiving his M.D. degree in 1812.

While Mitchell was still a student he helped organize the Columbian Chemical Society in 1811 and was the first president of the organization. He published material from his thesis on acidification and combustion in *Memoirs of the Columbian Chemical Society,* 1813. Only one volume of *Memoirs* appeared, and to this volume Mitchell contributed 9 of the 26 articles.

Mitchell practiced medicine in Philadelphia for almost two decades but did not stray far from chemistry. In 1819 he published a 131-page text, "Medical Chemistry, or a Compendious View of the Various Substances Employed in the Practice of Medicine, that Depend on Chemical Principles for their Formation." The following year Ohio University at Athens offered him the professorship of chemistry, but for reasons unknown Mitchell declined. He finally entered the education profession in 1831 when he accepted the professorship of chemistry at Miami University and then, before the year was over, transferred to Medical College of Ohio in Cincinnati. Here in 1832 he published "Elements of Chemical Philosophy," a text used in several midwestern institutions. In Cincinnati he also taught science at Lane Theological Seminary for a short time.

Mitchell moved to Kentucky in 1837 to accept the Chair of Chemistry at the Medical Institute of Louisville but left within a month to accept a similar position at Transylvania University in Lexington. At Transylvania he also later taught materia medica, therapeutics, and obstetrics. From 1849 onward he held a succession of medical professorships.

Mitchell was a prolific writer on medical and scientific subjects. He edited *Transylvania Journal of Medicine and the Associate Sciences* and was associate editor of *Western Medical Gazette.* His most influential book, "Materia Medica and Therapeutics" benefitted from his knowledge of chemistry. He died in Philadelphia, May 13, 1865.

Mitchell's career is reminiscent of that of a number of other chemists of his time. In his youth he demonstrated enthusiasm and competence in chemistry and presumably could have developed into an influential member of the profession, but his chemical growth was stunted by lack of opportunity. There were no academic, industrial, or governmental positions for him to occupy, and for 20 years he had to support himself by medicine. He was 40 years old before he became a professor of chemistry, and then the institution with which he was connected had little to offer him in the way of research facilities or advancement. We can only speculate what his career might have been had he been able to specialize in chemistry all of his life. Today Mitchell is remembered chiefly as one of the early chemistry teachers of the Midwest and as the author of one of the earliest chemistry texts published in that region.

Virgil F. Payne, "Thomas Duché Mitchell (1791-1865)," *Filson Club History Quart.* **31,** 349-357

(1957); "Dictionary of American Biography," **13**, 66, Chas. Scribner's Sons (1934); W. D. Miles, "The Columbian Chemical Society," *Chymia* **5**, 145-154 (1959).

WYNDHAM D. MILES

Samuel Latham Mitchill

1764-1831

The first chemist to sit in the United States Congress after the adoption of the Constitution was Samuel Latham Mitchill of New York. Mitchill taught chemistry and other subjects at Columbia University from 1792 until he went to the House of Representatives in 1801. He served as a representative from 1801 to 1804, as a senator from 1804 to 1809, and as a representative again from 1810 to 1813.

Mitchill was born in North Hempstead, Long Island, Aug. 20, 1764. He studied classics in a small school conducted by a clergyman, then studied medicine under Samuel Bard, one of New York's foremost physicians and a professor of chemistry at Columbia from 1770 to 1776, 1784 to 1785, and 1786 to 1787.

Mitchill traveled to University of Edinburgh Medical School in 1783. There he listened to the chemistry lectures delivered by Joseph Black and joined the Edinburgh University Chemical Society, before which he read a paper entitled "Magnesia." He received his M.D. degree in 1786, returned to New York, studied law for a brief period, was one of the commissioners appointed to purchase land in western New York from the Iroquois Indians at Fort Stanwix in 1788, and practiced medicine.

In 1792 Columbia University established a chair of "Natural History, Chemistry, Agriculture and the other Arts depending thereon," the professor of which was to teach chemistry, geology, meteorology, hydrology, mineralogy, botany, and zoology, including the applications of these sciences. Mitchill was elected to the professorship.

Lavoisier had recently published his influential treatise on chemistry. Mitchill had read and adopted Lavoisier's views, and he taught them to his initial and subsequent classes. Mitchill, years later, declared he was the first teacher in America to present Lavoisier's ideas to his pupils: "as early as July [1792] I taught the reformed chemistry of the French and unfurled the standard of Lavoisier sooner, I believe, than any other professor in the United States." There is not sufficient information about chemistry courses in American colleges around 1790 to verify or disprove Mitchill's statement, but if we assume that he was the first, then other teachers were close behind, for Joseph Priestley, in 1796, believed that the French chemistry was taught "in all the schools on this continent."

In 1799 the professorship with the long title was replaced by a professorship of chemistry and natural history, with Mitchill continuing as professor. But he had been caught up in politics, and in 1802 he resigned his post and went to Washington as a Jeffersonian Democrat.

Although Mitchill taught Lavoisier's system of chemistry, which destroyed the old notion of a substance called phlogiston, he did not agree with Lavoisier entirely. He transferred the name phlogiston to a hypothetical substance which he thought was the "basis" of hydrogen. To this American brand of phlogiston he attributed certain chemical properties, one of which was an ability to combine with "metallic matter" and form metals. Another was an ability to react with a small proportion of metal and form "a combination peculiarly favorable to ascend into the atmosphere, and, by inflammation, to furnish the materials of meteoric and atmospheric stones." The latter was about the most radical of Mitchill's theories concerning phlogiston. Generally, the reactions of Mitchill's phlogiston were the same as the reactions of hydrogen.

While at Columbia Mitchill published two pamphlets for his students; "Outline of the doctrines in natural history, chemistry and economics, which, under the patronage of the arts, are now delivering in the College of New York" (1792), and "Explanation of the synopsis of chemical nomenclature and arrangement; containing several important alterations of the plan originally reported by the French academicians" (1801). He also published several articles on chemistry.

While Mitchill was in Congress, a group of physicians organized the College of Physicians and Surgeons of New York. Mitchill was professor of chemistry, 1807-08, but was in Washington attending to his congressional duties while the college was in session and did not teach. He was professor of natural history from 1808 to 1820 and of botany and materia medica from 1820 to 1826.

Mitchill helped start *The New York Medical Repository* in 1797 and was one of its editors for a quarter of a century. It was America's leading scientific and medical periodical for a score of years and a place where Joseph Priestley, James Woodhouse, John Maclean, and other American chemists of that time published articles.

Chemistry was only one of Mitchill's interests. He was a genius who delved into biological sciences, earth sciences, medicine, and much else. He was a fair mineralogist and analyzed a few rocks, ores, mineral waters, and minerals. He helped the Navy Department ascertain the reason for the poor quality of gunpowder with which it was being supplied. For banks of New York he investigated the ink used by counterfeiters. He presided over the first convention for the United States Pharmacopeia in 1820. President Jefferson admiring and jokingly called Mitchill "the Congressional Dictionary." Senator John Randolph referred to him as "the Stalking Library" and poet Joseph Rodman Drake called him "the Fellow of 49 Societies." He died in New York City, Sept. 7, 1831.

Mitchill's books and articles; archives of Columbia University; Courtney R. Hall, "A scientist in the early Republic, Samuel Latham Mitchill, 1764-1831," Columbia University Press (1934), with portrait; Courtney R. Hall, "A Chemist of a Century Ago," *J. Chem. Educ.* **5**, 253-257 (1928); Wyndham D. Miles, "Washington's First Chemist-Congressman, Samuel Latham Mitchill," *Capital Chem.* **17**, 209, 211 (1967).

WYNDHAM D. MILES

Forris Jewett Moore

1867-1926

Moore, son of Forris Jewett and Ellen S. (Wightman) Moore, was born at Pittsfield, Mass. June 9, 1867. His father died during Moore's infancy, and he was raised in the home of his uncle, a physician, in Claremont, N. H.

He completed high school in Claremont and entered Amherst College, from which he received his bachelor of arts degree in 1889. At Amherst Moore was at first undecided between philosophy and chemistry, but the influence of his chemistry instructor, Elijah P. Harris, and perhaps his uncle, led him to choose chemistry. After graduation he continued at Amherst for a year as a laboratory assistant.

He then went to Europe for further training and worked under Victor Meyer and Ludwig Gatterman at University of Heidelberg. He was awarded his doctor of philosophy degree for work on isolating aromatic sulfonic acids in 1893. He returned to the U.S. in 1893, not only with a Ph.D., but also with a wife, the former Emma Tod of Edinburgh, Scotland, whom he had met in Heidelberg and married in 1892.

Upon his return, he served as instructor in chemistry at Cornell University from 1893 to 1894 and then went to Massachusetts Institute of Technology where he remained. Initially, Moore taught analytical chemistry to beginning classes of up to 125 students. This responsibility was shared with another instructor and one assistant. He himself carried out all the procedures that he taught and analyzed all student samples used. In 1893 he had carried out analyses at the Chicago World's Fair. It was in 1902 that he began to teach organic chemistry. Beginning at M.I.T. as an "assistant in chemistry" in 1894, he rose through the academic ranks (instructor 1895, assistant professor 1902, associate professor 1910) to professor of organic chemistry in 1912. During the years 1910-11, 1917-18, and 1918-19 he also lectured on organic chemistry at Harvard.

Moore's research dealt with sulfocinnamic acids, colored salts of Schiff's bases, the oxidation of uric acid by hydrogen peroxide, and the constitution of xanthogallol. He likened himself to an alchemist because of his preference for working in undeveloped fields and with new compounds. He was a successful teacher and the author of two textbooks—"Outlines of Organic Chemistry" (1910) and "Experiments in Organic Chemistry" (1911). However, he required students to use the German edition of Gatterman's text in his course. For many years, he was in charge of undergraduate instruction in organic chemistry at M.I.T. In 1925, heart disease forced him to give up active teaching although he still maintained a small group of research students. He died at his home in Cambridge, Mass., Nov. 20, 1926.

Perhaps Moore's most lasting memorial stems from his interest in history of chemistry. He was one of the founders of the Division of History of Chemistry of the American Chemical Society in 1921 and taught a course on history at M.I.T. For this course he wrote "A History of Chemistry" (1918), which enjoyed widespread popularity and which, in its third edition (1939) revised by William T. Hall, was in print until the 1950's.

Obituaries in *Technology Review,* January 1927, p. 150 with portrait; *Ind. Eng. Chem.* **19,** 1066 (1927), with portrait; *Ber. Deutsch. Chem. Ges.* **60,** 53 (1927); "A History of Chemistry," 3rd ed., McGraw-Hill (1939), with portrait and bibliography.

RUSSELL F. TRIMBLE

Gideon Emmet Moore

1842-1895

Moore, born in New York City, Aug. 21, 1842, studied chemistry under Samuel W. Johnson at Yale, graduated in 1861, and remained for a year as a graduate student. He conducted an analytical laboratory in San Francisco for a time, then moved to the primitive mining camp of Virginia City, Nev., as assayer for the Gould and Currie Mine.

In 1867 he traveled to Europe, studied under Fresenius at Wiesbaden for a year, under Bunsen, Kirchoff, and Kopp at Heidelberg where he earned his doctor's degree, and under Kolbe at Leipzig and Wichelhaus at Berlin. He married the daughter of Austrian Field Marshal Von Hildebrandt at Budapest in the autumn of 1871 and returned home.

Moore acted as chemist for the Passaic Zinc Co. and at the same time operated a private laboratory at 69 Liberty Street, New York City. He offered his services as an analyst of ores, minerals, fertilizers, waters, and products of the arts, as an expert witness before the courts, and as a consultant for metallurgical, chemical, and manufacturing industries. His researches, being directed toward the solution of problems for clients, were seldom publicized or published, but he is said to have "left a record marked with many triumphs." During his professional life he wrote at least 11 articles, prepared an account of American tobaccos for the special reports of the Tenth United States Census, was an original member of the American Chemical Society, and edited volume 2 of the society's journal.

Introspective, gentle, and courteous, Moore was a competent violinist and a poet. He died of pneumonia in New York, Apr. 13, 1895.

Charles F. McKenna, "Dr. Gideon E. Moore," *J. Amer. Chem. Soc.* **17,** 659-664 (1895), with portrait, and a list of 12 of Moore's writings; the quote is from this obituary. Information on Moore's consulting laboratory is from advertisements. Charles A. Browne, Mary E. Weeks, "A History of the American Chemical Society (A.C.S.)" (1952), with portrait.

WYNDHAM D. MILES

Richard Bishop Moore

1871-1931

The son of William Thomas and Mary (Bishop) Moore, Richard was born in Cincinnati, Ohio, May 6, 1871 and died in New York City, Jan. 20, 1931. Educated in the United States, England, and France, his professional life in chemistry was varied and distinguished in education, government service, and industry.

The family moved to England in 1878, and his early education was chiefly British, except for the year 1885-86, which was spent in Paris at the Institut Keller. He entered University College, London in 1886. During 4 years there he was greatly influenced by Sir William Ramsay. After teaching in British schools until 1895, he returned to the United

States where he was awarded his B.S. degree by University of Chicago in 1896.

His career as a teacher of chemistry really began in 1897 as an instructor in chemistry at University of Missouri. Remaining there until 1905, he became associated with Herman Schlundt. Together they initiated some of the early work in this country on radioactivity.

From 1905 until 1911 he was professor of chemistry at Butler University. A sabbatical leave in 1907–08 took him back to Ramsay's laboratory where his work with krypton and xenon generated in him a life-time interest in the noble gases. On returning to Butler, he resumed his investigations of radioactivity and began an unsuccessful search for a heavier noble gas (radon).

Then followed a period of 15 years of absence from academic work. First came an appointment at the Bureau of Soils in Washington, D. C. Because of his early work on radioactivity and of increasing interest in the use of radium for treatment of cancer, he was soon transferred to the Bureau of Mines in charge of investigating rare elements at Denver. In 1916 University of Colorado awarded him an D.Sc. degree in recognition of his contributions.

His proposal for using helium in lighter-than-air craft led to establishing experimental stations for recovering the element from natural gas in the Texas area. In 1919 he became chief chemist of the Bureau of Mines in Washington, where he established the Cryogenic Laboratory for additional studies on gases.

In 1923 he left government work to join the Dorr Co. in New York City as consulting engineer. Shortly thereafter he became general manager of the company.

The award, in 1926, of the Perkin Medal of the Society of Chemical Industry recognized his important industrial contributions. The Franklin Institute had already awarded him its Longstreth and Potts medals.

In 1926 he returned to academic life as dean of the School of Science and head of the Department of Chemistry at Purdue University. Here again his energy, vision, and leadership were soon exhibited. As dean of the School of Science he played an important part in establishing the Graduate School in 1928. Perhaps his greatest contribution was in the Department of Chemistry. For the first time the staff became research-oriented. A large new building was planned and adjacent space alloted, when needed, for a building equally as large. This addition was completed in 1972. The soundness of his foresight was proved by developments during the intervening four decades.

Dean Moore would have been gratified by the national reputation achieved by the Department of Chemistry. Unfortunately, he lived only long enough to move into his office in the first unit of the new building in 1930.

A. R. Middleton, *Proc. Ind. Acad. Sci.* **41,** 30 (1931); S. C. Lind, *Chem. Eng. News* **9,** 40 (1931), with portrait; personal recollections.

MELVIN G. MELLON

Robert Jerome Moore

1892-1947

Moore was born in Hoboken, N. J. June 4, 1892. His father was a song writer and musical publisher, head of Moore Publishing Co. located in Manhattan. Because of his musical background, in which both parents wrote and published music, he became a choir boy at an early age and was active in Trinity Church in Hoboken. This work led him to be proposed for a scholarship to Princeton Theological Seminary. An unfortunate accident by another student from Trinity at Princeton caused the scholarship to be abandoned, and Moore turned to chemistry and mathematics, in which he was an outstanding student. He attended Cooper Union night school and studied chemistry, receiving his B.A. degree in 1915. Concurrently he taught chemistry as a lecture assistant and then as an instructor at Columbia until 1920. The needs of a growing family caused him to set up a consulting business in New York. From 1920 to 1924 he was president and director of Fraser Laboratories.

Much of his undergraduate work and graduate studies were in the field of colloidal chemistry, particularly in oils, fats, and waxes. He became director of research of Pratt and Lambert Co. in Buffalo in 1924 and continued his work in the field of oils and resins. He is credited with having developed the first phenol-formaldehyde applications in quick drying floor varnishes, and his pioneer work in this field attracted the attention of Bakelite Corp. in Bloomfield, N. J., which was attempting to find more uses of their material in surface coatings.

In 1930 he was hired by Bakelite to head their development laboratories, and he stayed with Bakelite, later Union Carbide Corp. At the time of his death he was technical coordinator of all Union Carbide operations throughout the United States.

His earliest publications, many of which were with his brother-in-law, Jacque C. Morrell, and Gustav Egloff, had their origin in his research at Columbia University. His later articles covered a wide range of subjects in the field of paints, varnish, and resins. He was an expert in synthetic resin chemistry and contributed articles to almost all the technical journals and compendiums during the period from 1918 to 1947.

He was a founding member of the American Institute of Chemists and president of the institute from 1938 to 1940. He was honorary chairman of the American Section of the Society of Chemical Industry from 1934 to 1936. He was a national councilor and chairman of the North Jersey Section of the American Chemical Society. He was president of the Paint and Varnish Production Clubs of New York. He was a suburban vice president of the Chemists' Club of New York.

During World War I he was active at Columbia University in training army personnel on the special methods of handling toxic chemicals and gases. During World War II he applied his knowledge of protective coatings to many types of military vehicles as well as to corrosion problems of seagoing vessels. He spent many months during the period of 1941–45 working on specifications for coatings on locks and dams vital to national defense throughout the Eastern Seaboard and the Mississippi River. He was active with many governmental and scientific groups engaged in studying and testing the results of research and development programs throughout these areas of the United States, including the Army Corps of Engineers, Army Quartermaster Corps, Navy Department, American Society for Testing Materials, and Society of American Military Engineers.

Moore was sought after as a speaker on scientific subjects, and in the course of 20 years gave over 1,000 lectures and demonstrations to audiences throughout the United States. His best known talk was entitled, "There Are No Frontiers in the Laboratory—the Fourth Kingdom." The second session of the Herald Tribune Forum held Nov. 15, 1942, included six panelists, among them Moore; James F. Byrnes of South Carolina; Juan T. Trippe, founder of Pan American World Airways; Major Alexander de Seversky, famous airplane designer; Henry J. Kaiser, West Coast shipbuilder; and Wendell L. Willkie. The formidable stature of such men attest to the prominence Moore achieved in the field of chemistry and chemical engineering.

Moore was also greatly interested in the licensing of professional chemists and worked towards reaching this goal through the American Institute of Chemists and the American Chemical Society.

His career was cut short by death from a heart attack Jan. 6, 1947 at his home in Montclair, N. J.

Obituaries in *The Chemist*, February 1947; *The Percolator*, February 1947; *New York Times*, Jan. 8, 1947; *Montclair Times*, Jan. 9, 1947; *Chem. Eng. News*, Jan. 24, 1947; personal recollections.

RICHARD L. MOORE

John Motley Morehead

1870-1965

Morehead, one of the pioneers of Union Carbide Corp., was born into a wealthy, influential family at Spray (now Eden), N. C., Nov. 3, 1870. His grandfather, for whom he was named, had been governor of the state and his father, James Turner Morehead, was a prominent textile manufacturer.

After attending Bingham Military School, "Mot" Morehead entered University of North Carolina and graduated Phi Beta Kappa in 1891. He started his career in his home town as a chemist with Willson Aluminum Co., a firm that was financially backed by his father. Of about a dozen patents held by the company's president, Thomas L. Willson, one was for ore smelting in the electric arc furnace. Attempts by Willson to reduce lime with carbon led to the discovery, in May 1892, of the commercial process for making calcium carbide and acetylene gas. However, Willson was not successful in making aluminum in commercial quantities, and there was no market for calcium carbide, so the company went broke during the depression of 1893. James T. Morehead lost everything he had; he even had to pawn his pocket watch. Young Morehead got a job as an engineer with Westinghouse in Newark, N. J. and stayed there until 1895.

Meanwhile, the idea of using acetylene for lighting caught on, and Willson Aluminum Co. sold carbide manufacturing rights to a new firm, Electro Gas Co., which in turn sold rights worldwide. Morehead's knowledge of the carbide process was suddenly in great demand, and in the fall of 1895 he went to Europe to erect carbide plants in England and Germany. Returning to the U.S. in 1896, he built a carbide plant at Niagara Falls, N. Y. for Acetylene, Light, Heat, & Power Co. and afterward became the company's chief chemist and test engineer, located in Chicago. When this firm merged with others to form Union Carbide Co. in 1898, Morehead became chief engineer of the new corporation. In this capacity, he was granted a number of patents for improvements in the arc furnace. As the market for acetylene grew, in home, auto, and mine lighting and oxy-acetylene welding, Morehead became a leading authority on gas technology. In 1899 he developed an apparatus for analyzing illuminating gases which was standard in the industry for more than 20 years.

He was author of "Analysis of Industrial Gases" (1905), and contributed to the "Gas Chemist's Handbook" (1915). As a result of his research with gases under high pressure, Union Carbide became the first American producer of polyethylene.

In 1918 he was appointed chief of the Industrial Gases Section of the Chemical Division of the War Industries Board under Bernard Baruch. Commissioned a major in the U. S. Army, he was charged with solving the acute shortage of toluene, which he did through allocation and by building facilities to strip aromatic compounds from municipal gas supplies.

A member of a half dozen technical societies, he served as president of the International Acetylene Association and in 1922, established the James Turner Morehead Gold Medal Award of that organization. He was also vice-president of the American Gas Association and of the American Welding Society.

Morehead was active in Union Carbide Corp. as chief engineer and consultant all of his life—he never retired. His death on Jan. 7, 1965 at the age of 94 came as the result of a fall sustained while walking from his New York office to Grand Central Station to catch a commuter train.

During the Hoover Administration, he was Envoy and Minister Plenipotentiary to Sweden and was the first foreigner to receive the Gold Medal of the Royal Swedish Academy of Sciences. He maintained a life long interest in Scandinavian affairs and in 1932, advocated before the U. S. Congress adoption of the Swedish system of state control of alcoholic beverages to replace prohibition.

Noted for his philanthropy, Morehead's gifts to University of North Carolina over the years amounted to more than $30 million. Morehead Planetarium was the first major installation of its kind on a college campus. This building also houses a collection of paintings by Rembrandt, Gainsborough, and other 17th and 18th century artists.

Morehead was Mayor of Rye, N. Y. for three terms (1925–30). When in 1964 the city voted a $300,000 bond issue to build a new city hall which Morehead thought of poor design, he gave $500,000 to build one in the Federal style of which he approved. He was also benefactor of Morehead Hospital and Morehead High School in Eden, N. C.

Morehead was fascinated by time pieces and collected all sorts. Fine clocks were incorporated in the town hall at Rye and in the Morehead-Patterson Bell Tower at University of North Carolina. However, he liked to carry his own custom made pocket watch tied to a shoe string.

He had a great wit and once remarked that while money might not bring happiness, it helped to calm the nerves. On another occasion he said that he would never consider himself wealthy until he could write a check without worrying about the stub.

For many years after fashion styles dictated change, he dressed the part of a successful 1920's businessman, the foremost mark of which was the high "Hoover" collar, and justified his eccentricity by virtue of age and success.

He was married twice—in 1915 to Margaret Birkhoff, of Chicago (died, 1945) and in 1948 to Mrs. Lelia Duckworth Houghton (died, 1961). He had no children and at his death left no immediate survivors.

Williams Haynes, "American Chemical Industry—A History," D. Van Nostrand (1945) **1,** 134-135 (photographs, frontispiece and 36); C. J. Herrly, "The Acetylene Chemical Industry in America," *Chem. Eng. News,* **27,** 2026-2066 (July 18, 1949); obituary in *New York Times* (Jan. 8, 1965); Personal reminiscences and Correspondence, J. M. Morehead to H. T. Pratt; Marguerite Schumann, "The First State University—A Walking Guide," University of North Carolina Press (1972).

HERBERT T. PRATT

Agnes Fay Morgan

1884-1968

Morgan was born in Peoria, Ill., May 4, 1884. She obtained her B.S. (1904), M.S. (1905), and Ph.D. degrees (1914) in physical and organic chemistry at University of Chicago. She served as instructor at Hardin College from 1905 to 1907, teaching fellow at University of Montana in 1907, and instructor at University of Washington from 1910 to 1913. She was appointed assistant professor of nutrition at University of California in 1915 and was promoted to associate professor of household science in 1919 and to professorship in 1923. When the Department of Home Economics was established in 1938 she was appointed chairman. In the same year she was appointed biochemist in the Agricultural Experiment Station. She became emeritus in both of these positions in 1954. She died in Berkeley, Calif., July 20, 1968.

When Morgan embarked on her career in 1915 at University of California, the field of nutrition was in its infancy. Only two or three programs of study were being offered on the East Coast, only three books on nutrition in English were available, and there was a dearth of information. Morgan organized the available material into courses and began research projects. As a result of her efforts, her department grew and became world famous, and she contributed to the elevation of nutrition to a high scientific plane. She wrote a text for beginning students of nutrition with Irene Sanborn Hall, "Experimental Food Study" (1938), whose approach to problems dealing with food preparation was strictly chemical. Although her research dealt with a broad spectrum of problems of nutrition and home economics, most of her approximately 150 papers were concerned with vitamin analysis of processed foods, the effect of heat on the nutritional value of proteins, and the interrelationships of vitamins and hormones. She was the first to observe the effect of sulfur dioxide on ascorbic acid and thiamine, first to detect damage to adrenal glands due to pantothenic acid deficiency, and first to record heat damage to the nutritional value of proteins. During World War II Morgan spent 4 years working with the Office of Scientific Research and Development to improve the nutritive value of dehydrated foods. She was a member for 16 years on the Committee on Experiment Station Organization and Policy.

Because of her teaching, research, inspiration of young women, and role as consultant for government and private agencies, Morgan received many honors. She was awarded the Garvan medal by the American Chemical Society in 1949. In 1950 her colleagues at University of California chose her to deliver the Faculty Research Lectures, the first woman to do this. The Borden Award in Nutrition was presented to her in 1954 by the American Institute of Nutrition. In 1959 she was elected the first woman fellow of this institute. She was awarded an honorary doctorate by University of California in 1959. In 1961 the home economics building was renamed Agnes Fay Morgan Hall. She was the first recipient in 1962 of an award given by the Society of the Medical Friends of Wine. A commemorative symposium, "Landmarks of a Half Century of Nutrition Research," was dedicated to Morgan May 8, 1965, when the country's leading nutritionists presented papers and paid tribute to her significant role in the development of the field.

Archives, University of California, Berkeley; S. Edgar Farguhar, ed., "The Progress of Science, a Review of 1941," p. 260 with portrait, The Grolier Society (1941); "Garvan Medal to Agnes Morgan," *Chem. Eng. News* **27,** 905 (1949), with portrait on cover; "Report of Committee on Faculty Research Lecture, 1950-51," *University California (Berkeley) Faculty Bull.* **20,** no. 5, p. 41 (Nov. 1950); "Our Distinguished Faculty," *California Monthly* **44,** no. 9, 21 (May 1954); "Landmarks of a Half Century of Nutrition Research," *J. Nutrition* **91,** no. 2, supplement 1, part II, pp. 1-67 (Feb. 1967); *Chem. Eng. News* **46,** no. 33, p. 158 (1968); *Daily Californian,* July 23, 1968; *Oakland Tribune,* July 22, 1968, with portrait; *New York Times,* July 23, 1968.

MEL GORMAN

John Morgan

1735-1789

Morgan, born in Philadelphia, Oct. 16, 1735 studied medicine for several years under a preceptor, graduated from University of Pennsylvania, served as a regimental surgeon of Pennsylvania provincial troops during the French and Indian Wars, and sailed to Great Britain where he obtained his medical degree at University of Edinburgh in 1763. Among the courses that he took at Edinburgh was chemistry, taught by William Cullen. Returning to Philadelphia Morgan persuaded the trustees of the University to establish a medical school, the first in the British North American colonies, in 1765.

Morgan was professor of medicine in the new school. Because there was no one else in Pennsylvania or perhaps in North America with sufficient knowledge of chemistry to teach the subject, Morgan also acted as professor of chemistry. Thus he became the first person to present a complete course of lectures on chemistry in the colonies. One of his students, John Hodge, took notes of Morgan's lectures. Hodge's 225-page manuscript, entitled "A Compendium of Chemistry, Historical, Theoretical, and Practical. Being the Substance of a Course of Chemical Lectures, delivered by John Morgan, Professor of Medicine in Philadelphia, in the year 1766," is in the College of Physicians of Philadelphia. These notes show that Morgan's lectures were based on those of his teacher, William Cullen.

Morgan taught chemistry each session from 1765 until 1769 when Benjamin Rush took over. Thereafter Morgan confined himself to teaching and practicing medicine and acting as director general of the Medical Department of Washington's army through part of the Revolution. He died in Philadelphia, Oct. 15, 1789.

Whitfield Bell, Jr., "John Morgan, Continental Doctor," University of Pennsylvania Press (1965), has portrait and references to many articles about Morgan; J. S. Hepburn, "Smith and Morgan, our First Chemists," *General Magazine and Historical Chronicle* **35,** 491-510 (1933).

WYNDHAM D. MILES

Edward Williams Morley

1838-1923

Morley was the long-time (1868-1923) professor of chemistry at Western Reserve University. In the period from 1887 to 1889 he collaborated with Albert Michelson of Case School of Applied Science in making the famous Michelson-Morley experiment on the velocity of light. He did his most famous work in chemistry in the period from 1890 to 1895 when he determined with an accuracy of five significant figures the relative weights of oxygen and hydrogen, the most accurate such measurements made up to that time.

Morley was born Jan. 29, 1838, in Newark, N.J. His father, Sardis Morley, was a Congregational minister, so Edward was raised in an atmosphere of strict moral rules. He was tutored by his father until about 12 years of age and graduated from Williams College with highest honors in 1860. He spent a postgraduate year at Williams working with Albert Hopkins in the college observatory making observations of latitude which ended in his first published paper "On the Latitude of Williams College Observatory." He then entered Andover Theological Seminary to prepare for the Christian ministry and graduated after 3 years with a license as a minister in the Congregational Church.

After several years in which no pulpit was offered him, Morley took a job as a teacher. But in 1868 he accepted a pulpit at Twinsburg, Ohio. Within several months after arriving in Ohio he was approached by Western Reserve College to take a professorship of chemistry. Morley accepted, and after marrying Isabella Birdsall on Christmas Eve 1868, arrived at the college Dec. 31, 1868 to begin what was to be a 55-year association with Western Reserve.

Michelson and Morley made an ideal research team when they started their epoch-making experiment to measure the ether drift in 1887. Morley had a creative imagination and driving energy which was complementary to the quieter, more systematic Michelson. The result of their collaboration which must have been disappointing at the time, has been called the "greatest negative experiment of all time." The team was broken up after only 2 years when Michelson left Case to move to Clark University. Michael Pupin wrote in 1923 of the Michelson-Morley experiment: "The fame of . . . Michelson and Morley . . . rests upon an experimental demonstration, the importance of which was not until recently fully appreciated, . . . namely that there is no ether drift."

Morley's other great work, the measurement of the relative weights of oxygen and hydrogen, was carried out with an accuracy approaching one part in 300,000. His results were published by the Smithsonian Institution in 1896, and his final result he stated as follows: "If, then, no important source of error is detected . . . the atomic weight of oxygen referred to hydrogen as unity is very nearly: $0 = 15.879$." If oxygen is assumed to be 16.000, the atomic weight of hydrogen by Morley's results becomes 1.0076, which differs from the best modern value by only 4 parts in ten thousand.

Morley continued to be active in teaching and research until his retirement in 1906 and continued as emeritus professor until his death in 1923. He was elected to the board of directors of Dow Chemical Co. in 1898, and the stock which he acquired in that company was the source of much of the considerable fortune which Morley left at the time of his death on Feb. 24, 1923.

Morley was awarded the Davy Medal of the Royal Society in 1907. Other medals and honors included the Cresson Medal of the Franklin Institute in 1912 and the Gibbs Medal in 1917. He was elected president of the American Chemical Society in 1899.

Howard R. Williams, "Edward W. Morley; His Influence on Science in America," Chemical Education Pub. Co. (1957) includes a bibliography of Morley's writings; "Dictionary of American Biography," **13,** 192, Chas. Scribner's Sons (1934).

JOSEPH A. SCHUFLE

Harry Wheeler Morse

1873-1936

Morse was born in San Diego, Calif. Feb. 25, 1873. His father, Philip Morse, was a pioneer lumberman who had come to California from Arizona. Young Morse received his A.B. degree from Stanford University in 1897. He then went to Leipzig to study physical chemistry with Wilhelm Ostwald. He received his Ph.D. degree there in 1901 and served as Ostwald's laboratory assistant from 1900 to 1902. During this time he helped edit *Zeitschrift für physikalische Chemie* and contributed to Ostwald's "Handbuch." His relations with his teacher were close, and after his return to America he translated Ostwald's "Letters to a Painter on the Theory and Practice of Painting" (New York, 1907) and his "Fundamentals of Chemistry" (New York, 1909). In 1907

Ostwald and Morse published their "Elementary Modern Chemistry."

In 1902 Morse was offered an instructorship in physics at Harvard University. He accepted and remained there for 10 years, becoming an assistant professor in 1910. During these years he worked on photography, electrolytic conductivity, and storage batteries. In 1912 he published "The Chemistry and Physics of the Lead Accumulator."

In 1912 he was called to University of California as professor of chemistry. He was thus one of the relatively few men who held chairs in both physics and chemistry. However, during his year at Berkeley he became interested in the Cottrell process for precipitating smoke particles and recovering minerals from flue gases. He decided to leave academic life and accordingly resigned from the university in 1913 to take charge of the scientific work at Western Precipitation Co. in Los Angeles. A little later he also became technical manager of American Trona Corp. He continued in these positions until 1920. In that year he resigned to take up consulting work in chemistry and metallurgy. He established his home and office in Menlo Park, Calif. adjoining the campus of Stanford University. He was appointed an honorary research associate at Stanford.

During his years of consulting activity he prepared monographs on spectroscopy, fluorescence, diffusion, electrochemistry, metallurgy, and geological chemistry. Besides his scientific versatility, he was also interested in painting and amateur dramatics.

Morse died at his home in Menlo Park, Mar. 12, 1936.

A short obituary is in the *Palo Alto Times* of Mar. 13, 1936 and in a biography by George W. Pierce in *Proc. Amer. Acad. Arts Sci.* **72,** 372-374 (1938).

HENRY M. LEICESTER

Henry Morton

1836-1902

Henry Morton was born Dec. 11, 1836 in New York City where his great grandfather had been a flax merchant and his grandfather an attorney. His father, Henry J. Morton, was an Episcopalian rector in Philadelphia, and it was here that Henry grew up. He entered University of Pennsylvania at age 17 and graduated with the Class of 1857.

Following graduation he studied law for 2 years but gave it up for chemistry and physics, subjects which had always fascinated him. In 1859, he began teaching a course in general science at the Episcopal Academy which proved to be so popular that it was opened to the public the following year.

In 1863 Morton became professor of chemistry at Philadelphia Dental College. In 1864 he was appointed resident secretary of The Franklin Institute of the State of Pennsylvania and became editor of its *Journal* in 1867.

In 1865 under sponsorship of the institute, Morton commenced a series of science lectures at the Philadelphia Academy of Music that was to make him famous. He brought a rare combination of disciplines to his work —skill as an artist and illustrator, talent as a poet, a gift for oratory, mechanical aptitude, and a vivid imagination. Morton was a showman at heart, and in reality his lectures were magic lantern extravaganzas with detailed scenarios that kept 10 to 15 stagehands scurrying behind a giant 25 x 45 foot screen.

He perfected lighting and optical systems for projecting systems in motion such as a giant ticking watch, a 25-ft high locomotive, and a living hand that could seemingly reach off the screen and clutch the lecturer. With lecture billings like Phantasmagoria and Legion of Angels, it is no wonder that, more often than not, he packed the house's 3500 seats. The lectures continued through 1870 and were also given in Boston and New York. His improvements on the magic lantern and its use as a lecture tool were the basis for more than a dozen published articles.

In 1869, he organized and conducted the photographic division of the U.S. Eclipse Expedition to Iowa. Here he demonstrated that the bright line adjacent to the moon's edge, previously attributed to diffraction, was due entirely to chemical action during development of the photographic plate.

In 1870, Stevens Institute of Technology,

newly established under the will of Edwin A. Stevens with an endowment of a half million dollars, elected Morton as president. A course in mechanical engineering, the first such course in America, was organized and classes begun in 1871.

The most important of Morton's chemical researches were done in 1871 and 1872 when he prepared and examined the fluorescence and absorption spectra of more than 80 uranium salts. He also published work on the fluorescence of anthracene and pyrene and a new compound which he called thallene. His only textbook, "The Students Practical Chemistry," was coauthored with Albert R. Leeds in 1877. In 1877-78, he published a series of 10 papers on the chemistry, manufacture, and use of water gas. In 1879, he showed that of the isomeric purpurines, isopurpurine was only a mixture of antipurpurine and flavopurpurine.

Morton was frequently sought as an expert in patent litigation involving chemistry and physics and testified in suits such as those involving the use of dry calcium phosphate in baking powder, the manufacture of alizarine, and of celluloid.

He was appointed to the U.S. Lighthouse Board in 1878 to fill the vacancy created by the death of Joseph Henry. As chairman of the Committee on Scientific Tests, he investigated fire extinguishers, lighted buoys, and fog signals. Also, he published the first independent studies of the efficiency of the Edison electric lamp.

Morton's personal financial contributions to Stevens Institute were substantial. Over the years, he outfitted the machine shop, equipped the electrical laboratory, and built a new boiler plant. In addition, between the years 1889 and 1901 he gave more than $100,000 endowment for the Chair of Engineering Practice.

He was married to Clara Dodge of New York in 1863. The Mortons were social people and so frequently entertained guests that someone remarked that "they must be in a constant state of reception."

He was one of the men who founded the American Chemical Society and guided it through its early years, serving as vice president from 1876 to 1881. Honorary degrees were awarded him by Dickinson College, Princeton, and University of Pennsylvania.

He died in New York City, May 9, 1902.

"National Cyclopedia of American Biography"; James T. White Co., **24**, 374 (1935); Coleman Sellers and Albert R. Leeds, "Biographical Sketch of President Henry Morton," Engineering Press (1892), contains a portrait and bibliography of his publications, also some of his poetry and art.

HERBERT T. PRATT

Henry Moyes

1750-1807

Moyes, born in Scotland, 1750, was blinded by smallpox when he was 3 years old. Nevertheless he attended an academy and the University of Edinburgh where he heard Joseph Black lecture on chemistry. Searching for a means of supporting himself he tried public lecturing, first on music without much success, then on chemistry with great success. He lectured in Great Britain for several years, then turned toward the United States. He arrived in Boston, May 27, 1784 and presented a series of 21 lectures on "philosophical chemistry," illustrated by a variety of experiments.

Moyes next appeared in Providence where he was successful and then in New York where he attracted large audiences and was offered the professorship of natural philosophy at Columbia. Moyes had not yet earned enough money to support himself and his assistant on the salary offered by Columbia, so he asked the school to allow him 2 years' time in which to complete his tour. He also offered to accept the chair of chemistry and natural history, instead of natural philosophy, at a lower salary if the trustees would grant him leave of absence. The trustees agreed and elected Moyes professor of chemistry and natural history in January 1785.

Moyes then lectured in Philadelphia and in Baltimore. He re-visited New York and Philadelphia for a shorter series of lectures, some for the benefit of certain charities. He changed his mind about the Columbia professorship and wrote a letter of resignation in January 1786. Finally he headed for Charleston to end his American lecture tour and sailed for England in May 1786.

During his 2-year visit to the United States he lectured to what was probably the largest number of people to attend lectures on science up to this time. In Philadelphia 600, 900, and 1,000 turned out on different occasions; in Baltimore 200; in New York 300; and in Charleston 300. Noah Webster, the dictionary compiler, attended in Boston. Benjamin Rush, professor of chemistry in the University of Pennsylvania, listened to him in Philadelphia.

Moyes outlined his course in a little pamphlet, "Heads of a Course of Lectures on the Philosophy of Chemistry and Natural History." People could obtain copies without charge from ticket sellers. It served as a prospectus to draw people to the lectures and as a convenient outline for those who attended. A number of editions of this pamphlet are in existence, which is evidence that different printers ran off copies as Moyes journeyed from city to city.

Despite his handicap Moyes tried, with the help of an assistant, to experiment. He published at least two articles on electrolysis in Tilloch's *Philosophical Magazine.* He continued to lecture in Great Britain, and died at Doncaster, Dec. 11, 1807 while giving a course in natural philosophy.

E. V. Armstrong, C. K. Deischer, "Dr. Henry Moyes, Scotch Chemist, His Visit to America, 1785-1786," *J. Chem. Educ.* **24,** 169-174 (1947), port.; Archives, Columbia University; *Gentleman's Magazine and Historical Chronicle* 77, part 2, 1235 (1807).

WYNDHAM D. MILES

Samuel Parsons Mulliken

1864-1934

Mulliken, son of Moses J. and Sarah D. (Gibbs) Mulliken, was born at Newburyport, Mass., Dec. 19, 1864. The family can be traced back to the earliest settlers of Plymouth and Massachusetts Bay.

His interest in chemistry was aroused by reading Jane Marcet's "Conversation on Chemistry," the same book that introduced Michael Faraday to the subject. Together with his high school classmate, Arthur A. Noyes, he pursued the subject in a home laboratory, and after finishing high school he worked compounding prescriptions in a Newburyport drugstore. During this time he read Faraday's "Chemical Manipulations."

He entered Massachusetts Institute of Technology in 1883 and was excused from the freshman chemistry requirements. In 1887 he received his bachelor of science degree and went to University of Cincinnati where he taught chemistry for a year before going to Germany for advanced study. Originally intending to study in Munich under Baeyer, he found on arrival in Europe that there was no place available for him at Munich. After visiting several universities, he decided to go to Leipzig where he worked under Wisliscenus on the isomerism of α- and β-chlorocinnamic acids. He also measured ionization constants in Ostwald's new physical chemistry laboratory. He received his doctorate in 1890.

Mulliken was a fellow in chemistry at Clark University (1890-91), an assistant at Bryn Mawr (1891-92), and an instructor and acting head of the Chemistry Department at Clark University (1892-94). For a short time he worked in the private laboratory of Wolcott Gibbs in Newport, R.I. before going to Massachusetts Institute of Technology as an instructor of organic chemistry in 1895. He remained at MIT for the remaining 39 years of his life, rising through the academic ranks (assistant professor 1905, associate professor 1913, professor 1926), with time out to serve as a major in the Chemical Warfare Service during World War One. He died Oct. 24, 1934.

His major research interest was analytical organic chemistry and resulted in his four volume work, "A Method for the Identification of Pure Organic Compounds by a Systematic Analytical Procedure Based on Physical Properties and Chemical Reactions" (1904-22). Excluding the volume of dyestuffs, this work includes data on some 10,000 compounds. The amount of literature work involved was tremendous, as was the laboratory work of developing and testing the many experiments involved in locating compounds in the analytical scheme. He and Arthur Noyes introduced qualitative organic analysis into the graduate and undergraduate chemistry curriculum at MIT.

He made his home in Newburyport and commuted daily to Cambridge on the Boston and Maine Railroad as he had when a student. When it was calculated that he had travelled a million miles on the B&M, the railroad held a reception for him at North Station and presented him with a medal and scroll.

In 1893, Mulliken married Katherine W. Mulliken (her maiden name). They had three children, one of whom, Robert S. Mulliken, was the 1966 Nobel Prize winner in Chemistry.

"National Cyclopedia of American Biography," **25,** 73, James T. White & Co. (1936); obituaries in *Chem. Eng. News* **12,** 197-198 (1934), with portrait, and by T. L. Davis in *Proc. Amer. Acad. Arts Sci.* 70, 560-565 (1935-1936).

RUSSELL F. TRIMBLE

Charles Edward Munroe

1849-1938

Munroe, the last surviving charter member of the American Chemical Society, was born May 24, 1849 in East Cambridge, Mass., the third of six children of Enoch and Emaline (Russell) Munroe. His ancestors were accurately dated back to fifteenth century Scotland and included more than twenty Minutemen who fought at Lexington Apr. 19, 1775. His close family were persons of wealth and education and included a chemist, an instrument maker, the secretary of the Massachusetts Institute of Technology, and a prominent teacher-philanthropist. His early interest in chemistry led him to the local drug store to purchase potassium cyanide. The proprietor hired the inquisitive young man as an errand boy and permitted him to conduct experiments with other (presumably safer) chemicals from the store's inventory.

Fortunately young Munroe was able to attend Cambridge High School, one of the earliest in America to offer chemistry. Here his talent in the science was encouraged by his instructors, who made him a laboratory assistant. He attended Lawrence Scientific School of Harvard, where he studied under Wolcott Gibbs, Josiah Cooke, and Louis Agassiz (who encouraged him to collect minerals). He graduated *summa cum laude* in 1871 and remained at Harvard as assistant to Gibbs and later to Cooke, where he taught quantitative analysis to upperclassmen. Munroe routinely performed experiments before Cooke's classes and eventually began to lecture to students of Boston Dental College.

In 1872 against the advice of his friends who felt the subject was degrading, Munroe established a course in chemical technology at Harvard. He pioneered in establishing a summer school in chemistry in 1873 and 1874. From 1874 until 1886 he was professor of chemistry at the United States Naval Academy at Annapolis, Md. In 1883 he married Mary Louise Barker, daughter of chemist George F. Barker; they had two sons and three daughters. One of his acquaintances at Annapolis, Commander W. T. Sampson, convinced Munroe to follow him in 1886 to Newport, R.I. to help organize the Naval Torpedo Station and War College, where he spent six productive years. In 1892 he accepted the professorship of chemistry at Columbia College (later George Washington University) in Washington, D.C. where he remained until he retired in 1918 as professor emeritus. He also served as dean of the scientific school from 1892 to 1898 and dean of the graduate school (which he established) from 1893 until 1917. In recognition of his services the school awarded him the honorary degrees of Ph.D. in 1894 and LL.D. in 1912.

His intense patriotism led Munroe to offer his services to the government in 1898 and 1917. During the Spanish-American War he was concerned with mines and torpedoes and perfected an armor-piercing high explosive shell that would explode *within* an enemy battleship. This employed his shaped charge which saw use in World War II in the anti-tank weapon—the "bazooka." He was also a consulting expert on the question of the defense of Washington. Before World War I Munroe became a partner in a consulting firm and remained actvie in consulting work until his death at his home in Forest Glen, Md., Dec. 7, 1938. From 1871 on he had a reputation as a chemical expert in legal proceedings.

Munroe's first research interests were in analytical chemistry, a notable early (1871) invention of his being a clay filter cone used to replace paper filters in high temperature work. In 1872 he patented a device used for producing refrigeration by evaporating water from porous surfaces. While at Annapolis he studied nitroglycerine, gunpowder, guncotton, and other explosives; and until his death he served as explosives editor of *Chemical Abstracts.* At Newport he developed and patented "indurite," a smokeless powder. Long before 1914 he was aware of the need to develop the fixation of nitrogen as a basic chemical industry in America.

Munroe was among the group of eminent chemists who attended the August 1873 meeting of the American Association for the Advancement of Science in Portland, Maine. The group petitioned to form a subsection of chemistry, the petition was granted, and the subsection became Section C of the AAAS in 1881. In 1888 Munroe served as chairman of the section. He was one of the relatively small number of non-New York members of the American Chemical Society who remained loyal to the Society in the face of attempts of Frank W. Clarke and Harvey W. Wiley to organize a rival National Chemical Society. Munroe's letter to the ACS in 1889 suggested that if the Society was to succeed, meetings should be held outside of New York and local sections should be established. So persuasive was his letter that the August 1890 meeting was held in Munroe's home city of Newport, with Munroe serving as arrangements committee chairman. The first local section was formed in Providence June 4, 1891, under Munroe's leadership. In 1898 he served as president of the society.

The best sketch of Munroe's life is by Charles A. Browne in *J. Amer. Chem. Soc.* **61,** 1301-1310 (1939); bibliography by J. N. Taylor, *ibid.,* 1311-1316.

SHELDON J. KOPPERL

Eneas Munson

1734-1826

Eneas or Aeneas Munson was born in June 1734, in New Haven, Conn. He graduated from Yale in 1753, then studied divinity and was licensed to preach, but finally turned to medicine, learning the art from a physician. In 1756 he began to practice medicine, became one of the foremost physicians of New Haven, and later was the first professor of materia medica and botany at the establishment of the Yale Medical School although he was professor in name only and did not lecture. He served in the Connecticut legislature during the Revolution and died in his home town, June 16, 1826.

Munson may have learned chemistry by self-study, or he may have picked up the rudiments of the crude chemistry of the 1750's from his medical preceptor. According to Eli Ives, his pupil and a noted Yale medical professor, Munson "studied chemistry with zeal and made many chemical experiments. Previous to the introduction of the anti-phlogistic theory of chemistry he was looked upon as a master of the science, and no one in this vicinity was as well acquainted with mineralogy; he manufactured many of his medicinal chemical compounds. On the introduction of the discoveries of Lavoisier and Chaptal . . . he immediately adopted them and was the first in this country to use the new medicinal agents which were developed by those discoveries. He was looked up to by all his medical brethren on all subjects relating to chemistry and pharmacy."

Despite this praise, one wonders about Munson's understanding of chemistry because for a period of at least several years around 1790 he believed that metals could be transmuted; his biographer Bronson said he "was known as an experimental alchemist."

He was a founder and president of the Medical Society of Connecticut.

Eneas' son Elijah, born in 1766 and died in 1838, also a physician of New Haven, detested chemistry because, according to his own story, he was told to watch the fire during one of his father's experiments, he forgot his instructions, the apparatus blew up, and his father flogged him. "Ever after the sound of the word Chemistry cost him a sigh and a shudder."

Henry Bronson, "Medical History and Biography" (1876), a book reprinted from *Papers of New Haven Colony Historical Society* but with different pagination, has a sketch of Eneas, pp. 25-36, and of Elijah, pp. 135-139, and has the quotations given above; R. S. Wilkinson, "New England's Last Alchemists," *Ambix* **10,** 128-138 (1962); James Thacher, "American Medical Biography," DeCapo Press (1828, reprinted 1967) I, 401-403, with portrait, gives Eneas' birthdate as June 24, whereas Bronson gives June 13.

WYNDHAM D. MILES

George Moseley Murphy

1903-1968

Murphy was born in Wilmington, N.C., June 1, 1903. He attended public schools there and then enrolled at University of North Carolina, where he earned his B.S. degree in 1924 and his M.S. degree the following year. He taught at Clemson College for 2 years before he enrolled at University of Pennsylvania. He moved to Yale University when his thesis advisor, Herbert Harned, joined that university. He obtained his Ph.D. in physical chemistry from Yale in 1930.

Murphy's interests leaned toward the newly developed field of quantum mechanics, a subject not yet taught by chemical faculties at that time, so it was natural for Murphy to accept a post-doctoral appointment with Harold C. Urey, a co-author of a monograph on the subject, at Columbia University.

Urey had predicted the existence of isotopes of hydrogen with masses two and three. Urey's plan to prove their existence was to concentrate them by distilling liquid hydrogen. Urey and Murphy "worked out together the general formula for the difference in vapor pressure HD and H_2, which was essential for the concentration of deuterium," in Urey's own words.

Since Columbia University had no liquid hydrogen facilities, Urey enlisted the help of Ferdinand G. Brickwedde, then with the National Bureau of Standards. Brickwedde performed the distillations and sent the residual gases to Columbia, where Murphy, as Urey's assistant, analyzed the gases and found the intensity of the expected spectral lines to vary as predicted for an isotope of mass two.

Urey received the Nobel prize in Chemistry in 1934 for the discovery of deuterium, but many of Murphy's colleagues "asked themselves whether a more equitable distribution of the award might have been made." It is possible that Murphy's youth militated against his receiving part of the award, and secondly his natural modesty may have been a factor since he cared little for publicity and honors.

When the discovery of deuterium was announced Murphy stated that its discoverers "did not think the new hydrogen could be put to use commercially." At that time nobody could foresee the development of atomic energy. Later Murphy worked in the Manhattan project, where he was able to witness the importance of deuterium in developing nuclear reactions.

In 1936 Murphy joined Yale University as a faculty member, until he obtained a leave of absence during World War II to work as a division director of the Manhattan project. The wartime pressure on the scientists can be demonstrated by Murphy's remark to one of his cousins when he left New Haven: "Pray God we are in time." That we were on time was due to the devotion of scientists like Murphy.

In 1943 Murphy published a book with Henry Margenau, "The Mathematics of Physics and Chemistry," that reflected his interests in applying mathematics to chemistry, a preoccupation that remained with him in his long teaching career.

In 1948 Murphy joined New York University as professor of chemistry. Three years later he became chairman of the Chemistry Department. In 1958 he was appointed head of the all-university Chemistry Department, and in 1961 he became the first associate dean of Arts and Sciences and was acting executive dean for a while. His acceptance of the appointment in New York was influenced, no doubt, by the presence of the Metropolitan Opera House in that city as Murphy was an avid opera lover.

His health began to deteriorate in 1964, and he resigned his administrative posts to become distinguished professor.

Murphy was a slightly built person, capable of remaining cool under the New York summer heat. His teaching style was lively, sprinkled with anecdotes about his numerous consulting activities for private industries and for the Atomic Energy Commission. These vignettes made physical chemistry a relevant subject.

The church meant a great deal to Murphy, and he was a vestryman at Christ Church, New Brunswick, for many years. His hobbies were music and stamp collecting. He died after a long illness Dec. 6, 1968.

Personal communications from Mrs. Mary D. Murphy, Nicholas Moseley, F. G. Brickwedde, Harold C. Urey, Arthur M. Ross, John E. Vance, and E. J. McNelis; E. J. McNelis, "Memorial to Dr. George M. Murphy"; obituaries in *New York Times,* Dec. 7, 1968, with portrait; *Physics Today* **22,** 119 (1969).

JOE VIKIN

Walter Joseph Murphy

1899-1959

Walter Murphy was a chemist turned journalist who became one of the leading spokesmen for the chemical profession in the United States during the middle decades of the twentieth century.

Born in Brooklyn, N.Y., Aug. 20, 1899 he received his B.S. degree from Polytechnic Institute of Brooklyn in 1921 and then went to work as a research chemist for Air Reduction Có. In 1922 he became a sales engineer with American Cyanamid Co., for which he traveled extensively in Latin American countries, helping to apply cyanide insecticides and fumigants to those countries' agricultural problems. The use of calcium cyanide to control leaf-cutting ants, a serious pest in the citrus-growing industries of Cuba and Brazil, was one of many techniques he helped develop.

A pioneer worker in market research and development in the chemical industry, Murphy joined United States Rubber Co. in 1925 and conducted a market study of heavy industrial chemicals. He then was appointed vice-president of George Chemicals, Inc., a marketing organization in New York, and vice-president in charge of manufacturing of a subsidiary, Seaboard Crystal Co. In 1928 he became sales assistant to the president of Mutual Chemical Co. of America.

Murphy entered the editorial field in 1930 as managing editor of *Chemical Marketing,* later known as *Chemical Industries.* He became editor and general manager in 1939. In 1942 he began his 17-year career with the American Chemical Society when he was appointed editor of the semi-monthly *Chemical and Engineering News* and the monthly *Industrial and Engineering Chemistry,* then published in two editions—the Industrial Edition and the Analytical Edition. Under his guidance, *Chemical and Engineering News* was changed into a weekly and the Analytical Edition was made a separate monthly, *Analytical Chemistry.* This move did much to raise the morale and professional standing of the nation's analytical chemists. In 1953 a new publication, *Journal of Agricultural and Food Chemistry,* was established with Murphy as editor. Two years later he was named editorial director of the four publications, known as the ACS applied journals. He held this post until his death from cancer, Nov. 26, 1959.

Under his leadership the applied journals developed into the largest scientific publication program of its kind in the world with a total circulation of more than 165,000. Murphy's editorial achievements were recognized by his peers in 1954 when he served as president of the Society of Business Magazine Editors. One key to his accomplishments as an editor was his success in building an exceptionally competent and enthusiastically loyal staff.

Murphy believed strongly in the need for close, constructive academic-industrial relationships, and worked hard to bring them about. He, himself, maintained firm ties with the academic world as a member of the Corporation of Brooklyn Polytechnic Institute and of advisory bodies of Notre Dame and Fordham Universities.

Aware of the increasing importance of science and technology to the public, Murphy recognized the need to provide the public with accurate information about chemistry. He was the prime mover in the development of the Society's News Service, of which he was director, from a part-time publicity office into a well-rounded public relations organization which helped the public realize the importance of the work of chemists and chemical engineers.

After World War II Murphy was sent to Germany on a mission for the Technical Industrial Intelligence Committee, Joint Chiefs of Staff, to investigate wartime developments in the German chemical industry. On his return, he campaigned editorially for a more formal and continuing study of German wartime chemical technology and was instrumental in bringing about establishment of a government organization to disseminate the information thus obtained. Ultimately such an organization was set up in the Department of Commerce as the Office of Technical Services, and thousands of scientific and technical reports written in Germany during the war became available to scientists, technologists, and industry in the United States.

Murphy wrote about the investigation of the wartime German chemical industry in a book entitled "I Did Leave Home." He also was author of "The Lagoon of Decision," reporting the day-to-day developments of the Bikini atom bomb test, which he witnessed as a technical representative. He was coauthor of "Strategic Materials for Hemisphere Defense," wrote numerous magazine articles of a popular and semi-popular nature, and lectured widely on scientific and professional subjects.

A frequent visitor to Europe, Murphy numbered among his friends many of the leaders of chemical science and industry there as well as in the United States.

For his services to the chemical profession and his outstanding ability as a technical editor, Murphy received the Gold Medal of the American Institute of Chemists in 1950. He also received the honor scroll of the American Chemical Society's Division of Industrial and Engineering Chemistry in 1953, and an honor scroll from the American Section of the Societé de Chimie Industrielle in 1958. Centre College of Kentucky conferred the honorary D.Sc. degree on him in 1947.

Obituary, *Chemical and Engineering News*, p. 21, Dec. 7, 1959; personal recollections.

JAMES H. STACK

N

John Henry Nair, Jr.

1893-1971

Nair was born in Chicago, Ill., Feb. 20, 1893 son of John Henry and Isabel (Painter) Nair. He grew up in rural Wisconsin except for a period in North Carolina where he lived with an aunt while his mother obtained her medical degree after his father died. Later, when his mother was established in practice, John helped out as a technician in her office for a time.

John entered Beloit College to study chemistry and graduated with his B.S. degree *cum laude* in 1915, with a thesis (for honors) on electrolytes. He taught chemistry in Wausau, Wis., high school, 1915-16, coaching basketball and track on the side. He was encouraged by his Beloit major professor, who was a Syracuse graduate, to go to Syracuse University for graduate study. He was a teaching fellow there, 1916-17, studying the miscibility of organic and inorganic liquids under Herman C. Cooper. This study was terminated without a degree, however, when Nair enlisted in the Army Signal Corps in 1917; he entered as a private and returned from France as a company commander.

After the war Nair returned to Syracuse, and in 1919 he joined Merrell-Soule Co., manufacturers of dried milk, as a research chemist. He later became director of the control laboratory, and when Borden Co. acquired Merrell-Soule in 1928, he became assistant director of Borden's Syracuse research laboratories. His work involved many aspects of dairy research—oxidation of butter fat, lipase, dehydration of dairy products and of other foods. In 1938 he was made technical advisor to sales management for Borden's Dry Milk Division, headquartered in Atlanta; he helped solve many problems that ice cream companies were having in the South.

In 1942 Nair left Borden to become assistant director of research for Continental Foods, subsidiary of Thomas J. Lipton, Inc., which was then getting into dehydrated food products; he helped develop their soup mixes and tea extracts. At the same time he acted as food consultant in the U.S. for Unilever, Ltd., and continued working for Unilever for several years after retiring from Lipton in 1957.

Even more noteworthy than his technical services were his activities in, and contributions to, several professional societies. Nair joined the American Chemical Society in 1920 and was active in the Syracuse Section as secretary, 1924, as vice-chairman, 1925, and as chairman for two terms, 1926 and 1927, and as councilor 1929-30. He came to national prominence in the ACS when in 1933 he carried to the ACS council at the Washington meeting the proposal of the Syracuse Section to separate dues from subscriptions to journals. This was during the depression; dues payments had slowed, and the Society's subsidy from the Chemical Foundation had ceased. The dues, which had been $15 since 1921, included all of the Society's periodicals —*Journal of the American Chemical Society, Industrial & Engineering Chemistry,* including the *News Edition,* and *Chemical Abstracts.* Nair was appointed chairman of a Board-Council committee to study the plan. At the Chicago meeting that fall the Council voted to accept the plan, setting dues at $9.00, to include *News Edition,* with journal subscriptions at specified rates.

Nair also introduced the idea of student chapters of the American Chemical Society as sources of future strength. He was made chairman of the Committee to Establish Student Affiliate Chapters in 1934. The affiliate program was set up in 1937, and within 10 years, 3400 students were enrolled in 122 chapters.

Nair was active in the Division of Agricultural and Food Chemistry and served as secretary, 1930-34, as chairman, 1935, and as member of the executive committee, 1936-44. He was also active in the New York Section of ACS, which sent him to the Council, 1944-62. For a time during this second period on the ACS Council, he was a member of the powerful Council Policy Committee and was its vice chairman, 1954-56. He was chairman of the New York Section, 1950-51, and at the same time he was chairman of the organizing committee for the ACS Diamond Jubilee that was held in New York City in 1951.

When the ACS decided to erect a headquarters building in Washington, Nair was made head of the planning committee for the building fund campaign, 1957-60. Nair's services to the ACS were recognized by the Council which twice named him as one of the two candidates for the office of President-Elect of the Society—in 1953 and in 1960. In each of the elections, however, he lost. Finally, in 1963, he was elected a Director-at-Large of the ACS and served for three years on the ACS Board of Directors.

Nair was a strong supporter of the American Institute of Chemists, of which he was president, 1956-57, chairman of the board of directors for 5 years, and recipient of its Gold Medal, 1966. He was also active in the Institute of Food Technologists, which he helped found in 1939 and of which he was president in 1966. In 1958 his alma mater, Beloit College, recognized his many contributions to the profession of chemistry by awarding him an honorary D.Sc. degree.

In 1920 Nair married Claire L. Cook of Syracuse, and they had two children. His daughter, Janet, majored in medical technology in college and lives in North Carolina. His son, John H. III, followed his father into chemistry and was for years, technical director of the Microbiology Laboratory at Syracuse University, later regulatory affairs specialist with the Silicone Products Department of General Electric Co., Waterford, N. Y.

When he retired, Nair settled in North Carolina and served for a time as visiting professor at North Carolina State University. He was a director of Onyx Chemical Corp. and of Avi Publishing Co. Throughout most of his life he was interested in gardening, in philately, and in golf; he died on a golf course near Syracuse, N. Y., on July 26, 1971.

Personal recollections; information from John H. Nair III; files of *Chem. Eng. News;* records of the American Chemical Society.

ROBERT F. GOULD

Henry Bradford Nason

1831-1895

Henry B. Nason belongs to an early, formative stage of chemistry in nineteenth century America. An important aspect of this stage was the transfer of European skill and knowledge to the United States, which occurred in several ways. There was, first, the importation of European books, journals, and apparatus. Reinforcing this was the arrival of many European chemists who pursued their art in this country. There was, thirdly, the training of American students in European universities. Nason was one of these last, through whom the progress of American chemistry was advanced.

Nason was born in Foxboro, Mass., June 22, 1831, son of Elias and Susanna (Keith) Nason. His father was a merchant and member of the Massachusetts Legislature. He attended school in the state and graduated from Amherst College in 1855. In the same year he traveled to Europe and matriculated at Göttingen. Here he studied chemistry, mineralogy, and geology and received a Ph.D. degree in 1857. Frederick Wöhler was one of his teachers there as he was of many other American students.

In 1858 Nason was appointed professor of natural history at Rensselaer Polytechnic Institute where he remained until his death in Troy, Jan. 18, 1895. For the first few years he divided his academic services between Rensselaer and Beloit College in Wisconsin, where he started a student chemical society. Not until 1866 did Nason become exclusively the professor of chemistry and natural science at Rensselaer. The school acquired a new laboratory building at that time, and Nason made it a model for the country.

From the nature of Rensselaer as primarily

a school of engineering in which the sciences were largely secondary service studies, Nason had little opportunity to engage in original investigation. He was primarily a teacher of undergraduates, with whom he established considerable rapport. His professional interests were broad, including mineralogy and geology as well as chemistry. He traveled widely in Europe and America, visiting geological and mineral sites, often in charge of Rensselaer students. Aside from his doctor's "Dissertation on the Formation of Ether," his writings in chemistry were translations and editorial compilations. Chief among them were Wöhler's "Handbook of Mineral Analysis," "Table of Reactions for Qualitative Analysis," and two enlarged editions of William Elderhorst's "Manual of Blowpipe Analysis and Determinative Mineralogy."

On the practical side, Nason served for 10 years as a consultant to Standard Oil Co. and became interested in the abatement of smoke, odors, and other nuisances arising out of oil refining. He was named inspector of petroleum oils by the New York Board of Health in 1881. In 1884 he was a delegate to an international congress held in London to consider petroleum nuisances. Nason thus early anticipated, as it were, the modern concern with pollution of the environment.

Although not a Rensselaer graduate, Nason engaged actively in alumni affairs. He was long secretary of the alumni association and prepared "Proceedings of the Semi-Centennial Celebration" in 1874. Later he compiled and published a substantial volume, "Biographical Record of the Officers and Graduates of Rensselaer Polytechnic Institute" (1887).

Nason accumulated a considerable number of honors, both professional and social. In 1878 he was a juror representing the United States at the Paris Exposition where he was assigned to judge mineralogy and metallurgy. He received several honorary degrees, A.M. from Amherst, M.D. from Union University, and LL.D. from Beloit College. Certainly the most important of his honors was the attainment of the presidency of the American Chemical Society in 1889, in whose foundation a few years earlier he had participated.

"Appletons' Cyclopaedia of American Biography," **4,** 480, D. Appleton & Co. (1888); Henry B. Nason, "Biographical Record of Officers and Graduates of Rensselaer Polytechnic Institute," W. H. Young (1887); *J. Amer. Chem. Soc.* **17,** 339 (1895); "Dictionary of American Biography," **13,** 390, Chas. Scribner's Sons (1934).

SAMUEL REZNECK

John Ulric Nef

1862-1915

Nef was born June 14, 1862 at Herisau, Switzerland. His father, Johann Ulric Nef, was a textile mill foreman who came to the United States to study the textile industry and went to work in a mill in Housatonic, Mass. Johann brought his family over in 1866 to a small farmhouse 4 miles from the town.

The father taught Ulric to love books, sports, and music. He went to school in Great Barrington, walking 4 miles each way. He was an excellent swimmer and on one occasion saved his younger brother from drowning. He burst an eardrum swimming under water and was almost completely deaf in that ear for the remainder of his life.

When Ulric was 16 years old he was sent to New York to attend a prep school for 2 years, and then he enrolled at Harvard. He intended to be a physician, but after his first year in chemistry decided to become a chemist instead. In his senior year he was at the top of his class and received a travelling fellowship which enabled him to continue his education in Europe.

Nef decided to travel to Munich to study under Adolph von Baeyer. He received his Ph.D. degree after 2 years of research on phases of tautomerism that Baeyer had discovered. Before returning home he remained an additional year with Baeyer who rated him one of his most brilliant students.

In 1887 Nef accepted a position at Purdue. In 1889 Clark University was established as a graduate school, and Nef was hired as associate professor of chemistry. He became director in 1892 but resigned along with Albert A. Michelson and others because the source of funds had dried up. That same year University of Chicago opened with generous contributions from John D. Rockefeller. Nef and Michelson were selected for positions in the new school. Nef became head of the Chemistry Department in 1896 and remained at Chicago the rest of his life.

Nef was a strict disciplinarian in his department. He would not permit the slightest deviation from his own meticulous cleanliness and rigorous accuracy in every step of his laboratory procedures. He insisted on the most careful manipulation during the performance of experiments so that quantitative results could be obtained whenever possible. His insistence in adherence to that regimen sometimes resulted in angry correction of a careless student. Some students resented his manner, but most of those who worked for their Ph.D. degrees under him held him in high regard. Another mannerism that belied his friendly character was his harsh rebuking of anyone who intruded on him while he was in the midst of an experiment. He always worked alone and, with the exception of three papers at the beginning of his career, his was the only name on his publications.

Nef's contributions to theoretical and synthetic organic chemistry were profound. He had considerable influence on the teaching of and research in that branch of chemistry. He published only 34 papers, but they totaled some 1500 pages. Some of his articles in Liebig's *Annalen* ran to several hundred pages. He was primarily interested in the reactions of carbon in carbon compounds. Some of his work formed part of the foundation of the modern theory of radicals and the transition state theory.

Before the electronic theory of valence had been proposed, Nef held the opinion that the numerical value of bonds was changed during certain reactions of carbon compounds. He attacked the generally accepted opinion of Kekulé that the valence of carbon was always four. By difficult and exact experimentation, with the danger of poisoning and explosions from such compounds as carbon monoxide, isocyanates, and fulminates, he established the fact that carbon in those compounds had two bonds rather than four and that the valence of one carbon changed from two to four during the reactions. His theory of a transition state of bivalent carbon fragments foreshadowed the modern theory of a transition state. His theory of polymerization was essentially the same as that accepted today. He experimented at great length with mono- and dihalogen acetylenes, which led to a reaction vital to the commercial production of vitamin A.

During the period from 1904 to 1915, Nef studied the reactions of sugars in alkaline and acid solutions of varying strengths, hoping that he could unravel the molecular mechanism of sugar fermentation. He found that oxidation of an alkaline solution of an aldonic acid in an air stream results in the next lower aldose. The carbonyl group in sugars can move up or down the sugar chain through its entire length, producing many ketone and aldose structures.

Contrary to his expresed opinion that a chemist should be married to his profession, he married Louise Comstock in 1898. They had one son, John Ulric Nef, professor of history at University of Chicago.

In the 1880's Nef began climbing mountains in Switzerland. Thereafter he spent many summers hiking and mountain climbing in the Alps and in the Rockies of Colorado. He loved music, and was a subscriber to the weekly concerts of the Chicago Symphony Orchestra.

Nef was only 53 when he died Aug. 13, 1915 of a heart attack at Carmel, Calif.

Biographies by Lauder W. Jones in *Proc. Amer. Chem. Soc.* (1917), pp. 44-72; and by Melville Wolfrom in *Biog. Mem. Nat. Acad. Sci.* **34**, 204-227 (1960) and in Eduard Farber (ed.) "Great Chemists," Interscience (1961), pp. 1130-1143; personal recollections.

BERNARD E. SCHAAR

John Maurice Nelson

1876-1965

Nelson was born Oct. 19, 1876, near West Point, Nebr. of emigrant Scandinavian parents. His education included the local country school, high school, and 4 years at University of Nebraska. John White, who taught him analytical chemistry and had just returned from studying with Ostwald in Germany, inspired Nelson with his conversations about physical chemistry. Nelson was attracted to Columbia University by a fellow student, Hal T. Beans, from Nebraska. Both Nelson and Beans became professors of chemistry at Columbia.

Soon after receiving his Ph.D. degree in 1907, Nelson and K. George Falk, working at Columbia, proposed an electronic theory of valence which was an extension to organic structures of the theory initially proposed by J. J. Thomson in his *"Corpuscular Theory of Matter."* They suggested that the direction the electron moves would depend on the relative position of the elements in the periodic table. They diagrammed the direction of the electron shift through the use of arrows in place of bonds. Falk and Nelson recognized that carbon, sulfur, and halogens would either accept or give up electrons depending on the conditions and the nature of the reacting atom. This led them to suggest that one carbon of an ethane molecule should be more reactive than the other. This work and other papers on this subject attracted much attention for it offered an explanation for properties of organic compounds that had previously been a mystery. Nelson's promotion to the rank of professor was rapid.

Nelson and Harold Fales uncovered the effect of neutral salts on hydrogen ion activity which they found by direct electrometric measurement and by the effect on the rate of hydrolysis of sucrose.

Most of Nelson's research was devoted to investigating enzymes. For many years his attention was focused on the nature of invertase (sucrase) which catalyzes the hydrolysis of sucrose. Later he turned to the phenol oxidases, such as tyrosinase, which are copper proteins extracted from mushrooms.

Nelson was a superb lecturer and was persuaded to give the lectures in beginning organic chemistry. These lectures to 150–200 students annually were given for over 30 years, yet he kept them up-to-date.

Nelson's best known student was John H. Northrop (Nobel Laureate in 1946) who established the proteinous nature of enzymes by his use of a variety of different physicochemical procedures designed to separate mixtures of substances.

Nelson was a modest, quiet man who had an insatiable curiosity about how living cells carry out their many complex processes. After retirement and until his death at the age of 89, Nov. 15, 1965, he was active in reorganizing the Columbia Chemistry Museum.

Nelson was married and had one child, Mary, who became a practicing physician.

R. M. Herriott, *J. Chem. Educ.* **32,** 513-517 (1955), portrait; obituary, *New York Times,* Nov. 16, 1965.

ROGER M. HERRIOTT

William Henry Nichols

1852-1930

Nichols was born in Brooklyn, N.Y., Jan. 9, 1852, son of George Henry and Sarah Elizabeth (Harris) Nichols. He attended Brooklyn Polytechnic Institute when it was a prep school; he had 2 years of lab work there. He then enrolled in the science course at Cornell with the class of 1872 but only stayed a few months. He was expelled for supposed participation in a prank—putting a team of horses and a cart on a dormitory roof; he refused to divulge the names of others who might have been involved. He then went to New York University where he received his bachelors degree in 1870, M.S. in 1873. He attributed the choice of career to John W. Draper, under whom he chose to study as the most outstanding chemist of the time and with whom he was later to associate as a founder of the American Chemical Society.

With a friend, Charles W. Walter, Nichols started a business to make acids in 1870. Both men were underage, however, so Nichols' father assumed financial responsibility and became president of the company which was named G. H. Nichols & Co. with offices in downtown New York City. It continued under this name until 1890 when it became Nichols Chemical Co. As junior partner, Nichols looked after the plant and the laboratory while Walter took care of the office and business. Walter's accidental death about 1875, however, threw the entire responsibility for the firm on Nichols, and as a consequence he spent his mornings in the plant, afternoons getting orders, and evenings in the office. Tiring of this routine, he hired Francis J. B. Herreshoff as factory manager.

Nichols' competitive advantage was his technical knowledge in a field of manufacturing where the practical man had ruled. When he started making sulfuric acid, the standard product was 66°Bé although most of it was under strength. When he made his acid 66°, his competitors objected. When he insisted, they cut prices and took his business. But acid refining of petroleum was just getting started, and this required the stronger acid which Nichols supplied and on which his firm flourished.

Sicilian sulfur was the usual U.S. raw material for sulfuric acid in those days, but the price was high and erratic. To use pyrites, a cheaper source, Herreshoff developed a special burner that was installed in the Laurel Hill, Long Island, plant. When a Canadian mine owner showed Nichols a sample of his copper pyrites that nobody else would buy, Nichols bought the mine and began selling the copper matte that came off as a byproduct. When the copper refiners stopped buying his matte, Nichols commenced refining his own copper. To get a more exact method of determining copper, he developed an electrolysis method, and then he adapted this to producing electrolytic copper commercially. Here, the byproducts were gold and silver; the business prospered, and Nichols Copper Co. was spun off from the parent firm.

The sulfuric acid of the company's early days was chamber acid, adequate for fertilizer production and oil refining. New needs developed, such as coal tar dyes, that called for the stronger acid that could be made by the contact process, a European innovation. With Herreshoff, Nichols adapted the contact process to American conditions, and this gave him a commanding position in the industry. By this time, Nichols Chemical Co. had plants in Troy and Syracuse, N.Y., and in Capleton, Que., in addition to the original plant in Laurel Hill.

Nichols used this strong position to bring off what has been called "the first important American chemical merger"—12 companies with 19 plants into the General Chemical Co. in 1899. Nichols was made chairman and his son, William H. Jr., president. Nichols continued as chairman until 1920, thereafter as director. The company was formed as a defense against competition from German chemical imports. Later, Nichols went to Germany to seek rights to Badische's contact process, supposed to be the world's best. But Badische's price was too high so he put his own process into a new plant at Newtown Creek, Long Island. Badische sued, but the suit was settled out of court with exchange of patents and data.

When news of Haber's success with the ammonia synthesis plant at Oppau reached this country, Nichols set his company to work on a non-interfering process. The process was in production by 1916, and he offered it to the U.S. Government during World War I. In 1919 Nichols formed Atmospheric Nitrogen Co. with a plant in Syracuse, to follow up this work. It became the largest domestic producer of ammonia, and a larger plant was built at Hopewell, Va., in 1927.

In the spring of 1917 Nichols was named chairman of the Committee on Chemicals which cooperated with the Advisory Council to the cabinet-level Council of National Defense. It met every other week in the early summer of 1917, lining up supplies that would be needed for war. Nichols, who favored high prices to encourage production, ran into conflict with Bernard Baruch and Leland Summers of the Raw Materials Division who shied away from industry's making big profits in wartime. About this time sulfuric acid supplies were becoming critical, squeezed between rising demand and shortage of shipping for foreign pyrites. The Department of Commerce called together makers and users of sulfuric acid in July to consider imports of foreign pyrites; on August 1 the manufacturers met and formed the Chemical Alliance, which was to become the war board of the chemical industry. Nichols presided over the Alliance until December 1917.

Nichols also sat on two war service committees of the U.S. Chamber of Commerce—the Central Committee for Chemicals and the Foreign Pyrites Committee. During this period of wartime service Nichols lived in Washington.

The dye industry grew rapidly during the war; immediately after the Armistice there were 136 producers. This number dropped quickly, however, as firms went out of business and others merged. The largest merger formed National Aniline and Chemical Co. in 1917 from National Aniline, Schoellkopf Aniline and Chemical Works, Beckers Aniline and Chemical Works, and Benzol Products. The latter company was owned by General Chemical, Semet-Solvay, and Barrett. Nichols was made chairman of National Aniline and Chemical, but a year later he withdrew from active management while remaining as director. In 1920 Nichols brought together General Chemical, Barrett, National Aniline and Chemical, Semet-Solvay, and Solvay Process to form Allied Chemical & Dye Co., of which he became chairman.

Nichols played a role in founding the Chemists' Club (1898), which named a room after him. He was president of the Society of Chemical Industry (1904), of the Eighth International Congress of Applied Chemistry (1912), and of the American Chemical Society (1918-19). He was Commander of the Order of the Crown of Italy and a Knight of Saints Maurizio and Lassaro. In 1902 he funded the Nichols Medal, which the New York Section of ACS administers, and he gave the Nichols Chemical Laboratory and the School of Education Building to New York University. He received honorary degrees from Lafayette (LL.D., 1904), Columbia (Sc.D., 1904), New York University (LL.D., 1920), Pittsburgh (Sc.D., 1920), and Tufts (Sc.D., 1921). He was vice president of the board of Brooklyn Polytechnic Institute and was for many years a trustee of New York University, serving as acting chancellor of NYU in 1929 during the chancellor's illness. He was a director and vice president of Corn Exchange Bank and Title Guarantee and Trust.

Nichols married in 1873 Hannah Wright Bensel, who died in 1929. His son William H. Nichols, Jr., was president of General Chemical Co. and vice president of Allied Chemical Co. when he died in 1928. Another son, Charles Walter Nichols, was president of Nichols Copper Co. when that firm was sold to Phelps Dodge in 1930; thereafter he was a director of Phelps Dodge and then president of his Nichols Engineering and Research Corp. He died in 1950.

Nichols died Feb. 21, 1930, after a brief illness in Hawaii where he was visiting. He left over half of his fortune to charity and education, including $50,000 to the American Chemical Society.

Williams Haynes, "American Chemical Industry," Vols. 1-6 (1945-54), D. Van Nostrand; Allied Chemical Press Relations release; biographical sketches in: *Ind. Eng. Chem.* **15,** 424 (1923); **18,** 317 (1926); obituaries in: *Ind. Eng. Chem.* **22,** 394 (1930); *Science,* **71,** 528 (1930); *J. Amer. Chem. Soc.* **52,** 47 (1930); *New York Times,* Feb. 23, 1930; Feb. 24, 1930; Mar. 15, 1930; Dec. 1, 1930; Wm. H. Nichols, *J. Chem. Educ.* **5,** 448-51 (1928); Victor H. Peterson, "MCA 1872-1972: A Centennial History," Washington, D.C. (1972); New York and Brooklyn city directories, 1871-90.

DANIEL P. JONES
ROBERT F. GOULD

James Flack Norris

1871-1940

Norris was born in Baltimore, Md., Jan. 20, 1871. He received his elementary education in the schools of Baltimore and Washington. He attended Johns Hopkins where he joined the hiking club and debating society. He was fascinated by Ira Remsen's courses in organic chemistry and the history of chemistry, which he attended three times, and after receiving his A.B. degree in 1892 he stayed on at Hopkins for his Ph.D. in Remsen's laboratory. Norris' thesis topic was the chemistry of perhalide salts and selenium halide double salts of amines.

Norris received his Ph.D. degree in 1895 and joined the Chemistry Department at Massachusetts Institute of Technology as an instructor. He remained at MIT until 1904, when he became the first professor of chemistry at Simmons College, a college for women which was just opening. At Simmons he had general supervision of all courses in science and developed the chemistry curriculum and organized the laboratories. In the 1910-11 academic year, he took leave to study physical chemistry in Fritz Haber's laboratory in Karlsruhe. During summer vacations at his home in North Bridgton, Maine, Norris wrote a textbook, "Principles of Organic Chemistry," and a laboratory manual which were widely adopted. He left Simmons in 1915 for Vanderbilt University and returned in 1916 to MIT as professor of organic chemistry.

In 1917 Norris left MIT again, this time to use his talents as a chemist and an administrator in the service of his nation, which had entered the World War. He was in charge of chemical research on agents of offense (war gas) in the U.S. Bureau of Mines and in 1918 entered the army as lieutenant-colonel in the Chemical Warfare Service. He was in charge of the service in England, and in 1919 he led the investigation of the manufacture of war gas in Germany.

After the war Norris returned to MIT, and when the Research Laboratory of Organic Chemistry was organized in 1926, he was appointed director. His research interests included several areas of organic chemistry, such as condensation reactions and the Friedel-Crafts reaction, but his major work consisted of a series of 20 papers, "Reactivity of Atoms and Groups in Organic Compounds." In these investigations, Norris and his students measured the rates of esterification of substituted benzoyl and benzhydryl chlorides with various alcohols, the acidity of substituted benzoic acids, the reactivity of these acids with thionyl chloride, etc. Norris believed that relative rate constants for reactions of a series of structurally related compounds, under identical conditions, could provide fundamental information about the nature of bonding and how bonding is modified by changes within a molecule. In the 1920's, when Norris' work first was published, many chemists viewed the significance of rate constants with scepticism, but his careful measurements contributed to the development of physical organic chemistry, as practiced by Ingold and Hammett, in which kinetic measurements and structure-reactivity correlation assumed a major role.

Norris exerted a great influence in the development of chemistry and chemical education through his involvement with national and international scientific organizations. As early as 1904 he served as chairman of the Northeastern Section of the American Chemical Society and later as assistant editor of the society's *Journal.* He served as a member of the executive board of the National Research Council as well as chairman of its Division of Chemistry and Chemical Technology (1924-25). He was president of the American Chemical Society in 1925 and 1926 and was vice president of the International Union of Pure and Applied Chemistry from 1925 to 1928. He received many honors and awards in the United States and Europe and traveled widely to international chemistry meetings.

Norris was a distinguished educator who had a wide and deep knowledge of chemistry. In addition to his texts in organic chemistry he wrote books on inorganic chemistry, all written without the aid of reference books. He gave courses in several colleges in New England, and in the course of his career he taught virtually every chemistry course in the curriculum including the history of chemistry and food analysis.

In his later years, Norris spent his summers in walking trips through scenic areas of Europe. He planned to retire on reaching his seventieth birthday in 1941, but he died in Cambridge, Aug. 4, 1940.

T. L. Davis, *Chem. Eng. News* **14,** 325 (1936); Anon., *ibid.,* **18,** 730 (1940); "Current Biography," 620, H. W. Wilson Co., with portrait (1940); "Chemical Who's Who"; "National Cyclopedia of American Biography," **A,** 190, James T. White & Co. (1930).

MARTIN FELDMAN

John Pitkin Norton

1822-1852

Norton was a co-founder of the institution which evolved into the Sheffield Scientific School of Yale University. His ideas were also influential in founding the first agricultural experiment station in the nation, that in

Connecticut in 1875. In addition, he was one of the first trained chemists to apply the principles of his science to improve farming in the United States.

Norton was born July 19, 1822, in Albany, N.Y. His father, John Treadwell Norton, was a prosperous farmer who had served in the state legislature. He attended an academy in Brooklyn, N.Y. and in 1840 continued his education at Yale although he never enrolled as an undergraduate. He attended lectures on chemistry, mineralogy, and natural philosophy given by Benjamin Silliman, Sr., Denison Olmsted, and other teachers. He spent the academic year of 1841-42 in the private laboratory of Benjamin Silliman, Jr. where he was instructed in simple analyses of minerals and organic matter. In 1842-43, Norton studied in Boston, attending lectures in sciences given by distinguished members of the scientific community.

Encouraged by his understanding father, he decided to make agricultural science his life's work, and he re-entered the younger Silliman's laboratory to learn practical chemistry. Periods of intensive laboratory study in New Haven alternated with long summers during which he worked as a farmer. In 1844 Norton studied under James F. W. Johnston, head of the Agricultural Chemical Association laboratory in Scotland, who was one of the leading men in this field. He worked with Johnston for 2 years, during which time he completed an important investigation on the chemical constitution of the oat plant, for which he was awarded a prize of 50 pounds sterling by the Highland Agricultural Society of Scotland.

As his ability and zeal became apparent to his friends in Connecticut, he was offered teaching positions at Union College and Yale. In 1846 after considerable anxiety on his part and vacillation on the part of the Yale community, he was appointed "Professor of Agricultural Chemistry and Vegetable and Animal Physiology" at the same time that his former mentor, Benjamin Silliman, Jr. was elected "Professor of Chemistry and the kindred sciences as Applied to the Arts." Of money for these professorships, there was none. The new professors were expected to subsist on tuition charges and donations. In 1847 a college committee appointed to implement the new professorships recommended that a new department be constituted to embrace the practical sciences and certain other subjects although degrees were not conferred until 1852. The chemical curriculum was included in a "School for Applied Chemistry" which was headed by Norton and Silliman, Jr.

In the meantime Norton returned to Europe to study in the laboratory of Gerrit J. Mulder of Utrecht where for 9 months he worked intensively to prepare for the opening of the new school late in 1847. At this time, he was undoubtedly the best-trained and the most zealous American agricultural chemist.

Despite serious financial problems, the school commenced when an old residence was converted into a laboratory. Norton and Silliman, Jr. both supervised the chemistry which was mainly analytical in practice and such other sciences as geology and mineralogy which were thought to be essential in the preparation of scientific agriculturalists.

Norton proselytized assiduously for agricultural chemistry and scientific agriculture. Because he knew practical farming, he could unite theoretical science with practical farming, and he therefore received a wide hearing. He frequently spoke at agricultural fairs and other agricultural meetings in Connecticut, New York, and Massachusetts. He regularly contributed a column to a number of well-known agricultural periodicals. His textbook, "Elements of Scientific Agriculture," was well received and widely used. Norton was asked to join Louis Agassiz and James Dwight Dana in planning a national university in Albany, N.Y., and Norton gave several series of agricultural lectures at Albany to demonstrate his interest in this proposed university.

After his student years, Norton did not continue research even though he was an excellent chemist whose early research showed great promise. Instead he became a leader in the scientific farming movement in the United States and in the founding and improvement of the proto-Sheffield Scientific School. His students made important contributions in teaching science and in scientific research.

Norton died from tuberculosis in Farmington, Conn., Sept. 5, 1852, at the age of 30. Of him it was said:

"We feel that the whole country has suffered loss. . . Professor Norton was no ordinary man; he was one in whom the 'elements seemed so blended' that he commanded universal respect . . . we may safely say that to no man is this country more deeply indebted than to him, for the valuable truths which [he has] elicited on this subject [of the applications of science to agriculture]."

The literature on Norton is extensive although short and fragmentary. The most recent account is Louis I. Kuslan, "The Founding of the Yale School of Applied Chemistry," *J. Hist. Med. Allied Sci.* **24,** 430-431 (1969). The quote is from "Memorials of John Pitkin Norton," Yale College (1853).

LOUIS I. KUSLAN

Lewis Mills Norton

1855-1893

Norton was born in Athol, Mass., Dec. 26, 1855, the only son of Rev. John Foote and Ann Maria (Mann) Norton. He was the namesake of his grandfather who was a well-known inventor and manufacturer in the cheese and weaving industries. The grandfather erected the first cheese factory in the world in Goshen, Conn., invented the machinery and appliances used in the factory, and was the first to manufacture pineapple cheese in America. He also invented a power loom and made other contributions to the textile industry.

Norton studied chemistry at Massachusetts Institute of Technology from 1872 to 1875, at which time he was appointed assistant in analytical chemistry. After 2 years in this capacity, he went abroad and studied chemistry at Berlin, Paris and Göttingen. In 1879 he received his Ph.D. degree from the University of Göttingen.

Upon his return to America, Norton was employed as a chemist by the Amoskeag Manufacturing Co. of Manchester, N.H. In 1881 he returned to Massachusetts Institute of Technology as instructor in general chemistry. In 1883 he was promoted to assistant professor of organic chemistry and in 1885 to associate professor of organic and industrial chemistry. Owing to the growth of the department Norton gave up his organic chemistry teaching duties in 1891 and devoted all his energy to industrial chemistry, an area in which he had a deep interest. Norton was responsible for developing the course in chemical engineering which was begun at the institute in 1888 and had become firmly established by the time of his death 5 years later.

Norton contributed numerous papers to scientific and technical journals; his subject matter was varied, including organic problems, dyeing, bleaching, and the analysis of gases.

Norton wrote a book of experiments in general chemistry with William R. Nichols, which was used both at MIT and several other schools. His instruction in organic chemistry never lost sight of the industrial applications. But his main contribution was in teaching industrial processes where he excelled in the scope of topics covered and in his exhaustive treatment of these topics. His intimate acquaintance with the manufactures and manufacturers of the New England area was a great asset to this facet of his teaching career.

Norton died April 26, 1893.

J. Amer. Chem. Soc. **15,** 241-244 (1893); *Proc. Am. Acad. Arts Sci.* ns. **20,** 348-353 (1892-93); "National Cyclopaedia of American Biography," **4,** 301, James T. White & Co. (1891).

CHARLENE J. STEINBERG

Sidney Augustus Norton

1835-1918

Norton was born in Bloomfield, Ohio, Jan. 11, 1835, the eldest son of a pioneer family. He received his early education in the public schools of Cleveland, Ohio and was a member of the first class to be graduated from the high school of that city. He entered as a freshman at Western Reserve College, then located at Hudson, Ohio but left after one year to enroll at Union College, from which he graduated in 1856 as the first in a class of 90. He taught for 1 year at Bartlett School, Poughkeepsie, N. Y. but returned to Union College in 1857 for graduate study in chemistry under Charles Joy. He received his M.A. degree from Union in 1859. In 1858 he was appointed instructor in natural science at the high school in Cleveland and remained there until 1867 when he was appointed to a similar position at Mt. Auburn Young Ladies Seminary in Cincinnati. In 1868 he was appointed professor of chemistry at Miami Medical College of Cincinnati. Neither of these positions absorbed all his energy for he was concurrently a student at the medical college from which he received his M.D. degree in 1869. His formal education was completed by 16 months of study abroad while on leave of absence from his professorship at Miami Medical College. He studied chemistry under Engelbach at Bonn, under Kolbe at Leipzig, and under Bunsen at Heidelberg.

Norton returned to his position in Cincinnati in 1870 and remained there until 1872 when he was appointed acting professor of physics at Union College. In 1873 he was named professor of general and applied chemistry at newly established Ohio Agricultural and Mechanical College, which in 1878 became Ohio State University. Until 1893 Norton taught all the courses offered by the Chemistry Department of that institution. At the beginning he did this without any help but later had the services of a single assistant. In 1895, as the result of a dispute over the place of laboratory work in the course in general chemistry, he relinquished his position as head of the department. He retired as professor emeritus in 1899, and died of the infirmities of age Aug. 30, 1918.

He married Sarah Chamberlain of Cleveland in 1864. She died in Cincinnati in 1868 leaving one son, who died in 1899. In 1876 Norton married Jesse Carter of Columbus. To them were borne two sons and two daughters.

Norton was a prolific writer. His first large work, which appeared in 1863, was a revision of the widely used English grammar of Weld and Quackenbos. His "Elements of Natural Philosophy," published in 1870, had a large sale. In 1875 he published a textbook of physics and in 1878 one on inorganic chemistry, which became a textbook on general chemistry with the addition of a section on organic chemistry in 1884. He wrote many periodical articles, especially on methods of teaching science. Very few of his articles dealt with the technical aspects of chemistry itself.

Norton did little in the way of original investigation in chemistry. From the beginning of his career his chief interest was in teaching, and in this vocation he was eminent as is attested in various ways. For example, his valedictory lecture at the Miami Medical College so impressed the students of his audience that they had it printed and distributed at their own expense. Again, a series of testimonials by former students appeared as a tribute to him in the *Ohio State Monthly* in February 1914. These former students agreed that he excelled in systematic presentation of subject matter, in clarity of statement, and in the wealth of his lecture demonstrations. A number of his students later became prominent chemists, including two presidents of the American Chemical Society, William McPherson and William Lloyd Evans. His high reputation as a teacher and scholar spread beyond the confines of his university as shown by the honorary Ph.D. degree conferred on him by Kenyon in 1878 and the LL.D. degrees conferred on him by Wooster in 1881 and by Union in 1899.

J. H. Bownocker, C. A. Dye, and W. E. Henderson, "Report of a committee appointed to prepare a minute upon the death of Sidney A. Norton, in records of the faculty of The Ohio State University," November 1918; Earle R. Caley, "History of the Department of Chemistry," Ohio State University (1969).

EARLE R. CALEY

Thomas Herbert Norton

1851-1941

This versatile chemist, teacher, editor, diplomat, traveler, and dye expert lived for 90 interesting years. Norton was born June 30, 1851 at Rushford, N.Y., the son of a Presbyterian minister. From the age of 12 he showed great interest in science, which was fostered by his father. When he was a young man his father was called to a church in Saint Catharines, Ont., Canada; young Norton worked there as a reporter and later as a city editor.

He graduated from Hamilton College in 1873 as valedictorian and also won first prize in classics, mathematics, chemistry, and physics. Norton then went to Europe to study at University of Heidelberg. Here he obtained his Ph.D. degree in chemistry in 1875. The same year Hamilton conferred on him an Sc.D. degree *honoris causa*. The following year he studied with August W. Hofmann at University of Berlin; he then studied a year at University of Paris.

While in France Norton became manager of the dye works of Compagnie Generale des Cyanures at St. Denis. He managed the plant about 6 years before returning to the United States to become professor of chemistry at University of Cincinnati in 1883. Laboratory facilities were meager, but Norton did considerable research on cerium around 1886. A decade later he prepared a hydrated silica having the formula of orthosilicic acid, H_4SiO_4, by squeezing (about 100 kilos per sq in.) water out of silica gel and drying under constant conditions. Norton held the professorship of chemistry until 1900.

In 1900 President William McKinley appointed Norton consul at Harput in Turkey. It was a position of importance to American educational interests and of diplomatic rather than consular nature. During his stay in Turkey he and Ellsworth Huntington made a voyage down the Euphrates river on a native raft made from inflated goat skins. Norton hiked 12,000 miles through Europe and Asia and made newsworthy walks across Greece and Syria. During the winter of 1904-05 he was sent on a special mission to Persia by President Theodore Roosevelt. From 1905 to 1906 Norton was American consul at Smyrna. Finally, he was consul at Chimnitz, Germany from 1906 to 1914.

While in Germany Norton studied European chemical industries for the U.S. Department of Commerce. This investigation plus his dye making experience led to his being drawn into a political donny-brook over a protective tariff for American dye industry in 1916. I. Frank Stone, president of National Aniline Co., promoted the cause of American dye makers while U.S. Congressman Herman A. Metz, an importer, promoted the interests of importers and foreign manufacturers. The Secretary of Commerce, William C. Redfield, stated publicly that his department was very much interested in the matter and was doing everything possible to foster the development of the domestic dyestuff industry, but he would not advocate a higher tariff because it was against his philosophy and against the policy of the Wilson administration. Norton was engaged as a special representative of the Bureau of Foreign and Domestic Commerce of the Department of Commerce to prepare a report on dyes. He compiled the "Census of Artificial Dyestuffs used in the United States" in 1916. But his report was not received favorably, and a lecture he scheduled at College of the City of New York entitled "The Emancipation of American Chemical Industry," was cancelled by orders of the department. Protests from New York importers of German dyestuffs and German manufacturers caused Secretary Redfield to withdraw the proof sheets of the Census on Dyestuffs from public inspection. Finally the U.S. Tariff Commission was created, and the Commission revised Norton's Census and published it in January 1917.

Following this experience Norton joined E. I. du Pont de Nemours & Co. and later American Cyanamid Co. He was editor of *Chemical Engineer,* co-editor of *Chemical Color and Oil Daily,* and author of books on dyes, cottonseed products, potash production, the contributions of chemistry to methods of preventing and extinguishing fires, utilization of atmospheric nitrogen, and foreign markets for American chemicals.

He died of pneumonia at White Plains Hospital, N.Y., Dec. 2, 1941.

Obituary, *New York Times,* Dec. 3, 1941; *J. Chem. Educ.,* **9,** 241 (1932); *J. Ind. Eng. Chem.,* vols. 7-9 (1915-7); I. F. Stone, "The Aniline Color, Dyestuff, and Chemical Conditions from Aug. 1, 1914 to April 1, 1917" (1917).

DAVID H. WILCOX, JR.

Arthur Amos Noyes

1866-1936

Noyes was an outstanding chemist and educator who bridged the transition between the classical nineteenth century chemistry of Ostwald and the modern quantum chemistry and molecular biology of the twentieth century. He was born in the middle of the nineteenth century, Sept. 13, 1866, son of Amos and Anna Page (Andrews) Noyes in Newburyport, Mass. His father was a lawyer and had a local reputation as a philosopher. From his father he learned Latin, how to play chess, and more active sports such as swimming, rowing, and sailing.

He was a boyhood friend of Samuel Mulliken, and two youthful chemists, Sam and Arthur, carried on early chemical experiments in a backyard laboratory. They prepared phosphoretted hydrogen according to Joel Dorman Steele's "Fourteen Weeks of Chemistry," and many an experiment gone awry was overlooked with tolerance by indulgent parents.

Because of these studies made on his own, Noyes was admitted to Massachusetts Institute of Technology as a second year student in chemistry with high standing and graduated in 1886 with an S.B. degree. He continued to work in organic chemistry and earned the M.S. degree in 1887. At this period in his life he formed a close friendship with George E. Hale, who was to have a profound influence on the course of his later life.

In 1888 Noyes went to Leipzig to do research in organic chemistry with Johann Wislicenius. There he attended lectures by Wilhelm Ostwald and decided to transfer his efforts to Ostwald's field of interest, the newly developing field of physical chemistry.

In 1890 Noyes published several papers which resulted from his doctoral studies. One paper was concerned with deviations from van't Hoff's law of solutions. Another dealt with the validity of the solubility product principle which had just been proposed by Walter Nernst, a young privat-dozent in the laboratory. A third paper, written with Max LeBlanc, was concerned with freezing point determinations.

Noyes returned to MIT and within a few years proposed to establish a research laboratory of physical chemistry, the first such pure-science research laboratory established there.

During this early MIT period Noyes taught analytical, organic and physical chemistry in succession, and published a book on each subject: "Qualitative Analysis of Organic Substances," (1895); "Identification of Organic Substances," with Samuel Mulliken (1899); and "General Principles of Physical Science" (1902). Later on, he served a 2-year term, 1907-09, as acting president of MIT.

In 1915 George E. Hale was building up a staff of scientists in transforming Throop Polytechnical College into California Institute of Technology, to be a companion institution to Mt. Wilson Observatory. He contacted Noyes and invited him to move to California. Because of World War I, and Noyes' activities on the Nitrate Division of the National Research Council, Noyes continued his ties with MIT until the war ended. He then moved permanently to the West Coast and became director of Gates Chemical Laboratory at California Institute of Technology.

Noyes' early book on "Qualitative Analysis" received wide acceptance, and he continued to do research in the field. He devised a complete scheme of analysis which included many rare elements. Finally, with William Bray of University of California in 1929, Noyes published his monumental treatise entitled "A System of Qualitative Analysis for the Rare Elements," which he considered his most important contribution to chemistry.

In the preface he stated: "There has been worked out in detail . . . over the past thirty years . . . a system of qualitative analysis that includes nearly all of the metal-forming elements and makes possible their detection when present in quantity as small as one or two milligrams." Here, perhaps, classical solution chemistry, what we would call "wet chemistry," may have reached a zenith.

A teacher's accomplishments are often measured by his students, and Noyes played an important part in training many men who became leaders in science: W. R. Whitney,

W. D. Coolidge, H. M. Goodwin, G. N. Lewis, W. C. Bray, C. S. Hudson, C. A. Kraus, W. D. Harkins, to name but a few.

In his last years Noyes became interested in biochemistry and with Howard Estill in 1925, he published a paper on the effect of insulin on lactic fermentation."

In California Noyes exerted a profound influence on the development of his institution. The early educational policies of Cal. Tech were largely the result of his influence. Millikan said of him: "He spent more time than any other man on campus trying to create outstanding departments of physics, mathematics, humanities, geology, biology, . . . as well as chemistry . . . and what these departments are today they owe, more than they themselves know, to Arthur A. Noyes."

Noyes died on June 3, 1936. He never married.

Biographies by Miles Sherrill, *Ind. Eng. Chem.* **23,** 443-445 (1931), and Linus Pauling, *Biog. Mem. Nat. Acad. Sci.* **31,** 322-346 (1958) (port.); obituary by M. Sherrill, *Science* **84,** 217-220 (1936).

JOSEPH A. SCHUFLE

William Albert Noyes

1857-1941

Noyes was born on a farm near Independence, Iowa, Nov. 6, 1857, son of Spencer W. and Mary (Packard) Noyes. He attended a country school a few months each year but was self-taught to a considerable degree. Thus he learned the Greek alphabet while plowing corn, and his older sister helped him learn Latin.

Noyes entered the Academy at Grinnell, Iowa for one semester before enrolling in Grinnell College in 1875. He dropped out of college for the winter quarter each year to teach in country schools to earn funds for his education but graduated in the regular time, studying at night to learn material missed in the winter quarters. He took the classical course but also studied as much science as he could work into his schedule. As a result he received both his A.B. and B.S. degrees in 1879.

After graduation he taught Greek and chemistry in the college academy. In his spare time he trained himself in analytical chemistry. In the fall of 1880 he was given charge of chemistry courses in the college while the professor was on leave.

He entered Johns Hopkins University in 1881 and worked under Ira Remsen. He performed water analyses to pay his expenses. Yet he completed the work for his Ph.D. degree in 1½ years.

Noyes was instructor in chemistry at University of Minnesota in 1882, spending much time analyzing specimens for the Minnesota geological survey. In 1883 he became professor of chemistry at University of Tennessee and in 1886 professor at Rose Polytechnic Institute. While there he took a leave of absence from 1888 to 1889 to work under Adolf von Baeyer at Munich. As the only professor of chemistry at Rose, he taught all the chemistry courses and directed research for seniors.

In 1903 he accepted an appointment as the first chief chemist of National Bureau of Standards. His work in developing standard methods of analysis and standard specifications for chemicals established the base for the bureau's work in those areas. During this period he determined the ratio of hydrogen to oxygen in water and with H. C. P. Weber determined the atomic weight of chlorine.

In 1907 he was appointed head of the Department of Chemistry at University of Illinois, a position he held until he retired in 1926. His interest in teaching and research stimulated his colleagues, and together they laid the foundation for one of the eminent chemistry departments of the country. He was a great teacher and an excellent director of research for graduate students. He was always sympathetic and ready to help students in spite of other activities in which he was engaged.

While Noyes worked on atomic weights and analytical methods and standards, he was primarily an organic chemist. He investigated the structure of camphor and many rearrangements in the camphor series. He was one of the early chemists interested in the electronic theory of valence and published articles in the field. He was interested in the valence and character of nitrogen in nitrogen trichloride and worked with the dangerous chemical in spite of several explosions.

Noyes was a leader in the American Chemical Society. He edited the journal of the society from 1902 to 1917, helped establish *Chemical Abstracts* and was its first editor from 1907 to 1910, was first editor of *Chemical Reviews* from 1924 to 1926, and was first editor of the scientific series of ACS Monographs from 1919 to 1941. From 1904 to 1907 he was secretary and in 1920 president of the organization.

Noyes received many awards and honorary degrees. He and H. C. P. Weber received the Nichols Medal in 1908 for their work on the atomic weight of chlorine. His work in organic chemistry brought him the Willard Gibbs Medal in 1920. In 1935 he received the Priestley Medal.

No biography of Noyes would be complete without reference to his sincere interest in promoting peace. He had many friends in Germany and France and was particularly unhappy at the bitterness which arose between them during the first World War. He worked unsuccessfully to establish harmony between these leaders and to promote international understanding.

Noyes married Dec. 24, 1884 Flora Collier; they had three children—two girls who died young and W. Albert, Jr., who became a president of the American Chemical Society and professor of chemistry at University of Texas. Noyes' first wife died, and he married June 18, 1902 Mattie Elwell; they had a son, Charles Edward. Noyes' second wife died, and he married third in 1915 Katharine Macy; they had two sons, Henry Pierre and Richard Macy; the latter taught at Columbia and later became professor of chemistry at University of Oregon.

He was a deeply religious man and devoted much service to the Congregational Church. Two weeks before his eighty-fourth birthday he passed away on Oct. 24, 1941.

Biographies by R. Adams in *Biog. Mem. Nat. Acad. Sci.* **26-28,** 179 (1951-1954); by B S. Hopkins in *J. Amer. Chem. Soc.* **66,** 1045-1056 (1944), port.; by A. Patterson in *Ind. Eng. Chem.* **16,** 420 (1924). Charles A. Browne, Mary E. Weeks, "History of the American Chemical Society" (1952), port.; David H. Killeffer, "Eminent American Chemists" (1924), port.; personal recollections.

CARL S. MARVEL

O

George Oenslager

1873-1956

Oenslager was born in Harrisburg, Pa., Sept. 25, 1873. He graduated from Harvard University in 1894 and continued his study there under Theodore W. Richards, receiving his A.M. degree in 1896. He then went to work for S. D. Warren Co., Cumberland Mills, Maine. In 1906 at the invitation of Arthur H. Marks, chemically-trained general manager of Diamond Rubber Co., he went to Akron, Ohio. Oenslager later commented that when he arrived in Akron, already the center of the rubber industry, he was one of only three chemists engaged in the industry and that 25 years later the number had risen to 400.

Oenslager's first assignment at Diamond was, in his words, "the problem of the economic utilization of low-grade rubbers." These cheaper rubbers were slow-vulcanizing or, to use a more common term, slow-curing, and the cured rubbers were of low quality. The initial attack was to find a way to increase the rate of cure of low-grade rubbers with the expectation that the quality of the cured rubber would be improved at the same time. Marks had already shown that acetone extracted a material from high quality para rubber that accomplished this result. Such a material is now called an accelerator. It was also known that oxides of calcium, magnesium and lead are weak accelerators.

Oenslager first tried adding many elements, inorganic salts, and inorganic oxides to mixtures of rubber, sulfur, and zinc oxide. Mercuric iodide proved to be an excellent accelerator. In April 1906 tires using low-grade rubber cured with the aid of mercuric iodide were manufactured. Though initially apparently highly satisfactory, the tires soon disintegrated in use; the mercuric oxide also accelerated the destructive action of oxygen on rubber. Oenslager then turned to organic compounds. The basic character of the metal oxide accelerators suggested organic bases which he thought might have an added value because of their ready solubility in rubber.

Aniline, inexpensive and readily available, was the first organic base tried June 1, 1906. It proved an excellent accelerator as did many related aromatic amines. *p*-Nitrosodimethylaniline proved remarkably effective, but it had to be discarded because of the yellow color it imparts to whatever it touches. Aniline appeared the material of choice but presented toxicity problems. Oenslager therefore converted it to thiocarbanilide which proved an excellent solid accelerator with a much lower toxicity. In September 1906 experimental tires were made using thiocarbanilide in the tread rubber and aniline in the rubber which was calendered on the fabric for use in the tire carcass. Before tires were released, aging tests were run and a pleasant added value was discovered: the new organic accelerators were good preservatives of the vulcanized rubber. Beginning in 1907 all rubber compounds for tires manufactured by Diamond contained aniline or thiocarbanilide. The organic accelerators, of which there are many today, were thus established as important ingredients to make cured rubber compositions of high quality.

The discovery of organic accelerators and their development to commercial use was accomplished in an astonishingly short time, less than a year. Their impact on the industry was tremendous: improved rubber products, successful use of lower-cost rubbers for quality products, lower investment costs for curing machinery, lower labor costs because of the decreased curing times, better product control, and initially the successful utilization of certain low-cost natural rubbers.

Oenslager in 1907 designed and operated a plant to make 2,000,000 pounds of thiocarbanilide annually. In 1912 Diamond merged with B. F. Goodrich Co. and Oenslager continued with Goodrich until he retired in 1940. During this time he was instrumental in establishing carbon black as a reinforcing agent for rubber. From 1920 to 1922 he served as technical adviser for Yokohama Rubber Co. in Japan.

Oenslager was recipient of the Perkin Medal of the Society of Chemical Industry in 1933 in recognition of his "development of nitrogenous organic accelerators . . . and contributions to the development of carbon black tread for tires." In 1939 he was given an honorary D.Sc. degree by University of Akron. In 1948 he was awarded the Charles Goodyear Medal by the Division of Rubber Chemistry of the American Chemical Society.

On July 15, 1939 Oenslager was married to Ruth Alderfer. He died Feb. 5, 1956 while on board ship on the way to Europe. He was buried in Akron.

H. L. Trumbull, *Ind. Eng. Chem.* **25,** 230-32 (1933); *Rubber Age* **78,** 938 (1956); *Rubber World,* **135,** 849 (1956); the quotations above are from Oenslager's account of his work in *Ind. Eng. Chem.* **25,** 232-237 (1933).

GUIDO H. STEMPEL

Louis Atwell Olney

1874-1949

A distinguished descendant of sturdy New England stock extending back to the days of the founding of the colonies, Louis Olney was born in Providence, R.I., Apr. 21, 1874. His education included course work at the Bryant and Stratton business college and at Lehigh University, where he received his B.S. degree in chemistry in 1896 and his M.S. degree in 1908. In 1926 Lehigh conferred the Sc.D. degree on him.

A man of many talents, Olney distinguished himself as a scientist, educator, administrator, business man, and civic leader. He is perhaps best known as the founder of the American Association of Textile Chemists and Colorists, which filled a need in the United States for a society which would serve its members in the manner of the British Society of Dyers and Colourists. The first meeting of this society, in Boston Nov. 3, 1921, saw Olney elected as president, and he served in this capacity until 1927 when he became chairman emeritus. He remained as chairman of the research committee for many years.

Olney was an active participant in many scientific societies. His 45-year membership in the American Chemical Society included service as chairman of the Northeastern Section and as an assistant editor of *Chemical Abstracts.*

He was a director of Howes Publishing Co., which has honored him by establishing the Olney Medal as an award to be made by the AATCC for outstanding achievement in the field of textile chemistry. In addition to numerous technical articles, he was the author of several books including: "Elementary Organic Chemistry" and "Chemical Technology of Fibers." For many years he was editor of *American Dyestuff Reporter.*

As an educator Olney began his career as an instructor in chemistry at Brown University and later was professor of chemistry and dyeing and head of the department at Lowell Textile Institute from 1897 to 1944. Recognizing the importance of making chemistry a vital force in the service of mankind, he played a major role in developing a sound curriculum in textile chemistry; during his life there was probably at least one representative of his training in every dyeing and finishing plant in the country as well as in the laboratories of dyestuff manufacturers, synthetic yarn manufacturers, and testing and service laboratories.

As a business man and civic leader Olney also had a distinguished record. He was president of Wannalancit Textile Co. of Lowell, Lowell Institution for Savings, Lowell Morris Plan Co., Stirling Mills of Lowell, and Lowell Lingerie Co. He was a director and president of the Lowell Y.M.C.A., director of Isle of Shoals Congregational Corp., and treasurer of Northfield Conference of Religious Education.

Olney was a man who held to the highest standards, was endowed with a keen intelligence, and possessed excellent judgment. His vision, enthusiasm, and persuasive manner and his outstanding leadership in textiles characterized a noteworthy and distinguished career which was unexpectedly cut short. He died Feb. 11, 1949 at Jacksonville, N.C., following an automobile accident on Feb. 7th in which his wife, Bertha H. H., died immediately.

"American Contemporaries: Louis A. Olney," *Chem. Eng. News* **21,** 1620 (1943); "Louis Atwell Olney," *Am. Dyestuff Rep.* **38,** 156A-D (Feb. 21, 1949); Resolution by AATCC Council, *Am. Dyestuff Rep.* **38,** P331 (Apr. 18, 1949); personal recollections.

GEORGE R. GRIFFIN

John Morse Ordway

1823-1909

It was in 1884, during Tulane University's first year, that Ordway was called to this institution from Massachusetts Institute of Technology. He became the first professor of applied chemistry at the university and directed manual training courses, which he developed into a civil engineering curriculum. Later he offered the first lectures in academic biology in both the College of Arts and Sciences and Newcomb College, the women's division of Tulane University.

Ordway was born in Amesbury, Mass. in 1823, the son of John Morse and Hannah (Morse) Ordway. In 1830 the family moved to Lowell, Mass. He left school at age 13 to be apprenticed to a chemist and apothecary in Lowell, but it is assumed that he finished at Lowell High School because in 1844 he graduated with the degree of master of arts from Dartmouth College. He began to study medicine but was persuaded to go into industrial chemistry within the year. Three years later (1847) he became superintendent of Roxbury (Mass.) Color and Chemical Co.

In 1850 he moved, first to Illinois, then to St. Louis, and in the next year was given a professorship at Grand River College, Trenton, Mo. and at the same time was married. By 1854 the administration of the college was in his hands, but at this time the college burned so he took the position of superintendent of schools for the county. In the same year he was invited to return to Roxbury. The color company failed in 1857, and he joined Hughesdale Chemical Works at Johnston, R. I. where he stayed until the death of his wife in 1860. For the next 14 years he continued in industry, first as chemist, manager, and superintendent of the Manchester Print Works, then as chemist at Bayside Alkali Works, South Boston, alternating with Hughesdale. Although he was appointed professor of industrial chemistry and metallurgy at Massachusetts Institute of Technology in 1869, he continued in industry until 1874 when Bayside Chemical Works burned. After this he devoted his full time to academic work, an instructorship in biology having been added, and in 1877 became chairman of the faculty, serving in this capacity 4 years and carrying most of the duties of president. However, teaching and administration did not prevent his accepting the chairmanship of the Chemical Section of the American Association for the Advancement of Science, and publishing his "Plantarum Ordinum Indicator" (1881).

His second wife died in 1873, and in 1882 he married Evelyn M. Walton, a former assistant at Massachusetts Institute of Technology. They visited the Slojd (industrial) schools, and he translated from the Danish an account of the Swedish system of industrial education, published by Massachusetts Board of Education.

In 1884 he was called to Tulane University as professor of biology and industrial chemistry. His accomplishments in industry and education were acknowledged by his selection at this time as a member of a committee of three to call on Paul Tulane to interest him in establishing technical, i.e. engineering, education at the university.

Ordway was from 1884 on superintendent of the manual training courses at Tulane, a curriculum which developed into the College of Engineering, and from 1893 to 1896 he was acting professor of civil engineering. His publications in this field were carried in the *Transactions of the American Society of Mechanical Engineers.* He also published "A General Discussion of Cotton Production in the United States" in collaboration with E. W. Hilgard (1884). His articles on chemistry and biology appeared in *American Microscopical Journal, American Journal of Science* and *Journal für praktische Chemie.*

Ordway resigned from his positions in the College of Arts and Sciences in 1897, retaining until 1904 his professorship in biology at Newcomb College where his wife was professor of chemistry. After his retirement, he returned to Boston where he died July 4, 1909. Ordway bequeathed his personal library to the university, and it became the nucleus of Tulane's History of Science Collection. His journals, and the funds he left to continue them, became the base for the chemistry holdings of the university library.

Material in the archives of Howard Tilton Library of Tulane University, including: "The report to the Faculties of the Academic College of Tulane University of Nov. 20, 1909," and "Curriculum Vitae for John Morse Ordway" by Anne Ashby, Tulane University, Aug. 1956.

VIRGINIA MCCONNELL

Iwan Iwanowitch Ostromislensky

1880-1939

Ostromislensky, a Russian-American organic chemist and son of an officer of the Imperial Guard, was born in Moscow, Sept. 8, 1880. After graduating from the Moscow Military Academy in 1895 and studying at the Moscow Pilotekhnikum until 1899, he pursued graduate work at University of Zürich, receiving his Ph.D. degree in 1902 and M.D. in 1906. He completed his formal education by obtaining a chemical engineering degree in Karlsruhe in 1907. On his return to Russia he assumed the duties of assistant professor of chemistry at the Politekhnikum of Moscow University, teaching also at Moscow School of Dentistry.

From 1907 to 1912 his publications reflected a strong preoccupation with the physical properties of organic substances. He studied triboluminescence, the crystallography of stereoisomers, the structural origin of color, and the relationship of the structures of compound-pairs to their mutual solubility.

His technical writing, which tended strongly to broad generalizations, involved him in several priority controversies.

In 1912 he gave up teaching. During the preceding year he had established a private laboratory in Moscow in which he carried out research, particularly on the chemistry of rubber. He obtained a rubbery material by polymerizing vinyl halides and then dehalogenating with metallic zinc. His work came to be regarded as the principal support for the "head-to-head" theory of vinyl polymerization, a view definitively refuted and replaced by the "head-to-tail" theory only in 1939. He was the first to describe procedures for solution polymerization of vinyl chloride.

By 1914 his interests had shifted to medicinal chemistry. He worked on the composition of chaulmoogra oil and tried to offer a rationale for chemotherapy. His "dualistic theory" emphasized the independent significance of the binding and the chemically active portion of drugs, toxins, and antitoxins.

Under the pressure of war, Ostromislensky returned to work on rubber. In an extraordinary series of more than 20 journal articles in 1915 he described methods of preparing monomers, polymerization procedures, the basic physical chemistry of plastic and elastic systems, and methods of vulcanizing rubber without sulfur. The novel vulcanization systems used nitro, halo, or peroxy compounds. He also pointed out the value of aromatic amines in rubber formulation. In 1915 he developed the process for converting alcohol to butadiene which, with improvements by Lebedev in 1925 and Union Carbide in the early 1940's, was used to make tonnage quantities of butadiene during World War II.

He did not neglect medicinal chemistry. He developed improved organoarsenicals to replace the relatively toxic Salvarsan, and he developed medicinal applications for Pyridium, a phenylazodiaminopyridine, first synthesized by Chichibabin.

The course of World War I, revolution, and civil strife made life increasingly difficult for Ostromislensky, a staunch Czarist. He fled Russia in 1921. After a brief residence in Riga, Latvia, he was induced by the efforts of Ernest Hopkinson of U.S. Rubber Co. to come to America.

From 1921 to 1925 Ostromislensky worked on synthetic rubber, on vulcanization processes, on poly(vinyl chloride) and on polystyrene. He devised processes for manufacturing styrene monomer from ethylbenzene. He was the first to copolymerize a diene with styrene, a process basic to the modern synthetic rubber industry.

In 1925 he returned once more to medicinal chemistry. He founded Ostro Research Laboratories where he worked on methods of developing organoarsenicals and Pyridium. His "Scientific Basis of Chemotherapy," a small book, published in 1926, restated his "dualistic theory" and promoted the use of his medicinal products. A business association with Pyridium Corp. marked a further attempt to commercialize these compounds. Ostromislensky's efforts were not financially successful, and he terminated his connection with Pyridium Corp. in 1936.

In 1935 he developed several pyrazolone compounds which he claimed were useful for treating alcoholism and allergic diseases.

He had a brilliant and speculative mind. He attempted to work on problems for which the techniques and theories of his day were inadequate. A lack of meticulousness in his chemical experimentation exposed him to criticism which sometimes obscured the very substantial contributions he made, especially to the rubber and plastics industries.

A U.S. citizen since 1930, he retained his interest in Russian culture and traditions. He was president of the National League of Americans of Russian Origin, and founded, along with Count Ilya L. Tolstoy and Sergei Rachmaninoff, the Circle of Russian Culture.

He died at his home in New York City, Jan. 16, 1939.

Morris Kaufman, "The History of PVC," Gordon and Breach Science Pub. (1969); I. I. Ostromislensky, "Scientific Basis of Chemotherapy," Inter-American Medical Publishing Co. (1926); *Chem. Abstr.;* obituary in *New York Times,* Jan. 19, 1939; "Dictionary of American Biography," **22,** 505-506, Chas. Scribner's Sons (1958).

CHARLES H. FUCHSMAN

Emil Ott

1902-1963

For many years one of the world's leading authorities in the field of cellulose derivatives, plastics, and explosives, Emil Ott was known to his colleagues as one who could recognize important problems at once and then proceed to solve them.

Ott was born in Zurich, Switzerland, May 19, 1902 where he received his early education. He was the son of Emil and Anna (Goldfinger) Ott. Displaying a proficiency in chemistry, mathematics, and physics at an early age, he completed the Oberealschule in 1921 and became a diplomate of the Swiss Institute of Technology of Zurich in 1925. Remaining at the institute as an instructor in physics from 1925 till 1927, he completed his doctor of natural science degree there in 1927 in physics and chemistry.

Leaving Switzerland, Ott emigrated to the United States and accepted a position as a research and development chemist for Stauffer Chemical Co. in San Francisco during the year 1927-28. During the 1928-29 academic year, he continued his research activities at Johns Hopkins University as an American Petroleum Institute postdoctoral research fellow and then stayed on at Johns Hopkins as an assistant professor of chemistry for the next 4 years, where his ability as a teacher and research professor soon won him a large following.

Ott's unusual talent as an experimental physical chemist, especially in the then new field of applying x-rays and x-ray diffraction analysis to chemical systems, attracted the attention of Hercules Powder Co., which promptly invited him to become a consultant for them in 1933.

That same year he left Johns Hopkins to become a research chemist in the Development Department of Hercules in Wilmington, Del., and in 1937 he became head of the Research Division at Hercules Experiment Station, supervising the development of new processes and a variety of pilot plant operations.

In 1939 Ott was selected to be the first director of research at Hercules, a position he continued to hold until his retirement from the company in 1955. At Hercules he was responsible for coordinating the company's research program in its central research laboratory and its 12 plant laboratories and for organizing and developing a fundamental research program for the company.

The years at Hercules were fruitful ones for Ott, who led research groups into investigations involving explosives, cellulose and its derivatives, plastics and synthetic resins, rosin and terpene derivatives, commercial products and processes, reaction intermediates, and chemicals for paper making. His belief that the best research is that produced by a team effort was born while at Hercules.

During this period, Ott also served as editor of "Cellulose and Cellulose Derivatives," published in 1943, and presented lectures on high polymers at Western Reserve University, Wayne University, Polytechnic Institute of Brooklyn, and for the American Physical Society. His article entitled "The Team Approach in Research and Development" was published in *Chemical and Engineering News* in 1950 and another article "I Am Proud to Be a Chemist" was published in *The Chemist* the following year.

A strong supporter of all professional activities, Ott served in 1952 as president of the Association of Research Directors and in 1953 as president of the American Section of the Society of Chemical Industry as well as the secretary of the Association of the Directors of Industrial Research. During the latter year he also served as a member of the Committee on Macromolecules of the International Union of Pure and Applied Chemistry. For more than 20 years Ott was intimately connected with the activities of the American Chemical Society and the American Institute of Chemists as well as those of a number of other organizations.

Ott married Dorothy Aiken Wright Oct. 2, 1933, and they had five children: John Wright, David Emil, Peter Reynolds, Joan Nancy (Mrs. John Guyton), and Dorothy Ann (Mrs. George E. Darmstatter).

After he retired from Hercules in 1955, Ott became vice-president and director of chemical research for Food, Machinery, and Chemicals Corp. at Princeton, N.J. and in 1958 became vice-president for research and development of the Chemistry Division of the College of the City of New York. From 1960 to 1962 he served as a research specialist for Rutgers University, and from 1962 until his death on Sept. 29, 1963, he was a research professor at Stevens Institute of Technology.

Walter J. Murphy, "Emil Ott—An Appreciation of His Scientific and Professional Contributions," *The Chemist,* 1951; Harold Spurlin, "Emil Ott and Cellulose Chemistry," *The Chemist,* January, 1970.

JAMES C. COX, JR.

P

Howard Coon Parmelee

1874-1959

Parmelee, chemical engineer, editor, and educator, was born in Omaha, Neb., Dec. 4, 1874, the son of Edward A. and Sara (Coon) Parmelee. He was educated at University of Nebraska where he received his B.Sc. in chemistry in 1897 and his A.M. degree in 1899. Later when serving as president of Colorado School of Mines in 1916 he was awarded an honorary doctorate in science. He played a significant role in the pioneer development of the chemical engineering profession. He was an active member of a small group of western chemists and metallurgists responsible for the progress and world-wide adoption of basic processes and equipment. His contributions to the literature had important impact on chemical engineering education.

During his college days young Parmelee worked in the chemical and assay laboratories of the Union Pacific Railroad and also served as a graduate assistant. He moved to Denver in 1900 ultimately to become the chief chemist at the Globe plant of American Smelting and Refining Co. This association led indirectly to his editorial career, first as a contributor and later the editor of *Mining Reporter* in 1905 and of *Western Chemistry and Metallurgy* in 1907. Shortly thereafter he was appointed Western Editor of the national magazine, *Metallurgical & Chemical Engineering* which had been founded in 1902 by James H. McGraw and Joseph W. Richards of Lehigh as *Electrochemical Industry*. When its editor, Eugene F. Roeber, died in 1918, Parmelee moved to New York to begin his long and productive career with McGraw-Hill magazines and books. He was to become chief editor of "Chem. and Met." (now *Chemical Engineering*) and *Engineering & Mining Journal* as well as vice president and consulting editor of McGraw-Hill Book Co.

Shortly after he moved to New York Parmelee became an active member of the American Institute of Chemical Engineers and was appointed to its Committee of Chemical Engineering Education under the chairmanship of Arthur D. Little. Chemical engineering had experienced a mushroom growth since the first World War. More than 200 so-called chemical engineering subjects were being taught by almost 50 colleges and universities. Yet there was still no satisfactory answer to the question: what is really chemical engineering? It was only after several years of intensive study and exhaustive surveys that the Little committee succeeded in formulating a new concept and definition of chemical engineering and came up with a practical plan for its adoption by the major educational institutions of the country. Little drew heavily on his own association with Massachusetts Institute of Technology and particularly a new textbook, "Principles of Chemical Engineering" by Walker, Lewis, and McAdams of its Chemical Engineering Department.

In the June 1922 meeting of AIChE the report of the Little committee was officially adopted, including its still controversial definition. Most important was its recommendation that within 3 years the American Institute should come up with a list of "those institutions it believed were then offering satisfactory courses in chemical engineering." Only then did Little retire his chairmanship and was immediately succeeded by Parmelee who vigorously took over the crusade for the adoption of the new concept by universities and colleges. He first reorganized the committee to comprise an equal number of senior educators and major engineering executives.

After many stormy sessions, formal and informal conferences, editorials, and articles, it is to the lasting credit of Parmelee and his committee that the institute in its June 1925 meeting approved the accrediting of just 15 institutions. Eight years later when Harry A. Curtis had succeeded to the chairmanship, the list had grown to 24 . At the writing of this book it was 119. And most significantly for the entire engineering profession the AIChE plan was adopted in 1932 by the Engineering Council for Professional Development and became the basis for the accreditation of all undergraduate engineering courses in the United States and Canada.

Early in its work the AIChE committee foresaw the need for a new basic literature to serve as a lasting foundation for the growth of the new profession. After several meetings an informal editorial advisory committee submitted a report to the McGraw-Hill Book Co. outlining a coordinated series of more than a dozen texts and reference books. In all more than 50 such books had been published in the McGraw-Hill Chemical Engineering Series by 1972 and had become the basis for a world-wide profession—a lasting monument for Parmelee, who died in Bradenton, Fla., Nov. 17, 1959.

Personal recollections; obituary, *New York Times,* Nov. 18, 1959.

SIDNEY D. KIRKPATRICK

Samuel Wilson Parr

1857-1931

"The skyline of the chemist is to be seen in the advancing status of human welfare." Those words, taken from the speech Parr made when he became president of the American Chemical Society in 1928, reflect the motivation of his career in chemical technology, metallurgy, and teaching of applied chemistry. Parr is probably best known for his contributions to knoweldge of the chemistry of coal and his improvements in the art of fuel calorimetry.

Born Jan. 21, 1857 in Granville, Ill., Parr received his B.S. degree from Illinois Industrial University (now University of Illinois) in 1884 and his M.S. degree from Cornell University a year later. Subsequently, he did post-graduate studies at University of Berlin and the Polytechnic in Zurich.

Parr taught at Illinois College from 1885 to 1891 and then became professor of applied chemistry at University of Illinois. There he continued his laboratory studies in which he sought improved methods for testing the thermal properties of coal. The major result of his early work was development of an entirely new type of calorimeter, one which for the first time, used sodium peroxide rather than oxygen as the source of oxygen for combustion. Known as the Parr Standard Calorimeter, the apparatus was economical and convenient to operate and was therefore widely adopted for determining power plant heat balances and purchasing coal on specification.

In 1899 Parr founded Standard Calorimeter Co. to manufacture the scientific laboratory equipment he was designing. Known later as the Parr Instrument Co., it produced a wide range of scientific laboratory apparatus.

Although the Parr Standard Calorimeter found wide use for many purposes, a demand was growing for a more accurate calorimeter. In order to obtain increased accuracy, oxygen had to be used as the combustion medium. Oxygen bomb calorimeters were available, but they were expensive and were corroded by the acid products of combustion; this corrosion reduced the accuracy of the calorific tests.

Hoping to lessen or eliminate these undesirable effects and to develop a superior calorimeter, Parr turned his research to metallurgy. His objective was a corrosion-resistant alloy. His successful study led to his developing the first acid resistant alloy of its kind—a double ternary alloy of nickel/chromium/copper and tungsten/molybdenum/manganese. Named Illium after University of Illinois, Parr's alloy was one of the forerunners of stainless steel. In 1911 Parr began producing Illium solely for use in an improved, reasonably priced oxygen bomb calorimeter of his design.

Parr's 35-year-long teaching career paralleled his laboratory research and development efforts. He introduced the first chemical engineering curriculum at the University of Illinois. His textbook, "The Analysis of Fuel, Gas, Water and Lubricants," was used as a basic college text in industrial chemistry courses for many years. The book went through four editions; Parr revised it extensively during the year preceding his death.

In addition to his research and teaching, Parr wrote many articles for scientific journals and served in many capacities including director of the Illinois State Water Survey and consulting chemist of the State Geological Survey. He received many honorary degrees and awards, the most notable being the Chandler Medal in 1926.

Parr was known throughout the world for his contributions in coal chemistry, fuel calorimetry, and metallurgy. He excelled not only in research but also in making the results of his research intelligible and useful for a practical world. He died May 16, 1931.

Information from Parr's descendants; biography by E. Bartow in *Ind. Eng. Chem.* **17,** 985 (1925), with portrait; obituary by W. A. Noyes in *J. Amer. Chem. Soc. Proc.* pp. 1-7 (1932), with portrait.

JODY M. SMITH

Charles Lathrop Parsons

1867-1954

Parsons was born Mar. 23, 1867 at New Marboro, Mass., son of Benjamin Franklin and Leonora (Bartlett) Parsons. The family lived on a farm at Hopkinton, Mass. from 1869 to 1877 and then moved to Hawkinsville, Ga. Charles attended Cushing Academy in Massachusetts and in 1888 received his B.S. degree from Cornell.

Parsons' first job was that of assistant chemist at University of New Hampshire. In 1892 he was promoted to the professorship of general and analytical chemistry and to head of the Chemistry Department, for which he designed new laboratories. In 1903 he was named professor of inorganic chemistry. At New Hampshire Parsons developed the first method used by creameries in this country to analyze milk quickly. With Alfred J. Moses of Columbia University he wrote "Elements of Mineralogy, Crystallography, and Blowpipe Analysis from a Practical Standpoint," (1900). On occasions he acted as a consultant. He became an expert on the chemistry of beryllium, receiving the Nichols Medal for his work in 1905, and writing a monograph, "Beryllium: its chemistry and literature" (1909).

Parsons joined the American Chemical Society in Chicago in 1893 during the first national meeting he attended. In 1907 he was chosen as secretary of the society, a half-time position. Four years later he became chief mineral technologist of recently created U.S. Bureau of Mines and moved to Washington. He took the secretaryship with him; the bureau provided him space for it. In 1916 he became chief chemist of the bureau.

During his work at the bureau Parsons learned that Colorado and Utah carnotite, largest available source of radium in the world, was being shipped abroad for workup and its radium sold here for exhorbitant prices; he organized a project to develop an American process. Since government financing was not available, he found sponsors in James Douglas of Phelps Dodge Co., New York, and Howard A. Kelly, a Baltimore physician. The National Radium Institute was formed and was the operating body with which BuMines cooperated under Parsons' direction. Parsons contracted for the carnotite, supervised design of equipment and the plant that was built at Denver, contracted for disposal of residual vanadium and uranium values, and distributed the product. In its 3 years of operation starting in 1913, NRI produced over 4 grams of radium at a cost of about $40,000 per gram, one third the price of imported radium.

During World War I Parsons was influential in assisting in the establishment of munitions-related industries and in helping the country insure the best war-time employment of chemists. In 1916 he initiated work on nitric acid and nitrates at the Bureau and in industry, using a multiple platinum gauze catalyst that he developed. Sent to Europe by the Secretary of War to inspect nitrogen fixation plants, he returned with the recommendation that the United States adopt the Haber ammonia process. He directed the bureau's project to build a plant for making sodium cyanide by the Bucher process. Throughout the war he served on the Nitrate Commission, to which all plans for developing nitrogen fixation were first referred.

During his European tour, Parsons had been appalled at the waste of chemists as combatants, and he was determined not to let it happen here. On behalf of the ACS and the bureau he superintended a census of 16,000 chemists and metallurgists. This census formed the basis of selection of chemists who were assigned to wartime technical service with the government. As ACS secretary he knew all the prominent chemists in the country, and he recruited many of them for Navy and Army Ordnance and for the group which later became the Chemical Warfare Service.

Parsons resigned from the bureau Nov. 1, 1919 to give more of his attention to ACS affairs. Because the secretaryship was still only a half-time position, he again established himself as a part-time consultant, now as an authority on nitrogen fixation, uranium, and radium. He stopped consulting in 1931 to accept the additional job of ACS business manager and thereafter held both posts until he retired in 1946.

When Parsons joined the ACS it had 460 members; when he became secretary it had 3400; when he retired it had 43,000. During his long secretaryship he initiated several important changes within the organization. In 1909 he inaugurated corporation memberships, which for years were a substantial means of support for the society. In 1937 he carried off what he felt was the greatest coup of his career by obtaining a federal charter for the ACS, the only society of its size and character to have one. In the years 1942-44 when owners of Universal Oil Products Co. attempted to give the firm to the ACS, Parsons succeeded in directing the gift to a trusteeship under which the ACS administers the income for petroleum research. One of his close associates said of him; "He ruled the ACS with an iron hand, but it was just hand, and his aim and ends were always the welfare of the society, and through it the welfare of the country."

Parsons received many honors, including Officer of the French Legion of Honor (1922), Officer of the Crown of Italy (1926), Priestley Medal of the ACS (1932), and honorary doctorates from University of Maine (1911), University of Pittsburgh (1914), and University of New Hampshire (1944). In 1952 the ACS established a triennial award for public service named after Parsons and made him the first recipient.

Parsons married Alice D. Robertson in 1887, and they had five children—Charles Jr., and four daughters. He died Feb. 13, 1954, at Pocasset, Mass.

ACS official records; *Industrial and Engineering Chemistry; Chemical and Engineering News;* taped interviews with Parsons by Walter J. Murphy.

ROBERT F. GOULD

Austin McDowell Patterson

1876-1956

Patterson was born in Damascus, Syria, May 31, 1876 where his medical missionary father was located. Nine months later the family returned to Xenia, Ohio which became his home center for the remainder of his life. His mother helped him learn Arabic, and he was tutored in Latin and Greek. This familiarity with languages led him to enter Princeton University as a student of the classics. Here he first attended a chemistry course, without laboratory, for arts students. His interest in chemistry led him to experiment in a little laboratory in his dormitory bedroom. Upon graduation in 1897 he declined offers of a classical fellowship and a chemical assistantship at Princeton to work under Ira Remsen at Johns Hopkins University. He obtained his Ph.D. degree in chemistry in 1900.

Patterson taught at Center College from 1900 to 1901, and at Rose Polytechnic Institute from 1901 to 1903. At Rose he met William A. Noyes, Sr., who ushered him into the field of chemical documentation. In 1903 Patterson became chemistry editor of "Webster's New International Dictionary." He remained associated with the dictionary until 1953. This activity led to his selection as associate editor of *Chemical Abstracts* in 1908 and as editor in 1909.

An attack of tuberculosis interrupted his work on *Chemical Abstracts* in 1914, but he recovered by a change in climate. During his illness he began to compile special dictionaries, "A German-English Dictionary for Chemists" (1917) and "A French-English Dictionary for Chemists" (1921), reference books used by generations of chemists.

Patterson returned to *Chemical Abstracts* in 1916 to work out the indexing system for the first decennial index. In 1918 and 1919 he was a chemist with the U.S. Army Chemical Warfare Service.

Patterson was professor of chemistry at Antioch College from 1921 to 1941, vice president from 1930 to 1941. During these years he worked with committees of the American Chemical Society, National Research Council, and International Union of Pure and Applied Chemistry to compile an international report on nomenclature of organic chemistry. With Leonard T. Capell he spent 18 years in the painstaking task of listing and cataloguing all known ring systems of organic chemistry in a 600-page book, "The Ring Index" (1940). To help students find their way through journals and reference works, he and Evan J. Crane wrote "A Guide to the Literature of Chemistry" (1927).

Patterson served in Washington, D. C. from 1941 to 1943 with the Engineering, Science, and Management War Training Commission. Returning to Antioch as emeritus professor, he wrote for *Chemical Abstracts* and for "Webster's New International Dictionary" and was a welcome addition to the chemistry department as adviser and consultant.

In 1949 the Dayton Section of the American Chemical Society honored Patterson for his work in chemical documentation by establishing a biannual award in his name. For several years in the 1950s he wrote a column on chemical nomenclature, "Words About Words," in *Chemical & Engineering News*. The collected columns were later published as a book.

Patterson took an active part in the affairs of his home town. He was a member of the school board and the Charter Commission, chairman of the county Red Cross and the Boy Scouts, and editor of a county history. He once bought and operated a newspaper when a bad political situation needed and got "attention." He died Feb. 26, 1956, at his home in Xenia.

Biographies by E. J. Crane in *Chem. Eng. News* **19**, 743 (July 10, 1941), with portrait, and in "Words About Words: A Collection of Nomenclature Columns written by Austin M. Patterson" (1957), pp. vii-x, with portrait; obituary by J. F. Corwin, *J. Chem. Educ.* **33**, 396 (1956), with portrait; Charles A. Browne, Mary E. Weeks, "History of the American Chemical Society" (1952), with portrait; personal recollections.

JAMES F. CORWIN

Robert Patterson

1732-1824

Robert Maskell Patterson

1787-1854

Robert Patterson and Robert Maskell Patterson were one of the early father-son pairs of American scientists. Robert was born in Ireland, May 30, 1743, emigrated to Pennsylvania in 1768, was professor of natural philosophy in University of Pennsylvania from 1779 to 1814, and was appointed Director of the United States Mint by President Jefferson in 1805. He died in Philadelphia July 22, 1824. His greatest competence lay in mathematics, but he also taught the "principles of chymistry" in his natural philosophy course at the university. His chemistry lectures covered, presumably, the material in "A Brief Out-Line of Modern Chemistry" that he appended to his edition of George Adams' "Lectures on Natural and Experimental Philosophy," published in Philadelphia in 1806-07. He published other books in astronomy, mathematics, and physics.

Robert Maskell Patterson was born in Philadelphia, March 23, 1787. He graduated from University of Pennsylvania and the university medical school, then in 1809 he sailed to Europe for further study. He remained in Paris for 2 years, studying medicine, mineralogy, mathematics, physics, and chemistry and acting as temporary United

States consul. While there he attended the lectures of Louis Vauquelin, which he called "the most learned and complete course of chemistry given in Europe," and those of Louis Thenard, "the most popular lecturer on chemistry in Paris."

In 1811 he started home by way of London, attended Humphry Davy's lectures, purchased apparatus and chemicals that he could not obtain readily in Philadelphia, and sailed back across the Atlantic in 1812.

Robert substituted for his father, who was in ill health, at the university in 1812, was professor of natural history in 1813, professor of natural philosophy in the medical school the same year, and succeeded his father as professor of natural philosophy in the university in 1814. He remained at the university until 1828, then was professor of natural philosophy at University of Virginia until 1835. In the latter year President Jackson appointed him Director of the Mint. He superintended operations of the Philadelphia Mint and the three southern branches until he retired because of ill health in 1851.

In the first series of chemistry lectures Robert M. Patterson delivered at Pennsylvania in 1812, shortly after he returned from Europe, he gave an account of Dalton's atomic theory. This was the first time, according to Patterson, that the atomic theory was taught in America. During the years he was professor of natural philosophy at Pennsylvania and Virginia, Patterson included chemistry among the subjects he taught, a practice followed by many professors of natural philosophy at that time.

Patterson did little research, possibly because he was too busy with professional and civic duties. One of his friends said that he did not write much and wronged his memory by not publishing what he wrote. A conscientious teacher, an able administrator, and perhaps the first person to teach Dalton's atomic theory in America, Patterson died in Philadelphia, Sept. 5, 1854.

A biography of Robert Patterson, with references, is in "Dictionary of American Biography," **14,** 305, Chas. Scribner's Sons (1934); W. D. Miles, "Robert Maskell Patterson," *J. Chem. Educ.* **38,** 561-563 (1961); R. M. Patterson is depicted on a medal struck in his honor by associates at the Mint when he retired; copies may be purchased at a nominal sum from the Philadelphia Mint.

WYNDHAM D. MILES

Henry Clemens Pearson

1858-1936

Pearson was born in Le Roy, Minn., Feb. 13, 1858, the son of Charles H. and Emily (Clemens) Pearson. His mother was a favorably known novelist and biographer. Henry was educated in public schools but never had the opportunity of attending college.

In 1875 the Pearson family moved to Andover, Mass. where Henry went to work for Tyer Rubber Co. From 1881 to 1883 he was manager of Hayward Rubber Co. in Colchester, Conn. Then followed 3 years selling machinery and supplies to rubber manufacturers and 3 years in general newspaper work and service as a foreign correspondent.

In 1889 Pearson founded and became editor of *The India Rubber World,* the first journal of the rubber trade to be established in the United States. He continued as editor and publisher until 1926 when he sold *The India Rubber World,* since 1953 *Rubber World,* to the Bill Brothers Publishing Corp. After 3 more years as editor, he retired in 1929. He died in Pasadena, Calif., June 11, 1936.

Though not a scientist, Pearson was recognized as an expert in the rubber industry. He was a member of American Chemical Society, American Society for Testing Materials, Royal Geographical Society, and Institution of the Rubber Industry. He was a world traveler and visited all of the major sources of rubber in the tropics of both hemispheres. He served as an expert on rubber for the United States Government at the Paris Exposition in 1900. In the same year he founded the New England Rubber Club which later became the Rubber Manufacturers' Association, Inc. In 1904 he was secretary of the jury awards at the St. Louis Exposition and was vice-president of the International Rubber Exposition in New York in 1911.

Pearson was the author of several books: "What I Saw in the Tropics," 1906; "Rubber Country of the Amazon," 1911; "Rubber Machinery," 1920; "Pneumatic Tires," 1921; and "Crude Rubber and Compounding Ingredients," three editions, 1899, 1909, 1918, which has since been periodically expanded and brought up-to-date by *Rubber World.* Pearson was also an avid collector of books on rubber and in 1935 he gave his collection of over 1,000 titles to the Los Angeles Public Library.

In 1885 Pearson was married to Adelaide Ella French of Norway, Maine. They had an adopted daughter, Esther.

India Rubber J. **92,** 45 (July 11, 1936); *India Rubber World* **94** (4), 53 (1936).

GUIDO H. STEMPEL

Mary Engle Pennington

1872-1952

Mary Pennington ("Polly," to some of her family), pioneer in the chemistry, bacteriology, and refrigeration of perishable foods, remained active in food related industrial and governmental work until her sudden death in her eighty-first year.

The great-granddaughter of Joseph Engle, judge of Delaware County, Pennsylvania and chairman of a committee appointed by the governor of Pennsylvania to act as an escort to General Lafayette upon his second visit to America in 1825, Mary would have gladdened the Judge's heart could he have known of her colorful and successful career. She was born in Nashville, Tenn., Oct 8, 1872. Later the family resided in West Philadelphia, not too far from University of Pennsylvania where Mary was admitted to the Towne Scientific School and then to a course in biology, a rarity for a woman at that time. Graduate work with a major in chemistry and minors in botany and zoology led to her Ph.D. at the university in 1895. She was a fellow in chemical botany at the university for 2 years, then a 1-year fellow in physiological chemistry at Yale.

She became an expert in bacteriology and food chemistry, and during the period 1898-1906 she taught physiological chemistry at Women's Medical College where one of her students was a cousin, Georgiana Walter (another Engle descendant), with whom she later published a paper "A bacterial study of commercial ice cream," *New York Medical Journal,* November 1907. Her clinical laboratory, which she also maintained at this time, was patronized nationally by physicians, and she was consulted by the city of Philadelphia on the care of perishable foods during marketing, including the preservation of milk by refrigeration.

By 1908 Pennington had set up a Food Research Laboratory in Philadelphia for the U.S. Department of Agriculture to implement the recently passed federal food and drug act. Until 1919 she remained chief of this laboratory, which employed a staff of about 50 technicians and field men. The laboratory was charged by the government with the study of methods to establish quality of eggs, dressed poultry, and fish. The results led to a revolution in warehousing, packaging, and transportation of foods under refrigeration.

She was an official U.S. delegate in 1908 to the First International Congress of Refrigeration at Paris, to the second in 1910 at Vienna, and to the Third Congress at Chicago and Washington in 1913.

In 1919 she left government service to accept a position in New York with American Balsa Co. as manager of their research and development division. She remained in this post until 1923 when, because of her wide knowledge of the chemistry and bacteriology of perishable foods, she became a private consultant and investigator in this field. She established her office in New York City and maintained it until her death, which occurred Dec. 27, 1952.

From 1923 to 1931 she was director of the Household Refrigeration Bureau of the National Association of Ice Industries. Her pioneering work on food preservation continued into the area of frozen foods in which she did original research on frozen poultry.

She was called upon during World War II by the Office of the Quartermaster General as consultant to the Research and Development Branch of the Military Planning Division of the War Shipping Administration. Her work as a consultant extended to the design and construction of refrigerated warehouses and cooling rooms, commercial and household refrigerators, packaging, and transportation. She served as a U.S. delegate to the Eighth World Poultry Congress at Copenhagen in 1948. The Garvan Medal which honors annually a woman chemist of distinction was awarded to her by the American Chemical Society in 1940.

"Polly" Betts, Mary's namesake and the daughter of her sister, followed in Mary's footsteps by obtaining a degree in chemistry and marrying chemist Robert C. Elderfield, who became well-known in the field of organic chemistry.

In addition to contributions to technical journals, Mary Pennington wrote books, government bulletins, and magazine articles. She was the author of a handbook, "The Care of Perishable Food Aboard Ship." She made many addresses before technical and commercial groups. At the time of her death she was vice president of the American Institute of Refrigeration.

Information from Mary Betts Elderfield (Mrs. Robert C.); Alice C. Goff, "Women can be engineers," Edwards Brothers, Inc. (1946), pp. 183-214; obituary, with portrait, in *Ice and Refrigeration,* Feb. 1953, p. 58.

ETHEL ECHTERNACH BISHOP

John Howard Perry

1895-1953

Perry, chemical engineer and editor, was born in Hampden, Maine, Nov. 10, 1895, the son of Jeremiah David and Maude (Boober) Perry. He was educated at University of Maine, receiving his B.S. degree in chemistry in 1917, his M.S. at Northwestern in 1920, and his Ph.D. from Massachusetts Institute of Technology in 1922. Both of the latter degrees were in chemical engineering, and it was in the literature and practice of that profession that he was to make his major contributions.

When Perry began his career, chemical engineering was experiencing almost revolutionary growth and development, yet it was often regarded as an amorphous mixture of chemistry and engineering. His first job in industry was as a chemist for American Agricultural Chemical Co. where his work and interest in the technology of sulfuric acid were later to stand him in good stead at the U. S. Ordnance plant in Sheffield, Ala. and with the Bureau of Mines from 1923 to 1925. In the latter year he joined the Du Pont Co., where his first assignment was as a physical chemist for sulfuric acid research at the Experimental Station in Wilmington. He continued this work, studying the basic role of platinum in the contact process and developing improved catalysts using platinum deposited on magnesium sulfate carriers. In order to carry this work to full application, Perry was transferred in 1932 to the experimental laboratory of Grasselli Chemical Co. in Cleveland, then a Du Pont subsidiary. He returned to Wilmington in 1937 as a technical investigator for the Grasselli Chemicals Department and in 1942 was transferred to Du Pont's Central Development Department, where he continued until his death.

It is for his "extra-curricular" work however, that he is best remembered. This came about in the following way: the late Charles L. Reese, Du Pont's chemical director and vice president, was one of the great pioneers in chemical engineering. While president of the American Institute of Chemical Engineers from 1923 to 1925, he was among the first to recognize the need in education, as well as in industry, for a new and more adequate literature of chemical engineering. He sought the advice of Howard C. Parmelee of McGraw-Hill, and together they were instrumental in organizing an editorial advisory board made up of an equal number of prominent engineers and eminent educators. After many meetings and often controversial discussions this group of men succeeded in outlining a coordinated series of about a dozen texts and reference books that were to become the McGraw-Hill Chemical Engineering Series, which has since grown to more than 50 volumes. All agreed that the keystone of that structure should be a comprehensive engineering handbook, comparable with those of mechanical, civil, and electrical engineering.

To get the handbook project off the ground, Reese volunteered the cooperation of the Du Pont Co. in releasing some of its vast store of unpublished engineering information and data. The advisory committee felt that Fred Zeisberg, then chairman of the literature committee of A.I.Ch.E., was the logical man to head up such an undertaking, but for health and other reasons he was unable to accept the heavy responsibility. However, he and Reese were delighted when Harry A. Curtis, another member of the advisory committee, suggested Perry, who had been one of his graduate students at Northwestern. When Perry had accepted the assignment, it was agreed that William C. Calcott, head of Du Pont's Jackson Laboratory and long familiar with the company's technical resources and personnel, should serve as assistant editor. Then began years of hard work, much of it during the depression of the early thirties, until finally in January 1934, the first edition of Perry's "Chemical Engineers' Handbook" made its appearance. Reese's promise was realized when fully a fourth of its 2,610 pages reflected the contributions of Du Pont engineers.

Since 1934 there have been four completely revised editions. [A fifth issued in April 1974.—Ed.] There have been a half dozen translations of what has literally become the "Chemical Engineer's Bible" all over the world. It is an international monument to a great and dedicated engineering editor, whose life came suddenly to an end a few days after his 58th birthday. Fortunately, his son, Robert Howard Perry, a consulting engineer with wide experience in education and industry, was well qualified to accept the major responsibility for the fourth edition in 1958, in which Cecil H. Chilton and Sidney D. Kirkpatrick were to serve as co-editors.

A final chapter of Perry's short career resulted from his interest in the economic and business aspects of chemical industry. He recognized how important it was for chemical engineers to develop knowledge and skills beyond their technical specialties. He decided to do something about it, and in the last few years of his life he organized and edited another cooperative effort for the profession. With the help of 20 qualified section editors and more than 100 contributors, his "Chemical Business Handbook" was on the presses at the time of his death from lung cancer, Dec. 13, 1953.

Personal recollections, supplemented by communications from Thomas H. Chilton, Cecil H. Chilton, James R. Fair, and Robert H. Perry; obituary in *New York Times,* Dec. 14, 1953.

SIDNEY D. KIRKPATRICK

Charles Pfizer

1823-1906

About Charles Pfizer, founder in 1849 of the drug and chemical house which bears his name, relatively few biographical details are known. He appears to have been content to let his company's success and the quality of its products speak for themselves. The same is true of his son Charles Jr., who became the firm's first president when it was incorporated in 1900.

The elder Pfizer was born in Germany in 1823, educated in public schools in Ludwigsburg, Württemberg, and served as a chemist's apprentice for some years. In 1848 he and a cousin, Charles F. Erhart, who had been trained as a confectioner, emigrated to the United States. A year later they formed a partnership, bought a small building in Brooklyn, N. Y. (in which city the company would always have a center), and began producing pharmaceuticals and other chemicals.

From the start Pfizer and Erhart specialized in chemicals not produced in America or produced only in low purity which they could better. The first product they marketed was santonin, a sesquiterpenoid isolated from wormwood and other herbs (whose structure was at last completely elucidated as recently as 1955) and widely used as a vermifuge. Within a few years they added iodine, potassium iodide, iodoform, bismuth and mercury compounds, borax, boric acid, and camphor—all major items in the 19th century pharmacopeia.

The Civil War provided a tremendous boost to the Pfizer firm. Tartrates (tartaric acid; cream of tartar, or potassium acid tartrate; and Rochelle salt, or potassium sodium tartrate), important to the chemical and food industries, were no longer cheaply available from Europe. The company began to import argols (the deposits thrown down by wines as they age) from France and Italy and to recover tartrates from them. From this it was a short step to producing citric acid from Italian calcium citrate, prepared from cull citrus fruits. This product was first marketed in 1880. The uncertainty of the overseas source of raw material ultimately led the Pfizer company to set up a research project in fermentation chemistry, and in 1923 commercial production of citric acid by fermentation of ordinary sugar was begun. This was the first major production through fermentation of any pure chemical other than ethanol, and citric acid was soon followed by other organic acids. In the years that followed the United States became totally independent of foreign sources for citric acid, and the price of the chemical fell dramatically.

When the second World War brought massive demands for the new antibiotic penicillin, it was Pfizer's experience with fermentation technology which was principally responsible for successful production. After the war the company moved into terramycin production, and it now produces a number of antibiotics and organic acids by fermentation as well as a wide range of other fine chemicals and pharmaceuticals.

Charles Pfizer Sr. built his company and prepared his succession carefully. By 1868 he had purchased the building at 81 Maiden Lane, in lower Manhattan (a handsome specimen of the rare cast-iron frame construction which preceded structural steel) which would serve as international headquarters until the early 1950's. In 1873 John Anderson and in 1875 Franklin Black joined the firm; they would become officers when the company was incorporated in 1900 as Charles Pfizer & Co., and their sons would continue as executives after them. Charles Erhart died in 1891; his son William H. also became a corporate officer. Charles Pfizer Jr. took over as president in 1900, was succeeded by his brother Emile in 1907, and died in 1929; Emile, a bachelor, still served as president at his death, July 21, 1941.

Pfizer Inc. (the company's official name since 1970) has expanded more or less steadily since its founding. During the 19th century sales offices were established in Chicago and San Francisco. Within the present century the company has expanded both by construction and by acquisition and has major facilities at Groton, Conn., Maywood, N.J., and Terra Haute, Ind., as well as plants in 10 other states and 60 overseas factories on every inhabited continent. In 1950 the company entered the ethical drug market under its own name, beginning with terramycin sales. In addition to drugs the company produces agricultural and fine chemicals, specialty metals and oxides, cosmetics, and plastics.

Charles Pfizer Sr. and his wife Anna had five children: Charles Jr. and Emile, already mentioned, Helen, Gustav, and Alice. All of the last three returned to Germany or England to live as adults, Alice as Baroness von Echt. Charles Sr. died at the family's summer home in Newport, R. I., Oct. 19, 1906 of injuries sustained in a fall some weeks earlier.

New York Times, Oct. 21, 1906 (Charles Sr.), July 20, 1941 (Emile); "Pfizer: From Ludwigsburg to Brooklyn to Terramycin," *Resident and Staff Physician,* August 1970; W. Haynes, "American Chemical Industry," D. Van Nostrand Co., Vol. 6 (1949), pp. 335-37, as well as scattered references in Vol. 1 (1954); "Pfizer, A History of Growth" (company pamphlet, 1972). A superb account of the fermentation operations in Brooklyn can be found in Berton Roueché, "A Reporter at Large; Something Extraordinary," *New Yorker,* July 28, 1951, pp. 27 ff.

ROBERT M. HAWTHORNE JR.

Almira Hart Lincoln Phelps

1793-1884

Almira Hart, a pioneer in education of girls, was born in Berlin, Conn., July 15, 1793. She received her early education at home and in the private school of her sister Emma, later Emma Willard, a well-known teacher. She began to teach in Connecticut public schools when she was 16 years old and then taught in academies in Massachusetts, Connecticut, and New York.

In 1817 she married Simeon Lincoln, editor of the Hartford newspaper, *Connecticut Mirror.* He died in 1823, and she moved to Troy, N. Y. to teach in Troy Female Seminary. A very studious person, she learned mathematics, sciences, and several languages. Amos Eaton instructed her in chemistry at Rensselaer, and she passed her knowledge along to girls at the seminary. She used Eaton's method of teaching, having her students perform laboratory experiments.

At this time Eaton was considering compiling an elementary chemistry dictionary for his students. A copy of "Le Dictionnaire de Chimie" by three obscure French chemists named Brismontier, Boisduval, and Lecoq came across the Atlantic and into his hands. He used the book for several months, liked it, and persuaded Almira to translate it into English. In preparing the dictionary she abridged passages in the original, added material from the works of British and American chemists, prefaced the volume with a 14-page history of chemistry, and appended a 23-page explanation of elementary chemistry. Her "Dictionary of chemistry containing the principles and modern theories of the science, with its application to the arts, manufactures, and medicine, for the use of seminaries of learning and private students" was published in 1830.

In 1831 Mrs. Lincoln married John Phelps, a lawyer, and moved to his home in Vermont. While taking care of their household she also wrote elementary texts on botany, geology, physics, and chemistry. Her "Chemistry for beginners; designed for common schools, and the younger pupils of higher schools and academies," published in 1834, was a small volume written in a readable style. It was descriptive, with little mention of quantitative relationships and not much theory, but it emphasized lecture demonstrations and student experimentation at a time when these methods of teaching were seldom used in schools and academies. The text was used in schools for more than 30 years.

Mrs. Phelps' second chemistry text, "Familiar lectures on chemistry. For schools, families, and private students," was not as successful as her first, perhaps because other persons, less skillful as text writers, participated in the planning and writing. Emma Willard outlined the book and asked William F. Hopkins, chemistry teacher at the U.S. Military Academy, to fill it out. Hopkins began but resigned his post before completing the book and returned the manuscript to Willard. She asked her sister to complete it. The resulting book, a collaboration of three minds, was a too-long, mediocre text that did not appeal to the schools, families, and private students for whom it was intended.

In 1838 Mrs. Phelps resumed teaching, first in a girls seminary in West Chester, Pa., and then, in 1841, as principal of Patapsco Institute, Ellicott City, Md. She retired in 1856 and moved to Baltimore. In addition to her texts on various sciences, which sold well and widely for more than a generation, she wrote literary works and articles for religious, educational, and literary journals. She died in Baltimore, July 15, 1884, on her ninety-first birthday.

Biography in *Bernard's Amer. J. Educ.* **17,** 611-22 (1868); Emma L. Bolzau, "Almira Hart Lincoln Phelps," (a Ph.D. thesis) (1936); Mary E. Weeks and F. B. Dains, "Mrs. A. H. Lincoln Phelps and Her Services to Chemical Education," *J. Chem. Educ.* **14,** 53-57 (1937), portrait.

WYNDHAM D. MILES

Francis Clifford Phillips

1850-1920

The son of William Smith and Fredericka (Ingersoll) Phillips, Francis was born in Philadelphia, Apr. 2, 1850. Having received his early education from his mother, he studied at the Academy of the Protestant Episcopal Church in Philadelphia and entered the University of Pennsylvania in 1866. He left in his junior year without graduating. He spent a year as instructor of chemistry at Delaware College (Newark, Del.) but soon decided to further his education. From 1871 until 1873 he studied at Wiesbaden, Germany with Karl Regimus Fresenius, spending his last year as private assistant to this most respected analytical chemist. In 1874 he studied at Aachen under Hans H. Landolt, who investigated physical properties of organic compounds.

Because of his father's poor health Phillips was unable to complete his work for the doctorate. In 1875 he joined the staff of the Western University of Pennsylvania (now the University of Pittsburgh), where he remained until his retirement (as head of the department) 40 years later. He received his A.M. in 1879 *gratiae causa* and his Ph.D. in 1893, both from the University of Pennsylvania. His teaching interests included chemistry, geology, and mineralogy. For 1 year he taught chemistry at the Pittsburgh College of Pharmacy. During World War I he served with the Gas Warfare Service.

Phillips' research interests were in natural gas, an area in which he was an internationally recognized authority. Although he did not claim numerous publications, his work was skillful and he anticipated several principles later patented for commercial processes. His training with Fresenius showed itself in his interest in analytical methods. He edited the second edition of "Methods for the Analysis of Ores, Pig Iron, and Steel Used by the Chemists in the Pittsburgh Region" in 1904 and wrote "Chemical German" in 1913. When he died he had almost completed "Qualitative Gas Reactions." He used his knowledge of bacteriology to make extensive studies on drinking water leading to improvements in Pittsburgh's supply.

Phillips became an authority on Joseph Priestley. He had a large collection of Priestley items and had been working on a biography, "Life and Work of Joseph Priestley." He was active in the establishment of the American Chemical Society's Priestley Gold Medal. He married Sarah Ormsby Phillips in 1881 and became the father of two sons. He died Feb. 16, 1920 at his home in Ben Avon, Pa.

Biographies by Alexander Silverman in *Ind. Eng. Chem.* **12,** 399-400 (1920) and by Lyman C. Newell in "Dictionary of American Biography," **7,** 539-40, Chas. Scribner's Sons (1931), the latter sketch contains several errors and should be used cautiously.

SHELDON J. KOPPERL

Aaron Snowden Piggot

1822-1869

Piggot was born in Philadelphia in 1822, graduated from University of Maryland Medical School, practiced medicine, and taught anatomy and physiology at Washington University School of Medicine, Baltimore, from 1849 to 1851 and at Baltimore College of Dental Surgery from 1858 to 1861.

I do not know how he became proficient in chemistry. In the 1850's he began to act as an analytical chemist and a mining consultant. He analyzed guano for fertilizer companies in Baltimore and assisted firms seeking to develop copper and gold mines. He introduced into *American Journal of Dental Science* a section on dental chemistry and metallurgy, and wrote "Chemistry and Metallurgy, as Applied to the Study and Practice of Dental Surgery" (1854), the earliest American book on the subject. A few years later he wrote "The Chemistry and Metallurgy of Copper, including a description of the principal copper mines of the United States and other countries" (1858).

When the Civil War started, he took his family South, joined the Confederate forces as surgeon, and superintended the drug laboratory near Lincolnton, N.C. The war over, he returned to Maryland and in 1866 became professor of chemistry at Baltimore College of Dental Surgery. He continued his consulting work and died suddenly Feb. 13, 1869 while returning from an inspection of a mica mine in Spottsylvania County, Va.

Piggot wrote articles in dental and scientific journals, sections in medical reference works, and essays and reviews for literary magazines. Much of his work as a consultant went unpublicised although there is record of a 16-page pamphlet, "Geological and Mineralogical Reports of Mecklenberg Gold and Copper Mines" (1860). The anonymous writer of his obituary stated that "his great skill in his profession of analytical chemist, caused his opinions to be eagerly desired in many practical enterprises."

Obituary, *Amer. J. Dental Sci.* (3S) **2,** 549-552 (1868-1869); John R. Quinin, "Medical Annals of Baltimore," Press of Isaac Friedenwald (1884), pp. 145-147, has a long list of Piggot's writings.

Quinin's statement that Piggot graduated from Yale is erroneous for I cannot find him listed in *Catalogue of Yale Graduates.*

WYNDHAM D. MILES

George Gilbert Pond

1861-1920

Pond was born at Holliston, Mass., Mar. 29, 1861, son of Abel and Amelia (Robinson) Pond. His father was a merchant of high standing in the community. Pond prepared for college in the schools of Holliston and entered Amherst in 1877. He received a broad and deep cultural foundation and earned his A.B. in 1881.

Pond spent the succeeding year and a half at University of Göttingen where he studied under two great leaders of chemistry, Frederich Wöhler and Victor Meyer. He received the degree of Master of Synthetic Organic Chemistry. Upon returning to America he became an assistant professor of chemistry at Amherst where he received his Ph.D. degree in 1889.

In 1888 Pond became professor of chemistry at Pennsylvania State College (now University), and he spent the rest of his life there. When he went to Penn State, the faculty numbered 19 and the total student body 190. Under his direction the Chemistry Department became one of the strongest departments on the campus and was recognized nationally. He was affectionately known by Penn State chemistry graduates as "Swampy" Pond.

Pond was a good teacher. He had the ability to impart his knowledge in a vivid manner to the students, to develop in his students a desire to receive from their instructor all of which they might be capable, to instill in them an increasing purpose for trained and efficient work. His standards were high, and he was a constant inspiration and source of encouragement to his students.

In 1900 Pond wrote the first serious work in this country on acetylene: "The Application of Acetylene Illumination to Country Homes." The edition was soon exhausted and was revised and published in 1908 under the title, "Calcium Carbide and Acetylene." Over 100,000 copies of his work on acetylene were published. Pond was actively involved with the International Acetylene Association and frequently contributed new knowledge on the subject at its meetings. Pond also contributed to chemical journals and edited texts.

Upon the organization of schools at Penn State in 1896, he was made dean of the School of Natural Science. He carried into this new position the principles of integrity and loyalty that he had followed as head of the Chemistry Department. He commanded the loyal cooperation of his faculty, and he was able to coordinate his work with the other schools of the college. He became chairman of the Committee on Graduate Studies and Advanced Degrees in 1913, and he raised graduate standards and prepared plans for a more efficient organization and administration of graduate study.

The home of Joseph Priestley in Northumberland, Pa. was preserved as a shrine through Pond's efforts. Pond was interested in the history of chemistry and often visited the old Priestley homestead. He noticed the lamentable condition into which the house and grounds had fallen and drew up plans to purchase and to restore the property. With the help of the Chemical Alumni at Penn State, the property was purchased, and the George Gilbert Pond Memorial Association was created to own and operate it. It is now a National Historic Landmark and is operated by the Pennsylvania Historical and Museum Commission.

In 1920 the university trustees changed the name of the building known as Priestley Chemical Laboratory to George Gilbert Pond Chemical Laboratories.

Pond married Helen Palmer in 1888, and they had four children: Millicent, Clara, Gilbert, and Alfred. Pond contracted pneumonia while he was visiting one of his daughters in New Haven, Conn. and died there May 20, 1920.

Penn State Alumni News **6,** No. 11 (1920) and **13,** No. 1 (1926); *J. Chem. Educ.* **4,** 150-57 (1927); obituary by William E. Walker, *Ind. Eng. Chem.* **12,** 718 (1920) with portrait.

LESTER KIEFT

Frederick Belding Power

1853-1927

Power was born Mar. 4, 1853, in Hudson, N.Y., son of Thomas and Caroline (Belding) Power. He obtained his early education at a private school and at Hudson Academy. After working for 5 years in a local drug store and for a brief period in a Chicago pharmacy, he went to Philadelphia to work for pharmacist Edward Parrish. Although his employer died shortly thereafter, Power continued to work for the Parrish company and also attended Philadelphia College of Pharmacy. He graduated from the college in 1874 and remained with the Parrish firm for two more years before resuming his studies in Germany. At University of Strassburg, he studied chemistry with Rudolf Fittig and Friedrich Rose, pharmacology with Oswald Schmiedeberg, and pharmacognosy with Friedrich Flückiger. He served as private assistant to the latter in 1879-80. In 1880 he received his Ph.D. and returned to America.

After teaching analytical chemistry for 3 years at Philadelphia College of Pharmacy, Power went to University of Wisconsin in 1883 as first director of the newly created department of pharmacy. Through his teaching and research, he helped to place pharmaceutical education at Wisconsin on a firm scientific footing which was to make the university a leader in this field. In 1892 he resigned his position to become scientific director of the chemical laboratories of Fritzche Brothers in Passaic, N.J. Tragedy disrupted his life in the following year when his wife died 2 weeks after giving birth to their third child. The baby survived only a few days.

When his lifelong friend and fellow classmate at Philadelphia College of Pharmacy, Henry Wellcome, established the Wellcome Chemical Research Laboratories in London in 1896, Power was appointed as director. He held this position until 1914, when he returned to the United States. From 1916 until his death, he headed the Phytochemical Laboratory of the Bureau of Chemistry, United States Department of Agriculture.

His research interests were in the field of phytochemistry. He investigated many different plant substances, isolating and purifying numerous constituents and determining the structural formulas of several new compounds. In his years at the Wellcome laboratory, the most productive research period of his career, he studied the constituents of over 50 different plant products. For example, he isolated and identified 16 constituents from the essential oil of nutmeg. Experiments by Henry Dale of the Wellcome staff confirmed the view that the narcotic properties of nutmeg are due to myristicin. Another important research project, carried out while Power was at the Bureau of Chemistry, involved the investigation of the constituents of the cotton plant.

Probably Power's most widely known work was his research on chaulmoogra oil, a traditional remedy for leprosy in the Orient, which entered Western medicine in the nineteenth century. In the early years of the twentieth century, Power and his coworkers at the Wellcome laboratories isolated from the oil two new fatty acids, chaulmoogric and hydnocarpic acids. They also determined the structures of these acids, although the formulas assigned have since been modified.

At the time of his death Mar. 26, 1927, in Washington, D.C., Power was working on a comprehensive two-volume treatise on phytochemistry which was never completed.

Biographies by Max Phillips, *J. Chem. Educ.* **31,** 258-261 (1954), with portrait; C. A. Browne, *J. Assoc. Off. Agr. Chem.* **11,** iii-vi (1928), with portrait; Ivor Griffith, *Amer. J. Pharm.* **96,** 601-614 (1924), with portrait and bibliography; "Dictionary of American Biography," **15,** 154, Chas. Scribner's Sons (1935).

JOHN PARASCANDOLA

Albert Benjamin Prescott

1832-1905

Prescott was born in Hastings, N.Y., Dec. 12, 1832, the son of Benjamin and Experience (Huntley) Prescott. As a 9-year old youth, he had a severe fall which injured his right knee and which confined him to the house for the following 5 years, often to his bed, for months at a time and made him a cripple for life. While an invalid he learned Latin, French, German, and various branches of science, and took a keen interest in literature. In his formative years he became an active worker in the anti-slavery cause.

In 1854 he decided to devote himself to medicine. While preparing himself for the medical profession he taught school; chemistry and zoology became his favorite studies. In 1860 he went to Ann Arbor and entered the Department of Medicine and Surgery of University of Michigan, from which he graduated in 1864. He served in the army about a year and then entered upon his life work in 1865 on being appointed assistant professor of chemistry and lecturer in organic chemistry and metallurgy in University of Michigan.

He was made professor of organic and applied chemistry in 1870, dean of the School of Pharmacy in 1876, and director of the chemical laboratory in 1884. As a successful teacher he was widely known and highly esteemed. His interest in research was always keen, and he was constantly seeking opportunities to assist, in every possible way, the younger men who showed ability in this direction.

He was noted as a toxicologist and testified as an expert witness in poison cases. In the field of sanitary chemistry he examined drinking waters, harmful adulterants and preservatives in foods, and he investigated methods of detecting foreign fats in butter. He was influential in formulating Michigan laws concerning adulteration of foods and drugs. In pharmaceutical chemistry he carried out research for committees of revision of the U.S. Pharmacopeia.

Prescott published 36 original contributions to the chemical literature on a variety of subjects including the qualitative separation of arsenic from other elements, solubilities of alkaloids, the action of hydrochloric acid on metallic sulfates, and the composition of alkaline solutions of aluminum, zinc and silver, and a great number dealing with researches on alkaloids. He also published more than 120 articles and books including a monograph, "The Chemical Examination of Alcoholic Liquors," and books on qualitative analysis in collaboration with Silas H. Douglas.

He presided over the American Chemical Society in 1886, the American Association for the Advancement of Science in 1891, and the American Pharmaceutical Association in 1900. He chaired the section on analytical chemistry at the World's Congress of Chemists in Chicago in 1893 and the section on organic chemistry at the St. Louis Exposition in 1904. The University of Michigan awarded him a Ph.D. degree in 1886 and an LL.D. degree in 1896; Northwestern University awarded him an LL.D. degree in 1903.

He died Feb. 25, 1905.

Edward D. Campbell, "History of the Chemical Laboratory of University of Michigan," Univ. of Michigan (1916); "Albert Benjamin Prescott, In Memoriam," private printing by V. C. Vaughan (1906); biographies by M. Benjamin, *Sci. Amer.* **65**, Aug. 22, 1891; Edgar F. Smith, "Chemistry in America," D. Appleton & Co. (1914), pp. 272-275, taken largely from Benjamin; John H. Long, *Proc. Amer. Chem. Soc.* **27**, 76-78 (1905); V. C. Vaughan and others, *Physician Surg.* **27**, 103-108 (1905), portrait facing p. 97; J. O. Schlotterbeck, *Bull. Pharm.* **19**, 141-145 (1905), portrait; Oscar Oldberg, *Amer. J. Pharm.* **77**, 251-254 (1905).

LEIGH C. ANDERSON

Joseph Priestley

1733-1804

Priestley was born in Fieldhead, Yorkshire, England, Mar. 13, 1733. He attended Daventry Academy, a seminary for educating dissenting clergymen, from 1752 to 1755. He served as a pastor from 1755 to 1761, was tutor in Warrington Academy from 1761 to 1767, and a pastor again from 1767 to 1773. William Fitzmaurice, Earl of Shelburne, offered to finance Priestley's experiments and pay him an annual stipend, and from 1773 to 1780 Priestley was a companion to Shelburne, with a laboratory at Lansdowne House, Shelburne's home. In that laboratory in 1774 Priestley first isolated oxygen. He became the foremost "pneumatic" chemist, inventing apparatus for studying gases and isolating nitric oxide, nitrous oxide, hydrogen chloride, ammonia, sulfur dioxide, silicon fluoride, and hydrogen sulfide. Although he helped erect the foundation of chemistry, he sincerely believed in the old theory of phlogiston; none of the experiments and writings of his fellow chemists could persuade him to change his mind.

In 1780 Priestley became pastor of a church at Birmingham. His outspoken support of the French Revolution made him thoroughly unpopular in the neighborhood, and a mob sacked and burned his home in 1791. He fled to London where, through legal action, he recovered the value of his destroyed property, but he and Mrs. Priestley finally resolved to emigrate to the United States.

The Priestleys sailed from England Apr. 8, 1794 and landed in New York June 4 where they were warmly received by officials and citizens. They remained in New York for several days and then left for Philadelphia, where they arrived June 19. In mid-July they journeyed to Northumberland.

A few months after Priestley arrived in the United States he was offered the professorship of chemistry in the University of Pennsylvania's Medical School, chiefly through the influence of Benjamin Rush. Priestley vacillated, was elected Nov. 11, 1794, but, for several reasons, finally declined.

At Northumberland Priestley purchased land along the Susquehanna River in 1795. Construction of his house, with a laboratory, took more than 2 years, but in the interval he continued his theological and scientific studies.

Early in 1796 he read before the American Philosophical Society "Experiments and observations relating to the analysis of atmospherical air," and "Further experiments relating to the generation of air from water." In June 1796 he wrote a pamphlet, "Considerations on the doctrine of phlogiston, and the decomposition of water." Subsequently he published articles in *Transactions of the American Philosophical Society,* in *Medical Repository,* and in the British *Journal of Natural Philosophy, Chemistry, and the Arts.* In 1797 he published another pamphlet, "Observations on the doctrine of phlogiston and the decomposition of water," and in 1800 his last pamphlet, "The doctrine of phlogiston established, and that of the composition of water refuted." He readied a second edition of the latter shortly before his death, and it was published posthumously in 1803.

Priestley was the last major chemist to believe in phlogiston, and several of his Northumberland writings had as their purpose his defense of the theory of phlogiston. American chemists, among them Samuel L. Mitchill, James Woodhouse, and John Maclean engaged in mild controversies with Priestley over the subject, but after his death phlogiston was practically ignored by American chemists.

From Northumberland Priestley visited Philadelphia several times. He attended meetings of the American Philosophical Society, helped organize the first Unitarian church in the United States, visited George Washington and other leaders of the government and watched James Woodhouse experiment in his laboratory at the university.

Priestley's wife and one of his sons died in Northumberland and were buried there, as was Priestley after his death, Feb. 6, 1804.

Priestley's old home was purchased by the chemistry alumni of Penn State in 1919 to save it from possible demolition. Eventually it was acquired by the state of Pennsylvania and is now maintained by the Pennsylvania Historical and Museum Commission.

"Memoirs of Dr. Joseph Priestley" (1806). This autobiography has been reprinted, on occasion, perhaps the most recent being in I. V. Brown's "Joseph Priestley; selections from his writings," Pennsylvania State University Press (1962). A host of articles and books have been written about Priestley. An article on the Priestley house is in *J. Chem. Educ.* **4,** 145-199 (1927), and an article useful for its references is by Joseph Hepburn, *J. Franklin Inst.* **244,** 63-72, 95-107 (1947).

WYNDHAM D. MILES

Evan Pugh

1828-1864

Pugh was born Feb. 29, 1828 at Jordan Bank, Pa. His parents, Lewis and Mary (Hutton) Pugh, were Quaker farmers, and young Pugh grew up on the family farm until he was nineteen, when he entered a manual training school at Whitesboro, N.Y. He became heir to a small estate which included a private school for boys in Oxford, Pa. He conducted the school for 2 years and then decided to go to Europe to further his scientific education. In 1853 he entered University of Leipzig and took courses in chemistry and mathematics for a year and a half. He then went to University of Göttingen, becoming a student of Friedrich Wöhler. In 1856 he received a doctorate in chemistry, a distinction won by few Americans at that time.

Pugh spent 6 months in Heidelberg and Paris engaged in research on physiological chemistry. In 1857 the English agriculturalists Sir John Bennett Lawes and Sir Joseph Henry Gilbert invited him to their agricultural experiment station at Rothamsted, England. By 1859 Pugh had attained scientific eminence for his studies on plant growth, proving that plants do not assimilate free nitrogen and establishing the importance of soil ammonia in plant growth. For his investigations in agricultural chemistry he was elected a fellow of the Royal Society.

In 1859 Pugh returned to the United States to assume the presidency of Farmer's High School, which in 1862 became the Agricultural College of Pennsylvania (now Pennsylvania State University). For 5 years he laid the foundations of the new institution over the vigorous opposition of those who ridiculed the idea of a college education based on agriculture and mechanical arts. As the first president of Penn State he organized the system of instruction, planned the erection of buildings, and secured endowments. To this end he helped push through Congress the Merrill Land Grant College Act of 1862.

In addition to his work as administrator, Pugh gave lecture and laboratory instruction in chemistry, mineralogy, geology, and scientific agriculture. He published four articles in European journals during the 6-year period he was abroad and at least three on agricultural education after he returned to the U.S. He died of typhoid fever, Apr. 29, 1864 in Bellefonte, Pa., at the age of 36.

"Dictionary of American Biography," **15,** 257, Chas. Scribner's Sons (1935); "National Cyclopaedia of American Biography," **11,** 320, James T. White & Co. (1901); obituary in *Amer. J. Sci.* **38,** 301-302 (1864); biography, a portrait, and a selection of correspondence by Charles A. Browne in *J. Chem. Educ.* **7,** 499-517 (1930).

ALBERT B. COSTA

Q

John Francis Queeny

1859-1933

Edgar Monsanto Queeny

1897-1968

John Queeny and his son Edgar, were the founders of Monsanto Chemical Co. of St. Louis. John was born in Chicago, Aug. 17, 1859, the son of John and Sarah (Flaherty) Queeny. His father, a retired building contractor, was ruined by the Chicago fire of 1871, and John had to leave school in 1872. He started as office boy for the wholesale drug firm of Tolman and King, then from 1882 to 1891 worked for I. L. Lyons of New Orleans. He moved to St. Louis for 1 year with Meyer Bros. Drug Co., then to Merck and Co. in New York for 4 years, returning to St. Louis in 1897 as purchasing agent for Meyer Bros.

In 1899 Queeny's first chemical manufacturing venture ended in disaster when his plant to refine sulfur burned on its first day of operation. Queeny then decided to manufacture saccharin. He persuaded Louis Veillon of Switzerland to come to the United States to set up a production plant. The company was named the Monsanto Chemical Works, using his wife's maiden name in order not to embarrass Meyer Bros.

The German companies initiated a campaign to bankrupt the fledgling company, but Queeny added new products, phenacetin, chloral hydrate, phenolphthalein, and coumarin, and induced two additional Swiss chemists, Gaston DuBois and Jules Bebie, to join the firm. In 1905 Monsanto showed its first profit, and in 1907 John Queeny left his position with Powers-Weightman-Rosengarten Co. to give full–time to Monsanto. The war years brought problems in manufacturing intermediates but provided markets for American chemicals. Earnings rose from $81,000 in 1913 to $905,000 in 1916. The post-war years were characterized by purchases of other companies and creation of new plants in the United States and the British Isles.

John Queeny was president of Monsanto until 1919, when he was elected chairman of the board. In 1928 he learned that he was ill with cancer and resigned to turn over the management to his son, Edgar. John Queeny died in St. Louis, Mar. 19, 1933.

Edgar Monsanto Queeny was born in St. Louis, Sept. 29, 1897. His mother, Olga Mendez Monsanto, was married to John F. Queeny in 1896. Edgar attended St. Louis public schools and Pawling (N.Y.) School. In 1915 he entered Cornell University, majoring in chemistry and received his A.B. (war) degree in 1919. At Cornell he served as business manager of the school publication and on the student council.

Edgar enlisted in the U.S. Navy as seaman in 1917 and received his ensign's commission in December; he served at Plymouth, England, and Mare Island, Calif. In 1919 he was discharged from the Navy as a lieutenant (jg) and married Ethel Schneider of Washington, D.C.

Edgar joined Monsanto in 1919, instituted the company's first advertising and sales promotional activities, served as an apprentice in the export department, and later joined the general sales department. He became second vice–president and assistant general manager of sales in 1923. The following year, he was promoted to sales manager and subsequently to vice president.

Edgar was elected president in 1928 and served until 1943 when he was elected chairman of the board. He was elected a member of the executive committee in 1939 and the finance committee in 1947, of which he served as chairman from 1960 to 1965. When he stepped down from company chairman in

1960, Monsanto assets were one hundred times greater than when he assumed the presidency.

Under Edgar's leadership, Monsanto merged with Merrimac Chemical, Swann Corp., Resinox Corp., Thomas and Hochwalt Laboratories, Fiberloid Corp., and Lion Oil Co. The Central Research Laboratories were established. Chemstrand Corp. was founded, and the first sudsless detergent was produced. During the war the company operated Clinton National Laboratories in Oak Ridge and the Mound Laboratory in Ohio, and provided much needed materials for the war effort.

Edgar was interested in aviation and enjoyed hunting with gun and camera. He prepared nature films for the American Museum of Natural History and was the author of three books: "Cheechako (1941), "Spirit of Enterprise" (1943), a nonfiction best seller, and "Prairie Wings" (1946). In 1967 he received the St. Louis Award for outstanding community service in recognition of his services to Barnes Hospital. Edgar Queeny died of a heart ailment July 7, 1968.

William Haynes, "Chemical Pioneers," 225-242, D. Van Nostrand Co. (1939); *Monsanto Magazine* **30** (1951), and special supplement (1968); information from Monsanto Chemical Co. John Francis Queeny in "National Cyclopaedia of American Biography," **24,** 395, James T. White & Co. (1936); Edgar Monsanto Queeny, *ibid.,* **I,** 71 (1960).

JANE A. MILLER

R

George Washington Rains

1817-1898

Rains exemplifies the dedicated nineteenth century man of science. His unusually productive career began in 1838, the year of his entry at West Point, came to fruition during the Civil War, and continued practically until his death, March 21, 1898.

Rains was born in New Bern, N.C. in 1817 and was one of six children of Hester Ambrose and Gabriel (Manigault) Rains. At some time in his youth, he visited his brother, Gabriel, then a United States Army lieutenant stationed in Indian Territory (now eastern Oklahoma) for about a year. Returning, young George rode a dugout canoe 600 miles down the Arkansas River from Ft. Gibson to Little Rock. The frontier experience, climaxed by the dugout adventure, made an indelible impression and helped in establishing those qualities of self-confidence, curiosity, and willingness to undertake new ventures that typified Rain's mature career.

While at West Point (1838-42), Rains excelled in scholarship, finishing third in his class and first in scientific studies. He also became a proficient soldier, graduating as first captain of cadets.

Rains taught at the Military Academy from 1844 to 1846, as assistant professor of chemistry, mineralogy, and geology. In the Mexican War during which he served as dispatch bearer for General Winfield Scott, he was brevetted captain, then major, for gallantry. He resigned from the service in 1856 to become president of Washington Iron Works and Highland Iron Works of Newburgh, N.Y. where he remained until 1861. During this civilian interlude, he obtained several patents on steam boiler designs and published "Steam Portable Engines" (1860).

Soon after the Civil War erupted, Rains left Newburgh for Richmond, and there offered his services to Jefferson Davis in July 1861. At the close of the war, Rains became professor of chemistry and pharmacy, and later dean of the Medical College of Georgia in Augusta. He remained there as dean until 1883 and as emeritus professor until 1893, when he returned to Newburgh, N.Y.

Throughout his life he made numerous scientific contributions, and wrote textbooks and papers, always keeping abreast of scientific progress in his own and other fields. He published "Rudimentary Course of Analytical and Applied Chemistry" (1872), "Chemical Qualitative Analysis" (1879), and "Interesting Chemical Exercises in Qualitative Analysis for Ordinary Schools" (1880).

It was with the Confederate Ordnance Bureau that the scientific prowess of Rains had a most telling effect. Jefferson Davis presented Major Rains with carte blanche responsibility for the production of gunpowder and vague but flexible instructions that "the necessary works were to be erected as nearly central as practical; to be permanent structures, and of sufficient magnitude to supply the armies in the field and the artillery of the forts and defences."

With Herculean labors, the patience of Job, and Houdinian resourcefulness, Rains, starting from scratch, established the Confederate Powder Works at Augusta, Ga. It was an incredible plant in which black powder of the finest quality was produced. In 1861 he published "Notes on Making Saltpetre from the Earth of the Caves."

Commenting on his product after the war, Rains related how "the armor of the ironclads, though constructed expressly to withstand the heaviest charges and projectiles, gave way before its propelling force. Mr. Davis makes the statement that the engagement between the *Alabama* and *Kearsarge* would have resulted in a victory for the former, had Admiral Semmes been supplied with the powder from these works."

Summarizing the wartime contributions of Rains reveals his stature as a practical scientist in whom the characteristics of inspired imagination and indefatigable industry were most felicitously and synergistically combined.

Expertise—Rains built an efficient, high-capacity powder factory where none had existed before and, with all normal sources of raw material denied him, succeeded in manufacturing a product unmatched in quality.

Technology—Among Rains's many contributions to ordinance was an improved alloy for bronze field artillery castings made by fusing a small per cent of iron with the copper and tin.

Productive capacity—From April 10, 1862 to April 18, 1865, almost 3 million lbs. of gunpowder were made by the Augusta plant. At no time during the war did the plant have to work full–time.

Safety—During 3 years of continuous operation, the Augusta powder plant established an enviable safety record. Only a few explosions occurred and in only one did fatalities result.

Economics—In one year, the Augusta Powder Factory made 1 million lbs. of gunpowder, at a total cost of $1,080,000. Cost of the powder imported through the blockade would have been about $3 million.

Product quality—After the war, the remaining gunpowder produced at the Augusta works was regarded so highly by the Union Army that it was used in the School of Artillery Practice at Fort Monroe.

Papers and books on and by George Washington Rains in the Confederate Museum, Library of Congress, and National Archives; Rains, "History of the Confederate Powder Works," Chronicle & Constitutionalist Press (1882).

NORMAN P. GENTIEU

Lawrence Vincent Redman

1880-1946

Redman was born in Oil Springs, Ont., Canada, Sept. 1, 1880. He was the grandson of Edward Redman who emigrated from England in 1818 and settled in Quebec.

His early education was in Canada, culminating with an A.B. degree from University of Toronto in 1908. He accepted one of the original industrial chemical fellowships under Robert K. Duncan at University of Kansas in 1910. Redman stayed at the university until 1913, rising to the rank of associate professor of industrial chemistry. He then went to Mellon Institute in Pittsburgh to study phenol-formaldehyde condensation resins.

Redman's work at Mellon Institute was a continuation of his work at University of Kansas, which was in turn prompted by an earlier discovery by another investigator. In 1909, Leo Baekeland had announced the discovery of a phenol-aldehyde condensation resin which he called "Bakelite." Redman, at the time, was experimenting with a similar reaction as the basis of a superior furniture varnish. Subsequently, in 1914, with A. J. Weith and F. P. Brock, Redman formed the Redmanol Chemical Products Co., with himself as president, to market "Redmanol" and similar condensation plastics. The basic reaction involved was the interaction of anhydrous phenol with hexamethylenetetramine. The patents on which the company was founded were the first of more than 125 patents granted to Redman by 1931 in the field of resins.

The company flourished during World War I owing to the extensive use of phenolic condensate resins as insulators in automobile and airplane ignition systems and as a material for use in making airplane propellors. In 1922, after considerable litigation, Redman merged his company with its chief competitor, General Bakelite Co., to form Bakelite Corp. He became vice-president and director of research for the new company. The main office was in New York City, with the research and development plant at Bloomfield, N.J. erected in 1924. Later, branch offices were established in Chicago, Toronto, London, and Berlin. The principal manufacturing facilities were moved to Bound Brook, N.J. in 1931.

From the relatively few products in 1922, Bakelite Corp. expanded in the following decade into diverse fields, largely because of the inspiration and direction of Redman.

Among the new applications were protective coatings (varnishes and lacquers), radio tube bases, telephone speakers and receivers, table tops, and many other articles formerly made from wood or metal.

Lawrence Redman was not only exceptionally talented in the field of applied industrial research, he was an extremely gifted administrator, both in the laboratory and in the field of chemical economics. In recognition of this, he was awarded an honorary Ll.D. from University of Western Ontario in 1930 and an honorary Sc.D. from University of Toronto in 1931. In the latter year, Redman received the Grasselli Medal of the American section of the Society of Chemical Industry. His award speech on research as a fixed charge was still a novel idea in 1931. In the speech he said "Give us a monopoly of new knowledge and others may have their corner in raw materials." It was this guiding philosophy of Redman's which, carried into practice, accounted in large measure for the tremendous growth of the Bakelite Corp.

Redman was active for years in local and national affairs of the American Chemical Society. He was chairman of the Chicago Section of ACS, 1918-19, and when he moved to New Jersey he was a councilor of the North Jersey Section, 1922-29, vice chairman, 1928-29, and chairman, 1930-31. He was elected president of the American Chemical Society in 1932. He was chairman of the American Section of Society of Chemical Industry, 1926-27, and president of the Chemists' Club (New York), 1930.

In 1940, suffering from ill health, Redman retired to his farm in Burlington, Ont., where he pursued his life-long avocation of growing fruits and vegetables. He died Nov. 25, 1946 while visiting friends in Toronto, Canada.

Bull. Univ. Kansas **26,** 38 (1925); Redman's Grasselli Medal Award address, *Ind. Eng. Chem.* **24,** 112-115 (1932); communications from M. Corry, University of Toronto and J. M. Nugent, University of Kansas; Mary E. Weeks and C. A. Browne, "History of the American Chemical Society," (1952) pp. 486-87; obituaries in *New York Times,* Nov. 26, 1946, *Chem. Eng. News.* **24,** 3379 (1946), and *Oil, Paint, Drug Reporter* **150,** 39 (Dec. 2, 1946).

J. PAUL O'BRIEN

Charles Lee Reese

1862-1940

Charles Reese, the third son of John Smith and Olivia (Focke) Reese, was born in Baltimore, Nov. 4, 1862. His father was a successful merchant, and his mother came from a well-to-do Baltimore family. Reese was educated in private and public schools in Baltimore. After attending Johns Hopkins for the 1880-81 session he transferred to University of Virginia where he graduated in 1884. Then he went to Heidelberg, where he studied under Robert Bunsen, and received his Ph.D. degree in 1886. Before returning home he was a post-doctoral student of Victor Meyer at Göttingen.

Back in the United States he became an assistant in chemistry at Johns Hopkins until January 1888 when he was appointed instructor at Wake Forest. However, he only stayed for the spring semester and moved on to The Citadel in Charleston. He remained there as instructor of chemistry and physics until 1896. That year he returned to Johns Hopkins as an instructor and stayed there until 1900.

Before joining industry, Reese carried out research on the origin of Carolina phosphates, and later, in conjunction with Harmon N. Morse at Johns Hopkins, on manganese oxides.

On April 10, 1901 he married Harriet S. Bent; they had five sons: Charles Lee, John Smith, David Merideth, Eben Bent, and William Fessenden.

Reese became chief chemist for New Jersey Zinc Co. where he was largely concerned with the manufacture of sulfuric acid. His success with the contact process for sulfuric acid led Du Pont to engage him to take charge of its acid production.

Reese started with Du Pont in 1902, and his ability and chemical knowledge were soon recognized. His responsibility was first with the manufacture of acids and then with the improvement of commercial dynamites. He established the Eastern Laboratory at Gibbstown, N.J. Later four other research laboratories were built and placed under his guidance. His unusual capacity for organization and administration plus his uncanny ability to secure good men for research and administrative responsibilities helped him to acquire a group that had an outstanding influence on our chemical industry. In 1911 the Chemical Department was formed and placed under his administration. He remained in charge of the department until 1924 and served as consultant until his retirement. Reese was elected a member of Du Pont's board of directors Oct. 31, 1917 and retired Jan. 1, 1931.

As a distinguished industrial leader he received many honors. He served as president of the American Chemical Society in 1934; as president of the Manufacturing Chemists' Association in 1920-3; as president of the American Institute of Chemical Engineers in 1923-5; and as vice-president of the International Union of Pure and Applied Chemistry, to name the more important. Reese received honorary degrees from Colgate, Delaware, Heidelberg, Pennsylvania, and Wake Forest universities.

Reese published four research papers, and one general paper during his teaching career. His 13 papers published later were of general interest. Any innovations and ideas that Reese had were the property of Du Pont.

He died of a heart attack at Jacksonville, Fla., Apr. 12, 1940 and was buried at Wilmington, Del.

Obituary by Robert E. Curtin, Jr. in *J. Amer. Chem. Soc.* 62, 1889-1891 (1940); "Dictionary of American Biography," **22** (Supplement 2), 550, Chas. Scribner's Sons (1958).

DAVID H. WILCOX, JR.

Ira Remsen

1846-1927

Remsen, founder of the Chemistry Department and sometime president of Johns Hopkins University, was born in New York City Feb. 10, 1846. Except for an older sister who died in infancy, he was the only child of James Vanderbilt Remsen, a merchant, and Rosanna (Secor) Remsen; he was of the seventh generation of descendants of Rem Jansen Vanderbeeck, who settled on Long Island in 1642. His mother died during his childhood, and he grew up partly in the home of his maternal great-grandfather, James Demarest, a Dutch Reformed minister whose influence can be traced in Remsen's extensive knowledge of scripture, his absolute honesty, and his decidedly non-sectarian religious outlook.

Remsen entered the Free Academy in New York (later City College) at 14; although he did not finish, in 1892 he was granted an A.B. degree with the class of 1865. In 1867 he received his M.D. degree from College of Physicians and Surgeons, New York City, winning a prize for his graduating thesis, "The Fatty Degeneration of the Liver"—"a subject," he later noted, "on which I was and am profoundly ignorant." Remsen was briefly apprenticed to a physician in New York but finally turned to chemistry. He studied at Munich with Jacob Volhard, a privatdozent at the university, learning laboratory analytical chemistry from him while attending Liebig's lectures an organic and inorganic. After a year Remsen moved to Göttingen to work in Wöhler's laboratories under Rudolph Fittig. In 1870 he received his Ph.D. degree; Fittig was called to Tübingen, and Remsen accompanied him as his assistant, staying until 1872.

On his return to America he completed a translation of "Wöhler's Outlines of Organic Chemistry" which was published the next year and in the fall of 1872 accepted a position as professor of physics and chemistry at Williams College. There, despite an atmosphere indifferent to chemistry as he knew it, Remsen pursued his researches and began to develop the simple and lucid lecture style for which he became famous. He also embodied this lucidity in the first of a long series of exceedingly popular textbooks, "The Principles of Theoretical Chemistry."

In 1876 Remsen was offered the opportunity which absorbed and shaped the rest of his life. He went as professor of chemistry to the newly founded Johns Hopkins University in Baltimore, to help build a university on the continental model—a place for discovery, rather than just transmission of knowledge. The initial enrollment was about half graduate and half undergraduate. Graduate instruction was of a type new to

America: research work, a minimum of advanced lecture, and departmental "journal meetings" to keep everyone up-to-date in world chemical literature, a feat still possible in the 1870's. Remsen readily admitted that his system was German in origin, but he applied it with such a rare skill in teaching and directing students that by the turn of the century more than half the first-rank academic chemists in the country had trained at Johns Hopkins.

In 1879 Remsen founded *American Chemical Journal,* editing it through 50 volumes until it was absorbed into *Journal of the American Chemical Society* in January 1914. From 1910 to 1912 he served as president of Johns Hopkins; women were first admitted in these years, and the university acquired its present Homewood location in north Baltimore. During this period Remsen also served as a member of the Baltimore school board; as chairman of the Sewage Commission appointed to build a modern system following the Baltimore fire of 1904; and as chairman of President Theodore Roosevelt's advisory commission established under the Food and Drug Act of 1906. Dedicated and honest, he was nonetheless the target of much criticism, particularly in the federal position—some of it possibly occasioned by his impatience with those who lacked his habit of reasoning dispassionately from facts.

In 1913 Remsen, 67 years old, returned to the chemistry department to teach history of chemistry and edit late editions of his many famous textbooks. These included, with the one mentioned above, seven in all, with a total of 28 editions and 15 translations. One estimate places the number sold at half a million; their enormous popularity was well earned by their sound theory, logical organization, and extreme simplicity of style.

Remsen's research included oxidation of aromatic side chains and the protective effect of ortho-substituents (in the course of this work saccharin—the cyclic imide of benzene-1-carboxylic-2-sulfonic acid—was first prepared by Remsen and a post-doctoral student, Constantin Fahlberg; the latter produced the compound commercially); sulfonphthaleins; the "double halides" of transition and heavy metals with alkali halides (e.g., $PtCl_4 \cdot 2\,KCl$); and directive and kinetic effects of aromatic substituents. Although his research, together with that of his immediate graduate students, produced over 170 papers, and always tried to elucidate principles rather than just report compounds, the consensus is that it was relatively minor. His fame rests more on his brilliance as a teacher, text writer, builder of a university, and inspirer of students, than on his own direct efforts in chemistry.

Remsen was married in 1875 to Elizabeth H. Mallory of New York City, whose family summered in Williamstown, the site of Williams College. They had two children: Ira Mallory Remsen, an artist and playwright; and Charles Mallory Remsen, a surgeon who practiced in New York. Remsen was at one time or another president of the American Chemical Society, the American Association for the Advancement of Science, the National Academy of Sciences, and the Society of Chemical Industry, and was the first recipient of the Priestley Medal of the American Chemical Society, among other honors. He died of a cerebral hemorrhage in Carmel, Calif., Mar. 4, 1927.

The most complete, but least critical account of Remsen's life and work is the book, "The Life of Ira Remsen," Jour. of Chem. Ed. pub. (1940), by a former student, F. H. Getman; the biography, bibliography, and reminiscences by others which is usually given as Noyes and Norris, *Biog. Mem. Nat. Acad. Sci.* **14,** 207-57 (1932), is nearly as complete and contains some thoughtful dissenting opinions; Benjamin Harrow, "Eminent Chemists of Our Time," 2d ed., D. Van Nostrand (1927), pp. 197-215, 428-39, gives the best-organized summary of Remsen's research, together with an adequate account of his life; W. A. Noyes, *Science* **66,** 243-46, *J. Chem. Soc.* (1927), 3182-89, and J. F. Norris, *Proc. Am. Chem. Soc.* **50,** 67-86 (1928), are the sources from which their portion of the *Nat. Acad. of Science* reference, above, is drawn; Getman, *J. Chem. Educ.* **16,** 353-60 (1939), is a sketch for the book; a brief genealogy of the Remsen family, ending with Ira, may be found in Margherita A. Hamm, "Famous Families of New York," **2,** 71-80, Putnam (1901); *see also* "Dictionary of American Biography," **15,** 500-2, Chas. Scribner's Sons (1935); "National Cyclopedia of American Biography," **37,** 52, James T. White & Co. (1951); and *New York Times,* Mar. 6, 1927; all other sources have no more to offer than the sketch in "Encyclopedia Americana."

ROBERT M. HAWTHORNE JR.

Lloyd Hilton Reyerson

1893-1969

Reyerson was born in Dawson, Minn., May 1, 1893, to John Emil (of Norwegian descent) and Lydia [Hilton] (of Scotch-English descent) Reyerson. He married Nelle Nickell, Mar. 7, 1918. Their daughter, Jean Elizabeth, became Mrs. Albert H. Moseman. Their son, James Hilton, was killed in action in Europe, Mar. 5, 1945, in World War II. In World War I, Reyerson himself was a second lieutenant in the Chemical Warfare Service (1918-19) for whom he helped develop solid smoke producing agents and carried out catalytic researches. Reyerson died in the Yale-New Haven Hospital in New Haven, Conn., Sept. 7, 1969.

After completing his public school education in Dawson, Reyerson attended Carleton College, graduating in June 1915, with an A.B. degree (*cum laude*). Entering University of Illinois, he completed his M.S. degree in 1917 when his studies were interrupted by the war. His master's thesis advisor was Edward Washburn; while there and during his war service he was also influenced by Richard Tolman. After the war he resumed his studies at The Johns Hopkins University, receiving his Ph.D. degree in 1920. In 1956, Carleton conferred the honorary degree of D.Sc. upon him.

Joining the staff of the School of Chemistry at University of Minnesota as an instructor in 1921, he became an assistant professor in 1921, associate in 1926, professor of physical chemistry in 1930, and retired as emeritus in June 1961. From 1937 to 1954 he was administrative head of the School of Chemistry. From 1945 to 1954 he also served as an assistant dean and from 1934 until 1952 as director of the Northwest Research Institute at the university. He retired to Ridgefield, Conn., and there in 1962 joined the staff of the New England Institute for Medical Research, actively continuing interdisciplinary research until his death. While at the New England Institute, he served as chairman of the division of chemical sciences and also as the director of the summer student research programs for undergraduates.

In 1927 he was awarded a John Simon Guggenheim Memorial Foundation Fellowship and spent the academic year 1927-28 at Kaiser Wilhelm Institute in Berlin. A second such award in 1958 was taken at Carlsberg Laboratorium. During the tenure of these awards he came into contact with several eminent scientists such as Debye, Nernst, Haber, Einstein, Fruendlich, Linderstrom-Lang, and Svedberg. He formed lasting friendships with several of them, and these contributed to his growing interest in the history of science. He had many anecdotes about these researchers and frequently used them with good effect to stimulate and interest younger investigators and students. His autographed photo collection of these notables has been added to the Edgar Fahs Smith Memorial Collection on History of Chemistry at University of Pennsylvania.

He was a U.S. delegate to three IUPAC meetings: Warsaw, 1927; The Hague, 1928; and Stockholm, 1953. He was a Welsh Foundation lecturer in chemistry in 1962. He served as chairman of the scientific advisory commission to the Minnesota War Industries in 1942-1945, acted as a scientific consultant to the Royal Norwegian government in the summer of 1946, served as a chairman for the foreign research scientists program for the NAS, and was an assistant editor of *Journal of Physical Chemistry*, 1937-38. He was decorated with a Knight's Cross First Class of the Royal Order of St. Olaf by King Haakon of Norway in 1950, and received an Outstanding Alumni Achievement Award from Carleton in 1955.

Reyerson was an officer in several organizations, including American Institute of Chemists (counselor at large 1958-60, president 1965-66, chairman of the board 1966-67), and the American Chemical Society (chairman, Colloid Division 1939; chairman, Minnesota Section 1952). He joined the ACS in 1915 and often observed that at that time it was possible for one to know every other member of the society. He was also a member of the Chemists' Club (New York) and Cosmos Club.

Reyerson's major contributions (over 100 scientific papers) were in colloid chemistry, particularly in the areas of surface catalysis and behavior. He discovered the enhanced paramagnetism of fine palladium particles supported on silica gel, the apparently anomolous temperature sorption characteristics of NO radicals on silica and alumina gel, and the increased cross linking of nylon fibers when exposed to HCl vapor and then desorbed to produce a novel non-woven fabric (Cerex). His later researches were devoted to studies of vapor phase deuterium exchange in macromolecules of biological interest such as nucleotides and hemoglobins. Equally important with his research contributions, however, was his long career as a devoted educator, inspiring mentor, and dedicated teacher.

Personal recollections; obituary in *Chem. Eng. News*, Oct. 20, 1969, p. 74.

NELLE REYERSON
PHILIP J. KILLION
ALAN D. ADLER

Ellen Henrietta Swallow Richards

1842-1911

Ellen Swallow, daughter of schoolteachers, Peter and Fanny Gould (Taylor) Swallow, was born Dec. 3, 1842 in Dunstable, Mass. When she was 16 her father opened a store in Westford which enabled her to study at Westford Academy. In 1868 she entered the recently opened Vassar College and graduated with her A.B. degree in 1870. She was drawn to science, but there were few opportunities for women to gain a scientific education. Massachusetts Institute of Technology gave her special permission to enroll as a chemistry student, the first American woman enrolled as a full student in a scientific institution. In 1873 she became the first woman to receive a B.S. degree; Vassar awarded her a M.A. that same year. She remained at MIT until her death, serving as an assistant and later as instructor in sanitary chemistry. In 1875 she married Robert H. Richards, chairman of the Mining Engineering Department at MIT.

Ellen Swallow Richards directed her activities along three main lines: the use of chemistry to improve sanitary conditions, to broaden opportunities in science for women, and to develop home economics. She became an assistant in 1873 to John Ordway and William Nichols, experts in applied chemistry at MIT. Largely through her efforts, MIT established in 1876 a women's laboratory to afford better opportunities in scientific education for women. She guided its management, and the laboratory was so successful that it was abolished in 1883 in favor of admitting women on the same terms as men—at a time when women had no place in American science.

In 1883 MIT opened a laboratory in sanitary chemistry, the first in any institution, appointing Richards as an assistant under Nichols. Nichols was the world's foremost expert on water analysis, an area which Richards moved into when in 1887 the Massachusetts State Board of Health began a survey of the water supplies of the state. The success of the survey was largely due to Richards. She supervised a laboratory for water analysis, devised methods, and directed the examination of over 200,000 samples of water. She served as water analyst to the State Board of Health from 1887 to 1897.

In 1884 Richards became instructor in sanitary chemistry at MIT, a position she held until her death in 1911. She offered the first comprehensive course in sanitary chemistry in any school. For many years she directed the entire instruction in the chemistry of air, water, and foods for chemists, biologists, and sanitary engineers. She also served as a nutrition expert for the U.S. Department of Agriculture and conducted research on the danger of combustion in commercial oils for the Manufacturers Mutual Fire Insurance Co.

During the last 30 years of her life Richards was a leader in the home economics movement. She established the New England Diet Kitchen in Boston to improve the health of the poor. She worked for pure food laws, better lunches for school children, lectured throughout the country, and helped to found a nutrition department at University of Chicago. She was instrumental in founding the American Home Economics Association in 1908 and served as its first president. She sought to improve the material conditions of life and gave the name "euthenics" to the science which dealt with controlling the environment for right living.

As a leader in women's education Richards founded the Association of Collegiate Alumnae in 1882 (now the American Association of University Women). She was a charter member of the Naples Table Association for Promoting Laboratory Research by Women (1898), a director of the Women's Education Association, and trustee of Vassar College. For her achievements as a chemist and opening the doors of higher education for women, Smith College awarded her a doctor of science degree in 1890. She was stricken with angina pectoris and died at her home in Jamaica Plain, Mass., Mar. 30, 1911.

"Dictionary of American Biography," **15**, 553, Chas. Scribner's Sons (1935); Caroline L. Hunt, "The Life of Ellen H. Richards," Whitcomb & Barrows (1912), with portraits; Edna Yost, "American Women of Science," pp. 1-26, Lippincott (1943); obituary in *Technology Review*, 365-373 (July 1911), with portrait.

ALBERT B. COSTA

Joseph William Richards

1864-1921

Richards was born in Oldbury, Worcestershire, England, July 28, 1864. His grandfather, William Richards, had worked with Stevenson in England on the "Lion," one of the earliest steam locomotives. Richards was brought to America when he was 7 years old. His father, Joseph Richards, was a practical metallurgist in Philadelphia and was awarded the John Scott Medal of the Franklin Institute for the first successful solder for aluminum.

The boy was educated in the public schools of Philadelphia and received an A.B. degree from Central High School in 1882, being first in his class. Then he entered Lehigh University and majored in chemistry under William Chandler. Here, as a serious-minded student, he was labeled "Plug"—a title which stuck to him through life. Short of stature and of meager build, he was described in his class yearbook as "a hungry, lean-faced man—a mere anatomy."

Richards' senior thesis, under the direction of Chandler, was on the subject of "Aluminium" (the spelling to which he clung). His commencement address, June 24, 1886, was "Heroes of Science." Both subjects showed maturity, and no one was so impressed as Richards.

After graduation, he spent a year in industry as superintendent of Delaware Metal Refinery in Philadelphia. During this time he carefully revised his Lehigh thesis and had it published in 1887. Copies which he sent to his former professors, Chandler in chemistry and Benjamin Frazier in metallurgy, were well received. The result, as no doubt "Plug" hoped, was an invitation to return to his alma mater. His first degree had been A.C. (analytical chemist). Now, as an assistant instructor, he had time for graduate study and research. He earned his M.S. degree in 1891 and his Ph.D. degree in 1893. He was the first Lehigh student to receive the doctorate. His thesis was "A Calorimetric Study of Copper."

Richards' teaching was in metallurgy, mining, and blowpiping. In 1897 he went abroad to study in Heidelberg and Freiberg. Here he learned not only new techniques but also made contacts which were to aid him in his role as author and editor in the future.

Richards' success as a teacher stemmed from his quiet dignity and sincerity. He was meticulous in dress and had a humorously egotistical manner. Stories illustrated his lectures, whether on intricate calculations or profound description. He was an ardent champion of the metric system, for which he plead and debated. His interests were broad. In music he was an active supporter of the Bethlehem Bach Choir. He collected art works on his frequent travels. In religion he was a Unitarian although he was raised in the Methodist tradition.

He was the author of several texts which became internationally famous. His first was "Aluminium: Its History, Occurrence, Properties, Metallurgy, and Applications, Including Its Alloys," published in 1887, with later editions in 1890 and 1896. His second and more widely used text, "Metallurgical Calculations, Part I" (1906), grew out of his teaching practice. Its realm was general metallurgy. Part II, on iron and steel, was published in 1907, and Part III, dealing with non-ferrous metals, in 1908. These proved to be popular texts and were translated into five languages.

Richards, William Chandler, Edward Hart, and a dozen other men were charter members of the Lehigh Valley Section of the American Chemical Society in 1894. Richards' interest in the field of electrolysis led him to help organize the American Electrochemical Society, and he became its first president in 1902. Later, he was secretary of the society and editor of its publication. He died suddenly in Bethlehem, Oct. 21, 1921, at the height of his career, and his ashes were deposited in the chapel on the university campus where he did his life work.

Walter S. Landis, "Joseph W. Richards, the teacher—the industry," Memorial Lecture of the Electrochemical Society, 66th Convention, New York City, Sept. 28, 1934; Bradley Stoughton, "A Scholar and Aluminum," Lehigh Leaflets, No. 2, September 1923; R. D. Billinger, "America's Pioneer Press Agent for Aluminum—J. W. Richards," *J. Chem. Educ.* **14,** 253-55 (1937).

ROBERT D. BILLINGER

Theodore William Richards

1868-1928

Richards, the son of William, a noted painter of seascapes, and Anne (Matlack) Richards, was born in Germantown, Pa., Jan. 30, 1868. Because his mother felt that public education was geared to the slowest student in the class, Theodore received his elementary and secondary schooling at home. By the time he joined the sophomore class at Haverford College at the age of 14, his only formal education had been attendance at some chemistry lectures at University of Pennsylvania. In June 1885 he graduated at the head of his class with a degree in chemistry. Eager to study under Josiah Parsons Cooke, he entered the senior class at Harvard the following fall. He graduated with highest honors in chemistry in June 1886, the youngest member of the class.

As a graduate student at Harvard Richards undertook the difficult problem under Cooke's direction of accurately determining the composition of water to obtain the relative atomic weights of hydrogen and oxygen. He received his doctorate in 1888, and because of the merit of his dissertation he was awarded the Parker Fellowship, which enabled him to spend a year abroad making many important professional friendships.

On his return in the autumn of 1889 Richards joined the Harvard faculty as an assistant in analysis and stayed there for the rest of his life. He was chairman of the Chemistry Department from 1903 to 1911 and director of the Wolcott Gibbs Laboratory from its opening in 1912 until his death on Apr. 2, 1928.

Richards' best known studies were his determinations of the atomic weights of 25 elements, including those used to determine virtually all other atomic weights. For this work he was awarded the 1914 Nobel Prize in chemistry, the first American chemist to be so honored. About one-half of his nearly 300 published papers dealt with atomic weights.

When Richards began publishing his work in 1887, the accepted values of atomic weights were based upon those determined by Jean Stas in the 1860's. Stas's research was characterized by lengthy and careful procedures using large quantities of materials to achieve accuracy which far exceeded that of earlier workers. These values were so well received that until 1905 no investigator seriously questioned them nor attempted to check his work.

By 1905 Richards had become aware of serious errors in Stas's studies. Theory of precipitation had progressed far enough for him to see that his predecessor had neglected the slight but important solubility of silver chloride, had added solid silver nitrate to his solutions of metal halide salts, and had used such large quantities that impure samples dramatically increased errors. Consequently, the Harvard group determined the atomic weights of several major elements previously studied by Stas: silver, nitrogen, chlorine, sodium, and potassium. In all cases Richards' work resulted in significant changes in the accepted values.

Richards' other major contribution to the

field of atomic weights was a study of the atomic weight of radioactive lead from uranium minerals. The study by Richards and Max Lembert, published in 1914, was one of the first confirmations that lead from radioactive minerals has a different atomic weight from normal lead.

Richards also directed a vigorous research program in thermochemistry and electrochemistry. He became interested in these subjects in 1895 when, upon the death of Cooke, Harvard sent him to visit the laboratories of Wilhelm Ostwald at Leipzig and Walther Nernst at Göttingen to be suited better to teach physical chemistry. Many of his later investigations were a direct result of his theory of the compressible atom, an attempt to explain physically the variation of the constant "b" in van der Waals' equation of state. He proposed that an atom has a changeable volume, the magnitude of which depends upon its chemical state. Although the hypothesis was never adopted by other investigators, and Richards spent much of his last 10 years unsuccessfully trying to place the theory upon a firm mathematical foundation, his efforts led to the accurate determination of the physical constants of many elements and compounds. As a part of this study in 1902, while investigating the behavior of galvanic cells at low temperatures, he approached the discovery of the principles enunciated by Nernst in 1906 as the third law of thermodynamics.

While measuring thermodynamic values in his compressibility studies, Richards became aware of certain shortcomings in the calorimetric methods then in use, especially the need to apply a complex cooling correction to the calculation of his results because of heat transfer from the reaction vessel to the calorimeter jacket. In 1905, seeking to eliminate this problem, he devised an adiabatic calorimeter, in which the jacket temperature could be adjusted to that of the reaction vessel. Using continually improved versions of this device, he published 60 papers on thermochemistry; many contained data which are still listed among the standard values in handbooks. The Harvard group also investigated extensively the electrical and thermodynamic properties of amalgams.

During his years as a member of the faculty Richards created at Harvard a mecca for physical and analytical chemical research. Over 60 young men studied with him and became renowned chemists themselves.

Sheldon J. Kopperl, "The Scientific Work of Theodore William Richards," 1970, Madison, Wisc. (unpublished Ph.D. thesis); Harold Hartley, "Theodore William Richards Memorial Lecture," *J. Chem. Soc.*, pp. 1930-68 (1930); Aaron J. Ihde, "Theodore William Richards and the Atomic Weight Problem," *Science* **164**, 647-51 (1969).

SHELDON J. KOPPERL

John Leonard Riddell

1807-1865

Riddell was a man of many talents and accomplishments. He was in turn professor of botany, professor of chemistry, explorer, refiner of the New Orleans Branch of the United States Mint, and United States Postmaster there. In addition he was president of the Academy of Science in New Orleans, honorary member of the Microscopical Society of Great Britain, and an active member of the Louisiana State Medical Society. His services to the community are evidenced by his appointment by the governor as one of five Commissioners to devise a means of protecting New Orleans from flood (1844) and his election to the Louisiana State Legislature, a position which he held until his death, Oct. 7, 1865.

Riddell was born in Leyden, Mass., Feb. 20, 1807 the son of John and Zepha (Gates) Riddell. He obtained the degrees of A.B. and A.M. at Rensselaer, then lectured on chemistry, geology, and botany in Ogdensburg, Toronto, Erie, Pittsburg, and Cincinnati. In 1835 he graduated from Cincinnati Medical College, where he held the chair of botany and was adjunct professor of chemistry. In the same year he published his "Synopsis of the Flora of the Western States." His botanical research was well regarded, and the botanical genus "Riddelia" was named for him.

In the following year he wrote "Miasm and Contagion" in which he advocated the theory that organized living corpuscles were agents in contagious diseases.

Also in 1836 he was called to New Orleans as professor of chemistry at the newly founded Medical College of Louisiana, a post he held for nearly 30 years (Oct. 13, 1836 to Oct. 7, 1865). This institution became the Medical Department of University of Louisiana and later Medical College of Tulane University.

In 1838 Riddell was engaged by a company formed to purchase land for the mining of silver and gold in the "San Saba" region of Texas. Although no mines were developed, Riddell collected data by which to judge the general mineralogical character of the country, and received a share of the company, described as "equivalent to ten thousand acres of Texas lands." On his return to New Orleans he found that President Van Buren had appointed him melter and refiner of the local branch of the United States Mint, an office which he held from 1839 to 1849. His interest in numismatics led him to write several articles on the subject, notably his "Monograph of the American Dollar" (1845).

About 1844 Riddell began to devote attention to microscopy and, Oct. 2, 1852 at a meeting of the Physico-Medical Society of New Orleans, he demonstrated a binocular microscope, which he had devised and manufactured. The instrument was cumbersome, but Riddell used it to make numerous observations on infusoria, algae, and the microscopical characteristics of the blood and black vomit in yellow fever. Hence he was appointed to a commission to inquire into the causes of the yellow fever epidemic of 1853.

His articles on microscopy may be found in *London Microscopical Journal, New Orleans Medical and Surgical Journal,* and *American Journal of Science.* After Riddell's death, his widow presented his microscope to the Office of the Surgeon General in Washington (Apr. 14, 1879).

When Riddell was appointed United States Postmaster by President Buchanan, New Orleans printed its own postage stamps. After the secession of Louisiana, the Confederate Postmaster General ruled that all mail must be prepaid, but did not send stamps with which to carry out this policy, so Riddell printed stamps of his own design which were used until Apr. 26, 1862, when the city fell to the Federal forces.

Riddell was granted a pardon Aug. 14, 1865 by President Andrew Johnson for his activities on behalf of the Confederacy, conditional on his taking an oath of loyalty to the United States, although there is no evidence that he worked for the Confederate government except in a civil capacity. In fact, his views on secession were expressed forcibly a few weeks later when he called the State Democratic Convention to order, by saying, "Secession was worse than a crime, it was a blunder." The resultant controversy was so violent that it is claimed to have been responsible for Riddell's suffering a stroke, which brought on his death a few days later.

Biography by Joseph Jones, *Trans. Amer. Med. Ass.* **29**, 748-751 (1878); "Appleton's Cyclopaedia of American Biography," **5**, 248, D. Appleton & Co. (1888); notes in the archives of the Howard-Tilton Library, Tulane University; Riddell's articles in scientific and medical journals.

VIRGINIA MCCONNELL

Raymond Ronald Ridgway

1897-1947

Ridgway, who came to sign himself R³, was born in Morris, Ill., Aug. 27, 1897. He attended Lake Forest College and Massachusetts Institute of Technology, graduated from the latter in 1920 with the degree of B.S. in electrochemical engineering. For 1 year he worked for Aluminum Co. of America, mainly at Massena, N.Y. For another year, he worked with his father's firm, Ridgway Electric Co. in Freeport, Ill. In 1922, he joined the staff of Norton Co. in Niagara Falls, N.Y., which was then rebuilding after a drastic curtailment during the depression of 1921.

At Norton Co. he was active in the Research Department, eventually becoming associate director. The plant and laboratory of Norton Co. meanwhile were transferred to Chippawa, Ont., Canada less than 2 miles away as the crow flies, but Ridgway continued to live in Niagara Falls, N.Y., commuting daily across the river.

In 1921 he had married Margaret Lyman Longfellow of West Newton, Mass. Their four children were Stuart, Charlotte, Margaret, and Herbert.

R^3 was active in many local and national chemical and scientific societies. He was a leading spirit among the "Electrons" who met periodically for lunch. At various times secretary-treasurer and chairman of the Niagara Falls Section of the Electrochemical Society, he was elected president of the Electrochemical Society in 1941.

In 1943, he became the 13th recipient of the Jacob F. Schoellkopf Medal awarded by the Western New York Section of the American Chemical Society for his work in electrothermics, particularly the development of boron carbide.

The identification of boron carbide as B_4C, and establishing methods and equipment for commercially producing and fabricating this material were his most remarkable accomplishments, but many other contributions are also recorded to his credit. These included the design of electric furnaces, invention and production of a crystalline alumina abrasive (grown from a sulfide melt), processes for fusing non-conducting materials such as boric oxide glass, electrical grade magnesia for insulating metal wire heating elements, oxide resistor bars, thermocouples, ignitor tips, and lightning arrestors.

Ridgway and his staff were granted 26 U.S. patents as well as a number in other countries. They published 10 papers, 9 of which were contributions to the *Transactions of the Electrochemical Society*. Their studies of the temperature distribution in a silicon carbide resistance furnace are still the classical reference. The melting point of alpha alumina, the preparation of so-called "beta alumina," the hardness of abrasives, and the "hot-pressing" of ceramic materials were other fields of his interest.

His abundant energy overflowed from his work into social and club activities and sports. At M.I.T. he had learned to sail and had developed a taste for sea-food. He delighted in good food and introduced many of his friends to clam, oyster, and lobster dishes. He spent much time with his growing family, in swimming, dancing, figure skating, amateur dramatics, music, and puppets.

During World War II, he involved his laboratory group in two major projects. At the request of the Manhattan District, Corps of Engineers, it prepared various high boron alloys and produced commercially for the first time elemental boron of better than 98% purity. For these activities the laboratory received one of the awards for chemical engineering achievement given by the atomic bomb project. Rather less publicized was Ridgway's influence on and direction of the Bridgman Project, a cooperative effort among several interested companies to produce synthetic diamonds. During wartime this project suffered from lack of materials and money, for though the stakes were high, the probability of success was rated low. Later, the project was brought to successful fruition by General Electric Co.

In the post-war period, Ridgway was active in designing and building a new research laboratory for Norton Co. in Chippawa. He was never to enjoy this new facility. On June 12, 1947 after a family sailing party, he went back alone to collect some tools left aboard the sailboat. Attempting to return to shore, he drowned in Niagara River. A memorial plaque was placed in the entrance to the laboratory of Norton Research Corp.

Chem. Eng. News **21,** 858-62 (1943), portrait; "Who's Who in Engineering," Lewis Historical Publishing Co. (1941); obituary in *Double Bond* **20,** 7 (September 1947), portrait; obituary by Paul S. Brallier in *Trans. Electrochem. Soc.* **92,** 11-13 (1947); personal recollections.

GORDON R. FINLAY

Willard Bradley Rising

1839-1910

Rising was the first professor of chemistry in University of California. He was born in Mechlenburg, N.Y., Sept. 6, 1839. He began his college work at New York State Agricultural College, the germ of Cornell University, and continued his studies at Hamilton College, from which he graduated in 1864. In 1867 he obtained his M.A. degree from this institution and also the degree of M.E. from University of Michigan. He served for a time as instructor in chemistry in the latter university.

Soon after completing work for these degrees, he was appointed professor of natural science at College of California in Oakland, Calif. This was a private institution which had been organized by a group of men who hoped that it would become the state university. In 1869 when University of California was separately established, the college donated its lands to the university and closed its own doors. Robert Fisher, M.D. was at first appointed to the chair of chemistry, mining, and metallurgy in the new university, but because of disagreements he soon resigned and Rising was appointed to replace him in 1870.

Before assuming his new duties, Rising went to Heidelberg to study with Bunsen. He obtained his Ph.D. degree there at the end of 1871. In 1872 he returned to University of California as professor of chemistry and metallurgy. The original plan had been for him to organize a college of mines, but, reflecting his major interests, he organized a college of chemistry instead. A chemical laboratory was included in the first building on the university campus, but it was not until 1890 that a separate chemistry building was constructed. In 1876 Rising's title was changed to professor of chemistry, and he held this chair until his retirement in 1908. When the university opened a school of pharmacy, Rising taught chemistry in it for a short time.

Rising's chief chemical interests were in the fields of thermochemistry and explosives (he was a consultant for a powder company), but most of his activities were concerned with administering his department and with carrying on analytical work. After 1888 he served as state analyst for the state of California. He published technical bulletins on industrial analysis. He died in Berkeley, Feb. 9, 1910.

Most of the details of Rising's life are to be found in William Carey Jones "Illustrated History of the University of California," (F. H. Dukesmith) (1895); Rising's papers are deposited in the Bancroft Library, University of California, Berkeley; biographies in *Amer. Chem. J.* **43,** 385-386 (1910), and in "National Cyclopedia American Biography."

HENRY M. LEICESTER

Eugene Franz Roeber

1867-1917

Roeber, German-born electrochemical and engineering editor, was a pioneering proponent of scientific research in the early development of the American electrochemical and metallurgical industries. He was born in Torgau, Saxony, Oct. 7, 1867 and was barely 50 years old when he died Oct. 17, 1917. He received his education from German universities—Jena and Halle and at Berlin where he received his Ph.D. degree in 1892. Two years later, dissatisfied with job prospects under a repressive Prussian regime, he came to America and for a time was employed as a translater for Steiger Book Co. Through rigorous study and self discipline he was to become a gifted writer. But he never quite outgrew his German accent, manner, and appearance. One of his contemporaries, Howard C. Parmelee, described him as a "little round man, with a round chin, round glasses, and a big round hat, which on his head became *ein Professorenhut.*"

Roeber's professional career began in 1899 when he moved to Philadelphia to become an assistant to the eminent electrical engineer and chemist, Carl Hering. His editorial ability soon became evident as he took over a digest of foreign technical literature which his employer regularly contributed to *Electrical World* in New York. This also brought him to the favorable attention of its publisher, James H. McGraw.

Then came two significant developments. First was the burgeoning growth of the new electrochemical and metallurgical industries, stimulated by the abundant, cheap power then becoming available at Niagara Falls. Second was the beginnings in Philadelphia of a little group of electrochemists and engineers concerned with the problems involved in this new industrial development. This was to lead in September 1902 to the founding of the American Electrochemical Society, in which Hering and Roeber played an important part.

To these two men the time seemed right to publish a new magazine as an exponent of the science and technology of electrochemistry. They approached McGraw in New York, who shortly thereafter founded the Electrochemical Publishing Co. with Joseph W. Richards of Lehigh as president and Roeber as editor of *The Electrochemical Industry*. McGraw was publisher and business manager. The new magazine was successful from the start and in 1905 broadened its field to that of the *Electrochemical and Metallurgical Industry*. Five years later to recognize the growing importance of chemical engineering, it became *Metallurgical and Chemical Engineering*. This was changed during World War I to *Chemical & Metallurgical Engineering* and finally in 1946 to its present title, *Chemical Engineering*.

Fortunately for those who had trouble with these jaw-breaking titles, there developed the convenient nickname "Met & Chem" which later became "Chem & Met." But to many old timers it remained for 15 years as "Roeber's Journal." His virile and often controversial editorials were widely acclaimed. His sound advice provided guidelines often adopted by industry and education. When threats of war became pressing, he was among the first to urge that the chemical industries greatly increase their research and production for our Allies.

In an editorial on "Patriotism in Chemistry," which he wrote shortly after the United States entered World War I, he declared . . . "We must be ready to serve with our hearts and heads, our hands and fortunes. . . . The chemists of the United States have suddenly become, in a large measure, the trustees of our safety. Let us not fail." History records that the chemical profession and industries did not fail.

But in a very real sense Roeber was a tragedy of World War I. He was an extremely sensitive man and was sorely disturbed by the cruel criticism of his broken English and his German appearance. The burdens of war rested all to heavily on him. Much of the Germany he had loved and cherished was lost forever. Badly in need of rest and recuperation he gave up his active editorial work in June 1917 and retired to Battle Creek Sanitarium where he died. He was barely 50 years old, but in 15 years of what he chose to call "militant journalism" he had risen to a position of editorial leadership. He had established high standards for those who were to follow.

Editorial by Howard C. Parmelee and accompanying obituary and portrait in *Metallurgical Chem. Eng.* **17,** 511-515 (1917); personal tribute by Colin G. Fink with details and regulations of the E. F. Roeber Research Fund sponsored by the Electrochemical Society, and contributions by L. H. Baekeland, Charles F. Burgess, Wallace Cohoe, John V. N. Dorr, F. A. Lidbury, James H. McGraw, Malcolm Muir and Francis J. Tone, *Trans. Electrochem. Soc.* **64,** 6-7 (1933); Roger Burlingame, "Endless Frontiers, the Story of McGraw-Hill," pp. 228-38, McGraw-Hill (1959); obituary, *New York Times,* Oct. 19, 1917.

SIDNEY D. KIRKPATRICK

Allen Rogers

1876-1938

Rogers was born in Hampden, Maine, May 22, 1876. He received his B.S. degree in chemistry from University of Maine in 1897 and his Ph.D. degree from University of Pennsylvania in 1902. He was instructor in chemistry at University of Maine from 1897 to 1900 and of organic chemistry at University of Pennsylvania from 1903 to 1904. Oakes Manufacturing Co., Long Island, engaged him as chemist during the years 1904 and 1905.

In 1905 Rogers began teaching industrial chemistry at Pratt Institute. There in the School of Science and Technology he built laboratories in which students could learn practice, following the examples of the technical institutes of Europe. Thus, the laboratory had miniature soap plants, paint units, a tannery, a heavy chemical unit, and an organic dye unit.

The students were older than average and intense in their pursuits of vocations. They came from many lands, family backgrounds, and levels of society. Rogers became well known to these students through his development of sharkskin leather, his lectures on paints to the art students, his talks on the chemistry of foods to the nutritionists, and his lectures on other industrial topics.

Rogers' students had no holiday periods for he organized extended tours to chemical plants where the staffs answered the students' questions, no matter how prying these were. Students held him in high esteem; later, as entrepreneurs after graduation, they sought his counsel, particularly as an expert in litigation.

Rogers wrote a text, "Elements of Qualitative Analysis," early in his career, but his important books were on the industrial chemical processes of his time. His "Industrial Chemistry," written in collaboration with experts in various fields, first appeared in 1912 and reached a sixth edition in 1942. He published "Elements of Industrial Chemistry," an abridgement of his major work, in 1916 and 1926 and "Laboratory Guide of Industrial Chemistry" in 1908 and 1917. He wrote "Practical Tanning" in 1922 and "Manufacture of Leather" in 1929.

During World War I Rogers served as a major in the Chemical Warfare Service. In 1920 he received the Grasselli Medal for his development of a process for utilizing fish skins for leather. He died Nov. 4, 1938, after a fall down the steps of the chemical engineering building at Pratt Institute, and did not live to see the displacement of industrial chemistry by chemical engineering unit operations, both as subject matter and laboratory type.

Obituaries in *New York Times,* Nov. 5, 1938, and in *The Chemist* **15,** 390-91 (1938); personal recollections.

CHARLES L. MANTELL

Henry Darwin Rogers

1808-1866

Henry Darwin Rogers was born in Philadelphia, Pa., Aug. 1, 1808. He received much of his education from his father, Patrick Kerr Rogers, and at College of William and Mary. Around 1828 he taught school in Maryland and lectured on chemistry at Maryland Institute in Baltimore. He went to Dickinson College as professor of chemistry and natural philosophy in 1830, resigned after a moderate time, helped run a survey for a New England railroad, lectured on geology at Franklin Institute in 1833, and became professor of mineralogy and geology at University of Pennsylvania in 1835. While at the university he also directed the geological survey of New Jersey from 1835 to 1838 and of Pennsylvania from 1836 to 1842. In 1855 he became professor of natural history at University of Glasgow, Scotland and held this post until he died May 29, 1866.

Emma Rogers, "The Life and Letters of William Barton Rogers," Houghton, Mifflin & Co. (1896); *Amer. J. Sci.* (2s) **41,** 236-38 (1866); *Proc. Amer. Acad. Arts Sci.* **7,** 309-12 (1865-68).

WYNDHAM D. MILES

James Blythe Rogers

1802-1852

James Rogers was born in Philadelphia, Feb. 22 (or 11), 1802. His father, Patrick K. Rogers, was a chemist, and his three brothers, William B., Henry D., and Robert E., were well-known scientists.

James received his early education from his father, his college education at William and Mary, where his father was professor of chemistry and natural philosophy, and his medical education at University of Maryland, where he received his M.D. degree in 1822. He practiced medicine in Baltimore for several years, gradually came to dislike it, and in 1827 became superintendent of a chemical plant owned by Isaac Tyson.

From 1828 to 1835 he was professor of chemistry at Washington Medical College (a school no longer in existence) in Baltimore. During this period he held other jobs that overlapped. He taught chemistry in Maryland Institute around 1830, superintended Tyson's chemical factory for a time, and was partner in a financially unsuccessful apothecary shop.

In 1835 James was appointed professor

of chemistry in University of Cincinnati's Medical School and moved from Baltimore to Cincinnati. He taught chemistry to medical students for about 4 months during the winters, and from 1837 to 1840 he was a chemist with the geological survey of Virginia, conducted by his brother William, during the summers.

James moved to Philadelphia in 1840 to assist his brother Henry with the geological survey of Pennsylvania and worked on the survey for about 2 years. The faculty of University of Pennsylvania Medical Department elected him professor of chemistry in the department's summer school in 1841. Franklin Institute engaged him to instruct in 1844. He helped found Franklin Medical College (no longer in existence) in 1846 and taught chemistry there.

In 1847 he was elected professor of chemistry in University of Pennsylvania's Medical School and stopped teaching at other institutions. Earlier professors in the school had concentrated on inorganic chemistry; Rogers was one of the new order of chemists who emphasized organic chemistry.

Rogers published at least three articles. Many of his analyses appeared in the reports of the geological surveys of Virginia and Pennsylvania. In 1846 he and his brother Robert brought out an American edition of the combined texts of two noted British chemists under the title, "Elements of Chemistry, including the history of the imponderables and the inorganic chemistry of the late Edward Turner . . . and the outlines of organic chemistry of William Gregory."

James Rogers died in Philadelphia, June 15, 1852.

One of Rogers' colleagues said of him, "Dr. Rogers was a popular teacher; the full storehouse of his mind was drawn upon to instruct his pupils."

Emma Rogers, "Life and Letters of William Barton Rogers," Houghton, Mifflin & Co. (1896); Edgar F. Smith, "James Blythe Rogers, 1802-1852, Chemist" (1927), with portrait, private printing; Joseph Carson, "Memoir of the Life and Character of James B. Rogers, M.D.," T. K. & P. G. Collins (1852); biography by W. S. W. Ruschenberger in *Proc. Amer. Phil. Soc.* **23,** 104-146 (1888).

WYNDHAM D. MILES

Patrick Kerr Rogers

1776-1828

Patrick was born in Ireland in 1776. Emigrating to the United States in 1798, he settled in Philadelphia. He became a tutor at University of Pennsylvania, attended the university's Medical School, and received his M.D. degree in 1802.

He practiced medicine for a living, but apparently he would have preferred to work in chemistry had an opportunity been available. He was a member of the Chemical Society of Philadelphia, which existed from about 1790 to 1809, and he delivered courses of public lectures on chemistry, which Philadelphians could attend for a fee. His course in 1809 consisted of two or three evening lectures a week for 4 months in his laboratory in his house. Tickets cost $10.00, and Rogers showed "about 1500 interesting experiments." In 1810 Rogers published a 12-page "Syllabus of a course of lectures on natural philosophy and chemistry, with the application of the latter to several of the arts," for the ladies and gentlemen who attended his course. He applied for the professorship of chemistry in the university's Medical School after James Woodhouse died in 1809 but did not have sufficient support among the trustees to be elected.

Shortly after arriving in America Patrick married Hannah Blythe. The couple had four sons, James Blythe, William Barton, Henry Darwin, and Robert Empie Rogers, all of whom were notable scientists.

Rogers moved to Baltimore in 1812, hoping to find a wider scope for his talents. He practiced medicine, delivered public lectures, and opened an apothecary shop.

In 1819 he asked Thomas Jefferson to sponsor his application for the professorship of chemistry at University of Virginia, but the university visitors had already decided to offer the post to Thomas Cooper. The same year, College of William and Mary appointed Rogers professor of natural philosophy and chemistry.

Patrick moved to Williamsburg and taught at the college for the remainder of his life. He published a text, "Introduction to the mathematical principles of natural philosophy" (1822). Rogers had the reputation of being a profound scholar and was said to have been a careful teacher, singularly successful in his illustrative experiments before his class. He died Aug. 1, 1828 at Williamsburg.

Emma Rogers, "The Life and Letters of William Barton Rogers," Houghton, Mifflin Co. (1896); W. S. W. Ruschenberger, "A sketch of the life of Robert E. Rogers, with biographical notices of his father and brothers," *Proc. Amer. Philos. Soc.* Vol. 23 (1886); Galen W. Ewing, "Early teaching of science at the College of William and Mary," *J. Chem. Educ.* **15**, 3-13 (1938); Wyndham D. Miles, "Public lectures on chemistry in the United States," *Ambix* **15**, 129-53 (1968); Desmond Reilly, "Patrick Kerr Rogers (1776-1828) and early American chemistry," *Recorder; Bull. Amer. Irish Hist. Soc.* **17**, 22-25 (1955).

WYNDHAM D. MILES

Robert Empie Rogers

1813-1884

Robert Rogers was born in Baltimore, Mar. 29, 1813. He received his early education from his father, Patrick Kerr Rogers, who taught chemistry and physics at William and Mary College from 1819 to 1828, and, after his father's death, from his brothers, William, Henry, and James, noted chemists and geologists.

He attended University of Pennsylvania Medical School and received his M.D. degree in 1836. But instead of practicing medicine he served as chemist for the geological survey of Pennsylvania, directed by his brother Henry. He also taught chemistry at Philadelphia Medical Institute. He remained with the survey until 1842 when he became professor of chemistry at University of Virginia. He left Virginia in 1852 when his brother James, professor of chemistry at University of Pennsylvania Medical School, died; Robert went from Charlottesville to Philadelphia to succeed him.

Robert taught in the Medical School for 20 years. During the Civil War he also served as a surgeon in a military hospital in Philadelphia. He injured his right hand so badly in the hospital's laundry that it had to be amputated. He quickly learned to write with his left hand, and to carry out chemical manipulations without difficulty.

Robert's brother James, during his tenure from 1847 to 1852, had emphasized the trend away from inorganic chemistry toward organic chemistry. Robert went further and devoted much of his lectures to physiological chemistry, the logical extension of organic chemistry into medicine. For his classes he edited in 1855 one of the earliest texts on the subject printed in the United States, "Physiological Chemistry," which had been translated by the British chemist George Day, from Carl Lehmann's "Lehrbuch der Physiologischen Chemie."

Between 1865 and 1873 he gradually introduced chemistry laboratory instruction into the Medical School; previously chemistry had been taught by lecture and demonstration, with laboratory instruction being pursued voluntarily by a small proportion of interested students.

Robert Rogers resigned from the university in 1877 because of a disagreement among the faculty over changes in the Medical School. Later that year Jefferson Medical School engaged him to teach chemistry. At Jefferson he modernized the chemistry course and inaugurated laboratory instruction. He retired in 1884.

Rogers was chemist for the Gas Trust of Philadelphia from 1872 to 1884. In the 1870's he undertook for the Treasury Department a study of the wastage of silver in the refining process at the Philadelphia Mint and suggested improvements. He assisted the Treasury in planning the refinery for the San Francisco Mint when the building was under construction in the 1870's. From 1874 to 1879 he was a member of the U.S. Assay Commission.

He published a score of articles on a variety of topics in inorganic, organic, and analytical chemistry. With his brother James he edited "Elements of Chemistry, including the history of the imponderables and the inorganic chemistry of the late Edward Turner . . . and the outlines of organic chemistry by William Gregory," in 1846. He invented a number of devices, among them a steam boiler.

Robert Rogers died in Philadelphia, Sept. 6, 1884.

J. W. Holland, "A Eulogy on the Life and Character of Prof. Robert E. Rogers," Fell (1885); W. S. W. Ruschenberger, "A Sketch of the Life of Robert E. Rogers . . . with biographical notices of his Father and Brothers," *Proc. Amer. Phil. Soc.*, vol. 23 (1886); Emma Rogers, "Life and Letters of William Barton Rogers," Houghton, Mifflin and Co. (1896); biography by Edgar F. Smith in *Biog. Mem. Nat. Acad. Sci.* **5,** 291-309 (1905), with portrait, and in "Chemistry in America," D. Appleton and Co. (1914), pp. 236-241.

WYNDHAM D. MILES

William Barton Rogers

1804-1882

William Barton Rogers was born in Philadelphia, Dec. 7, 1804. He was educated largely by his father, Patrick Kerr Rogers, and at College of William and Mary. From 1827 to 1828 he taught chemistry at Maryland Institute, and in 1828 became professor of chemistry and physics at William and Mary, succeeding his father. He taught natural philosophy and geology at University of Virginia from 1835 to 1853 and concurrently was state geologist of Virginia from 1835 to 1842.

Rogers moved to Boston in 1853. A few years later he began to interest people in his concept of a technical college. This led to the chartering of Massachusetts Institute of Technology in 1862 and its opening in 1865 with Rogers as president. While speaking at graduation exercises of the institute on May 30, 1882, Rogers fell dead. During his life he published approximately 40 articles, of which at least 8 were on chemistry; and, in collaboration with his brothers, James, Henry, and Robert, at least 12 other articles on chemistry.

Emma Rogers, "The Life and Letters of William Barton Rogers," Houghton, Mifflin and Co. (1896); biography by Francis Walker in *Biog. Mem. Nat. Acad. Sci.* **3,** 1-13 (1895); W. S. W. Ruschenberger, "A sketch of the life of Robert E. Rogers, with biographical notices of his father and brothers," *Proc. Amer. Philos. Soc.* Vol. 23 (1886); Galen W. Ewing "Early Teaching of Science at the College of William and Mary," *J. Chem. Educ.* **15,** 3-13 (1938); obituary in *Proc. Amer. Acad. Arts Sci.* (n.s.) **10,** 428-38 (1882-83).

WYNDHAM D. MILES

Martin Andre Rosanoff

1874-1951

Rosanoff was born in Nicolaeff, Russia, Dec. 28, 1874. He came from a long line of scholars. His father, Abraham Rosanoff, taught history at the Imperial Classical Gymnasium and wrote a 10-volume encyclopedia of the Bible which was considered one of the best works on the subject. His younger brother, Aaron Joshua, became a renowned psychiatrist while his sister, Lillian, was a mathematician and head of the Department of Mathematics at Long Island University.

Young Rosanoff's early education was at the Gymnasium. In 1891 the family came to the United States. Martin attended New York University, receiving his Ph.B. degree in 1895. He studied chemistry at University of Berlin from 1895 to 1896, and then in Paris from 1896 to 1898, where he served as research assistant to Charles Friedel. On his return to the United States in 1898 he became an assistant to James M. Craft at Massachusetts Institute of Technology.

Rosanoff married Louise Place of New York City in 1901; they had three children. From 1900 to 1903 Rosanoff was editor for the exact sciences of "New International Encyclopedia." He then joined Thomas Edison in Edison's Orange, N.J., laboratory for 1 year. Rosanoff later wrote an account of his experiences with Edison: "Edison in His Laboratory" (1932).

In 1904 Rosanoff returned to New York University as an instructor in theoretical chemistry and advanced to assistant professor in 1905. He received his D.Sc. degree in 1908 from NYU. He accomplished most of his noteworthy researches while at Clark University between 1907 and 1914. At Clark he was head of the Graduate Department of Chemistry and director of the chemical laboratories.

In 1914 Mellon Institute of Industrial Research at University of Pittsburgh established a new department–the Department of Research in Pure Chemistry–and the Willard Gibbs Professorship of Research in Pure Chemistry. Rosanoff became the first scientist to hold this professorship and to serve as director of the new department. He served in these capacities until 1919.

The loss of his wife in an automobile accident in 1918 led him to retire from active research. For many years he was an industrial consultant. In 1932 he remarried and in 1933 became director of chemical research at Duquesne University in Pittsburgh. From 1934 to 1940 he was dean of the graduate school at Duquesne. During World War II he served as public panel chairman, War Labor Board, and received the U.S. Certificate of Merit.

Rosanoff's scientific investigations were mainly in the areas of physical and organic chemistry. His major work concerned partial vapor pressures and the theory of the distillation of mixed liquids. He held several patents for his processes and apparatus for fractional distillation. In papers published between 1906 and 1920 he developed processes for separating mixtures in a single distillation and methods for determining partial vapor pressures from the total vapor pressure of binary mixtures.

Another important area of research was catalysis, where he developed theories of the mechanism of action of different types of catalysts which enabled chemists to determine the best catalyst for different kinds of chemical reactions. His studies on the physico-chemical mechanisms of organic reactions formed an original contribution to chemical kinetics.

For his contributions to chemistry the New York Section of the American Chemical Society awarded Rosanoff its Nichols Medal for original research in 1910. The Franklin Institute awarded him its medal. Living in poor health for several years, he died in Mt. Lebanon, Pa., July 30, 1951.

"National Cyclopaedia of American Biography," **C,** 285, James T. White Co. (1930), with portrait; *New York Times,* July 31, 1951, with portrait.

Albert B. Costa

Robert Eustafieff Rose

1879-1946

Rose, an authority on the application of dyes to textiles, was born in Palermo, Sicily on June 2, 1879 where his father Robert, was acting British Consul. Privately educated as a child, he afterward attended Bowden House School and received his Ph.D. degree from Leipzig University in 1903. He was an instructor in chemistry at University of St. Andrews, Scotland from 1903 to 1905 and later lectured at University College, Nottingham, England from 1905 to 1907.

In 1907 he came to the United States and was associate professor of chemistry at University of Washington until 1917. After spending a year at Mellon Institute in Pittsburgh, he joined the Dyestuffs Department of Du Pont Co. in Wilmington, Del. Du Pont had started research in dye synthesis in 1916, began shipping indigo in 1918, and afterwards expanded its product line rapidly. In 1919 a new laboratory for research and development in dye application was built at the plant site on Deepwater Point, N. J., and in 1921 Rose was chosen to head it. He continued as director of the "Technical Laboratory" until his retirement in 1944.

Before joining Du Pont, Rose's chief research dealt with the composition of naturally occurring glucosides and sugars and wood decay. His work with wood decay was pioneering in the field. At Du Pont, he engaged primarily in the physical chemistry of dyeing, particularly of cellulosic textiles and paper. He held patents on baking powder, textile softeners, and oil soluble dyes. These dyes, which are still used in wood stains and leather finishes, were superior in colorfastness to any previously made. After managerial duties curtailed his active role in research, he published many papers of a popular and educational nature on dye usage and general chemistry.

A member of nine professional societies, he was quite active in several. A charter member of the American Association of Textile Chemists and Colorists, he at one time headed its research council and served as president from 1932 to 1935. Also, he was a director of Textile Research Institute at Princeton, N. J.

While at University of Washington, Rose married Glenola Behling, an assistant in the Chemistry Department. They had no children. He had a critical knowledge of classical music, was well known as a naturalist, and traveled extensively throughout the Americas and Europe. His collection of photographs taken from Alaska to the Balkans was considered one of the finest amateur travelogs in existence. He died in San Mateo, Calif. Sept. 23, 1946.

Amer. Dyestuffs Reporter, **35,** 485 (Oct. 7, 1946); "National Cyclopedia of American Biography," James T. White Co. (1954), **39,** 324; Walter J. Smith, "Chambers Works History" (1963), five volumes of unpublished typescript in Du Pont's Jackson Laboratory library.

Herbert T. Pratt

George David Rosengarten

1869-1936

Rosengarten, chemist, manufacturer, sometime president of the American Institute of Chemical Engineers and the American Chemical Society, was an example of the moral force and effect which can be generated when the best of the worlds of chemistry and commerce are combined.

George, son of Harry Bennett and Clara (Knorr) Rosengarten, was born Feb. 12, 1869 in Philadelphia, and virtually his whole personal and professional life was associated with that city. At the time of his birth the Rosengarten family, under the firm name of Rosengarten and Sons, were second generation chemical manufacturers. The company had begun the manufacture, in the 1820's, of quinine, morphine, and other alkaloids, silver salts, and ether, and had grown to be one of the city's two major suppliers of pharmaceutical and industrial chemicals—the other being their arch-rival Powers and Weightman.

George Rosengarten was accepted as a special student at Philadelphia College of Pharmacy and Science in 1884, when he was 15. From there he went to University of Pennsylvania, receiving his B.S. degree in 1890, and University of Jena, which awarded him his Ph.D. degree in 1892. He went directly into the family business in 1893. In 1898 his uncle Mitchell G. Rosengarten died, and two other uncles retired from the firm. This left George's father in charge, with George and three other family members of his generation, Adolph G., Joseph G. Jr., and Frederic Rosengarten. George became vice-president in 1901 when the business was incorporated under the name of Rosengarten and Sons, Inc. In 1905 the corporation merged with their former competitors to form the Powers-Weightman-Rosengarten Co., which later became Powers-Weightman-Rosengarten, Inc. George served as vice-president of this firm for its 22 years of existence, until it was succeeded by Merck and Co., Inc., in 1927. He thereupon served as a director for Merck until his retirement in 1929.

Rosengarten's life was far from being confined to his business activities, however, important as those were. He had a long-time concern for the support of basic research in the sciences, for the establishment and maintenance of standards in drug and chemical manufacture, and for the proper relation of "pure" and applied science in industry. These concerns were reflected in his many society memberships and in the honors which began to be conferred upon him in his middle years. He was president of the American Institute of Chemical Engineers from 1915 to 1917. He was extremely active in the American Chemical Society, serving as chairman of the Division of Industrial and Engineering Chemistry, councilor-at-large, member of the executive committee, chairman of the endowment committee, and as president in 1927. Rosengarten's presidential address spoke of some of his concerns as man and chemist, noting among other things that "We are plucking the fruit of trees of knowledge planted by our forebears . . . but we cannot much longer go on harvesting without planting."

During his ACS presidency Rosengarten founded and gained Chemical Foundation support for the ACS Institute of Chemistry, a 4-week period of brush-up courses and survey lectures on recent progress in chemistry. The first institute was held in July 1927 at Penn State College, the second in 1928 at Northwestern University.

Rosengarten was also a member of the 9th and 10th revision committees of the U.S. Pharmacopeia, as well as chairman of the subcommittee on organic chemicals, director of two Philadelphia banks and trustee of the Philadelphia College of Pharmacy and Science, and member of nearly a dozen and a half scientific and cultural organizations, including the Franklin Institute, of which he was a director. Rosengarten's relatively few technical publications were devoted either to organic synthesis or to definition of drug and chemical standards. In 1927 University of Pennsylvania awarded him an honorary Sc.D. degree.

Rosengarten married Susan Elizabeth Wright in 1895. They had no children. He died in Philadelphia, Feb. 24, 1936.

"National Cyclopedia of American Biography," **35**, 272, James T. White & Co. (1949); J. W. England, *Am. J. Pharm.* **99**, 21-22 (1927); *J. Ind. Eng. Chem.* **7**, 65 (1915), **19**, 329, 1076-77 (1927); *News Edition, I.&E.C.* **5**, Mar. 20-Oct. 10 (1927), **6**, Jan. 10-Sept. 10 (1928), **14**, 88 (1936); Williams Haynes, "Chemical Pioneers," D. Van Nostrand Co. (1939), pp. 26-41.

ROBERT M. HAWTHORNE JR.

Charles Ferdinand Roth

1886-1954

Charles Roth ("Mr. Chemical Show") was born in New York, Dec. 24, 1886, the son of Charles and Annie (Hart) Roth. He attended local schools and graduated with his B.S. degree in chemistry and the Cooper Medal from Cooper Union in 1912. While still in school he worked as control chemist with National Brewers Academy (1907-10), and as assistant, later chief, chemist with Standard Oil of New Jersey (1911-14).

For one year (1914-15) he was chemist in charge of manufacturing with Hartford Cast Stone Co., and one year he was a consultant with International Filtration Corp. This overlapped his organizing and managing of the first Chemical Exposition in 1915, which thenceforward became one of his major interests in life.

The Chemical Exposition was the inspiration of Charles H. Herty and an advisory committee comprised of Raymond F. Bacon, Arthur D. Little, Robert P. Perry, Theodore B. Wagner, and Henry B. Faber—all active members of the American Institute of Chemical Engineers—and this committee set high standards. Roth opened an office with, first, Adrian Nagelvoort, later Frederick W. Payne, as an assistant, and the Chemical Exposition became an annual affair of great importance, especially after the United States entered World War I. Later it was made biennial and was sponsored by the Salesmen's Association of the American Chemical Industry. This group (founded in 1921 by Williams Haynes) sponsored excellent educational programs during the expositions as well as exhibits of chemicals and equipment.

In 1918-19 Roth was in the Chemical Warfare Service as director of adjustment of personnel. From 1919 to 1925 he was director of Beckman and Linden Engineering Corp., but then he abandoned all work of this kind to concentrate on managing expositions.

His reputation as a good manager had spread. Starting in 1922 he managed the National Exposition of Power and Mechanical Engineering ("Power Show"). In 1923 he established the International Exposition Co., of which he served successively as secretary-treasurer, vice president, and (1936) as president. The number of expositions increased: Paper Industries, 1923-24; National Electrical Exposition (New York), 1929; International Heating and Ventilating, 1930; Power and Mechanical Engineering (Chicago), 1937; Pacific Heat and Air Conditioning (San Francisco), 1941; Electrical Engineering (New York), 1947. Altogether, he managed over 60 expositions; many of those named here became annual or biennial affairs.

Roth was an active member of the American Chemical Society; he served as secretary of the New York Section in 1916-17, as councilor in 1921 and on committees for award of the Nichols and the Perkin Medal. He was a member of several other scientific and engineering societies and of the Chemists'

Club in which he served as editor of *The Percolator* in 1917 and chairman of the house committee in 1921.

Roth was a well-rounded person with many varied interests. He was married in 1920 to Carol E. Thrall of Philadelphia, a talented amateur artist. They were always gracious hosts at their estate in Bedford Village, N.Y. (complete with spacious gardens, a lake, and a running stream). From his extensive library, the Chemists Club was given the unrestricted choice of 50 volumes.

He died June 24, 1954 as the result of an automobile accident, survived by his wife and a grandson, Charles McDonald.

Personal recollections; newspaper and journal clippings in possession of Florence E. Wall; obituaries in *The Chemist,* August 1954, and *New York Times,* June 25, 1954.

FLORENCE E. WALL

Edmund Ruffin

1794-1865

Ruffin sought to apply chemical knowledge to the restoration of the depleted soil of the family plantation in Prince George County, Virginia. The only child of George and Jane (Lucas) Ruffin, he was born Jan. 5, 1794. His mother died soon thereafter and his upbringing fell to George's next wife, Rebecca Cocke. There were several half brothers and sisters. The Ruffin's traced their American lineage back to 1666 when William Ruffin left England to settle on land in the valley of the James River. Over successive generations, the size of the estate grew, but when George Ruffin died in 1810 the soil was badly depleted from steady cropping of tobacco.

Edmund's early education was at the hands of private tutors. At 16 he entered William and Mary College but he appears to have had no talent for the academic life. He was soon dropped by the college, and he enlisted in a volunteer company preparing for duty in the War of 1812. He participated in no active fighting and, when mustered out in 1813, married Susan Travis of Williamsburg and settled down on the family plantation at Coggin's Point.

Although lacking experience for plantation management, Ruffin became an ardent student of contemporary writers on scientific agriculture. He found the answer to his land's failure to support clover in Humphry Davy's "Elements of Agricultural Chemistry" and became an advocate of lime. In 1818 he had his slaves dig marl from fossil beds on his farm and apply this to a test plot. Apparent success caused him to become a propagandist for marl on exhausted tidelands soil. He wrote extensively for the *American Farmer* and in 1832, expanded his articles in a book, "Essay on Calcareous Manures." His "Essay" was published in 1832, 1835, 1842, 1844, and 1852. The second and fourth were essentially reprints. The third and fifth were extensively expanded. The first edition was reprinted in 1961 with an introduction by J. C. Sitterson.

In 1833 he began publication of a monthly *Farmers' Register* to publicize the views gained from his reading and experimentation. He continued through 1842.

In his later years he became embittered at the lack of response to his ideas by Southern farmers. He was also caught up in the secessionist movement of his region. He attached himself to a guard unit and, according to some claims, set off the first artillery shell fired at Fort Sumter. In retaliation federal gunboats frequently took pot shots at his home, Beechwood, which he built near the James River in 1843 and which still stands. His age was against continued military service, however, and he bitterly watched the Southern cause deteriorate. Union troops pillaged his estate and vandalized his library. Following Appomattox he ended his life by shooting himself, June 18, 1865, because he couldn't bear to live under the federal government.

Avery O. Craven, "Edmund Ruffin, Southernor," Archon Books (1932, reprinted 1964); "Dictionary of American Biography," **16,** 214, Chas. Scribner's Sons (1935); Henry G. Ellis, "Edmund Ruffin and his Times," *John P. Branch History Papers of Randolph Macon College* **3,** 99-123 (1910); E. J. Dies, "Titans of the Soil," p. 55 ff., Univ. of North Carolina Press (1949); A. J. Ihde, "Edmund Ruffin, Soil Chemist of the Old South," *J. Chem. Educ.* **29,** 407-414 (1952); Emil Truog, "Putting Soil Science to Work," *J. Amer. Soc. Agronomy* **30,** 973-985 (1938).

AARON J. IHDE

Benjamin Rush

1745-1813

Rush, the outstanding early American chemist, was born Dec. 24, 1745 near Philadelphia. After graduating from Princeton in 1760, he studied medicine for 5 years in Philadelphia where he became friendly with John Morgan, M.D., founder of the University of Pennsylvania's Medical School and professor of medicine and teacher of chemistry there.

In 1766 Rush sailed to Scotland and attended University of Edinburgh Medical School. He paid close attention to chemistry, taught by Joseph Black. After receiving his M.D. degree in 1768 he visited France, where he met a number of the leading French chemists, and spent some time in England, learning about the empirical industrial chemistry of his time by visiting factories. Rush returned to Philadelphia in 1769 and, through the influence of Morgan, was elected professor of chemistry in University of Pennsylvania Medical School.

At that time medical school sessions lasted about 4 months of the year. Rush lectured almost every day, except Sunday. His lectures were based on those he had heard from Black. When Rush began teaching, there were no American textbooks on chemistry and few European works suitable for instruction. In 1770 he wrote "Syllabus of a course of lectures on chemistry" for his students. This was the first chemistry text published in this country. A few years later he issued another edition and in 1783 a third. These were small pamphlets, and are now extremely rare, only four copies being known. He showed his students a few, simple lecture demonstrations. There was no laboratory instruction; this method of teaching chemistry would not be started until many years later. Rush also taught his students elementary pharmaceutical chemistry; physicians of those days needed this knowledge to prepare their own medicines.

In 1773 Rush carried out one of his few chemical investigations, a crude qualitative and quantitative analysis of local spring waters. After giving a talk on the subject before the American Philosophical Society, he published his results in a pamphlet, "Experiments and observations on the mineral waters of Philadelphia, Abington, and Bristol, in the province of Pennsylvania" (1773).

Rush was one of the earliest persons to deliver courses of public lectures. Public lectures were ordinary events for more than a century in cities and villages of America. Delivered in all fields of learning by speakers ranging from prominent educators to charlatans, they provided an opportunity for citizens to broaden their education at a reasonable cost. In the winter of 1774-75 Rush advertised a series of evening lectures on chemistry. Tickets, costing a guinea, were sold at the London Coffee House. According to his ads, he lectured "on such parts of chemistry as abound with the greatest variety of the most useful and entertaining facts and experiments." This series of lectures may have been the first attempt in the colonies to provide an opportunity for private citizens to obtain a knowledge of chemistry.

The Medical School closed in 1774 because of the Revolution. For a time Rush was a physician in the Army. He sat in the Continental Congress and signed the Declaration of Independence. He knew the leaders of the nation. Apparently he violently disliked George Washington, but later in life he reversed his opinion.

At the outbreak of the Revolution there was a shortage of gunpowder and of its major ingredient, saltpetre. Rush wrote newspaper articles giving directions for preparing saltpetre from tobacco stalks and reprinting from the works of Cramer and Glauber old methods of obtaining saltpetre from earth. Rush's article later formed the main essay in a pamphlet, "Several methods of making salt-petre; recommended to the inhabitants of the United Colonies, by their representatives in Congress" (1775). Rush served on a committee to superintend the manufacture of saltpetre. This committee discussed its work in a pamphlet, "The process for extracting and refining salt-petre, according to the method practiced at the provincial works in Philadelphia."

The Medical School opened again in 1778, and Rush resumed his lectures in 1780. In

1787 Rush also designed a course for girls attending the Young Ladies Academy of Philadelphia. This was entirely different from his course at the university. He mentioned some of the important facts and theories of chemistry and physics, but his emphasis was on the chemical and physical basis of cooking and of household crafts such as dyeing and bleaching. This was the first course in the world, to the best of the writer's knowledge, on household chemistry and science, composed for and delivered to a class of girls. Rush published an outline of the course in a small pamphlet, "Syllabus of lectures, containing the application of the principles of natural philosophy, and chemistry, to domestic and culinary purposes."

Rush lectured on chemistry until 1789 and then moved to a professorship of medicine, which he held for the remainder of his life. But he still influenced chemical education in the medical school. He was largely responsible for having Joseph Priestley elected to the professorship of chemistry in 1794. After Priestley declined, Rush sponsored James Woodhouse, who became an excellent chemist. Upon the death of Woodhouse in 1809, he sponsored John Redman Coxe.

Rush's last appearance in chemistry was his honorary membership in the Columbian Chemical Society, shortly before his death on Apr. 19, 1813.

Many articles and books have been written about Rush, including biographies by Wyndham Miles in *Chymia* **4,** 37-77 (1953), and in Eduard Farber, ed., "Great Chemists," Interscience Publishers, 303-14 (1961).

WYNDHAM D. MILES

S

Charles Robert Sanger

1860-1912

Sanger, son of George Partridge and Elizabeth Sherburne (Thompson) Sanger, was born in Boston, Aug. 31, 1860. Entering Harvard in 1877, he showed an enthusiasm for chemistry and class activities (he was elected permanent class secretary). After his graduation in 1881, he remained for a year studying with Henry Barker Hill. In 1882-83 he studied at Munich and at Bonn under Richard Anschütz. Returning to Harvard, he continued his studies with Hill and received his Ph.D. degree in 1884.

He served as assistant at Harvard until 1886 when he was appointed professor of chemistry at the United States Naval Academy in Annapolis. In 1892 he became Eliot Professor of Chemistry at Washington University, St. Louis. In 1899 Hill was forced, because of failing health, to give up his analytical chemistry course. After a careful search to find an able successor, Sanger was appointed assistant professor at Harvard. Upon Hill's death in 1903, Sanger was appointed director of the chemical laboratory and promoted to professor. He also initiated a course in industrial chemistry although his other duties forced him to give it up. He remained active until his death in Cambridge on Feb. 25, 1912.

Sanger's early work with Hill was on the chemistry of pyromucic acid and its derivatives. Shortly after this study was completed Hill asked him to investigate the strange coincidence of a large number of obscure poisonings in the family of one of Hill's colleagues. Sanger ruled out carbon monoxide poisoning from a faulty furnace and observed that the wall paper was heavily coated with arsenic. Upon removal of the wall paper, the family's symptoms disappeared. This incident formed the basis for Sanger's most notable studies in arsenic poisoning. He modified the existing methods for quantitative testing arsenic by improving the Berzelius-Marsh test (in which arsenic is detected by a mirror on a capillary tube). Using this method, he studied the amount of arsenic in the excreta of persons living in different surroundings and found a definite correlation with the amount of arsenic in wallpaper. Significantly, even arsenic hidden beneath a coat of paint was absorbed by humans. Later he improved the Gutzeit method of arsenic detection—an even more sensitive test.

During his lifetime he published only 13 chemical papers partly because of his excess caution. He would not publish work done by his students until he repeated their work several times. His most important early paper was "The Quantitative Determination of Arsenic by the Berzelius-Marsh Process . . .," *American Chemical Journal,* **13,** 431 (1891). On Dec. 21, 1886, he married Almira Starkweather Horswell, with whom he had three children. She died in 1905; and on May 2, 1910, he married Eleanor Whitney Davis, who survived him.

Biographies by C. L. Jackson in *Proc. Amer. Acad. Arts Sci.* **48,** 813-822 (1912-13), and by Arthur B. Lamb in "Dictionary of American Biography," **16,** 350, Chas. Scribner's Sons (1935).

SHELDON J. KOPPERL

George Jackman Sargent

1885-1965

Sargent was born in Concord, N. H., Apr. 30, 1885. He studied at University of New Hampshire and received his B.S. degree there in 1909. His work with Charles L. Parsons on some organic compounds of beryllium was published in a joint paper [*J. Amer. Chem. Soc.* **31,** 1202 (1909)].

Sargent then studied at Cornell University from 1909 to 1914. His first work with Wilder D. Bancroft on the decomposition of bromoform was published in *Journal of Physical Chemistry* **16,** 407 (1912). In the summer of 1911 he began the work at Cornell on electrolytic chromium production for which he is noted and in 1912 was awarded his Ph.D. degree. He continued his investigation of chromium deposition for 2 years longer with the aid of a fellowship made possible by Hector Carveth, an earlier worker on chromium in the same laboratory and later president of Roessler and Hasslacher Chemical Co. The results were in some ways disappointing and were not published until 1920 [*Trans. Amer. Electrochem. Soc.* **37,** 479 (1920)].

Sargent narrowly missed the basic principles of the electrodeposition of chromium and ascribed the deposition of chromium metal to the trivalent chromium added to his chromic acid baths rather than to the sulfate or chloride ions accompanying the trivalent chromium additions. The principle of adding catalyst anions to chromic acid solutions to make chromium deposition possible was discovered by Colin C. Fink and Charles H. Eldridge in 1924. The distinction was so narrow that numerous workers later claimed to be using "Sargent's solution" in the hope of avoiding infringement of the Fink patents.

Upon leaving Cornell Sargent worked for Michigan Smelting and Refining Co. of Detroit, 1914-17; Connecticut Metal and Chemical Co., New Britain, 1917-20; and Dodge Brothers Motor Car Co., Detroit, 1920-21. For 2 years, 1921-23, Sargent did private research work in his own laboratory in Detroit on the utilization of nonferrous metal waste.

Returning to Concord, N. H., 1923-26, Sargent settled his mother's estate and worked part time for Cooks Lumber Co. of Laconia. In July 1926 he moved to Pittsfield, N. H. and married Lena Snow. In Pittsfield Sargent designed and built his own laboratory, a long-held desire. He was aided in this by being retained as a consultant from 1927 to 1937 by United Chromium Inc. in connection with their patent suits. His first purpose in his Pittsfield Laboratory was to work out a process for leaching zinc out of brass. This was not entirely successful, and the laboratory was later largely supported by pure electrolytic lead production until he retired in 1958.

Sargent was also connected with the Pittsfield National Bank for much of this period, becoming a director in 1945, and vice president for 9 years after that. The Sargents moved to Santa Barbara, Calif. in 1958, where he died June 13, 1965, survived by his wife, a sister, Mrs. Beulah S. Woods of Concord, and three nephews.

G. Dubpernell, *Plating* **47,** 35, 53 (1960); *Plating* **59,** 638-643 (1972).

GEORGE DUBPERNELL

Antoine Francois Saugrain de Vigni

1763-1820

Antoine Saugrain was born Feb. 17, 1763 in Paris (some sources say Versailles), the son of Antoine Claude and Marie (Brunet) Saugrain. His ancestors were booksellers and publishers. His uncle served as librarian for the Count d'Artois and saved the library from the revolutionaries at the fall of the Bastille. Saugrain's sister married Dr. Joseph Guillotin.

Saugrain was educated in chemistry, physics, and mineralogy and in 1784-85 and again in 1786 went on scientific expeditions to Mexico to study mines and minerals for the King of Spain.

During the autumn of 1787 Saugrain sailed for the United States with Picque, a botanist, and Raquet to explore the Ohio River. While in Philadelphia he visited Benjamin Franklin, and Jefferson gave him a letter of recommendation to General George Rogers Clark. He spent the winter of 1787-88 close to Fort Pitt, living in a cabin where he set up a crude laboratory and assayed ores from pits and mines in the neighborhood. In 1788 with an American companion the French group sailed down the Ohio. Opposite Big Miami they were attacked by Indians, Picque and Raquet were killed, and

Saugrain was captured. He escaped to Louisville where he spent some time recuperating from wounds and frostbite. While at Louisville he made a furnace and furnished fixed alkalies to local physicians.

After a short visit to France in 1790, Saugrain returned to the United States with a group of royalists to establish a settlement at Galliopolis, Gallia County, Ohio. Saugrain served as physician to the colony and was actively engaged in chemical research. He prepared and sold ink, barometers, thermometers, and glass tubes filled with phosphorus, which ignited when the tubes were broken. In 1793 he married Genevieve Rosalie Michau and in 1796 moved to Lexington to help establish a bar iron factory.

In 1800 Saugrain, his family, and in-laws moved to St. Louis, where he practiced medicine and served as the city's first postmaster. He depended primarily on vegetable compounds for treatment, many grown in his herb garden, and considered calomel a virulent poison. He continued to manufacture scientific instruments and to produce friction matches, the first in the United States. He supplied thermometers and matches to the Lewis and Clark expedition. He constructed an electric battery and startled the Indians with his experiments.

Saugrain introduced smallpox inoculation to St. Louis, offering the virus free to indigents and Indians. He served as Army Surgeon at Fort Bellefontaine on the Missouri River above St. Louis.

Saugrain died during the night of May 19-20, 1820 survived by his wife and six children. In 1928 the American Chemical Society, meeting in St. Louis, honored Saugrain as a pioneering American scientist.

"Dictionary of American Biography," **16,** 377, Chas. Scribner's Sons (1935); Henry Marie Brackenridge, "Recollections of Persons and Places in the West," J. B. Lippincott & Co. (1868); Edmond Meany, *Wash. Hist. Quart.* **22,** 3 (1931); N. P. Dandridge, *Ohio Arch. & Hist. Quart.* **15,** 192 (1906); W. V. Byars, "A Memoir of the Life and Work of Doctor Antoine Francois Saugrain," B. Van Phul (190-); Frederic L. Billon, "Annals of St. Louis in the Early Days under French & Spanish Dominations," G. I. Jones & Co. (1886); archives of Missouri Historical Society.

JANE A. MILLER

John Ahlum Schaeffer

1886-1941

Schaeffer, born in Kutztown, Pa., May 31, 1886, graduated in 1904 from Franklin and Marshall College, to which he returned in 1935 to become the college's sixth president.

Described as a happy combination of scientist, business executive, and scholar, he earned his A.M. degree from Franklin and Marshall in 1905 and his Ph.D. degree in chemistry from University of Pennsylvania in 1908. For 3 years he taught chemistry at Carnegie Institute of Technology and in 1911 was named chief chemist for Eagle-Picher Lead Co., Joplin, Mo. In 1921, he became the company's vice-president in charge of manufacture and 10 years later was appointed vice-president in charge of research. He was credited with a number of paint and ceramics inventions.

Schaeffer received an honorary degree of Doctor of Science from Franklin and Marshall in 1929. Somewhat reluctantly, he moved to the college campus 7 years later to assume the presidency of the college. Having made his decision, however, he moved with enthusiasm and devotion to build the institution in student enrollment, faculty membership, and endowment. He led the college in its celebration of its 150th anniversary in 1937. He instituted a scholarship program which assisted some 200 students and introduced comprehensive examinations and other measures aimed at improving scholarship at the institution.

During his presidency, two new buildings were erected: the Keiper Liberal Arts Building and the Fackenthal Library. His success was evidenced by the increases of the student body from 675 in 1935 to 983 in 1941, the teaching faculty from 38 to 48 in the same period, and the endowment from $968,000 to $1,334,000.

He received honorary LL.D.'s from Dickinson College (1937), Muhlenberg College (1938), and University of Pennsylvania (1939).

He collaborated in writing three books: "Analysis of Paints and Painting Materials" (with Henry A. Gardner in 1908); "Experiments in Chemistry for Engineering Students" (with Joseph H. James) in 1910; and "The Chemical Analysis of Lead and Its Components" (with Bernard S. White and John H. Calbeck) in 1912. He also contributed numerous articles on industrial and engineering chemistry to journals and trade magazines.

Schaeffer died suddenly at his home April 6, 1941 from a cerebral hemorrhage at the age of 54. He was the first president of the college to die while in office.

Archives, Fackenthal Library, Franklin and Marshall College.

BRUCE G. HOLRAN

William Jay Schieffelin

1866-1955

Schieffelin was born in New York City, Apr. 14, 1866. He graduated from Columbia School of Mines with his Ph.B. degree in 1887 and from University of Munich with his Ph.D. degree in 1889.

When Schieffelin was a young man his grandfather, John Jay, grandson of the first Chief Justice of the Supreme Court of the United States, used to say to him, "lend a hand." This was quoted to the writer, his eldest son, as a guiding stimulus to the patriotic, public, and charitable services to which he devoted the greater part of his life in addition to his contributions to the chemical, drug, and allied industries.

Schieffelin commenced his business life as chemist in the analytical laboratory of Schieffelin & Co. in 1889, assaying opium and coca leaves and standardizing nitrous ether. Later he assisted in developing products marketed by the firm. He was vice president from 1903 to 1906, president from 1906 to 1923, and chairman of the board from 1923 to 1929.

In the early 1900's he strongly supported Harvey Wiley in obtaining passage of the Pure Food and Drugs Act. As a result of his efforts, when that law first went into operation and industry guarantees were required, Schieffelin & Co. was awarded Guaranty No. 1.

In the late 1880's, Schieffelin was a member of Company K, 7th Regiment, New York National Guard. He used to tell his children that as the company marched west on 66th Street from the 7th Regiment Armory for maneuvers in Central Park, they passed the brownstone home of General Ulysses S. Grant. As Company K passed the windows they could see the old general, cigar in mouth, hands in pockets, inspecting them as they marched by.

In 1898 Schieffelin volunteered for war service and became first lieutenant and adjutant of the 12th N.Y. Infantry. General Peter Haines asked him to become his aide, in which capacity he accompanied the general in the brief Puerto Rican campaign of the Spanish American War. He participated in the attack which drove the Spanish troops into the hills shortly before the surrender and was breveted captain before being mustered out. Active all his life in black education, in World War I he was colonel of the Replacement Regiment of the 15th N.Y. Infantry in New York City.

From his earliest active life he participated in activities of many organizations. He was Civil Service Commissioner of New York City in 1896, a member of the Committee of Seventy in New York City, and chairman of the Citizens' Union for many decades, helping in the elections of mayors Seth Low, John Purroy Mitchell, and Fiorello La Guardia, with all of whom he was a good friend. In politics his efforts were constantly directed against Tammany Hall. The writer's earliest recollection as a small boy was the campaign slogan of Schieffelin, "Vote for Low and keep the grafters out."

He was president of the National Wholesale Druggists Association in 1910; president of The Chemists' Club in 1906; chairman of the Board of the N.Y. College of Pharmacy; vice president, American Bible Society; trustee, Hampton Institute; trustee, Tuskegee Institute; director, Maine Sea Coast Mission; and president, Huguenot Society. He was president of American Mission to Lepers; and the medical laboratory of the mission in Karigiri, South India, was named in his honor, "Wm. Jay Schieffelin, Ph.D., Leprosy Research Sanatorium."

Schieffelin died in New York City, Apr. 29, 1955.

Personal recollections; private papers of William Jay Schieffelin; "Who's Who in America," 1948-49; yearbooks of the National Wholesale Druggists' Association; obituary, *New York Times,* May 1, 1955.

WILLIAM J. SCHIEFFELIN, JR.

Hermann Irving Schlesinger

1882-1960

Schlesinger was born in Minneapolis, Oct. 11, 1882, son of Louis and Emily (Stern) Schlesinger. He entered University of Chicago as a freshman Oct. 1, 1900. Two and one-half years later, in March 1903, he was awarded his B.S. degree. In Aug. 1905, he received his Ph.D. degree. His research program for the doctoral degree was carried out under the supervision of Julius Stieglitz, who, through his kindly advice and interest was probably more influential than any other person in moulding Schlesinger's career in chemistry. His thesis was in the field of physical-organic chemistry: "Reaction Rates of Hydrolysis of Imido Esters."

At the time Schlesinger was a student at University of Chicago, he came under the tutelage of a number of stimulating teachers beside Stieglitz, among them Alexander Smith, John Ulric Nef, and Herbert McCoy in chemistry and Robert Millikan and Albert Michelson in physics. All of these men influenced Schlesinger's development as a teacher and scientific investigator.

After completing his doctoral work Schlesinger decided to spend a year or so abroad. He entered Walther Nernst's Laboratory in Berlin in October 1905 and remained there until the following August. In September 1906, Schlesinger moved to Johannes Thiele's Laboratory at Strassburg and remained there for about 5 months working on a problem in organic chemistry—the diazotation of dichlorostilbene. In February 1907, he accepted a position as research assistant to John Abel at Johns Hopkins Medical School. Schlesinger spent nearly a year there working on the extraction of the toxin of *amanita phalloides.* After some 6 months at the hospital he accepted an appointment as associate in chemistry at University of Chicago at an annual salary of $800.

His initial teaching position at Chicago was limited to general chemistry, qualitative analysis, and inorganic preparations. Since he was expected to teach courses which for the most part were inorganic, he shifted his research interests to inorganic chemistry in spite of the fact he had no special training in that field. Schlesinger, however, began his research program on the conductance of formates in anhydrous formic acid as a solvent, a project in physical chemistry. In 1910 he was appointed instructor in the Department of Chemistry, and it was this year he also was married to Edna Simpson. A year later he was promoted to the rank of assistant professor and in 1917 to associate professor. Five years later, in 1922, he was promoted to the rank of professor, and at this time he also accepted the task of secretary of the department.

For the next 23 years, until late 1945, Schlesinger was heavily involved in administrative work in the department. Following John U. Nef's death in 1915, Stieglitz took over the chairmanship of the department. In 1933 Stieglitz reached retirement age and although he continued to teach on a part-time basis until his death in January 1937, the administrative duties fell to Schlesinger who almost single-handed carried the heavy burden placed upon him until December 1945. This period of about 13 years probably was the most critical in the history of the department and, in turn, the most difficult and arduous for the chief administrative officer. The depression years and World War II created severe administrative problems, which Schlesinger met with painstaking planning. It was during these years that he initiated and developed an extensive research program on boron hydrides and borohydrides, which not only turned out to be his most important scientific contribution to the field of chemistry but also introduced an entirely new chemistry of wide influence and impact. His research with these compounds found applications in industry in the production of jet and rocket fuels, the synthesis of organic compounds, other inorganic hydrides, and compounds of biological interest.

Schlesinger received two of the top honors in American chemistry. In 1959 he received the American Chemical Society's Priestley Medal and the Willard Gibbs Medal awarded by the Society's Chicago Section. Also in 1959, he was awarded the U. S. Navy's top civilian medal, the Distinguished Public Service Award. He was the recipient of honorary degrees from Bradley University (1950) and from University of Chicago (1954). He was a member of the National Academy of Sciences and of the Bavarian Academy of Sciences. He was the first recipient of the Alfred Stock Memorial Prize awarded by the German Chemical Society.

Schlesinger was the author of a textbook in general chemistry, widely used over a period of 30 years, a laboratory manual, and more than 60 scientific papers. He died in Billings Hospital, University of Chicago, Oct. 3, 1960, after a short illness.

Records of University of Chicago; Warren C. Johnson, "Honor Scroll Award of Chicago Chapter of American Institute of Chemists to Schlesinger," *The Chemist,* November 1951; personal recollections; obituary in *New York Times,* Oct. 14, 1960.

WARREN C. JOHNSON

Rudolf Schoenheimer

1898-1941

Rudolf Schoenheimer was born May 10, 1898 in Berlin, Germany. After receiving his medical degree in 1922 from University of Berlin and serving as a resident pathologist for 1 year, he went to University of Leipzig as a research fellow under the sponsorship of Rockefeller Foundation. There he studied the role of cholesterol in arteriosclerosis. He continued to work on the metabolism of cholesterol at University of Freiburg from 1926 to 1933 and spent 1 year as a fellow at University of Chicago, 1930-31. Schoenheimer returned to the United States in 1933 and was appointed assistant professor of biochemistry at Columbia University. He remained at Columbia until his death as a result of poisoning on Sept. 11, 1941. The finding of the medical examiner was that he had committed suicide while suffering from mental depression.

Schoenheimer spoke of the significance of his work to biochemistry in his Harvey Lecture of 1937 and his Dunham Lectures of 1941. Chiefly as a result of his investigations on the metabolism of fats and proteins, researchers became convinced that there exists a constant turnover of these substances in mammalian tissues.

In 1934 Schoenheimer collaborated with David Rittenberg, a biochemist formerly associated with Harold C. Urey, in developing a technique for using compounds labeled with deuterium to study the intermediary metabolism of fatty acids in rats. In a series of papers in subsequent years Schoenheimer demonstrated the rapid turnover of stored fat in mammalian systems and established many of the chemical transformations which fatty acids undergo, such as reversible saturation and desaturation.

In 1938 Schoenheimer turned his attention to proteins. He administered amino acids labeled with ^{15}N and showed that proteins were also being constantly synthesized and degraded. Moreover he established that many transformations of amino acids were possible in mammalian systems.

Schoenheimer was a pioneer in the use of isotopic tracers in biochemical research in the years before radioactive isotopes were readily available. The development of this new experimental approach to biochemistry was his most significant contribution.

Rudolf Schoenheimer's papers can be found in *J. Biol. Chem.;* obituaries: Hans T. Clarke, *Science* **94**, 553-554 (1941); J. H. Quastel, *Nature* **149**, 15-16 (1942).

DANIEL P. JONES

Oswald Schreiner

1875-1965

Schreiner was born in Nassau, Germany, May 29, 1875. After immigrating to the United States with his parents in 1883 he attended Baltimore Polytechnic Institute, graduating in 1892. Two years later he received his Ph.G. degree from Maryland College of Pharmacy as an honor student who had won three gold medals in chemistry. In 1895 he pursued graduate studies at Johns Hopkins University and then transferred to University of Wisconsin where he received his B.S. degree in 1897, M.S. in 1899, and the Ph.D. in chemistry in 1902. While a student at Wisconsin he won the Ebert Prize for chemical investigations of sesquiterpene hydrocarbons and visited research laboratories in Europe. He served as a United States Pharmacopoeia fellow, 1895-96, assistant in pharmaceutical technique, 1896, instructor, 1897-1902, and instructor in physical chemistry, 1902-03. In 1902 Schreiner married and became a U.S. citizen.

Schreiner entered the Department of Agriculture in 1902 as an expert in physical chemistry and soon found his specialty in soil fertility work. He advanced to the rank of chemist in 1903, soil scientist in 1904, and chief of the Division of Soil Fertility Investigations in 1906. During these years he conducted numerous experiments on soil fertility, examining toxic plant secretions, oxidation, and minor soil elements. Many of his studies were published as department bulletins. Among his accomplishments was the development of the Schreiner colorimeter to determine water-soluble plant food constituents. In 1908 he made another tour of European laboratories and in 1912 was awarded the Longstreth medal for important research in agricultural chemistry.

During World War I Schreiner's laboratory worked on projects ranging from chemical warfare to the preparation of chemicals for medical use. His most important agricultural work during this period was on fertilizers, notably the use of byproducts of the Ammonium Nitrate Plant at Perryville, Md. and the development of special war fertilizers for sugar cane.

Between 1913 and 1921 the field work of the Division of Soil Fertility Investigations under Schreiner's charge grew from a single cooperative field station to 33 stations in different parts of the country with active fertilizer experiments on such diverse crops as cotton, corn, sorghum, and citrus fruits. These laboratories discovered over 50 new organic soil compounds. With Joshua J. Skinner, Schreiner established the triangular system for fertilizer experiments that was widely adopted by state experiment stations. He wrote some 90 articles on his investigations before 1920 and lectured before many universities and farm organizations.

In the 1920's and 30's Schreiner took part in many professional conferences. In 1921 he was a delegate to the Fourth Pacific Congress, and in 1921 and 1932 he attended the International Congress of Sugar Cane Technologists. In 1927 he served as chairman of the American Organizing Committee at the First International Soil Science Congress. While attending the Third International Soil Science Conference in 1935 he made a soil tour of Europe.

Schreiner's career in the Department of Agriculture assumed a more administrative character after his division was transferred to the Bureau of Plant Industry in 1915 and he was promoted to senior biochemist. In 1927 he was made principal chemist in the Bureau of Chemistry and Soils, returning to Plant Industry in 1935. His work at this time included acting as representative of USDA in all fertilizer grade reduction work at the request of the Secretary of Agriculture. In 1940 he was promoted to assistant to the chief of the Bureau of Plant Industry. After retiring in 1944 he remained an unpaid collaborator with USDA until 1964. Schreiner died in Fort Lauderdale, Fla., June 3, 1965.

Personnel File, Federal Records Center, St. Louis; A. S. Alexander, "The Inquiring Mind and the Seeing Eye," *Better Crops and Plant Food* **20,** 20-22, 41-45 (September-October 1935); obituary, *Washington Post,* June 4, 1965.

DOUGLAS E. BOWERS

Walter Cecil Schumb

1892-1967

Schumb was born in Boston, Mass., Sept. 10, 1892. He attended high school in Boston, receiving several prizes, and attended Harvard College. In 1914 he earned his A.B. degree and became a Sheldon Traveling

Fellow at Magdalene College, Oxford from 1914-15 where he studied with William H. Perkin, Jr. He returned to Harvard and obtained his A.M. degree in 1916 and his Ph.D. degree in 1918. At Harvard he worked with Grinnell Jones and Theodore W. Richards.

In 1918 Schumb served as an assistant gas chemist for the Chemical Warfare Service in Washington, D. C. He went to Vassar College in 1919 as an assistant professor and then to Massachusetts Institute of Technology in 1920. Beginning there as an assistant professor, he rose to associate professor in 1926 and to professor in 1934. In 1930 he was appointed director of the research laboratory of inorganic chemistry, which he had established. During World War II he was associated with the High Voltage Engineering Laboratory at MIT and received a Navy award for his work there. He also worked for the OSRD and ONR. After his retirement in 1959 he continued to work at MIT for Von Hippel's insulation research laboratory, one of the earliest material research centers. In addition to contributing syntheses, he served "Inorganic Syntheses" as an associate editor for volumes II to V (1946-57) and as a member of the advisory board for volumes VI to IX (1960-67). He died June 15, 1967, 2 months before his 75th birthday.

In his research Schumb showed a variety of interests, and although a majority of his papers dealt with silicon or fluorine chemistry, he cannot be said to have developed a "school" of students all working on some aspect of a common problem. I do not believe that he approved of this type of graduate training. He was very precise in manner as well as work, and a great deal of his research involved careful determinations of physico-chemical properties (refractive index, solubility, dissociation pressure, electrode potential, etc.). He was a contributing editor to the section of the "International Critical Tables" on the densities of aqueous solutions.

Some of Schumb's researches were carried out in collaboration with his MIT colleagues. In fact, the first contribution of the research laboratory for inorganic chemistry was a paper on the use of sodium peroxide–carbon fusions for the decomposition of refractories, written with George Marvin, an analytical chemist. He studied the partial hydrolysis and ammonolysis of silicon halides, the preparation of fluorine, and the reaction of fluorine with nitrides and carbides. During the war he investigated the possibility of using chlorine trifluoride and bromine trifluoride as incendiaries. It was also in connection with the war that he began a very prolonged study of concentrated hydrogen peroxide which resulted in his only book, "Hydrogen Peroxide" (1955), written with Charles Satterfield of the Chemical Engineering Department and Ralph Wentworth. He was the first chemist to investigate the use of electrodeless discharge to bring about chemical reactions and the synthesis of unusual compounds; he also studied the chemical effects of ultrasonic irradiation. In spite of the applied nature of some of his work, Schumb's primary interest was in pure science and basic research. His work was carried out in spite of his health, which was not robust (he suffered from a curvature of the spine).

A quiet, reserved man, Schumb was a dry lecturer, but his lectures were always orderly and carefully planned. His reserve extended even to close colleagues and his research students, whom he always addressed formally by title and last name. The post-war first name familiarity between teacher and student never established itself in Schumb's group. On the other hand, he did not continually press his students nor berate them. He was always available for consultation and encouraged his students to try out their own ideas.

Obituary in *Technology Review* **75**, No. 1, 69 (October/November 1967); personal recollections, and letters from L. F. Hamilton, R. R. Young, and E. L. Gamble.

RUSSELL F. TRIMBLE

Wilfred Welday Scott

1876-1932

Scott was born in Zanesville, Ohio, Aug. 13, 1876. He attended Ohio Wesleyan University where he obtained his B.A. degree in 1897 and his M.A. in 1901. He taught at Philander Smith College from 1898 to 1901 and at Morningside College from 1905 to 1909. He was chief chemist at Baldwin Locomotive Works from 1909 to 1910 and chemist for General Chemical Co., 1910 to 1921. He was professor of chemistry at Colorado School of Mines from 1921 to 1925. This institution bestowed on him an Sc.D. degree in 1923. He became head of the Chemistry Department at University of Southern California in 1925 and spent the rest of his career in this position until his death on May 3, 1932 in Los Angeles.

Scott's specialty was inorganic qualitative and quantitative analysis. He wrote seven texts and manuals on these subjects, and their contents demonstrate that he was interested in the aspiring student analyst as well as the practicing analytical chemist.

One of them, "Standard Methods of Chemical Analysis; a Manual of Analytical Methods and General Reference for the Analytical Chemist and Advanced Student," written and edited in collaboration with other specialists, is a classic in its field. The first edition appeared in 1917, and the book was in its fourth edition at the time of his death. It continued in print under succeeding editors and is now (1975) in a multivolume sixth edition.

Biography by Frank C. Touton in *Science* **76,** 93 (1932); obituaries in *San Francisco Chronicle,* May 4, 1932, and *The Daily Trojan,* May 4, 1932.

MEL GORMAN

Melville Amasa Scovell

1855-1912

Scovell, chemist by training, devoted nearly half his life to building and directing the Agricultural Experiment Station of the State University of Kentucky at Lexington. Scovell was born to Nathan and Hannah (Allen) Scovell Feb. 26, 1855 in Belvidere, N.J. in the seventh generation of an old American family of more than usual mobility; his father and mother were born in midstate and eastern New York, respectively; his family moved to Chicago, then to Champaign, Ill. when he was still quite young.

Scovell was educated in the local schools, and in 1871 entered University of Illinois (then called Illinois Industrial University), from which he received his B.S. degree in 1875 and M.S. degree in 1877. Thirty-one years later Illinois would grant him the Ph.D. for work done at the Experiment Station in Kentucky. He became an instructor in chemistry at Illinois in 1875, and in 1876-79 was private secretary to John Milton Gregory, president of the university, from whom he absorbed a clear view of the mission of an industrial university. In 1878 he was appointed instructor of agricultural chemistry, rising to professor in 1880. He married Nancy Davis, daughter of the Hon. Chester P. Davis of Monticello, Ill. in that year.

At Illinois Scovell interested himself in sugar production, patenting with Henry A. Weber, head of the Chemistry Department, a process for clarifying cane and sorghum juices by rapid superheating, and other processes for glucose, sugar, and syrup from sorghum. They established the Kansas Sugar Works at Sterling, Kans. and Scovell served as superintendent during 1883-84. For a year after this, he worked for the federal Department of Agriculture, with responsibility for the diffusion batteries used to extract cane and sorghum sugar in Kansas and Louisiana.

In 1885 when he was 30, Scovell was called to found and direct the Agricultural Experiment Station of the State University of Kentucky. The challenge was formidable: only a tiny budget of federal funds was available, the operation was housed in a basement, a few acres of university land were provided for experiment, and Scovell was the only strictly full-time staff member. The suspicion of working farmers for "book farmers" at the time is difficult to imagine today, and this was reflected in the attitude of the legislature, which provided no direct appropriations for the Experiment Station in its early years. Fortunately, Scovell had the solid backing of the university's administration and trustees. In 1886 state legislation regulating fertilizers was passed, with fees to be paid to the station for its services in testing. This would be the method of financing the station until shortly before Scovell's death, when he finally persuaded the legislature to provide a yearly appropriation.

Other testing services were added: food, feedstuffs, dairy products; and despite stringent budgetary restrictions the station grew, moving in 1889 and again in 1905 to its own building. By careful budgeting Scovell made research funds available from the beginning, and over the 27 years of his directorship the station turned out 168 bulletins covering all areas of agriculture—60 bore his name. The early involvement of the station with fertilizer analysis led Scovell to modify the standard Kjeldahl determination, by addition of zinc dust, to handle nitrate nitrogen; this was the work which he later submitted for his Ph.D. degree. He invented a sampling device for bulk milk and worked with the state and federal legislature for pure food and drug laws.

Scovell's outside activities were prodigious and reflected both his concern for civic affairs and his conviction of the value of professional organizations. He held office in the Association of Official Agricultural Chemists and the Association of American Agricultural Colleges and Experiment Stations. He was instrumental in starting a state fair, for agricultural purposes, in Kentucky. He was a member of Lexington's Park Commission and a trustee of the newly formed local telephone company. In 1910 Scovell was appointed dean of the College of Agriculture, and even in the short time remaining to him he left his mark by setting up a strong and efficient organization for the school.

Scovell died of heart disease Aug. 15, 1912, only 57 years old. He left no children.

23d Annual Report, Ky. Agr. Expt. Sta., 1-36 (1912); other sources add little, but a few have interestingly different views: *Mechanical and Electrical Engineering Record*, 4-10 (Nov. 1908); *Expt. Sta. Record, U.S. Dept. of Agr.* **27,** 401-405 (1912); *Breeder's Gazette,* 316-17 (Aug. 21, 1912); *see also* "Dictionary of American Biography," **16,** 512, Chas. Scribner's Sons (1935); H. Garman, *Proc. Soc. for Promotion of Ag. Sci.* **33,** 114-18 (1913); *Jersey Bulletin,* 1368 (Aug. 21, 1912); and H. W. Brainard, "A Survey of the Scovils or Scovills," 269-70, 349-50, 439-41, Hartford (1915).

ROBERT M. HAWTHORNE JR.

Atherton Seidell

1878-1961

Seidell was born Dec. 31, 1878, in Hartwell, Ga., the son of Charles and Emma (Roebuck) Seidell. He received his education in Atlanta public schools and gained his B.S. from University of Georgia in 1899, his masters degree from George Washington University in 1901, and his Ph.D. from Johns Hopkins University in 1903, the same year that he married Martha Adele Hooper of Baltimore.

Seidell was employed in the Bureau of Soils, Department of Agriculture in 1900. He transferred to the Bureau of Chemistry in 1905 and joined the U.S. Public Health Service in 1907, where he worked on thyroid studies with Reid Hunt. Also in 1907, he published the first volume of "Solubilities of Organic Compounds" (later, also "Inorganic"), a work which he updated every 10 years. In 1949 he gave this publication with its rights to the American Chemical Society which continued publication. His own words tell why he did this work; "Many years ago, as a struggling young chemist, I was given the problem of determining the solubility of calcium sulfate in aqueous solutions of sodium chloride. After completing the experiments I found that the results for the system had previously been published and my several weeks work had been wasted."

During World War I, Seidell was again with Reid Hunt, then professor of pharmacology at Harvard University Medical School, where they worked on studies pertinent to the war. After 1918 he devoted all of his research to the study of vitamins. In 1922 he went on inactive duty from the U.S. Public Health Service in order to pursue these studies in France. His experiments were conducted in private laboratories there until 1928 when he was given his own laboratory at L'Institut Pasteur. Here he worked until war came again. He was a pioneer investigator of the chemistry of vitamins, particularly of compounds from brewer's yeast. His concentration of the antineuritic substance was one of the major discoveries relating to vitamins. He closed his laboratory in 1939 and did not participate again in research. Fluent in French, he often assisted officers of the U.S. Public Health Service on their many tours of inspection in France.

His research now ended, he devoted his energies to furthering his ideas for disseminating information from scientific literature. In the late 1930's with his own money he established "Bibliofilm Service" in the Department of Agriculture library and shortly thereafter he did the same, again with his own money, at the Army Medical Library (now the National Library of Medicine) and called it "Medicofilm Service." After the war he established similar microfilm services in important scientific libraries of Paris and in many other libraries around the world. He made microfilm equipment and gave it to the libraries with the understanding that the library would give the microfilmed articles free of charge, thus serving the man at a distance, as well as the one able to come to the library. He very much disagreed with libraries' using such a service as a means of improving the budget. At a time when small viewers for reading microfilm were not obtainable commercially, he made them and dispensed them at less than cost, as he said, "for the advancement of science."

Realizing that such services called for a weekly index, he founded the "Current List of Medical Literature," published at the Army Medical Library; later it provided a basis for a system of classifying medical literature for computers.

Many awards came Seidell's way. Spain and Italy honored him for "promoting international good will in science." France made him a Commander of the Legion of Honor, "in recognition of [his] remarkable achievements in the scientific field." France honored him again in 1946, making him an officer of the Order of Public Health, "a tribute to the understanding which you have shown of the French problems and the aid which you have brought to our country under circumstances that were especially difficult." He was an honorary consultant of the Army Medical Library and received the highest awards of the American Documentation Institute and of the National Microfilm Association.

He died at his home in Washington, D.C., July 25, 1961.

Seidell's records on deposit at Smithsonian Institution, Washington, D.C.; obituaries in *Washington Post,* July 27, 1961, with portrait, and in *Chem. Eng. News* **39,** 91 (Aug. 7, 1961) with portrait; personal recollections.

ELIZABETH E. MEDINGER

Samuel Edward Sheppard

1882-1948

"More than any other single worker, Sheppard has been responsible for our present knowledge of the theory of the photographic process. He explored every section of the chemistry of that process, and everywhere his studies brought light." This was the evaluation of Sheppard's scientific career in 1949 by C. E. Kenneth Mees, himself a notable chemist in the field of photography.

Sheppard was born July 29, 1882, in London. After attending St. Dunstan's College, he went to University College, London, where he carried out research on the photographic process and received his B.Sc. degree in 1903. In college he struck up a lifelong friendship with Mees, with whom he published his first articles. After graduation he outfitted a laboratory in his home and continued research, for which he received his D.Sc. degree in 1906. He and Mees, who also carried out graduate research in photography, published their theses jointly in 1907 under the title, "Investigations on the theory of the photographic process."

In 1906 Sheppard received an 1851 Exhibition Scholarship which enabled him to study for 2 years on the continent, first with Karl Schaum at University of Marburg in photochemistry, and then with Victor Henri at the Sorbonne in colloid chemistry. Returning to England in 1908 he found himself at a loose end but finally joined the Agricultural Chemistry Department at Cambridge and studied the colloidal properties of bread dough.

George Eastman invited Mees to come to the United States to organize and direct the Kodak Research Laboratory at Rochester,

N.Y. In 1913 Mees called Sheppard to take charge of the sections of physical and colloid chemistry at the laboratory. In 1920 he was appointed chief of the Department of Physical, Inorganic, and Analytical Chemistry. In 1921 he was placed in charge of the development of x-ray screens, and in 1922 he organized the manufacturing department of this product. He directed research in 1922 and 1923 during Mees' absence and was appointed assistant director of research the following year. From then on he served as both assistant director of Kodak Research Laboratory and assistant superintendent of the Department of Physical and Organic Research until he retired Jan. 1, 1948. More than most workers in his field he covered all aspects, from theory to practice and from raw materials to the production of finished images.

Sheppard's early studies concerned the physicochemical properties of gelatin. This work was interrupted during World War I when the Submarine Defense Corp. asked Kodak to find a method of utilizing powdered coal (a waste product) as a fuel. Sheppard developed "colloidal fuel," a stable suspension of powdered coal in fuel oil, which was used on a fairly wide scale. The studies of gelatin continued for many years thereafter; one result was a method of manufacturing a standard de-ashed gelatin for use in chemical and biological laboratories.

An important study that Sheppard carried out was the frequency distribution of the various sizes of silver halide grains in an emulsion, and the mechanism of the precipitation of silver halide and its subsequent growth into grains, for the purpose of relating distribution and sensitivity. The results showed that, other things being equal, the larger the grain size the greater the sensitivity or speed.

Sheppard's most significant work was his search for the sensitizing substance in photographic gelatin. A painstaking series of analyses indicated that the sensitizer in natural gelatin was concentrated in the liquors obtained by the acid treatment of the raw material after liming. Eventually Sheppard found that the chemical properties of the sensitizer corresponded to those of allylthiourea and that the gelatin sensitizer was essentially one which could produce silver sulfide specks in silver bromide crystals.

Sheppard also studied the electrical response of silver halide to light, the colloidal structure of film base materials, the nature of dye sensitivity, the absorption of sensitizing dyes to silver halides, the absorption spectra of dyes in various solvents, and other areas of photographic chemistry.

Sheppard published, alone or with coworkers, over 165 scientific papers, nine books, and was granted over 90 patents (U.S. and foreign corresponding patents), dealing mainly with photographic theory, processes and products, and methods of manufacture.

Many honors recognized his achievements; the Royal Photographic Society awarded him the Hurter and Driffield Medal in 1928, as well as their Progress Medal. In 1929 he was awarded the Adelsköld Gold Medal by the Swedish Photographic Society, and in 1930 the New York Section of the American Chemical Society awarded him the William H. Nichols Medal. He was chairman of the Division of Physical and Inorganic Chemistry of the American Chemical Society, president of the Society of Rheology, and an associate editor of *Journal of Rheology*.

Sheppard died in Rochester, N.Y., Sept. 29, 1948.

Obituaries by C. E. K. Mees in *J. Chem. Soc.* Jan. 1949, pp. 261-263 (from which the quote above is taken), and in *J. Soc. Motion Picture Engrs.* **51,** 667-668 (1948) with portrait; also by J. Eggert in *Camera,* **27,** 380-381 (1948) with portrait; biography in *British J. Photogr.* **95,** No. 1581, March 5, 1945; archives of Eastman Kodak Co.

THOMAS T. HILL

Henry Clapp Sherman

1875-1955

Sherman was born in Ash Grove, Va., Oct. 16, 1875. His early education was in a rural, one-room school, complemented by home instruction. He then enrolled at Maryland Agricultural College (now University of Maryland), from which he received his B.S.

degree in 1893. For 2 years he served as an assistant in the laboratory of the State Chemist of Maryland. His work there resulted in a paper on the "Determination of Nitrogen in Fertilizers containing Nitrates." The quantitative nature of this paper set the tone for Sherman's future work.

In 1895 Sherman obtained a fellowship at Columbia University and received his M.S. degree in 1896, followed by his Ph.D. in 1897, with a thesis on "The Insoluble Carbohydrates of Wheat *(Triticum vulgare)*". After receiving his doctorate Sherman worked with Wilbur Atwater on energy metabolism and nutrition, and no doubt his association with Atwater influenced his choice of experimental field.

In 1899 Sherman joined Columbia University as a lecturer and was given charge of the quantitative organic analysis course being offered to students for the first time. In 1901 he became an instructor, and in 1905 an adjunct professor, the same year he published his first book, "Methods of Organic Analysis."

In 1903 Sherman married Cora Aldrich Bowen. Three children came from their union: two boys and a girl. One of the boys became a physician, and the daughter, Caroline (Sherman) Lanford, later collaborated with her father in writing their book, "Essentials of Nutrition."

One of Sherman's first areas of research was the investigation of the requirements for calcium, phosphorus, iron, and protein, using human subjects. From these experiments Sherman concluded that "of fuel food . . . the optimal intake is close to the actual need; of protein and some of the mineral elements . . . a margin of something of the order of 50 per cent is a desirable sort of insurance."

In 1907 Sherman became professor of chemistry. By this time he had begun to work on enzymes. He proved conclusively their proteineceous nature, in spite of the opinion to the contrary held by many European chemists.

During World War I Sherman served as a member of the American Red Cross mission to Russia, a public service action that showed he was willing to leave his ivory tower and apply his findings to improve the nutrition of human beings.

By this time Sherman had received an honorary doctorate from University of Maryland, and in 1919 he became executive officer of the Chemistry Department at Columbia, a post he held until 1939.

In 1920 Sherman began to pay attention to vitamins. As a result he developed biological assay methods (at that time the chemical structure of many vitamins was not known) for vitamin A, thiamine, ascorbic acid, and riboflavin. Charles G. King related that when Sherman was chided about the inordinate amount of time he was spending with animals, Sherman replied: "they are my burettes and balances. They give quantitative answers to many of man's greatest problems."

Sherman's patience and devotion to this line of research was recognized in 1934 when the New York Section of the American Chemical Society presented him with the Nichols medal. His contribution was described with the words: "So large a quantity of the literature on vitamins is valueless and there was needed a critically minded investigator who had the patience and capacity to review all that was published on the subject and eliminate what was worthless."

In 1924 Sherman was appointed Mitchill Professor of Chemistry, and 5 years later Columbia awarded him an honorary D.Sc. degree. In 1933 he was elected to the National Academy of Sciences; in the same year he received the medal of the American Institute of Chemists and in 1937 the Associated Grocery Manufacturers of America annual award.

In 1943 Sherman published his book "The Science of Nutrition," where he debunked spinach, which he found contained a "relatively large amount of oxalic acid," which renders calcium "unavailable and useless."

During World War II Sherman left Columbia to serve as the chief of the Bureau of Human Nutrition at the Department of Agriculture. In 1947 he received the Franklin medal of the Franklin Institute of Philadelphia, and the following year he became chairman of the Commission of Dietary Allowances of the National Research Council.

Another aspect of Sherman's work was on

the influence of diet on the life span. In his work with animals he established that old age could be postponed by a diet rich in "protective foods": fruits, vegetables, and milk. He also established that the internal chemistry of animals is not inflexible but can be influenced by the external environment.

Sherman's longevity was a good example of what good nutrition can do for individuals. He continued to publish until he was 75 and finally died just 9 days short of his 80th birthday Oct. 7, 1955 in Rensselaer, N.Y.

Communication from Charles G. King; biography by Edward C. Kendall in *J. Chem. Educ.* **32,** 510-513 (1955); *Current Biography,* 1949, p. 565-7, with portrait, H. W. Wilson Co.; obituary, *New York Times,* Dec. 8, 1955, with portrait; E. V. McCollum, "A History of Nutrition," Houghton Mifflin Co. (1957).

JOE VIKIN

Mary Lura Sherrill

1888-1968

Mary Sherrill was born in Salisbury, N. C., July 14, 1888 and died at High Point, Oct. 27, 1968. She received her bachelor of arts and master of arts degrees from Randolph-Macon Woman's College and continued teaching there until 1918 when she became associate professor of chemistry at the Woman's College of University of North Carolina. In 1921 after a year as a chemist at the Chemical Warfare Service's Edgewood Arsenal, she joined the chemistry faculty at Mount Holyoke College where she continued until her retirement in 1954. During the period 1946-54 she was chairman of the Chemistry Department.

During her first years at Mount Holyoke she completed her work for the doctor of philosophy degree at University of Chicago under Julius Stieglitz (1923); an honorary D.Sc. was also conferred upon her by University of North Carolina (1948). She held appointments to several fellowships that permitted residence in European laboratories during her sabbatical leaves of absence.

During the period 1925-40 Mary Sherrill participated in group research with her colleague Emma P. Carr and their students. In this work samples of the lower unsaturated hydrocarbons were prepared and their ultraviolet absorption spectra studied. An unusually high degree of compound purity was required in this work. The care with which it was carried out was demonstrated years later when standard samples, prepared by the American Petroleum Institute, were found to have spectra identical with those prepared at Mount Holyoke.

During World War II Miss Sherrill and her students engaged in the synthesis of novel antimalarial drugs under the auspices of the Office of Scientific Research and Development. Her syntheses of aminobenzothiazole derivatives developed renewed interest 30 years later during a search for prophylactic drugs active against drug-resistant malarias encountered in Southeast Asia.

Mary Sherrill believed that group research in her small department was interesting, stimulating, and more productive than any individual research project could have been. The group project made the best use of the limited research time for students and faculty alike. It was an outstanding success at Mount Holyoke College and gained for her and for Miss Carr an international reputation. She was awarded the Garvan Medal by the American Chemical Society at its meeting in New York in 1947.

Mary Sherrill's love for teaching was contagious. In the period 1937-46 there were 153 women who received the bachelors degree in chemistry at Mount Holyoke College. Of these 31% later received advanced degrees and 12% entered college teaching. In recognition of her 44 years of active teaching of chemistry, continued on a part-time basis even during her emeritus years, she was awarded in 1957 the James Flack Norris Award for outstanding achievement in the teaching of chemistry by the Northeastern Section of the American Chemical Society.

Biography by J. V. Crawford, *The Nucleus* **34,** No. 9 (1957) with several portraits.

EDWARD R. ATKINSON

Porter William Shimer

1857-1938

Shimer was born in Shimerville, Pa., Mar. 13, 1857, of colonial ancestry. His family moved to Easton in 1863, where he completed his education in the city's private and public schools. He entered Lafayette College in the same city, graduating in 1878 with a degree of engineer of mines. He spent 2 years of his college time as a special student in chemistry with Thomas M. Drown. After being employed for about 1 year as chief chemist with Thomas Iron Co. at its plant in Hokendauqua, Pa., Shimer returned to Lafayette. There he carried out research in Drown's laboratory. In 1885 Shimer opened a private chemical and metallurgical laboratory in his home, and continued it until he retired, 10 years before his death. From 1894 to 1902 he was resident lecturer on iron and steel at Lafayette. In 1898-99 he did graduate work at Lafayette, as a result of which a Ph.D. degree was conferred upon him in 1899. He died Dec. 7, 1938.

During his career Shimer served as consultant to Warren Foundry, Ingersoll-Rand Co., Bethlehem Steel Co., New Jersey Zinc Co., United States Steel Corp., National Bureau of Standards, and the Italian government.

Immediately after graduation, working during the summer in Drown's laboratory while the later was abroad, Shimer developed a method of determining silicon in steel which was not only more accurate but reduced the time necessary for the analysis from 2 days to 2 hours. Most of his published and unpublished research was connected with the perfection of analytical methods used in industry. The importance of his demonstrating the needs for analytical control in industry can not be overestimated. He discovered titanium carbide. He also invented a platinum crucible with a water-cooled stopper that greatly increased the speed and accuracy of the determination of carbon in iron and steel. His other contributions to the development of science and technology included; a new system of filtration in quantitative chemical analysis, a volumetric method for determining titanium in iron ores, a new process of case-hardening steel, molten baths for steel treatment, and a chaplet alloy used in iron foundries.

Among his publications were: "Methods and Apparatus used in Analytical Chemistry," "Application of Chemistry to Metallurgical Problems," "A New System of Filtration in Quantitative Analysis," "Metallurgy," and "Vanadium in Pig Iron."

He was awarded the John Scott medal and premium by the Franklin Institute. He also received a gold medal from the American Institute of Mining Engineers to mark 50 years of his membership in the institute. Because of his poor health the medal was presented to Shimer at a special testimonial dinner at the Hotel Easton in Easton, Pa., rather than at the regular meeting of the Institute in New York.

D. Arthur Hatch, ed., "Biographical Record of the Men of Lafayette," Lafayette College, 1948; David Bishop Skillman, "The Biography of a College," Lafayette College, 1932; newspaper clippings in Shimer's folder at the Alumni Office, Lafayette College.

GEORGE SIEMIENCOW

Benjamin Silliman

1779-1864

Daniel Webster once said to Mrs. Silliman: "Madam, if I were as rich as Mr. Astor, I tell you what I would do; I would pay your husband $20,000 to come and sit down by me and teach me, for I do not know anything." Webster was not alone in his admiration for Benjamin Silliman; tens of thousands of Americans heard him lecture and came away impressed.

Benjamin Silliman was born in Trumbull, Conn., Aug. 8, 1779, shortly after his father, Brigadier General Gold Selleck Silliman of the Revolutionary Army, had been captured and imprisoned by the British. Benjamin graduated from Yale in 1796, worked on his parents' farm, taught school, and then took up the study of law in New Haven. Timothy Dwight, president of Yale, offered him the newly-created professorship of chemistry in the college; and Silliman, who had never heard a chemistry lecture in his life, spent the winters of 1802-03, 1803-04 in Philadelphia learning chemistry from James Woodhouse.

He delivered his first lecture in April 1804. As the years passed he expanded his activities until he was teaching chemistry, pharmacy, geology, and mineralogy. His course in geology was said to have been the first in an American college. He relied at first upon William Henry's "Epitome of Chemistry" and "Elements of Chemistry," of which he brought out American editions, but in 1830-31 he published his own "Elements of Chemistry," a failure as a text because it was too advanced for college students of those days. For his geological classes he issued editions of Robert Bakewell's "Introduction to Geology." He resigned in 1853 after teaching for more than half a century. Yale then split his professorship into chemistry, given to his son Benjamin Silliman, Jr., and geology, to his son-in-law James Dwight Dana.

Silliman was well-known to literate Americans through his popular travel books, "A Journal of Travels in England, Holland and Scotland, and of two passages over the Atlantic, in the years 1805 and 1806" (1810, 1812, 1820); "A Tour to Quebec in the Autumn of 1819" (1820, 1821, London, 1822); and "A Visit to Europe in 1851" (five editions). He was equally well-known among scientists for the *American Journal of Science* founded by him in 1818 and still being published, the oldest scientific journal (with exception of society journals) in America. To the public at large he was famous for lectures that he delivered in scores of American cities, as far south as New Orleans and as far west as St. Louis, between 1834 and 1857. He was the greatest scientific lecturer of his time and through this medium did much to stimulate the study of chemistry and geology in the United States. "Through his public lectures, and by means of the journal," said Joseph Henry, "Professor Silliman became more widely known, and more highly appreciated, than any other man of science in the country."

Silliman wrote many articles, starting in 1808 with a paper on geology for the *Memoirs of the Connecticut Academy of Arts and Sciences,* but few of his contributions were of importance. His strength lay rather in the spoken word, in his ability to impart knowledge and to stimulate students. George Fisher called him "the most eminent of American teachers of natural science."

Benjamin Silliman died at New Haven on Thanksgiving, Nov. 24, 1864. After his death Samuel F. B. Morse wrote of him: "The course of science in the United States owes its progress to Professor Silliman more than to any other individual."

George Fisher, "Life of Benjamin Silliman . . . chiefly from his manuscripts, reminiscences, diaries, and correspondence," Charles Scribner and Co. (1866), 2 vols., with portrait, quotes are from this book; John Fulton and Elizabeth Thomson, "Benjamin Silliman," Henry Schuman, Inc. (1947), with portrait and bibliography; Edgar F. Smith, "Chemistry in America," D. Appleton and Co. (1914), with portrait; biography by W. D. Miles in Eduard Farber, editor, "Great Chemists," Interscience Publishers (1961), pp. 404-417; "Dictionary of American Biography," **17,** 160, Chas. Scribner's Sons (1935).

WYNDHAM D. MILES

Benjamin Silliman, Jr.

1816-1885

Silliman was born in New Haven, Conn., Dec. 4th, 1816, son of Benjamin Silliman, professor of chemistry and natural history at Yale College. Benjamin, Jr. at an early age was encouraged by his father to enjoy the sciences and practical studies. He was given a workshop by his father furnished for that purpose and quickly developed the useful mechanical and manipulative skills and interests so evident throughout his life. He was prepared in the New Haven schools, and after entering Yale College in 1833 he graduated with B.A. in 1837. While there was little opportunity for electives in sciences, science and mathematics at that time made up a large part of the required courses at Yale, and he received a sound theoretical training in them. In addition, young Silliman served as an assistant to his father during the course of his illustrated lectures which were famous for their interest and precision. During these college years he accompanied his father on several expeditions, such as a visit to certain Virginia gold mines on which the elder Silliman had been asked for his professional opinion.

After graduation, he continued to help his father as a paid assistant. In 1840 he received his A.M. degree from Yale, and was formally named an assistant in chemistry, mineralogy, and geology. In 1841 he spent 6 weeks in Charles T. Jackson's laboratory in Boston, where he studied analytical chemistry.

Benjamin, Jr. organized an informal teaching laboratory in cramped and inadequate quarters in the "old laboratory" of the college. His pupils included Yale students and also some promising young men without college training who sought to learn practical chemistry, mineralogy, and geology. In 1846 Silliman, Jr., together with John Pitkin Norton, conceived the idea of founding a special department at Yale to prepare scientists and scientific agriculturalists. Aided by the elder Silliman and by other influential friends, a department of "Philosophy and the Arts" was constituted in 1847. Young Silliman had earlier been named "Professor of Chemistry as applied to the Arts." Neither Silliman nor Norton, who also received a teaching appointment, was salaried. Their entire income was derived from students although the college did make available an old building at low rental for them. During the first half dozen years of this school, students came and went as they pleased—no examinations, no entrance requirements, no degrees. Silliman taught geology, mineralogy, and chemistry and continued to assist his father. Several of these early students, among them Samuel W. Johnson, George J. Brush, and William H. Brewer, became well known in the world of science.

In 1849 Silliman's financial needs grew more pressing as his family grew, and he accepted a post as professor of medical chemistry at University of Louisville, where he spent approximately 6 months each year.

In 1853 the elder Silliman retired after 50 years of service, and his son was called to fill the professorship of general and applied chemistry. He also took an active role on the faculty of Yale Scientific School and served as professor of medical chemistry in Yale Medical School.

Benjamin Silliman, Jr. had many interests besides teaching. He was frequently and profitably called on as a mining and oil consultant. His report of 1855 on Venango Oil is considered as one of the most penetrating and insightful studies in petroleum chemistry in that it laid the foundation for the petroleum cracking process and for many of the methods and uses associated with petroleum. He travelled widely through the United States and Canada, studying mining properties—gold, silver, coal—for promoters and interested financiers. He was criticized by some of his colleagues for sacrificing pure science in pursuit of money, and his professional ability was called to account on more than one occasion. Indeed, the vindictiveness of his accusers, notably Josiah Dwight Whitney, led to his resignation from his professorship at Yale College in 1870 and to an attempt to force him out from the National Academy of Sciences, of which he had been a charter member. The great majority of his professional judgments on these oil and mining controversies were ultimately vindicated although this vindication came too late in his career to be comforting.

Silliman was also a pioneer in the gas illumination industry, having served as one of the founders of New Haven Gas Co. in 1847, and much of his research was devoted to methods of improving gas lighting.

In addition to his long tenure as editor of *American Journal of Science,* which his father had founded in 1818 and on which he served from 1838 to the time of his death, he wrote "First Principles of Chemistry" in 1847, a textbook which remained in print for 20 years. He also wrote "First Principles of Physics or Natural Philosophy" in 1859. Historians of chemistry particularly honor him for his lengthy "American Contributions to Chemistry" which appeared in 1874 on the centennial of Priestley's discovery of oxygen.

Silliman was not an original scientific thinker. He was, however, an excellent technician. His interests were far ranging, and it was perhaps this combined with the constant need for money for his family which precluded the concentration on pure science that his critics demanded. He died Jan. 14, 1885 in New Haven.

Biography by Arthur W. Wright in *Biog. Mem. Nat. Acad. Sci.* **7,** 115-141 (1911), with portrait; biography in *Pop. Sci. Mon.* **16,** 550-553 (1880), with portrait; obituary by James D. Dana in *Am. J. Sci.* (3) **29,** 84-92 (1885); anonymous obituary in *Proc. Amer. Acad. Arts Sci.* **20,** 523-527 (1884-85); Gerald T. White, "Scientists in Conflict," Huntington Library (1968); Louis I. Kuslan, "The Founding of the Yale School of Applied Chemistry," *J. Hist. Med. Applied Sci.* **24,** 430-451 (1969); "Dictionary of American Biography," **17,** 160, Chas. Scribner's Sons (1935).

LOUIS I. KUSLAN

Alexander Silverman

1881-1962

Silverman was born in Pittsburgh, Pa., May 2, 1881. He received his Ph.B. degree in chemistry from the old Western University of Pennsylvania, now University of Pittsburgh, in 1902 and went to work as chemist for Macbeth-Evans Glass Co. of Charleroi, Pa. which in 1936 merged with Corning Glass Works. At that time there were only six other chemists employed in the American glass industry. He went back to school in 1905, attending Cornell University which granted him his A.B. degree in 1905. He then returned to his alma mater as a teaching fellow, receiving his M.S. degree from it in 1907.

He remained on the faculty for his entire career, becoming professor of chemistry and finally head of the department in 1918. Throughout this period he retained his interests in the chemistry of glass, carrying out and directing research in this area as well as serving as a consultant for a large number of companies manufacturing and using glass. He received honorary Sc.D. degrees from University of Pittsburgh in 1930 and from Alfred University in 1937.

Although Silverman maintained an active interest in the fundamental aspects of glass technology, he carried out extensive practical work concerned with the manufacture of colored and illuminating glasses. His early factory experience was gained in a plant in which all of the glasses were melted in 24 covered pots. His later extensive work included the conversion of old established compositions into new compositions for use in modern highly mechanized production processes. He contributed greatly to the development of alabaster and opal glasses essential for modern illuminants, and was awarded basic patents in these fields. He directed and inspired his students on research in these applied fields but did so even more extensively in such fundamental areas as the relations of bond energies and types to glass formation as well as to more precise and consistent correlations of compositions with the properties of the resulting glasses. This work resulted in his publishing more than 200 papers.

At the same time, he maintained an active interest in the broader aspects of inorganic chemistry and participated extensively in the activities of scientific and technical societies. In 1940 he was made an honorary member of the American Institute of Chemists for outstanding services as a teacher. During 1943 he served as a consultant for the War Production Board. He was a member of the Division of Chemistry and Chemical Technology of the National Research Council from 1947 to 1950 and served as chairman of its Committee on Ceramic Data. He served as U.S. Delegate to the meetings of the International Union of Pure and Applied Chemistry held in Liege in 1930, Madrid 1934, Lucerne 1936, Rome 1938, London 1947, Amsterdam 1949, New York 1951, Stockholm 1953, Paris 1957, and Munich in 1958, when he was Honorary President of its Committee on Inorganic Chemistry. He also maintained an active interest in the American Ceramic Society, serving as its vice president in 1931 as well as being elected a charter fellow. He was also a fellow of the British Society of Glass Technology and a member of Deutsche Glas-technische Gesellschaft.

He supplemented his activities in the scientific and technological aspects of glass by pursuing ardently the artistic and historical phases of this intriguing material. Over the years he amassed an outstanding collection of glass-ware ranging from those of archeological interest to modern craftmanship and works of art. His extensive foreign travel and the fact that he had married into the family of the oldest and largest glass and china store in Pittsburgh made possible the accumulation of a truly monumental collection, which he presented in 1951 to Alfred University, where it is on prominent display.

He received a number of recognitions for his contributions to chemistry and glass science, including the distinguished achievement award of the Pittsburgh Section of the American Chemical Society in 1940, the Francis Clifford Phillips Award in Chemistry from University of Pittsburgh, and the Albert Victor Bleininger Award of the American Ceramic Society in 1958. He died at the age of 83 at his home in Pittsburgh, Dec. 16, 1962.

Bull. Amer. Ceramic Soc. **9,** 216 (1930) portrait; **20,** 454 (1941) portrait; **36,** 246 (1951); **34,** 58 (1955) portrait; *News. Rev. British Soc. Glass Tech.* **35,** 56-60 (1951); personal recollections.

HENRY H. BLAU

Hezzleton Erastus Simmons

1885-1954

Simmons had the distinction of being the outstanding American educator of his time in the field of rubber chemistry. In this activity, as in many others, his lifetime career was indissolubly bound up with The University of Akron in Akron, Ohio, the "rubber capital of the world."

Simmons was born in 1885 in Lafayette, Ohio, and was reared on a Medina County farm. He entered Buchtel College (later The University of Akron) in 1903 to prepare for the study of medicine. However, chemistry exerted a strong appeal for him, and he ended up by obtaining his B.S. degree in that field in 1908. He continued his studies in chemistry in the graduate school of University of Pennsylvania, obtaining his M.S. degree in 1910. He then returned to his alma mater to accept an appointment as assistant professor of chemistry under Charles M. Knight, who in 1909 had already pioneered the first formal course in rubber chemistry ever offered—anywhere.

Appointed professor of chemistry in 1912, Simmons continued to develop the teaching program in rubber chemistry just at the time when Buchtel College, originally founded by the Ohio Universalist Convention in 1870, became the Municipal University of Akron in 1913.

With the enormous upsurge that occurred in the rubber tire industry between 1910 and 1920 (and the concomitant boom in Akron's population) the need for more technical education in rubber compounding became increasingly evident. Thus it was that the well-known trade journal, *Rubber Age,* which started publication in 1917, announced in that same year that it planned to publish Professor Simmons' lectures in rubber chemistry as a series of 27 serialized chapters. These covered the whole known area of the subject, from the chemical and physical characteristics of native *Hevea* rubber and its preparation to the complex chemical reactions of rubber compounding, vulcanization, testing, and analysis of the final product. This "correspondence course" obviously mirrored the rapid growth of a vigorous young industry.

Simmons' course in rubber chemistry at University of Akron was invaluable for senior undergraduate and graduate students in chemistry who were preparing for a career in the rubber industry. His published lectures on this topic were later incorporated in his book on rubber manufacturing, published in 1921; a description and discussion of this teaching program was published in 1922.

Simmons was active in research during this period, especially during the 1920's when the development and utilization of accelerators of vulcanization was in its heyday. He published on this topic jointly with the world-famous rubber chemist, G. Stafford Whitby. He was also active in his profession at the same time, acting as secretary-treasurer of the newly-organized Division of Rubber Chemistry of the American Chemical Society from 1926 to 1934.

By the end of the 1920's, however, his interests in higher education had broadened beyond the field of chemistry, and he became director of the Evening Division of the university in 1928, followed eventually by ascendancy to the presidency of the university in 1933. He held this post until his retirement in 1951, seeing the university through the dark days of the Great Depression and World War II. Although he could not personally devote himself to his own field of rubber chemistry during this period, he still was interested enough to play an important role in the building of the synthetic rubber industry by the U.S. government during the war. Thus, he held the post of associate chief of the Rubber Branch of the War Production Board from 1942 to 1944 and was especially instrumental in helping to organize the government-sponsored research program. One of the leading participants in this program was G. Stafford Whitby, the English rubber chemist whom Simmons recruited to come to The University of Akron to organize a research group—the forerunner of the present Institute of Polymer Science.

Simmons was honored in 1952 by being awarded the Goodyear Medal of the American Chemical Society's Division of Rubber Chemistry. He received honorary degrees from the universities of Toledo and Akron as well as from Wooster College. He died at his home in Akron on Dec. 30, 1954.

University of Akron Archives; *Rubber Age* **1,** 224 (June 11, 1917); H. E. Simmons, "Rubber Manufacture," D. Van Nostrand Co. (1921); *India Rubber Review* **22,** 41 (1922); *India Rubber World* **127,** 378 (1952).

MAURICE MORTON

William Simon

1844-1916

Simon, an industrial chemist and writer of a text widely used in pharmaceutical, dental, and medical colleges, was born in Eberstadt, Germany, Feb. 20, 1844. He attended school in Giessen, clerked in drug stores from 1860 to 1866, and studied chemistry at University of Giessen, obtaining his Ph.D. degree in 1869. He assisted Heinrich Will, noted analytical chemist, from 1869 to 1870, served in the sanitary corps of the German army during the Franco-Prussian War, and emigrated to the United States near the end of 1870. He lived in Baltimore for a few years, then in Catonsville, Md. for the rest of his life.

Simon was chemist for Baltimore Chrome Works from 1870 to about 1907 and concurrently taught chemistry at Maryland College of Pharmacy from 1871 to 1902, at College of Physicians and Surgeons of Baltimore from 1880 to 1916, and at Baltimore College of Dental Surgery from 1892 to 1916. An inspiring, sympathetic teacher in the classroom, he was frequently asked to speak before public and professional audiences whom he delighted with demonstrations employing liquid air, x-rays, photography, and other attractive media.

He wrote "Manual of Chemistry, A Guide to Lectures and Laboratory Work for Beginners in Chemistry," which passed through 13 editions between 1884 and 1927 and published many articles in chemical and pharmaceutical journals. He received several honorary degrees and served as a president of the Maryland Pharmaceutical Association.

Simon was a competent, amateur painter and an excellent photographer; his collection of color photographs was said to have been one of the finest in the country.

He died at his summer home in Eagles Mere, Pa. July 19, 1916.

Obituaries in *J. Amer. Pharm. Ass.* **5,** 886-87 (1916) and in *New York Times,* July 21, 1916.

WYNDHAM D. MILES

Lyndon Frederick Small

1897-1957

Small was born Aug. 16, 1897 in Allston, Mass., the second child of Frederick and Amanda (Corey) Small. A science teacher at Needham High School early recognized the young student's talents for physical sciences and encouraged Small in every possible way. This led to his enrollment at Dartmouth College where his intense desire to learn earned him a scholarship at the end of his first year, and this bursary supported him throughout the Dartmouth years. He graduated in 1920.

From 1920 to 1922, Small was a Henry Elijah Parker Fellow from Dartmouth College at Harvard where he studied under Prof. Kohler for his M.A. degree, which he received in 1922. After serving as an instructor in inorganic chemistry at MIT, Small spent 3 years at Harvard working with James B. Conant for his Ph.D. degree which was granted in 1926.

On Conant's recommendation, Small was awarded a Sheldon travelling fellowship for one year to Heinrich Wieland's laboratory in Munich. A second Munich year was funded by a National Research Council Fellowship. These 2 years in Munich proved decisive ones for Small's future activities. Of three research projects that Wieland suggested, Small chose the alkaloid problem, "The Ozonization of Thebaine." Within 2 years he acquainted himself thoroughly with the complex chemistry of the morphine alkaloids, and this subsequently became his life's work.

Although Conant invited him back to Harvard as his private assistant, Small accepted a position as research associate in the Chemistry Department at University of Virginia and continued his research in the morphine alkaloid area.

In 1929 the Division of Medical Sciences of the National Research Council established a committee whose purpose was to find and study means to cope with the increasing drug addiction problem. Based on the knowledge gained in replacing cocaine with the less hazardous novocaine, the possibility of separating the addiction liability of morphine from its beneficial analgetic properties by chemical means was envisioned. On Conant's recommendation, Small was appointed director of the so-called "Drug Addiction Laboratory" at University of Virginia. During the following 10 years Small and his students carried out a long series of painstaking researches on morphine and its cogeners. In addition he wrote, assisted by Robert E. Lutz of University of Virginia, "The Chemistry of The Opium Alkaloids" which, for the first time, marshalled together the whole chemistry of the opium alkaloids. This monograph was long considered to be the "Bible" on the subject for those working with morphine alkaloids. The hundreds of morphine derivatives produced at Charlottesville were pharmacologically evaluated by Nathan B. Eddy of the Department of Pharmacology, University of Michigan, which provided for the establishment of well-defined correlations between structure and analgetic activity.

Although it appeared that morphine chemistry had been fairly well exhausted by earlier researches of German and British chemists, Small, by virtue of tenacious skill, unearthed a wealth of new reactions as well as novel transformation and degradation products. One outstanding achievement was the synthesis of Metopon, which represented the first nuclear-alkylated morphine derivative. This substance, which was orally effective, showed promise of being superior to morphine as an analgesic in that its use was attended by lower addiction producing potential and elicited fewer undesirable side effects. Small was also particularly intrigued with the chemistry of thebaine (an opium constituent) and, in collaboration with his students, obtained unexpectedly interesting and fruitful results in this area.

In 1938, Small became editor-in-chief of *Journal of Organic Chemistry*. In this capacity, ably assisted by his wife, Marianne, he set high standards for the journal which earned him the lasting gratitude of American organic chemists.

With the outbreak of the second World War imminent in 1939 Lewis R. Thompson, director of the National Institutes of Health, invited Small to move his group from Charlottesville to NIH in Bethesda, Md., and to shift his energies to the development of new antimalarials (quinine substitutes). This transition occurred in late 1939. Small and his associates moved ahead energetically and successfully in this area of research and occupied an important position in the broad antimalarial program of the Office of Scientific Research and Development. Hundreds of novel compounds were synthesized, among them a dozen or more effective enough to replace quinine and atabrine. After the war Small returned to his first love, morphine chemistry.

In 1939, Small and Eddy received the first

Annual Scientific Award of the American Pharmaceutical Manufacturers Association, and in 1949 Small received the Hillebrand Prize of the Washington Section of the American Chemical Society. He died at his home in Bethesda, Md., June 15, 1957.

Personal recollections; records of the National Institutes of Health; biography by Erich Mosettig, *Biog. Mem. Nat. Acad. Sci.* **33,** 397-413 (1959), portrait; *Chem. Eng. News* **28,** 702 (1950), portrait.

LEWIS J. SARGENT

Alexander Smith

1865-1922

Smith was primarily a physical-inorganic chemist in his research work, but he made an even greater name for himself as a teacher and author of textbooks. His "Introduction to General Inorganic Chemistry," 1906, was one of the most important textbooks in the field during the first quarter of the twentieth century. It inaugurated a revolution in methods of instruction in America which can perhaps be compared only with that affected in Germany by Wilhelm Ostwald's "Lehrbuch der allgemeinen Chemie."

Smith's ideas on chemical education were put forth in an article entitled "The Training of Chemists," in which the following statements were made, quite unusual for their time, particularly from one who had received a classical education at Edinburgh and Munich:

"Listening to a lecture keeps the student in a *receptive* attitude of mind, whereas the attitude we desire to cultivate in him is the precise opposite of this. The student should begin by himself acquiring the ability to state simple ideas correctly, and later himself practice putting facts and ideas together and reaching conclusions. The conclusions are not new, but going through the operation of reaching them for himself is new to the student. . . .

"I am not proposing to abolish lecturing, . . . in the more advanced courses, lectures are of great value. . . . I am referring mainly to the elementary course for freshmen, where in the true sense, or has any knowledge of not one member in twenty has ever studied how to study. . . . Listening to lectures, in such a case, if the lectures are well constructed, only deludes him into thinking that he has fully grasped the subject, and *prevents* him from studying."

These are ideas that might sound revolutionary to some even today, a half-century later.

Smith was born in Edinburgh, Scotland, Sept. 11, 1865. His training was standard for his day, a B.S. degree in chemistry from Edinburgh (1886), followed by a Ph.D. degree at Munich in 1889. At Munich he studied organic chemistry with Adolph Ritter von Baeyer, and some of his first publications were in the field of organic chemistry (*e.g.,* "On Desylacetophenone," *J. Chem. Soc., (Proc.)* **57,** 643-652 (1890)). Actually, Smith first published four semi-popular articles on astronomy as an undergraduate at Edinburgh. After his studies at Munich Smith lectured for a year at Edinburgh and then came to the United States as professor of chemistry and mineralogy at Wabash College, where he stayed from 1890 to 1894.

In 1894 Smith moved to University of Chicago at the invitation of John Ulric Nef to take charge of work in inorganic chemistry. He spent the next 17 years there, perhaps the most fruitful years of his life. It was there that he formed his ideas on teaching which were published in 1902, with Edwin H. Hall, in "Teaching of Chemistry and Physics in Secondary Schools." Here he carried out his famous researches on the several forms of sulfur. Here also he and Alan W. C. Menzies developed the isoteniscope and with it measured the vapor pressures of a number of substances, among them mercury and calomel. The isoteniscope gave us a new method for measuring vapor pressures at high temperatures where previously measurements were difficult if not impossible.

In 1911 he became head of the Chemistry Department at Columbia University, where he remained until 1919 when he retired because of illness.

On Feb. 16, 1905 Smith married Sara (Bowles) Ludden, daughter of William Bowles of Memphis, Tenn. They had two children. He was elected president of the American Chemical Society in 1911. For his presidential address given in Washington, he chose to speak on the history of chemistry. He was elected to the National Academy of Sciences in 1915. The University of Edinburgh conferred the LL.D. degree on him in 1919. The citation reads:

"A most distinguished graduate of our own University, Professor Smith has risen to the rank of a super-chemist in the United States, head of a department embracing many specialized professorships, and director of one of the most important laboratories in the new world."

Smith died in Edinburgh, Sept. 9, 1922, after an illness lasting several years.

Alexander Smith, "The training of chemists," *Science* **43,** 619-629 (1916); Alexander Smith and E. H. Hall, "Teaching of chemistry and physics in secondary schools," Longmans, Green & Co. (1902); Alexander Smith, "An early physical chemist, M. W. Lomonossov," *J. Amer. Chem. Soc.* **34,** 109-119 (1912); biographies by James Kendall in *J. Amer. Chem. Soc. (Proc.)* **44,** 113-117 (1922), and *J. Chem. Educ.* **9,** 254-260 (1932), and by Ralph H. McKee in *J. Chem. Educ.* **9,** 246-253 (1932); obituary in *New York Times,* Sept. 10, 1922.

JOSEPH A. SCHUFLE

Daniel B. Smith

1792-1883

Smith was born in Philadelphia, July 14, 1792. He spent his childhood in Burlington, N. J. and learned chemistry from John Griscom at Griscom's academy in Burlington. Completing school, he went to Philadelphia and served an apprenticeship as a pharmacist, learning the practical side of chemistry.

Smith then entered the pharmaceutical profession and remained in it until 1853, building a large firm for manufacturing drugs and chemicals. He was one of the founders of the first pharmacy college in America, Philadelphia College of Pharmacy and Science, in 1821, and presided over the institution from 1829 to 1854. In 1825 he helped establish the first American pharmaceutical journal, *Journal of the Philadelphia College of Pharmacy,* later *American Journal of Pharmacy,* and wrote its first article, "On the Preparation of Glauber's and Epsom Salt and Magnesia, from Sea Water." Over the years he wrote several articles on pharmaceutical chemistry.

In 1834 Smith became professor of natural philosophy, English literature, and chemistry at Haverford College. In 1837 he published "The Principles of Chemistry, prepared for the use of schools, academies, and colleges," a text that appeared in revisions up to the 1860's. He remained at Haverford until 1846, when increasing business responsibilities caused him to resign. He died in Philadelphia, Mar. 29, 1883.

Joseph W. England, "First Century of the Philadelphia College of Pharmacy, 1821-1921" (pub. by the college) (1922); obituary by Charles Bullock, *Amer. J. Pharm. Allied Sci.* [4 S] **13,** 337-346 (1883).

WYNDHAM D. MILES

Edgar Fahs Smith

1854-1928

It is not an overstatement to say that any American chemist who seriously takes up the history of his profession will inevitably find his way to the Edgar Fahs Smith Memorial Collection in the History of Chemistry. And it is certainly true that any chemist who turns to the history of American chemistry will spend many hours reading Smith's "Chemistry in America," "James Woodhouse," "Robert Hare," "Chemistry in Old Philadelphia," and a score of his other works.

Smith was born in his father's grist mill near York, Pa., May 23, 1854. He attended York Academy and then Gettysburg College, where Samuel Sadtler influenced him to take up chemistry. He studied under Wöhler at University of Göttingen for 2 years and received his doctor's degree in chemistry in 1876.

Upon returning to America Smith became assistant in analytical chemistry at University of Pennsylvania, professor of chemistry at Muhlenberg College from 1881 to 1883, professor of chemistry at Wittenberg College from 1883 to 1888, and finally professor at Pennsylvania. In 1911 the trustees of the university elected him to the office of provost. When the results of the election were announced, Smith's popularity was so high that the students milled around Harrison Laboratory and built a bonfire in the street outside of his office. Smith resigned his offices of provost and professor in 1920 but remained at the university as emeritus professor until his death, May 3, 1928.

Smith's chief research interests were the complex inorganic acids, the rare earths, electrochemistry, and the revision of atomic weights of several elements. By himself and with graduate students, he carried out research reported in 169 articles. Between 1883 and 1918, 87 students completed their Ph.D. theses under his direction. He wrote, edited, and translated a dozen texts, which ran through at least 40 editions, on organic, inorganic, electrochemistry, and general chemistry.

Outside the lecture hall and laboratory Smith was busy with affairs of his city, state, and profession. He was, among other things, a member of the Electoral College for Pennsylvania in 1917 and 1925, of the Commission for Revision of the Constitution of Pennsylvania in 1919, of the U.S. Assay Commission in 1895 and from 1901 to 1905, and was appointed by President Harding to the Board of Technical Advisors of the Disarmament Conference of 1921. He presided over the American Chemical Society in 1895, 1920, and 1921. With Charles A. Browne he organized the Society's Division of History of Chemistry in 1921.

Smith's avocation was history of chemistry. As a young instructor at Pennsylvania in the 1870's he developed a series of lectures on history of chemistry, scrapped them because they bored the students, and worked up new lectures which the students enjoyed. He collected books, portraits, autographed letters, and memorabilia of famous chemists to exhibit to his classes and for research. This course in history lay close to Smith's heart, and he continued to teach it after he became emeritus professor and had passed his other subjects along to younger teachers.

Smith was primarily interested in the history of American chemistry, and almost all of his historical writings, which amounted to 32 articles, 15 brochures, and 7 books, were in this field. Most of his writings dealt with the lives and work of individual chemists. His preoccupation with people was not because of a lack of interest in the development of ideas, techniques, or technology, but it came about because Smith was a pioneer in the history of American science, and it was necessary for him to study the careers of early leaders.

Following Smith's death his widow presented his library, probably the finest private library of its kind in America, and a sum of money to endow it, to University of Pennsylvania, where it has grown and become internationally known as the Edgar Fahs Smith Memorial Collection in the History of Chemistry.

H. S. Klickstein, "Edgar Fahs Smith—His Contributions To The History of Chemistry," *Chymia* **5**, 11-30 (1959), has a list of Smith's publications and unpublished manuscripts in history; "The Edgar Fahs Smith Memorial Number," *J. Chem. Educ.* **9**, 607-750 (1933); biography by G. H. Meeker in *Biog. Mem. Nat. Acad. Sci.* **17**, 103-49 (1936), with portrait and list of Smith's scientific articles.

WYNDHAM D. MILES

James Smith

1740-1812

James Smith was the first professor of chemistry and materia medica in Columbia University Medical School from 1767 to 1770, and the first teacher in the United States to have the word "chemistry" in his title. He was not, however, the first to teach chemistry in this country. John Morgan had started to give complete courses on chemistry 2 years earlier in University of Pennsylvania Medical School.

Columbia's Medical School opened with public ceremonies in November 1767. The New York *Mercury* reported that "Dr. Smith, the Professor of Chymistry, give an introductory lecture on that branch, which for elegance and sublimity, met with universal approbation." Unfortunately, this is practically all we know of the chemistry career of this early American professor. We have not been able to learn anything of his course.

Little is known of Smith's life. He was born around 1740, attended Princeton University and University of Leyden (M.D., 1764). He learned chemistry as a medical student at Leyden. In 1770 Smith left New York for a time. The trustees thereupon dropped him from the faculty. For the remainder of his life Smith practiced medicine; he was a trustee of College of Physicians and Surgeons, 1807-11. He died in New York in 1812.

James Thacher, "American Medical Biography," De Capo Press (1828; reprint 1967) II, 95, has a very brief account of Smith; New York *Mercury,* Nov. 9, 1767; New York *Columbian,* Feb. 14, 1812; Milton H. Thomas, "Columbia University Officers and Alumni" (1936).

WYNDHAM D. MILES

John Lawrence Smith

1818-1883

Seated in the middle of the front row of the chemists who gathered at Northumberland, Pa., Aug. 1, 1874 for the Joseph Priestley Centennial was the doyen of American science, J. Lawrence Smith. Later, when the American Chemical Society became a reality, Smith was a vice-president the first year 1876 and president the second year 1877.

Smith was born near Charleston, S.C., Dec. 17, 1818. He was an exceptional child who could add and multiply at four, even before he could read, was doing algebra at eight, and was studying calculus at thirteen. He was sent to the best private schools in Charleston. At sixteen he entered University of Virginia where he studied civil engineering, natural philosophy and chemistry for 2 years. He began his career as a civil engineer for a railroad, but after a year he entered the Medical College of the State of South Carolina and graduated in 1840. The faculty awarded him a silver goblet for his thesis on "The Compound Nature of Nitrogen."

After receiving his M.D. degree Smith went to Europe where he attended lectures by well-known physicists, geologists, mineralogists, and chemists Dumas, Orfila, and Liebig. Van Klooster states that Smith was the first American to study at Giessen. Liebig suggested that he study spermaceti, and the results were published in *Annalen* **42,** 241 (1842).

In 1843 he returned to Charleston, began to practice medicine, and lectured on toxicology at the medical college. He and S. D. Sinkler in 1846 established *The Southern Journal of Medicine and Pharmacy,* which became *The Charleston Medical Journal and Review.* Smith soon abandoned his medical practice for science. He became the state assayer of gold bullion for Georgia, North Carolina, and South Carolina. Also, he studied the agricultural chemistry of the native materials around him. He was the first to show the presence of the large amounts of calcium phosphate in marl found under the soil along the South Atlantic coast. His investigations into the character of soils and the influence of soil components on the growth of crops resulted in Secretary of State James Buchanan's appointing him to a commission to teach Turkish farmers how to grow cotton.

Upon his arrival in Turkey in 1847 Smith lost interest in cotton and began a study of the mineral resources for the Sultan. His discovery of emery broke up the Greek monopoly on the mineral and greatly reduced the price. He also discovered chrome ores and coal which brought Turkey rich revenues. Because of his success as a mining engineer the Turkish government gave him many decorations and costly presents. While in Turkey he learned of Morse's telegraph and asked Benjamin Silliman to send him similar equipment, which he set up in the Sultan's palace so the latter could communicate with the port officials.

Smith returned to America in December 1850 and taught at Louisiana University until 1852. On June 24, 1852 he married Sarah Julia, daughter of James Guthrie, Secretary of the Treasury under President Franklin Pierce. He then taught at University of Virginia where he carried out extensive research on American minerals and devised a method for the determination of alkalies in silicate. In 1854 he moved to Louisville, Ky., where for the last 30 years of his life he performed much of his important analytical work in his private laboratory and taught chemistry at University of Louisville until 1866.

His interest in scientific endeavours was pandemic but mostly concerned meteorites and minerals. He invented the inverted microscope and reported it in *The American Journal of Science and Arts* **14**, 25 (1852). While working on samarskite of North Carolina he discovered what he thought was a new element which he named mosandrium after the chemist Mosander. His researches, which he carried out alone, filled more than 600 pages. He was a prolific writer with a total of 145 articles to his credit. In 1873 he published some of his more important papers in book form.

Smith was one of the first American chemists to be honored by European scientists. He was elected to membership in many learned societies. Edgar F. Smith, in "Chemistry in America," states "there could not be too much credit given Genth, Gibbs, and J. Lawrence Smith for the admirable contributions they made to the development of chemical science in the United States."

Smith and his wife had no children. Therefore, it is understandable that he founded and endowed the Baptist Orphans Home of Louisville. He died Oct. 12, 1883 in Louisville, Ky.

H. Hale, "The History of Chemical Education in the United States from 1870 to 1914," *J. Chem. Educ.* **9**, 729-44 (1932); J. R. Sampey, "J. Lawrence Smith," *J. Chem. Educ.* **5**, 123-8 (1928); B. Silliman, "Dr. John Lawrence Smith," *J. Amer. Chem. Soc.* **5**, 228-30 (1883), and *Biog. Mem. Nat. Acad. Sci.* **2**, 217-248 (1886); E. F. Smith, "Chemistry in America," D. Appleton and Co. (1914), 260-1, and "Mineral Chemistry," Golden Jubilee Number, *J. Amer. Chem. Soc.* 48, 71 (1926); "Dictionary of American Biography," **17**, 304-5, Chas. Scribner's Sons (1936).

DAVID H. WILCOX, JR.

Thomas Peters Smith

1776(?)-1802

Smith was born in Philadelphia during the chaotic days of the Revolution. As is the case with many other early American scientists, we know little about his personal life, and most of our knowledge about his career is derived from his writings or remarks made about him by contemporaries. He apparently did not receive much formal education and was largely self-taught; yet, according to an anonymous writer who knew him, "at eighteen he was a respectable mathematician, and at twenty an eminent chemist."

In the 1790's he joined the Chemical Society of Philadelphia, a lively organization where he associated with James Woodhouse, Robert Hare, Joseph Priestley, and other chemists. He served on a committee which analyzed ores and minerals as a free public service to increase knowledge of American mineralogy and geology and to encourage Americans to develop natural resources. In 1796 he presented a historial paper which the organization issued as a pamphlet, "A Sketch of the Revolutions in Chemistry" (1797). This is the earliest known publication of an American chemical society. Smith published at least seven articles in newspapers and magazines.

In 1797 the society organized a committee to collect information on the manufacture of niter. At this time the United States was apprehensive that war would break out with France, Great Britain, or the Mediterranean pirates, and if this happened there was a chance that a shortage of niter for gunpowder would develop as it had during the Revolution. Smith served on this committee and wrote an article, "On Niter, and the Best Means for Manufacturing it in the United States," for the Philadelphia *Weekly Magazine*.

In 1800 Smith sailed across the Atlantic to study European science and technology at first hand. He traveled through Germany, Denmark, Sweden, France, Switzerland, and Great Britain, inspecting factories, furnaces, and mines, listening to Vauquelin lecture in Paris, and visiting Gahn, James Watt, Charles Hatchett, and other scientists. After 2 years abroad he started home. During gunnery practice on board ship a cannon burst, wounding Smith mortally. He died at sea Sept. 22, 1802.

W. D. Miles, "Thomas Peters Smith, a Typical Early American Chemist," *J. Chem. Educ.* **30**, 184-188 (1953); Edgar F. Smith, "Chemistry in

America," D. Appleton and Co. (1914), pp. 13-41, reprinted Smith's "Sketch of the Revolutions"; the manuscript journal Smith kept in Europe is in the library of the American Philosophical Society.

WYNDHAM D. MILES

William Acheson Smith

1878-1933

Smith, son of William H. and Ellen (Acheson) Smith was born in Port Jervis, N.Y., Nov. 4, 1878. After graduation from Western University of Pennsylvania (now University of Pittsburgh) with a degree in mining engineering, he served Monongahela Coal and Coke Co. and Baggaley Mines of H. C. Frick Coke Co. until 1905. In that year he took a position at Niagara Falls with the graphite company founded by his uncle, Edward Goodrich Acheson. He remained with Acheson Graphite Co. for 23 years. His first assignments were in the experimental field connected with the electrothermal conversion of amorphous carbon to graphite. His success in this area soon led to his being placed in charge of the manufacture of new products, with the title of vice president. In 1916 he was made president of the Niagara enterprise.

The prosperity of Acheson Graphite Co. was in large part due to Smith's rare genius for organization, especially in the fields of business management and product distribution.

In 1928 when Acheson Graphite Co. was purchased by Union Carbide and Carbon Corp. and was fused with its National Carbon Co., Smith became a vice president of the latter company. His thorough knowledge of electrothermics, together with a magnetic personality and conversational charm, won him high regard in engineering and industrial circles.

Smith was a member of the American Institute of Mining and Metallurgical Engineers, the American Chemical Society, the Chemists' Club (New York), and the Union League Club of New York. He served a term as president of The Electrochemical Society and was on its board of managers. During World War I he was appointed to the War Industries Advisory Board. He was also a director of the Carborundum Co. and Power City Trust Co. of Niagara Falls, N.Y.

Death from a blood infection came unexpectedly, July 12, 1933, in the Johns Hopkins Hospital in Baltimore. Stricken in Paris, he was rushed to Baltimore directly upon his ship's arrival in New York. He was survived by his widow, the former Lucy Wood Carmack, and a daughter, Betty. He resided at time of death in Greenwich, Conn.

Raymond Szymanowitz, "Edward Goodrich Acheson—Inventor, Scientist, Industrialist," Acheson Industries (1965); obituaries in *New York Times*, July 13, 1933, and *J. Four Elec.*, October 1933.

RAYMOND SZYMANOWITZ

David Spence

1881-1957

Spence was born in Udny, Aberdeenshire, Scotland, Sept. 26, 1881. He was educated at Royal Technical College, Glasgow, University of Berlin, and University of Jena where he was awarded his Ph.D. degree in 1906. After 2 years as assistant biochemist at University of Liverpool, Spence came to Akron, Ohio, in 1909 to be director of the Research Laboratories of Diamond Rubber Co. Diamond was merged with B. F. Goodrich Co. in 1912, and Spence stayed on as director of the Research Laboratories for Goodrich until 1914.

In 1914 Spence was co-founder of Norwalk Tire & Rubber Co. and served as vice-president and general manager until 1925. In the meantime he offered his services to the United States Government during World War I and was put in charge of the Rubber Division of the National Research Council. In 1925 he became vice-president for development of International Rubber Co. in New York. In 1931 he retired to Pacific Grove, Calif. to concentrate on private research. During World War II he served as consultant to the Office of the Rubber Director, War Production Board. He returned to New York in 1952.

Spence was among the very early chemically trained investigators of the chemistry of rubber. He was particularly interested in the fundamental chemistry and kinetics of the vulcanization of rubber with sulfur, and he and his co-workers made the first definitive study of the reaction. They measured the temperature coefficient of vulcanization of rubber with sulfur and obtained values between 2.65 and 2.84, well within the limits for chemical reactions. They also showed that with time the reaction proceeds to complete consumption of the sulfur up to 32% sulfur, the maximum amount that will react with natural rubber. They also found that mastication of rubber has no effect on its rate of reaction with sulfur. These experimental data established sulfur-rubber vulcanization as a normal chemical reaction kinetically and stoichiometrically. The data also effectively disproved the Ostwald concept of vulcanization as a physical adsorption of sulfur on rubber, as well as a then current belief in a "trigger" temperature above which vulcanization suddenly proceeds at a rapidly increasing velocity.

Spence also contributed to the chemistry of rubber latex and was particularly interested in factors, such as enzymes, acids, bacteria, etc., in the latex which lead to tackiness in the latex coagulum. With M. L. Caldwell he developed a very accurate method for determining the rubber content of rubber-bearing plants. Spence was considered an expert on guayule latex and rubber.

In 1941 Spence was chosen the first Charles Goodyear Medalist by the Rubber Division of the American Chemical Society. However, illness prevented his giving the acceptance lecture, and the medal was finally awarded in 1948. In 1952 Spence donated his library, one of the most complete and extensive on rubber, to University of Southern California. In addition to ACS, he was a member of Society of Chemical Industry, Chemists' Club (New York), Cosmos Club (Washington), and life-fellow of Royal Institute of Chemistry.

Spence died in New York City, Sept. 24, 1957 and was buried in Scotland.

Rubber World, **137** (2), 289 (November 1957); *Rubber Age,* **82** (1), 142 (October 1957); K. Memmler, "The Science of Rubber," English translation edited by R. F. Dunbrook and V. N. Morris, Reinhold Publishing Co., New York, 1934, pp. 283-289.

GUIDO H. STEMPEL

Guilford Lawson Spencer

1858-1925

Spencer's interest in the technology and chemistry of sugar dated from his student days at Purdue University. Harvey W. Wiley, then professor of chemistry at Purdue, was already enthusiastic about beet sugar manufacture in the United States. He recognized in Spencer that chemistry was "his natural element" and referred to him after Spencer's death as "the most enthusiastic student in the laboratory." Wiley encouraged Spencer to specialize in sugar work and later to travel to France to study beet sugar technology under such eminent sugar specialists as Galois, Dupont, Pellet, and others.

Born in Lafayette, Ind., Dec. 20, 1858, Spencer entered Purdue in 1875 and graduated with the degree of analytical chemist in 1879. He did his graduate work at University of Michigan where he received his M.S. degree in 1882. His doctor's degree was granted by Purdue in 1893. Meanwhile Wiley had become chief of the Bureau of Chemistry in Washington where his interest in all phases of the sugar industry (beet, cane, and sorghum) became progressively greater. He gave Spencer a position as his assistant in the bureau when Spencer returned from his beet sugar studies in France. Spencer's first assignment as Wiley's assistant was at a small plant at Fort Scott, Kans., where he directed the study of sorghum sugar manufacture, using the diffusion and carbonation procedures common to beet sugar manufacture. Here he had the distinction of boiling the first strike of sorghum sugar ever discharged from a vacuum pan.

The next 40 years may be divided into two periods; the first from 1885 to 1905 when Spencer was with the Department of Agriculture where he served as Wiley's assistant and later as chief of the Sugar Laboratory in the Bureau of Chemistry. For the last 20 years of his life he was in commercial work in Cuba as general superintendent of manufacture of the Cuban-American Sugar Co., at that time the largest raw sugar producer in the world. During his years with the Department of Agriculture, Spencer kept in constant touch with commercial sugar manufacture and the practical problems of the industry.

The first of his work to attract wide attention took place at the Warmoth Plantation, "Magnolia," in Louisiana where he introduced the diffusion process to replace milling of cane for juice extraction. Five sugar crops were produced by this process from 1885 to 1889. Spencer published USDA bulletins covering each of these crops. The diffusion process gave much higher sucrose extractions than milling, but increased fuel costs because of difficulties in dewatering the extracted cane residue resulted in its abandonment at that time. The diffusion process for cane made a startling recovery in the mid-20th century, justifying Spencer's faith in the process he advocated in the 1880's. His publications about cane diffusion, long out of print, proved of value to later-day technologists who have adopted the continuous diffusers to compete with cane milling installations.

Spencer's name gained world-wide recognition with the publication in 1889 of his "Handbook for Cane Sugar Manufacturers and their Chemists." This early reference work on cane sugar factory analysis and control, now in its ninth edition as the Spencer-Meade "Cane Sugar Handbook," has been translated into Spanish and Japanese. His greatest achievement in technology came when he developed in 1910 the system by which factories under his direction produced only first grade sugar and molasses, eliminating the wasteful low grade sugars (89 test) that were a part of the output until then. This boiling system, first adopted in Cuba, spread with modifications throughout the entire cane-producing world. In connection with this work Spencer brought about the production of drier, cleaner raw sugars that would not deteriorate in storage or transit. Savings of millions of dollars resulted from these innovations.

One of the lesser known of Spencer's pioneering works was the establishment of the Central Control Laboratory at Cardenas, Cuba, in 1914. This was the first effort anywhere in the world to regulate the quality of the output of a number of related sugar factories as well as to check the methods of control employed in the factory laboratories. The general plan of a control laboratory along these lines was copied by the Hawaiian Sugar Planters Association, and by other sugar companies and associations in Cuba and Puerto Rico. Several important contributions to sugar analytical procedures were developed in the Cardenas Laboratory and published in American Chemical Society journals with Spencer's approval. He always encouraged members of his staff to publish their ideas independently. He himself was a frequent contributor in the Society's publications.

Spencer gave to the sugar industry the unusual combination of author, trained scientist, highly specialized technologist, and practical sugar-maker. He also had a flair for devising special apparatus to simplify the routine laboratory control which he had inaugurated and developed. His name is still associated with types of ovens, digesters, and extractors that he designed over 50 years ago.

He died of a heart attack in Cuba, Mar. 25, 1925, while travelling to one of the *ingenios* that he directed for the Cuban-American Co.

Biography in Spencer-Meade "Cane Sugar Handbook," 7th Ed., John Wiley (1929); obituaries in *Inter. Sugar J.*, 27-180, April 1925; *Planter and Sugar Manufacture*, April 9, 1925; *AOAC Journal* **9**, iii-vi (January 1926); "Reference Book of the Sugar Industry of the World," Planter & Sugar Manf. Co. (1925); Memorial tribute in Spanish, *Boletin Oficial de la Asociación de Tecnicos Azucareros de Cuba* **10**, No. 5, pp. 169-177 (July 1951).

GEORGE P. MEADE

Wendell Meredith Stanley

1904-1971

Stanley, who shared the Nobel Prize for Chemistry in 1946, for work in elucidating the nature of viruses, was born in Ridgeville, Ind., Aug. 16, 1904, the son of James and Claire (Plessinger) Stanley. His parents were publishers of the local newspaper.

When Wendell was in high school, the family moved to Richmond, Ind., where he later entered Earlham College.

He majored in mathematics and chemistry, but like many later-to-be-prominent men, he was interested mostly in football. (Christian Anfinsen, Chemistry Nobel Laureate for 1972, and Supreme Court Justice Byron White were also outstanding football players during their college days.) Stanley actually had an offer to remain at Earlham after graduation as assistant football coach, but in the spring of his senior year (1926) he visited Roger Adams at University of Illinois and became convinced that his real metier was chemistry.

Entering the University of Illinois Graduate School in 1926, he emerged in 1929 with his Ph.D. degree and 13 published papers. It was also at Illinois that he met his wife, the former Marion Jay, another graduate student under Adams. They were married in 1939 and had four children.

After obtaining his degree, Stanley remained a year at Illinois as a research assistant and instructor, then went to Munich for post-doctoral work with Heinrich Wieland, for whom he prepared the first optically resolvable disubstituted biphenyls.

In 1931 he returned to America, going to the New York branch of the Rockefeller Institute for Medical Research, where he assisted Osterhaut in preparing models of semipermeable membranes.

The following year he became scientifically independent when he moved to the Rockefeller Institute for Medical Research in Princeton, where he began the research on viruses which was to bring him the Nobel Prize.

By 1934 he had crystallized tobacco mosaic virus and shown it to be a nucleoprotein. This was earth-shaking. It had been believed that viruses were living organisms, because they were infectious, like bacteria, and because they were able to reproduce themselves prolifically. Since the 1950's it has of course become common knowledge that the nucleic acid portion of the virus serves as a template for reproduction of more virus by the macromolecule-synthesizing apparatus of the infected cell, but the virus itself is not alive. Some commentators have compared Stanley's breakthrough to Pasteur's work on the nature of bacteria. In any case, he shared half of the 1946 Nobel Prize with John Northrup, the other half going to James Sumner of Cornell, who had prepared the first crystalline enzyme, urease.

The remainder of Stanley's scientific career was devoted to work on viruses. He demonstrated that different viruses differ in their chemical structure, and he predicted that because of the chemical similarity between viruses and genes, and in view of the fact that he had been able to alter the properties of viruses by chemical means, the hereditary properties of organisms would some day be altered by chemical reagents. This has of course come to pass.

During World War II he interrupted his strictly scientific investigations to help the government develop influenza vaccines; these vaccines constitute the familiar "flu shots," which are administered yearly to millions.

In 1946 while flying to California to receive an honorary degree, Stanley met Robert G. Sproul, president of University of California, who asked him on the spot to create and direct a virus research laboratory at Berkeley. This he did, and he remained there from 1947 until his retirement in 1969. After his official retirement, he continued to work in the same place.

Stanley died of a heart attack June 15, 1971, while he was attending a conference at Salamanca, Spain.

Current Biography, 604-605, H. W. Wilson (1947); "McGraw-Hill Modern Men of Science," McGraw-Hill Book Co., 440-441 (1966); obituary, *New York Times,* June 16, 1971.

PETER OESPER

George Starkey

1627-1665

George Stirk, or as he called himself George Starkey, was born in Bermuda in 1627. He entered Harvard in 1643 and while there began to pursue alchemy independently. After graduating in 1646 he took up medicine, perhaps as a result of his interest in iatrochemistry, and soon was a practicing physician. He continued to study alchemy and iatrochemistry, borrowing books and perhaps glassware and chemicals from John Winthrop, Junior.

In 1650 he emigrated to England where he associated with scientists and physicians, among them Robert Boyle. Earning his living as a physician, he continued to experiment with the hope of finding new and better medicines and of transmuting metals, but none of his remedies could save him from the Great Plague that struck London in 1665. Several of his manuscripts are in the British Museum, and a small book, "Pyrotechny Asserted and Illustrated," (1658) contained his obscure ideas on chemical medicines.

R. S. Wilkinson, "George Starkey, Physician and Alchemist," *Ambix* **11,** 121-52 (1963); John L. Sibley, "Biographical Sketches of Graduates of Harvard University," Charles William Sever, University Bookstore, Cambridge (1873), I, 131-37; G. H. Turnbull, "George Stirk, Philosopher by Fire," *Publications Colonial Soc. Massachusetts* **38,** 219-51 (1949); Samuel E. Morison, "Harvard College in the Seventeenth Century," Harvard University Press (1936), 235.

WYNDHAM D. MILES

James Hervey Stebbins, Jr.

1857-1932

Stebbins was the American pioneer in the chemistry of azo dyes and the first chemist to publish articles on the subject in *Journal of the American Chemical Society.* In the 1870's fuchsin, a triphenylmethane compound, was the most important commercial synthetic dye. Synthetic alizarine and other kinds of synthetic dyes were just coming into use. Peter Griess' discovery of azo dyes from diazo compounds was exciting chemists. Stebbins' first paper, "Some New Azo Colors," was in volume 1 of the *Journal* (1879). In 1888 there were said to be 3000 German patent applications for azo dyes. By 1895 Stebbins had published 15 papers in the *Journal,* practically all on diazo compounds and azo dyes. During this period he obtained 22 American patents for coloring matters.

Stebbins worked for various companies from 1879 to 1886 and was an analytical and consulting chemist from 1880 until his death. The dye firm of T. & C. Holliday, in Brooklyn, may have been one of his employers since he stayed in New York City until 1894 at least. His work was known in Europe, the center of dye chemistry. *The Journal of the Society of Chemical Industry* abstracted his articles and published one of his papers, and *Berichte* published two papers.

In 1886 one or more unidentified German chemists published a humorous pamphlet, "Berichte der Durstigen Chemischen Gesellschaft." Two imaginary chemists mentioned were R. Butcher and S. C. Swine. The authors picked Stebbins out of all American chemists to be the butt of one of their jokes, ascribing to him a make-believe patent: "James N. (sic) Stebbins in New York, V.S.A. A Red Dye (American Patent 6813760 of March 1, 1886.) Diazo naphthalene is treated with an alkaline solution of picric acid."

Born in Brooklyn, N.Y., June 4, 1857 Stebbins graduated from Harvard with his B.S. degree in 1878, obtained his M.S. degree from Rutgers in 1880, and his Ph.D. degree from Omaha in 1893. He operated a consulting and analytical laboratory first in New York City, and later in Clayton, N.Y. In the latter town his interest turned to bio-chemistry and pathology. The cottage in Clayton which he used as a chemical laboratory still stands (1973).

Stebbins held many posts in the American Chemical Society ranging from recording secretary in 1881, to membership on the board of directors in 1892.

When August Wilhelm von Hofmann visited the United States, a banquet was given in his honor at the Hotel Brunswick, New York City, Oct. 16, 1883 by American chemists. A lone copy of the beautiful menu is in the rare book room at The Chemists Club. Along one edge is written: "Presented by Dr. J. H. Stebbins, Jr." Stebbins, in his will, bequeathed his scientific books, apparatus, and chemicals to The Chemists' Club.

Stebbins lived the latter half of his life in Clayton, N.Y., where he was well liked by the townspeople. He was senior warden of Christ Episcopal Church, in which there have been three organs; Stebbins gave the second, which is commemorated by a bronze plaque.

He was an authority on butterflies, and in summer he enjoyed gliding over the local waters in a little motor boat he named "The Microbe."

He was married twice, first to Alicia Ayeleen Radcliff, and some time after her death to Alzada M. Rees. He had no children.

Stebbins died of pleuro-pneumonia Apr. 22, 1932 at the age of 74 and was buried in Clayton Cemetery. His will indicated that he and his parents were financially well-to-do.

Personal communications from Gordon D. Cerow, Rev. Richmond N. Hutchins, and Mrs. Chester D. Bums, Clayton, N.Y., H. Ben Mitchell, Clerk, Jefferson Co., Watertown, N.Y., Mrs. Helethea R. Roy, *Thousand Islands Sun,* Alexandria Bay, N.Y., and Bernice M. Hetzner, Director, Library of Medicine, University of Nebraska. In the 1921 edition of "American Men of Science" Stebbins stated that he had received his Ph.D. degree from Omaha in 1893.

DAVID H. WILCOX, JR.

Joel Dorman Steele

1836-1886

Steele was an outstanding educator and an author of widely read texts. His books, a minimum of 15 (seven of which were co-authored by his wife), sold over one million copies. A few of his books were translated into Japanese. His astronomy book was published in Arabic. Steele's "Chemistry" was published in a braille edition.

The son of Reverend Allen and Sabra (Dorman) Steele, Steele was born May 14, 1836 in Lima, N.Y. His father was an itinerant Methodist minister, which resulted in an irregular early education for Joel. He attended Boys' Classical Institute in Albany, and Boys' Academy in Troy each for 2 years. In 1853 he worked in a publishing house in New York City. In 1854 he continued his studies at Genesee Wesleyan College in Lima. Wesleyan College later became Syracuse University. He was a diligent, industrious student. He worked summers on his father's farm and taught school to provide for his college education. He graduated high in his class in 1858.

Steele taught at Mexico Academy, Mexico, N.Y. in 1858 and became the principal in 1859. On July 7, 1859 he married Esther Baker, a music teacher at the academy. In 1861 he resigned from the academy and helped raise the 81st New York Volunteers, being appointed a captain. During the battle of Seven Pines he was badly wounded and after a lengthy recovery resigned his commission.

In 1862 he was appointed principal of Newark Union Free School at Newark, N.Y. Here he first became interested in the sciences. He created apparatus and methods of teaching, and soon became recognized as a gifted teacher and administrator.

The final school move came in 1866 when he assumed the duties of principal of Elmira Free Academy at Elmira, N.Y. The school was in a disorderly state, but he quickly brought it under control. He exerted great influence over the students and gained their respect and admiration. He taught from original, well-prepared, thoughtful notes and outlines in place of the current books. These notes led to a series of textbooks, the first of which, "Fourteen Weeks in Chemistry," he published in 1867. The Fourteen Weeks series continued with "Astronomy" (1868); "Natural Philosophy" (1869); "Geology" (1870); "Human Physiology" (1873); and "Zoology" (1875). The books were widely accepted by teachers as textbooks and by the public as general reading. They did much to make the study of the sciences a popular high school subject. One indication of the acceptance of his publications is apparent from the fact that in 1880 Steele's "Chemistry" was used in 60 out of the 122 public high schools in cities with populations of 7,500 or over. During the period 1871-84, Steele and his wife co-authored a series of history textbooks, the Barnes Brief History Series. Seven of his books were still in print in 1928.

He left Elmira Free Academy in 1872 and devoted the rest of his life to writing. His interest in education continued, and he regularly attended meetings of the Regents of the University of the State of New York.

Steele had a strong personality and a character in line with his deep Christian beliefs.

These beliefs were part of all of his writings. He led a life that was an inspiration and an example to his friends and acquaintances.

In 1870 he received a Ph.D. degree from the Regents of the State of New York, and in 1872 he was elected a trustee of Syracuse University.

Steele died May 25, 1886 at the age of fifty. He left $50,000 to Syracuse University to found a chair of theistic science, and his wife gave seed money ($5,000) to Syracuse to found Steele Hall of Physics in her name. She also helped erect Steele Memorial Library in Elmira.

Biography of Steele in Charles N. Sims, "One Hundredth Annual Report of the Regents of the University of the State of New York," 285-289 (1887); "Dictionary of American Biography," **17,** 556, Chas. Scribner's Sons (1935); "National Cyclopaedia of American Biography," **3,** 265, James T. White & Co. (1891); Paul Fay, "The History of Chemistry Teaching in American High Schools," *J. Chem. Educ.* **8,** 1545 (1931).

GEORGE J. VIRTES, JR.

Lewis Henry Steiner

1827-1892

Steiner, born in Frederick, Md., May 4, 1827, received his bachelor's degree from Marshall College, Mercersburg, Pa., in 1846, and his M.D. degree from University of Pennsylvania Medical School in 1849. He practiced medicine in Frederick from 1849 to 1852, then in Baltimore until the Civil War. When fighting began Steiner joined the U.S. Sanitary Commission, a non-government, volunteer organization which played the same role in the Civil War as Red Cross played in later wars but provided greater medical care for wounded soldiers. As Chief Inspector, U.S. Sanitary Commission, Army of the Potomac, Steiner treated the wounded at Bull Run, Antietam, Harrison's Landing, and other bloody fields. He was in Frederick when Lee's Army of Northern Virginia encamped there on its way to Antietam, and he later published an interesting diary of events mentioning that Stonewall Jackson had fallen asleep in church and that an "aged crone" (Barbara Frietchie?) had berated Confederate soldiers for dragging a U.S. flag in the dirt.

Sometime in his student years Steiner had become attracted to chemistry. He impressed his chemistry teacher at Marshall as a "bright, young man," and at Pennsylvania he attended Robert Hare and James B. Rogers, two of the best chemists in the country. Apparently he learned much of the subject, for a few years after graduation he began to lecture on chemistry in a medical preparatory school in Baltimore. In 1853 George Washington University elected him professor of chemistry and natural history in the college, and of chemistry and pharmacy in the medical school. At George Washington his annual, 4-month course included physics, inorganic and organic chemistry, and toxicology, with "proper demonstrations" to make it "most attractive to students." In 1855 he and Daniel Breed, Ph.D., a patent examiner who had studied under Liebig, translated Heinrich Will's "Outlines of Chemical Analysis," and had it published as a text for his classes.

Steiner taught at George Washington until 1856 and during this time established a reputation in Maryland as a good teacher. St. James College, Hagerstown, hired him to lecture on chemistry and physics from 1854 to 1859, and Maryland College of Pharmacy elected him professor of chemistry 1856 to 1861 and 1864 to 1865. Several articles on chemistry which he wrote during this period are in medical journals, the most interesting being reviews for physicians such as "Report on the recent contributions of chemistry to the medical profession" in 1857. He published at least two introductory lectures to his chemistry courses.

After the 1860's Steiner's other interests crowded out chemistry. He served on Frederick's school board several years, was political editor of the Frederick *Examiner,* sat in the Maryland State Senate in 1871, 1875, and 1879, and was a delegate to the Republican convention of 1876 which nominated Rutherford Hayes for the presidency. He was religious, serving on church committees which compiled a hymnal, a catechism, and an order of worship. He was a founder and a president of American Academy of Medicine and a vice president of American Public Health

Association. He wrote stories for children. When Enoch Pratt built the library which bears his name in Baltimore in 1886, he engaged Steiner to act as librarian.

Steiner "was a hard student, an eloquent speaker and wielded a facile pen." An avid reader, he "almost lived in his library, and he died surrounded by his books," of apoplexy, in his home in Baltimore, Feb. 18, 1892.

Obituary by Traill Green, *Bull. Amer. Acad. Med.* 1, 216-218 (1891-95); records, George Washington University; Steiner's writings; "Dictionary of American Biography," **17,** 562, Chas. Scribner's Sons (1935); quotes in the final paragraph are from obituary in *Baltimore Sun,* Feb. 19, 1892.

WYNDHAM D. MILES

David Stewart

1813-1899

Stewart was born at Port Penn, Del., Feb. 14, 1813. After attending New Castle Academy, he went to Baltimore and apprenticed himself to an apothecary. During a long apprenticeship he learned chemistry and pharmacy, and at about the age of 25 opened his own business as "chemist and pharmaceutist."

Civic-minded, he served as a member of the Baltimore City Council from 1835 to 1837, was school commissioner of Baltimore in 1836, and a state senator in 1840.

In 1840 he helped organize Maryland College of Pharmacy and delivered the first course of lectures on chemistry there in the winter of 1841-42. He became the first professor of theory and practice of pharmacy at the college in 1844 and taught the subject until 1846.

In the late 1830's Stewart's wife and infant son died; believing that their deaths resulted from inadequate medical care, he took up medicine. Busy with civic and professional duties, he nevertheless persevered with his studies and received his M.D. degree from University of Maryland in 1844. In 1847 he joined with several other physicians in establishing Maryland Medical Institute, a preparatory school where he lectured on chemistry and pharmacy until the school closed about 1853.

In 1850 the federal government appointed Stewart Inspector of Drugs for the Port of Baltimore. He ran assays and analyses of imported chemicals and drugs covered by the law, finally resigning in 1853.

Stewart accepted the professorship of chemistry and natural philosophy at St. John's College in 1855 and moved to the neighborhood of the school in Annapolis, Md. He taught there for 7 years and for a part of the time was also chemist for the State Agricultural Society. In the laboratory at St. John's he ran analyses of soils, fertilizers, and other materials as requested by farmers of the state.

The turmoil in Maryland caused by the Civil War led Stewart to return to his boyhood home, Port Penn, in 1862. There he practiced medicine until his death on Sept. 3, 1899.

Biography by George E. Osborne, *Amer. J. Pharm. Educ.* **23,** 219-230 (1959); Eugene F. Cordell, "Medical Annals of Maryland 1799-1899," Baltimore (1903), pp. 582-583.

WYNDHAM D. MILES

Julius Oscar* Stieglitz

1867-1937

Stieglitz was born in Hoboken, N. J., May 26, 1867, the son of Edward and Hedwig (Werner) Stieglitz. After he and his twin brother, Leopold, passed the entrance examinations to City College of New York at the age of 14, their father, who was born in Thuringia, took them to Karlsruhe where they attended the Realgymnasium until 1886. He was attracted to medicine, but when Leopold chose that field, Julius preferred to be different and chose chemistry. He obtained his Ph.D. in organic chemistry under J. C. W. Ferdinand Tiemann at University of Berlin and then spent a year with Victor Meyer at Göttingen.

Stieglitz returned to America in 1890 and

* The middle name "Oscar" is given in some directories, but Stieglitz never used it on his scientific papers.

worked a few months with John U. Nef at Clark University. He was toxicological analyst with Parke, Davis & Co. for 2 years, and in 1892 he joined the faculty of the new University of Chicago. In 1905 he became professor and was chairman of the Chemistry Department from 1915 to 1933. He was emeritus professor from 1933 to 1937 when he died on Jan. 10 from a heart attack.

Stieglitz married Anna Stieffel, Aug. 27, 1891 at Lake George, N.Y., where the Stieglitz family had a summer home. They had three children. His wife died in 1932, and Stieglitz married Mary M. Rising, an associate professor of chemistry at University of Chicago, Dec. 25, 1932.

During his many years at University of Chicago, Stieglitz carried out an extensive program of research in physical organic chemistry. Much of his early work dealt with the mechanisms of molecular rearrangments, including the Beckman rearrangement. The work of Stieglitz and his students in identifying intermediates and postulating mechanisms for these reactions served as models for research in this field. As the electronic theory of chemistry was developed, Stieglitz was a leader in applying this concept to organic reaction mechanisms. Stieglitz also conducted research on the mechanisms of catalytic reactions in which he directed his attention to the detection of intermediate compounds formed with catalytic agents. Another major area of his research was developing a general theory of indicators and color production in dyes. He had 118 doctoral students, but his name was on few of the papers published from their theses.

In addition to his research, Stieglitz was influential in setting the overall course of American chemistry. As president of the American Chemical Society in 1917, he was helpful in organizing American chemistry for World War I. Later, as advisor for the Chemical Foundation, he was an outspoken advocate for strong chemical industries which would insure American independence from Europe in the manufacture of fine chemicals. During the war Stieglitz served as chairman of the Committee on Synthetic Drugs of the National Research Council and was 19 years a member of the Council of Pharmacy and Chemistry of the American Medical Association. In these positions he was influential in the increased application of chemistry to medical research and thus helped promote the rise of American chemistry into prominence in this field in the period 1925–35.

Stieglitz was a member of the National Academy of Sciences; he received several honorary degrees and in 1923 the Gibbs Medal. He was associate editor of the *Journal of the American Chemical Society,* 1912–19, and on the board of editors of ACS Monographs, scientific series, 1919–36. Even in retirement he continued to be active, with an office in the new Jones Chemical Laboratory and a halftime position with Universal Oil Products Co. He played the cello well, and he was skillful in photography, the field in which his elder brother, Alfred Stieglitz, reached world renown.

William Albert Noyes, "Julius Stieglitz, 1867-1937," *Biog. Mem. Nat. Acad. Sci.* **21,** 275-314 (1941) contains a bibliography of his publications; Herbert N. McCoy, "Julius Stieglitz," *J. Amer. Chem. Soc.* **60,** 3-21 (Supplement to No. 11, Nov. 5, 1938); obituaries in *Proc. Amer. Acad. Arts Sci.* **72,** 387-391 (1937) and *Chem. Eng. News* **15,** 39 (1937).

DANIEL P. JONES

John Maxon Stillman

1852-1923

Stillman was the first professor of chemistry at Stanford University. He was born in New York City, Apr. 14, 1852, but his family moved to Sacramento while he was still very young and then settled in San Francisco. He graduated from University of California in 1874. He then spent 2 years in foreign study, at the universities of Strassbourg and Wurzburg. In 1876 he returned to University of California as an instructor in chemistry. He held this position until 1882, but in that year he received an offer of industrial work at a salary which the university would not meet. He therefore became superintendant and chief chemist at the plant of the Boston and American Sugar Refining Companies. However he was able to take his Ph.D. degree from California in 1885.

He had made the acquaintance of Senator Leland Stanford, and when the latter founded Stanford University he recommended Stillman to president David Starr Jordan. Stillman was appointed professor of chemistry at Stanford when the university opened in 1891. He was one of the three senior members in the original faculty of 15 when the school began. He served as head of the Chemistry Department until his retirement in 1917. In 1913 he became vice-president of the university and often served as acting president. His administrative skill was evident in these positions.

The original chemistry building which Stillman occupied was part of the main university quadrangle. It is said that one day Mrs. Stanford drove by the building in her carriage while a chlorination experiment was being conducted inside. She was enveloped in a cloud of chlorine. Stillman was at once given a large new chemistry building in a distant part of the campus.

Stillman carried out a variety of investigations during his career, paying particular attention to the chemistry of vegetable products and to the freezing points of organic substances. He was an art collector and had a fine collection of Japanese prints and carvings. However, the studies for which he is best known developed from a course which he gave in the history of chemistry. His interest in this field grew, and in 1912 he began to publish articles on the subject. At first he devoted much time and effort to the study of Paracelsus and his epoch. His last paper gave evidence that many alchemical writings were forgeries ascribed to earlier writers. After he retired he began his major work, the book, "The Story of Early Chemistry." In it he surveyed the major chemical developments, both practical and theoretical, down to the end of the eighteenth century. This work was for a long time one of the most authoritative books on its subject. It was in the hands of the publisher when Stillman died, Dec. 13, 1923.

Biographies of Stillman are by David Starr Jordan in *Science* **59,** 270 (1924), by S. W. Young in the foreword to Stillman's "The Story of Early Chemistry," Constable and Co. Ltd. (1924), and by H. M. Elsey in *J. Chem. Educ.* **6,** 466-472 (1929). The chlorination anecdote is a tradition at Stanford. *Sci. Mon.* **17,** 318 (1923).

HENRY M. LEICESTER

Thomas Bliss Stillman

1852-1915

Stillman was born May 24, 1852, at Plainfield, N.J. Thomas was the son of Charles Henry and Mary Elizabeth (Starr) Stillman. After receiving his early schooling in Plainfield, in Hamilton, N.Y. and at Alfred University, Stillman entered Rutgers University in 1870. Following his graduation in 1873 he remained at Rutgers as a teaching assistant in analytical chemistry.

From 1874 until 1876 he served as chemistry assistant to A. R. Leeds at Stevens Institute of Technology. He then went to Wiesbaden, Germany, where he studied analytical chemistry with Karl Remegius Fresenius. After 2 years he declined Fresenius' offer to become his assistant and returned to New York City where he opened an analytical laboratory in 1879. At the same time he served as an industrial consultant and associate editor of *Scientific American.* In 1881 he married Emma L. Pomplitz of Baltimore, with whom he had three children.

He returned to his old position as Leeds' assistant in 1882 and did graduate work at Stevens for which he received his Ph.D. degree in 1883. He became professor of analytical chemistry at Stevens in 1886 and head of the Chemistry Department (with the title professor of engineering chemistry) upon Leeds' death in 1902.

His interest in industrial chemistry led to his conducting a consulting practice and helping several local governments with water supply problems. He also became concerned about milk supplies and assisted the Medical Milk Commission of Newark for several years. He retired from his teaching position in 1909 and devoted his full time to his consulting and government work, becoming senior member of the New York firm of chemical consultants, Stillman and Van Siclen. From 1911 until his death, Aug. 10,

1915, he was city chemist of Jersey City and Bayonne.

Stillman's best known publication was "Engineering Chemistry, A Manual of Quantitative Chemical Analysis for the Use of Students, Chemists, and Engineers" (1897), a popular college textbook which went through six editions. His only other book was "Examination of Lubricating Oils" (1914). He published over 30 articles on analysis and engineering. He held patents on processes for manufacturing fertilizers and illuminating gas. In 1906 in New York he attracted attention to chemical processes by holding a "synthetic dinner," the food at which was prepared from synthetic products.

Obituary by F. J. Pond, *Ind. Eng. Chem.* **7**, 804-805 (1915), with portrait; *Stevens Indicator,* Oct. 1915; "Dictionary of American Biography," **18**, 27, Chas. Scribner's Sons (1936).

SHELDON J. KOPPERL

Charles Milton Altland Stine

1882-1954

Stine, an early advocate of fundamental research by industry, was the son of Milton Stine, a noted Lutheran clergyman and teacher, and Mary (Altland) Stine and great grandson of Charles Stein, who immigrated to America from Germany in 1798.

Stine was born in Norwich, Conn., Oct. 18, 1882. He attended grade school in Los Angeles, graduated from high school in Harrisburg, Pa., entered Gettysburg College at age 15 and earned four degrees: B.A. 1901 (first honors); B.S. 1903; M.A. 1904; and M.S. 1906. During 1904-05, he taught chemistry at Maryland College for Women. After one semester at University of Chicago, he entered Johns Hopkins University where he received a fellowship in 1906. After completing work for his Ph.D. degree in organic chemistry in 1907, he joined the Du Pont Co.'s Eastern Laboratory at Gibbstown, N.J. as a research chemist on nitration processes. His work over the next 2 years led to safer explosives for use in mining and to waterproofed and low freezing-point dynamites. From 1909 to 1916 he was in charge of all organic work at Eastern Laboratory. Under his direction were developed processes for making ammonia, aniline, nitric acid by the catalytic oxidation of ammonia, trinitrocresol, mononitrobenzene, and urea. The first commercial production of TNT in this country and a totally new process for making picric acid were under his direction. These developments, which resulted in 19 patents for Stine, radically affected the ability of the United States to produce high explosives during World War I.

With the coming of the war, the U.S. was cut off from its source of German dyes, and as a result, a number of companies jumped into dye synthesis. Stine headed Du Pont's efforts, which by 1918, resulted in processes for more than 150 dyes and intermediates.

Transferred to company headquarters in Wilmington, Del. in 1917, for the next 2 years he headed all organic research in the Chemical Department. In 1919 he was promoted to assistant director of the Chemical Department and from 1924 to 1930 he was director. In 1930 he was elected vice president in charge of research, and was made a director of the company and a member of its executive committee, positions which he held until he retired in 1948.

It was largely because of Stine's foresight that, in 1928, Du Pont embarked on a program of fundamental research, work that before had been done largely by universities. He convinced Lammot du Pont, then president, to set aside $750,000 to be used over five years in "a quest for facts about the properties and behavior of matter without regard to a specific application of the facts discovered." During the business depression of the 1930's, the large investments that were made in the study of high polymers led to neoprene synthetic rubber and to nylon fibers and plastics.

As a consultant to the United States Army Chemical Warfare Service, Stine was the first Du Pont official to be concerned with the atomic energy project; however, he refused a military commission.

As a means of testing his own theories of agriculture, Stine operated what was said to be one of the most scientifically controlled dairy farms in America. From studies of grassland management and drying grass artificially, he demonstrated that winter milk could be produced with virtually the same carotene content of summer milk. At Du Pont he pushed research in agriculture, and in 1952 the Stine Laboratory for the study of animal medicine and nutrition was dedicated in his honor.

Stine served one term as chairman of the Delaware Section of the American Chemical Society. He held a number of offices in the American Institute of Chemical Engineers and was president of that organization in 1947.

For several years he was advisor to the Departments of Chemical Engineering at Massachusetts Institute of Technology and at Princeton University. He was chairman of the trustees of Gettysburg College and on the executive committee of University of Delaware. A favorite topic was his insistence that chemical engineering is every bit as "cultural" as are the classics. His ideas on the role of the chemical industry in society were presented in more than 20 publications. In spite of ever pressing business affairs he still found time for active participation in many cultural and humanitarian causes. He was particularly interested in the ties between chemistry and medicine, and he fostered research in chemotherapy.

He married Martha Molly of Lebanon, Pa. in 1912 and had two daughters. Honorary doctorates were conferred on him by Gettysburg College (1926), Cumberland University (1932), Temple University (1941), and University of Delaware (1947). The Society of Chemical Industry awarded him the Perkin Medal in 1940. He died of a heart attack in Wilmington, Del., May 28, 1954.

Samuel Lenher, "Encourage the Seeker—and Find Nylon," *Nation's Business* **50,** 67-69 (January 1970); "The National Cyclopaedia of American Biography," James T. White Co. (1965) **47,** 168-169 (photograph); obituary, *Morning News,* Wilmington, Del. (May 29, 1954); Files of the Public Affairs Department, E. I. du Pont de Nemours & Co., Wilmington, Del.; "Who's Who in Delaware," National Biographical Society (1932) 193; C. M. A. Stine, "Rise of the Organic Chemical Industry in the United States," *Ind. Eng. Chem.* **32,** 137-144 (1940).

HERBERT T. PRATT

John Tappan Stoddard

1852-1919

Stoddard, first chairman of the Chemistry Department of Smith College in Northampton, Mass., the man for whom the college's chemistry building, Stoddard Hall, was named, was born in Northampton, Oct. 20, 1852, the son of William Henry and Helen (Humphreys) Stoddard. John Tappan was of the seventh generation of Stoddards in America, and his branch of the family had been prominent for six generations in Northampton—ministers, educators, and the like, after whom other buildings on the Smith campus were named.

Except for the years of his education, John Stoddard never left Northampton—though he was anything but provincial. He was educated in the public schools there, moved a few miles away to take his A.B. degree at Amherst College in 1874, and returned for a year as assistant principal of the Northampton high school. In 1876 he went to Germany to study under Hans Hübner at Göttingen, receiving his Ph.D. in 1877. His dissertation on anhydrobenzamidotoluic acid (2-(*p*-carboxyphenyl) benzimidazole), which appeared in the *Berichte* in 1878, proved to be one of only half a dozen research publications produced over his entire lifetime.

When Stoddard returned to Northampton in 1878 he accepted a position at Smith College, then in only its third year of operation. Initially he was appointed instructor in physics and mathematics, with Bessie T. Capen, from Wellesley, as chemistry teacher. In 1881 Stoddard became instructor of chemistry and physics; in 1892, when these two departments were divided (and academic ranks were established), he was made professor of chemistry and became the head of the department, both of which positions he occupied at his death 27 years later.

Stoddard was first and foremost a teacher; not merely an imparter of knowledge—of which he held a huge store—but a man of unusual clarity of thought, who could pass on some of that clarity to his students. Stoddard recognized the importance of research and of the seminar-type class, even at undergraduate level. To promote the first of these he helped design the chemistry building, built in 1898, as well as the addition in 1918; this is the building which was given his name after his death, and it was so well designed that it served for both research and teaching for a generation or more after that. Despite his own meager research output (or perhaps contributing to it), Stoddard encouraged research projects by students. Close contact among teachers and students, as well as an *esprit* necessary in an emerging women's college, was maintained in large measure by the colloquium Stoddard established in the middle '80's, one of the earliest and longest-lived clubs on the campus. In addition to the usual seminar function, the colloquium also served to acquaint girls outside the department with some aspects of the sciences, through occasional at-homes with displays in mechanics and chemistry, and optical trickeries, of a quality we usually associate with the best museums of science. Finally, Stoddard set forth his own understanding of chemistry for students in a series of texts: "Outline of Qualitative Analysis for Beginners" (1883); "Outline Lecture Notes on General Chemistry: Non-Metals" (1884), and "Metals" (1885); "Quantitative Experiments in General Chemistry" (1908); "Introduction to General Chemistry" (1910); and "Introduction to Organic Chemistry" (1914).

Stoddard's few research papers, aside from the thesis work mentioned above, dealt mainly with hydrocarbons and their flash points (his is not the name of "Stoddard solvent," which was named for the company which produced it, a decade or more after his death). In addition, his inquiring mind led him into some curious byways. In 1913 he published a book, "The Science of Billiards with Practical Applications," which one supposes must be the definitive work on the subject. He also became interested in "composite photography"—blending of features of many human subjects by multiple exposure—and published articles in *Science* and *Century Magazine*, as well as contributing the article on photography in Johnson's "Universal Cyclopedia."

Stoddard married Mary Grover Leavitt, of Northampton, June 26, 1879. They had three children: William L., James Leavitt, and Dorothy Stoddard Glascock. Stoddard died of heart disease Dec. 9, 1919.

Hampshire Gazette (Northampton, Mass.), Dec. 9 and Dec. 11, 1919; Henry Mather Tyler, "John Tappan Stoddard—A Tribute," in *Hampshire Gazette*, Dec. 18, 1919; "Dictionary of American Biography," **18**, 56, Chas. Scribner's Sons (1936); *New York Times*, Dec. 10, 1919; scattered references in L. Clark Seelye, "The Early History of Smith College, 1871-1910," Houghton (1923).

ROBERT M. HAWTHORNE JR.

Francis Humphreys Storer

1832-1914

Storer was born Mar. 27, 1832, in Boston, the son of David Humphreys and Abby Jane (Brewer) Storer. His physician father was a member of a group of Massachusetts naturalists whose scientific interests were an important influence on young Storer. In 1850 he entered Lawrence Scientific School of Harvard, where his ability in chemistry led to his serving as assistant to Josiah Parsons Cooke from 1851 to 1853. Simultaneously, he taught a private class in analysis at Harvard Medical School.

In 1853 he became chemist to the United States North Pacific Exploring Expedition, spending a year at sea. Upon his return he continued his education and received his B.S. degree from Harvard in 1855. He spent the next 2 years in Europe studying with Robert Bunsen at Heidelberg, Theodor Richter at Freiberg, Julius Stöckhardt at Tharand, and Emile Kopp at Paris. Returning to Boston in 1857, he opened a consulting laboratory and became chemist with the Boston Light Co., a position he held until 1871. In this position he made analyses of the gas used by customers, leading to some original research on the composition and illuminating power of coal and gas. He published "First Outlines of a Dictionary of the Solubilities of Chemical Substances" in 1864.

In 1865 Storer closed his consulting business and became professor of general and industrial chemistry at the newly organized Massachusetts Institute of Technology. Together with his close friend Charles W. Eliot, the professor of analytical chemistry, Storer wrote "A Manual of Inorganic Chemistry" (1867) and "The Compendious Manual of Qualitative Chemical Analysis" (1868). Both laboratory manuals were quite popular, the former being revised for nearly 50 years. During his tenure at MIT Storer spent several months in Paris observing the European chemical industry.

In 1870 the year after Eliot became president of Harvard, Storer was named professor of agricultural chemistry at Bussey Institution (the school of agriculture and horticulture) at Harvard. The following year he was named dean; and he held these offices until his retirement in 1907. In June 1871 he married Eliot's sister Catherine Atkins Eliot. As a result of a fire in 1872, the income of Bussey Institution was greatly reduced and Storer's research became restricted. He edited *Bulletin of the Bussey Institution* and personally contributed over 50 articles. His major work, "Agriculture In Some of Its Relations With Chemistry," appeared as two volumes in 1887. It subsequently ran through seven editions. With W. B. Lindsay he published "Elementary Manual of Chemistry" in 1894 and "Manual of Quantitative Analysis" in 1899. After his retirement he remained interested in chemical activities although a progressive deafness caused him to withdraw from active participation. He died July 30, 1914 in Boston.

Biographies by Charles W. Eliot in *Proc. Amer. Acad. Arts Sci.* **54,** 415-8 (1918-9), Charles A. Browne in "Dictionary of American Biography," **18,** 94-95, Chas. Scribner's Sons (1936); Tenney L. Davis, "Eliot and Storer," *J. Chem. Educ.* **6,** 868-79 (1929).

SHELDON J. KOPPERL

Bradley Stoughton

1873-1959

Bradley Stoughton was born Dec. 6, 1873 in New York City. He graduated from Yale University in 1893 with a major in civil engineering and continued his education in metallurgy under Robert Richards at Massachusetts Institute of Technology for 2½ years. His first position in the academic field was that of private assistant to Henry Howe, head of the Department of Metallurgy at Columbia School of Mines.

After a year at Columbia, he entered the steel and consulting fields and designed a new bessemer converter. From 1902 to 1908 he taught again at Columbia and was a partner in the firm of Howe and Stoughton, specializing in the design and improvement of steel furnaces. During this time he published "The Metallurgy of Iron & Steel," the leading book in its field for students and engineers for 40 years. He became the leading expert in the theory and operation of the practical steel foundry. In 1913 he was elected secretary of the American Institute of Mining Engineers and during World War I served on the National Research Council in Washington.

During 1923 Stoughton played an instrumental role in the change of 12-hour shifts to 8-hour shifts. President Harding urged the change, but steel executives opposed the move. Stoughton, as an authority and consultant in the steel business and yet not bound to an employing corporation, was chosen by the Federated Engineering Societies to make a survey and state his opinion. In his later reminiscences he recalled that his decision to reduce the working shift took more courage than anything else he had done. At first industry resisted the change but, with the pressure of public opinion, did agree in July 1923, to the 8-hour working shift.

In the fall of 1923, Stoughton again returned to academic life, this time as professor of metallurgy and later dean of engineering at Lehigh University until his retirement in 1939 at the age of 65. He was awarded an honorary degree upon his retirement. He was well-liked by the students and was often picked to be a speaker at class banquets. During the school year Sunday afternoon teas were held regularly at Stoughton's house. Humorous playlets were written by students and sometimes by Stoughton himself representing both students and faculty and were performed either at the Sunday teas or at the annual Metallurgy Christmas party.

In 1931 Stoughton was president of The Electrochemical Society. During World War

II, he served in Washington on the War Production Board and in 1943 accepted the post of corporate director of the Lukens Steel Co., Coatesville, Pa. He was still serving in this capacity at the time of his death on Dec. 30, 1959, at the age of 84. He published five books and 50 articles.

Gilbert E. Doan, "Bradley Stoughton," The American Society for Metals (1967).

MARY H. PERRY

Eugene Cornelius Sullivan

1872-1962

Sullivan was born in Elgin, Ill., Jan. 23, 1872. He received his bachelor's degree in chemistry from University of Michigan in 1894. While still an undergraduate, he worked part time as an assistant in the Department of Mines and Mining at the World's Columbian Exposition and after graduating he was employed briefly as a chemist by an explosives company and then by a baking powder company. These experiences convinced him of the need for further training, and he decided to seek this from the leading chemists of the day.

Accordingly, in the fall of 1896 he travelled to Germany and studied with Nernst at Göttingen and later with Ostwald at Leipzig where he obtained his doctorate in 1899. Following this, he returned to University of Michigan as instructor in analytical chemistry.

In 1903, Sullivan gave up teaching and entered the chemical laboratory of the United States Geological Survey. Here his work involved mineral analysis and silicate chemistry, and here he met Arthur Day, a leading authority on high temperature experimentation. It was Day who first interested Sullivan in glass chemistry and persuaded him, in 1908, to go to the Corning Glass Works to set up a research laboratory to study glass. This was the first such laboratory to be established in the United States.

One of the early accomplishments of Sullivan and his colleagues was the development of a heat-resistant borosilicate glass of good durability for the manufacture of chemical ware. The importance of this accomplishment is borne out by the fact that this glass quickly became the recognized standard for chemical equipment throughout the world and, with minor modifications, is still universally used. In recognition of this achievement, he was later awarded (along with W. C. Taylor) the Howard N. Potts Medal of the Franklin Institute.

Under Sullivan's guidance the Corning Laboratory grew rapidly in size and importance and attained world prominence. But his influence were by no means limited to the laboratory. By applying precise scientific methods he was in large measure responsible for the rapid growth and mechanization of the company's production facilities. In 1924 he was appointed to the post of manufacturing vice president and the following year was elected a director and also president. At various times he served as a director of many of Corning's affiliate and subsidiary companies.

Sullivan had long cherished the idea of developing materials that combined the desirable properties of the organic glasses, based on carbon chemistry, with those of the inorganic glasses, based on silicon chemistry. He initiated investigations along this line which ultimately led to the commercial production of the silicones and the founding of the Dow Corning Corp. of which he was, for many years, president.

In spite of the pressure of professional duties, Sullivan participated in many civic and philanthropic activities. He served 36 years on the Board of Education of the Corning schools, was a director of the Corning Chamber of Commerce, a member of the Corning Library Board, and a charter member of the Corning Museum of Glass.

For the American Chemical Society, Sullivan served many years on the editorial board of *Industrial and Engineering Chemistry*. For his country he served on the National Research Council and on chemical advisory committees to the armed services. He also served on the subcommittee on glass of the National Bureau of Standards.

Sullivan received the Perkin Medal of the Society of Chemical Industry, the Pioneer in Research Award of the National Association of Manufacturers, the Bleininger Award of the American Ceramic Society, and honorary doctorates from Michigan and Alfred Universities.

He remained remarkably active and mentally alert until a few weeks prior to his death on May 12, 1962 at the age of 90.

"Eugene Cornelius Sullivan—Glass Scientist," an illustrated booklet published by Corning Glass Works (1966); obituary, *Bull. Amer. Cer. Soc.* **41,** 436 (1962); biography, *Chem. Eng. News* **18,** 821 (1940).

ROBERT H. DALTON

James Batcheller Sumner

1887-1955

Sumner was a pioneer in the field of biochemistry and particularly in the study of enzymes. His isolation of a crystalline enzyme and his announcement that this enzyme was protein in nature were greeted with enthusiasm and support by some but with skepticism by most. Years of further work by Sumner and others were required before his discovery and views on the nature of enzymes became generally accepted.

Sumner was born Nov. 19, 1887 in Canton, Mass. where his family was active in farming and the cotton industry. He received his preparatory education at Roxbury Latin School (1900-06). Sumner entered Harvard University in 1906 and graduated 4 years later with his B.S. degree.

At the age of 17, Sumner suffered the loss of his left arm as the result of a hunting accident. Despite this handicap, he decided to study chemistry at Harvard. With his one arm and hand, Sumner was able to perform most of the manipulations required of a laboratory chemist, and he continued to do much of his own experimental work throughout his career. To further dispell any sense of inadequacy resulting from his loss, he participated in a number of strenuous sports including skiing, canoeing, and mountain climbing. He became particularly proficient at tennis and was still playing the game when he was in his sixties.

After his graduation from Harvard, Sumner spent a few months working in the family knitting mill. He was not reluctant to quit this type of work, and he eagerly accepted a one-term teaching assignment at Mt. Allison College, Sackville, New Brunswick. After his year at Mt. Allison, Sumner decided to continue his formal education in chemistry, and he accepted a research assistantship at Worcester Polytechnic Institute in 1911. He left Worcester a few months later to continue his graduate studies at Harvard under the guidance of Otto Folin. He was awarded his A.M. degree in 1913 and his Ph.D. degree in 1914.

Upon leaving Harvard, Sumner was appointed assistant professor of biochemistry at Cornell University Medical College. A heavy teaching load and a lack of laboratory help left Sumner with little time for research. It was this scarcity of time that caused him to select the "long shot" research problem of isolating an enzyme. He began in 1907 his attempts to isolate the enzyme urease from the jack bean. By 1926, Sumner had crystallized a protein he believed to be urease. This claim was attacked by several, but especially by a group of German workers who had already concluded that enzymes were not proteins. Those persons holding the same views as Sumner continued to be in the minority for several years. However, after a time other researchers, particularly John H. Northrop at Rockefeller Institute, began announcing findings on other enzymes which tended to substantiate Sumner's earlier findings with urease. Sumner himself had continued to work with urease, and during the decade 1926-36 he published about 20 papers concerned with the isolation and activity of this enzyme. Gradually, the fact that enzymes are proteins became generally accepted.

In 1929 Sumner was promoted to professor in the Department of Physiology and Bio-Chemistry of the Medical College. In 1938 he was appointed professor of biochemistry in the Department of Zoology of the College of Arts and Sciences. He was later associated with the Department of Biochemistry of the College of Agriculture. In 1947 Sumner was named director of the newly created Laboratory of Enzyme Chemistry. Sumner's activities at Cornell were interrupted by several periods of postgraduate work. He studied at the Medical School of the University of Brussels (1920-21), the University of Stockholm (1929), and at the University of Uppsala (1937).

Sumner was not a one-enzyme man. Other than urease, he isolated or investigated several enzymes including catalase and peroxidase. He was interested in protein isolations in general and was the first to prepare fibrinogen free from thromboplastic material. In addition to over 100 research papers, Sumner was author or co-author of several texts. "Textbook of Biological Chemistry" was published in 1927. With G. Fred Somers, he wrote "The Chemistry and Methods of Enzymes" (1943) and "Laboratory Experiments in Biological Chemistry" (1944). Sumner and Karl Myrbäck edited "The Enzymes, Chemistry and Mechanism of Action," a work which filled four volumes and which appeared over the period 1951-52. Sumner's contributions and leadership in biochemistry and enzyme research were recognized by the awarding in 1937 of the Scheele Medal of the Swedish Chemical Society and, with Northrop and Wendell M. Stanley in 1946, of the Nobel Prize in Chemistry.

In the spring of 1955, Sumner was preparing for his retirement from Cornell University and was making plans for a new and ambitious undertaking. He planned to go to Brazil after retirement to help organize an enzyme laboratory at the University of Minas Gerais, Belo Horizonte. Serious illness interfered, however, and he died Aug. 12, 1955 at Buffalo, N.Y.

Leonard A. Maynard, *Biog. Mem. Nat. Acad. Sci.* **31,** 376-96 (1958); "Current Biography Yearbook," 620-622, H. W. Wilson (1947); obituary, *New York Times,* Aug. 13, 1955.

GEORGE M. ATKINS, JR.

Gilbert Holm Swart

1903-1965

Swart was born in LaGrande, Oreg., Sept. 25, 1903. He received his B.S. degree in chemical engineering from the University of Washington in 1927.

Swart went to Akron, Ohio in 1927 to work as a rubber chemist with B. F. Goodrich Co. In 1936 he joined General Tire & Rubber Co. as chief chemist of their Wabash (Indiana) Mechanical Goods Division. In 1942 he returned to Akron to start a chemical research laboratory for General Tire. Here he became director of research in 1943, director of research and development in 1951, and assistant to the president for science and technology in 1962. He died Dec. 18, 1965.

Swart held about 20 patents, mostly in the fields of synthetic rubbers and resins. In 1943-44 he and his co-workers devised and reduced to plant practice a process for dispersing carbon black in synthetic rubber latex, followed by coagulation to give a solid rubber with black dispersed therein ready for factory processing. A variant of this process is still in use (1973). He discovered in 1944 that synthetic rubbers make excellent binders and fuels for rockets.

In 1949 Swart was co-inventor, with K. V. Weinstock and E. S. Pfau, of the process for making superior, but much less expensive, tire treads from very tough synthetic rubbers extended with more than 30 parts of inexpensive lube extract oils per hundred parts of rubber. The tread was compounded as if the oil plus rubber were 100 percent rubber; this saved about five cents a pound in the cost of tread rubber. U.S. Patent 2,964,083 for this oil-extended rubber process, finally issued Dec. 13, 1960, was the subject of the most extensive patent controversy in the United States and possibly in the world. In November 1973 the patent was upheld by the U.S. Court of Appeals, 6th Circuit.

Swart served as chairman of the Division of Rubber Chemistry of the American Chemical Society in 1962-63. In 1927 he married Florence McMeekin, and they had two sons: Gilbert H. and Alan.

G. H. Swart, E. S. Pfau, and K. V. Weinstock, *Indian Rubber World* **124,** 309-319 (1951); O. J. Scott, *Rubber World* **151,** No. 1, 33-39 (October 1964); R. F. Wolf, *Rubber Age* **102,** No. 8, 76-80 (August 1970); obituaries in *Rubber J.* **143,** No. 2, 42 (1966), *Rubber Age* **98,** No. 2, 130 (1966), and *Rubber World* **153,** No. 5, 18 (1966); personal recollections.

GUIDO H. STEMPEL

Magnus Swenson

1854-1936

Swenson ranked high among the leading chemical engineers of the nineteenth century at a time when that profession was scarcely recognized. He holds a similar rank as a chemist and as a mechanical engineer. During his last 30 years Swenson served with equal distinction as an administrator in public affairs and as a humanitarian on an international scale.

Swenson was born in Langeland, Norway, Apr. 12, 1854. When he was 2 years old his mother died, and the family moved to Larvik where his father owned a factory for manufacturing rope for sailing vessels. When the Swenson factory burned in 1868, Magnus left for America on the small sailing vessel *Victoria.* It took 11 weeks to reach the shores of New Foundland, 22 passengers died enroute due to starvation and disease. Swenson found a home with his uncle, a foreman of the Northwestern Railroad shops in Janesville, Wisc. After a few years in school Swenson worked with his uncle as a blacksmith helper, learning many skills in metal working. In 1876 Swenson enrolled at University of Wisconsin, receiving his B.S. degree in mining and metallurgy in 1880. During the summer of 1878 he worked as a surveyor for the Northwestern Railroad. In the summer of 1879 he worked with the U.S. Geological Survey. In the year 1880-81 he was appointed instructor in analytical chemistry at Wisconsin, and in 1881-82 was appointed assistant to W. A. Henry in the College of Agriculture. In this latter capacity Swenson constructed and operated a pilot plant for recovering sugar from sorghum cane. For his success Swenson received an unexpected check of $2500 from the U.S. Department of Agriculture. This was followed by a check of $10,000 from Colonel A. H. Cunningham for a year of service in operating a sugar refinery in Texas. In 1883 Swenson joined Judge W. L. Parkinson in operating a pilot plant in Ottawa, Kans. to produce sugar from sorghum cane followed by erecting and managing a commercial plant in Fort Scott, Kans. The U.S. Department of Agriculture was granted space in this plant to set up its own experiments in sugar refining under Guilford Spencer and Harvey Wiley of the USDA. Swenson was employed on the government project as well as by the industrial plant. At this time Swenson applied for a patent on preventing the inversion of sugar to glucose based upon his own work extending back 6 years. After much litigation this patent was granted Oct. 11, 1887, a pioneer case in establishing the conditions under which a federal employe may secure a patent on an invention related to his federal work. Swenson became known as the Eli Whitney of the sugar industry.

At Fort Scott Swenson helped organize American Foundry and Machine Works known later as Walburn-Swenson Co. in which Swenson owned half interest. With the rapid rise of the sugar cane industry Swenson shifted his activities to the invention, design, and fabrication of machinery for the sugar industry such as diffusion batteries, conveyors, cutting machinery, filter presses, evaporators, centrifuges, and strike pans. On Nov. 30, 1886 he received his first patent on a plate and frame filter press followed by several patents on multiple effect evaporators. In 1891 Swenson moved his headquarters to Chicago where he extended his activities to the fabrication of mining and chemical engineering machinery. The Swenson ore concentrator was used in all mining states and in Mexico, Chile, and Honduras. Over 20 patents in chemical engineering machinery were granted. In Chicago Swenson served as a consultant for many chemical industries engaged in the manufacture of soap, glycerine, pharmaceuticals and for the meat packing and tanning industries. For 6 years from 1896 on Swenson invented and improved machinery used in the cotton industry, especially in developing the cylindrical bale. In this he secured over 30 patents.

In 1902 Swenson retired from his consulting and manufacturing activities and turned his interests to public and humanitarian affairs. At the request of University of Wisconsin president C. F. Adams and of J. B. Johnson, dean of the College of Engineering, Swenson (beginning in 1899) promoted the formation of a Department of Chemical Engineering at Wisconsin with Charles F. Burgess as its first professor. In this behalf Swenson lectured about the state, appeared before the state legislature, and wrote several articles.

Swenson promoted the development of hydroelectric power from the Wisconsin River, serving as president of Southern Wisconsin Power Co. (1905-15). He became president of Norwegian-American Steamship Line, a director of Central Wisconsin Trust Co., director and vice-president of First National Bank of Madison, director of Union Refrigerator Transit Co., and member of the Great Lakes Waterways Commission. In 1906 he was appointed by Governor James Davidson as chairman of the Capitol Building Commission which he served for 10 years without compensation during construction of that magnificent building. In 1905 he was appointed by Governor Robert M. LaFollette as a member of the University Board of Regents.

During World War I Swenson was appointed by President Wilson as chairman of the Wisconsin Council of Defense and by Governor Emmanuel Phillipp as Federal Food Administrator for Wisconsin.

In December 1918 he was appointed by Federal Food Administrator Herbert Hoover to head agencies bringing relief to the hunger-stricken nations of Northern Europe. Swenson received honorary degrees from University of Wisconsin. In 1900 he received the John Scott Medal from Franklin Institute for his invention of machinery for producing the cylindrical cotton bale. For his food relief work he received the Gold Cross of St. Olav from King Hakon of Norway, the Order of the White Rose from President Mannerheim of Finland, and the United States Service Medal.

Swenson served as president of the Norwegian-American Historical Association. In his last years he lived gracefully and quietly with his children and grandchildren, occupied in gardening and painting landscapes of his native Norway. Swenson died Mar. 29, 1936, one year following the death of his wife.

Personal interviews with Magnus Swenson and his daughter, Mrs. Mary North; Swenson's personal correspondence and hundreds of newspaper clippings in possession of Mrs. North; Swenson's publications; biography by Olaf A. Hougen in *Norwegian-American Studies and Records* **10**, 152-175 (1938).

OLAF A. HOUGEN

T

Jokichi Takamine

1854-1922

Takamine's life spanned two centuries, two cultures, and two countries. He was born in Kanazawa, northern Japan, Nov. 3, 1854 to a Samurai family that practiced "medicine rather than war-science." Coincidentally, one year before his birth, Commodore Perry opened the doors of Japan to Western influence.

His father sent him to study in Osaka, and later he attended Imperial Engineering College, where he graduated as a chemical engineer in 1878. He was sent abroad by the Japanese government to complete his education and spent 3 years in Glasgow, attending Glasgow University, doing research on fermentation in the chemical laboratory of Andersonian College, and visiting industrial plants.

On his return to Japan he joined the Imperial Department of Agriculture and Commerce, where he was soon promoted director of the Government Chemical Laboratory. Later he became acting patent commissioner for 1 year. In his official capacity Takamine worked to improve "sake brewing, indigo ball manufacture, and paper." Each chemist under his employ was assigned a different task, but he reserved for himself the problems of sake fermentation.

He came to the United States in 1884 as chief Japanese commissioner of a delegation sent to visit the International Cotton Centennial Exposition of New Orleans. Here Takamine met Caroline Field Hitch, the daughter of a U.S. Army officer, whom he married the following year.

Takamine returned to his official duties in Japan until 1887, when he was given a leave of absence to establish the Tokyo Artificial Fertilizer Co., a factory for the manufacture of superphosphates, the first of its kind in the Orient.

During this time he worked out methods to extract cobalt oxide from manganese ore and also found new uses for manganese chloride, a material that was previously discarded as useless. The cobalt oxide Takamine used as a coloring agent for porcelain and the manganese chloride he used as a deodorizer and fireproofing agent for wood.

In 1890 Takamine returned to the United States with his wife and two sons, at the invitation of his father-in-law, to demonstrate his fermentation process that greatly shortened the fermentation time of malt. However, a series of events—illness, fire, and commercial competition—did not allow his enterprise to prosper, but his continuing researches led to the development of a starch-digesting enzyme, which he named Taka-Diastase. He sold his rights for manufacture and sale of the enzyme to Parke, Davis Co., for medicinal purposes.

He then moved from Chicago to Clifton, N.J. where he established a private laboratory. His researches there culminated in his isolation of adrenaline in 1901 from the suprarenal glands of sheep, the first hormone to be isolated in the pure state. Takamine first reported the isolation of adrenaline to the New York State Medical Society in January 1901. Lengthy litigation ensued after Takamine made his announcement, but after the U.S. Supreme Court heard the case, he was completely vindicated.

Another of Takamine's technical achievements was his development of a process to recover glycerin from waste printer's ink, after he tried unsuccessfully to find a substitute for this substance.

Takamine did not sever his links with his native land. He became a consultant to the Japanese government on matters related to industrial development. In this capacity he was instrumental in setting up a nitrogen fixation plant in Japan and also plants for extracting aluminum and manufacturing

Bakelite, alkalis, fertilizers, and dyes. He also helped to establish a large pharmaceutical plant. Through his efforts the Imperial Research Institute was organized by the Emperor in 1913.

Between 1912 and 1915 Takamine built a 5-story mansion on Riverside Drive in New York City; he had the first two floors decorated in ancient Japanese styles by Japanese artisans. It was destroyed by fire in 1927.

In his efforts to bridge American and Japanese cultures, Takamine helped to found the Japanese Association of New York, of which he became president. He was also a co-founder of the Nippon Club. The crowning achievement of Takamine's personal diplomacy was the donation of the cherry trees to the city of Washington which throngs of tourists come to admire every spring. Several of the cherry trees brought by Takamine were also planted in New York City and in Detroit.

Despite the fact that Takamine's work took place in the United States, that he lived in this country for a long time, and that he married an American woman, he could not become an American citizen because of the immigration laws in force at his time.

While Takamine was accumulating material success in the United States, his mother land did not forget him. It showered him with honors that included the doctor of engineering degree from University of Tokyo in 1899 and that of doctor of pharmacy in 1904, in addition to several awards from the Japanese government.

Takamine died in New York City, July 22, 1922, after a long illness.

Reliance was placed chiefly on Takamine's autobiography published in *The Saturday Evening Post,* Japanese Supplement, Dec. 30, 1916; "Dictionary of American Biography," **18,** 275-6, Chas. Scribner's Sons (1936); Henry George III, "Homage to Takamine," *Coronet,* September 1937, pp. 168-70; K. K. Kawakami, "Jokichi Takamine—A Record of his American Achievements," William Edwin Rudge (1928); Tom Mahoney, "The Merchants of Life," Harper & Bros. (1959); obituaries by F. O. Taylor, *Ind. Eng. Chem.* **14,** 990 (1922) and *New York Times,* July 23, 1922; personal communications from Wallace E. Davies, Joseph Head, and Tom Mahoney.

JOE VIKIN

Henry Paul Talbot

1864-1927

Talbot was born in Boston, May 15, 1864. His ancestors were among the earliest families to settle in Massachusetts. Upon graduation from high school he entered Massachusetts Institute of Technology in 1881. He received his B.S. degree in 1885 and remained as instructor of analytical chemistry for 3 years. He then went to Leipzig where he studied with Johannes Wislecenus and Wilhelm Ostwald. He received his Ph.D. degree *summa cum laude* in 1890.

Under Ostwald's influence, Talbot became aware of the value of the developing field of physical chemistry. Upon his return to MIT, he was appointed assistant professor and at once introduced physical chemistry into the curriculum. MIT was thus one of the first colleges in America to offer physical chemistry. In 1898 Talbot was promoted to professor and in 1902 was named chairman of the Department of Chemistry—a post he held for 20 years. From 1919 until 1921 he was chairman of the faculty, and he served as dean of students from 1921 until his death.

Although Talbot had little time to do research himself, as an administrator he encouraged it among his colleagues. He wrote a popular book "Quantitative Analysis" and co-authored "The Electrolytic Dissociation Theory" with Arthur A. Blanchard. He wrote numerous papers on chemical education. In 1921 he received an honorary D.Sc. from Dartmouth College.

Talbot was an active member of the American Chemical Society. From 1898 until his death he was a member of the Society's Council, and he served as one of the Society's directors. He was chairman of the Division of Inorganic and Physical Chemistry as well as associate editor of *Journal of the American Chemical Society.* He also served as president of the New England Chemistry Teachers' Association. During World War I Talbot served on an advisory board of the Bureau of Mines and the War Department. He also supervised some of the chemical warfare research. In 1891, he married Frances E. Dukehart, who survived him. They had one son who died as a child. Despite failing health Talbot remained active at MIT and died in Cambridge, June 18, 1927 during convalescence from major surgery.

Biographies by Arthur D. Little in *Ind. Eng. Chem.* **19,** 957 (1927) and Arthur A. Blanchard in "Dictionary of American Biography," **18,** 277-278, Chas. Scribner's Sons (1936).

SHELDON J. KOPPERL

Guy Baker Taylor

1888-1972

Taylor was born at Lexington, Ky., Oct. 5, 1888. He studied at University of Kentucky, obtaining his B.S. degree in 1908 and his M.S. degree in 1909, doing analytical chemical work in the summers at Kentucky Agricultural Experiment Station. He pursued graduate work at Princeton University under George Hulett, from 1911 to 1913, and received his Ph.D. degree in 1913.

Meanwhile he had worked for 2 years, 1910-11, at E. I. du Pont de Nemours' Eastern Laboratory, Gibbstown, N.J. From 1913 to 1919 he was with the U.S. Bureau of Mines, first in Pittsburgh and then in Washington. He came to Du Pont's Experimental Station in Wilmington, Del. early in 1919 and continued there until his retirement in 1953. He was a member of a mission sent in 1930 to Europe under the agreement with Imperial Chemical Industries, Ltd. for exchange of technical information with Du Pont and visited the important centers of research in physical chemistry in Great Britain and on the Continent.

Taylor was a prolific contributor to the chemical literature, with 55 entries in *Chemical Abstracts* (including 13 patents) on such diverse subjects as equilibrium, coal, explosives, oxidation of ammonia, manufacturing process development, and nylon. He will probably be best remembered for his direction of the work leading to the development of what became known as the Du Pont Pressure Process for manufacturing nitric acid (*see Ind. Eng. Chem.,* **23,** 860-65 (1931)). This work was a prime example of the successful application of the principles of physical chemistry to the design of an industrial process. The process became the prevailing method in the United States for producing this important chemical, with other installations world-wide.

Taylor was of a kindly and self-effacing disposition, but he was not one to be easily deterred from reaching the goals he set for himself.

He was married twice, first to Ione Stewart; they had a son and a daughter; and later to Charlotte Noyes. He died in Wilmington, Feb. 18, 1972, a few months after the death of his second wife.

Du Pont Company records; personal recollections; *Chemical Abstracts.*

THOMAS H. CHILTON

Ethel Mary Terry

1887-1963

Ethel Mary Terry was born in Hamilton N.Y., Feb. 10 in 1887, the daughter of a historian, Benjamin S. Terry and his wife Mary. Her father went to University of Chicago as professor of history when William R. Harper opened the university on Oct. 1, 1892. Ethel grew up on the campus of the university and received her A.B. degree in 1907 and her Ph.D. in chemistry in 1913. She was instructor in the Department of Chemistry from 1912 to 1918 and assistant professor from 1918 to 1927. On June 30, 1922 she married Herbert N. McCoy.

One of her first researches was with Julius Stieglitz: "The Coefficient of Saponification of Ethyl Acetate by Sodium Hydroxide."

In collaboration with her husband, she wrote and published their three-volume set of: 1. "Introduction to General Chemistry" 2. "Laboratory Outline of General Chemistry" and 3. "Teachers Manual and Notes." At the same time she carried out researches in Kent Chemical Laboratory and taught inorganic chemistry to large classes. During this period she published four papers, one on oxidation products from quinone, and three on the maleic acid-fumaric acid conversion (one of these on the Walden inversion, one on catalytic transmutation, and one on oxidation of both acids to tartaric acid).

Ethel remained at Chicago until Herbert decided he must live in California, and they moved to Los Angeles in 1927. Ethel had a number of research students who carried on her theories after she left the university.

With research ideas still in her mind she returned to University of Chicago to work every spring from 1927 to 1938 as a research associate. After 1938 she lived in retirement in Los Angeles until her death on May 23 1963.

Raised in the muted atmosphere of a professor's home, Ethel was indoctrinated in the Victorian ideals of taste for excellence. She liked quality in all things, and she liked good company. Her basic attitude was to strive for and accept only the first-rate. She resented and rejected everything common. And she herself was always good company. She was a lady of quality in blue velvet of the old school from which she strove strenuously to emerge, through her chemistry, in the image of the new professional woman of this age.

Her friends were beyond counting all over the country. She was revered as a most distinguished scientist, author, and teacher. Her encouragement to students to try for higher scientific achievements was superb. Her personal friendship and interest in any student never diminished. She helped graduate students work for extra money to support themselves.

Ethel Terry lived at a time when the progress of women in academic chemistry was slow, but working under a dedicated teacher like Julius Stieglitz she progressed in the Department of Chemistry at the University of Chicago. She was a handsome woman, dressed always in tailored suits commanding attention.

After her marriage her interests turned to color photography as a hobby. She disliked house work as something beneath her dignity. The only domestic thing she liked was a cat, the most sophisticated of pets.

Personal recollections; G. Ross Robertson, "Herbert Newby McCoy, 1870-1945" (privately printed, Los Angeles, 1964).

LILLIAN EICHELBERGER

John Webster Thomas

1880-1951

Thomas, a chemist who rose to distinction as president and chairman of The Firestone Tire & Rubber Co., was born on a farm near Akron, Ohio, Nov. 18, 1880. His father had been a coal miner in Wales before coming to the United States and settling in the coal district of Northeastern Ohio.

His youth was that of most farm boys. Chores at home cut into his school time with the result that he attended a one-room school until he was 17. In 1897 he had the highest marks in Summit County and won a scholarship to the academy of Buchtel College. His outstanding academic record at the academy resulted in a college scholarship to Buchtel College, now University of Akron.

As a student at Buchtel, he hiked 10 miles every day between school and home. He also helped out on the farm which left little time for other activities. But in his junior year he went out for the football team and was elected its captain. He was also elected class president. He received his Ph.B. degree in chemistry in 1904.

Thomas started as a chemist in the rubber compounding laboratory at B. F. Goodrich. In 1908 when quality control and improvement of rubber mixing and tire making depended more on guesswork than science, Harvey S. Firestone interviewed the young chemist. With his long-range viewpoint, Firestone realized that the bustling but immature tire business was no place for amateur experimenters. He wanted a professional scientist to set exact, scientific standards for tires and to work out formulas for their continuing improvement. The result of the interview was that Thomas went to work for Firestone at $100 a month.

He set up the company's first laboratory in a corner of the plant. Test tubes, beakers, testing equipment, and other scientific apparatus were moved in. There he worked on tire compounds determining specific formulas of rubber, sulfur, and other chemicals. He selected and tested fabric for use in tires.

As the company's production capacity grew, Thomas soon realized the need for additional technical and scientific personnel, and a second chemist was hired. Over the years Thomas' interest in science was the cornerstone for the development and growth of Firestone's vast research efforts. During his 38 years at Firestone the research and development staff grew from one chemist to more than 1,000 chemists, physicists, engineers, and other scientific and technical personnel.

After the laboratory was firmly established, Firestone put his first chemist in charge of wash lines, mills, and calenders on the night shift to test his executive ability. Working 12 hours a night and six nights a week, Thomas supervised factory operations for more than a year.

From that point on his upward progress in the Firestone organization was steady and, as he put it years later, "simply a matter of hard work and common sense."

Three years after he joined Firestone, Thomas was made general superintendent of Firestone plants. He was elected a director in 1916, vice president in 1919, president in 1932, and chairman and chief executive of the company in 1941. He retired in 1946 but remained on the board of directors for several years.

In the early 1930's Firestone undertook a major research program in synthetic rubber. By 1934 the company had built tires of synthetic rubber and was the first to supply the United States Army with synthetic rubber airplane tires. This early work permitted the company to build and put into operation the first government-owned synthetic rubber plant in the United States in Akron, Ohio. It began production in April 1942. Other plants built under the government's synthetic rubber program during World War II followed closely the standards established by Firestone. For his leadership in the successful synthetic rubber program, the American Institute of Chemists in 1945 awarded Thomas its Gold Medal.

Thomas was on the board of directors of University of Akron. His lifelong interest in the Boy Scouts led him to establish four university scholarships for them. At the time of his death, Nov. 26, 1951, he was moderator of the governing board of West Congregational Church in Akron.

Harvey S. Firestone, "Men and Rubber," Doubleday, Page & Co. (1926); A. Lief, "The Firestone story," Whittlesey House (1951); Karl H. Grismer, "Akron and Summit County," Summit Co. Historical Society (1952), p. 716; records of The Firestone Tire & Rubber Co.; information from R. L. Bebb.

JEAN SONNHALTER

Elihu Thomson

1853-1937

Elihu Thomson was born in Manchester, England, Mar. 29, 1853, at a moment in history most timely for a person who would later become an electrical inventor since the fundamental discoveries underlying that field were made in the decades immediately preceding his birth. His family migrated to America and settled in Philadelphia in 1858. Now 5 years of age, he was able to recite the alphabet, both forwards and backwards.

As a child he gained access to various factories, observing chemical and mechanical processes. He made small working models of cupola furnaces with fan blowers for air blasts and succeeded in melting iron. At the Philadelphia Navy Yard he was allowed during the noon lunch hour to operate a donkey engine used for boring holes in propellers for the iron-clad "Tonawanda" and the pursuit ship "Chattanooga," being readied for service in the Civil War. At 14 he was visiting plants where sulfuric, nitric, and hydrochloric acids, and paints and pigments were being made.

He was ready for enrollment at Central High School at the age of 11 but was denied admission because he was too young. He therefore spent 2 years reading, experimenting and making gadgets, the latter being a source of ridicule by his father until Elihu made a static machine from a wine bottle, which hurled the elder Thomson across a room and put an end to the ridicule.

Upon graduating, with honors, from high school he found employment in a chemical laboratory, where he did analytical work for a period of 6 months. At the age of about 18 he was appointed a teacher of chemistry and mechanics at Central High School. There he one day observed that when a nearby induction coil was in operation he could draw sparks from metallic objects such as brass door knobs and scientific instruments in the astronomical observatory at the top of the school building. He thus demonstrated the propagation of electrical waves through space, anticipating Hertz by 10 years and Marconi by 20, in wireless transmission. Likewise, he and his friend, William Greene, while still boys, constructed a telephone along the same principles employed by Alexander Bell a few years later.

Thomson remained at his teaching post from 1870 until 1880. During that time he published articles "On the Change of Color Produced in Certain Chemical Compounds by Heat," and "On the Inhalation of Nitrous Oxide, Nitrogen, Hydrogen" and began to make very important inventions—electric welding, the cream separator (from which the centrifuge descended), the three phase dynamo, improved arc-lighting to name but a few. In his teaching of chemistry he pointed out the close relation which elements bear to each other, thus constituting one large system, not "separate creations," and expressed the hope (fulfilled in his life-time), that the nature of this relationship would be someday understood and his theory accepted.

Thomson was the first to suggest using a mixture of helium and oxygen as an artificial atmosphere in caissons, as a preventive of "the bends," and he himself was probably the first to put oxygen gas to use in the treatment of pulmonary disturbances. He pioneered in a large way in the production and applications of fused quartz, his last great effort in this field being the attempt to make a two hundred inch quartz disk for the mirror of the giant reflecting telescope at Mount Palomar. His U.S. Patent 1,337,106 was for a process for providing an adequate supply of nitric acid for military explosives in war time.

When American Electric Co. was formed in New Britain in 1880 to control the Thomson-Houston patents, Thomson resigned from Central High School to go with the new firm. Two years later Charles A. Coffin of Lynn, Mass. entered the firm, which became Thomson-Houston Co. and in 1883 moved to Lynn. Coffin assumed management responsibilities which left Thomson free for research and development. The company grew rapidly from 184 workers in 1884 to 4,000 in 1892 when it merged with the Edison General Electric Co. of Schenectady.

Thomson guided the company into many new technical developments, but perhaps his greatest contribution was to demonstrate the value of industrial research. He won 700 U.S. patents (a number second only to that won by Thomas A. Edison), and he was awarded 17 American and foreign medals and many honorary degrees. Full of years and of honors, he passed away Mar. 13, 1937, at Swampscott, Mass.

David O. Woodbury, "Elihu Thomson, Beloved Scientist, 1853-1937," Whittlesey (1960); K. T. Compton, *Biog. Mem. Nat. Acad. Sci.* 21-22, 143-179 (1941-1943); Harold J. Abrahams and Marion B. Savin, "Selections from the Scientific Correspondence of Elihu Thomson," MIT Press (1971).

HAROLD J. ABRAHAMS

Edwin Ward Tillotson

1884-1956

Tillotson was born Feb. 28, 1884 in Farmington, Conn. where he attended grade and high schools. He received his B.A. degree from Yale University in 1906 and his Ph.D. in 1909. During his graduate career at Yale, he held both Loomis and Silliman fellowships. From 1909 he was a fellow in the Department of Industrial Research at University of Kansas under Robert Kennedy Duncan and then went with him to Pittsburgh as a founding member of the staff and an assistant director of Mellon Institute of Industrial Research.

His early investigations and publications were largely devoted to the relations of chemical compositions and the physical properties of silicate glasses. His work along these lines on densities and optical properties were highly regarded. His work and recognition of the importance of surface tension in glass processes and mechanisms were of a pioneering character and received international recognition. His functions at Mellon included the direction and supervision of all investigations on ceramics and related fields for a period of 43 years until his retirement in 1956 after having held the title of director of research.

Although his direct contributions to the glass and ceramic fields were monumental, his guidance and advisory inspiration were at least comparable in leaving their marks. His quiet and retiring manner tended to obscure the scope and number of suggestions and innovations which he so frequently contributed. He had a particular knack for recognizing and implementing research and developments involving over-lapping areas and companies of various fields.

He maintained active interests and participation in a number of scientific societies on both local and national levels. He joined the American Ceramic Society in its infancy in 1910 and was both a charter fellow and an emeritus member. He served as its vice president in 1922 and as president in 1925-26 after having been secretary of its Glass Division from 1919 to 1921 as well as associate editor of the society's *Journal* for several years. He was named honorary chairman of its annual meeting in Pittsburgh in 1943 and was recognized for his many achievements by receiving its Bleininger Award in 1949. In 1937 the British Society of Glass Technology awarded him the title of Fellow. He served as the American Ceramic Society representative on ASTM Committee D10 in 1922. He was also a member of the American Optical Society, the Society of Rheology, and the American Chemical Society, serving as the chairman of the ACS Pittsburgh Section in 1920.

On his retirement from Mellon Institute, a testimonial to Tillotson read: "He always had a great affection for the Institute, and all that it represents brought out his greatness of heart. Always generous in attitude and action, he has ever carried heavy administrative burdens most cheerfully. He has invariably been calm and helpful in different situations and has been unfaltering in his reliance on truth, in his adherence to settled rules of relations and others. . . . All these qualities have the esteem of all of his associates, the respectful opinions of many scientists and industrialists who have been associated with him."

He died July 31, 1956 at the age of 81 after suffering congestive heart failure.

Bull. Amer. Ceram. Soc. **1,** 120-121 (1922); **28,** 169-171 (1949); **35,** 384 (1956); **44,** 536 (1965).

HENRY H. BLAU

Richard Chace Tolman

1881-1948

Tolman was born in West Newton, Mass., Mar. 4, 1881. He received his bachelor's degree in chemical engineering from Massachusetts Institute of Technology in 1903, and then studied for a year in Germany at the Technische Hochschule, Charlottenberg, Berlin, and in an industrial laboratory at Crefeld. He returned to his alma mater in 1907 to a research position in chemistry, receiving his Ph.D. in 1910. He taught physical chemistry at the universities of Michigan (1910-11), Cincinnati (1911-12), California (1912-16), and Illinois (1916-18). During World War I Tolman was chief of the Dispersoid Section of the Chemical Warfare Service. After the conflict he held the posts of associate director (1919-20) and director (1920-22) in the Fixed Nitrogen Research Laboratory. In 1922 he accepted a professorship in physical chemistry and mathematical physics at California Institute of Technology.

During the next 18 years Tolman made contributions which were of fundamental and permanent significance in the clarification and extension of the interrelationships between physics and chemistry. During this period he completed his important early work (begun at Berkeley) of experimentally measuring the inertia of the electron. His interests broadened to a wide range of subjects including colloids, entropy of gases, quantum theory, partition of energy, general and special relativity, relativistic thermodynamics, statistical mechanics, chemical kinetics, and cosmology. He wrote four books, "The Theory of Relativity of Motion" (1917), "Statistical Mechanics with Applications to

Physics and Chemistry" (1927), "Relativity, Thermodynamics and Cosmology" (1934), and "The Principles of Statistical Mechanics" (1938). His monographs on statistical mechanics gave the most penetrating insights and extensions of the principles of this field since the works of Willard Gibbs. His papers on chemical kinetics were of particular value in clarifying the role of activation and the meaning of energy of activation. Making use of the astronomical data of the Mt. Wilson group, he enunciated a concept of the structure of the universe by combining the laws of thermodynamics with the theory of relativity. The key feature involved was a succession of identical expansions and contractions which avoided both the possibility of annihilation resulting from an irreversibly expanding universe and the degradation associated with maximum entropy. His publications of scientific papers totaled over 125 in 19 journals.

This period of extensive productivity was terminated by American involvement in the second World War, when Tolman went to Washington to offer his services. He played a major role in the coordination of the many scientific, industrial, and military activities. He was vice-chairman of the National Defense Research Committee where his advice on problems of armor and ordnance was significant. His expertise was utilized in the development of rockets and the proximity fuse. He served as scientific adviser to Lt. Gen. Leslie Groves in the Manhattan Project for the development of the atomic bomb. At the war's conclusion he contributed towards the achievement of international control of nuclear energy in his position as chief scientific adviser to Bernard Baruch, American representative to the United Nations Atomic Energy Commission. After his war duties, Tolman resumed his scientific studies at Pasadena in 1947, only to have his career cut short by death, Sept. 5, 1948.

Tolman was a person of wide humanistic interests, including psychology and the social sciences. He was interested in the philosophical and cultural impact of science on society. As a teacher he manifested concern for his students by presenting clear, precise, and challenging lectures. An honorary degree of doctor of science was conferred on him by Princeton University in 1942. For his war efforts he received the U.S. Medal for Merit and the rank of honorary Officer of the Order of the British Empire. The most outstanding characteristic of the man was his brilliantly versatile mind which enabled him to become a rare combination of scientist-technologist-statesman.

Biographies by J. G. Kirkwood, O. R. Wulf, and P. S. Epstein in *Biog. Mem. Nat. Acad. Sci.* **27,** 139-144 (1952), by V. Bush in *Science* **109,** 20-21 (1949), and by H. P. Robertson in "American Philosophical Society Yearbook," Amer. Philosophical Society (1948), pp. 295-299; Bernard Jaffe, "Outposts of science," Simon & Schuster (1935), pp. 506-514; Albert B. Christman, "History of the Naval Weapons Center, China Lake, Calif.," Vol. 1; obituaries in *New York Times* and *San Francisco Chronicle,* Sept. 6, 1948.

MEL GORMAN

John Torrey

1796-1873

Torrey, born to Capt. William and Margaret (Nichols) Torrey in New York City, Aug. 15, 1796, attended College of Physicians and Surgeons where he learned chemistry from William MacNeven. He received his M.D. degree in 1818 and practiced medicine for several years. In his spare time he wrote articles on chemistry, mineralogy, and botany and built a reputation as a competent scientist. He caught the eye of officials of the United States Military Academy and was hired to teach chemistry and mineralogy in 1824. He was professor at West Point until 1828, professor of chemistry and botany at College of Physicians and Surgeons, New York, 1827 to 1855, professor of chemistry at Princeton, 1830 to 1854, professor of chemistry at New York College of Pharmacy, 1829 to 1830, and professor of chemistry, botany, and mineralogy at New York University, 1832 to 1833.

In 1845 President Polk appointed Torrey Commissioner of the Mint to oversee the annual assay of coinage. Nine years later Torrey accepted the job of assayer of the U.S. Assay office in New York. He also acted as consultant for a California gold company, a New York-California petroleum firm, the Manhattan gas company, other firms, and for the Federal government, for whom he analyzed drinking water and building stone of Washington.

While the above accomplishments indicate ability in chemistry, Torrey was actually no more than an average American chemist of his time. His courses were typical of those taught in collegiate and medical schools, his dozen or so articles dealt with minerals and ores and demanded no talent above the ordinary, and his work for the Mint and private clients was no more than a competent chemist could have handled.

Torrey became famous in botany rather than chemistry. As a youth he was interested in natural history and turned out to be a prodigy in botany. "I must attend to chemistry," he wrote, "because I get my bread by it, and I love it very, very much. Yet I love botany more." His first book, "A Catalogue of Plants Growing Spontaneously within Thirty Miles of the City of New York" (1819), was followed by a long stream of botanical articles, monographs, and books. He prepared official reports on botanical specimens brought back by John Fremont and other western explorers, as well as by the naval expedition of Charles Wilkes. The governor of New York appointed him state botanist in 1836, and his two volumes on the flora of New York (1843) were a model of this type of publication. His herbarium formed the nucleus of the Columbia University herbarium, one of the great herbaria of the world.

Largely owing to Torrey, the Smithsonian Institution herbarium was organized. American naturalists memorialized his name in the genus *Torreya,* the "Torrey Pine," many plants, and Torrey's Peak in Colorado. One historian called him "the greatest early American man of science who sought to make the botany of the North American continent known."

If Torrey had concentrated on chemistry, would he have become as great as he did in botany? I find it hard to believe. He did not show any unusual aptitude in his chemical career, and I imagine that he would have remained an average chemist all of his days. It was well for science that chemistry provided Torrey with a livelihood and gave him the opportunity to unleash his great natural talent in botany.

One of Torrey's children, Herbert Gray Torrey (1841-1915), was a chemist and succeeded his father as assayer after the latter died in New York, Mar. 10, 1873.

Andrew D. Rodgers, "John Torrey, A Story of North American Botany," Princeton University Press (1942) has a portrait and list of Torrey's books and articles, quotes are from this book; biographies by A. Gray in *Amer. J. Sci.* 3S, **5,** 411-421 (1873), and in *Biog. Mem. Nat. Acad. Sci.* **1,** 265-276 (1877); Curt P. Wimmer, "The College of Pharmacy of the City of New York," (pub. by the college) (1929), pp. 23-31; "Dictionary of American Biography," **18,** 596, Chas. Scribner's Sons (1936). For Herbert Gray Torrey see "Who Was Who"; obituary in *New York Times,* Aug. 31, 1915.

WYNDHAM D. MILES

Olin Freeman Tower

1872-1945

Tower was born in Brooklyn, N.Y., Mar. 19, 1872. His ancestry goes back to early settlers at Plainfield and Canterbury, Conn. around 1700. He was related through his mother to Julia Ann Cleveland, to Moses Cleveland for whom Cleveland, Ohio was named. His youth was spent in Oregon, and he received most of his early schooling at home from his mother. It was in the wilds of Oregon that he learned to love camping and fishing. Later, when he and his wife (*nee* Elizabeth Williams) spent their 3-months' summer vacations at their Muskoka place in Canada near Gordon Bay, he often talked about his boyhood in Oregon where his father was a Methodist circuit rider. He certainly was at home in the bush—practically an Indian guide—would find a path through the undergrowth where, as far as one could determine, there wasn't a break. But, with a few scratches, scrapes, rips and pulls, he always ended up at the destination. Fishing was the same. If he trolled or cast in one area without any nibbles or bites, after about

10 minutes, he'd remark, "The fish aren't here today," and off he would row to another spot. Sure enough, before long there were fish.

At 16 he entered Wesleyan University from which he received his A.B. degree in 1892 and A.M. degree the following year. The next year he was assistant in chemistry to Wilbur O. Atwater at Wesleyan. He spent the period 1894-95 studying under Wilhelm Ostwald at Leipzig, Germany where he received his Ph.D. degree in 1895.

He returned to Wesleyan where he was assistant chemist in nutrition investigations until 1898. He was the first person to be tested in Atwater's human calorimeter. In the fall of 1898 he went to Western Reserve University as instructor in chemistry.

In 1901 he became assistant professor of chemistry. Upon the retirement of Edward W. Morley, he succeeded to the Hurlbut Professorship of Chemistry and became director of laboratories. He taught freshman chemistry and continued to do so until his retirement. He supervised the laboratory work and always graded the laboratory notebooks himself. He was very meticulous; the inferences to each experiment had to be written in his words.

The following courses were first offered and taught by him at Western Reserve: physiological chemistry in Adelbert College (now known as biochemistry and previously taught in the Medical School) in the fall of 1899; a course in the chemistry of food and nutrition in Mather College, spring of 1900; physical chemistry in the spring of 1901; colloid chemistry in the fall of 1925; radioactivity and the structure of the atom in the graduate school, spring of 1927.

He was an effective lecturer; his language was concise; he never wasted a word. He was brusque in manner, and the students were "scared to death" of him. Actually, his brusqueness covered a very tender feeling toward others. Some of the happy moments of his life were when he and his wife entertained for annual tea his second semester class of freshman girls. He brought out his snapshot albums and entertained with pictures and stories of Muskoka and some trips they had taken. He was fond of them and they of him. To those who really knew him, Tower was shy, modest, and kind.

He published several papers on dietary studies, but most of his publications stemmed from his training under Ostwald in physical chemistry. He published the first book on "Conductivity of Liquids" and one of the earliest books on "Qualitative Chemical Analysis" based on the ionic theory. In his later years his investigations were in the field of colloids.

He found recreation and relaxation in extensive reading and bridge. He and Mrs. Tower were an informed and skillful team in both. In the spring of 1944 he endowed the Olin Freeman Tower prize for excellence in the undergraduate course in physical chemistry. He was a member of many scientific societies, including the Chemical Society (London) and the American Philosophical Society.

In 1942 he became professor emeritus and retired to his new home in Mount Dora, Fla., where he passed away shortly after an automobile accident Dec. 21, 1945.

Archives of Case Western Reserve University; personal recollections; Case Western Reserve Chemistry Department publication, *The Reserve Chemist,* No. 9, February 1946.

FRANK HOVORKA

Benjamin Tucker

Around 1800 a teacher named Benjamin Tucker appeared in Philadelphia. Very little is known about him, but apparently he was a well-educated, scholarly, industrious man, and over the years he published several texts, among them "Sacred and Profane History Epitomized" in 1806, "Short Introduction to English Grammar" which reached a fourth edition by 1812, and "Epitome of Ancient and Modern History" in 1822. For a time he taught at Young Ladies Academy, a prestigious school to which girls came from all the states along the Atlantic coast and where, in the 1780's Benjamin Rush had introduced the first chemistry course for girls in the United States.

Public lectures on sciences and arts were a popular means of education and entertainment in Tucker's time, and in 1809 he offered a series of public lectures on chemistry for girls. The lectures, each an hour long and with "appropriate and brilliant experiments," were presented two evenings a week for several weeks at a fee of $8.00 for the course. They were so well attended that Tucker presented the lectures annually for several years, advertising them in the newspapers and by posters. For the convenience of his listeners he brought out an edition of a small (3½ in. wide, 5½ in. high) pocket-size text, "Grammar of Chemistry," originally published in London by Sir Richard Phillips under the pseudonym "Rev. D. Blair." A continual demand for the "Grammar" from beginning students led Tucker to issue editions in 1817, 1819, 1823, and 1827.

Tucker continued to live in Philadelphia until 1836, when, presumably, he moved or died. James Cutbush, who knew Tucker, said this about him: "For the introduction of popular chemistry, the citizens of Philadelphia are also indebted to . . . Mr. Benjamin Tucker, who . . . taught chemistry with much zeal and talents."

W. D. Miles, H. J. Abrahams, "The Public Chemistry Lectures of Benjamin Tucker," *J. Chem. Educ.* **34**, 450-451 (1957).

WYNDHAM D. MILES

Willis Gaylord Tucker

1849-1922

Tucker was an expert in water analysis, toxicology, the determination of adulterants in foods and drugs, and in other areas of chemistry connected with public health. He testified in many court cases involving these subjects, and his distinguished appearance and verbal delivery gave little doubt as to the accuracy of his testimony. On one occasion when questioned by the court about the authority on which he based his statements, it is said that he replied: "On the authority of Willis G. Tucker." The judge ruled that the testimony would stand as presented.

Tucker was born in Albany, N.Y., Oct. 3, 1849. He graduated from Albany Academy in 1866 and obtained his M.D. degree from Albany Medical College in 1870. He remained in Albany for the rest of his life where he was active as a chemist, educator, and administrator but never practiced medicine.

At Albany Medical College he held several posts on the faculty from 1871 until he retired in 1915 and taught inorganic, analytical, and organic chemistry, materia medica, and toxicology. One of the founders of the Albany College of Pharmacy in 1881, he was the first professor of chemistry and president of the faculty. He taught at the College of Pharmacy until 1918. For many years he lectured in chemistry at St. Agnes School for Girls, Albany High School, Albany Academy for Boys, and Albany Female Academy.

Tucker held many professional positions connected with the state of New York. He was an analyst for the New York State Board of Health from 1881 to 1891, a member of the Board of Medical Examiners of the University of the State of New York from 1882 to 1891, director of the State Board of Health Laboratory from 1891 to 1901, director of the State Department of Health from 1901 to 1907, and a member of the Medical Council and the Pharmacy Council of the University of the State of New York for many years. From 1882 to 1887 he was editor of *Albany Medical Annals*. He published many articles and at least three introductory lectures to his chemistry courses.

It is said that Tucker had a charming personality, was positive in his opinions, and was a great friend of students. He was well read, a lover of music and the arts, and possessed a rich library of musical biography and scores of operas in rare editions. He died in Albany, Apr. 21, 1922.

Albert Vander Veer, Frederick C. Curtis, "Biographical Sketch of Dr. Willis Gaylord Tucker," 31-page booklet (*ca.* 1922), with portrait, has list of talks and publications; F. C. Curtis, *Albany Med. Ann.* **43,** 296-299 (1922); "Who's Who in New York City and State" (1909); "National Cyclopedia of American Biography," **28,** 389, James T. White & Co. (1940); James J. Walsh, "History of Medicine in New York" (1919) IV, pp. 222-224; *New York Times,* April 23, 1922.

EGBERT K. BACON

John William Turrentine

1880-1966

Turrentine, an American potash chemist, was born in Burlington, N.C., July 5, 1880. After earning his Ph.B. (1901) and his M.S. degrees (1902) at University of North Carolina, he taught for 3 years at Lafayette College before beginning his doctoral studies at Cornell. His graduate research included studies on the electrochemistry of metals and on the properties of hydrazoic acid and hydrazine compounds. He received his Ph.D. degree in 1908 and returned to teaching, this time as instructor in physical chemistry at Wesleyan University.

In 1910 the unfavorable course of negotiations between American fertilizer manufacturers and German potash producers resulted in Congressional appropriations for developing domestic sources of potash. In 1911 Turrentine joined the U.S. Department of Agriculture's Bureau of Soils, one of the agencies thus funded, and began work on the possibility of utilizing Pacific coast kelps. These algae, when dried, were known to have a high potassium content. Turrentine also reported on the feasibility of using fish-cannery wastes as fertilizers.

By 1917 the World War I potash shortage was acute, and prices had risen to more than ten times their pre-war level. Turrentine was placed in charge of an experimental plant in Summerland, Calif. (near Santa Barbara) designed to produce potash from kelp.

Turrentine developed a process based on pyrolysis of dried kelp at temperatures low enough to prevent fusion of the potassium salts; followed by continuous counter-current leaching and recovery of the dissolved salts by crystallization. Turrentine abandoned the familiar vat-coolers for vacuum evaporator-crystallizers. He thus solved the problem of caking on conventional heat-transfer surfaces and converted the heat of crystallization from a disadvantage to an advantage. This method of crystallization later became general in solution-crystallization refineries in the U.S. and Europe.

With the end of the war, the kelp process became uneconomical although Turrentine tried to justify it by crediting it with the value of such by-products as iodine, "kelpchar" (an adsorptive carbonaceous pyrolysis residue), and organic distillate oils. In 1922 the plant ceased operation and Turrentine was assigned to other duties.

He remained interested in technical aspects of kelp and potash problems. He advocated kelp products to reduce dietary iodine deficiency. He devised a process for extracting potash from "greensand" (glauconite), whose economic feasibility depended heavily on the value accorded to an adsorptive silica byproduct.

For Turrentine 1926 was the watershed year. It was the year of his marriage to Katharine W. Bacon; the year of the publication of his book "Potash: a Review, Estimate and Forecast," and the year in which the great Permian basin potash deposits near Carlsbad, N. Mex. were discovered, ending America's quest for adequate sources. Although the kelp process was forgotten, its engineering innovations survived.

Turrentine remained with the Department of Agriculture until 1935, most of the last 13 years being devoted to gathering data on the potash industry. He attended European congresses, establishing personal contacts and broadening his understanding of the commercial aspects of potash production, distribution, and use.

He now saw the main opportunity for himself and for the potash industry in terms of increasing agricultural consumption rather than in improved production. In 1935 with the support of the American Potash and Chemical Corp. of Trona, Calif., of the recently established U.S. Potash Company of Carlsbad, N. Mex., and of the American agents for German and French potash producers, he organized the American Potash Institute, with himself as president. The Institute was to promote potash utilization through education and research. Increased potash consumption was promoted by the book, "Potash Deficiency Symptoms" published in 1937 in a German-French-English trilingual edition. The authors were O. Eckstein, A. Bruno, and Turrentine. In the same year, Turrentine received the gold medal of the Académie d'Agriculture of France for his work on potash.

The outbreak of World War II led to the withdrawal of the European representatives from the Institute. In his writings of 1942-43 and especially in his ACS monograph, "Potash in North America" (1943), Turrentine cited the potash industry's war production as a great national achievement.

Turrentine's interest in agricultural fertilizer's extended to phosphates. His proposed methods for treating phosphate rock with hydrochloric acid, nitric acid, and nitrogen oxides resulted in patents during 1938-40.

Turrentine retired Dec. 31, 1948 from the presidency of the Institute but remained as a consultant for several years thereafter. After the death of his wife Aug. 18, 1950, he continued to live in Washington, D.C. He died in a nursing home June 11, 1966 and was buried two days later in Burlington, N.C.

He was personally ambitious and valued professional status highly. He moved in socially prestigious circles. But he derived considerable satisfaction from being able to support "struggling young graduate students" through the research grant program of the institute.

Chem. Abstr., vols. 1-46; J. W. Turrentine, "Potash in North America," Reinhold (1943); obituary announcement, June 12, 1966, by American Potash Institute; S. W. Martin, "Dr. T.," *Better Crops with Plant Food* (1966); personal communication from J. O. Romaine.

CHARLES H. FUCHSMAN

U

Marvin J. Udy

1892-1959

Born in Farmington, Utah, Feb. 19, 1892, Marvin was the son of Mathias Cowley Udy and Emily Rebecca Hess. J. was part of his name; it was not an initial of a middle name. Trained as a chemical engineer, Marvin received his first degree from University of Utah in 1915 and followed this with his M.S. degree in metallurgy in 1916 from the same university.

Udy's professional career was divided into two nearly equal parts. The first two decades saw a succession of industrial appointments and the final two decades a broad diversity of consulting relationships and development activities in metallurgical processing. Upon completing his formal education, Udy first joined U.S. Mining and Smelting Corp. in Utah where he remained for 2 years as assistant research chemist. He then travelled east where he was associated with Hooker Electrochemical Co. of Niagara Falls, 1918, and Haynes-Stellite Co. as chief chemist in their Kokomo, Ind., laboratory, 1918-19, and as manager of their Leesburg, Idaho, cobalt mine, 1919-20.

Upon the merger of Haynes-Stellite into the Union Carbide Co., Udy returned to Niagara Falls where he pursued metallurgical research and development for UCC's Electrometallurgical Co. under Frederick M. Becket until 1931. From 1931 to 1934 he was associated with Swann Chemical Co. of Anniston, Ala. and from 1934 to 1937 with Oldbury Chemical Corp. of Niagara Falls. In these connections he was responsible for developing new processes for electroplating (Cd and Cr), electrowinning of metals from their ores (Co and Bi), formation of alloys (Stellite and ferro-alloys), and electric furnace production of various materials (calcium carbide, alumina abrasive, phosphorus, and ferrochrome). In nearly every one of these positions, Udy's work resulted in patenting new and valuable processes.

Udy's consulting career began in 1937 as an outgrowth of his studies for Oldbury Chemical Co. on the selective reduction of low-grade chromium ores. When small scale pilot studies proved successful, Canadian financiers set up a plant at Sault St. Marie, Ont. under the name of Chromium Mining and Smelting Corp. with Udy as consultant. This plant was very successful and, responding to wartime needs for chromium steels, grew to utilize 23,000 kw of energy; it was supplemented by other plants at Spokane and Chicago. Over the years, Udy held consulting relations with New Jersey Zinc, Vanadium Corp., Quebec Iron and Titanium, Ebasco Services, Carborundum Co., Pittsburgh Lectromelt Furnace Co., Battelle Memorial Institute, Bonneville Power Authority, the U.S. Bureau of Mines, and a host of others.

Several companies were formed to exploit developments originated by Udy. His cadmium plating process, developed at Haynes-Stellite, formed the basis of Udylite Co. of Detroit, privately financed by James C. Patton, vice-president of Haynes-Stellite. Denman Enterprises was a partnership established to develop processing of manganese ore. This and further ferro-smelting developments were later (1955) spun off as Strategic Materials Corp. which, in turn, subsequently formed Strategic-Udy Metallurgical and Chemical Processes Ltd. of Niagara Falls, Ont. and Strategic-Udy Processes of Niagara Falls, N.Y.

Udy's characteristics included a directness and clarity of thinking, patience and perseverance in the pursuit of an idea, and business acumen to see the economic importance of scientific and technological developments. Active in many professional societies, he received abundant recognition for his work: Schoelkoff Medal of the American Chemical Society's Western New York Section (1948), presidency of The Electrochemical Society (1954-55), and an honorary doctor of science degree from Alfred University (1956). He published 16 papers, edited vols. I and II of the book, "Chromium" and was awarded more than 76 patents.

Marvin Udy was married June 3, 1915 to Tessa McMurray by whom he had three sons, the eldest of whom, Murray Cowley, also pursued a metallurgical career and was associated at one time with him. The elder Udy died at his home in Niagara Falls, N.Y., April 11, 1959.

"Schoelkoff Award to Udy," *Chem. Eng. News* **26,** 1580 (1948); biography by M. C. Udy in Clifford A. Hampel, "Encyclopedia of Electrochemistry" (1964); obituaries in *J. Electrochem. Soc.* **106,** 167 (1959), and *New York Times* April 12, 1959.

J. H. WESTBROOK

V

Donald Dexter Van Slyke

1883-1971

Van Slyke was born in Pike, N. Y., Mar. 29, 1883, and received his Ph.D. degree in organic chemistry under Moses Gomberg at University of Michigan in 1907. After graduation he started work at Rockefeller Institute in 1909 and in 1914 was appointed director of the hospital of Rockefeller Institute. His work was directed to the field of biochemistry and clinical chemical studies. At this time biochemistry was in its infancy; methods for accurate blood study were just becoming available; proteins were not regarded as chemical entities; enzymes had not been isolated or chemically characterized; and hormones and vitamins were not identified as yet. For 65 years, Van Slyke, known as "Van," contributed to the knowledge of medicine and the young field of biochemistry with the publication of 317 papers and five books. He was instrumental in establishing medicine as a modern science based on data obtained by quantitative chemical procedures.

In the field of biochemistry, "Van" was among the first to study metabolic pathways by chemical means. In 1911 he developed a method for quantitatively determining aliphatic amino groups. With this tool he studied protein digestion and the role that the liver played in this digestion. His interest in proteins turned to enzyme kinetics, and he wrote several articles on urease. Both lipid and carbohydrate metabolism were studied in patients suffering from diabetes and nephritis, and the function of the kidney in ammonia production was established. Along with these metabolic studies he developed a urea clearance method which was useful in demonstrating kidney impairment.

Around 1925, Van Slyke studied the blood system as a physiochemical system, defining more fully the oxidation and reduction of hemoglobin. Again, he developed a gasometric manometer for determining many kinds of gas mixtures in blood and tissue. With this he determined the role played by hemoglobin, oxygen, and carbon dioxide in the distribution of water and electrolytes in the blood. He extended this work to the acid-base balance of the blood, including an exact mathematical definition of "buffer value." His discovery of the amino acid, hydroxylysine, was the result of 20 years of study; he first noted it as a discrepancy in the analytical data of gelatin hydrolysate.

In 1949, Van Slyke at the age of 66, became a member emeritus at Rockefeller Institute but then assumed a full time position of assistant director for biology and medicine at Brookhaven National Laboratory. While at Brookhaven he developed a sensitive method for determining carbon 14.

Van Slyke also served as managing editor of *Journal of Biological Chemistry* from 1914 to 1924 and was responsible to a large degree for the journal's high editorial standards. In 1931, he published his classic two-volume "Quantitative Clinical Chemistry," which served as a standard textbook for many years. His first publication was written with his father, and one of his later publications was written with his son. He held honorary degrees from seven American and European universities, received 14 scientific awards, was decorated by five foreign governments, was an honorary member of a score of scientific societies, and presided over several scientific groups.

Tennis was his principal extramural activity and he played the game to within a few months of his death on May 4, 1971 in Garden City, N. Y.

Biography in *Chem. Eng. News,* July 2, 1962, p. 7, with portrait on cover, and by A. Baird Hastings in *Fed. Proc.* **23,** 586-591 (1964); obituary by A. Baird Hastings in *J. Biol. Chem.* **247,** 1635-1640 (1972).

MARY E. PERRY

John Vaughan

1775-1807

In 1799-1800 the Philosophical Society of Delaware sponsored a series of public lectures on chemistry at Wilmington, one of the first such courses in America. The lecturer for the series was one of the Society's founders, John Vaughan, a 24-year old physician.

John Vaughan was born in Chester County, Pa., June 25, 1775, the son of Joshua Vaughan, a Baptist minister. He studied medicine in Philadelphia under William Currie and at the Pennsylvania Hospital. Besides building a fine practice during his 3 years in Wilmington, Vaughan also had edited a semi-weekly newspaper, helped to found the Abolition Society, and had been an active supporter of Jeffersonian politics and the Delaware Medical Society.

Vaughan was thoroughly French in his approach to chemistry and saw cumbersome nomenclature as one of the greatest impediments to teaching. Therefore, for his lectures he wrote a simple, up-to-date outline based on the works of Lavoisier. He was one of the first authors in this country to do so. It listed both the old and new chemical names and discussed chemistry in six divisions, *e.g.* salts, earths, metals, inflammables, waters, and airs. There were also brief sections on chemical affinity, fermentation, caloric, oxidation, amalgamation, and vegetable and animal substances. Entitled "Chemical Syllabus," the book contained 19 pages and sold for 25¢.

The course consisted of 15 lectures, most of which were illustrated. It was claimed that the society's pneumatical apparatus was the most complete in the country. The last lecture, a summary of the other 14, was published under the title "Valedictory Lecture," and gives ample evidence that the lectures were lively. The section on electricity ranges in content from a prediction of air conditioning to a warning to young girls to avoid negatively charged old bachelors. Properties of gases and a eulogy of Lavoisier comprise a major part of the section on chemistry. The last 4 pages champion the "mental emancipation" of women so that they might take rightful place beside men as teachers, lawyers, and senators. The course got good reviews from Samuel Mitchill in the *Medical Repository,* and Joseph Priestley wrote that he was elated that the effort had been made.

In 1801 Vaughan proposed to publish the lectures of Joseph Black but gave up his plans after Mitchill and Benjamin Rush convinced him that the work was out of date and would not sell. Of course, the lectures were published 2 years later by John Robison, Black's executor and went through several editions.

Vaughan viewed contagion and disease as resulting from chemical imbalance and published a paper on the use of potassium carbonate as a treatment for smallpox. He introduced vaccination for smallpox to Wilmington during the summer of 1802 and the following year distinguished himself by being the only physician to remain in the city during a yellow fever epidemic.

A deeply religious man, he was licensed to preach by the Baptist Church in 1806. He married Eliza Lewis in 1795 and had four children. He died Mar. 25, 1807 at the age of 32.

J. Thomas Scharf, "History of Delaware," L. J. Richards, **1,** 492-493 (1888); Herbert T. Pratt, "John Vaughan's Public Lectures On Chemical Philosophy (1799)," *J. Chem. Educ.* **39,** 42-43 (1962); D. I. Duveen and H. S. Klickstein, "The Introduction of Lavoisier's Chemical Nomenclature into America," *Isis* **45,** 373-374 (1954).

HERBERT T. PRATT

Victor Clarence Vaughan

1851-1929

Vaughan was born Oct. 27, 1851 near Mount Airy, Mo. The family went to Illinois during the latter part of the Civil War and started a new farm until the war was over and they could return to Missouri. His first training in "spelling, reading, writing, and figures" was given him by his mother; his first schooling was in the home of a neighboring physician. He attended Hazel Academy, Central College at Fayette, Mo., and Mount Pleasant College at Huntsville, Mo., where he earned his B.S. degree. In his first or second year at the latter school, at the age 18 or 19, he began teaching Latin and finally was put "in charge of Latin and chemistry." His connection with chemistry arose from finding a locked room in the college which had been equipped with chemicals and chemical equipment by some unknown teacher before the building was closed during the war. A copy of Silas Douglas and Albert Prescott's "Qualitative Analysis" came into his hands later on, and this book decided the question of whether his life work would be in the classics or science. He graduated in 1872 but continued teaching Latin and chemistry for another 2 years.

In 1874 on learning that there was a large and well equipped chemical laboratory at University of Michigan, he traveled to Ann Arbor to enroll for a master's degree in chemistry with geology and biology as minors. His master's thesis in June 1875 was on "The Separation of Arsenic and Antimony." He received his Ph.D. degree in 1876 and one of the three theses he presented for the degree was another paper on the separation of these same two elements. About this time a controversy arose over funds in the Chemistry Department, and Vaughan was appointed to succeed one of the men involved. He became instructor in physiological chemistry in 1876, even before receiving his M.D. degree which was granted in 1878. He was appointed lecturer in 1880, professor of hygiene and physiological chemistry in 1884, and dean of the Medical School in 1890. Among his students were such famous men as Frederick Novy, Donald Van Slyke, and Moses Gomberg.

While teaching, Vaughan engaged in the practice of medicine for 20 years. During his research he analyzed about 20 "opium cures" and found that all except one contained opium; the twentieth was a solution of sodium chloride. This led him to spend much time exposing fraudulent patent and proprietary medicines.

Vaughan was a member of the Michigan State Board of Health from 1883 to 1894, and 1901 to 1919. He served in the Santiago Campaign from 1898 to 1899 as major and surgeon, 33rd Michigan Volunteer Infantry and was recommended by the President for the brevet of lieutenant colonel. He was awarded the Distinguished Service medal in 1919. He was a member of the German Chemical Society and honorary member of the French and Hungarian Societies of Hygiene.

Twenty-seven original contributions were published by Vaughan while he was a member of the Chemistry Department, the first one dealing with the estimation of arsenic as sulfide and the remainder with the application of chemistry to medical problems. Thirty-six other articles not based on his research appeared with his name during this same period, 1879-1903. His research and publication record continued until his retirement from the School of Medicine in 1921. Altogether, his bibliography contained 324 entries.

He died Nov. 21, 1929 in Richmond, Va.

Charles A. Browne and M. E. Weeks, "A History of the American Chemical Society," ACS pub. (1952); Victor C. Vaughan, "A Doctor's Memories," Bobbs-Merrill Co. (1926); Victor C. Vaughan Memorial Issue, *J. Lab. Clin. Med.*, vol. XV, No. 9, June 1930; "Dictionary of American Biography," **19**, 236-237, Chas. Scribner's Sons (1936).

LEIGH C. ANDERSON

William Reed Veazey

1883-1958

Veazey was a teacher, consultant, coordinator, administrator, chemist, and inventor in the academic and industrial fields. He was also a devoted family man, churchman, and community service man. Born Dec. 29, 1883, in Chase City, Va., he lived for 10 years there, where his father (a United Presbyterian minister) was in charge of a mission school for blacks.

His family moved to New Wilmington, Pa., and there William earned his B.A. degree at Westminster College in 1903. In 1907 he was granted his Ph.D. degree in chemistry from Johns Hopkins University. He had taken time out in the 1904-05 school year to replenish his finances by instructing in chemistry at University of Oregon.

In September 1907 Veazey started his long

and successful association with Case School of Applied Sciences. He taught both chemistry and chemical engineering. He became head of the Department of Chemical Engineering in 1928 and held this position until 1936 when he resigned to accept a full time assignment at Dow Chemical Co. in Midland, Mich., as coordinator of research.

Veazey's association with Dow Chemical Co. had begun in 1916 when Herbert Dow, the company's founder, engaged him as a consultant. Dow's principal products were recovered from Michigan underground brine containing chlorides and bromides of sodium, calcium, and magnesium. World War I had cut off supplies of magnesium metal from its only source in Germany, and the many pyrotechnical requirements for the metal were not being met. Dow was producing magnesium chloride from brine, and Veazey's knowledge of physical and electrochemistry was needed to develop a cell and process for electrowinning metallic magnesium. By the end of the summer of 1916 Veazey and the small group of Dow people assigned to the project had devised a small prototype cell and a suitable molten electrolyte, and they produced over one hot summer night 1 lb of magnesium metal. This was the beginning of a process by which Dow Chemical Co. in 1972 turned out 245 million pounds of magnesium.

Veazey continued to spend his summers in Midland, advising Herbert Dow and later Herbert's son Willard on research policies and projects. In 1927, the value of Veazey's business acumen and his understanding of the company's processes were recognized when he was named to the board of directors of the Dow Chemical Co.

In spite of the financial depression of the 1930's, Dow's research organization grew to a size and at a pace that made it difficult for its president, Willard H. Dow, to maintain active direction, and he finally prevailed upon Veazey to give up his highly successful academic career and to undertake the coordination of research, development, and patents. Except for his position as a director of the company, Veazey had no official title. He seldom injected his own considerable knowledge of chemistry into conferences with Dow scientists but rather used his patient understanding, untiring optimism, and great good humor to the end of obtaining the best efforts of all those around him. To many of the younger scientists and engineers he was fondly called the "father confessor," and his guidance helped many young careers in Dow Chemical.

Veazey became a vice president and member of the board of Dowell, Inc. a wholly owned well-treating company arising from Dow research. He also served as a director of Dow Corning Corp. which was jointly owned by Dow Chemical and Corning Glass and manufactured a line of silicones.

After Dow's president, W. H. Dow, died in an airplane crash, Leland I. Doan, the new president of the company, appointed Veazey chairman of the Executive Research Committee. Veazey, now of retirement age, went about reorganizing Dow's large research and development operation to something more compatible with the very large corporation Dow had become. Finally, in 1952, Veazey's plans for Dow R&D culminated in naming one of his former Case students, Ray H. Boundy, to the office of director of research. Veazey had now weaned Dow's research, and, after a year of helping Boundy in his new duties, Veazey retired at age 70.

Five years later, on Dec. 20, 1958 in Youngstown, Ohio, Veazey died; he was buried in New Wilmington, Pa. He had, during his life, many honors bestowed on him by his church, college, community, and profession. He was on the board of elders of the Presbyterian Church, a member of the board of trustees of Westminster College, New Wilmington, Pa. and of Alma College, Alma, Mich. and was president of The Electrochemical Society in 1946. He was also on the board of trustees of the Midland Community Center, was president of Midland's Chamber of Commerce, and served on the Salvation Army board of directors.

The large Dow research center in Texas was dedicated to Veazey in October 1953.

Youngstown (Ohio) Vindicator, Mar. 24, 1957; biographical data from Dow Chemical Co.; *Midland (Mich.) Daily News,* editorial, June 3, 1953; *Bay City (Mich.) Times,* Aug. 13, 1953.

LEONARD C. CHAMBERLAIN

Francis Preston Venable

1856-1934

Venable was born in Prince Edward County, Virginia, Nov. 17, 1856 to Charles Scott and Margaret Cantcy (McDowell) Venable. He attended University of Virginia, where his father was professor of mathematics and where he became interested in chemistry through the stimulation of John Mallet. After graduating in 1879 he taught in a New Orleans high school for a short time and then went to Germany where he studied at the universities of Bonn, Berlin, and Göttingen under noted chemists, among them August Kekulé and Otto Wallach. He received his Ph.D. degree from Göttingen in 1881.

Venable accepted a position at University of North Carolina in 1880 and taught chemistry there for half a century. He also presided over the university from 1900 to 1914. He carried out research on zirconium from the 1880's into the 1920's. He began atomic weight determinations of the element in the 1890's, discontinued his research during the years when he presided over the university, and resumed the work in 1914. He and his assistant, James M. Bell, obtained atomic weights with variations larger than could be explained by experimental errors but which were explained later when Hevesy discovered hafnium and found that small quantities were present in zirconium materials. In 1922 Venable published a monograph, "Zirconium," that remained a standard work for many years.

Venable began to lecture on history of chemistry in the 1880's and continued until 1929, more than 40 years, perhaps longer than any other American who taught the subject. Of his approximately 110 published articles, about a dozen dealt with history. He used the historical approach in his books "The Development of the Periodic Law" (1896), and "The Study of the Atom; or the Foundations of Chemistry" (1904). In 1894 he published "A Short History of Chemistry," a book that passed through several editions, the last in 1922.

Other books that Venable published were "A Course in Qualitative Chemical Analysis," first published in 1883 with subsequent editions; "Inorganic Chemistry, according to the Periodic Law," with James L. Howe in 1898; and "A Brief Account of Radioactivity," in 1917.

Venable was a founder and in 1883 the first president of the Elisha Mitchell Scientific Society. He presided over the American Chemical Society in 1905 and over several Southern educational associations. Several honorary degrees were bestowed upon him, and Venable Hall at the university was named in his honor. He died at Chapel Hill, N.C., Mar. 17, 1934.

Biography in *J. Elisha Mitchell Sci. Soc.* **50,** 27-28 (1934), with list of publications, pp. 328-333, and frontispiece portrait; biographies by J. M. Bell in *Ind. Eng. Chem.* **16,** 755 (1924) and *J. Chem. Educ.* **7,** 1300-1304 (1930); "Dictionary of American Biography," **19,** 246, Chas. Scribner's Sons (1936).

Wyndham D. Miles

Louis Fenn Vogt

1880-1952

Vogt was born in Tallmadge, Ohio, Jan. 31, 1880, the son of William H. and Marietta Abigail (Fenn) Vogt. He attended Case School of Applied Science with the class of 1901. He was with Columbia Chemical Co. 11 years as analytical chemist, chemical engineer, and works chemist making soda ash, caustic soda, and other chemicals. For 4 years he was chief chemist and chemical engineer with Westinghouse Machine Co., East Pittsburgh, Pa., and afterwards superintendent of the reduction works for Vanadium Mines Co., Pittsburgh. He went with Standard Chemical Co., Pittsburgh which was formed in 1911 to produce radium from Colorado carnotite, and was sent to Cutter, N. Mex., in 1912 to manage the reduction plant there. After 1½ years he returned east as manager of the Canonsburg, Pa., plant. Later he became a director of the company.

He was vice president and a director of Radium Dial Co. of Pittsburgh and New York and was for a short time works manager of

American Cyanamid Co.'s Linden, N.J., heavy chemicals plant.

He held several basic patents for processes for extracting radium, uranium, and vanadium from carnotite, for treating carnotite, for extracting vanadium from ores with high phosphorus content, and for concentrating carnotite ores. He also held several patents for toughening aluminum by alloying it with uranium and three patents with R. M. Keeney for making ferro-uranium alloys with high uranium content (29%). With D. M. Goetschius he patented filters for acids and other liquids.

When Mme. Curie came to this country to receive the gift of a gram of radium that was presented to her by President Harding in 1921 on behalf of the women of the United States, she visited Pittsburgh where she was curious about how chemists there were able to treat low grade ores so successfully. She questioned Vogt closely on his refining techniques, and he recalled later that "She didn't disclose anything in return."

During World War II he volunteered his radiation knowledge to the Manhattan Project; he made a number of trips on its behalf but would never discuss with his family what he did or where he went. He was also chief of production service in the Newark, N.J., office of the War Production Board. He lived in Westfield, N.J., from about 1924, and in retirement he was an industrial real estate broker for David T. Houston Co., Newark. He died in Plainfield, N.J., Aug. 31, 1952.

Vogt married Lois Johnston in Tallmadge, Ohio, Aug. 27, 1901, and they had three children: Louis Fenn, Jr., Richard Johnston, and Lois Kathleen. He was a member of the American Chemical Society for a while and of The Electrochemical Society from 1917.

The Case Tech, Nov. 30, 1921, and Case Institute of Technology Alumni Records; Membership Application, American Chemical Society; Charles H. Viol, *Radium* (by Radium Chemical Corp.) **17,** No. 3, June 1921; *Chemical Abstracts* (for patents); *The Crucible* (published by Pittsburgh Section of ACS **51,** Special Issue, March, p. 14 (1966); obituaries in Plainfield, N.J., *Courier-News,* Sept. 2, 1952 and *New York Times,* Sept. 2, 1952; recollections of Mr. and Mrs. Louis F. Vogt, Jr.

ROBERT F. GOULD

Sabin Arnold von Sochocky

1882-1928

von Sochocky, a world authority on radium, was born in Lany (Bibrika), formerly within the Austro-Hungarian border but now in the Ukraine, Feb. 22, 1882, the son of Reverend Josef and Amelia (Petruschevych) von Sochocky. His uncle, Eugene Petruschevych, was (1918) president of the short-lived Republic of the West Ukraine.

He took his M.D. degree at University of Moscow but soon realized that he preferred science to medicine, so he obtained a Ph.D. degree in Vienna and other degrees in technical schools in Kiev, Berlin, Dresden, and Prague. He found his major interest when he met the Curies in Paris, and through continued study and research he soon learned just about all there was to know about radium at that time.

In 1906 von Sochocky came to the United States at the invitation of a leading university to teach in its medical school. He did not like our university methods, so he practiced medicine in New York for 10 years, utilizing, however, practically none of his knowledge of radium and radiotherapy.

In 1916 he met George S. Willis, M.D., of Morristown, N.J., who was looking for someone who knew about radium. von Sochocky had a radium-enlivened paint which he had introduced in 1913 (about 2000 watches with luminous dials were sold that year), so, thinking that this would be a good "pot-boiler" to finance their medical research, the two men founded Radium Luminous Material Corp. in Orange, N.J. in 1916.

Luminous paints were not new. They had been used in Germany before 1912, but those were not satisfactory. The new product was an immediate success, especially after the United States entered World War I, which brought demands for luminous dials, dashboards, all types of compasses. Thousands of watches arrived from everywhere to be touched up with the luminous paint; and it was in that summer of 1917 that the dial-painters started to point their brushes with their lips and were invaded by radium, thus bringing much notoriety to the company and doom to themselves.

"The doctor," as everyone called him, was also affected. When his left index finger showed signs of radium he hacked off half of the first phalanx. He was always careful about touching things that could transmit the "infection," but it was not until 1920 that he admitted his condition. He always denied in himself the necrosis of the bones that affected the women but accepted "aplastic anemia," and for the rest of his life he worked with private physicians, scientists, and public health authorities in efforts to study his condition and do something for it.

In 1921 the company was merged with others and von Sochocky left to become a consultant. He worked on some medicinal products, notably colloidal sulfur and colloidal lead, the use of vanadium in arteriosclerosis, and endocrine products; (he turned down the chance to introduce the first hormone cosmetic cream). A sample of the Sochocky-Willis Radioscope—an extremely sensitive device for determining radium—is preserved in the Smithsonian Institution.

The doctor was what some psychologists call a "B person"; that is, he usually started late but then worked until all hours. He was much interested in peace-time applications of radium and radium paint. Even before he left active duty with the company, however, it had discontinued the use of radium in favor of less dangerous mesothorium. He greatly enjoyed what home life he allowed himself. He was fond of good music; he was an excellent gourmet cook; and he was fluent in five languages: English, Czech, French, German, and Russian.

Within his last year his anemic condition became so bad that life seemed like one continuous transfusion. He finally died Nov. 14, 1928, survived by his wife, Marta Anna Lytwyn of Lvov, and a daughter, Stephanie.

Personal recollections; *American Mag.*, January 1921; notes from Stephanie von Sochocky Gury; obituaries in *New York Times*, Nov. 15, 1928 and *Brookln Eagle*, Nov. 25, 1928; *Med. Tribune*, May 11, 1966; *Chemistry*, April-May 1969.

FLORENCE E. WALL

W

William Henry Wahl

1848-1909

Wahl was born Dec. 14, 1848, in Philadelphia to John H. and Carolina R. Wahl. He received his A.B. from Dickinson in 1867 and his Ph.D. degree from University of Heidelberg in 1869.

Returning home he became secretary of Franklin Institute and an editor of *Journal of the Franklin Institute*. He was associated with the organization from 1870 to 1874. During the same period he taught science at Philadelphia Episcopal Academy from 1871 to 1873 and physics and physical geography at Philadelphia's Central High School from 1873 to 1874.

Work on the journal brought out Wahl's latent talent as a science editor and writer. From 1870 to 1874 he and Spencer F. Baird compiled "Annual Record of Science and Industry." He was an editor of *American Exchange and Review* from 1873 to 1876, editor of *Polytechnic Review* from 1876 to 1879, associate editor of *Engineering and Mining Journal* from 1878 to 1881, and an editor of *Manufacturer and Builder* from 1880 to 1896.

Wahl wrote, compiled, or edited several works, among them, "Report on the Light Petroleum Oils, considered as to their safety or danger in various domestic uses" in 1873; "Galvanoplastic Manipulations" in 1883; "Techno-Chemical Receipt Book" in 1885; "Hand-Book of Assaying," a translation from the German with additions by Wahl; and "The Franklin Institute of the State of Pennsylvania," a history of the organization, in 1895. He published a number of articles, most of them in the Franklin Institute's journal.

The discovery by Wahl and William H. Greene that metallic manganese could be obtained by heating a mixture of manganese oxide with metallic aluminum (or magnesium), and a flux of cryolite or of fluorspar, in a crucible lined with magnesite, the reaction releasing an enormous amount of heat, antedated Hans Goldschmidt's thermite process by a number of years. One application of this discovery was the use of aluminum for producing high temperatures (estimated at 3500°C.) in metallurgical processes.

Wahl again served Franklin Institute as secretary and editor of its journal from 1882 until 1909. During this period he organized several individual sections in applied sciences, these becoming a principal feature of the institute. He was the organizer and leader of the institute's Committee on Science and the Arts. During a trying period of several years, when the survival of the institute was in jeopardy, he maintained its smooth operation, despite small remuneration to himself. The library profited from his knowledge of applied science and from his efforts on its behalf. As principal director of the institute's school, his contribution to technical education was of lasting character. He died Mar. 23, 1909 near Doylestown, Pa.

Biography by Louis Levy and others in *J. Franklin Inst.* **167,** 473-478 (1909), portrait; obituary in *Philadelphia Public Ledger,* March 24, 1909; "Dictionary of American Biography," **19,** 314, Chas. Scribner's Sons (1936); J. W. Mellor, "Comprehensive Treatise on Inorganic and Theoretical Chemistry," **12,** 164, Longmans, Green & Co. (1932).

HAROLD J. ABRAHAMS

Leo Aloysius Wall

1918-1972

Wall was born May 26, 1918 in Washington, D. C. Having made a considerable reputation as a track athlete, he entered Catholic University intending to prepare for medicine.

Francis O. Rice and Karl Herzfeld succeeded in turning his interests toward physical chemistry and especially chemical kinetics. He received his Ph.D. degree in 1945. Except for a brief interval with California Research Corp., he was associated with the National Bureau of Standards where he ultimately became chief of the Polymer Chemistry Section. The complications encountered in thermal decomposition of polymers, at first considered as a possible analytical technique, led him to work out with Robert Simha the free radical chain treatment of polymer decomposition. As in hydrocarbon pyrolysis, a steady-state concentration of free radicals was envisioned, and a set of four elementary reactions was postulated:

1) Initiation: Polymer → Radicals
2) Propagation: Radical → $\text{Radical}_{(j-1)}$ + Monomer
3) Transfer: $\text{Radical}_{(a)}$ + $\text{Polymer}_{(j)}$ → Saturated $\text{Polymer}_{(a)}$ + Unsaturated $\text{Polymer}_{(j-k)}$ + $\text{Radical}_{(k)}$
4) Termination: 2 Radicals → 2 Polymers

A further essential was that the decomposing polymer is an open system from which not only monomer but any molecule can evaporate, once its size is reduced to less than L monomer units; L is governed by cohesive energy density.

The resulting set of integrodifferential equations became capable of solution by the computers developed during this period, allowing Wall and his associates to describe the course of conversion to volatiles and the changes of polymer molecular weight. They undertook an extensive program of experimental investigation, facilitated at first by the existence of extreme cases tractable by simpler mathematical methods. This work led naturally to efforts toward developing heat-resistant polymers, of which fluorocarbons received the most attention. The beginning of systematic perfluoroaromatic chemistry was one result. Related fields of activity were oxidative degradation, ultraviolet degradation, and radiation chemistry of polymers.

Wall held a Fulbright scholarship in 1951-52 and made many friends in France and England. He played an important part in founding the Division of Fluorine Chemistry of the American Chemical Society. He received a Department of Commerce Gold Medal. He satisfied his interest in teaching, first by presenting courses in polymer chemistry at the National Bureau of Standards and later by holding an adjunct professorship at American University.

His recreation, sailing, brought about his death. He was sailing alone on Chesapeake Bay toward Tangier Island when the boom apparently swung around, broke his neck, and knocked him overboard. His body was found on the shore of Smith Island, Sept. 19, 1972.

Personal recollections; Wall's articles, NBS publications, and chapters in his books.

ROLAND E. FLORIN

William George Waring

1847-1935

Waring, born Feb. 21, 1847, lived near Boalsburg, Pa., and from the age of seven attended Oak Hall School. In 1855 citizens of Bellefonte, Carlisle, and other towns agitated for a state school of agriculture, for which the Pennsylvania legislature provided funds. W. George's father, William Griffith Waring, was selected to supervise the erection of a suitable building. In 1858, the family moved into the partially completed structure and the lad attended a country school about one-half mile east of there. The Farmers High School, as it was first called, opened Feb. 16, 1859, with his father acting as superintendent. When the High School became the State Agricultural College (later, Pennsylvania State University) with Evan Pugh as the first president, William Griffith Waring remained on the faculty as professor of horticulture. W. George became a member of the first class and resided at the college as a student until the end of the term in 1863. The Civil War was at its height, and most of his classmates were taken into the army. On account of youth and small stature, he was not accepted.

Under Pugh's tutelage, W. George learned glass-blowing, and wishing to pursue this work he went to New York City where he obtained a job as glass-blower; but his health failed because of arsenic in the materials used. He was soon employed by a florist at Astoria, N.Y. Using the knowledge gained there, he set up a successful business of his own as a florist at Tyrone, Pa. He was also skillful in the use of Isaac Pitman shorthand and published a phonetic journal printed in shorthand. He became the official court stenographer for the central judicial district of Pennsylvania. While at Tyrone he did much reporting for newspapers; for this reason Henry C. Bolton asked him to record the proceedings of the Priestley Centennial at Northumberland, Pa., July 31-Aug. 1, 1874 for the N.Y. *Times, Sun,* and *Tribune.*

This meeting was a turning point in Waring's career; for, influenced by Thomas Egleston and other chemists he met at Northumberland he took courses in chemistry, blow-pipe analysis, and metallurgy at Columbia School of Mines and at Stevens Institute of Technology. In his memoirs written in 1932, he stated that the Priestley Centennial was for him the most important event in his entire mature life. Under Egleston he became such an expert with the mouth blowpipe that he was requested to demonstrate in the laboratories of universities in Ohio, Mississippi, Colorado, New Mexico, Arizona, and California.

His mastery of the methods of analysis of ores and minerals led him to move to Denver, Colo. in 1879 as a metallurgist with the U.S. Geological Survey. In 1882 he became superintendent and chemist for Rico Reduction Co., Rico, Colo. From 1884 to 1892 he was metallurgist for Flagler Reduction and Smelting Co., Silver City, N.M. He was consulting metallurgist and chemical engineer for the Hildago Mining Co. of Parral and Chihuahua and for the Montezuma Mining Co. of Lampozos, Sonora, Mexico from 1886 to 1898. From 1889 to 1914 Waring owned the Waring Research Laboratories in Joplin and Webb City, Mo. He was vice-president and president of Oronogo Mutual Mining Co.

When the New Mexico School of Mines, Socorro, was organized in 1891, Waring was appointed to the Board of Trustees. The Most Rev. John Baptist Lamy, Archbishop of Santa Fe, who has a pseudonym in Willa Cather's "Death Comes for the Archbishop," was a close friend of his. He was an accomplished violinist and flutist and published many papers of cultural value. In 1916 Pennsylvania State College recognized his work by conferring upon him a Master of Science degree. He belonged to American Chemical Society, AAAS, Society of Chemical Industry, and American Institute of Mining and Metallurgical Engineers.

In 1869 Waring married Mary Hull of Tyrone, Pa. They had two sons, William who died at the age of 12, and Guy who lived to work with his father on many projects, particularly in the Webb City, Mo., area. Guy was president and manager as well as part owner of New Year Trailing Mill, and manager of Oronogo Mutual Mining Co., Oronogo, Mo. Mary Hull Waring passed away in 1912 and W. George Waring later married her sister, Anne Hull. He died at Webb City, Mo. in 1935, leaving his son, Guy, as sole heir.

W. George Waring, Handwritten Memoirs, 1932; biography by Sister Mary Grace Waring, *J. Chem. Educ.,* **25**, 647-652 (1948).

SISTER MARY GRACE WARING

Lucien Calvin Warner

1841-1925

Warner founded a bewildering number of chemical companies which at one time made him the world's largest producer of electrolytic chlorine products and which were finally consolidated into the Westvaco Chlorine Products Corp. He became a chemical manufacturer only in middle life, but as he died in harness at 84, he put in well over a quarter century at this career.

Warner was born Oct. 26, 1841 at Cuyler, in mid-state New York, the second of only two children, both sons, of Alonzo Franklin and Lydia (Converse) Warner; he was of the ninth generation of descendants of Andrew Warner, who settled in Cambridge, Mass., in 1632. He was educated at De Ruyter Academy, then at Oberlin College, from which he received his A.B. degree in 1865 and his A.M. in 1870. In 1867 he completed his medical education and received his M.D. from University Medical College of New York University. Both his alma maters would later award him honorary LL.D. degrees; in addition he would serve on Oberlin's board of trustees for half a century and donate the college's gymnasium and music conservatory building.

Warner practiced medicine in New York state for a few years; toured with his brother, Ira De Ver Warner, giving popular medical lectures; and even wrote two medical books for laymen: "A Popular Treatise on the Functions and Diseases of Woman" (1873) and "Man in Health and Disease" (1873). Eventually his attraction to the world of business and manufacturing proved too great, however, and in 1874 the brothers set up a firm to manufacture corsets, which proved an immediate and great success. It was this success which allowed Lucien Warner to cast about in other directions, many of them money-losers, and eventually to become interested in the phosphate deposits of Grand Connetable, one of the groups of islands off the coast of French Guiana which include Devil's Island. In 1891 at the age of 50, he bought the interests which included the French government's original mineral concessions, and by 1898 he acquired sole ownership of the enterprise.

By the turn of the century the Warner Chemical Co. was manufacturing U.S.P. and commercial grade phosphates of various kinds in a new plant at Carteret, N.J. Soon they expanded into acid pyrophosphate and tartrate baking powders in the commercial and the home market, founding two new companies for the purpose. Almost by accident, because of a mistaken expansion into the vanillin market, the company acquired some electrolytic cells and began producing caustic and chlorine from brine. These were followed by phosphorus trichloride and oxychloride, acetyl chloride and acetic anhydride, and carbon tetrachloride. At the start of World War I, Warner Chemical was the only American firm producing these important materials.

With Ernest Klipstein, Warner organized yet another company and built an electrolytic plant in South Charlestown, W. Va., close to sources of salt and coal. During the war Warner also built a huge chlorine plant for the government at Edgewood Arsenal, Md., the largest of its kind in the world at the time. After World War I Warner consolidated his holdings further in the slack period of the early '20's. When he died in 1925 his son, Franklin Humphrey, continued as head of the firm, founding Westvaco Chlorine Products Corp. in 1928 to gather together the scattered Warner interests; in 1929 he formed United Chemicals Inc. to act as a holding corporation in the acquisition of small chemical companies.

In 1948 Westvaco Chlorine Products Corp. became the Westvaco Chemical Division of Food Machinery and Chemical Corp. Its nature changed later: for a time it produced caustics but eventually marketed only activated carbons, with another division, also called Westvaco, producing packaging materials of all kinds.

Lucien Warner was married in 1868 to Keren Osborne, who bore him four children: Agnes Eliza, Elizabeth Converse, Franklin Humphrey (mentioned above), and Lucien Thompson, who went on to head the corset firm. Warner died in New York City, July 30, 1925.

W. Haynes, "Chemical Pioneers," D. Van Nostrand Co. (1939; reprinted 1970), pp. 124-42; "National Cyclopedia of American Biography," **9,** 537, James T. White & Co. (1898); *New York Times,* July 31, 1925; for a history of the company to 1948 *see* W. Haynes, "American Chemical Industry," **6,** 478-80 (1949), D. Van Nostrand.

ROBERT M. HAWTHORNE JR.

Edward Wight Washburn

1881-1934

Washburn, born May 10, 1881, was one of four children of William Gilmor and Flora Ella (Wight) Washburn. Although his ancestors had settled in New England before 1626, Washburn's father had moved from Houlton, Maine to Edward's birthplace, Beatrice, Nebr. Despite his family's interest and success in commerce and politics, he became interested in chemistry even though his high school had no formal instruction in that subject. His home chemistry laboratory served as a substitute until the explosions he produced brought parental warnings to stop "all this foolishness."

He attended University of Nebraska for 2 years (1899-1901) and taught for a year at the McCook, Nebr. high school. Since advanced courses in chemistry were unavailable in that state, Washburn attended Massachusetts Institute of Technology, where he studied under Arthur A. Noyes. He received his B.S. degree in 1905 and Ph.D. in 1908.

After receiving his doctor's degree he became associated with University of Illinois as an assistant. Within 5 years he had been promoted to professor of physical chemistry. June 10, 1910, he married Sophie Wilhelmina de Veer, with whom he had four children. He was appointed head of the Ceramic Engineering Department in 1916. In 1926 Washburn was chosen head of the Chemistry Division of the National Bureau of Standards. In this position, which he held until his death, Feb. 6, 1934, he headed an active research program in the bureau.

Washburn's research interests covered a wide range of topics in physical chemistry. His doctoral studies led to a paper on "The Theory and Practice of the Iodometric Determination of Arsenious Acid." This prompted work on the thermodynamic treatment of buffer solutions. His early investigations also involved the first accurate measurement of true transference numbers and the relative hydration of aqueous ions.

At University of Illinois Washburn became interested in developing a simple system of thermodynamic chemistry "based upon an analogy with an imaginary perfect thermodynamic engine." Of more immediate value was his measurement of Faraday's constant with an iodine coulometer, his development of a highly precise viscosimeter, and his construction of precision apparatus for measuring the conductivity of aqueous solutions of electrolytes. While at NBS he became involved in petroleum research and was responsible for the isolation of the first crystals of rubber. In December 1931 he developed the fractional electrolysis of water on the basis of hydrogen isotopes, for which work he received the Hillebrand Prize of the Chemical Society of Washington.

In all Washburn published approximately 100 scientific papers. He authored a textbook in 1915, "An Introduction to the Principles of Physical Chemistry." In 1922 he became editor-in-chief of "International Critical Tables of Numerical Data: Physics, Chemistry, and Technology," and for 4 years worked on the first seven volumes which appeared between 1926 and 1930.

"Dictionary of American Biography," **19,** 498, Chas. Scribner's Sons (1936); biography by W. A. Noyes, *Trans. Illinois State Acad. Sci.,* **1,** 39-40 (1935); obituaries in: *Science,* Mar. 10, 1934; *Chem. Eng. News,* Feb. 10, 1934; *Washington Post,* Feb. 7, 1934.

SHELDON J. KOPPERL

Frank Sherman Washburn

1860-1922

Washburn, founder of American Cyanamid Co., was born in Centralia, Ill., Dec. 8, 1860, the son of Elmer and Elizabeth (Knight) Washburn. His father, of New England stock going back to the Massachusetts Bay Colony, was at various times president of the National Livestock Bank and of the Chicago Livestock Exchange, chief of police of Chicago, and chief of the United States Secret Service. Young Washburn was educated at Cornell University, where he earned his C.E. degree in 1883. For a year after that he worked for the Chicago and North Western Railroad but returned to Cornell in 1884 to do graduate work in economics, history, and political science. For the next two decades his professional activity would center about engineering, with chemical manufacturing coming only when he was nearly 50 years old.

By turns he built bridges for the Chicago and North Western, reorganized the Chicago Belt Line Railroad, studied the English railroad system for applications in this country, and organized an engineering firm with his father and Ira E. Shaler which built reservoirs and water supplies for New York City as well as the Third Avenue Cable Railway.

In 1897-98 Washburn spent several months in South America studying nitrate production for William R. Grace & Co. It was here that he first realized the wisdom of American independence of the Chilean nitrate monopoly and began to consider ways by which it might be achieved. For some years after this Washburn was absorbed in manufacturing mining equipment, but he gradually became involved in hydroelectric power installations and production after moving to Nashville, Tenn. in 1900; this provided the final link in the chain which led to the founding of American Cyanamid in 1907.

Well before that date Washburn realized that the Muscle Shoals section of the Tenessee River in extreme northern Alabama, where the river falls some 150 feet in a thirty-mile stretch, was ideal as the power source necessary for any nitrogen fixation process. It remained only to find a process which was not already tied up by other firms. This was the Frank-Caro process by which atmospheric nitrogen reacts with electrochemically produced calcium carbide to produce calcium cyanamid, from which a variety of further nitrogen chemicals can be made. The Muscle Shoals development proved to be impossible as first conceived so the first Cyanamid plant was opened in Niagara Falls, using power from the Canadian side of the Falls; the initial carload shipment went out in December 1909.

The first World War provided the necessary governmental impetus to complete the Muscle Shoals project, and in 1916 the Congress provided funds for power development for a Haber process ammonia plant to be operated by General Chemical Co. and a cyanamid process plant to be run by Washburn's company. The products of the company at this time were used principally in nitric acid for munitions and in nitrogen-phosphate fertilizers; for the latter purpose the company acquired Ammo-Phos Corp. of Warners, N.J. and Amalgamated Phosphate Co. of Brewster, Fla. After the war the Warners plant added a line of heavy chemicals, resins and varnishes, and other industrial products. This set the stage for the expansion of the company into the industrial giant we know today. In 1929 Calco Chemical Co., Bound Brook, N.J. with a total of seven plants, was acquired to introduce American Cyanamid into the dyes, pharmaceuticals, and coal-tar chemicals market. Later acquisitions were Selden Co., Pittsburgh; Kalbfleisch Corp., Brooklyn; Lederle Laboratories, now the company's ethical pharmaceutical division; and Chemical Construction Co., a fabricator of plant facilities, among many others. Today American Cyanamid markets a complete line of chemical products of all kinds and has branches or subsidiaries in all major countries overseas.

Washburn married Irene Russell of Atlanta, Ga. in 1890; they had two children, Frank S. Jr. and Elizabeth Washburn Irvine. The elder Washburn died in Rye, N.Y., Oct. 9, 1922. Frank Washburn Jr. was associated with American Cyanamid for his entire professional career, starting in sales in 1918, and heading the Canadian operation, North American Cyanamid, Ltd., and the Mexican division, as well as serving on the board of the American firm. He died on Apr. 7, 1963, ending the association of the Washburn name with the company, as he left only a daughter.

Williams Haynes, "Chemical Pioneers," 243-58, D. Van Nostrand Co. (1939; reprinted 1970); "National Cyclopedia of American Biography," **32,** 51, James T. White & Co. (1945); *New York Times,* Oct. 10, 1922, Oct. 12, 1922, Apr. 8, 1963, (Frank, Jr.). For a history of the company to 1948, *see* Williams Haynes, "American Chemical Industry," **6,** 21-25 (1949), D. Van Nostrand.

ROBERT M. HAWTHORNE JR.

Henry Stephens Washington

1867-1934

Washington was born in Newark, N.J., Jan. 15, 1867 the son of George and Eleanor Phoebe (Stephens) Washington. He entered Yale after having attended private school and received tutoring. He graduated in 1886 and spent the next 2 years at Yale as a fellow in physics. The following 6 years he travelled throughout the world and developed a permanent interest in archaeology. He spent part of this time as a student at the American

School of Classical Studies in Athens, where he learned chemical and petrographical methods to apply to his field. He studied the volcanic islands of the eastern Mediterranean area and published a series of papers on lavas. He also spent two semesters studying with the petrographer Ferdinand Zirkel at University of Leipzig. His research on volcanoes near Smyrna brought him his Ph.D. degree in 1893.

Washington returned to the United States in 1895 as an assistant in mineralogy at Yale. After a year he moved to his family's old home in Locust, N.J. and opened a laboratory, where for a decade he carried out research in petrology. From 1906 until 1912 he served as a geological consultant to the mining industry. In 1912 he became affiliated with the Geophysical Laboratory of the Carnegie Institution in Washington, D.C., where he remained for the rest of his career, except for a year (1918-1919) which he spent in Rome as scientific attaché to the American embassy. He died Jan. 7, 1934.

Washington's major research interests involved chemical aspects of geology. His collaboration beginning in 1899 with Joseph P. Iddings, Louis V. Pirsson, and Whitman Cross on chemical classification of igneous rocks led to the publication in 1903 of "Quantitative Classification of Igneous Rocks." The following year he published the first edition of "Manual of the Chemical Analysis of Rocks," which was used throughout the world as a standard handbook. Another important contribution was "Chemical Analyses of Igneous Rocks, Published from 1884 to 1913, Inclusive, with a Critical Discussion of the Character and Use of Analyses," published in 1917.

After joining the Carnegie Institution, Washington made several extended trips to study the characteristics of igneous rocks in the Pacific Ocean area. He published the results of his work as a five-part article, "Petrology of the Hawaiian Islands," in *American Journal of Science* (1923-1926). His other major publications were "The Chemistry of the Earth's Crust," in the *Journal of the Franklin Institute* (1920) and "The Composition of the Earth's Crust" (1924), written in collaboration with Frank Wigglesworth Clarke. Washington's work was known to petrologists throughout the world, and he received many honors, including the presidency of the Mineralogical Society of America in 1924. In 1918-19 he was scientific attache at the American Embassy in Rome and was decorated by the Italian government.

Biographies by Whitman Cross in "Dictionary of American Biography," **19,** 527-28, Chas. Scribner's Sons (1936); and by Albert Martin in *J. Chem. Educ.* **30,** 566-568 (1953).

SHELDON J. KOPPERL

George Watt

1820-1893

A grandnephew of the Scottish inventor James Watt, George was born on a frontier farm near Xenia, Ohio, Mar. 14, 1820. He attended primitive country schools, began teaching school when he was 17, studied medicine under a physician, practiced medicine, then attended Medical College of Ohio from which he graduated with his M.D. degree in 1848.

Becoming interested in dentistry, he attended Ohio College of Dental Surgery from 1852 to 1854. During the second term, he doubled as student of dentistry and professor of chemistry. After graduation he practiced dentistry, gave a series of public lectures on chemistry for girls, and went back to the college in 1855 as professor of chemistry and metallurgy. During the Civil War he joined an Ohio infantry regiment as surgeon, received an accidental injury from which he never fully recovered, and returned to teach chemistry and therapeutics at the dental college.

He co-edited *Dental Register of the West* from 1856 to 1864 and edited *Ohio Journal of Dental Science* from 1881 until he died at Xenia, Feb. 16, 1893. He published popular articles and lectures on chemistry in the *Register* and elsewhere; he was perhaps the most prolific writer of chemistry articles for dentists during his time. In 1867 he assembled several articles into a book entitled "Register Papers: a collection of chemical essays in reference to dental surgery" (1868).

Watt was not a practicing chemist. His knowledge of the science seems to have been rather shallow. He was primarily a dentist, secondarily a teacher. But he stands out as one of the few dentists of his time who showed an interest in chemistry. When ill health forced him to resign his professorship in 1871, the college catalog stated that as a teacher of dental chemistry he "never had a superior, if an equal." And one of his contemporaries credited him with being "the pioneer in directing the attention of the profession to chemistry."

Biography by J. Taft, *Dental Summary* **8,** 1-7 (1888), which contains the quote referring to Watt as the pioneer and was reprinted with additions in obituary by L. P. B., *Dental Summary* **13,** 105-13 (1893), with portrait; obituary, *Dental Cosmos* **35,** 326-27 (1893); biography and portrait in "Biographies of Pioneer American Dentists" by Burton L. Thorpe, being vol. 2 of "History of Dental Surgery," edited by Charles R. E. Koch, National Art Publishing Co., 1909, being also vol. 3 of the 1910 edition.

WYNDHAM D. MILES

Oliver Patterson Watts

1865-1953

Watts was born in Thomaston, Maine, July 16, 1865. Following his graduation from Bowdoin College and one year of graduate studies at Clark University, he was principal of the Thomaston grammar school for 2 years. He then taught chemistry and physics in Franklin Academy for 6 years and in the Waltham, Mass., high school for 5 years.

While at Waltham he became interested in the great developments in electrochemistry which were receiving attention from American industrial chemists. In 1902 he went to University of Wisconsin to study in the Department of Applied Electrochemistry that had been established 4 years earlier by Charles Burgess and in 1905 received the first doctor's degree granted by the University's College of Engineering. In the autumn of 1905 the Department of Applied Electrochemistry was merged with the new Department of Chemical Engineering, and Watts was placed in charge of the course in applied electrochemistry in the chemical engineering curriculum.

In 1914 Watts published "Laboratory Course in Electrochemistry." In his directions for each experiment he did not indicate the expected results but required the student to reach his own conclusions and interpretations.

Watts' scientific works, of which 31 papers appeared in the *Transactions of the Electrochemical Society,* include studies on the general theories and applications of electrochemistry, rates of corrosion, the rusting of iron, the effects of dissolved oxygen and water line on corrosion, the grading of corrosion resisting alloys, and the behavior of voltaic couples. In electroplating he worked on the problems of embrittlement, dezincification, structure of deposits, and the effect of addition agents. In electrometallurgy he investigated the production of calcium carbide, silicon carbide, alloys, and the borides, silicides and nitrides of metals.

In 1915 Kalmus, Harper, and Savelle published an article on cobalt plating in which they showed that higher current densities could be employed in concentrated baths. Watts then found that similar nickel baths were more efficient. Since that time almost all nickel plating has been done in "Watts Baths," including those that contain addition agents to yield bright deposits. Watts may well be called the Father of Modern Nickel Plating.

Watts' first wife was Mary J. Orton, daughter of Edward Orton, geologist and president of Ohio State University. After she died he married Estella N. Jones in 1948.

Watts was always fond of the out-of-doors. Until his retirement at the age of 70, he and Mrs. Watts continued to spend their summers sailing along the coast of Maine. After retirement the Watts toured the United States and Canada by car, with tent and camping outfit. Watts died in Madison, Wisc., Feb. 6, 1953.

Resolutions of the faculty of University of Wisconsin on the death of Watts; correspondence between Watts and Cleveland Nixon; personal recollections.

WILLIAM BLUM

Henry Adam Weber

1845-1912

Weber, the son of Frederick and Caroline (Tascher) Weber, was born July 12, 1845 in Clinton Township, Ohio, near Columbus. While a student at Otterbein College, he became interested in chemistry. He continued his education in Germany, studying at the Kaiserlautern from 1863 until 1866, and under Justus von Liebig at Munich from 1866 until 1868. In 1870, after his return to America, he married Rosa Ober, whom he had met in Germany. They had two daughters.

Analyses of the minerals and soils of Ohio occupied Weber's time as geological survey chemist for that state from 1869 until 1874. He joined the faculty as professor of chemistry at the University of Illinois in 1874, where his duties included the planning of a new university chemical laboratory. During this period he completed his formal education and received his Ph.D. degree from Ohio State University in 1879. He also served as chemist of the Illinois Board of Agriculture and the Illinois Board of Health, in which capacity he carefully studied the state's rivers to observe sanitation conditions.

In 1882 he left his positions to devote his time to the manufacture of sugar from sorghum. In conjunction with Melville A. Scovell he patented a clarification process that received considerable attention. Their work led them to found the Champaign Sugar and Glucose Co., which was established to prepare sucrose from sorghum juice and glucose from the starch of sorghum grain. Limitations on the use of sorghum and economic problems such as the reduction of the sugar tariff led to the collapse of the firm.

In 1884 Weber became professor of agricultural chemistry at Ohio State University, a position he held until his death. He also served as chief chemist of the Ohio Dairy and Food Commission (1884-1897) and in this capacity became recognized as a leading pioneer in Harvey W. Wiley's pure food movement. His interest in pure foods dated back to his work in Illinois. He was appointed as a member of the first American committee on food standards, established by Congress in 1903. His work, along with that of fellow members Scovell, Wiley, William Frear, and Edward H. Jenkins, led to the passage in 1906 of the Federal Food and Drugs Act.

Although he belonged to several professional societies and published numerous articles, his best known work was his committee work described above. He authored one textbook in 1875, "Select Course in Quantitative Analysis." He died at home in Columbus, June 14, 1912.

"Dictionary of American Biography," **19,** 582, Chas. Scribner's Sons (1936); "National Cyclopedia of American Biography," **19,** 277-8, James T. White & Co. (1926); "Semi-Centennial Alumni Record . . . University of Illinois" (1918).

SHELDON J. KOPPERL

Alfred Lambremont Webre

1881-1963

Webre was born into the sugar industry that he served all his life on the family plantation at St. James, La., Nov. 7, 1881. Indeed, the tradition is that Norbert Rillieu, the inventor of multiple-effect evaporation, had made his first commercial installation at the little Webre Sugar House some 25 years previously.

Webre studied mechanical engineering at Tulane University, obtaining his bachelor's degree "with distinction" in 1904 and his master's degree in 1906. He worked successively for John H. Murphy Iron Works, New Orleans, until 1916, and E. B. Badger & Son, Boston, until 1922. He then went with U.S. Pipe & Foundry Co., Burlington, N.J. under an agreement which allowed him to spend half his time privately as a consulting sugar technologist. In 1956 U.S. Pipe sold its sugar apparatus interests to Jackson Industries, Inc., Birmingham, Ala., and they retained Webre as their consulting sugar technologist until his death.

During the period from 1910 on, his role as a consulting sugar technologist became increasingly important. He started with the standard calandria evaporator, critically analyzed the function of several of the apparatus used in the process end of the cane sugar industry, both raw and refined, and made notable improvements in the fundamentals of design. He worked out standards for improving the performance of the boiling house which became so universal that their origin and uniqueness are largely forgotten.

In the 1920's he developed and patented a mechanical circulator for vacuum pans which, improved across the years, was widely accepted in both refineries and raw houses. His work on the comparison of the circulation in vacuum pans with and without mechanical stirring became a classic.

In 1926 in collaboration with Clark S. Robinson of MIT, he published a 500-page textbook, "Evaporation." During the same decade, he also published two highly practical booklets. One, published in 1924 and entitled "The Heat Balance in the Cane Sugar Factory" was a case study of each of the various cycles of heating and evaporation in use at the time. The other, published in 1930 and entitled "Masse Cuite, Molasses and Sugar," was a material balance for solids passing through the vacuum pans at different apparent purities for the two- and three-boiling systems. His subsequent writings consisted of a myriad of papers presented at meetings of various technical societies to which he belonged. He also wrote a chapter, "Circulation in vacuum pans" for Pieter B. Honig's treatise, "Principles of Sugar Technology," in 1958 and chapters "Evaporation" and "Crystallization of sugar" for the "Spencer-Meade Cane Sugar Handbook," 1948 and later editions.

Webre used the work of Eugene Gillette and others, to rejuvenate the abandoned two-boiling system; this, together with mechanical circulators, made it possible to produce superior raw sugar which easily met the standards desired by the refineries.

At the end of his career he collected the results of his investigations in a series of 16 articles published in *The Sugar Journal,* (La.), and *Sugar y Azucar* during the years 1960-62. These were practical papers, of value to equipment manufacturers as well as to factory operating personnel.

In 1938 Webre was made lifetime honorary president of the Asociacion de Tecnicos Azucareros de Cuba and in 1954 honorary professor of sugar technology by the Escuela de Quimica Azucarera de Sienfuegos de Cuba. In 1962 he received the annual citation from the Sugar Industry Technicians.

Webre gradually decreased his work load starting in 1960 but was still active until a few months before his death on Oct. 23, 1963. Essentially a humble man and a great teacher, he was remembered not only for his innovations but for his deep capacity to impart his knowledge to others.

Personal recollections.

ALFRED L. WEBRE, JR.

John White Webster

1793-1850

Webster was born in Boston, May 20, 1793, the son of an apothecary. After graduating from Harvard College in 1811 and Harvard Medical School in 1815 he travelled to England and studied at Guy's Hospital where the poet John Keats was his fellow student and friend. He seems to have been as much interested in science as in medicine, and after completing his studies at Guy's he took hikes in Scotland and England, studying the geology of the countryside. He visited noted mineralogists in Great Britain and France, studied their specimens, and brought back "a very select and considerably extensive cabinet of minerals."

He stopped at the Azores in 1817 on his way home and examined their geology, later publishing his observations in a book, "A description of the Island of St. Michael, comprising an account of its geological structure, with remarks on the other Azores or Western Islands" (1821). While there he fell in love with and married the daughter of the American consul.

John Gorham, M.D., a practicing physician and adjunct professor of chemistry at Harvard Medical School, had been Webster's preceptor in medicine. He was influential in having Webster appointed lecturer on chemistry, mineralogy, and geology in 1824.

In 1826 Webster advanced to the rank of adjunct professor of chemistry, and in 1827 he succeeded Gorham as Erving Professor of Chemistry and Mineralogy.

Webster was a conscientious, industrious teacher. He supplemented the apparatus at Harvard by purchasing, out of his own pocket, equipment from European and American instrument makers. He showed his students a variety of lecture demonstrations in chemistry and mineralogy as well as exhibits in geology and mineralogy. In 1826 he published a text, "A manual of chemistry on the basis of Professor Brande's; containing the principal facts of the science, arranged in the order in which they are discussed and illustrated in the lectures at Harvard University, N.E." Webster's "Manual" has an engraving showing a plan of the chemical lecture room and laboratory at Harvard at that time and is one of the few early illustrations of an American chemical laboratory. His text was used at Amherst, West Point, and other colleges and went through two later editions.

Webster was a kindly man, liked by his pupils. As students do with professors, they gave him a nickname, "Skyrocket Jack." For many years he sponsored a student chemistry society called the Davy Club. Among Webster's students were Josiah Parsons Cooke, John Bacon, Charles G. Page, Charles Thomas Jackson and other men who became teachers of chemistry or practicing chemists. A genial host, he entertained such celebrities as Henry W. Longfellow and Charles Dickens.

In 1818 and 1819 Webster presented courses of public lectures on geology and mineralogy for citizens of Boston. In 1826 and 1829 he offered public lectures on chemistry and mineralogy. It is probable that he offered public lectures in other years. In 1839 he appeared before the Cambridge Lyceum; again, it is quite probable he lectured before other lyceums and educational organizations.

At times Webster undertook chemical investigations for clients, and he gave expert testimony in trials involving poison—within a 3 month period in 1847-48 he ran analyses in four cases of suspected poisoning by arsenic.

Webster published articles and notes on geology, mineralogy, and chemistry. He was one of the founders of *Boston Journal of Philosophy and the Arts* in 1823 and one of the editors during the journal's existence from 1823 to 1826. In 1827 he brought out an edition of the British chemist Andrew Fyfe's text, "Elements of chemistry," which went through two later editions. He shepherded through the press American editions of two important books, Justus von Liebig's "Organic chemistry in its applications to agriculture and physiology" (1841), and "Animal chemistry or organic chemistry in its application to physiology and pathology" (1842, 1843).

Webster lived beyond his means and borrowed money; one of his creditors was George Parkman, M.D., a wealthy landlord of Boston. Parkman disappeared after pressing Webster for payment. Parts of a human skeleton, supposedly Parkman's, were found in Webster's laboratory in the Harvard Medical School. The resulting trial was one of the most famous in American history. Webster was convicted of the murder of Parkman and was hanged Aug. 30, 1850. Because of the advances that have been made since 1850 in protecting the rights of persons accused of committing crimes, it seems certain that Webster would not have been convicted today.

Perhaps more has been written about Webster than any other American chemist, but the writing has been done by persons intrigued over the drama of a professor convicted of murdering one of Boston's wealthiest citizens in the laboratory of Harvard Medical School. Little has been written about Webster as a teacher, chemist, and scientist. Edmund Pearson, "Murder at Smutty Nose and other murders," 94-114, Doubleday, Page & Co. (1926), interesting, illustrated, and with references, is one example of the many writings about the Parkman murder; but Pearson's book and almost all other pieces concerning the case are frivolous in comparison with the authoritative, analytical work by Judge Robert Sullivan, "The disappearance of Dr. Parkman," Little, Brown (1971), which shows that Webster did not receive a fair trial. "Quinquennial Catalogue of . . . Harvard University," (1900); Webster's articles and books; I. Bernard Cohen, "Some early tools of American science," Harvard Univ. Press (1950).

WYNDHAM D. MILES

Clarence Jay West

1886-1953

West, an internationally known bibliographer of literature in sciences associated with the pulp and paper industry, was born at Brighton, Mich., June 14, 1886. His education included a teaching degree from Michigan State Normal College at Ypsilanti and the A.B. and Ph.D. degrees from University of Michigan. Later he received an honorary master of education degree from Michigan State Normal College.

After earning his doctor's degree in physiological organic chemistry in 1912 he joined the staff of Rockefeller Institute for Medical Research. During World War I he served as a major in the Chemical Warfare Service and was later a lieutenant colonel in the Army Reserves. His work in chemical warfare led to literary collaboration with brigadier general Amos A. Fries, commanding officer of the Chemical Warfare Service, and they together wrote the first book on the subject, "Chemical Warfare" in 1921.

Following the war, West was employed by Arthur D. Little, Inc. until 1921 and then by the National Research Council in Washington, D.C. until 1936. In that year he joined the staff of The Institute of Paper Chemistry, a graduate school, research center, and center for the collection and permanent storage of the international scientific and technical information related to pulp and paper. West remained there until his death.

Though just beginning a major portion of his life's work with his position at the Institute as editor and research associate in technical bibliography, his contribution was already sufficiently outstanding that by 1937 he was awarded the Gold Medal of the Technical Association of the Pulp and Paper Industry.

His work had numerous aspects. He collaborated with Gustavus J. Esselen in translating Emil Heuser's "Cellulose Chemistry," one of the first standard German texts made available in English. While with the National Research Council he was editor of "International Critical Tables" and "Annual Survey of American Chemistry." For many years he was co-editor of the organic section of *Chemical Abstracts,* and from 1922 until his death was in charge of the section on cellulose and paper.

In his position at Institute of Paper Chemistry he was responsible for editing all technical and scientific publications of the organization and preparing all bibliographies. In addition, he was a member of the policy committee on all scientific articles emanating from the institute. His prominence in the editorial field led scientists and technologists in many other fields and from all parts of the world to seek his counsel.

He made an outstanding contribution to the technical progress of the pulp and paper industry. Because of his efforts the scientific literature of pulp and paper was probably better identified and organized than that of any other major industry. From 1920 until his death he was chairman of the TAPPI Committee on Abstracts and Bibliography. He served also as chairman of the editorial board of the Association's monthly publication, *Tappi.* His bibliographies in bound volumes cover the period from 1900 to 1953. He published more than 700 books, articles, translations, and book reviews. It has been estimated that he prepared more than 30,000 abstracts during his career, perhaps more than any other abstractor has compiled.

West was a member of TAPPI, American Chemical Society, a senior member of the Technical Section of the Canadian Pulp and Paper Association, and Sigma Xi.

West was active in Masonry and became a 32nd Degree Mason. His civic service included membership and presidency of service and social groups. He died in Appleton, Jan. 29, 1953. His body was interred in Arlington National Cemetery, Washington, D.C.

Biographies appeared in *Tappi* **36,** No. 2, February 1953, p. 108A, with port.; "A Tribute to C. J. West," by E. J. Crane, *Tappi* **37,** No. 9, September 1954, p. 196A, a review of West's work as an abstractor; Jack Weiner compiled a list of 158 publications by West in *Tappi* **37,** No. 9, September 1954, pp. 174A-189A; a biography also appeared in *Chemical Abstracts* **47,** No. 3 (Feb. 10, 1953), p. 1381.

ARNOLD E. GRUMMER

Charles Mayer Wetherill

1825-1871

Wetherill was born to Charles and Margaretta (Nayer) Wetherill in Philadelphia, Nov. 3, 1825. His family was well-known in the industrial and scientific world. The Wetherills had made the first white lead in America, in 1762 in Philadelphia. A cousin, Samuel Wetherill, developed a process of making zinc oxide directly from zinc ore in 1852.

Charles was a graduate of University of Pennsylvania and supplemented his chemical knowledge with training in the well-known laboratory of Booth and Boyé in Philadelphia. Then he went abroad to study for a year with Pelouze in Paris and 2 years with Liebig in Giessen. Here he received his doctor's degree in 1848.

Wetherill had an extensive career in public service. He was a public analyst, lecturer, and traveller. He was the first chemist of the U.S. Department of Agriculture from 1862 to 1863, and his "Analysis of Grapes" was the subject of the department's first official bulletin.

He was a special agent for President Lincoln in the investigation of gunpowder production during the Civil War. As chemist for the Smithsonian Institution from 1863 to 1865 he published researches on medical, agricultural, and mineralogical subjects. His interest in the field led New York Medical College to grant him an honorary M.D. degree.

Wetherill's longest treatise was a book published in 1860, "The Manufacture of Vinegar." One of his most important assignments in Washington was a study and plan for the ventilation of the United States Capitol.

In 1866 Wetherill became the first professor of chemistry at Lehigh University when the institution opened. He was one of the original faculty of five professors. All except Wetherill had to teach more than one subject. For his first class he prepared "A Syllabus of Lectures on Chemical Physics." Two years later he published his "Lecture Notes in Chemistry." These were only small text outlines, but they were milestones in the beginning of faculty publications.

At first the small university was completely housed in a Moravian Church which the founder, Asa Packer, had purchased together with extensive acreage on the north side of South Mountain. Two years later Wetherill moved into larger facilities in a new massive stone building, Packer Hall. Here he designed an excellent laboratory.

Wetherill was busy organizing courses, planning the laboratory, and teaching, but he found time for some investigation. A study of the mineral itacolumite (articulite) resulted in a paper which was commended by the well-known geologist James Dana.

Wetherill was asked to return to his alma mater, University of Pennsylvania, and planned to do so when he died suddenly Mar. 5, 1871. A brilliant career was terminated at the age of forty-five.

E. F. Smith, "Charles Mayer Wetherill," *J. Chem. Educ.* **6,** 1076-1089, 1215-1225, 1461-1477, 1668-1680, 1916-1927, 2160-2177 (1929), ill.; "Dictionary of American Biography," **20,** 22, Chas. Scribner's Sons (1936).

ROBERT D. BILLINGER

Samuel Wetherill

1821-1890

Samuel Wetherill, founder of Lehigh Zinc Co. which was later absorbed by New Jersey Zinc Co., prolific inventor and improver of mineral refining processes, came of a family involved for many generations in mineral recovery and chemical manufacture. Son of John Price and Maria Kane (Lawrence) Wetherill, he was the grandson of the Samuel Wetherill who in 1804 founded the paint pigment firm of Wetherill and Bro. in Philadelphia, where it operated continuously until its absorption by National Lead Co. in 1934. His first cousin was Charles Mayer Wetherill, professor of chemistry at Purdue and later Lehigh University; it was Charles' son Richard B. after whom Purdue's Wetherill Chemistry Laboratories were named. Samuel's sons John Price and Samuel Price Wetherill were destined to carry on in the same tradition, being closely associated with New Jersey Zinc throughout their careers.

Samuel Wetherill was born in Philadelphia, May 27, 1821. Educated in Philadelphia, he graduated in 1845 from University of Pennsylvania. He then entered his grandfather's firm, which manufactured white lead (the basic carbonate) and other paint pigments. Here he became interested in the possibility of substituting zinc oxide for the lead salt in paints and began to experiment with zinc-bearing ores to find a method of obtaining the oxide directly. In 1850 when he was twenty-nine, he moved to New Jersey Zinc Co., at Newark; and 2 years later he perfected and patented the Wetherill furnace, which was capable of producing zinc oxide directly from the rich but refractory ores of northern New Jersey, franklinite, zincite, and willemite. This was accomplished by a mixed carbon-reduction and air-oxidation process in which the ores were mixed with coal and heated, with a cold air blast operating at the same time. Wetherill also invented the "tower process" in which the volatile, powdery oxide was purified by being blown up a 70-foot tower, away from heavier ash materials; in a later version the product was passed through a film of water in the tower as well.

In 1853 Wetherill joined forces with Charles J. Gilbert to found Gilbert, Wetherill, Baxter, and Co. in South Bethlehem, Pa.; the name was shortly changed to Lehigh Zinc Co. Here zinc oxide was produced by Wetherill's process, and he immediately set to work to develop methods for producing zinc metal, or "spelter." This he did in 1854 by passing zinc oxide vapor through a bed of incandescent coal in a muffle furnace; later he used and patented vertical retorts. In 1857 he produced ingot from which the first zinc sheet from Pennsylvania ores was rolled.

At the outbreak of the Civil War Wetherill dropped his commercial activities and recruited two companies of cavalry in Bethlehem and was commissioned captain of the 11th Pennsylvania cavalry in August 1861. Three years later he was honorably discharged at the rank of major, and in 1865 he was brevetted lieutenant-colonel in the United States Volunteers for his services during the war. Wetherill returned to his business, but the rush of inventions, improvements, and patents was over. In 1881 two of his sons, John Price and Samuel Price, together with Richard and August Heckscher, purchased the Lehigh Zinc Works and operated it until 1897, when it was consolidated with New Jersey Zinc Co. John Price proved, after the merger, to be as prolific an inventor as his father had been, perfecting a new variety of Wetherill furnace and the Wetherill magnetic concentrating process for treating refractory ores.

Samuel Wetherill was married twice: first in 1844 to Sarah Maria Chattin, who bore him seven children and died in 1869, and then to Thyrza A. James, by whom he had three more children. Wetherill died June 24, 1890 in Oxford, Md., where he had retired some years earlier when the business was sold.

"Dictionary of American Biography," **20,** 23, Chas. Scribner's Sons (1936); biographies of Charles Mayer Wetherill and Samuel Wetherill (grandfather) may be found here also. "National Cyclopaedia of American Biography," **7,** 506, James T. White (1892); see also Vol. 27 for son John Price Wetherill; W. Haynes, "American Chemical Industry," Vol. I (1954), 356-58, D. Van Nostrand Co.; plus other scattered references.

ROBERT M. HAWTHORNE JR.

George Stafford Whitby

1887-1972

Stafford Whitby was born in Hull, England, May 26, 1887. He entered Royal College of Science in London in 1903 with the aid of a national scholarship in chemistry, one of five awarded that year. He graduated in 1907 as the top student of his class and continued as a demonstrator under Sir William Tilden, one of the pioneers in the synthesis of rubberlike materials from isoprene. At the suggestion of his mentor, Whitby accepted a post as chemist for Societé Financiere des Caoutchoucs and travelled to their plantations in Malaya in 1910. Out of this experience was to come his definitive and classic book "Plantation Rubber and the

Testing of Rubber," which he published, 1920.

He was one of the first chemists on a rubber plantation and turned his attention to a variety of problems from a better way to pack rubber for shipment to the variation of latex yield from the *Hevea brasiliensis* tree. By 1917 he had decided on further formal education and entered the graduate school at McGill University, obtaining his M.Sc. degree in 1918 and his Ph.D. in 1920, both in organic chemistry. He was retained at McGill as assistant professor of chemistry in 1920 and rose to the rank of professor only 3 years later.

A period of prolific research activity followed from 1920 to 1929 as he turned his attention to the composition and structure of Hevea rubber and latex as well as to the chemistry of the vulcanization reaction, especially the action of the accelerators, which were then being actively developed. He was responsible for developing several new "ultra-accelerators" and also showed the important role of fatty acids and zinc oxide in these reactions.

Whitby did not limit his interests during this period to the study of vulcanization but did extensive work on the chemical structure and physical properties of natural rubber. Furthermore he also began exploring the virgin territory of polymerization chemistry, with special reference to the problem of synthetic rubber. It should be remembered that substances such as rubber were still considered as "colloids" in those days since the idea of macromolecules was only beginning to be suggested by Staudinger.

In 1929 Whitby left academic life and accepted the post of director of the Chemistry Division of the National Research Council of Canada Laboratories. There he continued his studies of polymerization during the following decade with special reference to the diene monomers, which could lead to the formation of synthetic rubbers. His work was characterized by painstaking precision without the assistance of the many instruments now available for studying macromolecules.

In 1939 just before the outbreak of World War II, Whitby returned to England where he took the post of director of the Chemical Research Laboratories of the Department of Scientific and Industrial Research of Great Britain. But this was only a brief tenure since he emigrated to the United States in 1942 to join the newly-organized government-sponsored synthetic rubber research program. He became professor of rubber chemistry at University of Akron, where he stayed until his retirement in 1954. He became an American citizen in 1956. During this period he did extensive research on the well-known butadiene-styrene emulsion copolymerization reaction and was responsible for developing the very efficient hydroperoxide-polyamine ("peroxamine") redox initiators capable of yielding rapid rates at low temperatures.

Whitby was undoubtedly the most honored rubber chemist of all time. Besides a number of honorary degrees, he made a clean sweep of all available awards in his field. He received the first Colwyn Gold Medal of the Institution of the Rubber Industry of Great Britain in 1929, the Goodyear Medal of the American Chemical Society's Division of Rubber Chemistry in 1955, and the first Dunlop Lecture Award of the Chemical Institute of Canada in 1971. He wrote nearly 100 publications on polymer chemistry—from 1912 to 1970 he published at least one article each year.

His interests were not limited to chemistry or even science, but he showed a prodigious knowledge of and intense interest in historic events, philosophic thought, and literature of the English language. In a recent sketch, *Rubber World* aptly referred to him as "A Man for All Seasons."

He died at Delray Beach, Fla., Jan. 10, 1972. His citation in the International Rubber Science Hall of Fame at University of Akron reads as follows:

GEORGE STAFFORD WHITBY (1887-1972), English by birth and American by adoption, who, as teacher, scientist and administrator, introduced the science of chemistry into the growing and utilization of plantation rubber, with special emphasis on the development of organic accelerators of vulcanization; and whose accomplishments in synthetic rubber range from his early, pioneering studies of polymerization to his development of the efficient peroxamine initiators for modern emulsion polymerization of synthetic elastomers.

Chem. Abstr.; personal recollections; *Rubber World* **153,** 20 (December 1965); G. S. Whitby, "Synthetic Rubber," John Wiley (1954); *Ind. Eng. Chem.* **47,** 807 (1955).

MAURICE MORTON

Alfred Holmes White

1873-1953

White was born in Peoria, Ill., Apr. 29, 1873. He received his bachelor's degree in chemistry from University of Michigan in 1893. He served as an assistant in chemistry at University of Illinois for 3 years, then studied chemical technology for a year at the Polytechnicum, Zurich, Switzerland. In September 1897, he began his 46-year career at University of Michigan as instructor in chemical technology in the Engineering Department.

The young instructor was associated with Edward de Mille Campbell and Moses Gomberg in the chemical laboratory. With Campbell, he helped develop a curriculum in chemical engineering adopted in 1898, the second in the nation. His attention to this fledgling subject led to his elevation to the position of head of the Department of Chemical Engineering in 1914. Largely because of his efforts the East Engineering Building was erected (1923) to house the growing department. He assembled a fine team of teachers-researchers-consultants, including Walter L. Badger, Edwin M. Baker, George G. Brown, and Warren L. McCabe, who developed a vigorous graduate program leading to the Ph.D. degree. He guided this prominent Department of Chemical Engineering with a gentle and fatherly hand until 1942; he retired from teaching in 1943.

His early grounding in the tradition of chemistry inspired him to carry out much research. His 30-year study of Portland cement showed its growth by alternate wetting and drying. His work with the local gas industry resulted in the establishment of the Michigan Gas Association Fellowship in Chemical Engineering. He investigated methods of gas analysis, and his first book was "Technical Gas and Fuel Analysis" in 1913. Beyond these two areas, he made excursions into such topics as boiler water treatment and boiler scale, pyrolysis of wood and of coal, and manufacture of fertilizers. His 70 research papers and 14 patents covered a wide range of chemical technology.

White served as chairman of the Accrediting Committee of the American Institute of Chemical Engineers during its formative years—well before the ECPD was organized. He sponsored the first AIChE Student Chapter of Chemical Engineering at University of Michigan in 1922. Because of Edward de M. Campbell's interest in metallurgy, White had maintained a faculty with a metallurgical bent, and in 1935 he enlarged the department's name to Chemical and Metallurgical Engineering. His instruction in engineering materials and his widely used textbook by that name indicated his devotion to student instruction.

World War I provided an experience outside the university for White who petitioned the War Department for a commission and was appointed a captain in the Ordnance Department. During the War he was head of the Metallurgical Branch of the Inspection Division, technical advisor in the production of raw materials for explosives, and technical director of the Nitrate Division. He was influential in the development of processes at the Muscle Shoals Nitrate plants. This experience was a prelude to his developing a team of teachers interested in chemical industry. He was a strong supporter of Albert E. White of his department who inaugurated the Engineering Research Department of the college in 1920 to permit faculty and students to engage in solving industry's problems at the university.

He married Rebecca Mason Downey of Pueblo, Colo. in 1903. A daughter, Mary Julian White became a practicing physician. Their son, Alfred McLaren White, followed in his father's footsteps in a teaching career and in AIChE activities as chairman of the Student Chapters Committee until his death in 1936; the White name is honored by the A. McLaren White Award as the 1st Prize for the AIChE student contest problem.

Among the honors that White received were the presidency of the AIChE in 1928 and 1929 and of the American Society for Engineering Education in 1942. Northwestern University awarded him an honorary doctorate at the dedication of the Institute of Technology.

The Whites enjoyed 10 years of retirement together and celebrated their Golden Wedding Anniversary with many friends in the summer of 1953. White died at the age of 80 while he was working at his desk on a revision of his book on engineering materials on Aug. 25, 1953. He had been informed that University of Michigan was to award him an honorary doctorate, and it did so posthumously.

Personal recollections; biography in *Ind. Eng. Chem.* **17,** 607 (1939); records of University of Michigan.

DONALD L. KATZ

Frank Clifford Whitmore

1887-1947

Whitmore was born in North Attleboro, Mass., Oct. 1, 1887, son of Frank Hale and Lena Avila (Thomas) Whitmore. Rocky, as he was known to his friends, spent his early youth in Williamsport, Pa. and his high school years in Atlantic City, N.J. He took 4 years of Latin and 2 years of Greek and planned to be a professor of classics. At the same time he ran the quarter-mile on a relay team that twice won its class championship at the Penn Relays.

While in high school, Whitmore decided to go to Harvard. He had no Harvard connections; he simply decided that it was the best college there was. His family was not well off and while the family resources were marshaled to get him to college, he took a fifth year of high school, studying more Greek and running on the track team in those days of relaxed eligibility standards.

He entered Harvard in 1907. In the financial panic of that year his father lost his money. Cliff had to work his way through Harvard, which he did with a variety of jobs. One of these was guiding visitors through the Harvard yard in the summer, and this led to his meeting Marion Gertrude Mason, a Radcliffe student, whom he married in 1914. The summer of 1910, when he worked as a guide, was also important because it was then that he decided to be a chemist. He had taken chemistry courses during his sophomore and junior years and, having made his decision, he studied nothing but chemistry in his senior year. His work came to the attention of Charles Loring Jackson, and he became Jackson's laboratory assistant. Jackson retired in 1912, and Whitmore continued his graduate studies under Theodore William Richards. He received his Ph.D. degree in 1914.

Whitmore had decided on an academic career, but with a new wife and a growing family he postponed seeking a university position in favor of 2 years of tutoring at the "Widow" Nolan's school, one of several which devoted themselves to cramming well-to-do Harvard students for examinations. His first college teaching was at Williams College in 1916-17. In 1917 he took the big step to Houston, Tex., to teach at the newly established Rice Institute. He remained there only a year, leaving because of the president's reluctance to purchase a set of Liebig's "Annalen," which Whitmore regarded as indispensible to his research. With his wife, mother-in-law, and two little sons he drove in a 1918 Buick touring car from Houston to Minneapolis, where he became assistant professor at University of Minnesota. There he remained for 2 years and began publication of his papers on organic compounds of mercury. He was invited to lecture at Northwestern University and, as a result, was brought there in 1920 as a professor at the age of 32. He soon became head of the Chemistry Department and attracted the first of many graduate students in organic chemistry.

In 1929 he moved to Pennsylvania State College (now University) as Dean of the School of Chemistry and Physics. With the aid of Merrell R. Fenske and Dorothy Quiggle he built a chemical engineering curriculum and research program which, among other things, made contributions to developing cracking processes for petroleum. But his greatest pride was in his graduate program in organic chemistry. While he was dean, the graduate enrollment in chemistry increased from 18 to over 100 and the college awarded 871 bachelor's, 383 master's, and 215 doctor's degrees in chemistry and chemical engineering.

Through his years of teaching, administrative work, and society activity (he was president of the American Chemical Society in 1938), Whitmore continued his own research. Perhaps his best known work was in the field of molecular rearrangements, which took him into nearly every phase of synthetic aliphatic chemistry. His electronic conception of rearrangements was generally accepted. His friend Carl Marvel stated: "In particular he added to our knowledge of the Grignard reaction as a source for ketones from acid chlorides, for the preparation of tertiary hydrocarbons, and as a reducing agent. He devised improved procedures for the ozonolysis of unsaturated hydrocarbons in order to elucidate the structure of their rearrangement products. He developed the unusual chemistry of the neopentyl system which was essentially unknown before his researches in that field. He synthesized pure hydrocarbons, both saturated and unsaturated, to confirm their properties and structures. He related the rearrangements of olefins in acid solution to the rearrangement involved in the dehydrogenation of alcohols by various agents, and the rearrangements which occur in such reactions as the Hofmann degradation of amides, the Chugaev reaction, the transformation of tertiary halides, the pinacol and semi-pinacol rearrangements, action of nitrous acid on amines, the cationic polymerization of olefins and other similar transformations."

During World War II he was active in government research projects. He was proud of his work on penicillin, which contributed to its mass production, and of the synthesis of a variety of heterocyclic bases in the field of antimalarials. The first 100 lb. of the plastic explosive RDX were made in Whitmore's laboratory. His book, "Organic Chemistry," which came out in 1938, was one of the most comprehensive textbooks of descriptive organic chemistry that had been published to that time and was referred to as a "one-volume Beilstein."

Whitmore received many honors, including the Nichols and Willard Gibbs medals and in 1946, election to membership in the National Academy of Sciences. He died at his home in State College, Pa., June 24, 1947.

Biography by C. S. Marvel in *Biog. Mem. Nat. Acad. Sci.* **28,** 289-311 (1954); personal recollections.

FRANK C. WHITMORE, JR.

Josiah Dwight Whitney

1819-1896

The highest mountain in the United States south of Alaska, one of the smallest species of American owls, and a rare mineral are all named in honor of Josiah Dwight Whitney, Jr., chemist and geologist.

Born in Northampton, Mass., Nov. 23, 1819, to Josiah and Sarah (Williston) Whitney, he was as a young teen-ager, stimulated by one of Benjamin Silliman's public lectures on chemistry. He studied chemistry at Yale, graduated with a bachelor's degree in 1839, and then studied in Robert Hare's laboratory at University of Pennsylvania during the winter of 1839-1840.

With no prospect for a job in science, he was befriended by Charles Jackson, a consulting chemist and geologist of Boston, who took him as an unpaid assistant and later as a salaried assistant on the geological survey of New Hampshire.

Uncertain about the profession he wished to follow, Whitney began to study law in 1841, but he could not shake off the attraction he felt toward science. Accepting Jackson's advice he decided to stay with chemistry and geology and to seek further training in Europe. He went to France in 1842, attended lectures at Ecole des Mines and Collège de France, went on geological excursions with Elie de Beaumont, and studied chemical analysis with Rammelsberg in Berlin. He toured through France, Belgium, Germany, Sweden, Austria, Switzerland, went as far as Moscow in Russia, and spent a winter in Rome.

Called home because of business reverses suffered by his father, he prolonged his stay in Europe several months by translating for a Boston publisher Berzelius' "Use of the Blowpipe in Chemistry and Mineralogy," which was published in 1845. He arrived home in January 1845 and was again befriended by Jackson, who obtained for him a job as geologist for the Isle Royale Copper Co. Late in the year Jackson obtained for him two commissions that enabled him to return to Europe; one was to recruit German skilled workmen and furnace masters for American mining firms, and the other was to obtain information on German methods of dyeing calico for Cocheco calico print works in New Hampshire, a firm for which Jackson was a consulting chemist.

Whitney went to Heinrich Rose's laboratory in Berlin early in 1846. The reputation of another German chemist was growing and Whitney was listening to the talk about him: "I suppose I should be run after for a professorship, if I had studied at Giessen," he wrote his brother, "as it seems to be a settled point that no young man can be expected to know anything of chemistry, unless he has studied with Liebig." So Whitney went to Giessen in the autumn of 1846 and studied under Liebig until the spring of 1847 when he returned home.

Jackson appointed Whitney an assistant on the U.S. geological survey of the copper region of Michigan, and upon Jackson's resignation in 1849 he became one of the directors of the survey. The survey completed, he set up a private laboratory in Brookline, Mass., became a consultant for mining firms, and in 1854 published "The Metallic Wealth of the United States," the first comprehensive book on American ore deposits.

In 1855 Whitney was appointed professor of chemistry in University of Iowa, but, as the university existed largely on paper, his major activity for 3 years was as chemist of the state geological survey. He also made a geological survey of the lead regions of Wisconsin and Illinois, which adjoined the lead region of Iowa.

From 1860 to 1874 he was state geologist of California and undertook an elaborate survey of the state. Harvard University appointed him professor of geology in 1865, with leave of absence to direct the survey. On a field trip in 1864 his assistants spotted the towering peak which they named after their chief. Whitney served as a commissioner of Yosemite Park and was chairman of a committee to plan the state agricultural and mechanical college. In 1874 when the survey ceased because of a lack of funds, Whitney retired to Cambridge and thereafter taught at Harvard. He died at Lake Sunapee, N.H., Aug. 19, 1896.

Edwin T. Brewster, "Life and Letters of Josiah Dwight Whitney," Houghton Mifflin Co. (1909) has illustrations and a list of Whitney's publications; "Dictionary of American Biography," **20,** 161-163, Chas. Scribner's Sons (1936).

WYNDHAM D. MILES

Willis Rodney Whitney

1868-1958

Often considered the dean of industrial research, Whitney founded and directed for 28 years the General Electric Research Laboratory, one of the early, large industrial research laboratories of the United States.

Whitney was born in Jamestown, N.Y., Aug. 22, 1868. He graduated from Massachusetts Institute of Technology in 1890, remained on the teaching staff for 4 years, and then went to Europe to study chemistry. He received his Ph.D. degree from the University of Leipzig in 1896. After a short stay at the Sorbonne he returned to the institute to accept a position on the faculty. Here he was happy to settle down teaching and researching in the areas of colloids, corrosion of metals and modern theories of solutions. In 1900, however, he was persuaded by Edwin Rice, vice-president of the General Electric Co., to go to Schenectady, N.Y., to develop a research laboratory for the company.

The developmental laboratory started in Charles P. Steinmetz's barn, but when the barn burned shortly afterwards, the research operation was moved to a small building in the Schenectady Works. For 3 years Whitney commuted between Schenectady and Boston, but he found the work so interesting that he finally moved his family to Schenectady and devoted the rest of his active life to the development and expansion of this world renowned laboratory. He remained as director of research until 1928, was vice-president in charge of research for the whole company until 1932, then became honorary vice-president of the company.

Whitney had great love for experimentation. He often started his experiments on a hunch. To him no real experiment was useless. He liked the word "serendipity" which he defined as the art of profiting from unexpected discovery. Experimentation was fun, and in this spirit he developed a staff of scientists who could work as a group or as independent individuals such as Coolidge, Dushman, Hull, and Langmuir. Typical of the congenial atmosphere was the sign on the door of his office which read, "Come in—rain or shine." At all times he listened to ideas and gave encouragement and advice. While his own laboratory developments at General Electric, such as the metallized lamp filament and the inductotherm, he modestly accepted as his own, yet many of his ideas, such as the Calrod heating unit and hydrogen cooling in turbines that resulted in useful reality, he preferred to credit to the members of his staff.

He was a friendly person and liked to see other people happy. Before and after retirement he had many hobbies, all of which involved some sort of experimentation. On his "farm" he was occupied by such diverse interests as Indian arrowheads, the growing of corn under different conditions, and turtles.

Whitney was president of the American Chemical Society in 1909 and president of the Electrochemical Society in 1911. He received the Gibbs Medal, Chandler Medal, Perkin Medal, Gold Medal of the National Institute of Social Sciences, Franklin Medal, Edison Medal, Gold Key of the American College of Physical Therapy, Decoration of the French Legion of Honor, Public Welfare Medal of the National Academy of Sciences, John Fitz Medal, Award of the Industrial Research Institute, and honorary degrees from a number of universities. Medals and decorations received by Whitney and other members of the General Electric Research staff are on display in the reception room of the General Electric Research and Development Center in Schenectady, N.Y.

He died in Schenectady, Jan. 9, 1958.

Ind. Eng. Chem. **13,** 158-166 (1921) has Perkin Medal Award addresses by A. D. Little and C. F. Chandler and an autobiography by Whitney; contemporary sketch by L. A. Hawkins in *Ind. Eng. Chem.* **25,** 1058 (1933); C. A. Browne and M. E. Weeks, "A History of the American Chemical Society," ACS pub. (1952); G. Suits, *Biog. Mem. Nat. Acad. Sci.* **34,** 350-367 (1960), with portrait and list of publications; John T. Broderick, "Willis Rodney Whitney, Pioneer of Industrial Research," Fort Orange Press (1945); obituary: *New York Times,* Jan. 10, 1958.

EGBERT K. BACON

Ferdinand Gerhard Wiechmann

1858-1919

Wiechmann was born in Brooklyn, N.Y., Nov. 12, 1858. He attended Columbia University, obtaining his Ph.B. (1881) and Ph.D. (1882) degrees, and then spent a year at University of Berlin. After returning home he instructed in Columbia's Chemistry Department from 1883 to 1897 and concurrently was chemist for Brooklyn Sugar Refining Co., 1883-85, Havemeyer Refining Co., 1886, and Havemeyer and Elder Sugar Refinery Co. (later American Sugar Refining Co.), 1887-1909. During these years he wrote "Lecture Notes on Theoretical Chemistry" (1893; 2 ed. 1895), "Chemistry, its Evolution and Achievements" (1899), "Notes on Electrochemistry" (1906), and a novel "Maid of Montauk" under the pseudonym, Forest Monroe. He also wrote extensively on cane sugar and beet sugar, his best known publication being "Sugar Analysis," which appeared in 1890, and reached a third edition in 1914.

Unlike most of the other prominent figures in the industry Wiechmann never held an executive post either in the laboratory or in other branches of refining, so that his work in sugar technology was limited to a consulting capacity. His considerable reputation, both here and abroad, depended on several factors. His commanding presence, his fluency in French and German, and his wide acquaintances among scientists, especially those of high standing in the beet sugar industry in Germany, contributed to his ability as an organizer of various societies and associations.

In 1897 he met with Alexander Herzfeld, director of the Sugar Institute in Berlin, and other Austrian and German sugar specialists, to form the International Commission for Uniform Methods of Sugar Analysis (ICUMSA), with Herzfeld as president and Wiechmann as secretary. Wiechmann was active in the furtherance of the work of the Commission in subsequent sessions and served as chairman of the U. S. delegation at the New York meeting in 1912. The meetings of the commission discontinued because of World War I, but Wiechmann did a great service to the organization by compiling and publishing the resolutions and findings of the commission for the sessions through 1912.

Another project which he furthered and supported actively through its earlier years was the American Institute of Chemical Engineers, founded in 1908 with Wiechmann as a charter member. He was a strong proponent of the institute at a time when the American Chemical Society—or at least a powerful segment of it—considered the new society as a divisive factor among chemists and chemical engineers. Although his efforts to increase the membership of the institute had no marked effect until World War I brought chemical engineering into prominence, he should receive due credit for calling the name and aims of the institute to public attention during that period.

In 1892 he received a patent on the use of kieselguhr in connection with Taylor bag-filtration of refinery liquors but this proved of no practical value. The kieselguhr available at that time was high in iron salts which caused color formation in the treated liquor.

Wiechmann spent 6 months at the Colonial Sugar Company's refinery in Gramercy, La., as a consultant. For several years before his death, April 24, 1919, he did similar work for the Warner Brothers Refinery in Edgewater, N. J.

H. C. S. de Whalley, "ICUMSA Methods of Sugar Analysis," (London, 1964); Charles A. Browne and M. E. Weeks, "History of the American Chemical Society" (Washington, 1952); "The First 50 Years of AICHE" (New York, 1958); "Dictionary of American Biography"; "National Cyclopedia of American Biography," vol. 14, p. 343; obituaries in *Louisiana Sugar Planter,* May 19, 1919, and *Sugar,* June 1919; personal recollections.

GEORGE P. MEADE

Harvey Washington Wiley

1844-1930

Wiley was born Oct. 18, 1844, in a log farmhouse near Madison, Ind. to parents of Scotch-Irish descent, Preston Pritchard and Lucinda (Maxwell) Wiley. He graduated from Hanover College in 1867, his education interrupted by service with the 137th Indiana Volunteers. He received his M.D. degree at Indiana Medical College in 1871. Not sure that he wanted to practice medicine, he studied chemistry at Harvard from 1872 to 1873, and after passing special examinations, received his bachelor of science degree.

The following year he taught chemistry at Northwestern Christian (now Butler) University and in 1874 became professor of chemistry at fledgling Purdue University. There he instituted laboratory work as the principal form of instruction. He became interested in analytical chemistry and drew attention for his studies of sugar and its adulterants.

In 1883 the Commissioner of Agriculture brought Wiley to Washington as chief of the Division of Chemistry. A scientifically sound and practical sugar expert, he headed for two decades the technical side of the politically potent drive to establish a domestic sugar industry. The effort to develop sorghum as a sugar source failed, but the work of Wiley and his lieutenants to improve the plant made it more valuable as cattle feed. Their lasting contribution to the cane-sugar industry was to introduce chemical control. Perhaps their greatest achievement was to advance the cultivation of sugar beets. In cooperation with state experiment stations, they studied the principal determinants of beet production and published maps illustrating their finding that temperature was the most potent environmental factor.

Concurrently, Wiley concentrated on developing analytical methods to investigate the adulteration of foods. The Division of Chemistry studies were reported in "Foods and Food Adulterants," a landmark bulletin which appeared in nine parts between 1887 and 1898. It detailed the normal characteristics of food products, described the best methods of analysis, and reported common adulterations. It became the scientific foundation of the pure-food movement. To provide a national base for this work, Wiley led in the establishment of the Association of Official Agricultural Chemists. His three-volume "Principles and Practice of Agricultural Analysis" (1895-97) won recognition as the best English-language treatise on methods.

As director of the principal chemical laboratory of the government, Wiley became a national leader in his profession. Soon after arriving in Washington, Wiley teamed with Frank W. Clarke, chief chemist of the U.S. Geological Survey, in the activities of the Chemical Society of Washington. Both had resigned from the American Chemical Society as it degenerated into a local New York group, and they started taking steps toward a national, or continental, chemical society, Wiley working through AOAC and Clark working through Section C of the American Association for the Advancement of Science. The leaders of the American Chemical Society met these moves with steps in 1890-91 that transformed it into a truly national body. The rifts were healed when Wiley was elected president of the ACS in 1892; he served two terms—1893 and 1894, during which local sections were formed, and Wiley and others brought their groups into the ACS; membership zoomed from 322 to 859, and the society's journal was put on a sound editorial and financial footing.

It was as leader of the fight for federal pure food and drugs legislation that Wiley made his greatest mark. From 1898 on he was caught up in the fight to achieve the tools for federal regulation. Circumstances, a strong sense of social obligation, and a compelling drive for personal achievement combined to put him at the fore.

With a shrewd sense for publicity, Wiley tested the effects of chemical food preservatives on the health of a panel of volunteers, the "Poison Squad." He spoke at countless meetings, recruited workers and allies, sought to consolidate support for specific proposals, and worked closely with Congressional leaders, supplying information and helping to draft legislation. He deserves the lion's share of the credit for the Pure Food and Drugs Act of 1906.

The new law placed responsibility for enforcing the general prohibition against adulterating and misbranding in the Department of Agriculture. Wiley at once became involved in a series of controversies over misleading terminology and the use of preservatives. He took positions that struck Presidents Roosevelt and Taft and Secretary of Agriculture James Wilson as doctrinaire and unreasonable. When they devised administrative machinery to control his excesses, he concluded that they were abandoning the consumer to the rapacity of mercenary manufacturers. The fight simmered behind the scenes until 1911, when Wiley survived an inept attempt to force him out. In 1912 faced with another unjustified assault on his integrity, he resigned to campaign for Woodrow Wilson.

The Wilson administration and the Republican administrations of the twenties disappointed Wiley in their failure to enforce the law as he believed it should be. As director of the *Good Housekeeping* Bureau of Foods, Sanitation, and Health, he continued to fight old battles and to propagandize for purer food and better nutrition. A sense of defeat embittered these years, but this was leavened by a happy family life with the former Anna Kelton, whom he married in 1911, and his two sons. Active to the last, he died June 30, 1930.

In 1956 on the 50th anniversary of the passage of the Pure Food and Drug Law the United States issued a commemorative postage stamp, and on it was pictured, fittingly, Wiley.

Harvey W. Wiley Papers, Library of Congress; Records of the Bureau of Agricultural and Industrial Chemistry and Records of the Office of the Secretary of Agriculture, National Archives; "Harvey W. Wiley; An Autobiography," Bobbs-Merrill (1930); Oscar E. Anderson, "The Health of a Nation; Harvey W. Wiley and the Fight for Pure Food," Univ. of Chicago Press

(1958); C. A. Browne and M. E. Weeks, "A History of the American Chemical Society," ACS (1952).

OSCAR E. ANDERSON

Thomas Leopold Willson

1860-1915

Although Willson was a Canadian citizen, his pioneering work with the electrothermal furnace during his 13-year stay in the United States merits for him a place in this book. He was born in Princeton, Ont., Mar. 14, 1860, grandson of John Willson, Speaker of the United Canadian Assembly. Willson entered Hamilton Collegiate Institute in 1876, but after his father died in 1879, he had to drop out to support himself. He set up a small laboratory in the Hamilton blacksmith shop where he worked and soon perfected an arc lighting system, the first seen in the Hamilton area.

At age 22, Willson moved to the United States and for the next 5 years worked at various jobs in the mechanical and electrical trades before he finally settled in Brooklyn, N.Y. in 1887. His work over the next 3 years resulted in six patents, the most important of which secured for him the rights in the United States for the use of the electric arc furnace for ore smelting. In December 1890, the Willson Aluminum Co. was formed with stock of $30,000 to manufacture aluminum and other metals and alloys. An experimental plant was built in the fall of 1891 at Spray (now Eden), N.C. on land owned by one of the backers, Major James Turner Morehead.

On Apr. 20, 1892 Willson applied for a patent for making aluminum in the electric furnace by reducing alumina with tar. Although the theory was right, in practice only a few globules of aluminum were recovered. Then, Willson reasoned that if he could use his process to make a more chemically active metal such as calcium, he could, in turn, use the calcium to reduce alumina. Accordingly, on May 2, 1892, a mixture of lime and tar was subjected to the heat of the arc. When the furnace was tapped and the resulting product tossed into water, it produced a flammable gas as calcium was expected to do. However, unlike hydrogen, the gas burned with a sooty flame for which there was no ready explanation. Francis P. Venable of University of North Carolina was brought in as a consultant, and during the summer and fall of 1892 he proved that the furnace product was calcium carbide and that the gas evolved with water was acetylene. Willson filed for a patent on this process Aug. 9, 1892 and sent samples of calcium carbide to Lord Kelvin for examination on Oct. 3. Henri Moissan of France also prepared metallic carbides in the arc furnace in 1892 but did not communicate his findings until Dec. 12 of that year.

Failure of the stock market in 1893 and the ensuing depression soon bankrupted Willson Aluminum Co. At one point, the whole company, its patents, plant, and process were offered for sale for $5,000 but there were no takers. The first sale of calcium carbide was made in January 1894 to Eimer and Amend, a chemical supply house in New York. The Willson Laboratory Co., formed to develop chemical uses for acetylene, outfitted a lab at Eimer and Amend. Chloroform and aldehydes were produced, and during experiments to liquify compressed acetylene, an explosion destroyed the laboratory. Willson narrowly escaped unhurt.

In February 1894 Willson applied for a patent for manufacturing hydrocarbon gas in which he cited acetylene as a raw material from which "many chemical products of great value in the arts may be derived. Because of its cheapness, it promises to supersede many of the present processes for the manufacture of various hydrocarbon products."

When it was realized that acetylene emitted more than three times as many candle power per dollar as did water gas, its use as an illuminant developed rapidly. In August 1894 Willson Aluminum Co. sold its rights to calcium carbide and acetylene for lighting purposes in the U.S. to a new company, the Electrogas Co. but retained the chemical rights. Willson retained all rights for Canada. He returned to Canada in 1895 and by 1896 had a carbide plant under construction at Merritton, Ont. As he sold rights for carbide manufacture to others, he developed many interests, forming new companies as he went: water power, acetylene lighted marine bouys, fertilizer, cement, ammonia, phosphoric acid, and paper. He died of a heart attack in New York, Dec. 21, 1915 while raising money for yet another project.

Willson was married to Mary Parks of Marysville, Calif. in 1895; they had four children. He loved art and music and was active in sports, particularly bicycling and pole vaulting in his younger days. Toronto University awarded him the first MacCharles Prize in 1905 for his discoveries.

Alfred B. Garrett, "The Flash of Genius," D. Van Nostrand Co. (1963), pp. 88-92; Lyman B. Jackes, "Canadians and World Progress," *The Canadian Magazine* (January 1926); S. A. Miller, "Acetylene—Its Properties, Manufacture, and Use," Academic Press (1964); C. J. S. Warrington, and R. V. V. Nicholls, "A History of Chemistry in Canada," Pitman (1949), pp. 166-171.

HERBERT T. PRATT

Robert Erastus Wilson

1893-1964

Robert E. Wilson, a distinguished scientist, engineer, inventor, industrialist, and government administrator, was born Mar. 19, 1893 in Beaver Falls, Pa., the son of William Hyatt Wilson and Madge (Cunningham) Wilson. He was graduated *magna cum laude* in 1914 from College of Wooster and in 1916 received his degree in chemical engineering from Massachusetts Institute of Technology.

Wilson served briefly as a research associate at MIT, but when the United States entered World War I, he went to Washington to work as a chemical engineer in the Bureau of Mines. In 1918 he entered the Army as a captain and later served as a major in the Chemical Warfare Service, where he directed research on gas mask absorbents and protective clothing.

In 1919 Wilson returned to MIT as associate professor of chemistry and director of the research laboratory of applied chemistry. After 3 years he left MIT to begin a long career with Standard Oil Co. of Indiana. There he advanced from assistant director of research to director and vice-president in charge of research, which position he held until 1934. From 1935 to 1958 he served successively as vice chairman, president, and chairman of the board of the Standard Oil subsidiary, Pan American Petroleum and Transport Co., and its successor, American Oil Co. He also served as chief executive officer and chairman of the board of the parent company from 1945 to 1958, with offices in Chicago.

Throughout this period Wilson was continuously involved in directing and carrying out scientific research to improve company products and production. He held individually or jointly more than 90 patents for inventions in chemical engineering. He also published more than 120 technical papers on subjects in which he had a special competency, such as petroleum refining and hydrocarbon synthesis. In recognition of his research contributions, he was awarded the Chemical Industry Medal in 1939, the Perkin Medal of the Society of Chemical Industry in 1943, and the third Cadman Memorial Medal of the Institute of Petroleum (England) in 1951. He also received several awards in other fields of interest, such as the Northwestern University Centennial Award in 1951. He served the American Chemical Society as chairman of two divisions (Physical & Inorganic, 1922-23, and Industrial & Engineering, 1929-31) and as member of the board of directors for 14 years (1931-39 and 1941-45).

Wilson retired from Standard Oil at age 65 in 1958, but that was not the end of his productive activity. He had been interested in atomic energy ever since that new field of research became generally known after World War II. In 1956 President Dwight Eisenhower named him to the General Advisory Committee to the Atomic Energy Commission. Four years later he gave up this advisory post to become a member of the Atomic Energy Commission.

As a Commissioner from 1960 to 1964, Wilson was an energetic leader in the effort to develop nuclear reactors for electric power production. He was the chief architect of the Commission's program to allow private ownership of special nuclear materials. In a speech December 1961 before the Japanese Atomic Industrial Forum in Tokyo, he announced the U.S. policy to supply enriched uranium fuel to other countries on a long-term basis.

In October 1963 Wilson suffered a mild coronary attack and four months later resigned from the Commission because of difficulty in coping with the heavy workload. He was nearly 71. On Sept. 1, 1964, he died of a cerebral hemorrhage in Geneva, Switzerland, where he was attending the Third United Nations Conference on Peaceful Uses of Atomic Energy. His widow, the former Pearl Rockfellow, and three married daughters survived him.

Throughout his career, Wilson maintained a strong interest in education. He was a life member of the MIT Corporation and a trustee of College of Wooster and Carnegie Institution of Washington. Seventeen honorary degrees from colleges and universities were awarded him. He was a director of Chase Manhattan Bank in New York and First National Bank of Chicago. He was a member of the National Academy of Sciences, American Philosophical Society, and the Newcomen Society and a fellow of the Royal Society of Arts (London).

"Leaders in American Science," **1**, 686, Who's Who in American Education, Inc. (1953-54); "Who's Who in Atoms," **2**, 1692, Vallency Press (1959); biographical information from files of the Atomic Energy Commission.

L. ROBERT DAVIDS

Saul Winstein

1912-1969

Winstein was born in Montreal, P.Q., Canada, Oct. 8, 1912. His family came to the United States when he was 11 years old, and he became a naturalized citizen in 1929. In 1930 he graduated from high school in Los Angeles and entered University of California at Los Angeles. As an undergraduate he was deeply interested in chemistry and did research under the supervision of William G. Young which led to a paper on allylic Grignard reagents in 1933. After Winstein's graduation in 1934 he remained at UCLA for his M.A. degree with Young in 1935. Winstein published eight papers with Young based on his undergraduate and master's degree work. Winstein's interest in allylic rearrangements lasted throughout his career, and he collaborated with Young in this area for 36 years.

Winstein continued his work in physical organic chemistry at Caltech, where his doctoral research was supervised by Howard J. Lucas. At Caltech, Winstein established the intermediacy of cyclic bromonium ions in substitution reactions, and this led to the concept of neighboring group participation which he exquisitely developed throughout his career. After his Ph.D. degree was awarded in 1938, he spent a postdoctoral year at Caltech and another as a National Research Fellow at Harvard with Paul Bartlett. He spent a year as an instructor at Illinois Institute of Technology and then returned to UCLA where he remained for the rest of his life.

Except for his research in World War II on antimalarials, almost all of Winstein's work could be traced back conceptually to the studies he began with Young and Lucas. The major area of Winstein's interest was the nature of cationic intermediates in organic reactions. From the early investigations of bridged ion formation with neighboring halogen. Winstein generalized the phenomenon to other neighboring groups containing non-bonded electron pairs, olefinic and aromatic groups, and finally, carbon-carbon and carbon-hydrogen single bonds. Studies of the bridged and "non-classical" ions led to the concepts of homoallylic and homoaromatic stabilization, which were illustrated in many elegant experiments.

Aside from the important structural concepts Winstein developed, he and his students scrutinized the mechanisms of organic reactions, especially substitution and elimination reactions. He emphasized the importance of ion pairs as intermediates in organic reactions and distinguished their behavior from covalent and dissociated ionic species. He provided quantitative treatments of the ionizing power of solvents and elucidated the nature of salt effects in solvolysis reactions.

Winstein was interested in all areas of chemistry, and he critically analyzed the experimental results and theoretical interpretations of eminent chemists as well as of his students. He used techniques from various disciplines to solve problems, and he remained at the forefront of chemical research by acquiring expertise in new techniques. Often the development of new instrumental methods allowed Winstein to return to unsolved aspects of central problems. Winstein used and developed techniques in spectroscopy, kinetics, stereochemistry, isotopic labeling, molecular orbital calculations, and synthesis in his research, and his approach to problems in physical organic chemistry served as a standard throughout the world. Prominent and young chemists alike came from many countries to work in Winstein's laboratory.

In the 1960's Winstein received many honors and traveled widely, but he still followed his students' work closely. He always demanded precise experimental work and logic from his students and was a superb teacher of practical science. A total of 72 students received their Ph.D. degrees under his supervision, and 86 postdoctoral fellows collaborated with him.

Winstein died suddenly Nov. 23, 1969, at the height of his career. He was honored posthumously with the award of the National Medal of Science in 1970.

W. G. Young and D. J. Cram, *Int. J. Chem. Kinetics* **2**, 167 (1970); P. D. Bartlett, *J. Amer. Chem. Soc.* **94**, 2161 (1972); A. Streitwieser, Jr., *Prog. Phys. Org. Chem.* **9**, 1 (1972); personal recollections.

MARTIN FELDMAN

John Winthrop, Jr.

1605-1676

John Winthrop the Younger, as he is called to distinguish him from his notable father, was born in England, Feb. 12, 1605. He attended Trinity College, Dublin, studied law in London, and was admitted to the bar. He sailed on a British naval expedition and then made an extensive tour of Europe.

His father immigrated to New England in 1630, and John followed in 1631. In 1634 he returned to England on colonial business. His father's friends undertook to establish a plantation in Connecticut with Winthrop as governor. In 1635 he was commissioned governor and returned to America.

Winthrop sailed to England again in 1641 and remained there for 2 years. In 1645, after his return to America, he was granted extensive property in eastern Connecticut where he founded the town of New London. In 1657 he was elected governor of Connecticut and, with the exception of 1658, was reelected each term until his death.

In 1661 he visited England for a third time and returned with the charter for Connecticut. In the autumn of 1675 he journeyed to Boston to discuss with other colonial officials the worsening relations between Indians and settlers which later deteriorated into King Philip's War. During the bitter winter he contracted a fatal illness and died in Boston, Apr. 5, 1676.

Perhaps while Winthrop was attending Trinity College or between the time he left Trinity to sail to America, he became interested in alchemy and chemistry and began to purchase books, apparatus, and chemicals. After he settled in New England, he imported books and materials. Eventually he accumulated the first large scientific library in the colonies, part of which is still in existence. He became friendly and corresponded with British scientists, among them Robert Boyle and Kenelm Digby. He loaned books and materials to, and otherwise encouraged, other New England chemists, among them George Starkey and Jonathan Brewster.

Winthrop prepared chemical medicines and promoted several industries. In 1638 he built a saltworks on the shore of Massachusetts Bay. He began to manufacture iron in 1643 at Braintree. In 1650 he planned a chemical stock company, the first in the colonies, to manufacture saltpetre. Connecticut granted him a monopoly in 1651 to work lead, tin, copper, vitriol, alum, and other materials.

He was elected to the Royal Society in

1662, while in England, and he read before it the first scientific paper presented by an American to a scientific organization, "Of the Manner of Making Tar and Pitch in New England." He also contributed papers on the preparation of potash, black lead, brewing beer from American corn, and other topics.

Winthrop was not successful financially in his chemical enterprises, but his failures were not owing to his lack of persistence, industry, or technical ability. He was defeated by circumstances beyond his control. The mineral deposits that he sought to exploit were not rich enough for profitable operation. Markets were too remote or too small to support his chemical enterprises permanently and profitably. He was a pioneer chemical industrialist on the frontier of the British empire, and in his dream of establishing chemical industries he was a century or more ahead of his time.

R. S. Wilkinson, "The Alchemical Library of John Winthrop, Jr. (1606-1676) and His Descendants in America," *Ambix* **11,** 33-51 (1963), and **13,** 139-188 (1965-66), with portrait; C. A. Browne, "An Old Colonial Manuscript Volume Relating to Alchemy," *J. Chem. Educ.* **5,** 1583-90 (1928); C. A. Browne, "Scientific Notes from the Books and Letters of John Winthrop, Jr. (1606-1676)," *Isis* **11,** 325-342 (1928); Williams Haynes, "Chemical Pioneers," 13-25, D. Van Nostrand & Co. (1939), with portrait.

WYNDHAM D. MILES

James Renwick Withrow

1878-1953

Withrow was born in Philadelphia, Pa., Aug. 29, 1878. He attended the public schools of that city and entered University of Pennsylvania with the intention of studying mine engineering. However, this subject was deleted from the university offerings after he enrolled, and he majored in chemistry as a second choice. During all his undergraduate years at the university he held the City of Philadelphia Competitive Scholarship. He graduated with his B.S. degree in 1899. On graduation he became a chemist for the Barrett Co. of Philadelphia and after 1 year was placed in charge of the laboratory. In 1903 he began graduate work in chemistry at University of Pennsylvania under Edgar Fahs Smith. He received his Ph.D. degree in 1905.

After a year at University of Illinois, first as an assistant instructor and then as an instructor, Withrow was appointed assistant professor in the Department of Chemistry at The Ohio State University, where he taught courses in applied chemistry and chemical engineering. He was promoted to associate professor in 1907 and to professor in 1912. In 1924 all the courses in applied chemistry and chemical engineering were transferred to the newly created Department of Chemical Engineering, and Withrow was named chairman of that department, a position which he held until his retirement in 1948. Withrow died March 20, 1953 after an active retirement. He was survived by his wife (*nee* Alice Steelman), one daughter, and two sons.

The development of chemical engineering as a distinct profession occurred during the first half of Withrow's teaching career, and he was a major contributor to this development. In 1908 he advocated the teaching of chemical engineering by the study of the unit operations common to many chemical industries. This method was later generally followed. He advocated other educational procedures in the teaching of chemical engineering by extensive writing and speaking. He was the first chairman of the Chemical Engineering Education Committee of the American Institute of Chemical Engineers and also chairman at other times.

About 1500 industrial chemists and chemical engineers received their training under Withrow. Many of these rose to prominence as presidents, vice presidents, and directors of research in industrial concerns. Some became administrators or professors of chemical engineering at other universities. Withrow was unusual in that he kept in touch with a large proportion of his students long after their graduation by means of frequent correspondence and personal visits.

Throughout his university career and even in retirement, Withrow was frequently called on as a consultant to industrial concerns covering almost the entire spectrum of the chemical industry. At various times he was a consultant to departments of the federal government and to the state of Ohio. He was, for example, a member of the U.S. Naval Consulting Board for Ohio in 1915-17, and a consulting chemical engineer to the Chemical Warfare Service in World War I. He was also called on as an expert witness in patent litigation and in civil suits involving damages from disasters and industrial accidents. These outside activities helped make him an exceptionally capable teacher in his field.

Withrow was not only a member or fellow of a large number of scientific, technical, engineering societies and other organizations but a very active member of many of them. For example, he served a term as a councilor of the American Chemical Society, was elected a director of the American Institute of Chemical Engineers four times, was chairman of the Chemical Engineering Section of the Ohio Society of Professional Engineers for four consecutive years, and was the chairman or a member of at least six committees of the American Society for Testing and Materials.

His reputation as a speaker and writer on technical subjects was high. He spoke before local sections of the American Chemical Society in 15 states and gave various invited lectures both in the United States and foreign countries. He was the author or co-author of about 150 technical papers on his many research interests, which included sugar and petroleum refining, electrochemistry, wood distillation, insecticides, and paints.

His position as a leader in his field was recognized by various honors such as an honorary D.Sc. degree from Geneva College. What Withrow probably appreciated most was an elaborate scroll presented to him in 1950 by his former students in grateful acknowledgement of his devoted teaching and his continued interest in their welfare through the years.

A. M. Brant, D. J. Demorest, and J. H. Koffolt, Memorial to Professor James R. Withrow, Minutes of the College of Engineering Faculty, Ohio State University, 1953; Earle R. Caley, "History of the Department of Chemistry," Ohio State University (1969); "American Men of Science," 5th Ed. (1933), Science Press, New York; personal recollections.

EARLE R. CALEY

Rudolph August Witthaus

1846-1915

Witthaus, chemical toxicologist and poison expert, was born in New York City, Aug. 30, 1846. His parents, Rudolph August and Marie Antoinette (Dunbar) Witthaus, had come to this country from Germany 13 years earlier; his father had established himself as a successful merchant in New York. Young Rudolph was educated at the Redfield and Charlier schools in New York and received his A.B. and M.A. degrees from Columbia University in 1867 and 1870. He spent the next year at Columbia law school but finally decided that medicine was preferable and took a year's course at Bellevue Medical College. In 1873 he attended lectures at the Sorbonne and Collège de France, in Paris. Returning to New York, he completed his studies at University Medical College, New York University, receiving his M.D. degree in 1875.

Witthaus was immediately appointed associate professor of chemistry and physiology at University Medical College, and the remainder of his life can very nearly be summed up in a recital of his professional positions, achievements and publications. A small, spare, sandy man who married in 1882 but separated soon after, Witthaus was always withdrawn, and in his later years became exceedingly cynical, particularly in matters of religion, and irascible. Nonetheless, his friendship was valued by the few who enjoyed it.

Witthaus's learning and accomplishments were such that he came to occupy as many as four posts simultaneously. From 1878 to 1898 he was professor of chemistry and toxicology at University of Vermont; from 1882 to 1888 he occupied a like position at University of Buffalo, as well as being Buffalo's City Chemist, a post which was created for him. In 1882 University Medical College, New York University, made him professor of physiologic chemistry, and in 1886 he accepted the chair of chemistry and physics at the same institution, which he held until 1898. In that year, he was made professor of chemistry, physics, and toxicology at

Cornell University Medical College, New York City, from which he retired in 1911 as professor emeritus.

Witthaus's professional output during these 30 years was prodigious, particularly considering his disinclination for collaboration. In addition to many articles in the medical and chemical literature on homicide by morphine, detection of quinine, post-mortem imbibition of poisons, researches at the Loomis Laboratories, etc., Witthaus was author, contributor, or editor of a number of books. In 1879 he published "Essentials of Chemistry, Inorganic and Organic," which by 1908 had gone through six editions. In 1881 he produced "General Medical Chemistry for the Use of Practitioners of Medicine," and in 1886, "A Laboratory Guide in Urinalysis and Toxicology." All of these texts were much in demand and passed through a number of editions. Witthaus contributed articles on poisoning by hydrocyanic acid, oxalic acid, opium, and strychnine, and on the ptomaines, to A. H. Buck's "A Reference Handbook of the Medical Sciences" (1885-93). His major publishing effort was the editorship of the four-volume "Medical Jurisprudence, Forensic Medicine and Toxicology" (1894-96), which came to be known as "Witthaus and Becker's Medical Jurisprudence," the standard work in the field, and which appeared in a second edition, 1906-11. Witthaus wrote the introduction and the entire fourth volume, which dealt with toxicology.

In addition to his eminence through these publications, Witthaus was known world-wide as a poison expert who gave testimony in a number of the notable murder trials which were a form of public entertainment in the decades about the turn of the century. He died Dec. 20, 1915, leaving an estate valued at nearly $250,000, including a collection of first editions and *objects d'art*. In his will he left $20,000 to one Jennie Cowan, who was said to have lived with Witthaus in his final years, and he directed that the remainder of his estate be put in trust to provide her an income of $300 per month, with his scientific books and the remainder of the income to go to the New York Academy of Medicine. He further directed that on the death of his beneficiary the residue of the trust was to be turned over to the New York Academy of Medicine for a library that would be open to the public.

Jennie Cowan turned out to be Mrs. Jennie Shore, said to have been married in 1908 without Witthaus' knowledge. Members of his family contested the will, but the judge of the surrogate court ruled that Witthaus was competent and not under restraint when he made the will (1912) and its codicil (1915) and that it should be duly executed.

Obituary, *New York Times,* Dec. 21, 1915; (July 16, Sept. 22, Sept. 23, 1916) (will and contest); *Science* **43,** 527-28 (1916); "National Cyclopedia of American Biography," **11,** 60-61, James T. White & Co. (1901); "Cyclopedia of American Biography," **20,** 439, Chas. Scribner's Sons (1936); Howard Kelly and Walter Burrage, "Dictionary of American Medical Biography," 1320-21, Norman, Remington Co. (1928); also, notices in *J. Am. Med. Assn.* **66,** 132 (1916); *Medical Record,* p. 1100 (Dec. 25, 1915); Record of Wills, Surrogate Court for the County of New York, Book 1026, pp. 342-6.

ROBERT M. HAWTHORNE JR.

Richard Leopold Wolfgang

1928-1971

Richard Wolfgang was born in Frankfurt am Main, Germany, July 24, 1928, the son of Julius and Wilhelmina Wolfgang. He emigrated to England in 1939 and to the United States in 1945. His graduate and undergraduate work were both completed at University of Chicago.

Wolfgang made many contributions to several areas of physical chemistry. His Ph.D. research, under the direction of Willard F. Libby, was in the area of nuclear chemistry. He continued this research at Brookhaven, Florida State University, and Yale. One of his projects concerned nuclear stripping reactions in which two nuclei "brush up against each other" and a few nucleons are exchanged.

His greatest contribution was the development of hot-atom chemistry, the study of reactions at higher than thermal energies. As an example, hot tritium (H^3) atoms can be produced by the nuclear reaction He^3 + $n \rightarrow T + p$ in a reactor. These tritium atoms slow down on collision with molecules in the system and many react before reaching thermal energies. One observes reactions like $T^* + CH_4 \rightarrow HT, CH_3T$. Wolfgang characterized many of these reactions and derived a quantitative theory—a hot-atom Arrhenius equation—to treat the results. He also studied the reactions of free carbon atoms produced by nuclear reactions. One gets reactions like $C^{11}(^1D$ or $^1S) + C_2H_4 \rightarrow CH_2C^{11}CH_2$ as well as C-H bond insertion. In 1968 he won the American Chemical Society's award for Nuclear Applications in Chemistry.

In the 1960's he used molecular beams to provide better control over reagent conditions and provide more information about the products than the earlier nuclear recoil methods. One beam machine, ADAM, was built to study reactions of hot tritium atoms. Tritium ions are produced in a discharge, accelerated to the desired energy, and then neutralized by charge exchange, thus producing a beam of tritium at a known and variable energy.

A second machine, EVA, was originally constructed for the same purpose but in preliminary trials was used to study ion-molecule reactions. It proved so successful in this mode of operation that it has not, to this time, been converted to study neutral reactions. In EVA a mass-selected ion beam of known and variable energy is crossed with a molecular beam. The angular, energy, and mass distributions of the products of the ion-molecule reaction are measured. One such reaction was $H^+ + D_2 \rightarrow D_2^+ + H$, $HD^+ + D$, $D^+ + HD$, $D^+ + H + D$. The reactions were studied in detail and, simultaneously, the angular and energy distributions were calculated using *ab initio* potential energy surfaces. The two agree, thus providing the first comparison between an *ab initio* theory and *detailed* chemical kinetics experiments.

Wolfgang was also a dedicated teacher. He directed the research of 25 graduate students and 25 postdoctoral fellows in his 16-year academic career. He was active in the field of chemical education and served as chairman of the panel of Chemistry for Citizens at the International Conference on Education in Chemistry 1970 (*see J. Chem. Educ.* **48,** 23 (1971).

Wolfgang disappeared June 19, 1971 while sailing alone near New Haven. The Coast Guard found his capsized boat that day, but his body was not recovered until June 27.

Personal recollections.

R. JAMES CROSS, JR.

Melville Lawrence Wolfrom

1900-1969

Wolfrom, of German and Swiss ancestry, was born Apr. 2, 1900 at Bellevue, Ohio. All his early education was received in the public schools of that town. His initial interest in chemistry was aroused by William A. Hammond, his high school science teacher. This grew to such an extent that he resolved to become a chemist or chemical engineer. Unfortunately, his family lacked sufficient funds to send him to any college or university. Soon after leaving high school in 1917 he obtained a position in the works laboratory of National Carbon Co. at Fremont, Ohio, a concern making electric batteries. In the autumn of 1918 he enrolled at Western Reserve University in Cleveland but left at Christmas, in part because of dissatisfaction with the course offerings which did not include chemical engineering. He then went to New York City to assist a brother in sales work. In the autumn of 1919 he entered Washington Square College of New York University, but again dissatisfaction with the courses available caused him to leave in a short time. After supporting himself with various odd jobs, including that of a factory hand and a railroad laborer, he enrolled at Ohio State University in the autumn of 1920 and finally started on a curriculum leading to a degree in chemical engineering. He followed this curriculum until his senior year when he transferred to the arts curriculum with a major in chemistry. This shift of interest was largely caused by the stimulating influence of William Lloyd Evans, under whom he worked as a student assistant in a research program in carbohydrate chemistry. He graduated with his B.A. degree *cum laude* in 1924.

At the suggestion of Evans he went to Northwestern University for graduate study under W. Lee Lewis. There he supported himself by teaching but spent the major part of his time in research on carbohydrates. In his final year there he married Agnes Louise Thompson of Auburn, Ind. In the course of their marriage five children were born, four of whom survived infancy, a son and three daughters. He received his M.S. degree in 1925 and his Ph.D. degree in 1927.

After graduation in 1927 Wolfrom was awarded a National Research Council Fellowship which enabled him to pursue his research for an additional 2 years. He first studied with Claude S. Hudson of the National Bureau of Standards, then the leading American authority on carbohydrates, next with Phoebus Levene of Rockefeller Institute, and finally with Evans of Ohio State, where he was engaged in original research on carbohydrates that was largely independent.

In the autumn of 1929 Wolfrom was appointed instructor in chemistry at Ohio State. He was promoted to the rank of assistant professor in 1930, associate professor in 1936, and professor in 1940. In 1948 Wolfrom was made head of the division of organic chemistry at Ohio State, a position that he held until 1960, when he was named research professor. In 1965 he was appointed regents professor, the highest teaching rank in the state university system of Ohio. He died unexpectedly June 20, 1969.

Although Wolfrom did his share of undergraduate teaching, especially in his earlier years at Ohio State, his chief interest was in graduate education and in the direction of research. He was the preceptor for more than 100 M.S. and Ph.D. candidates. Many postdoctoral fellows, both from the United States and foreign countries, came to work in his laboratory. The research work on carbohydrates and other classes of organic compounds done in this laboratory is remarkable for both its high quality and great volume.

His research publications, mostly with graduate students, postdoctoral fellows, and colleagues as co-authors, totaled about 500. Most of these appeared as papers in scientific journals, but over 50 appeared as articles, chapters, or large sections in books. Wolfrom also did an enormous amount of editorial work acting as editor or co-editor for 22 volumes of "Advances in Carbohydrate Chemistry," co-editor or consulting editor for "Methods in Carbohydrate Chemistry," and editor for several years of the section on carbohydrates in *Chemical Abstracts.* Furthermore, he was at various times the chairman or a member of national and international committees on chemical nomenclature and documentation.

The achievements of Wolfrom were recognized by numerous honors, such as the Honor Award (now the Hudson Award) of the Division of Carbohydrate Chemistry of the American Chemical Society in 1952, and a citation from the Austrian Government in 1959. A special number of the journal *Carbohydrate Research* was published as the Wolfrom Memorial Issue in April 1970.

Earle R. Caley, "History of the Department of Chemistry, The Ohio State University" (1969); biography by D. Horton in "Advances in Carbohydrate Chemistry and Biochemistry," **26,** 1-47 (1971); personal recollections.

Earle R. Caley

Edward Stickney Wood

1846-1905

Wood was born Apr. 28, 1846 in Cambridge, Mass. to Alfred and Laura (Stickney) Wood, descendents of early seventeenth-century settlers in Massachusetts. He entered Harvard in 1863 and soon decided to specialize in chemistry to prepare for a career in medicine. Following graduation in 1867 he served as house pupil in both the Marine Hospital in Chelsea and Massachusetts General Hospital in Boston. He received his M.D. degree from Harvard Medical School in 1871.

The resignation of James Clarke White from the Chemistry Department at Harvard Medical School created a vacancy, to which Wood was appointed. To increase his knowledge of biological chemistry he spent 6 months in Berlin and Vienna. Upon his return he became assistant professor, developing one of the best systematic medical chemistry courses in the United States. In 1876 he was promoted to professor, a position he retained until his death, July 11, 1905, at Pocasset, Mass. He was a very successful teacher, his lectures being described as "eminently practicable and intelligible, developing the enthusiasm which always attend the work of a popular lecturer."

Wood's activities in chemistry were varied. For several years he served on the commission which revised the U. S. Pharmacopoeia. He revised the standard monograph by Neubauer and Vogel, "A Guide to the Qualitative and Quantitative Analysis of the Urine" (1879). He contributed chapters to books on medical jurisprudence, and he co-edited the fourth edition of the Wharton and Stille "Treatise on Medical Jurisprudence" (1882).

Among Wood's professional activities were service as chemist to Massachusetts General Hospital and as a member of sanitary commissions for Boston and Massachusetts, concerned with the water supply and facilities for gas lighting in Boston. He cooperated with the Massachusetts State Board of Health in matters involving chemistry, and his activities are noted in the board's annual reports. He published scores of articles, most of them in *Boston Medical and Surgical Journal,* many of them on facets of legal chemistry and on urinalysis as a diagnostic aid. During the 1870s he ran a series of review articles under the general title, "Recent progress on medical chemistry," in that journal.

He was the best known toxicologist in New England; it was said that there was hardly a trial for a capital crime in which his knowledge of chemistry, and especially his skill in identifying bloodstains of various kinds, was not called upon.

In 1872 Wood married Irene Eldridge Hills, who died in 1881. In 1883 he married Elizabeth A. Richardson, who survived him.

Biographies by J. Collins Warren, in *Proc. Amer. Acad. Arts and Sci.* **51,** 929-30 (1915-16), and Henry R. Viets in "Dictionary of American Biography," **20,** 455-56, Chas. Scribner's Sons (1936).

SHELDON J. KOPPERL

Richard George Woodbridge, Jr.

1886-1946

Woodbridge was born in Iowa City, Iowa, Mar. 18, 1886, the son of the Rev. Richard G. and Anna A. (Rode) Woodbridge. He had two older brothers and two younger sisters. Graduating from MIT in 1907 with a B.S. degree in chemistry, he returned to MIT during the year 1907-08 as the holder of a fellowship for the study of chemistry of cellulose established by Arthur D. Little of Boston. This resulted in Woodbridge's first publication (with Forris Moore) on "Colored Salts of Schiff's Bases" (1908). His MIT work on cellulose resulted in his thesis "Researches on Cellulose" and later in a paper "Notes on Cellulose Esters."

In 1908, Woodbridge joined E. I. du Pont de Nemours and Co. as a research chemist in the newly organized Smokeless Powder Division of the Experimental Station near Wilmington, Del. Thus he began a career with Du Pont which lasted 38 years and resulted in 26 patents, many articles, and in recognition as the world's expert in military smokeless powders.

Woodbridge married in 1911 in Waltham, Mass. Ethel Lytle (Whall) Chandler, and they had two children, Richard George III and Margaret Lytle.

With Alfred L. Broadbent, his assistant for many years, Woodbridge rapidly brought the Du Pont Co. to supremacy in the manufacture of military smokeless powders with a sequence of patents and advances in the art. Key patents (Broadbent and Woodbridge, 1,312,463 and 1,313,459) covered the hot water treatment of nitrocellulose powder grains which had been previously coated with dinitrotoluene. This treatment greatly improved the ballistics of smokeless powders for rifles and cannons.

Shortly after the outbreak of World War I, with orders pouring in for cannon and rifle powder, Du Pont, after preliminary tests of Woodbridge's new powder, Improved Military Rifle (I.M.R.) Powder as it was to be named, decided to offer it to the British. It was an immediate military success. Difficulties in making the new powder (but not in

Woodbridge's portion of the process) led to Woodbridge's going to the smokeless powder plant at Carney's Point, N.J. where in August 1915, the output per powder press was almost at a standstill at about 600 pounds per day. When Woodbridge left Carney's Point in June 1916, production was over 3,000 lbs. per press per day and was to reach 4,500. He turned Du Pont's failure into success!

It was under Woodbridge's patents that Du Pont produced during World War I the vast amount of rifle and cannon powder which contributed to Allied victory. The wartime profits accruing to Du Pont in large part permitted Du Pont the means to embark on the vast program of expansion and diversification to become a giant chemical company. These patents of Woodbridge were referred to as the most valuable patents in the explosives industry, second only to those of Alfred Nobel of Sweden. In 1940 Woodbridge (jointly with his assistant Broadbent) received from the National Association of Manufacturers the "Modern Pioneer Award" for his development of I.M.R. Powder.

At the close of World War I, Du Pont had on its hands large amounts of leftover cannon powder. Woodbridge developed a process for removing the stabilizer (diphenylamine) from this powder and reducing its viscosity so that it could be used in the new nitrocellulose lacquers and plastics (patent 1,439,656). Under this patent over 50 million lbs. of this excess explosive were processed for peacetime use in Du Pont's coated fabric "Fabrikoid" and in Du Pont's "Pyroxylin" lacquers for the burgeoning automobile industry.

In Du Pont, Woodbridge held many positions. In 1921 he became the first director of the Brandywine Laboratory, Smokeless Powder Department, and in 1927, chemical director, Smokeless Powder Department.

After World War I Woodbridge continued his work on improving smokeless powder. This involved the development of a flashless powder that could not be seen when it was fired at night, the development of greatly improved powders for sportsmen, the use of wood pulp as a raw material for smokeless powders, and many improvements in equipment and technique.

During World War II Du Pont and Hercules Powder alone manufactured more than 700 million lbs. of rifle powder and more than 3.5 billion lbs. of cannon powder under Woodbridge's patents. His work in World War II was recognized by the War Department with a Certificate of Appreciation, "As a member of the Joint Army-Navy Smokeless Powder Board, for patriotic service in a position of trust and responsibility."

His heavy responsibilities and arduous schedule during World War II hastened his death, Nov. 7, 1946, in Wilmington, Del.

Manuscript, "History of the development of improved military rifle powder at the Experimental Station" by Woodbridge; information from registrar, M.I.T.; records of Du Pont Co.; patents from Richard C. Woodbridge, U.S. Patent Office; Woodbridge's publications; personal recollections.

RICHARD G. WOODBRIDGE, III

James Woodhouse

1770-1809

Woodhouse, born in Philadelphia, Nov. 17, 1770, attended University of Pennsylvania and the University's Medical School. While still a medical student in 1791 he became a surgeon's mate in General Arthur St. Clair's campaign against the Indians of the Northwest Territory.

Woodhouse studied chemistry in medical school, wrote a thesis "On the Chemical and Medical Properties of the Persimmon Tree, and the Analysis of Astringent Vegetables," and followed chemistry on his own. In 1796 he was recommended for the professorship of chemistry in the Medical School by Benjamin Rush. Woodhouse's background seems meager for a professor of chemistry, yet it was equal to or better than that of other 18th century American chemistry teachers.

As soon as Woodhouse was elected, he plunged into the science. Charles Caldwell, an occasional teacher of chemistry, left the following reminiscences about Woodhouse:

"He became, in a short time, so expert and successful an experimenter, as to receive from Dr. Priestley, who had just arrived in the

United States, very flattering compliments on his dexterity and skill. That distinguished gentleman, on seeing him engaged in the business of his laboratory, did not hesitate to pronounce him equal, as an experimenter, to anyone he had seen in either England or France. . . . At times, his devotion to [chemistry] and the labor he sustained in the cultivation of it, were positively marvelous—not to say preternatural. To the young men who attended his lectures, he . . . recommended "Miss Chemistry as their only mistress." . . . During an entire summer (one of the hottest I have ever experienced), he literally lived in his laboratory, and clung to his experiments with an enthusiasm and persistency which at length threw him into a paroxysm of mental derangement, marked by the most extravagant hallucinations and fancies. He even believed, and, on one occasion, proclaimed, in a company of ladies and gentlemen, that, by chemical agency alone, he could produce a human being."

Woodhouse became nationally known as a teacher. When Benjamin Silliman wanted to learn chemistry in 1802, he picked Woodhouse over other American teachers and journeyed from Connecticut to attend him. According to Silliman, Woodhouse "was short, with a florid face. He always dressed with care; generally he wore a blue broadcloth coat with metal buttons; his hair was powdered, and his appearance was gentlemanly." But Silliman criticized him as "wanting in personal dignity," and for being "out of lecture hours, sometimes jocose with the students." Woodhouse's informality did not seem quite proper to young men whose model was the dignified, rhetorical type lecturer, and one of them criticized their teacher thus:

"Professor Woodhouse was a plain, unpretending man, though a learned chemist, and an accurate experimenter, and bold enough too. Little can be said of his style of lecturing. He did not attend to its effect; and he did himself great injustice thereby, and lost much of his influence over his pupils, by neglecting the manner of communicating instruction."

Woodhouse wrote perhaps a score of articles in old science and medical journals. For students he wrote "The Young Chemist's Pocket-Companion" (1797), and edited three American editions of European texts: James Parkinson's "Chemical Pocket-Book" (1802), Samuel Parkes' "Chymical Catechism" (1807), and Chaptal's "Elements of Chemistry" (1807).

When Woodhouse began to teach, the Chemical Society of Philadelphia, organized a few years earlier, was moribund. He revived the society and spurred it into activity. Within a few years it had its own laboratory and had issued several pamphlets. Woodhouse remained the chief support of the society, and it collapsed after his death.

As with almost all early American chemists, we know little of Woodhouse's personal life. He remained a bachelor. In 1802 he visited England and France where he met Davy and other European chemists. He died of apoplexy in Philadelphia, June 4, 1809, at the age of 39.

James Thacher, "American Medical Biography," **II,** 220-222, De Capo Press (1828, reprint 1967); "Autobiography of Charles Caldwell," Harriot Warner, ed., 173-176, Lippincott, Grambo & Co. (1855, reprint 1968); Edgar F. Smith, "James Woodhouse A Pioneer in Chemistry," a 298-page book with portrait, fairly complete but unfortunately lacking references to all sources of information, John C. Winston Co. (1918); the quote, "Professor Woodhouse was a plain, unpretending man," is from anonymous article, "Reminiscences of Olden Times," *Maryland Med. Surg. J.* **1,** 75-76 (1839-40); George P. Fisher, "Life of Benjamin Silliman," Charles Scribner and Co., 1, 100-103 (1866).

WYNDHAM D. MILES

Lincoln Thomas Work

1898-1968

Work, a chemical engineer, contributed notably to education, to chemical engineering technology, and to his profession during the era of great technical and industrial changes that encompassed pre- and post-World War II.

Born in Hartford, Conn., Dec. 28, 1898, son of Norman Porter and Hattie (Lincoln) Work, he grew up in New York City where his closely-knit family encouraged scholastic achievement. He obtained the degrees of

A.B. (1918), Ch.E. (1921), A.M. (1928), and Ph.D. in chemical engineering (1929), all from Columbia University. He married Sept. 7, 1922, Clara Radcliff, and they had two daughters, Lillian and Dorothy Lincoln.

In 1921 he joined the Columbia faculty as instructor in chemical engineering. There he helped to create a new, much-needed course in process development which was added to the curriculum and adopted by other institutions. He helped to administer the "extension" courses in chemical engineering, designed for those in industry who needed to keep up with technical progress.

Columbia encouraged its faculty to do consulting work, both to make the academic world more relevant to industrial progress and to supplement modest faculty salaries. Work concentrated his consulting endeavors on such specialties as cement, pigments, paints, grinding, surface chemistry, particle-size, and heterogeneous systems in chemical and metallurgical engineering. He thoroughly explored and aided the development of the emerging field of powder metallurgy, which during the war provided substitutes for scarce strategic metals and also supplied unique materials suitable, among other uses, for special filters and self-lubricating bearings.

Work advanced through the academic ranks of instructor, assistant professor, and associate professor while his consulting activities became recognized for thorough research, dependability, and results. In 1940 he decided to join Metal & Thermit Corp., Rahway, N.J., as director of research and development. He built up the firm's scientific staff and constructed laboratory and pilot plant facilities to carry out its research.

It was not until 1959 that the public was told about Work's significant contribution to World War II. Then John R. Dunning revealed that:

"Work and his Metal & Thermit group gave tremendous help to the Manhattan Project. Following the work on fission of U-235, and our invention of the Gas Diffusion Separation Process, it was necessary to develop porous separation barriers, fine enough to serve as atomic sieves. Work's fundamental studies on fine metal powders and his company's experience in production . . . provided tremendous help in making possible the huge atomic plants at Oak Ridge, Paducah, and Portsmouth, which gave and maintain our atomic leadership. The Office of Scientific Research & Development, through the National Defense Committee, awarded Work a certificate of appreciation for his contributions."

In 1949 Work set up his own consulting chemical engineering practice in New York city, most recently at The Chemists' Club Building. Clients consulted him on laboratory research, development, and design, on research management, on process development, and on a range of individual subjects, including abrasives, adhesives, cements, ceramics, cosmetics, fertilizers, fuels, and textiles.

The rapid growth of industry and government unfavorably affected the professional status of scientists and engineers, of whom 98% had become salaried employees, with consequent loss of individual recognition and with a diminished sense of professional consciousness. Work, humanitarian by nature and devoted to his profession, lectured constructively on this problem to many societies. His talents for leadership and management also guided several professional societies through difficult internal transition periods.

He served as president of Phi Lambda Upsilon (1930-35), of The American Institute of Chemists (1952-54), of the Association of Consulting Chemists and Chemical Engineers (1957-58), and of The Chemists' Club (1960-62). His competence kept him in demand for committee assignments for these and other associations, including the American Institute of Chemical Engineers, American Institute of Mining & Metallurgical Engineers, American Society for Testing & Materials (vice chairman 1952; honorary vice chairman 1968), Society of Chemical Industry, and the Society for the Promotion of Engineering Education.

He was the author of well over 100 publications and patents in addition to private reports pertaining to his consulting practice. He received many honors in recognition for specific services, for guidance of chemical engineering students, and for high professional and ethical standards.

He died Nov. 3, 1968, exactly 3 weeks after the death of his wife.

The Percolator, June 1962, Dec. 1964; biography by John R. Dunning, in *The Chemist,* January 1960; obituary in *The Chemist,* December 1968; personal recollections and correspondence.

VERA KIMBALL CASTLES

Theodore George Wormley

1826-1897

Wormley, a noted toxicologist, was born to David and Isabella Wormley in Wormleysburg (a town named for his ancestors), Pa., Apr. 1, 1826. He attended Dickinson College for 3 years and left before completing his collegiate education to study medicine at Philadelphia College of Medicine, where he received his M.D. degree in 1849. He practiced medicine for 3 years, first in Carlisle, Pa., then in Chillicothe, Ohio, and finally in Columbus, Ohio.

Wormley had studied science as much as possible at Dickinson and Philadelphia Medical. In Columbus he became friendly with William Sullivant, a botanist, under whose guidance he became an expert microscopist. Within a short time he impressed the educational community sufficiently so that he was offered the instructorship in chemistry at Esther Institute, a seminary for girls, and the professorship of chemistry and natural science at Capital University, in 1852. He stopped practicing medicine and taught at Esther Institute until 1854, at Capital until 1865, and also was professor of chemistry and toxicology at Starling Medical College from 1854 to 1877. In 1877 he moved to Philadelphia to become professor of chemistry and toxicology in the University of Pennsylvania's Medical School and held this position until his death.

Wormley began to research and publish on urinalysis and toxicology in the early 1850's. Gradually toxicology became his favorite study. He tried all the chemical tests for poisons that he could find in the literature, he picked out the best, ascertained the effect of interfering substances, and determined the limits of tests under as nearly actual conditions as a toxicologist would meet in practice. With his microscope he studied the crystals formed by reagents acting on poisons. After years of experimenting, interrupted by Wormley's service with the Sanitary Commission during the Civil War, he published in 1867 "Micro-chemistry of Poisons, including their physiological, pathological, and legal relations: adapted to the use of the medical jurist, physician, and general chemist." Wormley's wife learned the art of steel engraving to produce the fine illustrations of crystals after the publisher protested that he could not find an engraver competent to make the plates. The book was accepted immediately as an authority and remained in use for half a century until it was displaced by more modern works.

Wormley's reputation as a toxicologist caused him to be sought as an expert witness and he appeared in a number of murder cases, including those of Culbertson, Freet-Converse, Dresbach, Wharton, Carlyle W. Harris, and Herbert W. Hayden. He seems gradually to have come to abhor courtroom battles and toward the end of his life was loathe to testify.

Aside from toxicology, Wormley was business manager and one of the editors of *Ohio Medical and Surgical Journal* from 1862 to 1864, Gas Commissioner of Ohio from 1867 to 1875, and chemist to the geological survey of Ohio from 1869 to 1874. His articles, which number more than 50, appeared in *Chemical News* and several American chemical, medical, pharmaceutical, and other scientific journals.

He was a vice president of the Centennial of Chemistry at Northumberland, Pa. in 1874, a vice president of the American Chemical Society, and a recipient of three honorary degrees.

Wormley was a punctual, conscientious teacher. In private life he was quiet and reserved, a skillful musician who could play several instruments, and during the summer liked to work in the garden. He died of Bright's disease in Philadelphia, Jan. 3, 1897.

Obituaries by John Ashhurst, Jr., in *Trans. College Physicians Phila.* [3 S] **19,** lxxix-lxxxviii (1897), and E. F. Smith in *J. Amer. Chem. Soc.* **19,** 275-279 (1897); W. M. MacNeven, "Theodore

George Wormley—First American Microchemist, 1826-1897," *J. Chem. Educ.* **25,** 182-186 (1948), portrait; "Dictionary of American Biography," **20,** 535, Chas. Scribner's Sons (1936).

WYNDHAM D. MILES

Ralph Garrigue Wright

1875-1954

Wright, long-time head of the Department of Chemistry at Rutgers, after whom the university's present chemistry laboratories are named, was born in St. Louis, Mo., Apr. 29, 1875, the son of Thomas Wright, a merchant and banker, and Emily (Garrigue) Wright. He was educated in the St. Louis public schools and at Columbia Grammar School in New York; going on to Columbia University, he received his B.S. degree in 1899. Thereafter he went to Switzerland to finish his education, taking his M.A. degree from the Polytechnic Institute in Zürich in 1902, and his Ph.D. from University at Basel in the same year.

Initially Wright returned to St. Louis after completing his graduate work, serving as a chemist for the brand-new (1901) Monsanto Chemical Co. John Queeny, Monsanto's founder, was engaged in a price-cutting competition with the Germans to produce vanillin, and Wright was able to recommend his former Swiss roommate, Gaston DuBois, as an interpreter and later a chemist; DuBois eventually became a vice-president of the company.

In 1904 Wright returned to the academic world to stay. He was assistant in chemistry at Columbia for a year, then professor of chemistry at Washington and Jefferson College, from 1905 to 1907. In 1907 he was appointed professor and head of the Department of Chemistry at Rutgers University, which positions he would hold for the remainder of his professional life. Rutgers, an early land-grant state university, was interested in building its academic programs to the level of those in agriculture and technology, and Wright attacked this problem with enthusiasm, in other departments as well as his own. He insisted from the start upon research as an adjunct to teaching and an indispensable activity of the faculty in his department. Wright's own research interests were on the structure of dyestuffs, oxidation, and fluorescence in organic compounds, although his output of publications was meager. He was the moving force in the approval and construction of chemistry laboratories built in 1910.

Always deeply interested in the arts, in his later years with the university Wright concerned himself with improvement of the departments of art, music, and literature, as well as chemistry. He retired from teaching in 1930, occupying himself thereafter with extensive civic activity. At times before or after his retirement he was a member of the local board of education, the board of the Middlesex County General Hospital, and the New Brunswick Board of Trade (later Chamber of Commerce). In 1938 he became a trustee of Rutgers University, a position for which his broad background and acquaintance in business and the university fitted him to a degree unusual among university trustees.

Wright was married in 1911 to Margaret Bevier, daughter of Dean Louis Bevier of Rutgers. Mrs. Wright was prominent in the 1930's and 40's through her activities in the national Girl Scout movement. The Wrights had two children, Margaret (Wright) Buckingham and Ernest Bevier Wright, a faculty member at Rochester University's Medical College.

Wright died at his summer home in Cherry Valley, N.Y., June 21, 1954, less than a month after Rutgers University dedicated the present University Heights chemistry laboratories in his name.

New York Times, June 23, 1954; "National Cyclopedia of American Biography," **40,** 432, James T. White & Co. (1950); *Chem. Eng. News* **30,** 1130 (1952).

ROBERT M. HAWTHORNE JR.

Y

Edward Livingston Youmans

1821-1887

Youmans, son of Vincent and Catherine (Scofield) Youmans, would have been lost in a chemical laboratory; yet by arduous study, he developed himself into one of the best writers and lecturers on chemistry that this country has produced. Born in Coeymans, N. Y., June 3, 1821, he worked on a farm in summer and attended school in winter. When he was thirteen, he and his teacher learned chemistry together by reading John Comstock's "Elements of Chemistry."

An eye disease attacked Youmans when he was fourteen, rendering him partially, and at times totally, blind for the remainder of his life. For years he underwent medical treatment, improving then relapsing. During this period he continued his education by having his sister Eliza and his friends read to him. Hoping to find a physician who could aid him, he moved to New York City. He helped support himself by writing reviews and essays for the press, devising a writing board to hold paper and guide his pen. Eliza took chemistry lessons in Thomas Antisell's private laboratory to explain the science better to Edward.

Youmans' first venture into chemical education was a wall chart he designed for classrooms. This chart, 5 by 6 feet, showed reactions as taking place between atoms. The atoms were drawn as squares with areas proportional to atomic weights, and painted in colors characteristic of the elements, as yellow for sulfur. This chart was widely used as a teaching aid in schools in the 1850's.

The popularity of the chemical chart encouraged him to attempt an elementary text. He attended John Draper's lectures at New York University, had reference books read to him, and then laboriously wrote "Class-Book of Chemistry." He made the book interesting and useful, rather than detailed and exhaustive, by emphasizing the chemistry of everyday life—of farming, housekeeping, and manufacturing. Published in 1851, the book sold at least 150,000 copies, an astonishing number at a time when chemistry was not studied as widely as it is today.

In 1854 Youmans brought out "Chemical Atlas," designed to assist students by means of diagrams. Perhaps the most unusual text published in this country, it had folio size plates painted in different colors picturing compounds, isomers, homologous series, respiration, combustion, the organic balance of nature, and other topics.

Youmans attempted science lecturing in New York in 1851 with a lecture on chemistry. He was successful and became one of the leading lecturers of his time. For two decades he spent several months each year travelling thousands of miles by train to towns in the Midwest and Canada to speak on science topics, generally related to chemistry. He stopped only when rheumatism and ill-health forced him to lead a sedentary life.

In 1869 Youmans accepted the editorship of *Appleton's Journal of Popular Literature, Science, and Art.* Unhappy because it carried so little sicence, he resigned in 1870 and 2 years later founded *Popular Science Monthly.* In 1871 he conceived the idea of a series of books written by leading American and European scientists, the volume appearing simultaneously in several languages. This was the origin of the "International Scientific Series," popular a century ago for placing first-class science works within the reach of lay readers. Youmans edited more than 50 volumes of the series.

Youmans died Jan. 18, 1887, in New York. He was a remarkable man. Handicapped by eye disease and lack of formal education, he struggled forward to become a notable interpreter and diffuser of science.

Edward's brother, William Jay Youmans, born Oct. 14, 1838, at Milton, N. Y., studied chemistry under Charles Joy at Columbia, attended Sheffield School at Yale, and finally took a degree in medicine at New York University. His chief interests were chemistry and physiology. He edited *Popular Science Monthly* after Edward's death until 1900 and contributed major reviews of the year's advances in chemistry, physiology, mineralogy, and meteorology to "Appletons' Annual Cyclopedia" from 1880 to 1900. He died at Mount Vernon, N. Y., Apr. 10, 1901.

John Fiske, "Edward Livingston Youmans . . . with selections from his published writings and extracts from his correspondence," D. Appleton and Co. (1894), illus.; John Fiske, "A Century of Science," Houghton Mifflin Co. (1899), pp. 64-99; biographies by W. D. Miles in *Sch. Sci. Math.* **64,** 491-498 (1964); by R. Burlingame in *Pop. Sci. Mon.* **150,** 136-141 (May 1947); "Dictionary of American Biography," **20,** 615, Chas. Scribner's Sons (1936); obituary, *New York Times,* Jan. 19, 1887.

For William Jay Youmans see "Dictionary of American Biography," ibid., p. 616; "Appletons' Annual Cyclopedia," with portrait, 3rd series, **6,** D. Appleton & Co. (1901).

WYNDHAM D. MILES

Abram Van Eps Young

1853-1921

A paternal ancestor of Young, one John Youngs, a dissenting clergyman of Hingham, England, came with his flock to America in 1640 and founded one of the earliest settlements on Long Island. In these early days the minister was a magistrate as well and the chief man of the parish. Accordingly, we find the Young family holding high office in the municipality for many generations. Almost 200 years later, a member of the family migrated to central New York and joined his fortunes with the substantial Dutch family of Van Eps. In 1840 Abram Young moved to Sheboygan, Wisc. There his son, the object of this sketch, was born June 5, 1853. He prepared for college in the Grand Rapids, Mich. high school and entered University of Michigan in 1870. At the end of his sophomore year he went abroad for a year's study in Paris. Returning to Ann Arbor, he graduated with the degree of Ph.B. in 1875 and also qualified as a pharmaceutical chemist.

Young stayed in Ann Arbor for more than a year as assistant to the professor of general chemistry and physics. In late 1877 he began graduate study at Johns Hopkins University. In 1881 he moved to Harvard University for further graduate study and private instruction. He spent another year abroad and then returned to Harvard as an assistant to Josiah Cooke. Among his duties was supervision of the summer school of chemistry. In the summer of 1885 he became professor of chemistry and head of the department at Northwestern University, a position he filled for 35 years.

Before Young took charge, the department had no laboratory; instruction had been given by lectures and textbook study. Young immediately fixed up temporary laboratory space in an academic building and began plans for a proper chemistry laboratory. He succeeded in getting the laboratory built. A contemporary wrote of it; "The new chemical laboratory was planned by Professor Young after an extended tour throughout the East, and few, if any, laboratories anywhere are more completely or conveniently equipped. Many ingenious arrangements of the Professor's own invention served to make the laboratory a model of convenience and efficiency." The building, Fayerweather Hall, housed both the chemistry and physics departments until 1941 when new and larger laboratories for both subjects were constructed in the new Technological Institute. The Young laboratory was then razed.

Young published a number of scientific papers in the field that has come to be known as physical chemistry.

Near the close of Young's career the head of Northwestern University's Department of German, James Hatfield, wrote of him; "With all his interest in science, Professor Young has always found some time for the cultivation of the humanities. He is especially fond of music, was for several years president of the Evanston Browning Club, and is a member of the Chicago Literary Club. Professor Young is known and admired in the village for his breadth of culture and refinement of manners, respected and loved by his students for his clear and forcible style as a lecturer, his skill and tact as an instructor, combined with the genial kindness and patience of the true teacher, and is honored by his colleagues on the faculty and his co-workers in science as a patient investigator and accurate scientist."

Young retired in 1920 and died on his farm in Hendersonville, N.C. in 1921.

Northwestern University Syllabus, 1892; *Northwestern Univ. Alumni J.*, March 1918.

CHARLES D. LOWRY

Z

Frederick Clemens Zeisberg

1888-1938

Fred Zeisberg was born in Jefferson City, Mo., May 13, 1888, the son of Francis and Clara (Hugershoff) Zeisberg. Francis Zeisberg had immigrated from Germany in 1881 and was professor of music at several colleges for women.

Fred Zeisberg was educated in private and public schools in Virginia and Tennessee. He studied at King College, Bristol, Tenn. (1904), Abington (Va.) Male Academy (1906), and the University of Virginia (1909).

During 1907-08, he was employed as a chemist by Mathieson Alkali Works and in July 1909 joined the Du Pont Co.'s Eastern Laboratory as a research chemist in explosives. In 1912 he was transferred to the Chemical Department as a specialist in acid manufacture. He became manager of Elementary Process Design in 1918 and from 1919 to 1921 headed the Intelligence Division of the Chemical Department. Here he was concerned with systematizing research reporting, information retrieval, and building an extensive library of chemical technology.

From 1921 to 1927 he headed Chemical Engineering Investigations for the Development Department, and from 1927 until his death from a heart attack Nov. 12, 1938 he was a special technical investigator.

Zeisberg was best known for his thorough knowledge and experience on the manufacture of sulfuric and nitric acids, on which he held a dozen patents and contributed numerous technical articles. He supplied charts included in the International Critical Tables, published in 1928, on vapor pressures, boiling points, and vapor compositions of the system H_2O — H_2SO_4 — HNO_3; also tables of vapor pressures of aqueous solutions of HCl. His development of methods of supplying heat by means of Duriron heaters to furnish steam for countercurrent contacting with sulfuric acid containing nitric acid and so obtaining concentrated nitric acid overhead, caused this process to become known as the Zeisberg process. He was also an authority on the oxidation of ammonia and high pressure gas techniques. His expertise contributed greatly to the establishment of a dye and intermediate industry in the United States.

He was active in the American Institute of Chemical Engineers and was chairman of the Publications Committee, reading for some years every paper submitted for presentation and publication. He served as director, vice president, and president during 1937-38, and had just been elected to the second term a few days before his death. He was chairman of the Delaware Section of the American Chemical Society in 1921, served on the editorial board of the Technologic Series of ACS Monographs, and also served as editor of the section on "Acid, Alkalies, Salts, and Sundries" of *Chemical Abstracts* from 1921 to 1928.

He was married in 1915 to Madeline Davis, and they had three children, Frederick Clemens, Jr., Margaret Elizabeth, and Millard Davis.

Mr. Zeisberg, as he was known to his younger associates, was called "Fritz" by his contemporaries, or by closer friends "Zip," for alliteration. He kept the most meticulous notebooks, always written in green ink. He was interested in report writing and was the author of the section of that subject in the early editions of Perry's "Chemical Engineers' Handbook."

The Delaware Valley Section of the American Institute of Chemical Engineers, wishing to honor his memory, established shortly after his death the F. C. Zeisberg Award, given annually to a student in chemical engineering at any one of the colleges in the area for the best report submitted in the course in unit operations, the prizes taking the form of books on chemical engineering subjects.

He took a humane interest in those working with him, on one occasion taking into his own home a newly recruited young engineer for whom nursing care was needed and not otherwise available. His interests included stamp and book collecting and horticulture; he planted hundreds of trees in his neighborhood.

He was remembered professionally as an outstanding technologist, especially in the field of inorganic acid manufacture, and personally as a wise and genial friend and associate.

Personal recollections; *Chem. Abstr.;* portrait, *Chem. Eng. News* **15,** 525 (1937); obituaries in *Chem. Eng. News* **16,** 616 (1938); *Chem. Met. Eng.* **45,** 667 (1938); Wilmington *Morning News,* and Wilmington *Evening Journal,* Nov. 14, 1938.

THOMAS H. CHILTON
HERBERT T. PRATT

Frederick William Zerban

1880-1956

Zerban was born in Oppenheim, Germany, Oct. 20, 1880. He received his education in Germany and graduated *summa cum laude* at Munich in 1903 with a Ph.D. degree. His thesis was on radioactive thorium. He taught 1 year at Munich, then came to this country as a Carnegie Assistant at College of the City of New York, where he stayed for 2 years. Thereafter his entire career was in sugar research and the study of sugar analytical procedures. This work took him to several countries; chemist, Louisiana Sugar Experiment Station, 1906-08; director, Sugar Experiment Station, Lima, Peru, 1909; asst. director, Agricultural Experiment Station, Tacuman, Argentine Republic 1909-10; research chemist, Puerto Rican Sugar Planters' Experiment Station, 1911-17. Then in 1917 he returned to Louisiana as research chemist of the Sugar Experiment Station in New Orleans.

The 3 years from 1920 to 1923 covered his only venture into industrial work when he was director of research for Penick & Ford, dealers in molasses products in New Orleans.

Zerban's appointment as chemist-in-charge, New York Sugar Trade Laboratory, in 1923, in which position he succeeded Charles A. Browne, gave him free play for his three major interests—research, writing, and activities in sugar associations and scientific societies around the world. He continued as director of the Sugar Trade Laboratory until he retired in 1955. This laboratory served as referee in the testing of all raw sugars that entered the U. S. market.

Zerban's facility with languages enabled him to learn Latin, Greek, English, French, Spanish, Dutch, Polish, Russian, Czech, Portuguese and Italian. His fluency in several of these languages made his work as assistant editor of *Chemical Abstracts* especially valuable during World War II.

Zerban was a prolific writer. His best known publication was the monumental Browne-Zerban "Sugar Analysis," the third edition of which (1941) was long one of the leading reference authorities in its field. This edition was almost entirely Zerban's work as Browne was in failing health during the rewriting and revision. Another important set of methods compiled and edited by Zerban was the "System of Sugar Factory Control" of the International Society of Sugar Cane Technologists. Two editions (1941 and 1946) were trans ated into Spanish and Portuguese. His monograph, "The Color Problem in Sucrose Manufacture" (1947) gave a careful review and extensive bibliography of the whole subject about which Zerban had previously published several studies. His findings and publications covered the whole spectrum of cane sugar analytical procedures from the filterability of raw sugar solutions to the determination of various physical constants of sucrose and other sugars. His many publications on the characteristics and properties of cane molasses gave results that were valued by distillers and molasses users.

In spite of the many demands on his time, Zerban carried on a voluminous correspondence with inquiries from all over the world. Any question on analytical methods received a prompt answer, generally with supporting data. Many honors were conferred on him by scientific societies in various countries. Among those so honoring him, the more unusual ones were La Academia de Ciencias of Lima, Peru; the Sugar Technologists Association of India; La Societé de Zymologie and Assoc. Chim. de Sucrerie et de Dist. de France.

His work with the Association of Official Agricultural Chemists for many years involved his position as associate referee for the chapter on "Sugar and Sugar Products" as well as papers he presented at annual meetings. These contributions contained numerous methods developed by him and his co-workers. In the American Chemical Society he served several terms as chairman of the Division of Sugar Chemistry and Technology, and in 1949 received the C. S. Hudson Award of the division. In the American Institute of Chemists he was a fellow and chairman of the New York Section. The International Society of Sugar Cane Technologists elected him president for 1929-32, and he headed the committee on the system of factory control and standard methods of analyses that resulted in his publication of the codified results in 1941 and 1946.

Possibly Zerban's work with the International Commission for Uniform Methods of Sugar Analysis had greater impact than any of his other activities. He served as international referee and associate referee on committees dealing with varied subjects and was especially influential in standardizing procedures for the polarization of raw sugars, the figure which determined the price of raw sugars in all world markets. Because of his active participation in the work of the commission for so many years he was awarded a scroll of honor and named Life Honorary Vice President of the commission at its Paris Session in 1954. He died in Boston on Aug. 30, 1956.

Biographical Note—Sugar Research Foundation Technical Report No. 2; obituaries in *Intern. Sug. J.* **58,** 321 (1956); *J. Ass. Offic. Ag. Chem.* **40,** 101-103 (1957); *Percolator* **32** (3), 39, 40 (1956); personal recollections.

GEORGE P. MEADE

INDEX

INDEX